PLANETARY NEBULAE

INTERNATIONAL ASTRONOMICAL UNION

UNION ASTRONOMIQUE INTERNATIONALE

PLANETARY NEBULAE

PROCEEDINGS OF THE 131ST SYMPOSIUM OF THE
INTERNATIONAL ASTRONOMICAL UNION,
HELD IN MEXICO CITY, MEXICO, OCTOBER 5-9, 1987

EDITED BY

SILVIA TORRES-PEIMBERT

*Instituto de Astronomía,
Universidad Nacional Autónoma de México*

KLUWER ACADEMIC PUBLISHERS
DORDRECHT / BOSTON / LONDON

Library of Congress Cataloging in Publication Data

International Astronomical Union. Symposium (131st :
1987 : Mexico City, Mexico)
   Planetary nebulae.

   At head of title: International Astronomical Union,
Union Astronomique Internationale.
   Includes indexes.
   1. Planetary nebulae--Congresses.   I. Torres-Peimbert,
Silvia. II. Title.
QB855.5.I67 1987   523.8   88-26615

ISBN 0-7923-0002-5

Published on behalf of
the International Astronomical Union
by
Kluwer Academic Publishers, P.O. Box 17, 3300 AA Dordrecht, The Netherlands.

Kluwer Academic Publishers incorporates
the publishing programmes of
D. Reidel, Martinus Nijhoff, Dr W. Junk and MTP Press.

Sold and distributed in the U.S.A. and Canada
by Kluwer Academic Publishers,
101 Philip Drive, Norwell, MA 02061, U.S.A.

In all other countries, sold and distributed
by Kluwer Academic Publishers Group,
P.O. Box 322, 3300 AH Dordrecht, The Netherlands.

All Rights Reserved
© *1989 by the International Astronomical Union*

No part of the material protected by this copyright notice may be reproduced or
utilized in any form or by any means, electronic or mechanical including photo-
copying, recording or by any information storage and retrieval system, without
written permission from the publisher.

Printed in The Netherlands

TABLE OF CONTENTS

The Organizing Committees	xii
Preface	xiii
Group Photograph	xv
List of Participants	xix

I. OBSERVATIONAL DATA ON NEBULAE

Invited presentations

Recent UV and Optical Observations of Planetary Nebulae S. Torres-Peimbert	1
Infrared Observations of Galactic Planetary Nebulae A. Preite-Martinez	9
Radio Images of Planetary Nebulae Y. Terzian	17
New and Misclassified Planetary Nebulae L. Kohoutek	29
Catalogues of Planetary Nebulae A. Acker	39

Contributed presentations

Narrow Band Photometry and Mapping of the Planetary Nebulae NGC 6210 and NGC 7009 E. E. Mendoza, C. Chavarría-K., and V. M. Arévalo	49
New Observations of Planetary Nebulae S22 and YM29 V. P. Arkhipova, T. A. Lozinskaya, E. I. Moskalenko, and T. G. Sitnik	50
Two Southern Low Excitation Planetary Nebulae H. Moreno, A. Gutiérrez-Moreno, and G. Cortés	51
Spectroscopic Survey of Planetary Nebulae A. Acker, J. Köppen, and B. Stenholm	52
The Probable Low Excitation Planetary Nebula M3-44 A. Gutiérrez-Moreno, H. Moreno, and G. Cortés	53
Optical and Infrared Observations of the Peculiar Planetary Nebula He2-442 V. P. Arkhipova, V. F. Yesipov, and B. F. Yudin	54
The Photometric (UBV) Study of the Planetary Nebula Variability in 1968 - 1987 E. B. Kostyakova	55
The Photometric and Spectral Variability of the Planetary Nebula IC 4997 E. B. Kostyakova	56
Planetary Nebula He 2-467 Turned out to be the Symbiotic Star with a Period about 500 Days. V. P. Arkhipova and R. I. Noskova	57
Infrared Energy Distribution of Selected Planetary Nebulae P. Persi, A. Preite-Martinez, and M. Ferrari-Toniolo	57
Near Infrared Photometry of Compact Planetary Nebulae M. Peña and S. Torres-Peimbert	58
The Near Infrared Spectrum of NGC 7027 J.-P. Baluteau and D. Pequignot	59
A VLA Radio Continuum Survey of Planetary Nebulae A. Zijlstra, S. R. Pottasch, and C. Bignell	60
A New Planetary Nebula E. Capellaro, M. Turatto, and F. Sabbadin	61
Is NGC 2242 a New Planetary Nebula? L. Jiying, H. Yongwei, and F. Xingchun	62

Planetary Nebulae Near the Galactic Center I: Method of Discovery and Preliminary Results
    S. R. Pottasch, C. Bignell, R. Olling, and A. A. Zijlstra   63
An Infrared Search for New Planetary Nebulae   S. R. Pottasch, C. Esteban, A. Manchado,
    and A. Mampaso   63

II. NEBULAR PROPERTIES

Invited papers

Distances to Planetary Nebulae   J. H. Lutz   65
Galactic Distribution, Radial Velocities and Masses of PN   W. J. Maciel   73
The Shapes and Shaping of Planetary Nebulae   B. Balick   83
Expansion Velocities and Characteristics of Galactic Planetary Nebulae   R. Weinberger   93
Multiple Shell Planetary Nebulae   Y.-H. Chu   105
Dust in Planetary Nebulae   P. F. Roche   117
Molecules and Neutral Hydrogen in Planetary Nebulae   L. F. Rodríguez   129
Abundances in Planetary Nebulae   R. E. S. Clegg   139
Photoionization Models   J. P. Harrington   157

Contributed papers

Planetary Nebulae and the Galactic Bulge   M. W. Feast, T. D. Kinman, and B. S. Lasker   167
Spectroscopic Distances to Central Stars of Planetary Nebulae   R. H. Méndez,
    R. P. Kudritzki, A. Herrero, D. Husfeld, and H. G. Groth   168
Investigation of two Planetary Nebulae and their Angular Vicinity in Cygnus   W. Saurer
    and R. Weinberger   168
Structure and Morphology of NGC 6369   T. K. Chatterjee and J. Campos   169
H_2 and H I Emission Line Imaging of the Ring Nebula NGC 6720   M. A. Greenhouse,
    T. L. Hayward, and H. A. Thronson Jr.   170
Deep Narrow Band Interference Filter Photographs of Selected Extended Planetary
    Nebulae   M. Rosado and M. Moreno   171
CCD Images of Selected Planetary Nebulae   M. Turatto, E. Cappellaro, and F. Sabbadin   172
CCD Images of Soutern Hemisphere Planetary Nebulae   J. Lutz, N. J. Lame, and B. Balick   173
IRAS Observations of Extended Planetary Nebulae   A. Leene and S. R. Pottasch   174
Nebular Density Distributions; A Critical Look   R. C. Kirkpatrick   175
The Temperature Structure of NGC 7027   C. T. Daub and J. P. Basart   176
The Display and Manipulation of PN Images on a IBM PC or Compatible   M. J. Hoey and
    D. Whelan   177
Infrared Images and Line Profiles of Planetary Nebulae   M. G. Smith, T. R. Geballe,
    C. Aspin, I. S. McLean, and P. F. Roche   178
NGC 2899: An Evolved Bipolar Planetary Nebula   J. A. López, L. H. Falcón, M. T. Ruiz,
    and M. Roth   179
The Structure and Kinematics of Bipolar Planetary Nebulae   W. G. Weller and
    S. R. Heathcote   180
Collimated Outflows in Planetary Nebulae   B. Balick, H. L. Preston, and V. Icke   181
Kinematical Properties of Planetary Nebulae   L. Bianchi, M. Grewing, J. Barnstedt, and
    Chr. Diesch   182
Kinematics of Abell 30   G. H. Jacoby and Y.-H. Chu   183
High Resolution Long Slit Spectroscopy of A78   A. Manchado, S. R. Pottasch, and
    A. Mampaso   184
The Structure and Velocity Field of A78   P. Pişmiş, and M. A. Moreno   185
Spatial Deconvolution of IRAS Observations of Planetaries   G. Hawkins and
    B. Zuckerman   186

Evidence of Expansion Velocity in the Central Region of NGC 2346   *D. P. K. Banerjee,*
   B. G. Anandarao, J. N. Desai, S. K. Jain, and D. C. V. Mallik ... 187
Unusual Emission Line Profiles of M 1-1   *K. Shibata and S. Tamura* ... 188
Emission Line Profiles in the Planetary Nebulae IC 4593 and NGC 6153   *D. P. K. Banerjee*
   and B. G. Anandarao ... 189
A New Study of Some Galactic Planetary Nebulae   *S. J. Meatheringham, P. R. Wood,*
   and D. J. Faulkner ... 189
Expansion Velocities of [N II] and [O III] from Compact Planetary Nebulae   *K. Shibata*
   and S. Tamura ... 190
High and Low Resolution Spectra of Selected Planetary Nebulae   *S. Cristiani,*
   F. Sabbadin, and S. Ortolani ... 191
The Planetary ESO 166 - PN21   *M. T. Ruiz, S. R. Heathcote, and W. G. Weller* ... 192
The Structure of NGC 2392   *M. J. Hoey* ... 193
High Velocity Outflows in Post-Main Sequence Nebulae   *J. P. Phillips and A. Mampaso* ... 194
Long-Slit 2-Dimensional Spectra of the Giant Halos Around NGC 6543 and NGC 6826
   D. Middlemass, R. E. S. Clegg, and J. R. Walsh ... 195
The Halos of NGC 6543 and NGC 6826   *A. Manchado, S. R. Pottasch, and A. Mampaso* ... 196
Expansion Velocities of Southern Planetary Nebulae   *K. C. Sahu and S. R. Pottasch* ... 196
Detection of an Extended Optical Halo around IC 418   *D. J. Monk, M. J. Barlow,*
   and R. E. S. Clegg ... 197
Internal Motions of Faint PN Halos   *Y.-H. Chu and G. H. Jacoby* ... 198
Spectroscopic Investigations of Halos of Planetary Nebulae   *M. Bässgen, G. Bässgen,*
   M. Grewing, S. Cerrato, and L. Bianchi ... 199
Kinematic Structure and Chemical Composition of the Double Shell PN NGC 3242
   K. C. Sahu, S. R. Pottasch, B. G. Anandarao, and J. N. Desai ... 200
Thermal Infrared Emission by Dust in the Planetary Nebula NGC 3918   *J. P. Harrington,*
   D. J. Monk, and R. E. S. Clegg ... 201
The Dust Content of Planetary Nebulae with Neutral Halos   *M. G. Hoare* ... 202
Spatially Resolved Observations of the Unidentified Dust Features in BD+30°3639
   C. H. Smith, D. K. Aitken, and P. F. Roche ... 203
Observations of CO and HCN (J=1-0) in NGC 2346 and NGC 7293 with the Nobeyama
   45m Telescope   *J. R. Walsh, R. E. S. Clegg, and N. Ukita* ... 204
A Young Planetary Nebula with OH Molecules: NGC 6302   *H. E. Payne, J. A. Phillips,*
   and Y. Terzian ... 205
CO in the Bipolar Nebula NGC 2346   *A. P. Healy and P. J. Huggins* ... 205
Fluorescent H_2 Emission in the Planetary Nebulae BD+30 3639 and Hb 12
   H. L. Dinerstein, J. S. Carr, P. M. Harvey, and D. F. Lester ... 206
Molecular Hydrogen Emission from Cold Condensations in NGC 2440   *N. K. Reay,*
   N. A. Walton, and P. D. Atherton ... 207
The Systematics and Distribution of Molecular Hydrogen in Planetary Nebulae
   P. W. Payne, J. W. V. Storey, B. L. Webster, M. A. Dopita, and S. J. Meatheringham ... 208
Detection of OH Maser Emission at 1667 MHz from IC 4997   *S. Tamura and I. Kazes* ... 209
OH Maser Emission from Young Planetary Nebulae   *A. Zijlstra, S. R. Pottasch,*
   P. te Lintel, and C. Bignell ... 210
Helium Abundances in Gaseous Nebulae   *R. E. S. Clegg and J. P. Harrington* ... 211
Collisional Excitation of the 10830 He I Line and the Population of the 2^3S He I State in
   Gaseous Nebulae   *M. Peimbert and S. Torres-Peimbert* ... 212
Planetary Nebulae and the Pregalactic Helium Abundance   *W. J. Maciel* ... 213
N/O Abundances in Planetary Nebulae from Far-Infrared Line Observations
   H. L. Dinerstein and M. W. Werner ... 214
Investigations of DDDM-1; The Fourth Halo Planetary Nebula   *A. Yu. Shchelkanova* ... 215

The Peculiar Planetary Nebula M1-78 S. R. Pottasch, A. A. Zijlstra, N. Ukita, A. Manchado, and M. Ratag	216
Oxygen Depletion Variations in Planetary Nebulae and Shells Ejected from Luminous Population I Stars R. J. Dufour	216
Magnesium Abundances in Planetary Nebulae and Interstellar Absorption of Mg II $\lambda 2800$ Å D. Middlemass	217
Abundances of C, N and O in Planetary Nebulae A. A. Nikitin, A. F. Kholtygin, A. A. Sapar, and T. H. Feklistova	218
Some Statistics of Nebular Chemical Compositions L. H. Aller and C. D. Keyes	219
Electron Densities in Planetary Nebulae L. Stanghellini and J. B. Kaler	220
Low Resolution Spectroscopy of 13 Low Surface Brightness PN's A. Manchado, S. R. Pottasch, and A. Mampaso	220
Effect of Density Variations on Elemental Abundance Determinations in Gaseous Nebulae R. H. Rubin	221
A Possible Subdivision of Type II Planetary Nebulae M. Faúndez-Abans and W. J. Maciel	222
Observations and Models of the "Helix" Nebula NGC 7293 R. E. S. Clegg and J. R. Walsh	223
A Photoionization Model Study of NGC 7027 R. B. Gruenwald and D. Pequignot	224
Emission Lines of C I and N II in Planetary Nebulae V. Escalante	225
The Continuum Emission from Planetary Nebulae S. M. Viegas-Aldrovandi	226
Comparison of Different Treatments of the Radiation Transfer in Model Calculations of Planetary Nebulae G. Bässgen, M. Bässgen, and M. Grewing	227
The Spatial Distribution of Line Ratio O III/O II in High Excitation Planetary Nebulae A. Noriega-Crespo and M. McCall	228

III. CENTRAL STARS

Invited papers

Magnitudes, Spectra and Temperatures of Planetary Nuclei J. B. Kaler	229
Wind Features and Wind Velocities M. Grewing	241
Close-Binary and Pulsating Central Stars H. E. Bond	251
Binarity and Intrinsic Variability in Central Stars of PN R. H. Méndez	261
Model Atmospheres and Quantitative Spectroscopy of Central Stars of Planetary Nebulae R. P. Kudritzki and R. H. Méndez	273
Mass Loss Rates in Central Stars of Planetary Nebulae M. Perinotto	293

Contributed papers

The Central Star of NGC 7027 N. A. Walton, S. R. Pottasch, N. K. Reay, and T. Spoelstra	301
Magnitude Measurements of Central Stars of Planetary Nebulae R. Gathier and S. R. Pottasch	302
Stromgren Photometry of the Central Stars of Planetary Nebulae R. Costero and J. Echevarría	302
New Identifications of Faint Central Stars in Extended PN K. B. Kwitter and G. H. Jacoby	303
Einstein X-ray Observations of Planetary Nebulae and their Implications S. P. Tarafdar and K. M. V. Apparao	304
The Origin of the Zanstra Discrepancy: UV Excess in the Central Star Continuum? R. B. C. Henry and H. L. Shipman	305
Are Zanstra Temperatures Always Real? L. H. Aller	306
Temperatures and Luminosities of Planetary Nebulae Nuclei L. Bianchi, E. Recillas, and M. Grewing	307
Broad Baseline Flux Distributions of Planetary Nuclei S. Heap and A. V. Torres	308

Photometric and Spectroscopic Observations of Peculiar Nuclei of Planetary Nebulae
    G. Jasniewicz and A. Acker   309
HFG1: A Planetary Nebula with a Close-Binary Nucleus   H. E. Bond, R. Ciardullo,
    T. A. Fleming, and A. D. Grauer   310
Nonradial Pulsational Analyses of the Pulsating Central Stars of Planetary Nebulae
    S. Starrfield and A. N. Cox   311
A Search for Cool Companions of Planetary Nebula Nuclei   A. F. Bentley   312
A Case Study of a WC Nucleus   J. B. Kaler, R. A. Shaw, and W. A. Feibelman   313
The Nature of the Hot Companion of the G8 IV Nucleus of Abell 35   M. Grewing
    and L. Bianchi   314
V605 Aquilae - The Most Extreme Hydrogen-Poor Object   W. C. Seitter   315
Echelle Spectra of a Large Sample of Planetary Nebula Nuclei   J. K. McCarthy   316
The Metal-Line Spectra of Central Stars of Planetary Nebulae   M. Roth, A. Herrero,
    R. H. Méndez, R. P. Kudritzki, K. Butler, and H. G. Groth   317
Revisited Mass-Loss Rates of Planetary Nebula Nuclei Observed with IUE
    D. Hutsemékers and J. Surdej   317
Quantitative Investigations on Mass Outflow from Planetary Nebulae Nuclei   S. Cerrato,
    L. Bianchi, M. Grewing, M. Bässgen, and G. Bässgen   318

IV. EXTRAGALACTIC NEBULAE

Invited papers

Planetary Nebulae in the Magellanic Clouds   M. J. Barlow   319
Planetary Nebulae in Galaxies Beyond the Local Group   H. C. Ford, R. Ciardullo,
    G. H. Jacoby, and X. Hui   335

Contributed papers

A Survey of Planetary Nebulae in the SMC and M31   N. Meyssonnier, M. Azzopardi,
    J. Lequeux, and R. Gathier   351
The Kinematics of the Planetary Nebulae in the Large Magellanic Cloud
    S. J. Meatheringham, M. A. Dopita, H. C. Ford, and B. L. Webster   352
N 66: A High Excitation N Rich Planetary Nebula in the LMC   M. Peña and M. T. Ruiz   353
Chemical Abundances in Magellanic Cloud Planetary Nebulae   D. J. Monk, M. J. Barlow,
    and R. E. S. Clegg   354
Central Star and Nebular Masses for Magellanic Cloud Planetary Nebulae.   D. J. Monk,
    M. J. Barlow and R. E. S. Clegg   355
Evolution of Magellanic Cloud Planetary Nebulae   S. J. Meatheringham, M. A. Dopita,
    P. R. Wood, B. L. Webster, D. H. Morgan, and H. C. Ford   356
Planetary Nebulae as Standard Candles for Extragalactic Distances   G. Jacoby, H. Ford,
    and R. Ciardullo   357

V. ORIGIN OF PLANETARIES

Invited papers

OH/IR Stars and other IRAS Point Sources and Progenitors of Planetary Nebulae
    H. J. Habing, P. te Lintel Hekkert, and W. E. C. J. van der Veen   359
Carbon Stars as Planetary Nebula Progenitors   G. R. Knapp   381
Thermal Pulses and the Formation of Planetary Nebula Shells   A. Renzini   391
Progenitors of Planetary Nebulae   S. Kwok   401
Models of Planetary Nebulae: Generalisation of the Multiple Winds Model   F. D. Kahn   411

Stellar Evolution and the Planetary Nebulae Formation Rate   J. P. Phillips   425

Contributed papers

The Proto-Planetary Nebula Vy 2-2   R. E. S. Clegg, M. G. Hoare, and J. R. Walsh   443
The Optically Resolved Planetary Nebula/OH Maser Vy 2-2   R. Falomo and F. Sabbadin   444
New IR-Observations of Post AGB Stars and Proto-Planetary Nebulae   W. E. van der Veen,
 H. J. Habing, and T. R. Geballe   445
IRAS 17516-2525: The Birth of a Planetary Nebula?   W. E. van der Veen, H. J. Habing,
 and T. R. Geballe   446
The Shocking Truth About some "Proto-PN"   R. W. Goodrich and L. Bianchi   447
New OH/IR Stars: Proto-Planetary Nebulae?   J. Eder, B. M Lewis, and Y. Terzian   448
Infrared Photometry of OH/IR Stars   M. Peña and J. Fierro   449
The Presence of Water Masers in Color Selected IRAS Sources   B. M. Lewis and
 D. Engels   450
Some Dependences for Long-Period Variables and a Possible Scheme of their Evolution.
 I. L. Andronov, L. S. Kudashkina, and G. M. Rudnitskij   451
Proto-Planetary Nebulae: Models and IRAS Observations   K. Volk and S. Kwok   452
HCN The First Strong Maser in Carbon-Rich Stars   A. Omont, S. Guilloteau, and R. Lucas   453
Lower Limit for NPN's Masses   A. Harpaz   454
Some Hypothesized Observational Aspects of Magnetic Fields in Protoplanetary Nebulae
 G. Pascoli   455
Stalled Winds: Interactions Between Nebulae and Stellar Winds   J. B. Kaler,
 W. A. Feibelman, R. A. Shaw, and H. Henrichs   456
Optically Thick Wind from Post-AGB Stars and Formation of Planetary Nebulae   M. Kato   457
Direct Evidence for a Bipolar Stellar Wind in NGC 2392   C. R. O'Dell   458
Two-Dimensional Hydrodynamical Models of Planetary Nebulae (PNe)
 I. V. Igumenshchev, B. M. Shustov, and A. V. Tutukov   459
The Spatial Structure of Planetary Nebulae with Binary Central Stars   I. G. Kolesnik
 and L. S. Pilyugin   460
Morphologies of Planetary Nebulae with Close-Binary Nuclei   H. E. Bond, M. Livio, and
 M. Meakes   461
CCD Images of Three Planetary Nebulae with Binary Nuclei   J. Lutz and N. J. Lame   462

VI. EVOLUTION

Invited papers

Evolutionary Tracks for Central Stars of Planetary Nebulae   D. Schönberner   463
The Distribution of Planetary Nebula Nuclei in the log L - log T Plane: Inferences from
 Theory   R. E. Shaw   473
The Position of the Central Stars of PN on the HR Diagram   S. R. Pottasch   481
Initial Masses   D. C. V. Mallik   493
Binary Stars and Planetary Nebulae   I. Iben Jr. and A. V. Tutukov   505
The Evolution of the Common Planetary Nebula   J. Köppen   523
Planetary Nebulae with Massive Central Stars   R. Tylenda   531

Contributed papers

A New Method for Observational Testing of the Planetary Nebulae Nuclei Evolution
 R. Szczerba   539
Mass Distribution and Birth Rate of Central Stars of Planetary Nebulae: Comparison with
 White Dwarfs, and Influence of Selection Effects   V. Weidemann   540

Common Envelope Evolutions of Binary System and Formation of Planetary Nebulae
    I. Hachisu and M. Kato   541
Theoretical Models for the Evolution of Planetary Nebulae Nuclei Tested by Observations
    R. Tylenda and G. Stasińska   542
Snapshots of Evolving Model Planetary Nebulae   *G. Stasińska*   542
Evolution of Planetary Nebulae: A comparison with Observed Central Stars
    M. Schmidt-Voigt   543

VII. PLANETARIES AND WHITE DWARFS

Invited papers

White Dwarfs and Planetary Central Stars   *J. Liebert*   545
Properties and Evolution of White Dwarf Stars   *H. L. Shipman*   555

VIII. SUMMARIES AND A VIEW TO THE FUTURE

Invited papers

Summarizing Remarks on the Structure and Evolution of Planetary Nebulae and Properties
    of their Central Stars   *L. H. Aller*   567
Comments on the Applications of Planetary Nebulae Research   *M. Peimbert*   577

Author Index   589
Object Index   593
Subject Index   603

ORGANIZING COMMITTEES

-Scientific Organizing Committee

A. Acker
M. J. Barlow
J. B. Kaler (Chairman)
G. S. Khromov
R. P. Kudritzki
S. Kwok
J. H. Lutz

M. Perinotto
S. R. Pottasch
A. Renzini
L. F. Rodríguez
Y. Terzian
S. Torres-Peimbert
R. Tylenda

-Local Organizing Committee

J. Cantó
R. Costero
J. Fierro (Secretary)
M. Peimbert (Chairman)

M. Peña (Treasurer)
A. Sarmiento
A. Serrano

PREFACE

Every 5 years since 1967 a meeting has been held to discuss the subject of planetary nebulae and their central stars. Previous meetings have been held in Tatranská Lomnica (Czechoslovakia); Liege (Belgium); Ithaca, New York (U. S. A.); and London (Great Britain). IAU Symposium 131 was sponsored by IAU Commision 34, on Interstellar Matter and co-sponsored by IAU Commisions 35 and 36 on Stellar Constitution and Theory of Stellar Atmospheres.

The symposium was held at the Universidad Nacional Autónoma de México in Mexico City, October 5-9, 1987. It took place in one of the old buildings of the University of Mexico in the downtown area. The inner patio of the building provided very pleasant surroundings for the poster sessions and for extensive discussions among the participants. The meeting was attended by 160 scientists from 22 countries. The Scientific Organizing Committee, under the chairmanship of J.B. Kaler, prepared a comprehensive scientific program based on a set of invited presentations. All contributed papers were presented in poster form.

The Scientific Organizing Committee would like to thank the staff of the University of Illinois Department of Astronomy: Dr. Ron Allen for granting financial support; Carol Stickrod, Louise Browning, Deana Griffin and Sandie Osterbur for their help with the organization. IAU provided economic assistance to a group of young astronomers.

The Local Organizing Committee acknowledges the Universidad Nacional Autónoma de México for the general support provided for the meeting, mainly through the following departments: Instituto de Astronomía, Coordinación de la Investigación Científica, and Rectoría. Many other branches of the University, collaborated for the successful development of the meeting. Additional financial assistance provided by the Consejo Nacional de Ciencia y Tecnología is gratefully acknowledged.

The present proceedings include the 39 invited papers and 138 abstracts of contributed papers. The invited papers were delivered in their final form by the authors and are their sole responsibility. Many of the abstracts were re-typed to follow the required format. To facilitate the use of the material we include author index, object index and subject index. The organization of these proceedings follows very closely the order of presentations of the meeting. The editor is grateful to Elizabeth Themsel for typing the abstracts.

J. B. Kaler
Chairman
Scientific Organizing Committee

M. Peimbert
Chairman
Local Organizing Committee

S. Torres-Peimbert
Editor of the Proceedings

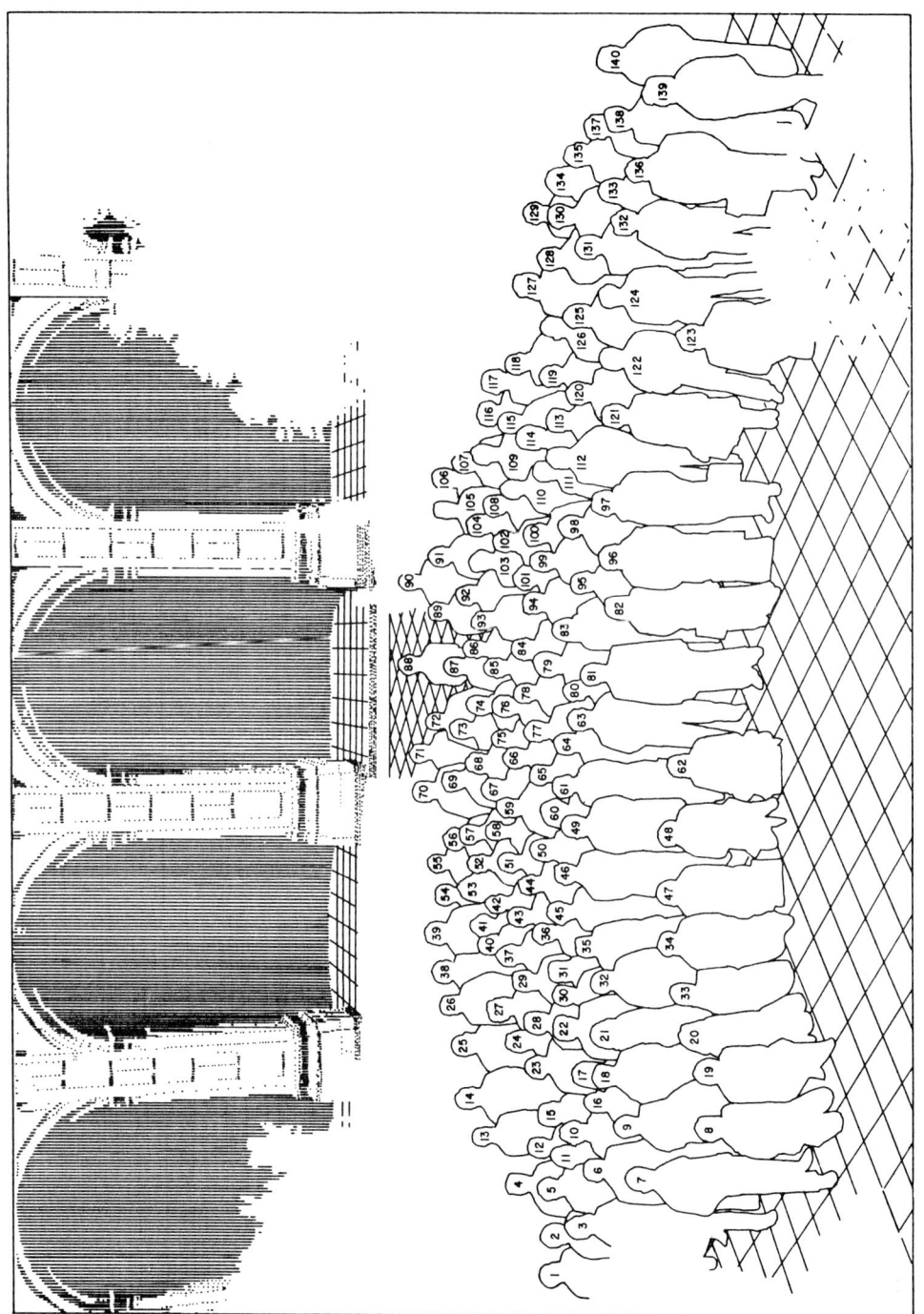

GROUP PHOTOGRAPH

1. C. Chavarría-K.
2. E. Moreno
3. M. W. Feast
4. M. Grewing
5. R. P. Kudritzki
6. C. T. Daub
7. M. G. Smith
8. L. Hernández Falcón
9. J. Belley
10. W. J. Maciel
11. H. J. Habing
13. S. R. Heathcote
14. J. W. Liebert
15. G. Bässgen
16. R. C. Kirkpatrick
17. S. Cerrato
18. J. A. García-Barreto
19. R. Carrillo
20. Y. Gómez
21. M. Peña
22. M. Peimbert
23. M. Bässgen
24. W. G. Weller
25. B. M. Lewis
26. E. E. Mendoza V.
27. A. Zijlstra
28. G. Hawkins
29. J. Köppen
30. R. B. C. Henry
31. E. Recillas
32. H. Dinerstein
33. A. Acker
34. S. R. Pottasch
35. P. Pişmis,
36. L. Bianchi
37. C. Smith
38. L. Carrasco
39. A. Serrano
40. N. A. Walton
41. S. Viegas-Aldrovandi
42. D. Middlemass
43. D. Monk
44. K. C. Sahu
45. L. Maupomé
46. G. Jacoby
47. A. González
48. J. M. Alcalá
49. K. B. Kwitter
50. L. Stanghellini
51. M. G. Hoare
52. R. Goodrich
53. S. Navarro
54. L. J. Corral
55. S. Curiel
56. J. Guichard
57. L. Neri
58. K. Volk
59. R. A. Shaw
60. J. H. Lutz
61. Y.-H. Chu
62. V. Robledo
63. R. J. Dufour
64. Y. Terzian
65. S. Tamura
66. J. B. Kaler
67. R. R. Robbins
68. A. Bentley
69. J. P. Phillips
70. J. K. McCarthy
71. H. E. Payne
72. H. Ford
73. H. E. Bond
74. M. Schmidt-Voigt
75. I. Iben
76. A. V. Tutukov
77. R. Rubin
78. K. M. V. Apparao
79. A. Gutiérrez-Moreno
80. D. C. V. Mallik
81. T. K. Chatterjee
82. J. Fierro
83. M. T. Ruiz
84. H. Moreno
85. D. Schönberner
86. M. Perinotto
87. C. R. O. O'Dell
88. D. Hutsemekers
89. V. Weidemann
90. C. Bignell
91. S. Heap
92. W. J. Schuster
93. R. Tylenda
94. S. González-B.
95. M. Greenhouse
96. R. Costero
97. R. H. Méndez
98. A. Herrero
99. S. J. Meatheringham
100. A. Chelli
101. A. Arrellano-Ferro
102. P. Patriarchi
103. R. Szczerba
104. C. Allen
105. R. Weinberger
106. J. P. Harrington
107. L. F. Rodríguez
108. M. Chávez
109. D. Pequignot
110. A. Preite-Martinez
111. M. Azzopardi
112. M. Roth
113. S. Kwok
114. V. Caloi
115. F. D. Kahn
116. A. Poveda
117. P. F. Roche
118. A. Renzini
119. P. Persi
120. R. E. S. Clegg
121. R. Peniche
122. J. H. Peña
123. L. H. Aller
124. M. J. Barlow
125. J. Galindo
126. P. J. Huggins
127. M. Ferrari-Toniolo
128. E. Capellaro
129. S. Torres-Peimbert
130. M. A. Moreno
131. J. Cantó
132. G. Koenigsberger
133. G. Stasińska
134. J. Espresate
135. M. A. Hobart
136. A. López
137. V. Escalante
138. W. van der Veen
139. A. Manchado
140. A. Harpaz

LIST OF PARTICIPANTS

Agnes Acker, Observatoire de Strasbourg, 11 rue de l'Université, F- 67000 Strasbourg, France
Juan M. Alcalá Estrada, Instituto de Astronomía, UNAM, México 04510, D. F., México
Christine Allen, Instituto de Astronomía, UNAM, México 04510, D. F., México
Lawrence H. Aller, Astronomy Department, Univ. of California, Los Angeles, CA 90024, U.S.A.
Krishna M. V. Apparao, Tata Institute of Fundamental Research, Dr. Homi Bhabba Rd., Colaba, Bombay 400 005, India
Armando Arellano-Ferro, Instituto de Astronomía, UNAM, México 04510, D. F., México
Marc Azzopardi, Observatoire de Meudon, F-92195 Meudon, Principal Cedex, France
Bruce Balick, Department of Astronomy, University of Washington, Seattle, WA 98195, U.S.A.
Michael J. Barlow, Deptartment of Physics & Astronomy, University College London, Gower St. London WC1E 6BT, U. K.
Gabriele Bässgen, Astronomisches Institut der Universität, Waldhauserstr. 64, D-7400 Tubingen, Fed. Rep. Germany
Martin Bässgen, Astronomisches Institut der Universität, Waldhauserstr. 64, D-7400 Tubingen, Fed. Rep. Germany
Julien Belley, Department of Physics, University of Laval, Quebec PQ G1K 7P4, Canada
Alan Bentley, Dept. of Physical Sciences, Eastern Montana College, Billings, MT 59101, U.S.A.
Luciana Bianchi, Osservatorio Astronomico, Strada Osservatorio 20, I-10025 Pino Torinese, Italy
Carl Bignell, National Radio Astronomy Observatory, P.O. Box 0, Socorro, NM 87801, U.S.A.
Howard E. Bond, Space Telescope Science Institute, Homewood Campus, Baltimore, MD 21218, U.S.A.
Vittoria Caloi, Istituto di Astrofisica Spaziale, C.P. 67, I-00044 Frascati, Italy
Jorge Cantó, Instituto de Astronomía, UNAM, México 04510, D. F., México
Enrico Cappelaro, Osservatorio Astronomico di Asiago, I-36012 Asiago (Vicenza) Italy
Luis Carrasco, Instituto de Astronomía, UNAM, Ensenada 877, B. C., México
René Carrillo Moreno, Instituto de Astronomía, UNAM, México 04510, D. F., México
Simona Cerrato, Astronomisches Institut der Universität, Waldhauserstr. 64, D-7400 Tubingen, Fed. Rep. Germany
Tappan K. Chatterjee, Instituto Nacional Astrofísica, Optica y Electrońica, Apartado Postal 51, Tonantzintla, Puebla, México
Carlos Chavarria-K., Instituto de Astronomía, UNAM, México 04510, D. F., México
Miguel Chávez Dagostino, Instituto de Astronomía, UNAM, México 04510, D. F., México
Alain Chelli, Instituto de Astronomía, UNAM, México 04510, D. F., México
You-Hua Chu, Astronomy Department, Univ., of Illinois, 1011 W. Spri, Urbana, IL 61801, U.S.A.
Robin E. S. Clegg, Deptartment of Physics & Astronomy, University College London, Gower St. London WC1E 6BT, U. K.
Pedro Colín, Instituto de Astronomía, UNAM, México 04510, D. F., México
Cecilia Colomé Canales, Instituto de Astronomía, UNAM, México 04510, D. F., México
Luis J. Corral, Instituto de Astronomía, UNAM, México 04510, D. F., México
Rafael Costero, Instituto de Astronomía, UNAM, México 04510, D. F., México

Salvador Curiel, Instituto de Astronomía, UNAM, México 04510, D. F., México
Clarence T. Daub, Dept. of Astronomy, San Diego State Univ., San Diego, CA 92115, U.S.A.
Harriet Dinerstein, Astronomy Department, Univ. of Texas, Austin, TX 78712, U.S.A.
Reginald J. Dufour, Deptartment of Space Physics & Astronomy, Rice University, Houston, TX 77001, U.S.A.
Vladimir Escalante, Center for Astrophysics, 60 Garden St., Cambridge, MA 02138, U.S.A.
Julia Espresate, Instituto de Astronomía, UNAM, México 04510, D. F., México
Michael W. Feast, South African Astronomical Obs., P.O. Box 9, Observatory, South Africa
Marco Ferrari-Toniolo, Istituto di Astrofisica Spaziale, C.P. 67, I-00044 Frascati, Italy
Julieta Fierro, Instituto de Astronomía, UNAM, México 04510, D. F., México
Holland Ford, Space Telescope Science Inst., Homewood Campus, Baltimore, MD 21218, U.S.A.
Jesus Galindo, Instituto de Astronomía, UNAM, México 04510, D. F., México
J. Antonio García-Barreto, Instituto de Astronomía, UNAM, México 04510, D. F., México
Yolanda Gómez, Instituto de Astronomía, UNAM, México 04510, D. F., México
Javier González, Instituto Nacional de Astrofísica, Optica y Electrónica, Apartado Postal 51, Tonantzintla, Puebla, México
Salvador González-Bedolla, Instituto de Astronomía, UNAM, México 04510, D. F., México
Bob Goodrich, Lick Observatory, University of California, Santa Cruz, CA 95064, U.S.A.
Matthew A. Greenhouse, Wyoming Infrared Observatory, University of Wyoming, Laramie, WY 82071, U.S.A.
Michael Grewing, Astronomisches Institut der Universität, Waldhauserstr. 64, D-7400 Tubingen, Fed. Rep. Germany
José Guichard, Instituto de Astronomía, UNAM, México 04510, D. F., México
Adelina Gutiérrez-Moreno, Departamento de Astronomía, Universidad de Chile, Casilla 36-D, Santiago, Chile
Harm J. Habing, Sterrewacht Leiden, Postbus 9513, 2300 RA Leiden, The Netherlands
Izumi Hachisu, Department of Physics & Astronomy, University of Louisiana, Baton Rouge, LA 70803-4001, U.S.A.
Amos Harpaz, Deptartment of Physics, The Technion, Haifa, Israel
J. Patrick Harrington, Astronomy Program, Univ. of Maryland, College Park, MD 20742, U.S.A.
Herbert Hartl, Institut für Astronomie, Universitat Innsbruck, Technikerstr. 25, Austria
Ilse Hasse, Instituto de Astronomía, UNAM, México 04510, D. F., México
George Hawkins, Department of Astronomy, Univ. of California, Los Angeles, CA 90024, U.S.A.
Sarah Heap, NASA, Goddard Space Flight Center, Greenbelt, MD 20771, U.S.A.
Steve R. Heathcote, Cerro Tololo Inter-American Observatory, Casilla 603, La Serena, Chile
Richard B. C. Henry, Department of Physics & Astronomy, University of Oklahoma, Norman, OK 73019, U.S.A.
Luis Hernández Falcón, Instituto de Astronomía, UNAM, México 04510, D. F., México
Artemio Herrero Davo, Institut für Astronomie und Astrophysik, Universität München, D-8000 München, Fed. Rep. Germany
Melvin G. Hoare, Department of Physics & Astronomy, University College London, Gower St., London WC1E 6BT, U. K.
Marco A. Hobart, Instituto de Astronomía, UNAM, México 04510, D. F., México
Michael J. Hoey, Physics Department, University College Dublin, Belfield, Dublin 4, Ireland
Patrick J. Huggins, Physics Deptartment, New York Univ., New York, NY 10003, U.S.A.
Damien Hutsemékers, Institut d'Astrophysique, Avenue de Cointe 5, 4200 Cointe-Liege, Belgium
Icko Iben Jr., Department of Astronomy, University of Illinois, Urbana, IL 61801, U.S.A.
George H. Jacoby, Kitt Peak National Observatory, Tucson, AZ 85726- 6732, U.S.A.
Franz D. Kahn, Department of Astronomy, The University, Manchester M13 9PL, U. K.
James B. Kaler, Department of Astronomy, University of Illinois, Urbana, IL 71801, U.S.A.
Mariko Kato, Department of Physics, Keio University, 4-1-1 Hiyoshi Kouhuku-ku, Japan

Ronald C. Kirkpatrick, Los Alamos National Laboratory, Los Alamos, NM 87545, U.S.A.
Gillian R. Knapp, Dept. of Astrophysical Sciences, Princeton Univ., Princeton, NJ 08544, U.S.A.
Gloria Koenigsberger, Instituto de Astronomía, UNAM, México 04510, D. F., México
Joachim Köppen, Institut für Theoretisches Astrophysik, Im Neuenheimer Feld 561, D-6900 Heidelberg, Fed. Rep. Germany
Rolf P. Kudritzki, Observatory München, Scheinerstr. 1, D- 8000 München, Fed. Rep. Germany
Karen B. Kwitter, Williams College, Hopkins Observatory, Williamstown, MA 01267, U.S.A.
Sun Kwok, Deparment of Physics, University of Calgary, Alberta T2N 1N4, Canada
Brian M. Lewis, Arecibo Obsevatory, P. O. Box 995, Arecibo, PR 00612, Puerto Rico
James W. Liebert, Steward Observatory, University of Arizona, Tucson, AZ 85721, U.S.A.
Alberto López García, Instituto de Astronomía, UNAM, Ensenada 877, B. C., México
Julie Lutz, Program in Astronomy, Washington University, Pullman, WA 99164-2930, U.S.A.
Walter J. Maciel, Instituto Astronômico e Geofísico, Universidade de São Paulo, Caixa Postal 30.627, São Paulo SP, Brasil
Dipankar C. V. Mallik, Indian Institute of Astrophysics, Bangalore 560 034 India
Arturo Manchado, Instituto de Astrofsíca de Canarias, La Laguna, Tenerife, Spain
Lucrecia Maupomé, Instituto de Astronomía, UNAM, México 04510, D. F., México
David Mayer, Instituto de Astronomía, UNAM, México 04510, D. F., México
James K. McCarthy, Astronomy Dept., California Institute of Tech., Pasadena, CA 91125, U.S.A.
Stephen J. Meatheringham, Mount Stromlo Observatory, Private Bag P. O. Woden, ACT 2606, Australia
Roberto H. Méndez, Instituto de Astronom'ia y Física del Espacio, C. C. 67 Suc. 28, 1428 Buenos Aires, Argentina
Eugenio E. Mendoza V., Instituto de Astronomía, UNAM, México 04510, D. F., México
Desmond Middlemass, Department of Physics & Astronomy, University College London, Gower St. London WC1E 6BT, U. K.
Andrea Miranda, Instituto de Astronomía, UNAM, México 04510, D. F., México
Víctor M. Mondragón A., Instituto de Astronomía, UNAM, México 04510, D. F., México
David Monk, Department of Physics & Astronomy, University College London, Gower St. London WC1E 6BT, U. K.
Edmundo Moreno, Instituto de Astronomía, UNAM, México 04510, D. F., México
Hugo Moreno, Departamento de Astronomía, Universidad de Chile, Casilla 36-D, Santiago, Chile
Marco A. Moreno, Instituto de Astronomía, UNAM, México 04510, D. F., México
Silvana Navarro, Instituto de Astronomía, UNAM, México 04510, D. F., México
Luis Neri, Instituto de Astronomía, UNAM, México 04510, D. F., México
Alberto Noriega-Crespo, Canadian Institute of Theoretical Astronomy, University of Toronto, Mc Lennan Labs., Canada
C. Robert O'Dell, Dept. of Space Physics & Astronomy, Rice Univ., Houston, TX 77001, U.S.A.
Laura Parrao, Instituto de Astronomía, UNAM, México 04510, D. F., México
Patrizio Patriarchi, Osservatorio Astrofisico di Arcetri, Largo E. Fermi 5, I- 50125 Firenze, Italy
Harry Payne, National Radio Astronomy Observatory, P.O. Box 2, Greenbank, WV 24944, U.S.A.
Manuel Peimbert, Instituto de Astronomía, UNAM, México 04510, D. F., México
José H. Peña, Instituto de Astronomía, UNAM, México 04510, D. F., México
Miriam Peña, Instituto de Astronomía, UNAM, México 04510, D. F., México
Rosario Peniche, Instituto de Astronomía, UNAM, México 04510, D. F., México
Daniel Pequignot, Observatoire de Meudon, F-92195-Meudon, Principal Cedex, France
Mario Perinotto, Istituto di Astronomia, Úniversita di Firenze, I- 50125 Firenze, Italy
Paolo Persi, Istituto di Astrofisica Spaziale, C.P. 67, I-00044 Frascati, Italy
J. Anthony Phillips, National Astronomy & Ionosphere Center, Cornell University, Ithaca, NY 14853, U.S.A.
John P. Phillips, Physics Department, Queen Mary College, London N14 NS, U. K.

Paris Pişmiş, Instituto de Astronomía, UNAM, México 04510, D. F., México
Stuart R. Pottasch, Kapteyn Astronomical Institute, Postbus 800, The Netherlands
Arcadio Poveda, Instituto de Astronomía, UNAM, México 04510, D. F., México
Andrea Preite-Martinez, Istituto di Astrofisica Spaziale, C.P. 67, I- 00044 Frascati, Italy
Elsa Recillas, Instituto de Astronomía, UNAM, México 04510, D. F., México
Alvio Renzini, Dipartamento di Astronomia, Universita di Bologna, I-40126 Bologna, Italy
R. Robert Robbins, Department of Astronomy, University of Texas, Austin, TX 78712, U.S.A.
Víctor F. Robledo Rella, Instituto de Astronomía, UNAM, México 04510, D. F., México
Patrick F. Roche, Royal Observatory, Blackford Hill, Edinburgh, EH9 3HJ, Scotland, U. K.
Luis F. Rodríguez, Instituto de Astronomía, UNAM, México 04510, D. F., México
Margarita Rosado, Instituto de Astronomía, UNAM, México 04510, D. F., México
Martin Roth, University Observatory München, Scheinerstr. 1, D-8000 München, Fed. Rep. Germany
Robert Rubin, NASA, Ames Research Center, Moffet Field, CA 94035, U.S.A.
Ma. Teresa Ruiz, Depto. de Astronomía, Universidad de Chile, Casilla 36-D, Santiago, Chile
Franco Sabbadin, Padua Observatory, Vicolo Osservatorio 5, I- 35100 Padova, Italy
K. C. Sahu, Kapteyn Laboratorium, Postbus 800, The Netherlands
Antonio Sarmiento G., Instituto de Astronomía, UNAM, México 04510, D. F., México
M. Schmidt-Voigt, Max-Planck Institut für Physik und Astrophysik, Karl Schwarzschild Str. 1, D-8046 Garching b. München, Fed. Rep. Germany
Detlef Schönberner, Institut fr Theoretisches Physik und Sternwarte, Universität Kiel, D-2300 Kiel, Fed. Rep. Germany
William J. B. Schuster, Instituto de Astronomía, UNAM, México 04510, D. F., México
Alfonso Serrano, Instituto de Astronomía, UNAM, México 04510, D. F., México
Richard A. Shaw, Lick Observatory, University of California, Santa Cruz, CA 95064, U.S.A.
Gregory Shields, Department of Astronomy, University of Texas, Austin, TX 78712, U.S.A.
Harry L. Shipman, Physics & Astronomy Dept., Univ. of Delaware, Newark, DE 19716, U.S.A.
Craig H. Smith, Department of Physics, University College, Australia
Malcolm G. Smith, U K Infrared Telescope U., 665 Komohana St., Hilo, HA 96720, U.S.A.
Letizia Stanghellini, Department of Astronomy, University of Illinois, Urbana, IL 61801, U.S.A.
Grazyna Stasińska, Obs. de Paris-Meudon, F-92195-Meudon, Principal Cedex, France
Ryszard Szczerba, Copernicus Astronomical C, ul. Chopina 12/18, 87-100 Torun, Poland
Shin'ichi Tamura, Astronomical Institute, Tohoku University, Aobayama, Sendai, Japan
Yervant Terzian, National Astronomy & Ionosphere Center, Cornell University, Ithaca, NY 14853, U.S.A.
Silvia Torres-Peimbert, Instituto de Astronomía, UNAM, México 04510, D. F., México
Alexander V. Tutukov, Astronomical Council, USSR Academy of Sciences, Pyatnitskaya ul. 109017, Moscow, U. S. S. R.
Romuald Tylenda, Copernicus Astronomical Center, ul. Chopina 12/18, 87-100, 87-100 Torun, Poland
Wil E. C. J. van der Veen, Sterrewacht Leiden, Postbus 9513, NL-2300 RA Leiden, The Netherlands
Sueli Viegas-Aldrovandi, Dept. of Physics, Ohio State University, Columbus, OH 43210, U.S.A.
Kevin Volk, NASA, Ames Research Center, Moffet Field, CA 94035, U.S.A.
Nicholas A. Walton, Kapteyn Laboratorium, Postbus 800, 9700 AV Groningen, The Netherlands
Volker Weidemann, Institut fr Theoretisches Physik und Sternwarte, Universität Kiel, D-2300 Kiel, Fed. Rep. Germany
Ronald Weinberger, Institut für Astronomie, Universitat Innsbruck, Technikerstr. 25, Austria
William G. Weller, Cerro Tololo Inter-American Observatory, Casilla 603, La Serena, Chile
Albert Zijlstra, National Radio Astronomy Observatory/VLA, P.O. BOX 0, Socorro, NM 87801, U.S.A.

RECENT UV AND OPTICAL OBSERVATIONS OF PLANETARY NEBULAE

Silvia Torres-Peimbert
Instituto de Astronomía
Universidad Nacional Autónoma de México
Apartado Postal 70-264, México 04510 D.F., México

ABSTRACT. This review contains a brief survey of the issues that have been the concern of optical and ultraviolet studies of planetary nebulae since the last IAU Symposium on this subject in 1982.
 The nature of this review is such that it is not possible to do justice to the wealth of work that has taken place in this period, I will just point out some characteristic examples of the different aspects of the work done.

1. ULTRAVIOLET OBSERVATIONS

The last five years have seen great advances in UV research on PN. Firstly because of the extended lifetime of IUE that has allowed the increase in the number and length of exposures allowing fainter targets to be observed, and secondly the IUE Data Bank has been steadily growing and it allows revisions of large samples of essentially homogeneous data. Other reviews on this subject are by Koppen and Aller (1987) and Perinotto (1987).
 Up to now more than 130 PN have been observed in the low dispersion mode. An atlas of merged short and long wavelength spectra is being prepared by Feibelman, Oliversen and Nichols-Bolhin (1986). In most instances the aperture was centered on the PNN, although a few offset obervations have been obtained. There is also a considerable data bank of high dispersion observations that includes more than 50 PN.

1.1 Nebular Data

a) Abundances

The study of UV spectrum of PN has allowed a better understanding of the chemical composition of these objects. The UV nebular spectrum includes intercombination lines of C II], C III], N III], N IV], O IV], Si II], and Si III]; resonance lines of C II, C IV, N V, Mg I, Mg II, Si II and Si IV; as well as forbidden lines from the levels which also produce the optical lines of [N II], [O II], [O III], [Ne III], [Ne IV], [Ar III], [Ar IV], [Ar V], and [Mg V]. A number of ions can only be observed in UV, or their interpretation in terms of ionic abundances is more straightforward. Ultraviolet observations have thus made possible to derive abundances of additional elements, namely carbon which plays such a crucial role in stellar evolution, as well as of Mg and Si and improve our information on nitrogen, oxygen and neon.
 The complete derivation of physical parameters and chemical abundances in the nebulae are usually carried out by optical spectrophotometry in conjunction with IUE spectra and very often are supplemented with model ionization structures that allow for unobserved ionic stages; normally from

the stellar UV continuum and nebular line ratios an improved value of the interstellar reddening is also derived. UV data limit the possible nebular models since more severe constraints on physical processes, excitation conditions, and stellar fluxes below 912 A can be set.

Usually IUE spectra have been acquired for individual objects that have been investigated in detail for abundance determinations. Aller and Czyzak (1983) presented extensive optical observations for 41 PN and used available IUE data for one third of their objects. They derived C, N, O, Ne, S, Cl, and Ar values whenever possible. In a recent compilation, Zuckerman and Aller (1986) list C, N, O, Ne, Ar, and S abundances for 44 nebulae, where most have C abundances derived from IUE data.

The same analysis can be carried out for extragalactic planetary nebulae. Maran et al. (1982) performed UV observations on 3 high excitation PN in the Magellanic Clouds, (LMC P40, SMC N2, and SMC N5) and determined their chemical abundance where they found that C is greatly enhanced relative to the interstellar medium. Maran et al. (1984) observed the only known PN in the Fornax galaxy, probably the most distant PN that can be observed with IUE. They combined it with optical data and found that it resembles the three PN in the Magellanic Clouds; it is deficient in N, O, Ne, S and Ar, but not in C, relative to the planetaries in the Milky Way.

b) Physical Processes

IUE observations have also given the opportunity to carry out a more thorough examination of physical processes, in particular dielectronic recombinantion, and charge exchange processes whose effects are more pronounced in the UV. For example, Clegg, Harrington and Storey (1986) reported the detection of the 2600 A triplet and quintet Ne III lines from high dispersion spectra in NGC 3918 produced by charge transfer reactions between Ne^{3+} and H^o.

1.2 Data on Central Stars

a) Stellar Parameters

For most of the planetary nebulae nuclei (PNN) observed in the optical range only the Rayleigh-Jeans tail of the energy distribution can be detected. The extension of the stellar continuum and line spectrum to shorter wavelengths improves considerably our understanding of the nature of hot stellar atmospheres. Traditional optical spectroscopy for determining stellar temperatures can be improved with IUE data. The Zanstra method can be applied to the stronger stellar continuum and the brighter He II 1640 A line.

Kaler and Feibelman (1985) analized the spectra of the central stars of 32 extended PN (larger than 0.2 pc). From ultraviolet and UV- to-optical flux ratios they derived color temperatures. Their values of the color temperatures are far in excess to those derived from the Zanstra method. They found that in general, the stars where the continuum contribution is indistinguishable from a Rayleigh-Jeans distribution have lower intrinsic luminosities.

b) Stellar Wind Properties

Although it had been known that PNN of the W-R type have expanding atmospheres, definite proof of mass loss in PNN was presented only through observations of P Cygni profiles in UV lines (Heap et al. 1978). The great breadth of their P Cygni lines permit fast winds to be recognized even in the low spectral resolution mode. PNN show these mass loss manifestations mainly in the resonance lines of N V, Si IV, and C IV. The edge velocities in these lines are of the order 1000 to 3000 km s^{-1}. Little is known about the ionization mechanisms and stratification in the winds.

Mass loss rates for these winds have been determined from fitting high dispersion spectra of NGC 1535 (Adam and Koppen 1985). Cerruti- Sola and Perinotto (1985) analyzed low dispersion IUE spectra of 60 central stars; they found that 22 out of 42 spectra with recognizable stellar continuum display P-Cygni profiles. All low excitation PN (without nebular He II) have winds but for some high excitation PN no wind has been detected. Cerruti-Sola and Perinotto were able to determine that for PNN with surface gravity less than about log g = 5.2, there is strong mass loss present in the star. They determined mass loss rates in the 10^{-10} to 10^{-7} $M_\odot$/yr range.

2. OPTICAL OBSERVATIONS

We have seen during this period both the increase in number of scientists devoting their energy to the research of PN as well as widespread operation of better instrumentation. In general it can be said that there are more data, better quality measurements and new observational and data reduction techniques that have broadened our scope.

The major instrumentation advances for optical studies have dramatically affected the subject; very sensitive 2-dimensional detectors have given us a more detailed picture of structure of the PN as well as spatial information on line intensities and velocity fields.

2.1 Search for New Objects

The discovery of new PN continues to take place. At present the number of identified galactic PN is of about 1600 objects; for comparison it should be pointed out that the number of entries in Perek and Kohoutek's (1969) catalogue is of 1067. Most of the discovery work is done in optical wavelengths.

In our galaxy we know only about 10% of all PN, and the rest are heavily reddened. The northern sky has been carefully searched on different sets of material, the southern sky has not been studied so thoroughly, nevertheless new objects are still being found in both hemispheres. For example, Hartl and Tritton (1985) reported 14 new objects from deep J and R plates taken at the UK 1.2-m Schmidt telescope and Shaw and Wirth (1985) reported 7 new PN in Baade's window, while Ellis, Grayson and Bond (1984) report 6 new possible PN from a new revision of the Palomar Survey prints.

As is to be expected, there has been a dramatic increase in the identification of new extragalactic PN. Morgan (1984) classified 134 PN in the Magellanic Clouds and estimated [O II]/Hβ ratios for 29 nebulae in the SMC and 85 in the LMC. Morgan and Good (1985) reported 10 new possible PN in the SMC. Meysonnier, Azzopardi and Lequeux (1988) reported a survey covering 2/3 of M31 and found about 1200 objects showing emission lines between 4350 and 5300 Å, believed to be PN.

2.2 Nebular Research

a) Direct Imaging

The possibility of obtaining deep CCD images of PN has allowed the search of faint extended nebulosities beyond the dense nebulae. Jewitt, Danielson, and Kupferman (1986) from CCD images in Hα of 44 objects of 10 to 100" diameter found halos previously undetected in 29 of them; many of the halos exhibit filamentary structure; they also determined that their masses are comparable to the masses of the denser shells.

Chu, Jacoby, and Arendt (1987) from direct images through narrow band filters plus available data, found that the frequency of multiple shell events is larger than 50%. Balick (1987) from direct images of 51 PN in the light of Hα, [O III], [N II], He II, and [O I] defined 3 morphological categories: 'round', 'elliptical' and 'butterfly'. He proposed that nearly all PN fall in these general categories suggesting a common history.

Also the possibility of direct imaging through narrow band filters has allowed detailed studies to determine temperature, density and ionization structures of extended PN. Images in the light of Hα, [N II], [O II], [O III], [S II], in NGC 40 and NGC 6826 were obtained by Jacoby, Quigley and Africano (1987), from these data they derived detailed physical conditions and ionic abundances at each position.

b) Speckle Interferometry

Wood, Bessell and Dopita (1986) have obtained speckle interferometry in the light of [O III] 5007 A and have derived angular diameters of Magellanic Clouds PN (two objects in the SMC and nine objects in the LMC). The mass of ionized gas is derived in each nebula from the angular diameter and published Hβ line fluxes; the derived masses range from 0.005 to 0.19 $M_\odot$ with a mean value of 0.08 $M_\odot$; all the PN observed are the brightest objects in the clouds. They concluded that the nebulae are most certainly partially ionized, and that the masses derived at the ionized part of the nebulae are lower limits to the total nebular mass. Barlow *et al.* (1986) obtained speckle interferometry in [O III] of N2 in SMC; they found the nebula to be distributed in two shells, an inner one of 0.22" and an outer one of 0.38", which correspond to sizes of 0.06 and 0.10 pc, and to masses of 0.09 and 0.27 $M_\odot$, respectively.

c) Spectrophotometry

The accumulation of high quality line intensities has allowed detailed analysis of nebular parameters for a large number of individual PN. This work has been coupled with ionization structure models. It has been possible to determine the He, C, N, O, S, Ar, Cl, and Mg abundances relative to H for large number of objects. This research has been pursued by many authors that have carried out detailed analysis of individual nebulae (see references in compilation by Zuckerman and Aller 1986). Abundaces for a large number of objects have been determined by Aller and Keyes (1987) for 51 objects, by Aller and Czyzak (1983) for 41 objects, by Kaler (1985) for 12 nebulae, and by Peimbert and Torres-Peimbert (1987) for 16 PN of Type I. It is of interest to note that Acker and Koppen (1988) present a paper in this Symposium where they report to have obtained homogeneous spectrophotometric data for 900 PN.

It has also been possible to derive chemical abundances of PN in other galaxies. Jacoby and Ford (1986) analyzed 3 PN in M31. Spectrophotometry of PN is at present the only means of measuring chemical abundances in individual stars at the distance of M31. The oxygen abundances in a disk PN and in an H II region appear to be higher in the disk of M31 than in the disk of the Galaxy. They also found from the abundance of 2 of the PN, which are halo objects and which have different O/H ratio, that the halo of M31 is chemically inhomogeneous. Monk, Barlow and Clegg (1988) have derived chemical abundances for 71 PN in the Magellanic Clouds.

Barlow (1987) studied PN in the SMC and LMC; he derived the electron density from [O II] 3727/3729 and from Hβ determined their mass. In 18 PN of the SMC he derived a nebular mass of 0.02 to 0.45 $M_\odot$; in his sample, he found Type I objects, that is, PN with N/O > 1. In 14 objects in the LMC he derived nebular masses from 0.0076 to 0.69 $M_\odot$.

d) Radial Velocity Measurements

The number of PN for which there are available radial velocity determinations is large; for example, the catalogue by Schneider *et al.* (1983) comprises 524 objects; it lists 287 objects, with an accuracy better than 10 km s^{-1}. Most of this work has been done in the optical lines.

e) Nebular Kinematics

Many individual nebulae have been studied in detail through Fabry-Perot interferometry, high dispersion coude and echelle spectroscopy. The increase in data has been considerable, and more details of their velocity fields are known about them.

The catalogue of expansion motions by Sabbadin (1984) comprises 165 objects; an updated compilation is being prepared by Weinberger (1988). From a comparison of expansion velocity *vs.* linear size, Sabbadin concludes that there are systematic differences between planetary nebulae of class C and B morphological types of Greig (1971), and that the expansion models are consistent with the class C objects having been ejected and excited by post asymptotic giant branch stars having a final mass ~ 0.60 $M_\odot$, and a nebular mass ~ 0.20 $M_\odot$, while the class B objects having been ejected and excited by central stars of final mass > 0.60 $M_\odot$ and nebular mass from 0.3 to 1.0 $M_\odot$.

Anomalies in the expansion motions of nebular material have been found. High velocity jet-like bipolar mass flow in NGC 2392 have been found from high dispersion Hα and [N II] images; the flows are of 200 km s^{-1} (O'Dell and Ball 1985; Gieseking, Becker, and Solf 1985). Walsh and Meaburn (1987) studied the Helix Nebula, NGC 7293, and found a faint external filament that shows a radial velocity of 50 km s^{-1}, as compared to 25 km s^{-1} in the bulk of the nebula; they attributed it to an ionized remnant of an early ejection event of the central star, where the high velocity expansion is an indicator of a superwind stage. Reay and Atherton (1985) have determined the kinematics of NGC 7009 from observations at [O I] 6300 A; they found condensations symmetrically disposed in velocity about the central star.

High expansion velocities have been found in extragalactic objects as well; for example, in a few PN of the LMC expansion velocities of 100 km s^{-1} have been observed (Dopita, Ford and Webster 1985).

2.3 Data on Central Stars of PN

a) Magnitudes of Central Stars

Walton *et al.* (1986) obtained narrow band continuum CCD images of 21 PN where the central stars had been poorly dectected or undetected (of magnitude fainter than 14). They were able to measure the visual magnitude of the central stars in 19 of them. Three objects in the sample have temperature greater than 2×10^5 K (NGC 3918, NGC 3211, and NGC 2440); their position in the H-R diagram can be explained if they have masses greater than 0.8 $M_\odot$. This technique is particularly useful for those nebulae whose central stars are faint or below the visual threshold.

b) Variability of Central Stars

Studies of photometric variability of PNN have established the presence of close binary systems not previously known. For example, Grauer *et al.* (1987) reported the discovery of sinusoidal variations with a 13.96 hr period and amplitude of 1.1 mag in the B band of the central star of HFG1. Reviews on the presently available data on binary systems are given by Bond (1988) and Mendez (1988).

Observational evidence was presented for the first time that shows the central star of a PN to be a pulsating variable (Grauer and Bond 1984). K1-16 has a main period of 28.3 min with a semiamplitude of about 0.01 mag. Spectroscopically and photometrically, the central star of K1-16 closely resembles the previously known hot pulsator PG 1159- 035; these two objects exhibit a new pulsational instabilty mechanism for extremely hot degenerate or predegenerate stars. It demonstrates pulsational instability present in a new region of the H-R diagram and provides an opportunity of direct observation of stellar evolution.

c) Temperatures and Gravities

There has been a considerable effort to derive more dependable physical parameters of PNN. Mendez et al. (1985, 1987) from spectroscopic data of central stars, and by fitting stellar H and He absorption lines with theoretical profiles, have been able to derive accurate stellar temperatures (± 5000 K), log g values (± 0.2 dex), and photospheric He/H ratios (± 0.02) for a sample of 28 objects.

3. CONCLUSIONS

The advances in observational material for the study of PN have been outstanding. High quality data have been secured for an increasing number of objects, including the fainter ones.

Our understanding of the properties of the central stars has increased. Stellar winds, magnitudes and colors have been measured from observations; also effective temperatures and gravities have been derived. It will be very important to determine stellar wind characteristics on different positions of the H-R diagram.

Our knowledge on the nebular parameters has improved; especially in the abundance determinations, kinematics and morphology. Increased attention is being paid to morphology and kinematics since it is hoped that thet can help us to a better understanding of the ejection mechanisms of the nebulae.

Ultraviolet work on the objects will depend on the continued operation of IUE and on the succesful launch of the Space Telescope. Optical work on fainter objects and of higher angular resolution from larger telescopes that are under construction will be possible.

The Space Telescope operation will allow for better studies of Magellanic Clouds PN, where nebular sizes, masses, chemical composition, and central stars magnitudes and colors will allow direct comparison of evolutionary tracks with observations.

REFERENCES

Acker, A. and Koppen, J. 1988, in this volume.
Adam, J. and Koppen, J. 1985, *Astron. Astrophys.*, **142**, 461.
Aller, L. H. and Czyzak, S. J. 1983, *Ap. J. Suppl.*, **51**, 211
Aller, L. H. and Keyes C. D. 1987, *Ap.J. Suppl.*, **65**, 405.
Balick, B. 1987, *A. J.*, **94**, 671.
Barlow, M. J., Morgan, B. L, Stanley, C., and Vine, H. 1986, *M. N. R. A. S.*, **223**, 11.
Barlow, M. J. 1987, *M. N. R. A. S.*, **227**, 161.
Bond, H. E. 1988 in this volume.
Cerruti-Sola, M. and Perinotto, M. 1985, *Ap. J.*, **291**, 237.
Chu, H.-Y., Jacoby, G. H., and Arendt, R. 1987, *Ap.J. Suppl.*, **64**, 529.
Clegg, R. E. S., Harrington J. P., and Storey, P. J. 1986, *M. N. R. A. S.*, **221**, 61p.
Dopita, M. A., Ford, H. C., and Webster, B. L. 1985, *Ap. J.*, **197**, 593.
Ellis, G. L., Grayson E. T., and Bond, H. E. 1984, *P. A. S. P.*, **96**, 283.
Feibelman, W. A., Oliversen, N. and Nichols-Bohlin, 1986 in *New Insights in Astrophysics: 8 Years of UV Astronomy with IUE*, ed. E. J. Rolfe, ESA SP-263, p. 299.
Gieseking, F., Becker, I. and Solf J. 1985, *Ap. J.*, **295**, L17.
Grauer, A. D. and Bond, H. E. 1984, *Ap. J.*, **277**, 211.
Grauer, A. D., Bond, H. E., Ciardullo, R. and Fleming, T. A. 1987, *Bull. A. A. S.*, **19**, 643.
Greig, W. E. 1971, *Astron. Astrophys.*, **10**, 161.
Hartl, H. and Tritton, S. B. 1985, *Astron. Astrophys.*, **145**, 41.
Heap, S. et al. 1978, *Nature*, **275**, 385.
Jacoby, G. H. and Ford, H. C. 1986, *Ap. J.*, **304**, 490.

Jacoby, G., Quigley, and Africano, J. 1987, *P. A. S. P.*, **99**, 672.
Jewitt, D. C., Danielson, G. E., and Kupferman, P. N. 1986, *Ap. J.*, **302**, 727.
Kaler, J. B. 1985, *Ap. J.*, **290**, 531.
Kaler, J. B. and Feibelman, W. A. 1985, *Ap. J.*, **297**, 724.
Koppen, J. and Aller, L.H. 1987, in *Scientific Accomplishments of the IUE*, ed. Y. Kondo, (Dordrecht: Reidel), p. 589.
Maran, S. P., Gull, T. H., Stecher, T. P., Aller, L. H., and Keyes, C. D. 1984, *Ap. J.*, **280**, 615.
Maran, S. P., Aller, L. H., Gull, T. R. Stecker, T. P. 1982, *Ap. J.*, **253**, L43.
Mendez, R. H. 1988 in this volume.
Mendez, R. H., Kudritzki, R. P. and Simon, K. P. 1985, *Astron. Astrophys.*, **142**, 289.
Mendez, R. H., Kudritzki, R. P., Herrero, A., Husfeld, D. and Groth, H. G. 1988, *Astron. Astrophys.*, in press.
Meysonnier, N., Azzopardi, M., and Lequeux, J. 1988 in this volume.
Monk, D. J., Barlow, M. J., and Clegg, R. E. S. 1988 in this volume.
Morgan, D. H. 1984, *M. N. R. A. S.*, **208**, 633.
Morgan, D. H. and Good, A. R. 1985, *M. N. R. A. S.*, **213**, 419.
O'Dell, C.R. and Ball, M. E. 1985, *Ap. J.*, **289**, 526.
Peimbert, M. and Torres-Peimbert, S. 1987, *Rev. Mexicana Astron. Astrofis.*, **14**, 540.
Perek, L. and Kohoutek, L. 1969 *Catalogue of Galactic Planetary Nebulae*, (Prague: Czechoslovakian Acad. Sci.).
Perinotto, M. 1987 in *Planetary and Proto-Planetary Nebulae: From IRAS to ISO*, ed. A. Preite-Martinez, (Dordrecht: Reidel), p. 459.
Reay, N. K. and Atherton, P. D. 1985, *M. N. R. A. S.*, **215**, 233.
Sabbadin, F. 1984, *Astron. Astrophys. Suppl.*, **58**, 273.
Schneider, S. E. Terzian, Y., Purgathofer, A., and Perinotto, M. 1983, *Ap.J. Suppl.*, **52**, 399.
Shaw, R. A. and Wirth, A. 1985, *Pub. A. S. P.*, **97**, 1071.
Walsh, J. R. and Meaburn, J. 1987, *M. N. R. A. S.*, **224**, 885.
Walton, N.A., Reay, N. K., Pottasch, S. R. and Atherton, P. D. 1986, in *New Insights in Astrophysics: 8 Years of UV Astronmy with IUE*, ed. R. J. Rolfe, ESA SP-263, p. 497.
Weinberger, R. 1988, in preparation.
Wood, P. R., Bessell, M. S., and Dopita, M. A. 1986, *Ap. J.*, **311**, 632.
Zuckerman, B. and Aller, L. H. 1986, *Ap. J.*, **301**, 772.

First Row: Stuart Pottasch, Manuel Peimbert and Michael Feast. Second Row: Detlef Schönberner and Volker Weidemann.

INFRARED OBSERVATIONS OF GALACTIC PLANETARY NEBULAE

A. Preite-Martinez
Istituto di Astrofisica Spaziale
C.P. 67
00044 Frascati, Italy

1. INTRODUCTION

Five months after the 1982 Symposium in London the InfraRed Astronomical Satellite (IRAS) was launched. Undoubtedly this has been the major event of the last five years in the infrared world, with a great impact in the field of Planetary Nebulae (PN) research (^).

In the following, IRAS observations of PN will be presented and discussed, as well as ground based observations. The presentation and discussion of the data will highligt the general properties of the infrared emission of PN, with particular emphasis on the spatial distribution of the IR emission, on the energy distribution over the range 1 to 100 micron, and on some of the relevant properties of dust and ionized gas emission features.

The frequency range covered by IR observations is rather broad, and so is the number of questions we can address ourselves and try to answer with the help of IR data. A comprehensive list of all possible problems is out of the scope of this review, and is covered also by other reviews in this volume. Nonetheless it is worth stressing two points connected mainly with the evolutionary scenario : with infrared observations we can shed light on the link between asymptotic giant branch stars and nuclei of planetary nebulae, and whether dust properties are related to basic parameters of the progenitor star.

IRAS photometric and spectral data, and ground-based photometric observations of various kind will be presented and discussed in sections 2 and 3, respectively.

2. IRAS OBSERVATIONS

We know that planetary nebulae are strong sources of emission in the far infrared (FIR) since 20 years (Gillet et al. 1967), but until 1983 only some of the brightest PN were observed in the near IR and 10 micron windows, and very few in the FIR. With IRAS we had the first survey in the FIR covering 98% of the sky, but missing one of the most interesting and brightest objects: NGC 7027.

The three onboard instruments were: (i) a Survey Array operating in four bands centered at 12, 25, 60, and 100 μm with a rectangular field of view (FOV) ranging from 0.75x4.5 to 3x5 arcmin; (ii) a Chopped Photometric Channel operating at 50 and 100 μm with circular FOV of 1.2 arcmin; and (iii) a Low Resolution Spectrometer operating between 7.5 and 23 μm with FOV in the range 5 to 7.5 arcmin and resolution between 14 and 35.

(^) The Infrared Astronomical Satellite (IRAS) was developed and was operated by NIRV, NASA, and SERC.

Three main data products are now available: (i) the Survey data on point sources, collected in the Point Source Catalogue (PSC), (ii) an Atlas of low resolution spectra (Olnon and Raimond, 1986), and (iii) Skyflux maps. Other products, as the Small Scale Structure Catalogue (SSSC), are also available. Further details can be found in the IRAS Explanatory Supplement (Beichman et al.,1985).

IRAS observations of planetary nebulae are also discussed in Pottasch et al.(1984a), Pottasch (1986, 1987), Kwok et al.(1986), and Iyengar (1986).

2.1. Maps

One of the point to bear in mind is that flux densities for objects larger than about 2 x 4 arcmin are severely underestimated (if present at all) in the PSC. Leene and Pottasch (1987a) derived maps of NGC 7293, the Helix nebula, in order to estimate the total energy emitted by the nebula and give an indication of the origin of the total emission. The four panels of Figure 1 show the infrared data of NGC 7293 for 12, 25, 60, and 100 µm. The maps at different wavelengths show a totally different structure. It is clear that emission from dust is unable to explain the different morphologies observed. They show that ion line emission plays an important role at 12 and 25 µm. Indeed, the 12 µm emission can be entirely attributed to the S IV 10.52 µm and possibly to Ne II 12.81 µm, while the centrally peaked 25 µm emission is attributed to the O IV 25.87 µm line. Only the 60 and 100 µm bands are dominated by dust emission.

The effect of line emission on the IRAS data of planetaries has been also pointed out by Preite-Martinez and Pottasch (1987) and Leene and Pottasch (1987b).

Another important result is the determination of integrated flux densities. In order to show how standard IRAS products can underestimate fluxes for large objects, in Table 1 we compare the integrated flux densities for NGC 7293 as derived by Leene and Pottasch (1987a) with flux densities listed in IRAS catalogues (PSC and SSSC).

Table 1.

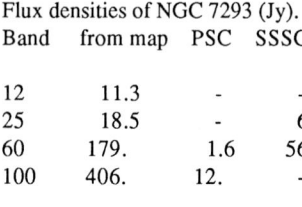

Flux densities of NGC 7293 (Jy).

Band	from map	PSC	SSSC
12	11.3	-	-
25	18.5	-	6
60	179.	1.6	56
100	406.	12.	-

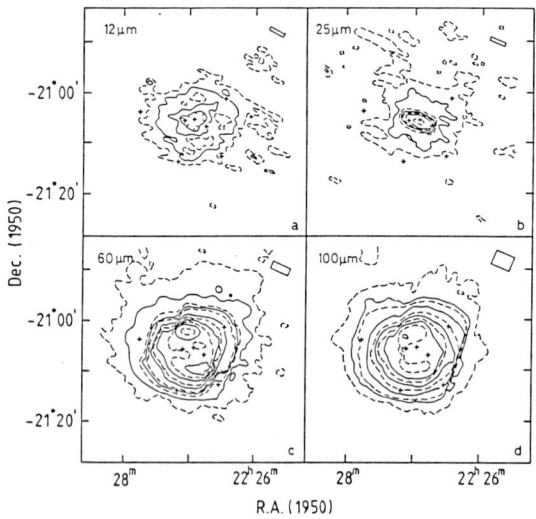

Figure 1. IR maps of NGC 7293 from Leene and Pottasch (1987a).

A similar analysis has been done by Zhang et al.(1987) for NGC 6853 and it is in progress on other extended nebulae, as A 21.

2.2. Search for IR halos

During the IRAS mission a sample of 67 PN were observed in the pointing mode by Leene et al. (1987). The nebulae were selected for their large optical diameter and/or the presence of multiple rings and halos. Also objects smaller than the beam size were selected to search for the presence of weak FIR halos. It is important to know if a source is extended or has some kind of halo showing up in the IR, because the spatial distribution of IR emission can shed light on the origin and evolution of the dust shell, as discussed by Kwok (1980), Natta and Panagia (1981), and Lenzuni et al.(1987).

From the analysis of their sample of PN Leene et al. (1987) find that FIR emission usually originates in the ionized region. This is particularly true for old, optically thin nebulae. On the contrary, for BD+303639 and NGC 6543, two relatively young nebulae, there is clear evidence for dust emission well outside the ionized region. It should be mentioned though that far-IR slit scans of NGC 7027 at 50 and 100 microns taken by Lester et al. (1986) from the Kuiper Airborne Observatory show that the spatial distribution of the far-IR emission corresponds very closely with the distribution of ionized gas.

2.3. IRAS spectra

There are several advantages over the optical and UV lines in using infrared lines: (i) some of the ions are only observed in the IR; therefore they are very useful in determining "total" abundances; (ii) intensities are less sensitive to electron temperature because of the small excitation energy involved, and they are also less influenced by extinction; (iii) we can derive total line intensities for PN with sizes up to about 40", while smaller diaphragms are used from the ground in order to minimize sky background.

On the other hand we should bear in mind that the reliability of IRAS spectral data is rather uneven, both for line intensities and continuum levels. Line fluxes and continuum intensities can vary up to a factor of 2 from one spectrum to another. Sometimes features present in the averaged spectrum appear in only one of the individual spectra. Moreover, as the sources become more extended than 40", the resolution of the spectrometer is degraded: this is why some bright, but large, nebulae have not been measured.

Recently Pottasch et al.(1984b, 1985, 1986) presented and discussed more than 60 LRS spectra of PN. They found that the relative importance of ionic lines and continuum emission is strongly variable from object to object, and that many lines of Ne, S, and Ar are strong enough to be detected. Seven lines were measured, Ne III 15.5 and Ne VI 7.65μm for the first time. The ratio of IR to optical lines of NeIII, NeV and S III were used to derive electron temperatures and densities. The determination of Te and ne from the Ne III and S III ratios agrees very well with values found from optical line ratios. On the other hand the values of the density found from the Ne V ratio are an order of magnitude higher, indicating a higher density in the higly ionized regions near the centre of the nebula. More reliable total abundances of Ne and S were derived, in particular for Ne, where in many nebulae no correction for missing stages of ionization had to be made.

2.4. The IRAS Point Source Catalogue

More than 900 PSC entries are associated with objects listed in the P-K catalog of planetary nebulae (Perek and Kohoutek, 1976), but not all PN known today were listed in the P-K catalogue, and not all the sources in P-K are PN (see Acker et al. 1987). In Table 2 we list the number of PSC object associated with planetary nebulae, with additional information on the quality of the detection and galactic distribution. At point 4, "NoVar" and "NoConf" mean that the variability index is lower than 20%, and that the confusion flag is set to 0 if flux quality is >1, respectively. The number of misclassified PN is also indicated ("noPN").

Table 2: PSC objects associated with P-K sources

Sample	Tot.	$|b|\leq 15°$	noPN
1) Associated	928	868	86/928
2) Detected (Flux Quality $\geq 2,2,2,2$)	168	148	
3) Detected (Flux Quality $\geq 1,2,2,1$)	685	642	39/685
" " " F12/F25 < 1		577	
4) Detected, NoVar, NoConf	479	445	14/479
" F12/F25 < 1		404	
" $|l| \leq 15°$	162		
" $|l| > 15°$	317		

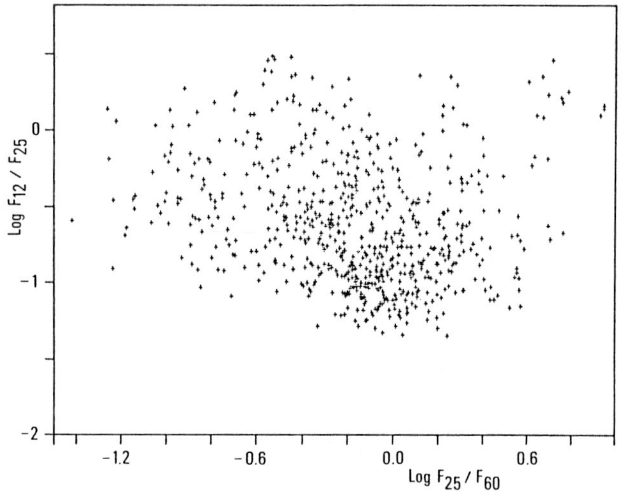

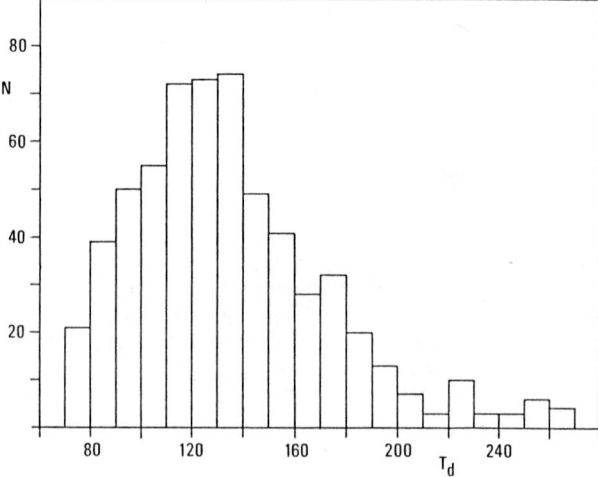

Figure 2. Colour-colour diagram for 685 planetaries with good IRAS detection (Flux Quality $\geq$ 1, 2, 2, 1).

Figure 3. Distribution of dust temperature for the same sample of 685 PN shown in Figure 2.

The 685 PK nebulae detected at least at 25 and 60 microns are shown in Figure 2 in the colour-colour plot involving the first three photometric IRAS bands. In Figure 3 the distribution of dust temperature for the same sample is also shown.

All nebulae show continuum emission in the far-IR, in excess of expected nebular continuum emission, due to dust heated to about 100-150 K. The dust temperature Td is the parameter used to describe the distribution of emission. It is calculated between 25 and 60 microns, because the nebular flux usually peaks in this range.

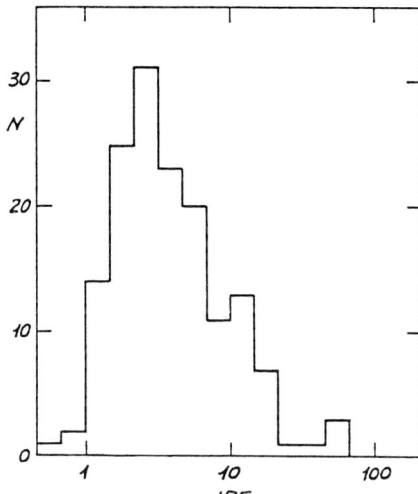

Figure 4. Distribution of InfraRed Excess (IRE) for a sample of 159 PN selected from Pottasch et al. (1984a, 1987).

What is the origin of dust heating? As suggested by Krishna Swamy and O'Dell (1968) an important heating mechanism is scattering of Lyman-α photons generated in the ionized nebular gas, if the far-IR radiation comes from dust well mixed with the ionized gas. It is possible to calculate the IR flux expected if all Ly-α photons produced by recombinations in the nebula heat the dust, using the observed radio continuum emission to estimate the number of recombinations (Pottasch, 1984). We can then compare the expected IR flux F(IR) with the total observed IR flux and define as infrared excess (IRE) the ratio of observed total infrared emission to the energy available in Ly-α photons. In Figure 4 the distribution of the IRE is shown for a sample of 159 PN that includes galactic centre nebulae. Many authors (Pottasch, 1986, 1987; Pottasch et al. 1984a; Kwok et al. 1986; Iyengar 1986) have shown that the IRE is related to properties of the nebula: Ly-α heating is able to explain the amount of observed IR radiation, in particular in the case of large, old nebulae. Values of F(IR) slightly larger than expected could be explained by absorption of other strong nebular lines, as the CIV UV doublet. High values of the IRE (>10) are found in compact, probably very young, nebulae. In these cases, heating by direct starlight is the most plausible cause of dust heating.

An interesting point is whether stellar radiation is absorbed directly by dust in the ionizing continuum, thus competing with the gas, or longward of the Lyman limit. Since large values of the IRE are usually found in nebulae excited by low temperature stars, probably the stellar radiation is absorbed by dust longward of the Lyman limit. This can be the case for ionization bounded nebulae surrounded by IR thermal emission due to heated dust. If ionizing photons are absorbed by dust, the determination of the stellar temperature can be affected, in particular when the Zanstra method is used: photons not absorbed by the gas will not be counted, thus the resulting temperature of the central star will be underestimated. Because of its different nature, the Energy-Balance method is much less affected.

3. GROUND-BASED OBSERVATIONS

The success of the IRAS mission stimulated a reniewed interest in ground-based observations. A great number of planetary nebulae was observed in the near-, mid-IR from ground-based telescopes. In the following only photometric observations and spatial studies of selected nebulae will be discussed.

3.1. Photometry

More than 200 PN have been observed in the near-IR in recent years by Whitelock (1985), Kwok et al. (1986), Persi et al. (1987), and Preite-Martinez et al. (1987, private communication).

The importance of near-IR photometry is manyfold. First of all, it completes the information on the IR energy distribution. Secondly, it gives us a powerful tool for the classification of nebulae according to the principal source of emission. Furthermore, the presence of hot dust can be investigated.

Whitelock (1985) first pointed out that the near-IR two-colour diagram (J-H, H-K) can be used in order to classify PN according to their main source of emission. Three different sources can play a role. (i) *Nebular emission* : there can be a relevant contribution from thermal continuum due to H and He free-free and bound-free emission; two-photon emission can also contribute al low electron densities. Recombination lines of H (as Pβ, Pγ, and Brγ) and He (as He I 1.083μm, 2.058μm, and He II 1.162μm) are expected to be strong features in this range. PN dominated by nebular (continuum and line) emission are classified as type N. (ii) *Dust emission* : cool dust, while dominating the emission in the far-IR, is not expected to contribute significantly in this range. Dust with temperatures >800K can be seen in the near-IR, if present. PN showing a substancial contribution of thermal emission by hot dust are classified as type D. (iii) *Stellar continua* : the Rayleigh-Jeans tail of the thermal emission from hot, bright central stars is sometimes seen, especially in the J band. A contribution from a cool companion of the central star can also show up in a few PN. In these cases nebulae are classified as type S.

The colours (J-H, H-K) of two large samples of PN are shown in the diagrams of Whitelock (1985, her Figure 1) and Persi et al. (1987, their Figure 1, here reproduced as Figure 5). It is clear from these diagrams that PN can be well separated and classified according to their near-IR colours. About half of

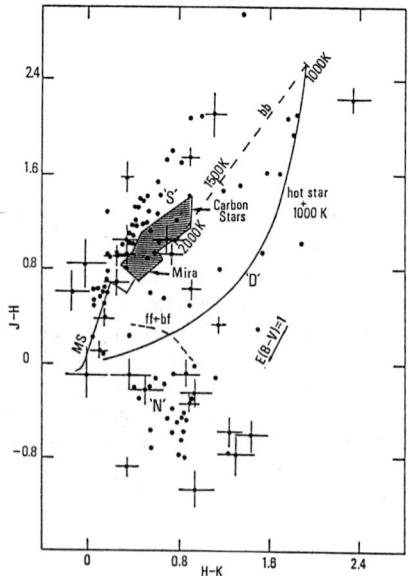

Figure 5. Near-IR two-colour diagram from Persi et al. (1987). Regions where different types of PN concentrate are indicated by the letters N, D, and S.

the nebulae of the total sample fall in a region called "nebular box" by Whitelock (1985). It can be roughly defined as the region of the two-colour diagram delimited by values of the observed J-H < 0.2 and observed H-K > 0.3 mag.

Type D nebulae can be found above the nebular box and to the right of the cooler end of the main sequence. Type S nebulae mostly fall in the region of Mira's and carbon stars. A significative fraction of the objects showing in Figure 5 the near-IR energy distribution of a cool main sequence star have been carefully reobserved. I have found that the majority of them were field stars and that the correct position for the objects is either in the N or D regions.

One of the most important point worth stressing about this near-IR classification scheme is that we can build up the two-colour diagram using *observed* magnitudes, regardless of extinction, with little loss in classification effectiveness. This is because the direction of the reddening correction is almost parallel to the black-body line and to the locus of S type objects. In other words, it is impossible to transform a reddened D or N type object into a dereddened S type object, while dereddening S types only makes them hotter. Dereddened N sources do not change type, too. Such a feature of the (J-H, H-K) colours can also have practical applications in IR observing runs of planetaries.

If hot dust is present in the nebula, it is rather easy to recognize its signature in the two-colour diagram (H-K, K-L), although the L magnitude has been measured only for a sub-sample (<30%) of nebulae. Objects of type S, N, and D are well separated also in this diagram (see Persi et al., 1987, their Figure 2). Sources of type D, as expected, and also many type N, show a distinct K-L excess with respect to stellar or nebular emission. It is not possible to ascertain whether the contribution to the L band is due to thermal emission of hot dust or to emission features (or both) from photometric data alone. In order to give a more reliable indication on the nature of the emission processes involved, the spectrum and a thorough study of the IR energy distribution are required.

3.2. Spatial studies

Near-IR scanning (Phillips et al. 1984) and mapping techniques (Bentley et al. 1984; Lester and Dinerstein, 1984) were used in high angular resolution studies of a relatively small number ($\leq$20) of PN. Phillips et al. (1984) found in most cases that near-IR sizes were smaller than optical sizes. Only in NGC 6826, 6891, 7027, and J900 dust and gas are probably well mixed. The most interesting result comes from IC 418: emission in band H originates from within the ionized zone, while K and L emissions extend outside the H II zone. From a detailed analysis of the spatial distribution of K and L emissions they inferred the presence of a circumnebular shell of dust with rather high temperatures and a strong temperature gradient. On the other hand Bentley et al. (1984) found that the observations of BD+303639 can be explained by the presence of two distinct species of dust, one of which is responsible for the emission features at 8.6 and 11.3 μm, possibly concentrated toward the edges of the nebula. Also in this case the observations are consistent with the dust being well mixed with the ionized gas.

Using a different but very promising technique, Arens et al. (1984) have performed observations of NGC 7027 with a two-dimensional charge-injection-device (CID) array operating at 10 μm. The 32x32 pixel CID array permits the mapping of extended sources at angular resolutions close to diffraction limits. The main result is that with a resolution <2" radio maps and mid-IR maps are very similar, not only in their general appearance, but also in their details. This is the best evidence to date that the dust responsible for the emission in the 10 μm window is coextensive and well mixed with the ionized gas.

Now that more sensitive and larger two-dimensional IR arrays are becoming available for ground-based as well as for the next generation of infrared satellites, the acquisition of high spatial and spectral images of PN will greatly help in the determination of the basic parameters of dust and gas IR emission, broadening our knowledge of the origin and properties of planetary nebulae.

References

Acker.A., Chopinet,M., Pottasch,S.R., Stenholm,B.: 1987, Astron.Astrophys.Suppl.Ser. **71**, 163
Arens,J.F., Lamb,G.M., Peck,M.C., Moseley,H., Hoffmann,W.F.,Tresch-Fienberg,R., Fazio,G.: 1984, Astrophys.J. **279**, 685
Beichman,C.A., Neugebauer,G., Habing,H.J., Clegg,P.E., Chester,T.J.,eds. 1985, "Explanatory Supplement to the IRAS Catalogs and Atlases" (Washington:GPO)
Bentley,A.F., Hackwell,J.A., Grasdalen,G.L., Gehrz,R.D.: 1984, Astrophys.J. **278**, 665
Gillet,F.C., Low,F.J., Stein,W.A.: 1967, Astrophys.J. Letters **149**, L97
Iyengar,K.V.K.: 1986, Astron.Astrophys. **158**, 89
Krishna Swamy,K.S., O'Dell,C.R.: 1968, Astrophys.J. Letters **151**, L61
Kwok,S.: 1980, Astrophys.J. **236**, 592
Kwok,S., Hrivnak,B.J., Milone,E.F.: 1986, Astrophys.J. **303**, 451
Leene,A., Pottasch,S.R.: 1987a, Astron.Astrophys. **173**, 145
Leene,A., Pottasch,S.R.: 1987b, "Planetary and Protoplanetary Nebulae: from IRAS to ISO", ed. A.Preite-Martinez (Reidel:Dordrecht),p.233
Leene,A., Zhang,C.Y., Pottasch,S.R.: 1987, Astron.Astrophys, in press
Lenzuni,P., Natta,A., Panagia,N.: 1987, "Planetary and Protoplanetary Nebulae: from IRAS to ISO", ed. A.Preite-Martinez (Reidel:Dordrecht), p.249
Lester,D.F., Dinerstein,H.L.: 1984, Astrophys.J. Letters **281**, L67
Lester,D.F., Harvey,P.M., Joy,M.: 1986, Astrophys.J. **304**, 623
Natta,A., Panagia,N.: 1981, Astrophys.J. **248**, 189
Olnon,F.M., Raimond,E.: 1986, Astron.Astrophys.Suppl.Ser. **65**, 607
Perek,L., Kohoutek,L.: 1967, "Catalogue of galactic planetary nebulae", Academ. Publish. House of the Czechoslovak Acad.of Sciences
Persi,P., Preite-Martinez,A., Ferrari-Toniolo,M., Spinoglio,L.: 1987, "Planetary and Protoplanetary Nebulae: from IRAS to ISO", ed. A.Preite-Martinez (Reidel: Dordrecht), p.221
Phillips,J.P., Sanchez Magro,C., Martinez Roger,C.: 1984, Astron.Astrophys. **133**, 395
Pottasch,S.R.: 1984, "Planetary Nebulae", Reidel: Dordrecht
Pottasch,S.R.: 1986, "Light on dark matter", ed. F.P.Israel (Reidel:Dordrecht), p.131
Pottasch,S.R.: 1987, "Planetary and Protoplanetary Nebulae: from IRAS to ISO", ed. A.Preite-Martinez (Reidel:Dordrecht), p.1
Pottasch,S.R., Baud,B., Beintema,D.A., Emerson,J., Habing,H.J., Harris,S., Houck,J.R., Jenning, R.E., Marsden,P.: 1984a, Astron.Astrophys.**138**, 10
Pottasch,S.R., Beintema,D.A., Raimond,E., Baud,B., van Duinen,R., Habing,H.J., Houck,J.R., de Jong,T., Jenning,R.E., Olnon,F.M., Wesselius,P.R.: 1984b, Astrophys.J.Letters **278**, L33
Pottasch,S.R., Preite-Martinez,A., Olnon,F.M., Raimond,E., Beintema,D.A., Habing,H.J.: 1985, Astron.Astrophys. **143**, L11
Pottasch,S.R., Preite-Martinez,A., Olnon,F.M., Jing-Er,M., Kingma,S.: 1986, Astron.Astrophys. **161**,363
Pottasch,S.R., Bignell,C., Olling,R., Zijlstra,A.A.: 1987, Astron.Astrophys. , in press
Preite-Martinez,A., Pottasch,S.R.: 1987, "Planetary and Protoplanetary Nebulae: from IRAS to ISO", ed. A.Preite-Martinez (Reidel:Dordrecht), p.197
Whitelock,P.A.: 1985, Mon. Not. Roy. Astron. Soc. **213**, 59
Zhang,C.Y., Leene,A., Pottasch,S.R.: 1987, Astron.Astrophys. **178**, 247

RADIO IMAGES OF PLANETARY NEBULAE

Yervant Terzian
Cornell University
NAIC
Ithaca, NY
USA

ABSTRACT. The continuum radio spectra of planetary nebulae are discussed, and the structure of these objects is examined from the observed aperture synthesis brightness distributions determined with the Very Large Array. The use of radio observations in determining distances to planetary nebulae is examined. The detection of atomic neutral hydrogen at $\lambda 21$ cm associated with planetary nebulae, as well as the associated CO and OH components are discussed. An upper limit, of the nebular magnetic field associated with the neutral material, of 1mG is reported for NGC 6302.

1. INTRODUCTION

Observations of planetary nebulae at radio wavelengths continue to make some of the most important discoveries about the nature and evolution of these objects. The aperture synthesis techniques using the Very Large Array (VLA) have made possible the detailed study of the brightness temperature distributions of many nebulae with angular resolutions ten times better than that obtained at optical wavelengths, and without suffering from any interstellar extinction. The $\lambda 3$ mm CO observations, the $\lambda 18$ cm OH observations, and the $\lambda 21$ cm HI observations of planetary nebulae have made possible the study of the neutral envelopes of young nebulae and proto-planetary nebulae.

It appears that we are beginning to understand the relatively fast and interesting transition from red giant stars to proto-planetary nebulae and to the normal planetary nebulae. Such progress has been made possible by the identification and observations of several objects caught in the fast transition from red giant stars to normal planetary nebulae. The recent observations of the OH/IR stars promise to add more objects in these interesting and important phases of the late stellar evolution.

2. CONTINUUM RADIO EMISSION

2.1. Flux Density Measurements and Radio Spectra

Most of the flux density measurements of planetary nebulae at radio wavelengths during the last few years have been made as a by-product of the two dimensional aperture synthesis studies of these objects, or of the radio line observations of HI, OH and CO. Taylor et.al. (1987) have presented an analysis of the radio continuum spectra of 18 compact planetary nebulae based on published flux densities plus new observations at 327 MHz made with the Westerbork Synthesis Radio Telescope. They show that the radio spectra are well represented by a model in which the radio emission arises in a photon-limited, ionized shell of a stellar wind type envelope. Theoretical models seem to fit the observed spectra best when a radial density power law in the shell of the form r^{-2} is assumed.

When the central star of a planetary nebula is hot enough ($\geq 2 \times 10^4$K) to ionize the neutral material surrounding it, then the free-free transitions in the ionized gas begin to produce observable radio emission. Such young nebulae have small radio angular sizes, and have relatively high electron densities approaching 10^6 cm^{-3}. The ionization front rapidly moves outwards, and as the central star evolves and reaches surface temperatures exceeding 5×10^4K, the surrounding nebula also undergoes rapid changes, and its radio emission increases. Recently Gathier (1986a) has computed theoretical radio continuum spectra as a function of time for a uniform density, spherical, and isothermal (electron temperature $T_e = 10^4$K) nebula with a total mass of 0.5 M$_\odot$. Figure 1 shows Gathier's results, which were computed for a central star of 0.6 M$_\odot$ and a surface temperature, at t = 0 yr, of 2.8×10^4K at a distance of 1 kpc. The initial electron density was 10^6 cm^{-3}, and the expansion velocity was 20 km/sec. It is clear from these results that the radio continuum spectrum should undergo significant changes as a function of the age of the nebula. Figure 2 shows the radio and infrared spectrum of the well observed planetary nebula NGC 7027 where the radio emission is due to free-free radiation from the ionized gas, and the infrared emission is due to thermal radiation from the warm dust.

Measurements of flux densities of compact nebulae have been made by Turner and Terzian (1984) at 2380 MHz with the Arecibo radio telescope, by Kwok (1985) at 5 GHz using the VLA, and by Schneider et.al. (1987) at 2380 MHz at Arecibo. The lower limit of the sensitivity of these observations is about 10 mJy. It should be possible to detect even weaker sources using the VLA or the Arecibo instruments with longer integration times. This might be useful in identifying the extremely young nebulae when the ionization is just beginning and the flux densities are still very low.

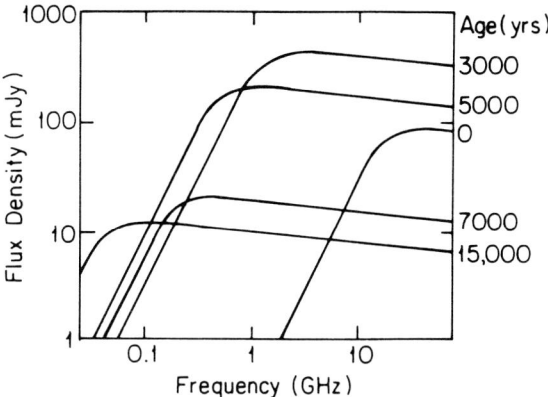

Figure 1. The evolution of the radio spectrum of a thermal spherical nebula with uniform density and temperature as described by Gathier (1986a).

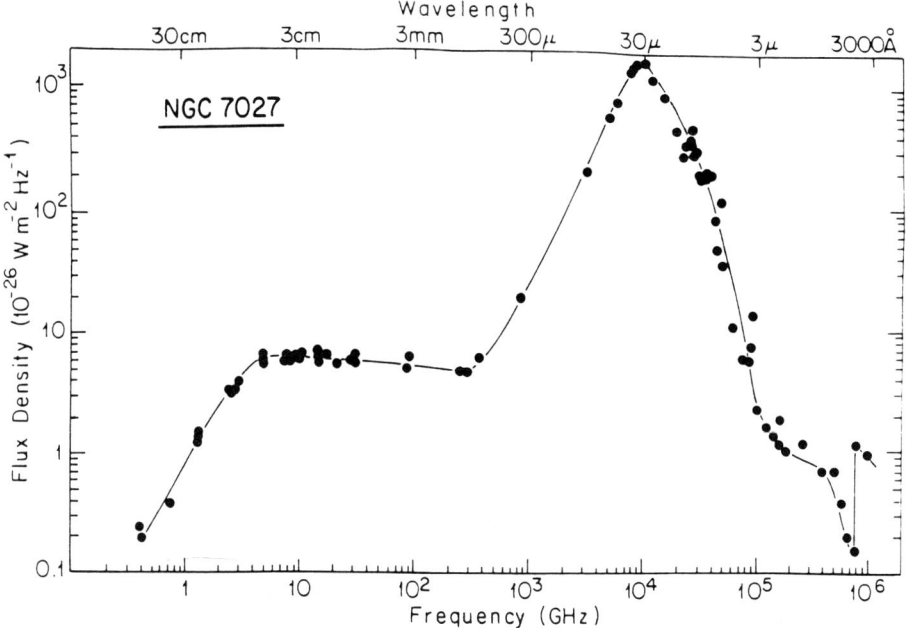

Figure 2. The radio, infrared and optical spectral regions of NGC 7027 adapted from Terzian (1977). The radio spectrum indicates a thermal origin of the emission with the nebula being optically thin at frequencies higher than ~ 5 GHz, and optically thick at lower frequencies. The infrared spectrum indicates the presence of a warm dust envelope. The submillimeter fluxes at 370, 780 and 1090 μm are from Gee et.al. (1984).

2.2. Aperture Synthesis Brightness Distributions

One of the most important contributions of the radio observations of planetary nebulae has been the high angular resolution mapping of these objects, and unaffected by interstellar extinction. Aperture synthesis results of planetary nebulae were first reported by Balick et.al. (1973) and Scott (1973) for NGC 7027, and by Terzian et.al. (1974) for 14 nebulae. These early observations achieved resolutions of 2 arc seconds and were performed with the use of the Green Bank 3-element interferometer, and the Cambridge 5-km radio telescope. The Very Large Array during the last several years has made possible such high resolution mapping with much higher efficiency and accuracy. A review of the early VLA work has been given by Bignell (1983). Gathier et.al. (1983) used the VLA to observe many nebulae in the direction of the galactic center, and Kwok (1985) has reported VLA observations of ten compact nebulae with angular resolutions of ~ 0.4 arc second. Similar additional measurements have been reported by Balick et.al. (1987), and Terzian (1987). It is now possible to achieve a resolution of up to 0.06 arc second with the VLA at a frequency of 22 GHz. Kwok (1987) states that approximately 100 nebulae have been observed to date with the VLA and that most of them are resolved with an angular resolution of 0.4 arc second at a frequency of 5 GHz. Figure 3 shows examples of radio maps of compact nebulae observed by Kwok (1985). These results show a clear shell structure with abrupt boundaries expected from ionization fronts. It can be seen that the nebula II 5117, which has a very high electron density of ~ 10^5 cm^{-3}, is resolved into a bipolar structure with an overall angular size of 1.6 arc seconds. Such nebulae represent the very young stages of planetary nebulae and may be only slightly more evolved than the extremely young and compact objects like Vy 2-2 and CRL 618.

Many planetary nebulae also show outer extended faint envelopes at optical wavelengths (see for example Terzian 1983). If these represent low density ionized gases then these envelopes should also be detectable at radio waves with full-synthesis VLA observations. The current work by Aaquist and Kwok (reported by Kwok 1987) shows that such extended haloes are indeed detectable with the VLA in some cases.

Very recently Basart and Daub (1987) have also presented VLA observations of BD30°3639, NGC 6572, NGC 6590 and NGC 7027 at 1.4 and 5.0 GHz. In addition to presenting brightness temperature distributions for these objects they also construct models and show the two dimensional optical depth, emission measure, and electron temperature distributions. Three of these nebulae, BD30°3639, NGC 6572 and NGC 7027 have also been observed with the VLA by Terzian et.al. (as reported by Terzian 1987), together with the nebulae NGC 3242, NGC 2392 and NGC 6210 in a continuing program to measure the angular expansions of these objects. In order to test the response of the different observing instruments and techniques, Figure 4 shows three

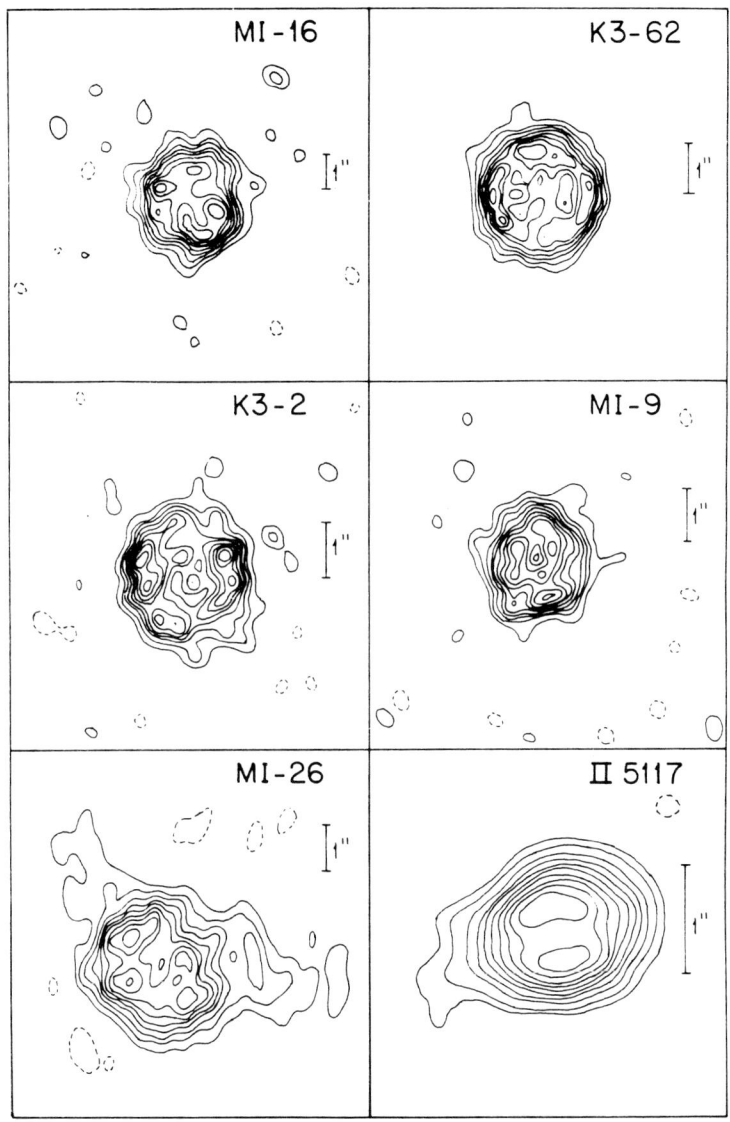

Figure 3. Six representative radio images of planetary nebulae made with the VLA at a frequency of 5 GHz, and a resolution of ~ 0.4 arc sec., selected from the work by Kwok (1985).

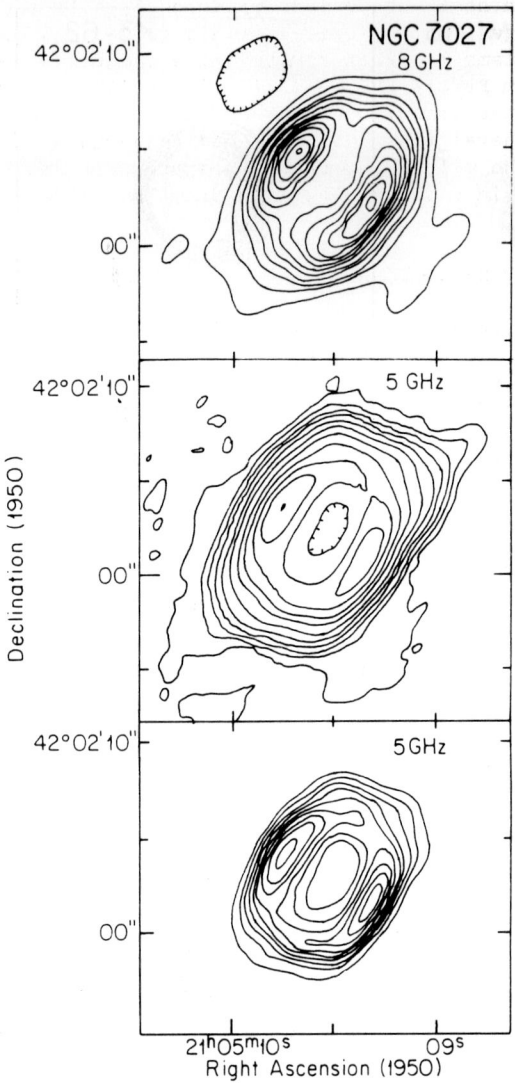

Figure 4. A comparison of the observations of the radio brightness temperature distribution of NGC 7027. The top figure shows the results obtained by Terzian, Balick and Bignell in 1974 using the NRAO Green Bank 3-element interferometer. The middle figure is from Basart and Daub (1987), and the lower figure is from Masson (1986), both of which were made with the VLA.

independent aperture synthesis radio maps produced by different authors. The first image is from Terzian et.al. (1974) and was made with the Green Bank 3-element interferometer at a frequency of 8 GHz, the second image is from Basart and Daub (1987), and the third image is from Masson (1986), the last two were made with the VLA at a frequency of 5 GHz. In general there is very good agreement between these radio maps, and the minor differences are probably due to somewhat different criteria in cleaning the data from the noise.

3. DISTANCES FROM RADIO MEASUREMENTS

The problem of distance determinations of planetary nebulae has remained very difficult, inspite of several studies that have examined this issue very carefully. It is now clear that the assumption of a constant nebular mass in deriving distances is not adequate even in a statistical manner, as discussed by Gathier et.al. (1983) in their study of planetary nebulae at the Galactic Center. Schneider et.al. (1987) have also examined the distance scale problem of planetary nebulae. They use the results of Schneider and Terzian (1983), who showed that for a large sample of nebulae in order for them to be consistent with Galactic rotation they must have a distance scale of 1.4 times Acker's (1978) statistical value. Schneider et.al. (1987) then use the results by Milne (1982) who had scaled his radio distances by a factor of 1.45 times Acker's, and by applying a small correction (to be consistent with the above factor of 1.40) derive a radio distance relation for young, optically thick nebulae as follows:

$$D = 27(S_{6cm})^{-1/2},$$

and for optically thin nebulae

$$D = 24(S_{6cm})^{-1/5}(\theta)^{-3/5},$$

where the distance D is in kpc; the flux density at the wavelength of 6 cm, S_{6cm}, is in mJy; and θ, the angular radius is in arc seconds. The last expression, for example, gives a distance of 1.4 kpc for NGC 7027 assuming that this object is optically thin. This value agrees well with that of Pottasch et.al. (1982) who derived a distance between 1.0 and 1.5 kpc for this object using various methods. However, assuming the nebula to be optically thick, which is likely given the molecular envelope surrounding NGC 7027, one obtains a distance of only 330 pc. Clearly significant uncertainties exist in the method used by Milne (1982) and Schneider et.al. (1987) since the method assumes a constant luminosity for the nebula, which is not realistic (Gathier 1986b).

The high angular resolution provided by the VLA presents new possibilities in deriving accurate distances by measuring the angular expansion of the planetary nebulae as a function of time. Masson (1986) has used this method and has derived a distance of 940 ± 200 pc for NGC 7027. Terzian et.al. (as reported by Terzian 1987) have also begun such a program at the VLA and the first epoch observations, performed during 1983-84, include the objects BD30°3639, NGC 7027, NGC 3242, NGC 2392, NGC 6572 and NGC 6210. The second epoch observations will be made in 1988-89.

Neutral hydrogen $\lambda 21$ cm absorption observations have been used for 24 nebulae by Gathier et.al. (1986), using the Westerbork Synthesis Radio Telescope to determine kinematic distances to these nebulae. For 12 nebulae the derived distances have an uncertainty of ~ 50%, and such methods provide only lower limits to the distances of the nebulae.

The problem of determining accurate distances of planetary nebulae has remained very difficult, but significant new efforts and some progress have been made.

4. ATOMIC NEUTRAL HYDROGEN IN PLANETARY NEBULAE

We are beginning to understand that only a few thousand years separate the end of the Asymptotic-Giant-Branch evolution and the onset of the ionization of planetary nebulae. During this transition period the surface temperature of the central star increases to the point where the neutral gas surrounding it begins to get ionized. Such proto-planetary nebulae are composed of a compact very dense ionized core surrounded by neutral material. Since part of this neutral gas is in the form of atomic hydrogen radio astronomers have tried to detect the $\lambda 21$ cm line emission from it. Only recently such attempts have given positive results when Rodriguez and Moran (1982) detected an HI absorption feature in the direction of the young planetary nebula NGC 6302. Rodriguez et.al. (1985) showed that the HI arises from the dark lane perpendicular to the bipolar ionized nebular components.

More recently Altschuler et.al. (1986) have reported HI absorption towards the young nebula IC 4997, and Taylor and Pottasch (1987) have detected absorption and emission components associated with the compact nebula IC 418. Figure 5 shows the neutral hydrogen spectra of NGC 6302, IC 4997 and IC 418. In all three cases the HI gas is blue-shifted with respect to the radial velocity of the ionized nebula as indicated in Table 1. In addition to the above observations Gathier et.al. (1986) have reported the marginal detection of HI absorption in the direction of NGC 6790. The HI mass detected in these objects ranges from 0.002 to 0.07 $M_\odot$ and represents only a very small percentage of the total nebular mass.

A few dozen other nebulae have been searched for associated neutral hydrogen with negative results, and it appears that in most

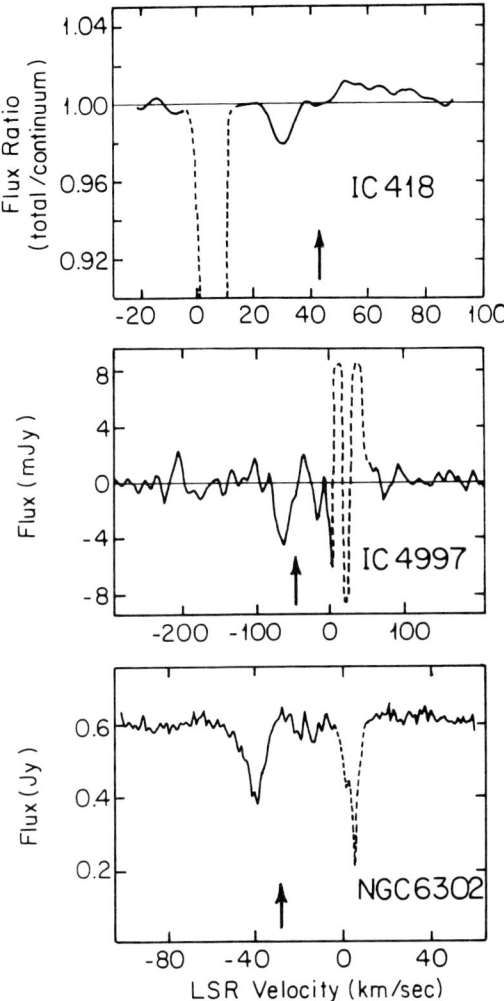

Figure 5. The radio spectra of the three planetary nebulae showing the HI λ21 cm absorption lines due to the neutral nebular envelopes surrounding the ionized clouds. The arrows indicate the radial velocities of the ionized hot nebulae. The dashed profiles indicate HI due to galactic absorption and emission not associated with the nebulae. Note that in all cases the nebular HI absorption is blue-shifted from the radial velocities of the ionized clouds. (Spectra adapted from Taylor and Pottasch 1987, Altschuler et.al. 1986, and Rodriguez et.al. 1985).

cases the ionization of a circumstellar envelope takes place during a relatively short period of time.

TABLE 1. HI MEASUREMENTS

Nebula	Radial Velocity* km/sec	HI Absorption Velocity* km/sec	HI Mass** $M_\odot$
IC 418	+ 43.4	+ 30	0.07
IC 4997	− 49.8	− 64	0.002
NGC 6302	− 31.4	− 40	0.06
NGC 6790	+ 56.8	+ 40	0.01

*Velocities are with respect to the Local Standard of Rest.

**Masses are uncertain since they depend on parameters like nebular distances, geometry and HI excitation temperature, all of which are not well known.

5. MOLECULES IN PLANETARY NEBULAE

A significant fraction of the information on proto-planetary nebulae and the early stages of their evolution has come from infrared and molecular observations. Both of these subjects are treated in detail by other authors in these proceedings. At radio waves CO, CN, and OH have been detected from young and compact nebulae indicating mass loss rates in the range of 10^{-6} to 10^{-4} $M_\odot$/yr. Knapp (1986) has listed ten nebulae with detected molecular envelopes of CO, OH, and H_2. Some of these nebulae, like NGC 7293 and NGC 6720, are well established planetary nebulae, yet molecular emission is still detectable and it must come largely from the outer regions of the nebular envelopes. It is also possible that the molecules survive because they may be shielded from the stellar radiation by dust particles.

Until recently mostly CO and H_2 were observed in association with planetary nebulae. OH was only seen in Vy 2-2, and CN only in NGC 7027. However, Payne et.al. (1987) have now detected OH from NGC 6302, and Pottasch et.al. (1987) have detected continuum radio emission from two OH/IR stars which make them probable objects in transition from OH/IR stars to planetary nebulae.

6. NEBULAR MAGNETIC FIELDS

Very little is known about magnetic fields in planetary nebulae. Gurzadyan (1962) and very recently Pascoli (1987) have discussed the possible influence of a dipole magnetic field as the origin of the bipolarity in the observed morphology of planetary nebulae. Rodriguez et.al. (1985) have tried to measure the intensity of the magnetic field in the HI gas near NGC 6302 by the Zeeman effect in the $\lambda 21$ cm line. They used the VLA and recorded the right and left circular polarizations of the $\lambda 21$ cm line. The difference in the right and left circular polarization spectra produce the characteristic "S" shaped Zeeman pattern which results from a substantial magnetic field. No significant magnetic field signature was detected and a 3σ upper limit of ~ 1 mG was derived for the magnetic field strength in the neutral gas of NGC 6302. However, it should be possible to extend these observations with longer integration times and even with better frequency resolutions, to examine this important issue further.

7. CONCLUSIONS

Significant advances in our knowledge of planetary nebulae have taken place during the last few years through observations of these objects at radio wavelengths. High angular resolution radio images of many nebulae have revealed their structure, neutral hydrogen $\lambda 21$ cm and OH $\lambda 18$ cm measurements have indicated the presence of neutral envelopes around these objects, and CO $\lambda 3$ mm observations have given us realistic estimates to the mass loss from the progenitors of planetary nebulae.

Future work will certainly include the examination of the neutral envelopes, the derivation of accurate physical parameters from radio synthesis observations and the possibility of establishing a more accurate distance scale for the planetary nebulae population by the nebular expansion method using high resolution radio images.

The author wishes to thank R. Gathier for providing information in advance of publication. This work was supported in part by the National Astronomy and Ionosphere Center, which is operated by Cornell University under a management agreement with the National Science Foundation.

REFERENCES

Acker, A. 1978, <u>Astr. Ap. Suppl.</u>, <u>33</u>, 367.
Altschuler, D.R., Schneider, S.E., Giovanardi, C. and Silverglate, P.R. 1986, <u>Ap. J. (Letters)</u>, <u>305</u>, L85.
Balick, B., Bignell, C. and Terzian, Y. 1973, <u>Ap. J. (Letters)</u>, 182, L117.
Balick, B., Bignell, C., Hjellming, R.M. and Owen, R. 1987, <u>Astron. J.</u>, (in press).
Basart, J.P. and Daub, C.T. 1987, <u>Ap. J.</u>, <u>317</u>, 412.
Bignell, C. 1983, IAU Symposium 103, ed. D.R. Flower, (D. Reidel Publ. Co.), 69.
Gathier, R., Pottasch, S.R., Goss, W.M. and van Gorkom, J.H. 1983, <u>Astr. Ap.</u>, <u>128</u>, 325.

Gathier, R. 1986a, <u>Late Stages of Stellar Evolution</u>, ed. S. Kwok and S.R. Pottasch, (D. Reidel Publ. Co.), 371.
_____ 1986b, <u>Astro. Ap.</u>, (in press).
Gathier, R., Pottasch, S.R. and Goss, W.M. 1986, <u>Astr. Ap.</u>, 157, 191.
Gee, G., Emerson, J.P., Ade, P.A.R., Robson, E.I. and Nolt, I.G. 1984, <u>M.N.R.A.S.</u>, <u>208</u>, 517.
Gurzadyan, G.A. 1962, <u>Vistas Astron.</u>, <u>5</u>, 40.
Knapp, G.R. 1986, "Mitteilungen der Astronomischen Gesellschaft", (in press).
Kwok, S. 1985, <u>Astron. J.</u>, <u>90</u>, 49.
_____ 1987, <u>Late Stages of Stellar Evolution</u>, ed. S. Kwok and S.R. Pottasch, (D. Reidel Publ. Co.), 321.
Masson, C.R., 1986, <u>Ap. J. (Letters)</u>, <u>302</u>, L27.
Milne, D.K., 1982, <u>M.N.R.A.S.</u>, <u>200</u>, 51P.
Pascoli, G., 1987, <u>Astr. Ap.</u>, <u>180</u>, 191.
Payne, H.E., Phillips, J.A. and Terzian, Y. 1987, <u>Ap. J.</u>, (in press).
Pottasch, S.R., Goss, W.M., Arnal, E.M. and Gathier, R. 1982, <u>Astr. Ap.</u>, 106, 229.
Pottasch, S.R., Bignell, C. and Ziljlstra, A. 1987, <u>Astr. Ap.</u>, <u>177</u>, L49.
Rodriguez, L.F. and Moran, J.M. 1982, <u>Nature</u>, <u>299</u>, 323.
Rodriguez, L.F., Garcia-Barreto, J.A., Canto, J., Moreno, M.A., Torres-Peimbert, S., Costero, R., Serrano, A., Moran, J.M. and Garay, G. 1985, <u>M.N.R.A.S.</u>, <u>215</u>, 353.
Schneider, S.E. and Terzian, Y. 1983, <u>Ap. J. (Letters)</u>, <u>274</u>, L61.
Schneider, S.E., Silverglate, P.R., Altschuler, D.R. and Giovanardi, C. 1987, <u>Ap. J.</u>, (in press).
Scott, P.F. 1973, <u>M.N.R.A.S.</u>, <u>161</u>, 35P.
Taylor, A.R. and Pottasch, S.R. 1987, <u>Astr. Ap.</u>, 176, L5.
Taylor, A.R., Pottasch, S.R. and Zhang, C.Y. 1987, <u>Astr. Ap.</u>, <u>171</u>, 178.
Terzian, Y. 1977, <u>Sky and Telescope</u>, December, 459.
Terzian, Y., Balick, B. and Bignell, C. 1974, <u>Ap. J.</u>, <u>188</u>, 257.
Terzian, Y. 1983, <u>Planetary Nebulae</u>, ed. D.R. Flower, (Reidel Publ. Co.), 487.
_____, 1987, <u>Asteroids to Quasars</u>, ed. P. Lugger, (Cambridge University Press), (in press).
_____, 1987, <u>Sky and Telescope</u>, February, 459.
Turner, K.C. and Terzian, Y. 1984, <u>Astron. J.</u>, <u>89</u>, 501.

NEW AND MISCLASSIFIED PLANETARY NEBULAE

L. Kohoutek
Hamburg Observatory
Hamburg-Bergedorf, W.Germany

ABSTRACT. 75 objects have been classified as new planetary nebulae since 1982. They are summarized in Table 1 which gives the designations, names, coordinates and references to the discovery. In the list of misclassified PN (Table 2) 41 objects have been included; Table 3 presents objects with incorrect identification in CGPN. The main properties of a PN and of its nucleus are given in a summary which can be useful for a correct classification of planetaries.

This third supplementary list to the "Catalogue of Galactic Planetary Nebulae" (CGPN - Perek,Kohoutek,1967)contains 75 discoveries which were published mainly between 1982 and 1986. As in the previous lists the designations, names, coordinates and references to the discovery as PN are given in Table 1. An asterisk affixed to the galactic number means an uncertain classification (suspected,possible or probable PN).

It is suggested to remove 41 objects (Table 2) from the CGPN or from the previous supplementary lists (Kohoutek,1978 - Paper I, 1983 - Paper II):they are mostly M stars without H_α emission, galaxies, reflection nebulae or plate faults. There are numerous further objects the classification of which as PN is still questionable: it is possible to find them in both the list of planetaries and the list of other emission-line objects. The catalogue of symbiotic stars given by Allen (1984) contains 40 stars also classified as planetaries. We are rather reserved concerning their reclassification alone on the basis of symbiotic behaviours: let us mention the symbiotic object 330+4.1 (Cn 1-1) which was removed from the list of PN (Paper I) whereas the recent papers of Lutz (1984) and Bhatt,Mallik (1986) classify it again as a PN.

It is useful to present a list of planetary nebulae with incorrect identification in CGPN (Table 3). Besides

TABLE 1 NEW PLANETARY NEBULAE (1982-1986)

Design.	Name	R.A. (1950) Decl.		Discovery	Rem.
124+10.1*	ELO103+73	1ʰ03ᵐ.6	+73°17'	Ellis,al. 1984	R
128 -4.1*	S 22	1 27.4	+58 07	Arkhipova,Lozinskaya 1978	R
148-48.1*	GRO155+10	1 55.3	+10 43	Ellis,al. 1984	
136 +5.1	HEFE 1	2 59.53	+64 43.0	Heckathorn,al. 1982	
138 +4.1*	HtDe 2	3 06.9	+62 37	F Hartl,al. 1983	
149 -9.1*	HtDe 3	3 23.8	+45 13	Hartl,al. 1983	
137+16.1*	ELO419+72	4 19.4	+72 42	Ellis,al. 1984	
166 -6.1*	CRL 618	4 39.56	+36 01.2	Proto-PN in Paper I	
158 +0.1*	Sh 2-216	4 41.3	+46 44	Reynolds 1985	R
205-26.1*	MaC 2-1	5 01.25	-6 13.7	F MacConnell 1982	
203-18.1*	MaC 2-2	5 26.42	-0 43.1	MacConnell 1982	
156+12.1*	HtDe 4	5 33.8	+55 30	Hartl,al. 1983	
204-16.1*	MaC 2-3	5 35.71	+0 12.8	MacConnell 1982	
173 +2.1*	PP 40	5 37.53	+35 41.0	Turner,Terzian 1985	
204-13.1*	MaC 2-4	5 45.02	+0 37.7	F MacConnell 1982	
197 -6.1	WeDe 1	5 56.64	+10 41.5	F Weinberger,al. 1983	
218-10.1*	HtDe 5	6 21.2	-10 11	Hartl,al. 1983	
192 +7.1*	HtDe 6	6 37.2	+21 28	Hartl,al. 1983	R
231 -8.1*	Y-C 34	6 49.64	-20 10.8	Cesco,al. 1984	
221 +4.1*	Y-C 36	7 19.73	-5 50.0	Cesco,al. 1984	
221 +5.2*	Y-C 37	7 23.99	-5 16.0	Cesco,al. 1984	
223 +4.1*	Y-C 39	7 24.46	-7 26.8	Cesco,al. 1984	R
219 +7.1*	RWT 152	7 27.4	-2 00	Pritchet 1984	
247 -4.1*	FEGU 248-5	7 40.48	-32 40.7	F Fesen,al. 1983	
235 +4.1*	Y-C 40	7 47.80	-17 44.4	Cesco,al. 1984	
211+18.1*	HtDe 7	7 52.2	+9 43	Hartl,al. 1983	
211+22.1*	BN0808+11	8 08.5	+11 06	F Ellis,al. 1984	
271 -8.1*	Y-C 41	8 33.46	-54 53.3	Cesco,al. 1984	
214+31.1*	Y-C 42	8 43.24	+12 48.2	Cesco,al. 1984	
221+46.1	BN0950+13	9 50.3	+13 59	F Ellis,al. 1984	

Design.	Name	R.A. (1950)	Decl.		Discovery	Rem.
273 +6.1*	HBDS 1	9h50m.8	-46°03'		Heber,Drilling 1984	R
299 -4.1*	HtTr 1	12 13.82	-66 29.0		Hartl,Tritton 1983	R
315+59.1*	Y-C 43	13 14.26	-2 47.9		Cesco,al. 1984	
321 -3.1	HtTr 2	15 26.17	-60 51.4		Hartl,Tritton 1983	R
335+12.1*	DS 2	15 39.7	-39 10	F	Drilling 1983	R
336 -1.1*	VERA 90	16 34.40	-48 36.8	F	Vega,al. 1980	
333 -4.1	HtTr 3	16 35.69	-52 43.4		Hartl,Tritton 1983	
336 -2.1*	VERA 104	16 38.31	-49 39.4	F	Vega,al. 1980	
335 -3.1	HtTr 4	16 41.13	-51 06.8		Hartl,Tritton 1983	R
94+38.1*	EL1647+64	16 47.2	+64 18		Ellis,al. 1984	
343 -0.1*	HtTr 5	16 57.90	-43 01.6		Hartl,Tritton 1983	R
11+17.1*	DeHt 1	17 04.17	-9 43.1	F	Dengel,al. 1979	R
75+35.1	Sa 4-1	17 12.5	+49 19		Sanduleak 1983	
358 +2.5*	HtDe 8	17 28.7	-28 39		Hartl,al. 1983	
36+21.1*	Y-C 44	17 36.12	+12 42.6	F	Cesco,al. 1984	
36+20.1*	Y-C 45	17 40.75	+12 21.9		Cesco,al. 1984	R
332-16.1	HtTr 6	17 47.36	-60 22.6		Hartl,Tritton 1983	R
332-16.2*	HtTr 7	17 49.60	-60 49.4		Hartl,Tritton 1983	R
6 +1.1*	HtTr 8	17 52.90	-22 58.6		Hartl,Tritton 1983	P.
1 -3.3*	SAWI 1	17 59.27	-29 25.2	F	Shaw,Wirth 1985	
1 -3.4*	SAWI 2	17 59.85	-29 46.1	F	Shaw,Wirth 1985	
1 -3.5*	SAWI 3	18 00.08	-29 50.7	F	Shaw,Wirth 1985	
1 -3.6*	SAWI 4	18 00.45	-29 46.0	F	Shaw,Wirth 1985	
1 -3.7*	SAWI 5	18 00.70	-29 51.6	F	Shaw,Wirth 1985	
1 -3.8*	SAWI 6	18 00.80	-29 27.0	F	Shaw,Wirth 1985	
1 -3.9*	SAWI 7	18 01.88	-29 19.7	F	Shaw,Wirth 1985	
351-10.2	HtTr 9	18 05.41	-41 48.9		Hartl,Tritton 1983	R
22 +4.1*	MA 2	18 12.52	-6 58.2	F	Maehara 1982	
23 +4.1*	MA 3	18 15.13	-6 49.6	F	Maehara 1982	
30 +6.1*	Sh 2-68	18 22.43	+0 49.9	F	Fesen,al. 1983	R
23 +1.1*	MA 13	18 27.80	-7 29.8	F	Maehara 1982	

32

Design.	Name	R.A. (1950)	Decl.	Discovery	Rem.
40 +7.1*	Y-C 46	18^{h}34^{m}04	+10°16.2	Cesco,al. 1984	
9 -8.1*	Y-C 47	18 34.37	-24 29.2	Cesco,al. 1984	
31 -0.2*	HtTr 10	18 47.82	-1 43.7	Hartl,Tritton 1983	R
68+14.1*	SP 4-1	18 58.75	+38 17.1 F	Stephenson 1985	
36 -1.2*	HtTr 11	19 00.48	+2 57.9	Hartl,Tritton 1983	R
38 -0.1*	HtTr 12	19 01.32	+5 05.2	Hartl,Tritton 1983	R
11-14.1*	HtDe 10	19 02.63	-25 28.5	Hartl,al. 1983	R
36 -2.1*	HtTr 13	19 05.52	+2 16.6	Hartl,Tritton 1983	R
41 -0.1*	HtTr 14	19 06.79	+7 00.8	Hartl,Tritton 1983	R
51 +2.1*	IRAS1912+172P09	19 12.77	+17 17.5	Whitelock,Menzies 1986	R
34-10.1*	HtDe 11	19 28.7	-3 49	Hartl,al. 1983	
14-25.1*	HtDe 12	19 55.1	-26 31	Hartl,al. 1983	
75 +5.1*	V1016 Cyg	19 55.33	+39 41.5	Proto-PN in Paper I	
99 -8.1*	HtDe 13	22 28.1	+47 15	Hartl,al. 1983	

* Possible planetary nebula. F Finding chart. R Remarks.

REMARKS

6 +1.1   F Hartl,Tritton (1985).
11+17.1   Coord.Weinberger (priv.comm.)
11-14.1   ESO 524-G?06 (Lauberts,1982).
30 +6.1   HtDe 9,discov.indep.by Hartl,al. in Johnson (1955); Sh 2-188
          (1983); Simeiz 291 (Gaze,Shajn, (Sharpless,1959).
          1954); YM 15 (Johnson,1955).   YM 22 (Johnson,1955);Simeiz 288
                                          (Gaze,Shajn,1954).
31 -0.2   F Hartl,Tritton (1985).   Confirmed by Pasachoff,al.(1984);
                                          F Rubin,al.(1974).
36 -1.2   F Hartl,Tritton (1985).   ESO 558-G01 (Lauberts,1982).
36 -2.1   F Hartl,Tritton (1985).   LSS 1362
38 -0.1   F Hartl,Tritton (1985).   F Hartl,Tritton (1985).
41 -0.1   F Hartl,Tritton (1985).   F Hartl,Tritton (1985).
51 +2.1   F Hrivnak,al.(1985).   F Hartl,Tritton (1985).
124+10.1 HtDe 1,discov.indep.by Hartl,al. F Hartl,Tritton (1985).
          (1983) No.1; see Lynds (1965). LSE 125
128 -4.1 PN (Rosado,Kwitter,1982); No.91 F Hartl,Tritton (1985).
                                          F Hartl,Tritton (1985).
                                          F Hartl,Tritton (1985).

REFERENCES TO TABLE 1

Y-C	Cesco C.U.,Sanguin J.G.,Sanchez G.,Mira H.,Cesco M.R.,Vicentenla J.A.,1984,Bol.Asoc.Argent.Astron. No.28,159.
S	Arkhipova V.P.,Lozinskaya T.A.,1978,Sov.Astron.Lett. 4(1),7.
DeHt	Dengel H.,Hartl H.,Weinberger R.,1979,Mitt.AG 45,182.
DS	Drilling J.S.,1983,Astrophys.J.270,L13.
EL,GR BN	Ellis G.L.,Grayson E.T.,Bond H.E.,1984,Publ.Astron. Soc.Pacific 96,283.
Sh 2 FEGU	Fesen R.A.,Gull T.R.,Heckathorn J.N.,1983,Publ. Astron.Soc.Pacific 95,614.
--	Gaze V.F.,Shajn G.A.,1954,Izv.Krym.AO 11,39.
HtDe	Hartl H.,Dengel J.,Weinberger R.,1983,Mitt.AG 60,325.
HtTr	Hartl H.,Tritton S.B.,1983,Mitt.AG 60,328.
--	Hartl H.,Tritton S.B.,1985,Astron.Astrophys.145,41.
HBDS	Heber U.,Drilling J.S.,1984,Mitt.AG 62,252.
HEFE	Heckathorn J.N.,Fesen R.A.,Gull T.R.,1982,Astron. Astrophys.114,414.
--	Hrivnak B.J.,Kwok S.,Boreiko R.T.,1985,Astrophys.J. 294,L113.
--	Johnson H.M.,1955,Astrophys.J.121,604.
--	Kohoutek L.,1978,IAU Symp.No.76 (ed.Y.Terzian),D. Reidel Publ.Comp.,p.47 (Paper I)
--	Lauberts A.,1982,The ESO/Uppsala Survey of the ESO(B) Atlas,European Southern Observatory.
--	Lynds B.T.,1965,Astrophys.J.Suppl.12,163.
MaC 2	MacConnell D.J.,1982,Astron.Astrophys.Suppl.48,355.
MA	Maehara H.,1982,Contr.Bosscha Obs.71,1.
--	Pasachoff J.M.,Kwitter K.B.,Massey P.,1984,Bull.AA Soc.16,994.
RWT	Pritchet C.,1984,Astron.Astrophys.139,230.
Sh 2	Reynolds R.J.,1985,Astrophys.J.288,622.
--	Rosado M.,Kwitter K.B.,1982,Rev.Mexicana Astron. Astrof.5,217.
--	Rubin V.C.,Westpfahl Jr.D.,Tuve M.,1974,Astron.J. 79,1406.
Sa 4	Sanduleak N.,1983,Publ.Astron.Soc.Pacific 95,619.
--	Sharpless S.,1959,Astrophys.J.Suppl.4,257.
SAWI	Shaw R.A.,Wirth A.,1985,Publ.Astron.Soc.Pacific 97, 1071.
SP 4	Stephenson C.B.,1985,Publ.Astron.Soc.Pacific 97,930.
PP	Turner K.C.,Terzian Y.,1985,Astron.J.90,59.
VERA	Vega E.I.,Rabolli M.,Muzzio J.C.,Feinstein A.,1980, Astron.J.85,1207.
WeDe	Weinberger R.,Dengel J.,Hartl H.,Sabbadin F.,1983, Astrophys.J.265,249.
IRAS	Whitelock P.A.,Menzies J.W.,1986,Monthly Notices Roy.Astron.Soc.223,497.

TABLE 2   MISCLASSIFIED PLANETARY NEBULAE

Design.	Name	Remarks and references
0 +2.2	ESO-520-13	Plate fault (Fredrick,West,1984)
0 -6.1	ESO-456-73	Plate fault (Fredrick,West,1984)
3 -3.1	Sa 3-119	M star without H_α emission (MacConnell,1983)
3 -4.10	ESO-456-64	Plate fault (Fredrick,West,1984)
9 -6.1	ESO-522-29	Plate fault (Fredrick,West,1984)
28 -4.2	Th 1-I	M star without H_α emission (MacConnell,1983)
35 -2.1	K 4-14	M star without H_α emission (MacConnell,1983)
37 -2.1	Ap 3-1	M star without H_α emission (MacConnell,1983)
37 -3.1	K 4-18	M star without H_α emission (MacConnell,1983)
43 +1.1	K 4-13	M star without H_α emission (MacConnell,1983)
97 +3.1	A 77	Compact HII region (Sabbadin,al.,1986)
196-12.1	A 11	No em. lines, very probably a reflection nebula (Lutz,Kaler,1983)
227+33.1	A 32	A galaxy or a plate deffect on POSS (Lutz,Kaler,1983)
239-18.1	ESO-426-13	Galaxy (Fredrick,West,1984)
241 -7.1	M 4-1	Em.-line galaxy (Kohoutek,Pauls,1985)
242 -3.1	ESO-429-04	Galaxy (Fredrick,West,1984)
245 -3.1	ESO-429-17	Galaxy (Fredrick,West,1984)
247-21.1	K 2-13	Plate fault (Kohoutek,Pauls,1985) Plate fault (West,Kohoutek,1985)
248-12.1	ESO-367-03	Bar galaxy (Fredrick,West,1984)
249-22.1	ESO-308-08	Plate fault (Fredrick,West,1984)
251 -4.1	ESO-369-01	Plate fault (Fredrick,West,1984)
265 +5.1	ESO-314-12	Probably a galaxy (Fredrick,West,1984)
266 +2.1	Pe 2-3	M star without H_α emission (MacConnell,1983)
274 -0.1	ESO-212-08	Plate fault (Fredrick,West,1984)
284-39.1	Lo 2	Probably a galaxy (West,Kohoutek,1985)
292 -3.1	SP 2-14	M star without H_α emission (MacConnell,1983)
308 -1.1	ESO-097-03	Plate fault (Fredrick,West,1984)
309 +6.1	Sm 2	M star without H_α emission (MacConnell,1983)
310 +2.1	Sm 3	M star without H_α emission (MacConnell,1983)
327+14.1	ESO-328-04	Among galaxies (Fredrick,West,1984)

Design.	Name	Remarks and references
329+12.1	ESO-328-40	A number of galaxies around the given position (Fredrick,West,1984)
336 -8.1	ESO-180-05	Plate fault (Fredrick,West,1984)
341+17.1	ESO-450-16	Probably a galaxy (Fredrick,West,1984)
341-15.1	ESO-182-04	Galaxy (West,Kohoutek,1985)
343+16.1	ESO-451-03	Late-type star plus nebula (Fredrick, West,1984)
345+10.1	ESO-390-05	Em.-line galaxy (Fredrick,West,1984)
346+19.1	ESO-515-19	Late-type star (Fredrick,West,1984)
347 +7.1	ESO-391-02	Plate fault (Fredrick,West,1984)
349-10.1	ESO-280-02	Either a galaxy or a reflection nebula (West,Kohoutek,1985)
353-55.1	ESO-289-19	Probably a galaxy (West,Kohoutek,1985)
358 -3.2	H 2-30	M star without H_α emission (MacConnell,1983)

REFERENCES TO TABLE 2

Fredrick L.W.,West R.M.,1984,Astron.Astrophys.Suppl.56,325.
Kohoutek L.,Pauls R.,1985,Astron.Astrophys.Suppl.60,87.
Lutz J.H.,Kaler J.B.,1983,Publ.Astron.Soc.Pacific 95,739.
MacConnell D.J.,1983,Rev.Mexicana Astron.Astrof.8,39 (T.2).
Sabbadin F.,Strafella F.,Bianchini A.,1986,Astron.Astrophys. Suppl.65,259.

those objects given in this table we would like to point out 1-3.1 (H 1-47) and 1-3.2 (Ap 1-7): the identification of both objects is identical whereas the respective coordinates differ. Therefore the chart(s) must be incorrect.

In order to answer the question "what is not a PN" with the consequence to remove such an object from the respective list it is necessary to indicate what is a normal PN. We have therefore collected (using the current review literature) the main properties which correspond to the general conception of PN - in the following summary the typical values are given, but they are changing very much during the evolution of the nebulae and of their nuclei (the nebular parameters correspond to the main nebular structure). We believe that only a thorough discussion of <u>all</u> properties can lead to the decision whether or not the respective object belongs (with a certain probability) to the class of PN. For such a discussion not only extensive observations but also a comprehensive theory are necessary.

REFERENCES

Allen D.A.,1984,Proc.ASA 5(3),369.
Bhatt H.C.,Mallik D.C.V.,1986,Astron.Astrophys.168,248.

TABLE 3 MISIDENTIFIED PLANETARY NEBULAE

Design.	Name Remarks and references
1 -0.1	Bl 3-11 The CGPN identif.chart incorrect (Sanduleak,1976). Misclassified PN (Paper II).
2 +1.1	H 2-20 Probably incorrect identification,the object is a A0 star without em.lines (Lutz,Kaler,1983).
59 -1.1	He 1-3 Misidentified in CGPN,correct chart given by Sabbadin,Bianchini (1979).
164+31.1	NGC 2474/75 Not identical with NGC 2474/75 (Barbieri,Sulentic,1977). New name JnEr 1 proposed.
324 -1.1	He 2-133 Finding chart incorrect,the PN is located 1.5mm to the north and 1.2mm to the east of the star (A0V) indicated in the CGPN (Lutz,Kaler,1983).
353 +8.1	MyCn 26 A- or early F-type star,no em.lines.A misidentification or not a PN (Lutz,Kaler,1983).
355 +2.3	Th 3-11 Not correctly identified on the CGPN chart (Sanduleak,1976).
356 -0.1	Th 3-34 Lies Sp object marked on CGPN chart (Allen,1979).
358 +1.4	Bl B Lies Np object marked on CGPN chart (Allen,1979).
358 -2.2	Bl 3-6 Sanduleak (priv.comm.)finds the object 30"E of the identification given in CGPN (Allen,1979). Misclassified PN (Paper II).

REFERENCES TO TABLE 3
Allen D.A.,1979,Obs.99,83.
Barbieri C.,Sulentic J.W.,1977,Publ.Astron.Soc.Pac.89,261.
Lutz J.H.,Kaler J.B.,1983,Publ.Astron.Soc.Pacific 95,739.
Sabbadin F.,Bianchini A.,1979,Publ.Astron.Soc.Pac.91,65.
Sanduleak N.,1976,Publ.Warner Swasey Obs.2,No.3,55.

Kohoutek L.,1978,IAU Symp.No.76 (ed.Y.Terzian), D.Reidel
   Publ.Comp.,Dordrecht,Boston,p.47   (Paper I).
Kohoutek L.,1983,IAU Symp.No.103 (ed.D.R.Flower),D.Reidel
   Publ.Comp.,Dordrecht,Boston,London,p.17 (Paper II).
Lutz J.H.,1984,Astrophys.J.279,714.
Perek L.,Kohoutek L.,1967,Catalogue of Galactic Planetary
   Nebulae,Academia Praha.

PLANETARY NEBULA:

MORPHOLOGY: symmetrical shape (mostly circular or elliptical disc or ring, sometimes bipolar structure) with apparently sharp outer boundary, often multiple shells (main nebula + faint outer structure or halo)
- depending on: wavelength (stratification)
   intrinsic absorption
- reflecting orientation in space

DIMENSION: diam. 0.1pc - 0.2pc (limits ~0.005pc, ~1pc)
- depending on wavelength (stratification)

EL.DENSITY: $10^3 cm^{-3} - 10^4 cm^{-3}$ (but $<10^3 cm^{-3}$ and $>10^4 cm^{-3}$ possible for large and small nebulae)

EL.TEMPERATURE: $9000°K - 15000°K$ (limits $8000°K, 23000°K$)

TOTAL MASS: $0.1 M_\odot - 0.2 M_\odot$ (limits $\sim 0.001 M_\odot, \sim 1 M_\odot$)

EXP.VELOCITY: non-isotropic, ~25km/s (limits 4km/s, 60km/s)

SPECTRUM: Em.lines:
recombination lines mostly of H and He
collisionally excited (forbidden) lines of C,N,O,Ne,Mg,Si,S,Cl,Ar
fluorescent lines (rare) of OIII and NIII
- depending on: exc.conditions (exc.class)
   stratification
   chemical composition

$I([OIII]5007+4959)/I(H_\beta) \approx 1$ to 15
≈ 0 to 1 for very low-exc.nebulae

Continuum emission:
free-bound, free-free, two-quantum processes, emission from grains (dust)

CENTRAL STAR:

TEMPERATURE: $50000°K - 100000°K$ (limits $25000°K, \sim 200000°K$)

LUMINOSITY: $\sim 5 \times 10^3 L_\odot$ (limits $\sim 10^4 L_\odot, \sim 10^1 L_\odot$)

RADIUS: limits $\sim 0.005 R_\odot, \sim 1.5 R_\odot$

MASS: $\sim 0.6 M_\odot$ (progenitors from $0.8 M_\odot$ up to $6-8 M_\odot$)

MASS LOSS: $\sim 10^{-10} M_\odot/yr - 10^{-7} M_\odot/yr$

GRAVITY: $\log g \sim 4.5 - 7.0$

SPECTRUM: WR, Of, WR+Of, OVI, cont.
O, sdO, peculiar

CATALOGUES OF PLANETARY NEBULAE

Agnès Acker
Observatoire, Centre de Données de Strasbourg
11, Rue de l'Université
67000 Strasbourg
France

ABSTRACT. Firstly, the general requirements concerning catalogues are studied for planetary nebulae, in particular concerning the objects to be included in a catalogue of PN, their denominations, followed by reflexions about the afterlife and computerized versions of a catalogue. Then, the basic elements constituting a catalogue of PN are analyzed, and the available data are looked at each time.

A. INTRODUCTION

The principal aim of a catalogue is to identify a large number of objects having the same nature, and to present and preserve all observational data available for these objects. Planetary nebulae (PN) are particularly interesting to catalogue. The sample of observable PN constitutes a highly representative fraction of the PN system in the Galaxy; as the detection and analysis of PN is done essentially using emission lines, the survey can be done at great distance, enabling the group to be studied as representing a population of the old disk of the Galaxy.

Planetary nebulae are fascinating objects, varied, not easily classifiable, interesting in all spectral regions, each area bringing some specific information (dust in the infrared, molecules in the radio, the central star in the ultraviolet and the visible, ...), from all of which the object called "planetary nebula" can be perceived and described.

As of today, two main catalogues concerning PN exist: "The Catalogue of Galactic Planetary Nebulae" (CGPN), by Perek and Kohoutek (1967), containing 1036 objects and 477 bibliographical references, and "The Catalogue of the Central Stars of True and Possible PN" by Acker et al (1982), containing 480 stars of which 95 have been discovered after 1967, and 894 references. The other existing catalogues give only specific data of samples of PN observed in various spectral ranges. Before listing the catalogues presently available, I shall go over general characteristics defining a catalogue.

given as "PN" in various lists in fact turn out to be symbiotic stars or galaxies or other kinds of object. Other errors may be typographical or clerical, misidentification (BD-12°133 instead of BD-12°134 for the nucleus of NGC 246), or wrong data (the visual magnitude of the central star of NGC 2440, accepted as near to 14 up until 1986, is in fact much fainter, near to 19). As noted by Jaschek, once the error is detected, it usually has a long, happy life, because scientific journals do not like to publish errata, and because, even if published, errata are largely ignored by users. Thus, we still find the denominational error attributing to PN 164+31°1 = VV 47 the name NGC 2474-75 which refers to a nearby pair of galaxies.

As observations are mainly taken with the help of a catalogue, a catalogue becomes out of date all the faster according to how long it has been awaited and how good it is ! This explains how, since the publication of Perek and Kohoutek's catalogue, the annual number of publications devoted to planetary nebulae rose from 35 in 1965 - 67 to 102 in 1968-70 (see fig.1). The large amount of relevant data is more and more difficult to present in its entirety in a catalogue; especially when it is a case of complex images in two dimensions; in this case it is therefore impossible to give in a catalogue more than the title ("label") of a work, and its bibliographical reference.

3. Updating and error correction are facilitated in **computerized versions** of a printed catalogue, and such versions also give rapid access to stored data and sampling facilities. However, the lifetime of a magnetic tape seems to be about one order of magnitude shorter than that of a printed catalogue, and another disadvantage pointed out by Jaschek is that the computer version escapes the control of its author, being very easily changeable by anyone who has access to the tape; the replacement of tapes by videodisks should overcome some of these difficulties. Even so, it is never possible to give the same information in a computer version as in printed versions : for instance, only printed versions can contain finding charts, spectroscopic recordings, detailed qualitative notes and introductions, ...

Figure 1 -

Evolution of the annual number of papers devoted to planetary nebulae

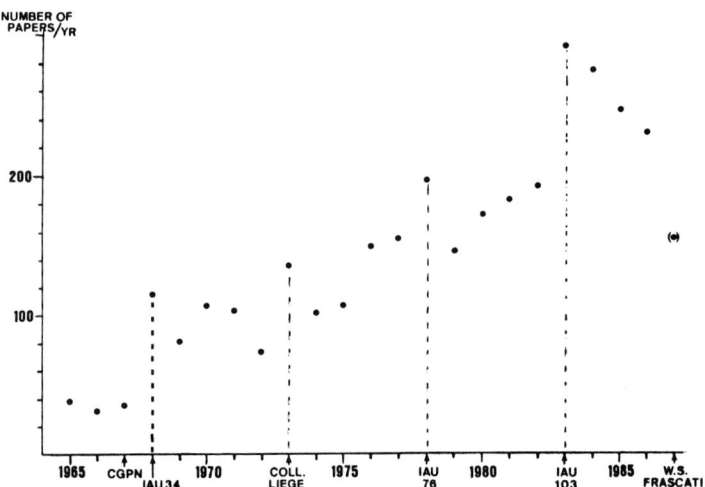

C. BASIC ELEMENTS OF A CATALOGUE OF PLANETARY NEBULAE

The essential observational data which should appear in a catalogue of PN concern the position and the structure, velocities, fluxes in various spectral ranges, and the central stars.

1. Positions and identification; dimensions and morphology

. **Optical coordinates** of a large number of PN have been determined since 1969 by various authors (Table 1), by measuring the positions of the PN on Palomar or ESO sky survey prints or plates, the calibration being done using known stars in each field. The accuracy is of about 0.1 arcmn.

Table 1 - OPTICAL COORDINATES OF PLANETARY NEBULAE

Number of objects	References
32	Higgs L.A., 1969, J. Roy. Astron. Soc. Can. **63**, 200-202
153	Milne D.K., 1973, Astron. J. **78**, 239-242
345	Milne D.K., 1976, Astron. J. **81**, 753-758
84	Blackwell S.R., Purton C.R., 1981, Astron. Astrophys. Suppl. **46**, 181-183
722	Lauberts A., 1982, The ESO/Uppsala Survey of the ESO (B) Atlas.

The positions are also determined from the radio maps for about 450 objects by the authors cited in Table 4. The VLA position is given at $\pm$ 0.5 arcsec; these coordinates can be quite different from the optical ones, in particular because the maximum emission is not the same in various spectral ranges (in particular, in the case of IC 2120). The same remark can be applied to the coordinates supplied by IRAS (divergence with the optical positions for 4-2°1, 254 + 5°1, 299-1°1).

. The **identification** of the PN on "finding charts" is unclear or inexistent in 230 cases (16% of the whole sample), in particular for the crowded area in the Milky Way and for the objects without an identification chart in the discovery papers. This situation will be clarified for most of the objects observed by Acker and Stenholm (spectroscopic survey); besides this, A. Gutierrez-Moreno, B.M. Lasker and T.D. Kinman have completed 43 sets of photographs in the direction of the galactic center, and are preparing a second program of 52 sets.

New finding charts will be prepared by C. Schohn for the Strasbourg - ESO Catalogue of PN; southern and equatorial PN will be identified on ESO and SRJ plates; but a homogenous solution has not yet been found for the northern PN (about 400), as the new Palomar Sky Survey is not yet ready.

. **The dimensions and structure** of PN are poorly known, for two main reasons:
- the nebular contours are badly defined, as the dimension and the morphology differs from one ion to the other, and therefore depend on the spectral area of the observation and on the exposure time. In addition, radio pictures are often much more symetric than optical ones, because the extinction in the visible is quite strong (case of NGC 7027).
- PN are distant objects; almost half of the known PN are unresolved, starlike

objects, obviating the study of any properties that require knowledge of their angular sizes; these include the absolute fluxes, the mass, the distance, ...

An effort has been made in this direction, thanks to **radio interferometric observations**, which have been able to resolve very compact PN and to determine their angular size with an angular resolution of 0.1 to 0.5 arcsec (see Terzian's work since 1979; later: Isaacman, 1984; Kwok, 1981 and 1985 - references in Table 4).

Other substantial progress concerning the dimensions and the morphology of PN has been made in the optical field with the help of CCD images; about 100 PN have been measured up to now, using this technique, in particular by Jewitt, Chu, Grewing and their collaborators. It has been found that about 20% of the well-studied PN have two or even more shells, and are very often surrounded by large halos.

For each value given for the dimension in a catalogue, it is indispensable to specify whether it concerns a radio or optical measurement. Morphological information can be specified by a summary description (no definitive classification exists up to now); but above all by a high quality image given in the "Finding Charts".

2. Radial and expansion velocities

Contrary to the shape and size of PN, velocities are objective data, which can be determined without ambiguity. The radial velocities of 532 PN have been presented in the catalogue of Schneider et al (1983), who have made a critical study of all the values published in the literature and have calculated mean values. New radial velocities have since been published for about fifteen PN.

Expansion velocities have been compiled for 165 PN in the catalogue of Sabbadin (1984), and more recently by Weinberger who has now listed 237 objects in his unpublished catalogue.

3. Fluxes and line intensities

These very important data have been increasing over the last ten years, due to more rapid and luminous instruments, and due to the amount of spatial results (IUE, IRAS).

. **Absolute Hβ fluxes** are known today for about 450 objects (see Table 2); Kaler and Cahn have developed an unpublished catalogue of data relative to the distances and temperatures of PN, listing in particular absolute Hβ fluxes, with errors, corrected to the modern photometric standard.

. **Relative line intensities in the optical range**
The reference work is Kaler's catalogue (1976), giving line strengths for 335 PN, as observed by various workers; Kaler is still updating this compilation, and has plans to publish a second edition of his catalogue. Spectroscopic surveys are conducted by several authors, in particular by Aller and Keyes (results

Table 2 - ABSOLUTE Hβ FLUXES

Number of objects	References
116	Perek L., 1971, B.A.C. **22**, 103-107
≈ 300	Cahn J., Kaler J.B., 1971, Astrophys. J. Suppl. **22**, 319-368
25	Kaler J.B., 1976, Astrophys. J., **210**, 113-119
33	Torres-Peimbert S., Peimbert M., 1977, Rev. Mex. Astron. **2**, 181-207
44	Barker T., 1978, Astrophys. J., **219**, 914-930
19	Kaler J.B., 1980; Astrophys. J. **239**, 78-88
30	Kohoutek L., Martin W., 1981, Astron. Astrophys. Suppl. **44**, 325-328
55	Carrasco L., Serrano A., Costero R., 1983, Rev. Mex. Astron. **8**, 187
72	Webster B.L., 1983, P.A.S.P. **95**, 610-613
31	Kaler J.B., 1983, Astrophys. J. **264**, 594-598
57	Kaler J.B., 1983, Astrophys. J. **271**, 188-220
49	Cahn J., 1984, Astrophys. J. **279**, 304-309
51	Shaw R.A., Kaler J.B., 1985, Astrophys. J., **295**, 537-546
134	Shaw R.A., 1985, Thesis, University of Illinois at Urbana-Champaign
14	Gutierrez-Moreno A., Moreno H., Cortes C., 1985, P.A.S.P. **97**, 397-403
13	Viadana L., de Freitas Pacheco J.A., 1985, Publ. Observatorio Nacional CNPq - Brazil, N° 10
102	Acker A., Stenholm B., 1987, Spectroscopic survey of planetary nebulae, private communication

For 450 objects, the absolute Hβ flux has been measured

Table 3 - INFRARED DATA

Number of objects (observed at various near - IR bands)		References
80	(H, K, L)	Allen D.A., 1973, M.N.R.A.S **161**, 145-166
81	(J, H, K, L)	Allen D.A., Glass I.S., 1974, M.N.R.A.S. **167**, 337-350
51	(upper limit at K)	
42	(J, H, K, L)	Allen D.A., 1974, M.N.R.A.S. **168**, 1-13
95	(upper limit at K)	
113	(H, K, L)	Cohen M. Barlow M.J., 1974, Astrophys. J. **193**, 401-418
16	(J, H, K, L)	Allen D.A., Glass I.S., 1975, M.N.R.A.S. **170**, 579-587
40	(upper limit at K)	
17	(J, H, K, L, M)	Phillips J.P., Sanchez Magro C., Martinez Roger C., 1984, Astron. Astrophys. **133**, 395-402
80	(J, H, K)	Whitelock P.A., 1985, M.N.R.A.S. **213**, 59-69
97	(J, H, K, L)	Persi A., Preite Martinez A., Ferrari Toniono M., Spinoglio L., 1987, "Planetary and proto-planetary nebulae : from IRAS to ISO", Preite Martinez A., Reidel Dordrecht, p. 221
		Preite Martinez A., 1987, Near IR photometry of planetary nebulae, private communication

For 418 objects, near - IR data are known (186 of these objects with only upper limit of K)

published for 51 PN up to now), and Acker and Stenholm (results so far obtained for 750 PN at La Silla, ESO, Chile, and 180 PN at the Observatoire de Haute-Provence, France).

. Ultraviolet fluxes

In 1985, more than 150 PN have been observed with IUE; results have appeared in various papers, but an "atlas of IUE spectra of PN" has not yet appeared. In the forthcoming PN catalogue, only the IUE Merged Log will be given, specifying high- or low-resolution spectra.

. Infrared fluxes

As shown on Table 3, more than 400 objects have up to now been observed in the **near-IR**; but reliable data are known for only about half of these objects. In addition, from 1981 to 1986, several authors (Aitken, Roche, Whitmore) have done $8 - 13\,\mu m$ spectral observations for about 30 PN.

Since 1983, however, the most important data in the IR are those collected by **IRAS**; up to now, about 950 IRAS sources have been identified as being PN. According to a sample of 100 PN, fluxes are given with the following qualities:

Quality	$12\,\mu m$	$25\,\mu m$	$60\,\mu m$	$100\,\mu m$
1 (upper limit)	55%	9%	15%	79%
2	4%	6%	7%	6%
3	41%	85%	78%	15%

Besides this, several hundred IRAS sources without optical counterparts presented IRAS colors typical of PN, and should be possible PN.

Table 4	Number of objects	Frequency 4.9 GHz (6 cm)	14.7 GHz (2 cm)	Others	References
RADIO DATA	256		X		Rubin R.H., 1970, Astron. Astrophys. **8**, 171-180
FOR PLANETARY	121			X	Higgs L.A., 1971, M.N.R.A.S. **153**, 315-336
NEBULAE	74			X	Aller L.H., Milne D.K., 1972, Austr. J. Phys. **25**, 91-98
	165	X			Milne D.K., Aller L.H., 1975, Astron. Astrophys. **38**, 183-196
	167	X			Milne D.K., 1979, Astron. Astrophys. Suppl. **36**, 227-235
	98			X	Milne D.K., Webster B.L., 1979, Astron. Astrophys. Suppl. **36**, 169-171
	40	X		X	Kwok S., 1981, Astrophys. J. **250**, 232-239
	397		X		Milne D.K., Aller L.H., 1982, Astron. Astrophys. Suppl. **50**, 209-215
	42	X	X	X	Burton C.R., Feldman P.A., Marsh K.A., Allen D.A., Wright A.E., 1982, M.N.R.A.S. **198**, 321-338
	196			X	Calabretta M.R., 1982, M.N.R.A.S. **199**, 141-150
	42	X			Gathier R., Pottasch S.R., Gross W.M., Gorkom J.H., 1983, Astron. Astrophys. **128**, 325-334
	62	X		X	Isaacman R., 1984, M.N.R.A.S. **208**, 399-408
	10	X		X	Kwok S., 1985, Astron. J. **90**, 49-58
	≈ 400	X			Zijlstra A., 1987, A VLA Survey of planetary nebulae, Private communication

For 439 objects, the radio flux is known, essentially at 5 GHz

. **Radio fluxes**

The first important catalogue of radio data was done by Higgs (1971) for 557 PN. Accurate radio fluxes are known for about 450 objects (see Table 4). A VLA survey is being conducted by A.Zijlstra since 1986 and intends to cover the whole population of PN; the observations and reductions should be finished at the end of 1987; the results (total fluxes accurate to 5%) will appear in the forthcoming Strasbourg-ESO catalogue of PN.

Low frequency (< 600 MHz) observations of PN have been reported in particular by Calabretta (1982, reference on Table 4, with references to previous works therein).

4. Central stars

A compilation of data for 480 nuclei of PN was published by Acker et al (1982), as finding charts for all these stars; "new" nuclei can now be seen on good quality ESO charts in particular (see the case of 298-4°1 = NGC 4071). On the other hand, the true central star of very extended PN often remains unknown (as in the case of 27+16°1 = DeHt 2). One cannot say that the most probable candidate is the bluest star in the central region, because in several large PN, the visible central star is a cold giant star (see the case of A 35, LoTr 5, ...).

A lot of spectroscopic observations have been done, by Mendez in particular. Kaler completed an "Atlas of PN central star spectra", containing data on 76 stars from 72 references.

5. The PN in Local Group Galaxies

are catalogued by Jacoby and Ford and their collaborators. A catalogue of 315 PN in M31 appeared in 1978 (ApJ Suppl.38), and was supplemented by 19 new PN in 1982 (ApJ 256) and by 37 PN in 1987 (ApJ 317), bringing the total number of PN catalogued in M31 to 371. In addition, several papers giving finding charts and positions for PN in M 32, NGC 147, NGC 185, and NGC 205 have been published. The same authors plan to publish a catalogue of the positions and magnitudes of 185 PN in M 81 and of 200 PN in NGC 5128.

D. THE STRASBOURG - ESO CATALOGUE OF GALACTIC PLANETARY NEBULAE

The new "Strasbourg - ESO Catalogue of Galactic Planetary Nebulae" which is in preparation since 1984 and which should appear in 1988, will be presented in book form; excerpts from the preliminary draft containing 102 PN are available now, at Symposium 103 here in Mexico. The content of the forthcoming book is described in the "Announcement" published at the Vulcano Workshop (Acker, 1987). For those who wish, a computer version will be available, in the form of twelve specific "files", presenting the various data for PN ordered by their galactic PK number.

The data contained in the book (and in the files) are specified by bibliographical references collected at the **"Centre de Données de Strasbourg"** (CDS). From 1965 to mid-1987, the papers concerning individual PN or lists of PN represent a total of 3324 (excluding papers reporting only theoretical studies). As shown on figure 1, the annual number of publications is marked by the astronomical events devoted to PN, and reflects the publication of the Proceedings of the IAU Symposiums on PN. Since 1970, thanks to the advent of "SIMBAD" at the CDS, it is possible to search the astronomical literature by object; this is how we have been able to see that the object with the largest number of bibliographical references in the whole Stellar Data Center which, on October 1st 1987, contained 688 520 objects, of which 90 000 are non-stellar objects, of which 70 000 are galaxies, is NGC 7027 with 730 references from 1965 to now.

Another interesting fact is the large number of papers devoted to the **determination of distances.** Since the IAU Symposium 76 held at Ithaca in 1977, individual distances of PN have been given in 17 papers, and 9 statistical distance scales have been published; thus, the individual distance has been published for 111 PN, and one or several statistical values have been given for 679 PN: the distance is known for more objects than is the exact radius (even though it is necessary for most distance scales) or the spectrum ! This does not prevent the distances from being extremely imprecise; improvements will only come from Hipparcos.

A good data catalogue will help towards the fruitful preparation of future science missions: Hipparcos (Ariane), HST (Shuttle) and ISO (Ariane), scheduled respectively in April 1989, mid 1989 and 1992-93.

A catalogue should set the record straight for current knowledge, but should also be a new departure point, for complementary observations and for research into new properties of PN.

REFERENCES

Acker A. : 1987, "Planetary and Proto-Planetary Nebulae : from IRAS to ISO", p. 35-38, A. Preite-Martinez (ed.), D. Reidel Publishing Company

Acker A., Gleizes F., Chopinet M., Marcout J., Ochsenbein F., Roques J.M. : 1982, "Catalogue of the central stars of true and possible planetary nebulae", Publication Spéciale du C.D.S. N° 3 - Observatoire de Strasbourg, and Complements I (1982), II (1983), III (1984) - (480 stars)

Acker A., Marcout J., Ochsenbein F, Lortet M.C.: **1983,** "Index and Cross-identification of PN ", Astron. Astrophys. Suppl. **54,** 315-364

Fernandez A., Lortet M.C., Spite F. : 1983, Astron. Astrophys. Suppl. **52,** N° 4

Higgs L.A. : 1971, "Catalog of radio observations of PN and related optical data", P.A.B. Vol. 1, N° 1

Jaschek C. : 1984, Q. J. astr. Soc., **25,** 259-266

Kaler J.B. : 1976, "Catalog of relative emission line intensities observed in PN and diffuse nebulae", Astrophys. J. Suppl. **31,** 517-688

Perek L., Kohoutek L. : 1967, "Catalogue of galactic planetary nebulae", Praha, Academia Press, CSSR

Sabbadin F. : 1984, "Catalogue of expansion velocities of PN", Astron. Astrophys. Suppl. **58,** 273-285

Schneider S.E., Terzian Y., Purgathofer A., Perinotto M. : 1983, "Catalog of radial velocities of PN", Astrophys. J. Suppl. **52,** 399-423

Wray J.D. : 1966, "Study of $H\alpha$ emission objects in the southern milky way", Thesis - Ph.D. - Northwestern University

NARROW BAND PHOTOMETRY AND MAPPING OF THE PLANETARY NEBULAE
NGC 6210 AND NGC 7009

E.E. Mendoza, C. Chavarría-K., and V.M. Arévalo
Instituto de Astronomía, Universidad Nal. Autónoma de México
P.O. Box 20-158
México 20, D.F. 01000

This paper is based upon observations carried out at the Observatorio Astronómico Nacional, San Pedro Mártir, B.C.N. México. They are of two types: (1) $\alpha(16)\Lambda(9)$-photometry, obtained with the 1.5-m Johnson telescope, and (2) maps at the wavelengths of Hα and $\lambda7751$ A [Ar III] emission lines secured with the 2.1-m telescope.

The results clearly indicate that planetary nebulae are extremely well separated from stellar objects in the $\alpha(16)\Lambda(9)$-plane: they fall above and to the right of the extreme Herbig Ae/Be stars (Mendoza 1987: Proceedings IAU Coll. 92, "Physics of Be Stars"). This is so because the Hα total emission line in NGC 6210 and NGC 7009 is stronger than in stellar objects (Mendoza 1987: Revista Mexicana Astron. Astrof., 14, 310 and references therein), and because the $\lambda7751$ A [Ar III] falls in the short wavelength filter that defines the "blue" continuum of the $\Lambda(9)$-index. These characteristics easily distinguish planetary nebulae from stellar objects.

We have also developed a subroutine to obtain maps of extended sources with one dimensional detectors at any wavelengths, such as the He I $\lambda10830$ A. We have secured maps of Hα and $\lambda7751$ A [Ar III] emission lines to test our technique, which by the way, allows us to measure linearly ($\sim 1\%$) in a flux range of six decades.

The results of the maps yield that: (1) the flux at the center of NGC 6210 and NGC 7009 is higher by a factor of about 10^5 than at the edges; (2) structure in the surface brightness is easily detected, and (3) the $\alpha(16)$ and a new index defined for the [Ar III] emission line are weaker in those pixels that contain the central star.

The above indicates that the outlined procedure is satisfactory, thus, we plan to obtain maps of several planetary nebulae at different wavelengths, including the He I $\lambda10830$ A line.

NEW OBSERVATIONS OF PLANETARY NEBULAE S22 AND YM29

V.P.Arkhipova, T.A. Lozinskaya, E.I. Moskalenko, and
T.G. Sitnik
Sternberg Astronomical Institute
Moscow, USSR

ABSTRACT. Spectral and monochromatic observations of two thin-filamentary nebulae S22 and YM29 have been carried out in the lines [O III] 5007A, [N II] 6584A and [S II] 6717+6732A. Radial stratification of the emission typical for photoionization excitation has been found: displacement between [O III] - bright regions and [S II] - and [N II] - bright regions is equal $\Delta R \sim 0.1R$ (about 0.01 to 0.2 pc), the first ones being closer to the nuclei. Both nebulae are characterized by more diffuse morphology in the [O III] line and by thin-filamentary one in the [N II] and [S II] lines.

Both objects seem to be old planetary nebulae of PNI type probably having stellar wind from their nuclei.

Optical 4000-7000A and UV 1150-3150A spectra of YM29 nucleus have been analyzed in looking for stellar wind indicators. The spectra seem to be similar to sdO spectra with black-body temperature of 58000 ± 6000 K. Overall spectral structure may probably show some hints of PCyg-type lines, but one needs to obtain spectra with better resolution and statistics. The nebula YM29 and its nucleus have been also observed in X-ray range 0.2 - 3.5 KeV, but clear indications on strong wind have not been found. There are neither compact nor diffuse sources with the 3σ upper limit for 0.2 - 3.5 KeV flux: $F_{nucl} \lesssim 1.4 \cdot 10^{-13}$ ergs/cm^{2}s and $F_{YM29} \lesssim 8 \cdot 10^{-14}$ ergs/cm^{2}s under "standard" spectrum $J(E) \sim E^{-1.5}$ photons/cm^{2}s and $N_H = 6 \cdot 10^{19}$ cm^{-2}.

The detailed results are given in Arkhipova, V.P., Lozinskaya, T.A., Moskalenko, E.I., 1986, Pisma Astron. Zh. (Sov. Astron. J. Lett.), v. 12, p. 890; Lozinskaya, T.A., Sitnik, T.G., Toropova, M.S., 1984, Pisma Astron. Zh. (Sov. Astron. J. Lett.,), v. 10, p. 122; Lozinskaya, T.A., Sitnik, T.G., Toropova, M.S., 1986, Astron. Zh. (Sov. Astron. J.), v. 63, p. 255.

TWO SOUTHERN LOW EXCITATION PLANETARY NEBULAE

H. Moreno, A. Gutiérrez-Moreno, and G. Cortés
Departamento de Astronomía
Universidad de Chile

ABSTRACT: Within a spectroscopic study of some southern planetary nebulae, we have observed 32 objects. Some of them are symbiotic or suspected symbiotic stars, and one (He 2-61) is evidently not an emission object.

We discuss here two nebulae with similar characteristics: He 2-138 and He 2-151. They are both classified by Stenholm and Acker (1987) as possible PN? with high density. Our observations cover the wavelength range $\lambda\lambda 3400$ to 8600 A. The main characteristics of the spectra are as follows: the continua are blue, reaching a maximum at about $\lambda 3650$ A; [O III] is not observed; [O II] $\lambda 3727$ is conspicuous; [N II] $\lambda 6584$ is comparable to Hα, though fainter in He 2-151 than in He 2-138; helium lines are not detected; the [S II] doublet at $\lambda\lambda 6717, 6731$ is clearly seen, being fairly well separated, with I(6731) > I(6717) for He 2-138; these lines are more blended in He 2-151.

He 2-138 has already been recognized as a low excitation PN; consequently, we may assume that He 2-151 falls in the same category.

A detailed study of both nebulae will be published elsewhere.

REFERENCES

Stenholm, B. and Acker, A. 1987, *Astron. Astrophys. Suppl. Series*, 68, 51.

SPECTROSCOPIC SURVEY OF PLANETARY NEBULAE

A. Acker and J. Köppen
Observatoire de Strasbourg, France
B. Stenholm
Lund Observatory, Sweden

ABSTRACT. Since 1984, we have undertaken a spectroscopic survey of planetary nebulae, both at La Silla (E.S.O.)., and at the Observatoire de Haute Provence (France). Up to now, the spectra of about 900 PN have been obtained; all 723 spectra observed at La Silla have been reduced and the line intensities have been measured for 250 of them; about one fourth of the 182 OHP spectra have been reduced.

1. Misclassified Planetary Nebulae

Through our survey and IRAS data, and following comments in the literature, we have shown that 202 objects are surely (and 60 others possibly) misclassified planetary nebulae.
About one third of these objects are in fact symbiotic or late-type stars; others are galaxies (19), H II regions (22), plate faults (10),...

2. Determination of Physical Properties of the Nebulae

A computer code has been developed which automatically analyses emission line spectra, using a single zone, constant temperature and density (model). For the measured spectrum, we deduced: interstellar extinction, electronic temperature and density, the abundances of all observed ions and elements, and Zanstra temperatures and luminosities of the central star. At each stage of the analysis the quality is assessed, as measured against the ideal case if all relevant lines were observed.
From the first sample of about 200 spectra, it seems that for 27 objects the abundances are well determined; for 46, the data could be better, and for other spectra, the parameters are poorly determined. Say, one third of all observed objects can be used for further work.
This very homogeneous and reliable material will thus be treated statistically, with the collaboration of G. Jasniewicz, regarding galactic gradients and problems of stellar evolution.

THE PROBABLE LOW EXCITATION PLANETARY NEBULA M3-44

A. Gutiérrez-Moreno, H. Moreno, and G. Cortés
Departamento de Astronomía
Universidad de Chile

ABSTRACT: The Hα emission object M3-44 was observed photographically and spectroscopically. Photographs were obtained in the visual region, in [O III] λ5007, and in Hα. The visual image is faint and has almost stellar aspect; the object is not visible in [O III], but the Hα image is very intense, with a diameter of the order of 5".

A spectrum was obtained in the region λλ4300-7000 A. The continuum is practically 0 at λ4300, reaching a value of the order of 1.5×10^{-15} at λ7000; this can be attributed to reddening. The only lines visible in this part of the spectrum are the hydrogen lines, the [N II] lines λλ6548 and 6584 (with λ5755 almost lost in the noise), and the [S II] lines λλ6717 and 6731. The Balmer decrement is very steep, implying a high reddening (in agreement with the aspect of the continuum).

The characteristics of the observed spectrum are very similar to those of He 2-138 and He 2-151, two very low excitation PN. This preliminary classification will be confirmed by the observations of other portions of the visual spectrum (λλ3400 to 4300 and 7000 to 8500 A), to study the presence and intensity of other relevant lines.

OPTICAL AND INFRARED OBSERVATIONS OF THE PECULIAR PLANETARY NEBULA
HE2-442

V.P. Arkhipova, V.F. Yesipov, and B.F. Yudin
Sternberg Astronomical Institute
117234 Moscow, USSR

ABSTRACT. Infrared observations of He2-442 carried out at Crimean Station of Sternberg Institute during 1984-1986 revealed that in the past ten years the 1-10 μm flux has decreased by factor ~ 2.4, the colour temperature determined from (K-L) index has decreased by ~ 300 K, and the infrared energy distribution has been changed also essentially. However, the definite cool star features, for example, Mira-like variability was not yet recorded.

In the optical spectra of He2-442 the low and high ionization emission lines of FeII, HI, HeII, [ArV], [FeVII] and forbidden lines with different critical density from $\sim 5 \times 10^3$ cm^{-3} for [S II] 6716 A to $\sim 5 \times 10^6$ cm^{-3} for [ArV] 7005 were identified. Thus in the gaseous envelope of He2-442 the regions of low and high density exist. There were also observed the small brightness variations of the star ($\Delta V = 0\overset{m}{.}25$ between two nights in 1983).

It was found that He2-442 contains two emission line sources: a large ($\lesssim 7"$) nebula of a low surface brightness and a stellar source inside of the latter. However spatial structure of He2-442 is not yet investigated in detail.

THE PHOTOMETRIC (UBV) STUDY OF THE PLANETARY NEBULAE VARIABILITY IN 1968-1987

E.B. Kostyakova
Sternberg State Astronomical Institute, Moscow, USSR

ABSTRACT. The photoelectric UBV-observations of planetary nebulae variability, begun in 1968 at Crimean Station of Sternberg State Astronomical Institute (USSR) and Skalnate Pleso Observatory (Czechoslovakia), were carried out during 20 years, 1987 inclusive.

The results of our UBV-observations (1968-87) present the total UBV-magnitudes (the nebula plus the central star) of 6 nebulae: NGC 6572, Hu 2-1, NGC 6891, IC 3568, NGC 6720 and NGC 6543. The values Δv, Δb, and Δu, estimated relative to the comparison stars, give the average data for each observation season: usually several observation nights during one moonless period; each nebula was observed several times per night. The accuracy of the given data is generally several thousandths of one stellar magnitude. The data concerning the well-known nebula IC 4997, showed the most marked variability in brightness and spectrum, are presented separately, in the next contributed paper.

The obtained results permit to draw the following conclusions:

1) The nebula NGC 6572 after 1968 became clearly brighter by $0^{m}.3 - 0^{m}.5$ in visual light. However, from about 1978-79 its V-brightness either stabilized, or even began to weaken.

2) The nebula Hu 2-1 and 6891 showed fluctuations (of similar character) of the UBV-brightness in the range $0^{m}.2 - 0^{m}.4$.

3) The nebulae IC 3568 and NGC 6720, suspected earlier in variability, during our observations showed no changes of total UBV-brightness $> 0^{m}.1$.

4) The nebula NGC 6543 was observed only from 1979; therefore conclusions on its variability would be yet premature.

Besides, from 1972 we have studied systematically the changes in spectra of the same objects for comparing then with the changes in UBV-brightness and analysing the causes of revealed variability.

THE PHOTOMETRIC AND SPECTRAL VARIABILITY OF THE PLANETARY NEBULA IC 4997

E.B. Kostyakova
Sternberg State Astronomical Institute, Moscow, USSR

ABSTRACT. Among the planetary nebulae showing noticeable long-time variations of brightness (see our foregoing contributed paper; *Astron. Circ. USSR*, No. 1430, 3, 1986, and earlier publications) the young stellar planetary IC 4997 is the most prominent. Our photoelectric observations revealed its rather surprising behaviour. During 1968-85 its total UBV-brightness was monotonously decreasing, especially in filter V: the reduction of the value Δv amounted $\sim 0^m.5-0^m.6$. The colour indices of the planetary showed that during the observation period the object as a whole, became definitely bluer. In 1985-86 our observations revealed an unexpected stop of mentioned brightness decrease, but in 1987 the object appeared to start brightening.

Beginning from 1972, we have systematically studied the spectrum of IC 4997, using the 50-cm Maksutov telescope with objective prism. These observations allowed to reveal noticeable changes both in certain nebular lines and in continuous spectrum of the central star. In particular there was found a systematical change of the line intensity relation:

$$R = \frac{F(\lambda 4363[O\ III])}{F(H\gamma)},$$

indicating the excitation degree of the nebular spectrum. There were detected a noticeable increase of the absolute energy flux in the emission line $\lambda 4363[O\ III]$, a marked growth of the relation:

$$\frac{F(\lambda 4363)}{F(\lambda 4959)} [O\ III],$$

and a rise of the relation:

$$\frac{F(\lambda 4363\ [O\ III])}{F(\lambda 3727\ [O\ II])}$$

which indicates a growth of the ionization degree in the nebula.

Besides, we observed an increase of the central star radiation flux in $\lambda_{ef} \cong 4220$ A. Noteworthy, that during the whole observation period the intensity of hydrogen lines has almost not changed, but the intensity of helium line $\lambda 4471$ has somewhat increased.

The results of photoelectric and spectral observations permit to conclude that the revealed changes in brightness and spectrum of the nebula IC 4997 evidently reflect the changes in the central star radiation field, namely, the relative increase of ultraviolet radiation in its spectrum, which can indicate a growth of its effective temperature.

PLANETARY NEBULA HE 2-467 TURNED OUT TO BE THE SYMBIOTIC STAR WITH A PERIOD ABOUT 500 DAYS

V.P. Arkhipova and R.I. Noskova
Sternberg State Astronomical Institute
117234 Moscow, USSR

ABSTRACT. By means of broad and narrow-band photometry in UBV spectral diapason the variability of the object He 2-467 earlier classified as peculiar central star of planetary nebula has been revealed. The light amplitude significantly decreases with the wavelength, from $1^{m}_{.}7$ in U-band to $0^{m}_{.}3$ in V. The brightness variations were found to be periodic, with P $\approx$ 500 days. The observations of He 2-467 have been interpreted using the model of binary consisting of very hot subdwarf and G5 II giant. The parameters of both components have been derived. The hot star is probably the evolved low mass nucleus of planetary nebula already dissipated. The periodic variations in U-band may be the result of the reflection effect due to the presence of hot extended region on the side of cold star facing the subdwarf. The subdwarf UV-flux can heat and ionize the upper atmosphere of the giant giving birth to the emission lines and Balmer continuum. The yellow symbiotics to which He 2-467 belongs may be predecessors of red symbiotics with M-giants.

INFRARED ENERGY DISTRIBUTION OF SELECTED PLANETARY NEBULAE

P. Persi, A. Preite-Martinez, M. Ferrari-Toniolo
Istituto Astrofisica Spaziale, CNR,
C.P. 67, 00044 Frascati, Italy

ABSTRACT. Using the IRAS and near-IR photometry obtained at the 1-m, 3.6-m ESO telescopes, and 1.5-m TIRGO telescope, we will discuss the 1 - 100 microns energy distribution of selected PNe.
　　In addition, 8 - 13 microns spectrophotometry and around the Br γ line (2.168 µm) taken with CVF is also reported for the selected PNe.

NEAR INFRARED PHOTOMETRY OF COMPACT PLANETARY NEBULAE

M. Peña and S. Torres-Peimbert
Instituto de Astronomía
Universidad Nacional Autónoma de México
México

ABSTRACT. The study of young planetary nebulae can provide information about the ejection process of the nebular material as well as on nebular evolution and the enrichment of the interstellar medium. The youngest PN are expected to be compact and very dense, they usually show signs of variability and IR excess due to the presence of warm dust which, in some cases, seems to be mixed with the ionized gas.

We have started an extensive program to try to obtain IR data of a large sample of PN. We are interested in investigating a set of young, compact, high density PN in order to study their general characteristics that well help us understand better the infrared emission mechanism and their evolution.

In this work, we present JHK photometry for more than 30 compact ($\phi < 10"$) planetary nebulae, which have been obtained with the infrared photometric system at the 2.1-m telescope of the Observatorio Astronómico Nacional in Baja California.

The unreddened colors $(J-H)_o$ and $(H-K)_o$ have been determined for all the objects with E(B-V) available.

From these data we find:
1) In the color-color diagram the PN are grouped in an area about $(J-H)_o = -0.51$ and $(H-K)_o = +0.68$.
2) High density nebulae ($\log N_e \geq 4$) have an excess in the mean value of $(H-K)_o$ over low density objects.
3) Most high density nebulae show and IR excess in H and K filters over the expected free-free emission that is not present in low density objects. This excess is interpreted as due to warm dust in the high density (younger) objects.
4) A good correlation of the $(J-H)_o$ color with the He^+ ionic abundance is found. This is due to the He^+ 1.08 μm emission line which is included in the J band.

THE NEAR INFRARED SPECTRUM OF NGC7027

Jean-Paul BALUTEAU
Observatoire de Haute-Provence
04870-Saint Michel l'Observatoire, FRANCE

Daniel PEQUIGNOT
Observatoire de Paris-Meudon
92195 Meudon Principal Cédex, FRANCE

NGC7027 has been observed at the 193cm telescope of OHP in the range 6800 - 10500 Å, using the CARELEC spectrograph equiped with a RCA CCD. The dispersion corresponded to a resolution 1 Å/px.

More than 100 recombination lines belonging to the (3,n) series of HI, the (5,n) and (6,n) series of HeII and to 7 out of the 10 (3ℓ, nℓ') series of HeI could be identified. Using case B recombination theory (Hummer, Storey, 1987) and the nebular physical conditions from a recent model (Gruenwald, Péquignot, 1987, these proceedings), it is found that the observed Paschen decrement agrees with theory within calibration and atmospheric correction uncertainties ($\sim$ 3% for n = 7-19 and $\sim$ 7% for n = 20-40). However unidentified blends must be postulated with n = 13, 26, 34-36. There is no indication of the high electron densities found by Kaler et al (1976) from optical decrements. The Paschen series provides calibration of the near infrared to better than 5%. The HeII series are consistent with an unreddened ratio :

I(HeII 4686) / I(Hβ) = 0.50 ± .04

in agreement with the optical data of Kaler et al (1976). The HeI series are identified for the first time and are consistent with a pure case B recombination (Storey, 1987, private communication) unreddened ratio :

I(HeI 4471) / I(Hβ) = (3.35 ± .30) x 10^{-2},

indicating that λ4471 is not collisionally enhanced by more than 10%. The continuum agrees with recombination theory within 3% for wavelengths $\geqslant$ 7650 Å but an excess $\sim$ 10% is found for λ < 7600 Å.

Many forbidden lines are measured for the first time. All multiplet line ratios available (CI, NI, SII, ClII, SIII, ArIII, ArIV, ClIV) are consistent with theory within observational uncertainty (often $\leqslant$ 3%) except that [SII]10370 may be $\sim$ 10% weaker than predicted. [NiII]7378 and most lines of multiplets [FeII]1F, 13F, 14F are observed, providing the first detection of these ions in this nebula. Several strong lines are lacking convincing identification.

A VLA RADIO CONTINUUM SURVEY OF PLANETARY NEBULAE

A. Zijlstra[1,2], S.R. Pottasch[1], and C. Bignell[2]
1. Kapteyn Astron. Inst., Groningen
2. NRAO (VLA) Socorro, NM. USA

ABSTRACT. With the Very Large Array it is now possible to make high resolution radio continuum maps with sensitivity less than a milliJansky in an observation of only 5 minutes. We have used this so-called snapshot capability to measure about 400 PN north of declination -35. Most of the measurements were carried out at 6 cm. Some of the stronger sources were observed at several frequencies. Most sources were detected, however many nebulae were too weak to map in detail. The resolution ranges from 1.5" to 1', depending on the size of the PN. The selected PN have sizes in the range from 4" to 6'.

From this sample we will get the following results: 1) accurate values for the total flux densities will be obtained. Most published radio flux densities of planetary nebulae were obtained with single dish radio telescopes which compared to VLA measurements have a much lower sensitivity and suffur from confusion. 2) The total radio flux density in combination with the angular size of the nebulae allows physical quantities to be derived, such as ionized mass and density as function of the distance. 3) A comparison with the total far infrared flux yields the infrared excess (IRE). The IRE is an indication of the evolutionary stage of the PN. For nebulae which are optically thick to the stellar radiation, the IRE also gives an estimate for the temperature of the central star. 4) The morphology of the nebulae can be used to model the density distribution. In many cases the morphology also distinguishes H II regions from PN.

All observations have been completed, but data reduction is still in progress. As a preliminary result we found that pk 132-0.1, 169-0.1 (Ic2120), 195-0.1, 35-0.1, 223-2.1, 118+2.1 and 176+0.1, all of which are classified as planetary nebulae, probably are H II regions. We plan to publish the final results in the form of a catalog. This catalog will contain total radio flux densities or upper limits for all of the measured nebulae, and maps for those for which we had sufficient resolution and adequate signal to noise. Preliminary results will be presented at the conference.

A NEW PLANETARY NEBULA

E. Capellaro, M. Turatto and F. Sabbadin
Asiago Astrophysical Observatory, Asiago (VI) Italy

ABSTRACT. The object ($\alpha_{1950} = 18^h04^m.3$; $\delta_{1950} = -8°56!4$) was discovered in a 103a-E+RG 1 objective prism plate taken with the 92/67-cm Schmidt telescope of the Astrophysical Observatory of Asiago (Italy). It presents only the Hα emission and no stellar continuum; following Kohoutek (1965, 1969, 1972) it is a *bona fide* planetary nebula. This classification is confirmed by the appearance of the object in the red and infrared plates of the Near Infrared Photographic Survey of the galactic plane (Sabbadin, 1986): it is quite bright in the red plate and almost invisible in the infrared one.

Figure 1 is a Hα + [N II] interference filter CCD frame of the new planetary nebula obtained at the Cassegrain focus of the 182-cm telescope of Asiago Observatory at Cima Ekar. The non-stellar nature of the object is confirmed by its FWHM = 3.5 arcsec, to be compared with FWHM = 2.2 arcsec of the field stars. Moreover, the object appears slightly elongated in P.A. ≅ 145°.

A detailed spectroscopic study of this compact planetary nebula is in progress at the Astrophysical Observatory of Asiago.

REFERENCES

Kohoutek, L. 1965, *Bull. Astron. Inst. Czech.*, 16, 221.
Kohoutek, L. 1969, *Bull. Astron. Inst. Czech.*, 20, 307.
Kohoutek, L. 1972, *Astron. Astrophs.*, 16, 291.
Sabbadin, F. 1986, *Astron. Astrophs. Suppl. Series*, 65, 301.

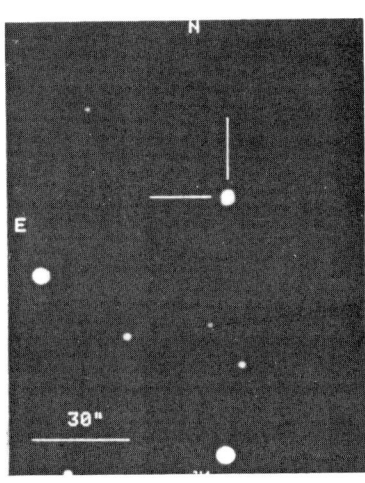

Fig. 1. Hα + [N II] CCD frame of the new planetary nebula.

IS NGC 2242 A NEW PLANETARY NEBULA?

Liu Jiying, Huang Yongwei, and Feng Xingchun
Beijing Observatory, Academia Sinica
China

ABSTRACT. An interesting object was found on the IIIaJ objective prism plate SP1270, taken on Jan. 12.6 UT, 1983 with the Beijing Observatory 60/90-cm Schmidt telescope plus $5°.3$ objective prism with a dispersion of 580 A/mm at Hγ. The emission lines [O III]$\lambda\lambda$4959+5007 and Hβ were extremely strong. The lines Hγ---Hδ and [O II]λ3727, all in emission, were broad and conspicuous. Two prominent emissions were tentatively identified as He II λ4686 and He II λ4542. Of all these lines none showed any noticeable redshift. It belongs to the Galaxy. The overall spectrum looked like that of a planetary nebula. But on the POSS overlay this object was designated as RNGC 2242 and ZWG 204.005. It was listed as a galaxy either in ZWG, or in A Master List of Nonstellar Optical Astronomical Objects, but it was absent from any previous catalogues of the planetary nebula.

The precise position of the central image is:
$\alpha = 6^h30^m28^s.06$ (1950)
$\delta = 44°48'58".2$ (1950)

The integrated photographic magnitude was $14^m.5$ given in the RNGC.

According to the above informations, we suppose that NGC 2242 may be a planetary nebula. Then we informed about it to Prof. He Xiang-Tao, Prof. He and H. Maehara *et al.* have made follow-up observations of this object with the Kiso 105-cm Schmidt telescope and the Okayama 188-cm and 91-cm telescope. Luminosity and color distributions and a small heliocentric velocity (-30 km/s) are all inconsistent with previous classification as a galaxy. NGC 2242 is probably a planetary nebula located at $\sim$ 2 kpc from the sun and at $\sim$ 500 pc above the galactic plane.

PLANETARY NEBULAE NEAR THE GALACTIC CENTER I: METHOD OF DISCOVERY AND
PRELIMINARY RESULTS

S.R. Pottasch[1], C. Bignell[2], R. Olling[1], and A.A. Zijlstra[1,2]
1. Kapteyn Astron. Inst., Groningen, The Netherlands
2. NRAO (VLA) Socorro, NM, USA

ABSTRACT. A method is described for finding planetary nebulae. Use is made of the far infrared IRAS colors and radio continuum measurements. The method is applied here to a region within 15° of the galactic center. The first results are given, including 36 new PN. The characteristics of the nebulae are described. While they are generally similar to known nebulae, the method of selection gives an emphasis to younger objects. A substantial number of the new nebulae may be in the transition phase between OH/IR stars and PN.

AN INFRARED SEARCH FOR NEW PLANETARY NEBULAE

S.R. Pottasch[1], C. Esteban[2], A. Manchado[2], and A. Mampaso[2]
1. Kapteyn Astron. Institute, Groningen, The Netherlands
2. Instituto de Astrofísica de Canarias, Tenerife, Spain

ABSTRACT. We present the first results of a large scale infrared search for new planetary nebulae. More than 1000 unidentified sources were selected from the IRS PSC having infrared colours similar to those of known PN's. Subsequent near-infrared photometry and optical spectroscopy will be made to investigate their precise nature. We report here one to five microns photometry of 30 such sources obtained with the 1.5-m CSM telescope (Tenerife, Spain). The preliminary results indicate that many of the observed sources have near infrared colours of heavily reddened PN's (A_V greater than 30 magnitudes in some cases), while a smaller fraction could represent obscured normal stars or giants surrounded by circumstellar envelopes.

Foreground: Reginald Dufour and Rafael Costero. Background: Malcolm Smith.

DISTANCES TO PLANETARY NEBULAE

Julie H. Lutz
Program in Astronomy
Washington State University
Pullman, WA 99164-2930
USA

ABSTRACT. Finding distances to planetary nebulae remains a frustrating undertaking, but significant progress has been made over the past several years. This review covers primarily work done on distances since 1980, with some references to earlier papers. Some interesting new methods have been tried recently and some methods that have been used for years have been refined. Missions such as the Hubble Space Telescope and Hipparcos may provide new data on distances. Advances in ground-based telescopes and instruments will make possible new studies of distances.

1. INTRODUCTION

Distances to planetary nebulae (PN) are a fundamental quantity but accurate methods for determining distances have remained elusive. Much progress has been made both in determining distances for individual objects and for methods that are applied to large numbers of PN. However, because PN are so diverse in terms of their physical parameters (optically thick or thin, morphology and filling factors, nebular mass, helium content, etc.), it is unlikely that there will ever be a single method for finding distances that can be applied to all PN. We will have to be content with applying appropriate methods to particular PN and remaining cognizant of the limitations and selection effects associated with each method.

The disagreement among investigators comes in the assessments of which methods are appropriate for which nebulae and the associated errors and limitations. Table 1, adapted from Gathier, Pottasch and Pel (1986), illustrates that the amount of disagreement between various distance determinations can be substantial.

In this review I will discuss the principles and limitations of various distance methods, particularly those that have appeared in the literature since 1980. To set the context, I will begin with a brief discussion of some earlier work and then proceed to more recent studies, grouped according to the techniques employed.

Table 1. Comparisons of Distances

Object	Distances (Kpc)
NGC 3918	1.60(1); 1.77(2); 0.80(3); 0.92(4); 0.58(5); 2.24(6)
NGC 5315	2.36(1); 4.05(2); 2.80(3), 0.69(5); 2.62(6)
NGC 6567	3.49(1); 3.55(2); 1.20(3); 2.08(4); 1.48(5); 1.68(6)

References

(1) Cahn and Kaler (1971)
(2) Milne and Aller (1975)
(3) Acker (1978)
(4) Maciel and Pottasch (1980)
(5) Daub (1982)
(6) Gathier et al. (1986)

2. STUDIES PRIOR TO 1980

Since the early 1970's major efforts have been made to determine distances to large numbers of PN. Cahn and Kaler (1971) used the photometric distance method to estimate distances (or upper limits to the distances) for more than 600 PN. The photometric distance method is based upon the assumptions that the mass of the nebular shell is the same for all PN and the nebulae are optically thin. The scale is calibrated by using one or more PN with distances determined by some "reliable" method such as trigonometric or spectroscopic parallaxes. Once the scale is calibrated, the only observations that are needed are the nebular H-beta flux, interstellar extinction and angular radius.

Cudworth (1974) used proper motions to determine statistical parallaxes for 62 PN. He divided the nebulae into two groups according to their classification as B ("binebulous") or C ("centric") morphologies as defined by Grieg (1971, 1972). The kinematics of the two groups came out slightly different, but the numbers of PN in each group and the errors of observation made Cudworth cautious about his general conclusions and the validity of the individual distances.

Acker (1978) combined all of the distance determinations available (both individual and statistical) into a synthetic scale that included 330 PN. Because of the heterogenity of the data, it is not particularly meaningful to compare the distances obtained by this method with those found in other studies. The paper provides a complete set of references for all of the individual and statistical distances published up to that time.

3. EXTINCTION DISTANCES

The first studies of extinction distances began in 1970's (Lutz 1973) and have continued with refinements to the present. The idea is that a plot can be made of interstellar extinction versus distance for the region of the PN and then the extinction of the nebula, as determined from Balmer line ratios, radio/H-beta measurements or other techniques, can be used to find the PN distance. Limitations of this method are severe. The interstellar medium may be non-uniform along the line-of-sight to the nebula and/or insufficient stars may be available in the region of the PN so that patchiness across the region becomes a problem.

Kaler and Lutz (1985) redid the objects done earlier by Lutz (1973) with improved data for both the PN extinctions and the stars that comprised the extinction diagram. They revised the techniques used to fit the nebular extinction to the stellar data. From their error analysis it has become clear that a photometry system that is more refined (i.e., narrower bands, more filters) than UBV is desirable for obtaining reasonable extinction diagrams.

Maciel (1985) and Maciel, Faundez-Abans and de Olivera (1986) have measured extinction distances for He2-131, NGC 6565 and NGC 5979 by using stars within 2 degrees of the nebulae. UBV photometry was obtained for stars chosen because their spectral types were known already.

Gathier, Pottasch and Pel (1986) determined the distances to 12 PN by using extinction diagrams. They used Walraven VBLUW photometry in conjunction with model atmospheres and stellar evolution models. In addition, they made estimates of internal reddening from IRAS colors of the program nebulae. The fields around the PN were restricted to angular radii of less than 0.3 degree so the patchiness of the interstellar medium was less of a problem than in other studies. Their results showed that there are still problems with extinction distances even when the greatest care is taken to minimize the errors. Some fields showed a large amount of scatter in the extinction diagram, while other fields had an extinction diagram that allowed quite reasonable estimates of distance to be made. The estimated errors of the distance determinations ranged from 10% to 40%.

4. H I ABSORPTION OBSERVATIONS

Gathier, Pottasch and Goss (1986) have made H I absorption measurements at 21 cm for 24 PN with the Westerbork Synthesis Radio Telescope. They observed a background source located within 1 degree of the PN and the 21 cm emission spectrum in the direction of the PN. They used these three observations in conjunction with either a galactic rotation curve or by using data on H II regions (distances and radial velocities) to determine the galactic structure in the regions of the PN. For half of the PN, the observations could not be used to derived distances, but for the other 12, distances with accuracies of 25% to 50% were derived. The H I absorption technique was used previously to estimate a distance for NGC 7027 by Pottasch et al. (1982).

The paper of Gathier, Pottasch and Goss (1986) points out some limitations of the H I technique. In particular, a mean rotation curve is used for the analysis but it is well known that there are substantial deviations from circular motion and substantial random motions associated with particular H I clouds.

5. WIND DISTANCES

Kaler, Mo and Pottasch (1985) used P Cygni profiles that appear in the ultraviolet spectra of some central stars in combination with stellar atmosphere and evolution models to derive "wind distances" to 16 PN. The terminal wind velocity, which can be measured from the P Cygni profiles, is presumed to be related to the escape velocity which in turn is related to the temperature, mass and luminosity of the central star. Temperature is found from the Zanstra method. Stellar evolution provides a second relation between temperature, mass and luminosity, so all three parameters can be calculated. Distances can then be derived from apparent magnitudes.

This method depends upon the assumption that the theory derived for the winds of Population I stars can be applied to the central stars of PN. As the authors point out, Hubble Space Telescope observations would be desirable for deriving wind distances to a much larger number of PN.

6. METHODS INVOLVING A MASS-RADIUS RELATION

Maciel and Pottasch (1980) and Pottasch (1980) developed a method that would allow distance determinations to optically thick PN by establishing a relation between nebular ionized mass and nebular radius. This relation is calibrated by using PN of "known distance", viz. those that have trigonometric, spectroscopic or expansion parallaxes or some other type of individual distance determination. Also needed to establish the relation are electron densities, filling factors and helium abundances. Maciel and Pottasch (1980) used electron densities which were derived from forbidden line ratios. They assumed that the filling factor and the helium abundance were the same for all PN (0.65 and 0.11 respectively). They applied the mass-radius relation to and derived distances for 121 PN that had 5 GHz radio flux measurements (Milne and Aller 1975). Major uncertainties come from the electron density determinations, size measurements and assumption of a single value for the filling factor.

Maciel (1981a,b) extended this technique to 81 additional PN and used all the distances to estimate the birthrate of PN and the total number in the Galaxy. Maciel (1984) used the same method to find distances for all PN in the catalogue of Cahn and Kaler (1971).

Milne (1982), in response to the work on the mass-radius relation, suggested an alternative way to find distances given the 5 GHz flux and the angular size of the PN. His proposed technique was a variation of the methods which assume constant luminosity during the optically thick stage.

Daub (1982) used the basic idea of a mass-radius relation, but chose to make a calibration of a quantity involving the nebular mass, temperature and filling factor versus a quantity involving the nebular angular size and the 5 GHz flux. He used the relation to derive distances for 299 PN. Phillips and Pottasch (1984) used another variant of a mass-radius relation combined with parameters observed directly or derived from observations (electron density, 5 or 14 GHz radio flux, angular diameter) to derive distances to 55 PN. Amnuel et al. (1984) calculated distances to 335 PN after calibrating the mass-radius relation by using several different considerations including individual distances, 2.7 GHz fluxes and electron densities. One of their results is that the filling factor decreases as nebular radius increases.

Kwok (1985) considered the evolution of the ionization structure of PN based upon evolutionary models of the central star. He found that the observed parameters of the nebula are a function of both the nebular evolutionary state and the central star evolution. He derived a non-linear mass-radius relation, as contrasted with the linear relation used by previous investigators. He concluded that the relationship between the radio flux density and the angular size as a result of expansion had the same form as the flux-angular size relation for PN at different distances. If his conclusion is correct, the distances for optically thick PN could not be determined uniquely from the measurements of flux (radio or optical) and angular size alone.

7. RECENT DIRECTIONS

Sabbadin (1986) evaluated distances that were found by various methods for 81 PN. His study included consideration of nebular morphologies (B and C nebulae).

Mendez et al. (1987) have developed a new method for finding distances that looks promising. They determine the properties of the PN central star atmosphere from high dispersion spectra and then use the derived properties (surface gravity, model atmosphere flux at 5840 A, visual magnitude, and extinction) to determine a distance. This method is similar to the "wind distances" in being limited to PN for which model atmosphere analyses are obtainable, but there is the prospect of using this method on many more PN as more high dispersion central star spectra become available.

Pottasch (1987a, b) has considered the problem of finding nebular masses for PN that have "reliable" distances. Instead of taking the PN that are usually used for such studies, he has calibrated the mass-radius relation by using PN in the galactic bulge, PN in the Magellanic Clouds and PN that have distances determined from model atmospheres (Mendez et al. 1987). He concluded at another parameter is necessary to explain the scatter in the mass-radius relation and proposed that the central star luminosity (or mass) might be the missing parameter.

8. PRESENT PROBLEMS AND FUTURE WORK

The work of Kwok (1985) needs a response. A great deal of effort has gone into the distance method which involves a mass-radius relation and, if the method is ambiguous in its present form, perhaps there are ways to work around the ambiguity. In addition, there are some ways in which the basic data available for distance determinations can be improved. For example, PN imaging programs can provide better information about nebular sizes and filling factors. Indeed, one thing that is bothersome about the method based upon the mass-radius calibration is that the morphology exhibited by many nebulae in the low ionization potential lines that are used for electron density determinations is very different from the morphology exhibited in the light of hydrogen. Hence, the electron densities may not be representative of the regions where hydrogen radiates and the filling factors may be very different. Another puzzling aspect of the mass-radius method is that the investigators appear to be finding that most PN are optically thick. It is difficult to understand why so many of the PN show multiple shells and faint outer structures (Jewitt, Danielson and Kupferman 1986, Balick 1987, Louise et al. 1987).

The extinction method for finding distances is showing a lot of promise and should be extended by using appropriate CCD detectors in combination with filters and detailed atmospheres/evolution analyses such as those done by Gathier, Pottasch and Pel (1986).

Two methods from the past should be tried again. Expansion parallaxes (Liller, Welther and Liller 1966) are an excellent concept but in the past the measurements have been unreliable because of the uncertainties in measuring the positions of nebular features. Given modern CCD detectors and data reduction software, "first epoch" images should be obtained for a project on nebular expansions. In addition, the advent of multifiber feeds for spectroscopy should make it possible to get a good expansion velocity for particular nebular features. However, care must be taken to assure that real matter rather than expanding ionization fronts are being observed. Statistical parallaxes for PN derived from proper motions and radial velocities should be redone and the numbers of nebulae included in such studies increased. Cudworth (1974) mentions the desirability of doing such a study "in another decade" and 1984 has gone by already. It would be interesting to rework the statistical parallax studies by including more recent studies of nebular morphologies to divide PN into categories.

Two space missions have the possibility of giving considerable additional information on PN distances. Hipparcos will be launched in the late 1980's and has among its many targets some central stars of PN. The Hubble Space Telescope could provide us with excellent data (fluxes, sizes, terminal wind velocities) that can be used in a variety of distance determinations, (including "wind" and "stellar atmospheres" distances), as well as for ionization models and central star atmosphere and evolutionary models. In addition, the Magellanic Cloud PN that we struggle to observe with ground-based techniques will be easy to study. Instead of observing only the brighter part of the distribution of the Magellanic Cloud PN, observers will be able to study the properties (such as nebular mass) of even faint PN. However, since these missions will be spending only a small amount of observing time on PN, the major hope for improving distances remains with ground-based optical, radio and infrared studies.

REFERENCES

Acker, A. 1978, Astr. Ap. Suppl., 33, 367.
Amnuel, P. R., Guseinov, O. H., Novruzova, H. I., and Rustamov, Yu. S. 1984, Ap. Space Sci., 107, 19.
Balick, B. 1987, A. J., 94, 671.
Cahn, J. H., and Kaler, J. B. 1971, Ap. J. Suppl., 22, 319.
Cudworth, K. M. 1974, A. J., 79, 1384.
Daub, C. T. 1982, Ap. J., 260, 612.
Gathier, R., Pottasch, S. R., and Goss, W. M. 1986, Astr, Ap., 157, 191.
Gathier, R. Pottasch, S. R., and Pel, J. W. 1986, Astr. Ap., 157, 171.
Grieg, W. E. 1971, Astr. Ap., 10, 161.
_____. 1972, Astr. Ap., 18, 70.
Jewitt, D. C., Danielson, G. E., and Kupferman, P. N. 1986, Ap. J., 302, 727.
Kaler, J. B., and Lutz, J. H. 1985, Pub. A.S.P., 97, 700.
Kaler, J. B., Mo, J. E., and Pottasch, S. R. 1985, Ap. J., 288, 305.
Kwok, S. 1985, Ap. J., 290, 568.
Liller, M. H., Welther, B. L., and Liller, W. 1966, Ap. J., 144, 280.
Louise, R., Macron, A., Pascoli, G. and Maurice, E. 1987, Astr. Ap. Suppl., 70, 201.
Lutz, J. H. 1973, Ap. J., 181, 135.
Maciel, W. J. 1981a, Astr. Ap., 98, 406.
_____. 1981b, Astr. Ap. Suppl. 44, 123.
_____. 1984, Astr. Ap. Suppl., 55, 253.
_____. 1985, Rev. Mexicana Astr. Ap., 10, 199.
Maciel, W. J., Faundez-Abans, M., and de Olivera, M. 1986, Rev. Mexicana Astr. Ap., 12, 233.
Maciel, W. J., and Pottasch, S. R. 1980, Astr. Ap., 88, 1.
Mendez, R. H., Kudritzki, R. P., Herrero, A., Husfeld, D., and Groth, H. G. 1987, Astr. Ap., in press.
Milne, D. K. 1982, M.N.R.A.S., 200, 51P.
Milne, D. K., and Aller, L. H. 1975, Astr. Ap., 38, 183.
Phillips, J. P., and Pottasch, S. R. 1984, Astr. Ap., 130, 91.
Pottasch, S. R. 1980, Astr. Ap., 89, 336.
_____. 1987a, ESO Workshop, in press.
_____. 1987b, Torino Workshop, in press.
Pottasch, S. R., Goss, W. M., Arnal, E. M., and Gathier, R. 1982, Astr. Ap., 106, 229.
Sabbadin, F. 1986, Astr. Ap. Suppl., 64, 579.

GALACTIC DISTRIBUTION, RADIAL VELOCITIES AND MASSES OF PN

Walter J. Maciel
Instituto Astronômico e Geofísico da USP
Caixa Postal 30.627
01051 São Paulo SP, Brasil

1. INTRODUCTION

   Since the discovery in 1779 of the first planetary nebula in the Lyra constellation by Antoine Darquier, the total number of objects known as PN has been steadily increasing. About 1600 PN have been identified in the Galaxy, although 10-20% are probably misclassified objects, which include H II regions, reflection nebulae, symbiotic stars, etc. (see for example Kohoutek, 1987; Acker et al., 1987; Stenholm and Acker, 1987).
   Several hundred nebulae have been discovered in other systems, especially in Andromeda and in the Magellanic Clouds. In the Galaxy, however, the higher resolution achieved permits more detailed analysis and classification, so that PN are becoming increasingly useful as tools for the study of galactic structure.
   Recent work show that 4-6 different types of PN can be identified, regarding their space distribution, kinematics and chemical composition (Peimbert, 1978; 1983; Peimbert and Serrano, 1980; Peimbert and Torres-Peimbert, 1983; Maciel and Faúndez-Abans, 1985; Faúndez-Abans and Maciel, 1987). A possible scheme of PN classification is given below, where the original types introduced by Peimbert (1978) have been extended to include a subdivision of the type II PN and the galactic centre objects as well.

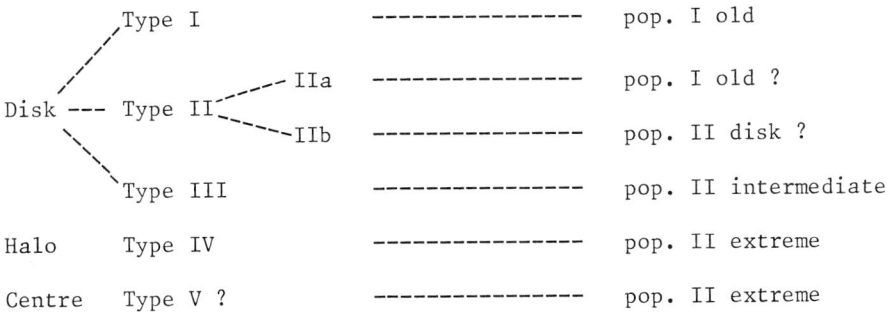

Other somewhat different schemes may be found in the literature (cf. Greig, 1971; 1972; Kaler, 1983; Heap and Augensen, 1987). However, the classification proposed by Peimbert seems more suitable, as it contains criteria stemming from the three main sources of data, namely, spatial distribution, kinematics, and chemical composition. Generally speaking, type I PN belong to class B (Greig, 1971), whereas type II PN are of classes B and C.

In the present work, an attempt is made to review the recent results concerning the galactic distribution, radial velocities, and masses of galactic planetary nebulae. Especially related reviews appearing in this conference are presented by Kohoutek (new and reclassified PN), Lutz (distances), Clegg (compositions) and Phillips (formation rate). Previous reviews on the subject of galactic distribution of PN include Minkowski (1965), Perek (1968), Cahn (1968), Cahn and Wyatt (1978), Terzian (1980), and the more general works of Pottasch (1984) and Aller (1984). Radial velocities have been reviewed by Schneider et al. (1983), and the determination of masses of planetary nebulae has been treated recently by Pottasch (1980; 1983; 1984; 1987).

2. STATISTICS

Most known PN are within 5 kpc from the Sun, and correspond to a small fraction of the total number in the Galaxy. The estimate of statistical parameters concerning PN strongly depends on two different facts: (1) most observationally determined parameters are derived for the solar neighbourhood, and have to be extrapolated for the whole Galaxy; (2) the main conclusions depend heavily on the adopted distances, which remain poorly known. Although recent distance determinations using individual methods may reach accuracies of 30-50% (Pottasch, 1983; 1984), most statistical studies on PN are based in statistical methods, which may have uncertainties of a factor 2 for individual objects (Maciel, 1981a; 1984).

Table I shows a selection of the main results obtained in the past 15 years. Here n_v is the space density of PN in the solar neighbourhood, n_s is the density projected on the galactic plane, h is the scale height, χ is the PN formation rate, t is the lifetime of the PN stage, and n is the total number of PN in the Galaxy.

Most recent work on the statistics of PN tend to agree with each other, except for the much larger values of the space density and formation rate given by Ishida and Weinberger (1987), which are based on PN closer to the Sun than 500 pc. However, it seems that this result depends rather strongly on the distances adopted, which are uncertain in many cases. An average increase of 50% in the distances would be sufficient to bring the formation rate close to the remaining values of Table I. On the other hand, the results by Ishida and Weinberger (1987) are interesting in the sense that they call attention to the large, evolved nebulae, which contribute to the galactic population of PN, but are difficult to detect at large distances, due to their low surface brightness.

Table I - Statistics of planetary nebulae

	n_v	n_s	h	$10^3 \chi$	t	n
	(kpc^{-3})	(kpc^{-2})	(pc)	(kpc^{-3} yr^{-1})	(yr)	
Cahn and Kaler (1971)	50	13	90	3.2	16000	300000
Cahn and Wyatt (1976)	80	19	115	5.0	16000	38000
Alloin et al. (1976)						
Cahn and Kaler	50	13		3.1	16000	9000
Cudworth	15	6		0.6	24000	4000
Smith (1976)	150	30	100	11.0	14000	
Weidemann (1977)	36	11	150	1.8	20000	20000
Acker (1978)	48	13		3.0	16000	25000
Khromov (1979b)	430	170	200			190000
Maciel (1981a)	41	12	144	2.0	20000	31000
Daub (1982)	53	13	125	5.0		28000
Mallik (1982, 1983)	44	15	160	2.4	16000	28000
Amnuel et al. (1984)	117	26	130	4.6	25000	40000
Pottasch (1984)	50	25	250	2.0	30000	
Aller (1984)	50	12	120	2.0	30000	15000
Ishida and Weinberger (1987)	326	81	100	8.3		140000

The number of planetary nebulae in the Galaxy is not expected to be lower than $2 \pi R^2 h n_v \simeq 10000$, where we have used R = 15 kpc, h = 140 pc and n_v = 50 kpc^{-3}. An upper limit is more difficult to establish, and may be in excess of 10^5, depending on the adopted space density in the solar neighbourhood. Observations of PN in other galaxies may help solving this problem. Data on PN in the Local Group suggest that the specific number of PN is $k \simeq 1\text{-}4\ 10^{-7}$ M$\odot^{-1}$ (see for example Pottasch, 1984), which would imply n = 15000-60000, if the mass of the Galaxy is M = 1.5 10^{11} M$\odot$.

From stellar evolution theory, the PN formation rate is expected to be similar to the rate at which stars in the mass range 1-5 M$\odot$ leave the main sequence, and to the white dwarf formation rate. As seen in Table I, this seems to be true for most recent estimates of the formation rate (cf. Phillips, 1987). On the other hand, the formation rate of cool giants, comprising Miras, OH/IR stars and other cool evolved giants is a factor of 3 lower than the PN rate. However, such rates are particularly uncertain, and a significant fraction of the white dwarfs may originate from AGB stars which do not produce PN, or even from stars not massive enough to climb the AGB (Drilling and Schonberner, 1985).

The rate at which stars leave the main sequence depends on the mass range of stars producing PN. Generally assumed to be 0.8-1.0 M$\odot$, the lower limit may reach 1.2-1.3 M$\odot$ (Wood and Cahn, 1977) or even higher values (Mallik, 1983; 1985). As for the upper limit, stars having masses between 5 and 8 M$\odot$ may also form PN (cf. Terzian, 1983), although such stars do not strongly affect the formation rate, as the IMF decreases sharply towards higher masses (Tinsley, 1978).

The mass ejected per year in a column perpendicular to the galactic plane is $(dM/dt)s = 2\text{-}4\ 10^{-10}$ M☉ $pc^{-2}\ yr^{-1}$ (Maciel, 1981a; Pottasch, 1984). Therefore, the PN, Miras, OH/IR stars, and other cool giants are the main responsible for the mass returned to the interstellar medium, corresponding about 30% to the PN alone (Alloin et al., 1976; Pottasch, 1984; Knapp and Wilcots, 1987). Explosive events such as supernovae and novae are comparatively less important, due to their lower frequency and higher volume of ocurrence. On the other hand, the disk PN have no influence on the injection of kinetic energy in the interstellar medium, which is essentially due to the supernovae, novae and hot stars.

3. GALACTIC DISTRIBUTION

The space distribution of planetary nebulae shows a pronounced concentration to the galactic plane (GP), although not as intense as in the case of H II regions (Figure 1). A general concentration towards the galactic centre (GC) is also observed, which is particularly true for the PN having small apparent sizes (angular diameters smaller than about 10 seconds of arc), which constitute the majority of the observed objects. PN with larger apparent sizes, typically having diameters greater than 50 seconds of arc, are not concentrated in the direction of the GC, and reach relatively high galactic latitudes, indicating that they are nearby objects. As reviewed by Pottasch (1984), most of the nebulae in the direction of the galactic centre lie actually close at it. Although a selection effect may artificially increase the number of PN in such direction, a reduction of discoveries in the direction of the anticentre is observed, so that a density gradient $dn/dr = 5\ 10^{-9}\ pc^{-3}\ kpc^{-1}$ can be estimated.

The latitude distribution of PN close to the GC shows relatively few objects with b = 0. Moreover, some differences exist between the number of objects above and below the GP, especially for l = 0, 1 and 2 degrees (Pottasch, 1984). Both effects are attributed to the inhomogeneity of the extinction near the galactic plane (see also Terzian, 1980).

Adopting a distance scale (Maciel, 1984), a distribution perpendicular to the GP can be obtained for about 600 PN, 47% of which have $|z| < 200$ pc. To avoid incompleteness effects, one could take into account only PN within 1 kpc from the Sun, which imply that 2/3 of the objects are within 250 pc from the plane (Pottasch, 1984).

Planetary nebulae of types I and II have a shorter scale height as compared to the remaining types, in agreement with the classification scheme presented in section 1. Given the relatively small sample of well classified objects, type II nebulae of both subtypes seem to have a similar distribution above the galactic plane, so that the main differences between them lie in the heavy-element abundances and progenitor masses.

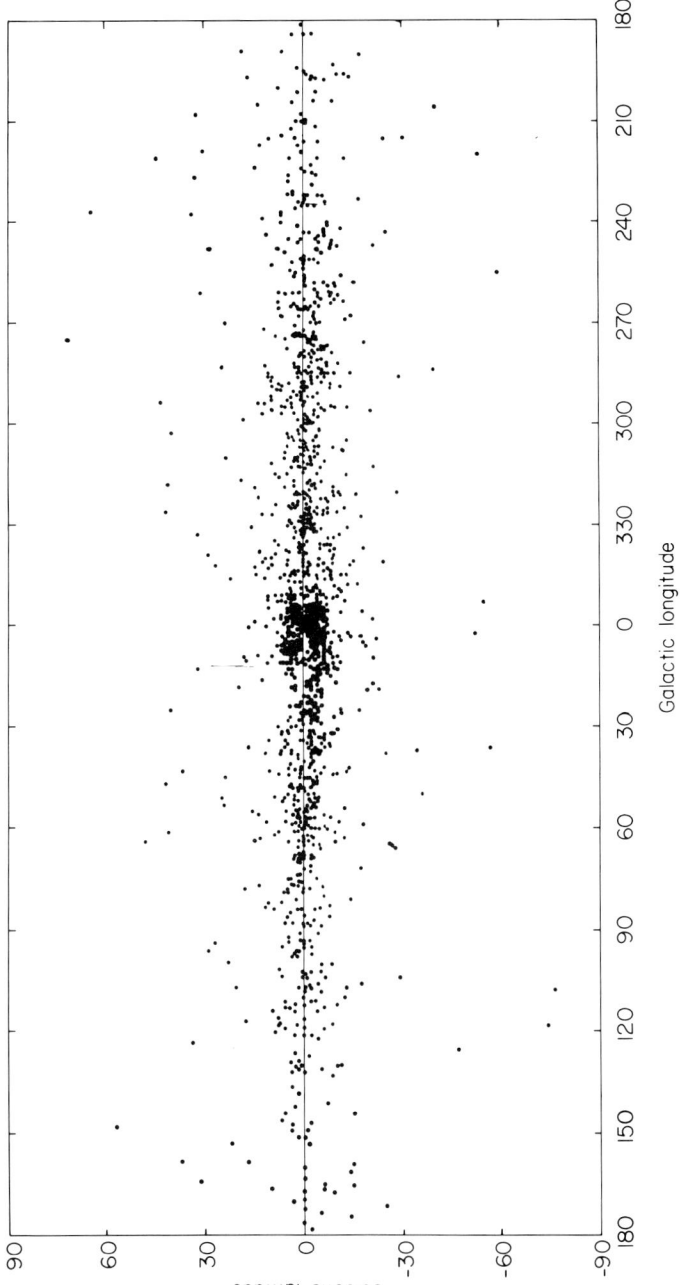

Figure 1. Galactic distribution of planetary nebulae.

4. RADIAL VELOCITIES

The main progress in kinematical studies of PN, apart from their internal motions, lies in the determination of radial velocities. Accurate measurements of proper motions remain difficult, basically due to the large distances involved (Kaler, 1985). About a hundred objects have known proper motions (Khromov, 1979a; Kiosa and Khromov, 1979; Cudworth, 1974). Most old values are uncertain, and the new measurements imply a considerably large discrepancy in the distance scale (cf. Maciel, 1981a; Kaler, 1985).

The first determinations of radial velocities of PN were made by Keeler (1894), for 13 nebulae at a dispersion of 20 A/mm. The classical work of Campbell and Moore (1918) includes a list of radial velocities for 99 PN, which has been considerably extended by N. U. Mayall, R. Minkowski and others (see for example Minkowski, 1965).

Presently, some 550 PN have measured radial velocities (Acker, 1985), corresponding to about 35% of the known PN in the Galaxy. The main source of reference is the catalogue by Schneider et al. (1983), which lists heliocentric as well as LSR velocities for 524 PN. The catalogue is based on 102 different sources, including 920 velocity measurements and new observations for 19 nebulae. Most recent work is based on (1) spectroscopic techniques, especially using image tubes, (2) Fabry-Perot interferometry, and (3) observations of radio recombination lines.

The errors in the determination of radial velocities have been discussed by Schneider et al. (1983) and Acker (1985), and are lower than or equal to 10 km/s for half of the objects in the catalogue. For reciprocal dispersions up to 150 A/mm, the errors increase approximately linearly with the dispersion. Dispersions better than 100 A/mm produce, in average, errors smaller than 10 km/s (Schneider et al., 1983). As a consequence, variations of the radial velocities with periods from a few hours to a few days have been detected, and are used to investigate the binary nature of nuclei of some planetary nebulae (Augensen, 1985).

The galactic distribution of the radial velocities of PN relative to the LSR is shown in Figure 2. As discussed more than 20 years ago by Minkowski (1965), important differences exist between the distribution of the PN and the distribution of the galactic plane objects which participate in the circular rotation motion around the galactic centre. The curves show the limits for the radial velocities compatible with a recent galactic rotation curve (Clemens, 1985).

A large dispersion of about 140 km/s exists near the galactic centre (Pottasch, 1984), and several objects lie outside the curves, probably indicating non circular orbits. Apart from the galactic centre objects, the position of most nebulae in Figure 2 is in agreement with the curves of circular rotation for distances to the Sun smaller than 12 kpc, and an average dispersion of about 40 km/s. The galactic centre objects display a characteristic behaviour typical of population II systems, such as the globular clusters and RR Lyrae variables. The distribution of the OH/IR maser sources is also remarkably similar to the PN in the direction of the galactic centre, which is clearly seen by considering in Figure 2 only the PN with small angular dimensions (cf. Pottasch, 1984).

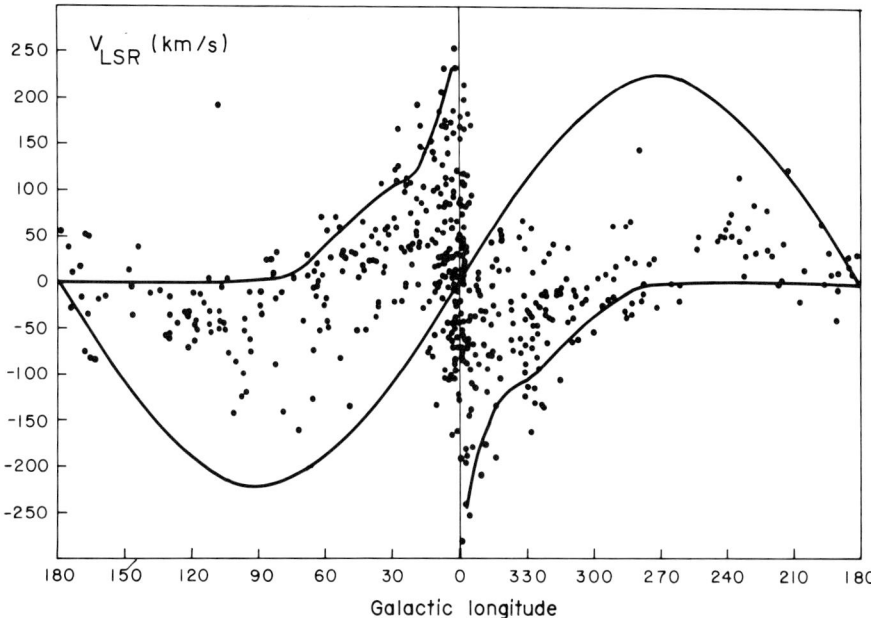

Figure 2. Distribution of radial velocities of PN according to their galactic longitudes.

The general interpretation of these distributions is that both PN and OH/IR objects in the galactic centre form a spherical population with high dispersion, superimposed on which there is a younger, flattened population, which approximately obeys the galactic rotation curve. According to the classification scheme given in section 1, PN of types I and II are expected to show smaller differences relative to the galactic rotation curve, which was found to be true for the more numerous type II objects (Maciel, 1987). This places the interesting possibility of determining the galactic rotation curve from a selected sample of PN. An obvious advantage of such procedure lies in the large number of known PN, their brightness, and accuracy of the radial velocity measurements. On the other hand, uncertainties in the distance scale and the lack of a large number of carefully classified objects have limited the applicability of this method. A tentative investigation along these lines has been made by Schneider and Terzian (1983), who found that the rotation curve does not fall after the solar circle, in agreement with independent determinations of the rotation curve based on the CO molecule (Blitz et al., 1980).

Also referring to the given classification scheme, it is interesting to comment on the halo type IV nebulae. Having radial velocities in excess of 100 km/s, these nebulae have less massive progenitors, do not participate in the galactic rotation motion, and present a strong underabundance in heavy-elements, showing therefore a distinctively population II behaviour.

5. MASSES

The problem of mass determination in planetary nebulae is closely related to the difficult problem of establishing a distance scale. To this date, no direct mass determinations exist for PN, so that one has to rely on indirect estimates which usually involve the distance. According to the Shklovsky method, popularly used through the seventies, the derived distances are $d \propto M^{2/5} \theta^{-3/5} F^{-1/5}$, where M is the ionized mass, θ is the angular radius, and F is the observed flux. Therefore, from this method one would obtain that $M \propto d^{5/2}$, so that an uncertainty in the distance scale of a factor two would imply an uncertainty of a factor 6 in the ionized mass.

The total masses of planetary nebulae are essentially in the range 0.1-0.5 M☉ (Pottasch, 1980; 1984; 1987; Maciel, 1981b; Mallik, 1982; Wood, 1987; Gathier, 1987; Barlow, 1987). The ionized mass can be a small fraction of the total mass for optically thin nebulae. Most determinations refer to the ionized masses, whereas the neutral material - essentially atomic and molecular H - has masses similar to the average ionized masses (Phillips and Pottasch, 1984; Kaler, 1985; Pottasch, 1987).

Due to the dependence with distance, few determinations of masses of PN have been made until recently. Early estimates featured Magellanic Cloud objects, whereas Perinotto (1975) used optical and radio data to estimate the masses of 40 galactic PN. However, several of the distances implicitly assumed by him are unnaceptable today (cf. Pottasch, 1980; Maciel, 1981b). More recently, Pottasch (1980) determined masses for 25 objects having independent distance estimates. It was shown that the ionized masses could vary over about 3 orders of magnitude, being strongly correlated with the electron density and the nebular size, suggesting that most objects are optically thick (see also Pottasch, 1983). Maciel (1981b) determined ionized masses and distances for 202 nebulae based on the empirical mass radius relationship developed by Maciel and Pottasch (1980), a procedure that was followed and modified later by others (see for example Daub, 1982). A larger set of distances, and hence masses, has been given by Maciel (1984). For these objects, an average uncertainty of a factor 3 was estimated, based on the masses of selected objects given by Pottasch (1980).

Recent work by Pottasch (1987) shows that the strong correlation between mass and size holds for PN in the galactic bulge and Magellanic Clouds, apart from a selected sample of nearby PN whose distances are well known. Although selection effects may play a role, it seems clear that a real correlation exists, supporting the conclusion that most objects are optically thick or do not deviate very much from this condition. The correlation shows some real scatter, however, which has been interpreted as to indicate the need of an additional parameter, such as the mass of the central star. In fact, the different tracks in the mass-radius plane can be interpreted as due to central stars with different luminosities.

The objects of the galactic centre and Magellanic Clouds are from the sample of Gathier (1987) and Wood (1987), respectively, who have presented similar mass-radius correlations. It is interesting to notice that, in the radius interval 0.01-0.4 pc, the average masses of both samples do not deviate very much from the masses derived with the empirical relationship of Maciel and Pottasch (1980).

Acknowledgements. I am indebted to Dr. S. R. Pottasch and Dr. A. Acker for some papers sent in advance of publication. This paper was partly supported by CNPq (Brasil).

References

Acker, A. 1978, Astron. Astrophys. Suppl. 33, 367
Acker, A. 1985, Bull. Inf. CDS Strasbourg n. 28, 5
Acker, A., Chopinet, M., Pottasch, S. R., Stenholm, B. 1987, Astron. Astrophys. Suppl. (in press)
Aller, L. H. 1984, Physics of thermal gaseous nebulae, Reidel
Alloin, D., Cruz-Gonzalez, C., Peimbert, M. 1976, Astrophys. J. 205, 74
Amnuel, P. R., Guseinov, O. H., Novruzova, H. I., Rustamov, Yu. S. 1984, Astrophys. Space Sci. 107, 19
Augensen, H. J. 1985, Monthly Notices Roy. Astron. Soc. 213, 399
Barlow, M. J. 1987, Monthly Notices Roy. Astron. Soc. 227, 161
Blitz, L., Fich, M., Stark, A. A. 1980, IAU Symp. 87, ed. B. H. Andrew, Reidel, p. 213
Cahn, J. H. 1968, IAU Symp. 34, ed. D. E. Osterbrock, C. R. O'Dell, Reidel, p. 44
Cahn, J. H., Kaler, J. B. 1971, Astrophys. J. Suppl. 22, 319
Cahn, J. H., Wyatt, S. P. 1976, Astrophys. J. 210, 508
Cahn, J. H., Wyatt, S. P. 1978, IAU Symp. 76, ed. Y. Terzian, Reidel, p. 3
Campbell, W. W., Moore, J. H. 1918, Publ. Lick Obs. 13, 75
Clegg, R. E. S. 1987, IAU Symp. 131, ed. S. Torres-Peimbert, Reidel
Clemens, D. P. 1985, Astrophys. J. 295, 422
Cudworth, K. M. 1974, Astron. J. 79, 1384
Daub, C. T. 1982, Astrophys. J. 260, 612
Drilling, J. S., Schonberner, D. 1985, Astron. Astrophys. 146, L23
Faúndez-Abans, M., Maciel, W. J. 1987, Astron. Astrophys. (in press)
Gathier, R. 1987, Late Stages of Stellar Evolution, eds. S. Kwok, S. R. Pottasch, Reidel, p. 371
Greig, W. E. 1971, Astron. Astrophys. 10, 161
Greig, W. E. 1972, Astron. Astrophys. 18, 70

Heap, S. R., Augensen, H. J. 1987, Astrophys. J. 313, 268
Ishida, K., Weinberger, R. 1987, Astron. Astrophys. 178, 227
Kaler, J. B. 1983, IAU Symp. 103, ed. D. R. Flower, Reidel, p. 245
Kaler, J. B. 1985, Ann. Rev. Astron. Astrophys. 23, 89
Keeler, J. E. 1894, Publ. Lick Obs. 3, 161
Khromov, G. S. 1979a, Astrofizika 15, 269
Khromov, G. S. 1979b, Astrofizika 15, 445
Kiosa, M. N., Khromov, G. S. 1979, Astrofizika 15, 105
Knapp, G. R., Wilcots, E. M. 1987, Late Stages of Stellar Evolution,
    eds. S. Kwok, S. R. Pottasch, Reidel, p. 171
Kohoutek, L. 1987, IAU Symp. 131, ed. S. Torres-Peimbert, Reidel
Lutz, J. H. 1987, IAU Symp. 131, ed. S. Torres-Peimbert, Reidel
Maciel, W. J. 1981a, Astron. Astrophys. 98, 123
Maciel, W. J. 1981b, Astron. Astrophys. Suppl. 44, 123
Maciel, W. J. 1984, Astron. Astrophys. Suppl. 55, 253
Maciel, W. J. 1987, Late Stages of Stellar Evolution, eds. S. Kwok,
    S. R. Pottasch, Reidel, p. 391
Maciel, W. J., Faúndez-Abans, M. 1985, Astron. Astrophys. 149, 365
Maciel, W. J., Pottasch, S. R. 1980, Astron. Astrophys. 88, 1
Mallik, D. C. V. 1982, Bull. Astron. Soc. India 10, 73
Mallik, D. C. V. 1983, IAU Symp. 103, ed. D. R. Flower, Reidel, p. 424
Mallik, D. C. V. 1985, Astrophys. Lett. 24, 173
Minkowski, R. 1965, Galactic Structure, ed. A. Blaauw, M. Schmidt,
    Un. of Chicago, p. 321
Peimbert, M. 1978, IAU Symp. 76, ed. Y. Terzian, Reidel, p. 215
Peimbert, M. 1983, II Advanced School of Astrophysics, IAGUSP
Peimbert, M., Serrano, A. 1980, Rev. Mexicana Astron. Astrofís. 5, 9
Peimbert, M., Torres-Peimbert, S. 1983, IAU Symp. 103, ed. D. R.
    Flower, Reidel, p. 233
Perek, L. 1968, IAU Symp. 34, ed. D. E. Osterbrock, C. R. O'Dell,
    Reidel, p. 9
Perinotto, M. 1975, Astron. Astrophys. 39, 383
Phillips, P. 1987, IAU Symp. 131, ed. S. Torres-Peimbert, Reidel
Phillips, J. P., Pottasch, S. R. 1984, Astron. Astrophys. 130, 91
Pottasch, S. R. 1980, Astron. Astrophys. 89, 336
Pottasch, S. R. 1983, IAU Symp. 103, ed. D. R. Flower, Reidel, p. 391
Pottasch, S. R. 1984, Planetary Nebulae, Reidel
Pottasch, S. R. 1987, Torino Workshop
Schneider, S. E., Terzian, Y. 1983, Astrophys. J. 274, L61
Schneider, S. E., Terzian, Y., Purgathofer, A., Perinotto, M. 1983,
    Astrophys. J. Suppl. 52, 399
Smith, H. 1976, Astron. Astrophys. 53, 333
Stenholm, B., Acker, A. 1987, Astron. Astrophys. Suppl. 68, 51
Terzian, Y. 1980, Quart. J. Roy. Astron. Soc. 21, 82
Terzian, Y. 1983, IAU Symp. 103, ed. D. R. Flower, Reidel, p. 487
Tinsley, B. M. 1978, IAU Symp. 76, ed. Y. Terzian, Reidel, p. 341
Weidemann, V. 1977, Astron. Astrophys. 61, L27
Wood, P. R. 1987, Late Stages of Stellar Evolution, eds. S. Kwok,
    S. R. Pottasch, Reidel, p. 197
Wood, P. R., Cahn, J. H. 1977, Astrophys. J. 211, 499

THE SHAPES AND SHAPING OF PLANETARY NEBULAE

Bruce Balick[1]
University of Washington
Dept. of Astronomy, FM-20
Seattle, Washington 98195
U.S.A.

ABSTRACT. One of the most striking attributes of planetary nebulae are their complex, yet highly symmetric shapes. The process(es) which shape planetaries are only beginning to be understood. It is proposed that the morphologies of most PNs can be understood within the context of the "interacting winds" wherein a fast but light wind driven by the nucleus rams into an older, slower, and more massive wind, or red giant envelope ("RGE") ejected earlier.

In order to explain the shapes of noncircular PNs, it seems necessary to hypothesize that the slow wind was originally ejected with an enhanced density along an equatorial plane. The morphologies of nearly all PNs can be understood through two basic parameters: the equatorial density contrast in their RGEs, and the degree of interaction between the fast and slow winds.

1. INTRODUCTION

Winds are now recognized as important in the shaping of many types of nebulae (e.g. Pikel'ner 1968, 1973; Kwok et al. 1978; Kwok 1980, 1982; Kahn 1983, Okorokov et al. 1985; Volk and Kwok 1985; Kahn and West 1985). Kwok and his collaborators have been instrumental in developing the "interacting winds" model for planetary nebulae (hereafter PNs) wherein fast, light winds from a central star interact hydrodynamically with the former outer layers of the star ejected as the star evolved from the red-giant stage. The "fast wind" is characterized by mass flows of $\sim 10^{-7}$ to $\sim 10^{-8}$ at radial velocities of about 10^3 km s^{-1}.

According to one-dimensional model calculations, the hydrodynamic interaction generates two stable shocks. The outermost of the two is the region where a hot, relatively high pressure interior bubble energized by the winds interfaces to the as-yet unaffected RGE. In times short compared to the dynamical age of the RGE the outer shock becomes associated with a snowplow, which is observable as a dense and bright "rim" between the hot interior bubble and the RGE. The snowplow is sustained by the negative radial pressure gradient caused by the wind. The second

---

[1] Visiting Astronomer, Kitt Peak National Observatory, National Optical Astronomy Observatories which is operated by the Association of Universities for Research in Astronomy, Inc. under contract with the National Science Foundation.

shock is reverse shock which settles into place near the star at the points where the pressure of the radially streaming wind is matched by the pressure of the hot bubble upstream.

The simple models are in qualitative accord with the morphologies of the relatively rare circular PNs such as NGC 1535, NGC 6894, and IC 3568. Even for simple (i.e. round) PNs, detailed agreement between observation and theory must await knowledge of the changes in the properties of the fast wind as the PN nucleus evolves, and such knowledge is elusive (see Kwok 1987 and Schoenberner 1987 for discussions).

However, the typical PN is not round. Clearly either the RGE or the fast winds (or both) are nonaxisymmetric. Can the precepts of the interacting wind model be applied to explain the shapes of non-round PNs? This is the underlying scientific question which motivates the work described here.

The scientific question is a large one, and one which requires computation of full two- or three-dimensional time-dependent hydrodynamic model calculations. Such calculations are difficult and unavailable, and they rely on a knowledge of initial and boundary conditions (i.e. gas density and motion distributions) which are yet to be supplied by observers. Hence, the question is not yet answerable.

2. THE DATA BASE

Observations reported here were made with the 2.1-m telescope of the Kitt Peak National Observatory through a variety of narrow ($\sim$10-20Å bandpass) emission line filters. The detector was a Texas Instruments CCD whose large dynamic range makes possible the study of the interrelations of both subtle background features (such as the RGE) and bright highlights (e.g. the bright rim) in a single deep exposure. The data are described and presented by Balick (1987a,b). Selected images shall be presented later.

3. MORPHOLOGICAL SEQUENCE

Based on these data, Balick (1987b) proposed a morphological sequence which is illustrated in fig. 1. All arrows in the figure should be ignored for the present. From left to right there are **classes** of PNs which are based on the symmetry of the of bright parts of the nebula. The simplest PNs are round (R) and are shown in the left column. PNs in the center column are designated as "ellipticals" (E), and those on the right as "butterflies" (B). In each column, from top to bottom, PNs are designated by **types**, i.e. "early" (e), "middle" (m), and "late" (l). PNs can be designated as eE (early elliptical), mB (middle butterfly), etc. Obviously, there will be PNs of intermediate classes and types.

In order to be useful, a morphological sequence must encompass a major fraction of known objects. Balick (1987b) argues that most, though not all, PNs are accommodated by the sequence in fig. 1. Of Balick's 51 PNs, 29 of the best examples which fit into the classification scheme are shown in figs. 2, 3, and 4. Another eleven apparently fit into the sequence but are sufficiently small in angular size or faint that they are not shown here.

Another eleven of Balick's 51 PNs fail to fit into the sequence. "Lumpy" PNs such as NGC 2452, NGC 6210, IC 4593, and J 320 seem to consist of many knots,

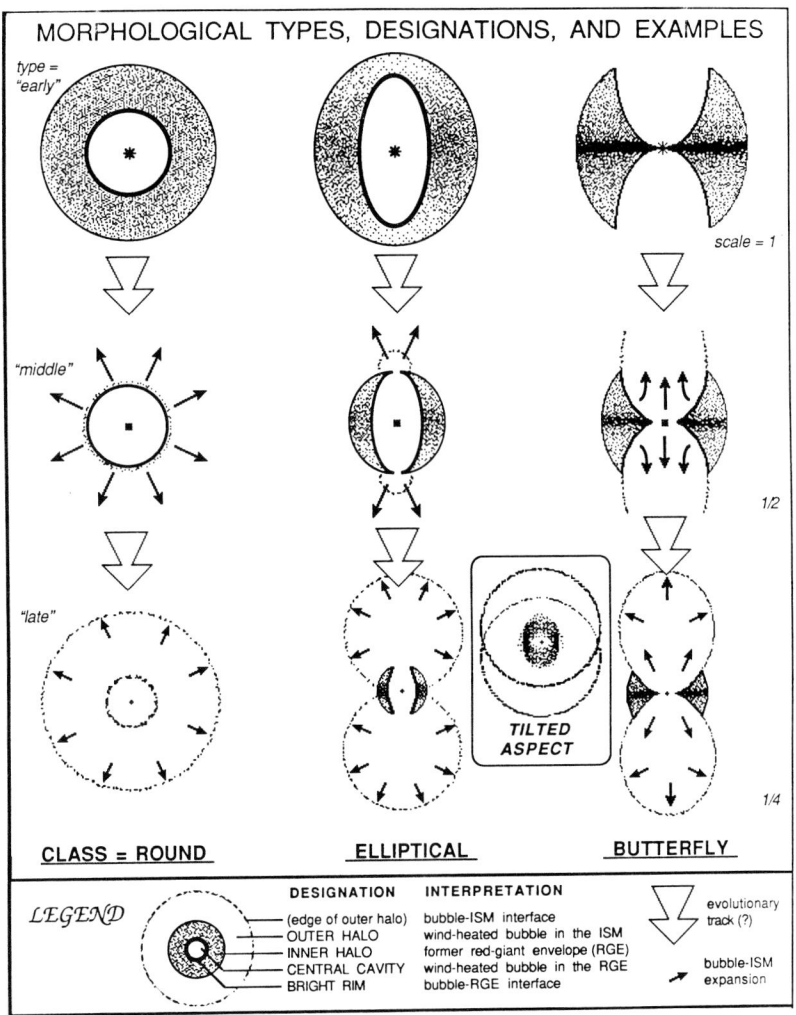

Fig. 1 – Schematic illustrations of two-dimensional cuts through the centers of various morphological archetypes of PNs. Grey regions are remaining regions of the RGE (inner halos), where grey tone is related to gas density. Heavy lines indicate regions where the stellar wind has pushed portions of the RGE into snowplows (rims). The morphological types are defined by the shapes (after projection onto the sky) of inner halos and rims.

Light, broken lines indicate the leading edge of the interface between winds and the interstellar medium which can be identified with outer halos. Wide, white arrows indicate evolutionary pathways through the diagram, and thin black arrows indicate the directions of the advancing wind-ISM interface. Scale changes among the figures are indicated by the size of the icon representing the central star.

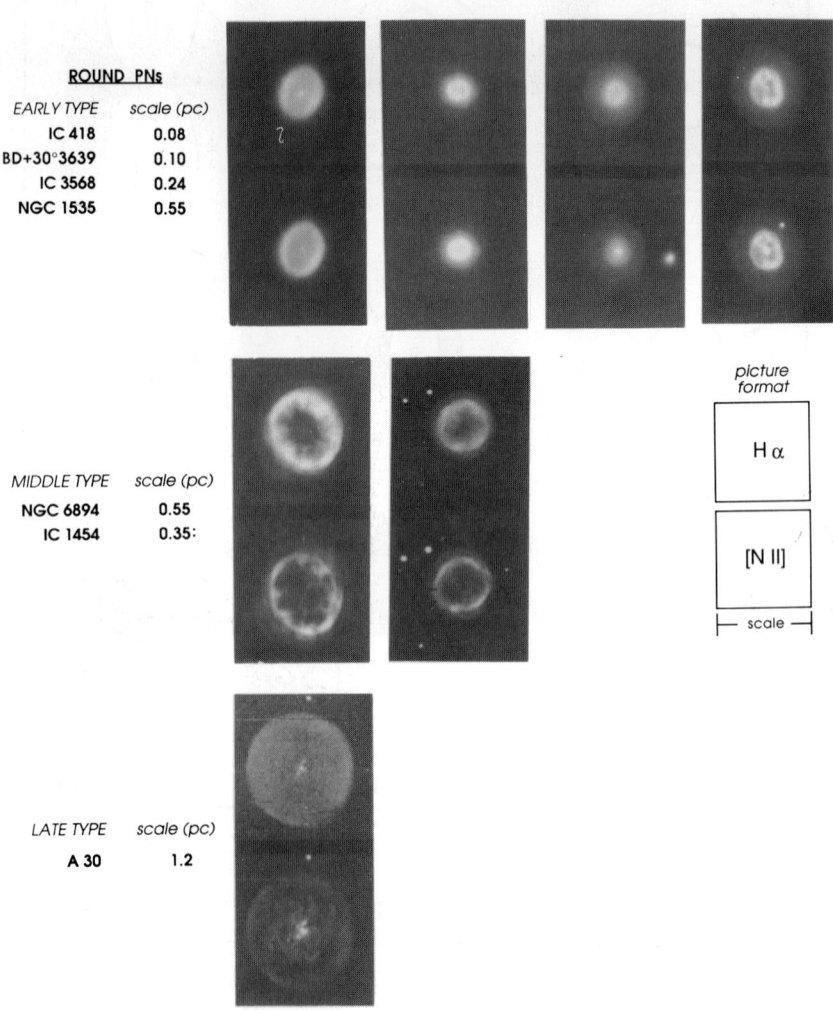

Fig. 2 – Examples of Round PNs of various types. The pictures are from CCD images through Hα (top panels) and [N II] (bottom panels). The contrast has been altered to emphasize faint features while not suppressing highlights in the images.

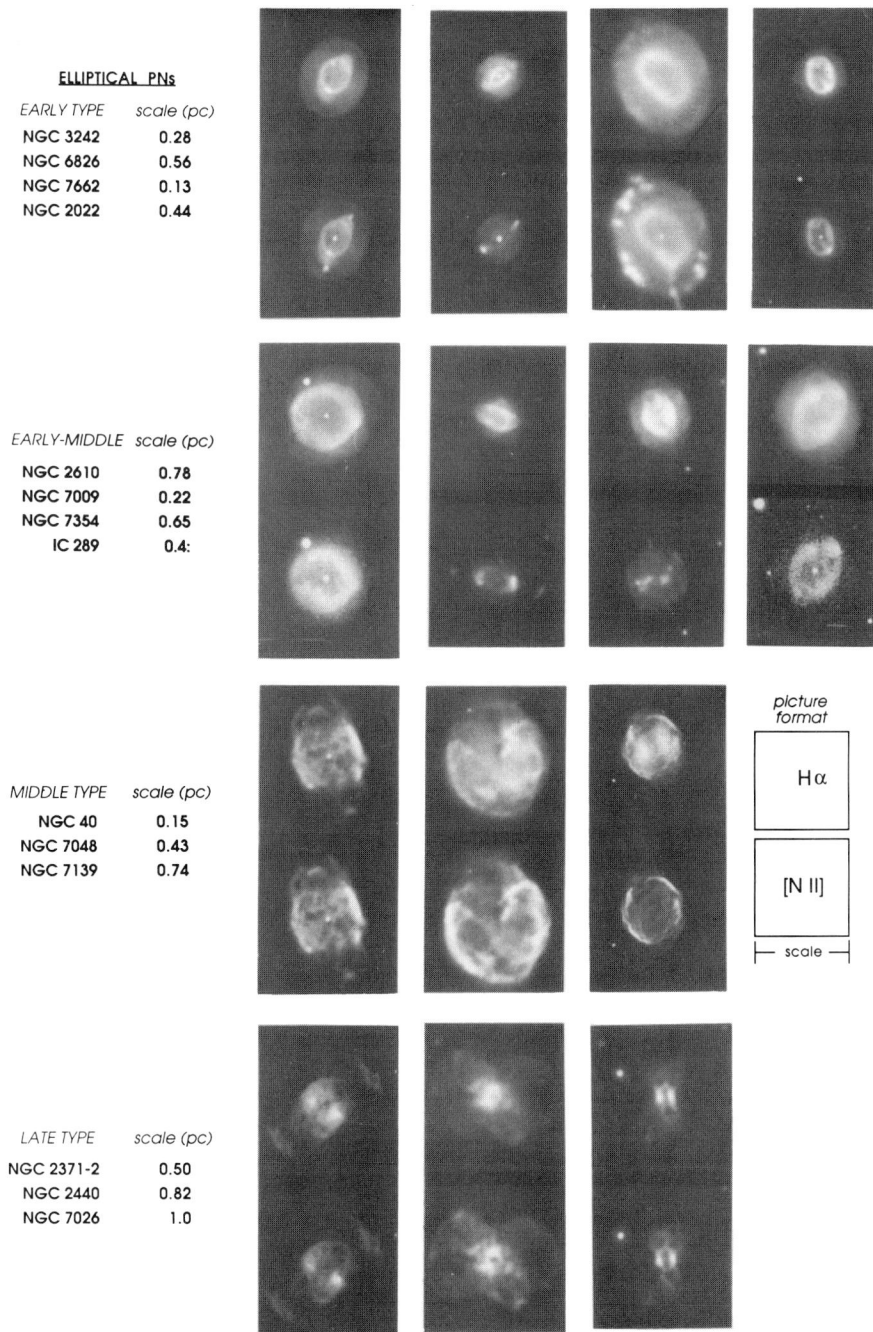

Fig. 3 – Examples of Elliptical PNs of various types.

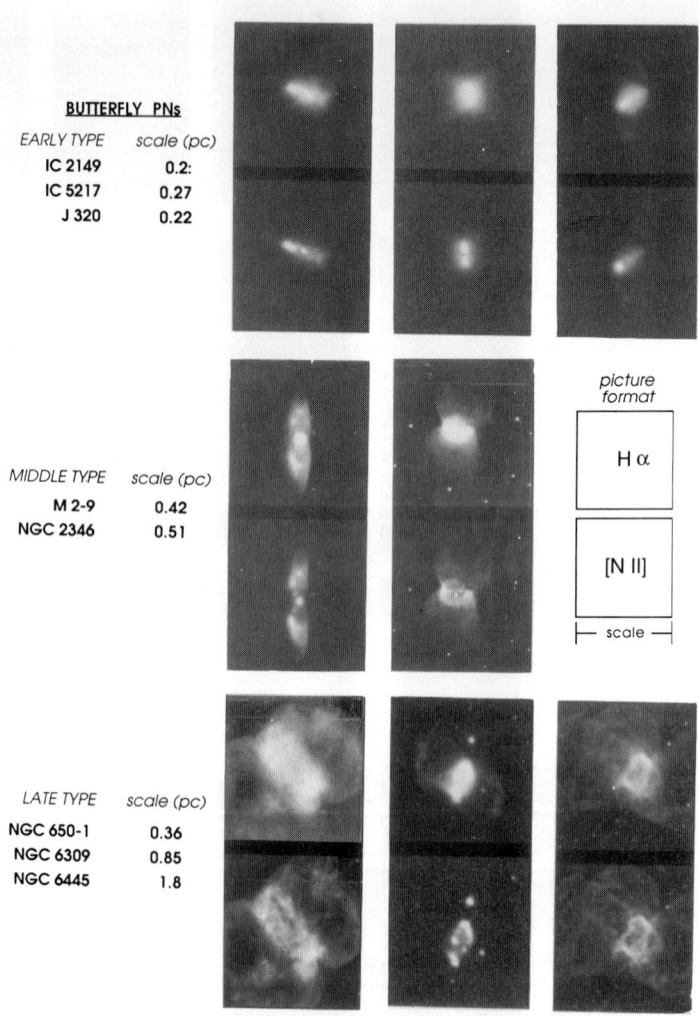

Fig. 4 – Examples of Butterfly PNs of various types.

often in pairs on opposite sides of the central star. NGC 1514 and NGC 3587 are "amorphous" – they contain none of the bright fine structure which a critical defining characteristic of the present sequence.

NGC 6543, NGC 7008, and NGC 7293 are called "peculiar". Even though peculiar PNs are extremely symmetric, their morphologies are apparently singular among all PNs. Finally, NGC 7027 is an example of an anomalous PN whose intrinsic structure (as observed at IR and radio wavelengths) fits into the morphological scheme, but whose optical appearance is heavily affected by strong and patchy foreground extinction. Finally, it must be noted that PNs are classified according to their *projected* appearance on the sky. Some PNs may be misclassified as a result. NGC 6720, NGC 6781, and IC 1747 are likely improperly classified candidates owing to projection effects.

4. PHYSICAL SIGNIFICANCE

The goal for defining a morphological sequence, whether in biology or astronomy, is to attempt to provide an organizational framework within which the patterns of morphologies might find an interpretation. The immediate goal here is to determine whether the shapes of PNs are *consistent* with a hydrodynamical model of interacting winds. Although consistency is not to be confused with proof, the establishment of consistency between theoretical expectations and observations is a very important motivation for possible future research.

Since no detailed multi-dimensional hydrodynamical models based on realistic initial and boundary conditions are available, it is important to emphasize that the attempt to show consistency between data and theory is based on theoretical *expectations*, not calculations. Such an approach is treacherous, and in the long term the proposed morphological sequence might be found to have little relevance to the manner in which PNs evolve!

4.1. The Observed Environments

As seen in nearly all of the early-type PNs of figs. 2 – 4, the slow wind is taken to have a higher density, and thus higher pressure, along an equatorial axis that along the symmetry axis of the equator (i.e. along the polar axis). The equatorial-polar density ratio, or density contrast, is unity for circular PNs and increases regularly for elliptical and butterfly PNs. The equatorial density enhancement is particularly obvious in pictures of e-m E and B PNs such as NGC 650-1, NGC 2346, NGC 2610, NGC 3242, NGC 6826, NGC 7354, and NGC 7662.

The equatorial zones of butterfly nebulae may be sufficiently dense to be optically thick to ionizing radiation. Such RGEs might be described as dense disks, portions of which are largely neutral. IR and radio molecular studies are showing the existence of these disks (e.g. Zuckerman and Gatley 1988; Bieging and Nguyen-Quang-Rieu, preprint; and Balick, Gatley, and Zuckerman, private communication).

The process that ejects the RGE with is not clearly understood. Poe and Friend (1986), Friend and Abbott 1986 and their coworkers have argued that the massive stars are likely to contain similar bands.

4.2. Early Evolution

In the subsequent discussions we adopt the general conditions of the interacting winds model (e.g. Kwok 1982 and Volk and Kwok 1985). The RGEs expand homologously until overtaken by the fast winds. We assume that the fast wind is isotropic as it leaves the stellar photosphere.

Round PNs show no evidence of density variations other than radial. Thus one expects the one-dimensional models to describe the nebular structure. For such PNs the snowplow associated with the outer shock can be seen as a circular bright rim along the inner edge of the RGE. NGC 1535, IC 3568 (fig. 1) and Shapley 1 (Pottasch 1984, p. 3) are archetypical eR PNs. Inside the bright rim is a hot bubble which is too sparse and highly ionized to be detected at visible wavelengths. Ultraviolet absorption line studies (e.g. Kaler et al.1987) confirm that there exists a transition region between the nebula and the hotter, more highly ionized interior of some PNs.

In elliptical and butterfly PNs the outer shock pushes more quickly towards the low-density polar regions than the equator. In eE-type PNs the snowplow is prolate ellipsoidal or even peanut-shaped, and is seen as a closed bright rim when projected on the sky. Simplified models of such nebulae have been computed by Kahn and West (1985). Balick et al.(1987) have investigated how eE PNs produce the collimated gas flows seen frequently in this class of PNs. Their results are summarized elsewhere in this volume (Balick et al.1988).

4.3. Subsequent Development

If the stellar wind continues to maintain a higher pressure in the bubble interior than in the RGE, the outer shock will drive outward. Eventually the outer shock will reach the edge of the RGE and, if the density near the RGE edge decreases, instabilities may develop (see, e.g., NGC 40, NGC 6894, NGC 7048, and the halo of NGC 6543 in Millikan 1974). Most PNs in this evolutionary state are of the middle type, and are drawn along the center row of fig. 1. mE PNs are characterized by a pair of polar protruberances, as seen in fig. 3. Other examples of mE PNs might be Abell 43 and 72 (see Pottasch 1984, pp 12 and 13) and NGC 6905.

In cases where the density contrast of the RGE was originally large, the development of rapidly expanding bipolar lobes is likely to begin very quickly. Models of the development of bipolar lobes in disk-dominated environments have been developed by Icke and Choe (1987), Icke et al.(1988), and Pudritz and Norman (1986), among others. The dynamics of such systems is particularly easy to study in PNs of the butterfly class such as NGC 2346 and M 2-9 owing to their relatively high surface brightness and low extinction.

Once pierced, the RGE can no longer contain the bubble. The hot material expands nearly adiabatically until eventually confined by the ISM or material ejected earlier in the lifetime of the star. Except for round PNs, two large lobes will develop. These lobes will be limb brightened as the shocks along their leading edge again form snowplows in the ambient medium. Often the lobes will be faint and very difficult to detect (Chu, Jacoby, and Arendt, 1987; Jewitt, Danielson, and Kupferman 1986). Their expansion motions are likely to be highly supersonic. Heathcote and Weller (1987) and Balick et al.(1988) report that the two outer lobes of NGC 2440 are expanding at ~ 75 km s^{-1}. Other late-type PNs which might be in a similar stage of evolution are shown in the bottom rows of figs. 3 and 4.

5. CONCLUSIONS

If one adopts the hypothesis that PNs are shaped by fast and isotropic stellar winds which interact with a red giant envelope expelled earlier with some degree of equatorial density enhancement, then a sequence of evolutionary states for PNs can be predicted. The shapes of most PNs fit nicely into the expected morphological categories (with some exceptions, of course). While such an exercise does not prove that PNs are shaped by winds, the morphological agreement between expectation and reality justifies further research.

There are several important research directions that are required before the process of wind-shaping is ready for serious credibility. On the theoretical side, two-dimensional hydrodynamic model calculations with both photon and collisional heating must be developed. It is up to the observers to determine appropriate initial and boundary conditions for these calculations. Among the missing data are the mass and velocity distributions of RGEs in PNs and their evolutionary predecessors, and the kinematics of the gas in the rims, RGEs, and other important (and readily observable) nebular structural components. Such studies are underway; however, the enormity of the task invites the participation of everyone.

6. ACKNOWLEDGEMENTS

It is a pleasure to thank many cooperative people who have been working to understand the structure of PNs. My collaborations with Heather Preston and Vincent Icke have been especially rewarding, both personally and scientifically. George Jacoby and You-Hua Chu have enthusiastically shared their latest results with us on many occasions. Important discussions and encouragement has been freely offered by Lawrence Aller, Ian Gatley, Sun Kwok, Stuart Pottasch, Yervant Terzian, and Ben Zuckerman, to name only a few. Financial support was received through grants AST 82-08041, 83-10552 (for astronomical image processing at the University of Washington), and 86-12228 from the National Science Foundation.

REFERENCES

Balick, B. (1987a). Sky and Tel. **73**, 125.
Balick, B. (1987b). Astron. J. **94**, 671.
Balick, B., Preston, H.L., and Icke, V. (1987). Astron. J. (December 1987 issue).
Balick, B., Preston, H.L., and Icke, V. (1988). This volume.
Bieging, J.H. and Nguyen-Quang-Rieu (1987). Preprint.
Chu, Y.-H., Jacoby, G.H., and Arendt, R. (1987). Preprint.
Friend, D.B., and Abbott, D.C. (1986). Astrophys J. **311**, 701.
Icke, V., and Choe, S.-U. (1987). Submitted to Astron. Astrophys.
Icke, V., Preston, H.L., and Balick, B. (1988). In preparation.
Jewitt, D.C., Danielson, G.E., and Kupferman, P.N. (1986). Astrophys. J. **302**, 727.
Kahn, F.D. (1983). In *Planetary Nebulae*, IAU Symposium No. 103, edited by D. R. Flower (Reidel, Dordrecht), p. 305.
Kahn, F.D., and West, K.A. (1985). Mon. Not. R. Astron. Soc. **212**, 837.
Kaler, J.B. Feibelman, W.A., and Henrichs H.F. (1987). Preprint.

Kwok, S. (1980). J. R. Astron. Soc. Canada **74**, 216.
Kwok, S. (1982). Astrophys. J. **258**, 280.
Kwok, S. (1987). In *Late Stages of Stellar Evolution*, edited by S. Kwok and S.R. Pottasch (Reidel, Dordrecht), p. 321.
Kwok, S., Purton, C.R., and Fitzgerald, M.P. (1978). Astrophys. J. Lett. **219**, L125.
Millikan, A.G. (1974). Astron. J. **79**, 1259.
Okorokov, V.A., Shustov, B.M., Tutukov, A.V., and Yorke, H.W. (1985). Astron. Astrophys. **142**, 441.
Pikel'ner, S.B. (1968). Astrophys. Lett. **2**, 97.
Pikel'ner, S.B. (1973). Astrophys. Lett. **15**, 91.
Poe, C.H., and Friend, D.B. (1986). Astrophys. J. **311**, 317.
Pottasch, S.R. (1984). *Planetary Nebulae* (Reidel, Dordrecht).
Pudritz, R.E., and Norman, C.A. (1986). Can. J. Phys. **64**, 501.
Schoenberner D. (1986). In *Late Stages of Stellar Evolution*, edited by S. Kwok and S.R. Pottasch (Reidel, Dordrecht), p. 337.
Volk, K.M. and Kwok, S. (1985). Astron. Astrophys. **153**, 79.
Weller, W.G. and Heathcote, S.R. (1987). In *Late Stages of Stellar Evolution*, edited by S. Kwok and S.R. Pottasch (Reidel, Dordrecht), p. 409.
Zuckerman, B. and Gatley, I. Astrophys. J. (Jan. 1988 issue).

EXPANSION VELOCITIES AND CHARACTERISTICS OF GALACTIC
PLANETARY NEBULAE

Ronald Weinberger
Universität Innsbruck
Institut für Astronomie
Technikerstrasse 25, A-6020 Austria

ABSTRACT. An up-to-date account of expansion data and their interpretation is provided.- The importance of spatiokinematical models is underlined; bipolar outflows appear to be characteristic of very young objects and prolate spheroids (including toroids and rings) seem to be a useful approximation for many PN.- High-velocity features are much more frequent than previously assumed.- Expansion velocities vs. linear radii diagrams for [OIII], HI, and [NII], based on a new catalogue show a broad, rather homogeneous distribution and thus are of limited value for studies of the dynamical evolution or formation mechanisms of PN - future investigations of these relations should be pursued separately for groups of PN with similar physical properties; a difference in the kinematical properties of B and C nebulae could, however, be confirmed.- Several recommendations for work in this area are added.

1. INTRODUCTION

A knowledge of expansion in PN is the basis for an understanding of the kinematical and dynamical processes in these objects. Expansion velocities and their correlation with various nebular and stellar parameters are, generally, of great value with regard to the conceptions on the evolution of the nebulae and their central stars.
   Campell and Moore (1918) were the first who noticed the double, bowed appearance of emission lines. The first interpretation as expansion was given by Wilson (1950); he also noticed that the smallest expansion velocities (V_{exp}) are observed for the highest ionization degrees, and the largest V_{exp} for the lowest ones; also, V_{exp} was found to increase with distance from the nebular center. Weedman (1968), by deriving simple spatiokinematical models, supposed most PN to be prolate spheroids (i.e., an ellipse

rotated about its major axis). The earliest fully hydrodynamical models of expanding nebulae were calculated by Mathews (1966).

In recent years, an enormous increase of expansion data took place, by use of long slit high-dispersion coudé and echelle spectrographs, and single aperture Fabry-Perot interferometers.

Within the last 5 years, with regard to V_{exp}, the by far most data were obtained by Sabbadin and his collaborators (19 papers - for the majority of references see, e.g., Sabbadin 1984), on compact and extended PN. Numerous (i.e., >10) PN were also observed by: Robinson, Reay, Atherton (1982; mostly compact PN); Welty (1983; mostly extended PN); Gieseking, Hippelein, Weinberger (1986; very extended PN); Mendez et al. (1987; mostly compact PN); Hippelein, Weinberger (1987; very extended PN); Chu and Jacoby (1987; multiple shell PN).

A catalogue of all V_{exp} was presented by Sabbadin (1984). A new catalogue of V_{exp}, complete up to July 1987, was prepared by Weinberger (1987).

As to spatiokinematical models, most of them were derived by Sabbadin and his co-workers (for most references see, e.g., Sabbadin 1984). Further references are given below and a complete list of references can be found in Weinberger's catalogue.

Recent interpretations of internal motions in context with nebular and/or stellar evolution are contained in: Sabbadin et al. (1984); Phillips (1984); Okorokov et al. (1985); Volk and Kwok (1985; Schmidt-Voigt and Köppen (1987).

In this review the emphasis is laid on spatiokinematical models and their common properties, PN with high-velocity features, the new compilation of expansion velocities and its various implications, and some cautionary remarks.

Neither expansion characteristics of the shells of multiple shell PN nor those of extragalactic PN will be discussed, since they are parts of other reviews in this symposium (Chu 1987, and Barlow 1987, respectively).

2. SPATIOKINEMATICAL MODELS

Spatiokinematical models are of crucial importance for the understanding of the 3-dimensional form and the kinematics and dynamics of PN. One of the main advantages is the possibility to derive true (=deprojected) expansion velocities, due to the non-sphericity of the vast majority of PN. The construction of detailed, reliable models is, however, rather time consuming and difficult and, as a consequence, only few reliable spatiokinematical models exist.

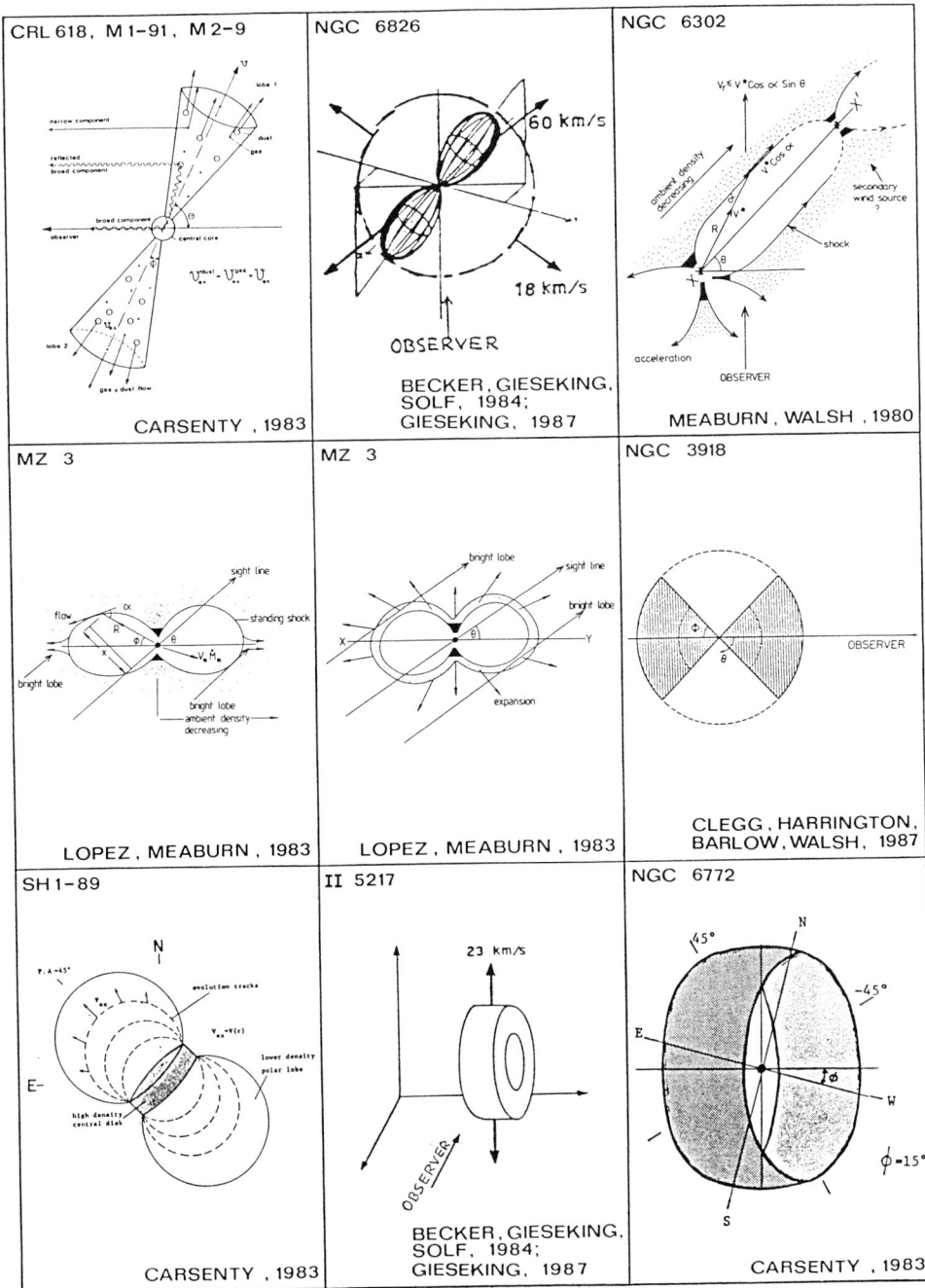

Figure 1. Sketches of typical spatiokinematical models; the drawings shown here comprise 1/3 of all published sketches.

At present, spatiokinematical models for 42 PN are known; for a few objects, more than one model exists.
In the literature, altogether 27 sketches of such models (comprising 19 PN) could be found. In Fig. 1 one third of these drawings, a representative sample, is shown. From the 27 sketches and the residual descriptions on all 42 PN as taken from the literature, we draw the following, preliminary and necessarily rough conclusions (by taking the evolutionary status of the PN into account):

Very young PN consist of compact (disklike/toroidal) central nebulae and distinct bipolar fast outflows; young and middle-aged PN show bright, thick toroids or rings and faint bipolar outflows, i.e., they are prolate (often truncated) spheroids, expanding more rapid along their major axes. For old PN too few models exist to form analogous opinions about them. In addition, a rich variety of faint (usually outer) features (ansae, filaments, halos etc.) with heterogeneous expansion characteristics exist. High dispersion spectra also show that numerous condensations are quite typical.

One may surmise that an increase of detailed spatiokinematical models will lead to a modification of the above tentative conclusions, e.g., by introducing new conceptions like that of Recillas-Cruz and Pismis (1984) on NGC 650-1 for some other objects.

3. HIGH-VELOCITY FEATURES

We will first shortly discuss the most spectacular high-velocity features and then try to find some common properties.
NGC 6302: Meaburn and Walsh (1980a),and Barral et al. (1982) report on a central disk and a wind produced ionized elongated main cavity and additional cavities. The ionized walls of the main cavity flow at ca. 300 km/sec away from the dense core. In the core region,the [NeV] 3426Å line is observed to be 800 km/sec in extent and is interpreted as direct emission from radiatively ionized gas, provided the distance is not much larger than D = 150 pc.
Mz-3: This object appears to be spatiokinematically similar to NGC 6302. The Hα line is 2460 km/sec broad and is observed in the core region; it is interpreted as produced by electron-scattering, provided $D \approx 1$ kpc (Lopez and Meaburn 1983).
M2-9: Swings and Andrillat (1979) and Walsh (1981) report on a very broad Hα line stemming from the core region. According to the former authors, its extent is 1000 km/sec.
He2-111: a range in velocity of about 400 km/sec between two parts near to the nebular borders in the southeast and northwest is reported. If the nebula is a cylindrical shell

(the major axis appears to be close to the plane of the sky), then the true outward velocity must be appreciably larger (Webster 1978).

NGC 2392: Besides high-velocity components, the main body of this pole-on object has the largest V_{exp} (true V_{exp} 53 - 93 km/sec) of all PN. O'Dell and Ball (1985) found a high-velocity stream with a true V_{exp}([OIII]) = 190 km/sec away from the center. Gieseking, Becker and Solf (1985) also detected a jetlike multiknot bipolar mass flow starting from the center, with a true V_{exp}([NII]) = 200 km/sec. These features are best reproduced in the annual report 1986 of the Max-Planck-Institut für Astronomie, in Mitt. Astron. Ges. 69, p. 166.

NGC 6537: Becker and Solf (1983) suppose the nebular center to be a compact, possibly ringlike structure; in addition, a bipolar flow appears to be present: hollow, elongated shells expand away from the center with true V_{exp}([NII]) ≈ 230 km/sec.

NGC 6543: A bipolar jet with a true V_{exp}([NII]) ≲ 50 km/sec was discovered by Solf (for a reproduction see annual report 1986 of the Max-Planck-Institut für Astronomie, in Mitt. Astron. Ges. 69, p. 166).

NGC 6826: Becker, Gieseking and Solf (1984) suppose this object to consist of 3 main components, one being an "hourglass" with a true V_{exp}([NII]) of ca. 60 km/sec along the major axis.

NGC 7293: According to Meaburn and Walsh (1980b), a small (15") region projected near the inner rim of the bright helix expands with V_{exp}([OIII]) = +66 km/sec with respect to the mean velocity of the PN. Meaburn and White (1982) noted that this small region is a part of a much larger very faint feature. Walsh and Meaburn (1987) found a filament, 11' west of center, that moves with V_{exp}([NII]) = +50 km/sec with respect to the mean velocity; it may be part of a radially expanding ellipsoidal shell.

He2-36: Feibelman (1985) discussed a possible high-velocity jet with a true V_{exp} of ca. 600 km/sec, but later Lutz et al. (1986) found no evidence for the jet.

Provided that the results on high-velocitiy features are representative, we conclude:
1) the frequency of such features is much larger than previously assumed; 2) young and/or proto-PN (particularly "butterflies") show distinct high-velocity flows in their lobes and very broad line components in their centers; 3) in evolved PN high-velocity jets, knots, filaments etc. can be present, but their number or other details are largely unknown; 4) the physical nature of high-velocity features is poorly or not at all understood (anyway, the stellar winds appear to be of importance).

4. THE NEW V_{exp} CATALOGUE AND SOME CONSEQUENCES

Up to now, compilations of V_{exp} in PN were mainly used for $V_{exp}([OIII])$ vs. linear radii (r) diagrams, as basis for understanding the dynamical evolution and formation mechanisms of these nebulae.

The most recent catalogue of expansion velocities in PN was compiled by Sabbadin (1984) and contains 165 objects. The present catalogue (Weinberger 1987) contains expansion velocities observed in 237 planetaries, as derived from about 100 papers. Most linear radii in the new catalogue are based on an enlarged set of "reliable" (i.e., ≤±50%) distances, with the enlargement to a large part due to distances taken from Mendez et al. (1987). In case of several nebular components, the data listed refer to the main, usually centrally located nebula and generally are the expansion velocities at or near the centers in extended objects. The catalogue comprises P&K designations, names, observed expansion velocity data (for all lines), references, notes, the adopted $2V_{exp}$ for [OIII], HI, and [NII], distances, linear radii, and classes (B or C).

In the following, we shall concentrate on [OIII], HI, and [NII] expansion velocities. There are:
202 PN with $V_{exp}([OIII])$, including 41 objects with less reliable data and limits,
101 PN with $V_{exp}(HI)$, including 12 with smaller reliability and 77 PN with $V_{exp}([NII])$, including 12 with smaller reliability.

Some mean values (the less reliable ones are not taken into account):
mean $V_{exp}([OIII])$ = 20.0 km/sec,
mean $V_{exp}(HI)$ = 20.0 km/sec and
mean $V_{exp}([NII])$ = 24.3 km/sec.

For mean velocity differences between these lines possible observational selection is reduced by only taking PN with "reliable" measurements in [OIII] AND HI (54 PN), [OIII] AND [NII] (42 PN), and HI AND [NII] (48 PN):
mean $V_{exp}([OIII])$ − HI) = 1.0 km/sec,
mean $V_{exp}([NII]$ − [OIII]) = 4.1 km/sec and
mean $V_{exp}([NII]$ − HI) = 4.2 km/sec.

The higher velocity of [NII] is due to the wellknown increase of V_{exp} with radial distance from the star and ionization stratification, i.e. [NII] emission predominates in the outermost ionized layers.

In retrospect, the most frequent use of V_{exp} data was in correlation with linear nebular radii (r); only [OIII] velocities were taken. It was suggested that:
a) the majority of compact nebulae show V_{exp} proportional to r and a slow decrease of r for larger radii (e.g., Sabbadin et al. 1984),
b) a high velocity sequence and a low velocity sequence

seem to exist (Robinson, Reay and Atherton 1982),
c) most nebulae conform with the relation $V_{exp} \propto r^{-0.22}$ (Phillips 1984), etc.

The observed V_{exp}([OIII]) vs. r data were frequently used to study the dynamical evolution and formation mechanisms of PN, in part also by including the evolution of central stars (Sabbadin et al. 1984; Okorokov et al. 1985; Volk and Kwok 1985; Schmidt-Voigt and Köppen 1987).

Unfortunately, the data from the new catalogue do neither support a) nor b) nor c) and will be of limited value for the just mentioned kinds of studies. In Fig. 2 we show $2V_{exp}$ vs. r relations for [OIII] (Fig. 2a), HI (2b), and [NII] (2c); the small symbols denote less reliable expansion values. In Fig. 3 the true V_{exp}([OIII]) as derived from the spatiokinematical models vs. r is presented. Both quantities refer to the minor axis in case of prolate spheroids, elliptical rings etc. by assuming a linear dependence of V_{exp} with r; the largest symbols correspond to the most reliable data.

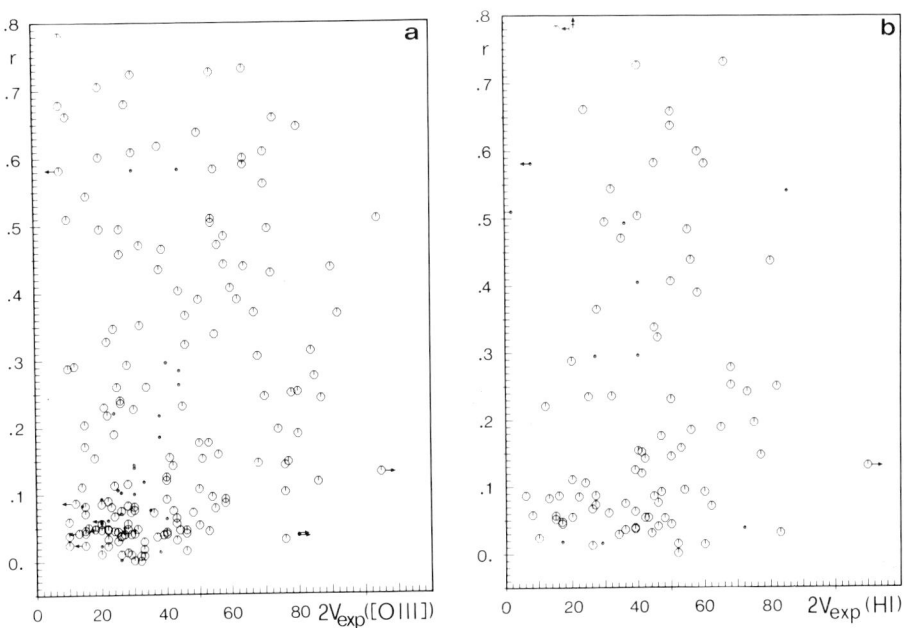

Figure 2. Observed expansion velocities vs. linear radii. 2a) [OIII] data, 2b) HI data, 2c) [NII] data. Small symbols correspond to less reliable velocities.

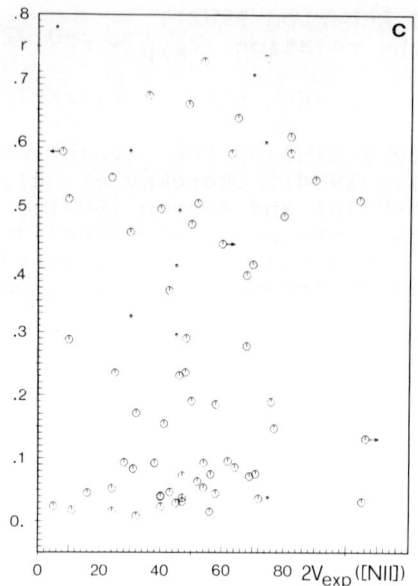

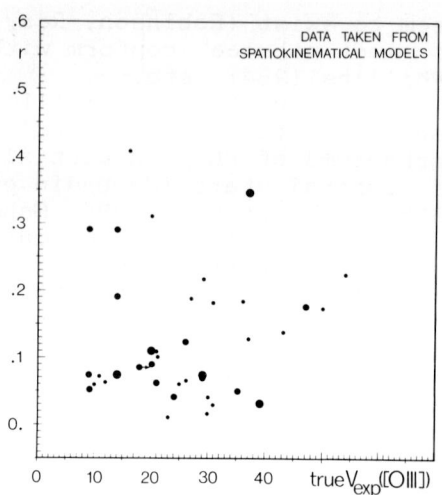

Figure 2. – Continued.

Figure 3. True (=deprojected) expansion velocities derived from spatiokinematical models versus linear radii.

Figs. 2 and 3 lead us to the following conclusions: even by considering the large effects of the various error sources, a very broad and rather homogeneous distribution (for $r \gtrsim 0.1$ pc) is obvious that appears to reflect — at least in part — the heterogeneity in the PN population.

The scientific content of such diagrams may be better worked out by applying them to physical similar groups of PN, e.g., with respect to nebular excitation, stellar spectral types etc.; we made two such attempts in order to find out i) whether B and C nebulae can be discriminated, and ii) whether highly evolved nebulae in the galactic plane might be influenced by the interstellar medium.

The classes "B" and "C" (=non-B) were proposed by Greig (1971, 1972): B's are younger, more concentrated to the plane, have higher progenitor, central star and nebular masses, and are stronger in [NII], [OII], and [OI] emission compared to C's; these results are, in part, in connexion with the kinematically different behaviour, as Sabbadin and co-workers had suggested in many papers (e.g., Sabbadin 1984). Gieseking, Hippelein and Weinberger (1986) suggested that highly diluted, i.e., very evolved PN in the plane suffer from a deceleration of their expansion.

In Fig. 4 the $2V_{exp}$([OIII]) vs. r relation is shown for B (open symbols) and C nebulae (crosses). In Fig. 5 the $2V_{exp}$([NII]) vs. r relation for PN with $0 \leq |z| \leq 0.15$ kpc (filled symbols), $0.15 < |z| < 0.50$ kpc (open symbols), and $|z| \geq 0.50$ kpc (crosses) is shown.

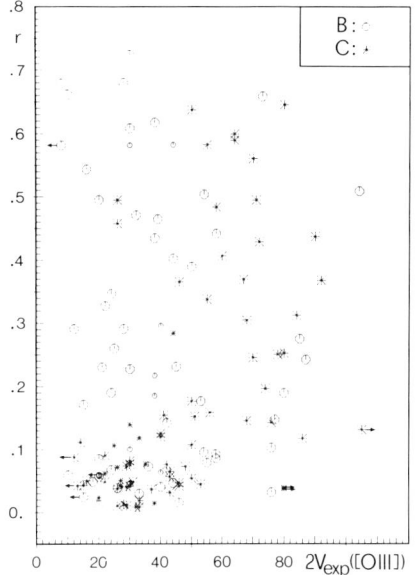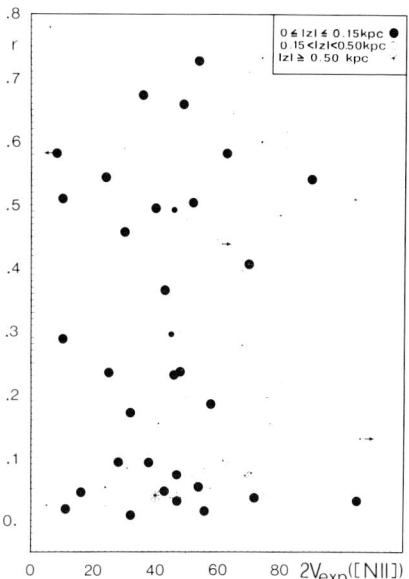

Figure 4. The $2V_{exp}([OIII])$ vs. r relation for nebulae of type B (circles) and C (crosses).

Figure 5. The $2V_{exp}([NII])$ vs. r relation for nebulae in three different $|z|$ intervals.

From both figures we conclude: it seems that both B nebulae are kinematically different (=slower) from C nebulae and that a deceleration of evolved PN in the galactic plane can occur.

5. WARNINGS AND RECOMMENDATIONS

In all above-mentioned measurements and investigations a variety of shortcomings exist - not surprising, if it is considered that this area of research is, to some extent, still in its infancy. For newcomers in this field and/or those working mainly with V_{exp} vs. r relations it is advisable to read the paper by Chu et al. (1984). Several suggestions from their article and a few own ones are presented in the following in a condensed form, as general recommendations, which may help to reduce the shortcomings in the field of expansion velocities and characteristics of planetary nebulae.
1) <u>You should try to derive detailed spatiokinematical models.</u> We need more quality instead of quantity. Concerning their importance, there are by far too few such models. Tentative models are of limited value; measure at many positions and include also the faintest possible

brightness levels.
2) <u>Bear in mind the untrustworthiness of Vexp for kinematically unresolved PN.</u> Usually, V_{exp} is deduced from FWHM/2 under assumption of simple expanding shells. This assumption can be quite wrong.
3) <u>You should not measure in one line only.</u> Remember that at least [OIII] AND [NII] (or [OII]) should be observed in order to uncover the kinematics not only in the interior but also at the surface of the nebula.
4) <u>Relate appropriate nebular dimensions to the expansion data.</u> In several cases it is forbidden to use, say, Hα + [NII] dimensions (from the Palomar Sky Survey, for example) and to relate them to [OIII] expansion velocities. Also, mean dimensions of a nebula should only be taken, if the V_{exp} data really refer to them.
5) <u>When working with Fabry Perot interferometers, bear their free spectral range and the possible influence of a finite aperture in mind.</u> These instruments usually cover a limited velocity range; high-velocity gas will escape detection. Apertures approaching the size of the nebular diameter lead to definite lower limits in V_{exp}.
6) <u>For investigations of Vexp vs. r relations you should not treat all PN alike.</u> Future investigations of V_{exp} vs. r relations should be pursued separately for groups of PN with similar physical properties.

Acknowledgements. This work was supported in part by the "Fonds zur Förderung der wissenschaftlichen Forschung", project no. 5708. The author acknowledges the allotment of observing time at the Calar Alto Observatory and the kind hospitality both there and at the Max Planck Institut für Astronomie in Heidelberg. Thanks are due to P.J. Teutsch for his assistance during preparation of this article.

REFERENCES

Barlow, M.J.: 1987, in <u>Planetary Nebulae</u>, IAU Symp. No. **131**
Barral, J.F., Canto, J., Meaburn, J., Walsh, J.R.: 1982, <u>Mon.Not.R.Astron.Soc.</u> **199**, 817
Becker, I., Gieseking, F., Solf, J.: 1984, <u>Mitt.Astron. Ges.</u> **62**, 253
Becker, I., Solf, J.: 1983, <u>Mitt.Astron.Ges.</u> **60**, 319
Campbell, W.W., Moore, J.H.: 1918, <u>Publ.Lick Obs.</u> **13**, 75
Carsenty, U.: 1983, <u>thesis</u>, Ruprecht-Karls-Universität Heidelberg
Chu, Y-H.: 1987, in <u>Planetary Nebulae</u>, IAU Symp. No. **131**
Chu, Y-H., Kwitter, K.B., Kaler, J.B., Jacoby, G.H.: 1984, <u>Publ.Astron.Soc.Pac.</u> **96**, 598
Chu, Y-H., Jacoby, G.H.: 1987, priv. comm.
Clegg, R.E.S., Harrington, J.P., Barlow, M.J., Walsh, J.R.:

1987, Astrophys.J. **314**, 551
Feibelman, W.A.: 1985, Astron.J. **90**, 2550
Gieseking, F.: 1987, priv. comm.
Gieseking, F., Becker, I., Solf, J.:1985, Astrophys.J.Lett. **295**, L17
Gieseking, F., Hippelein, H., Weinberger, R.: 1986, Astron.Astrophys. **156**, 101
Greig, W.E.: 1971, Astron.Astrophys. **10**, 161
Greig, W.E.: 1972, Astron.Astrophys. **18**, 70
Hippelein, H., Weinberger, R.: 1987, in prep.
Lopez, J.A., Meaburn, J.: 1983, Mon.Not.R.Astron.Soc. **204**, 203
Lutz, J., Balick, B., Kaler, J., Shaw, R., Heathcote, S., Weller, W.: 1986, Bull.Amer.Astron.Soc. **18**, 951
Mathews, W.G.: 1966, Astrophys.J. **143**, 173
Meaburn, J., Walsh, J.R.: 1980a, Mon.Not.R.Astron.Soc. **193**, 631
Meaburn, J., Walsh, J.R.: 1980b, Astrophys.Lett. **21**, 53
Meaburn, J., White, N.J.: 1982, Astrophys.Space Sci. **82**,423
Mendez, R.H., Kudritzki, R.P., Herrero, A., Husfeld, D., Groth, H.G.: 1987, Astron.Astrophys., in press
O'Dell, C.R., Ball, M.E.: 1985, Astrophys.J. **289**, 526
Okorokov, V.A:, Shustov, B.M., Tutukov, A.V., Yorke, H.W.: 1985, Astron.Astrophys. **142**, 441
Phillips, J.P.: 1984, Astron.Astrophys. **137**, 92
Recillas-Cruz, E., Pismis, P.: 1981, Astron.Astrophys. **97**, 398
Robinson, G.J., Reay, N.K., Atherton, P.D.: 1982, Mon.Not.R.Astron.Soc. **199**, 649
Sabbadin, F.: 1984, Astron.Astrophys.Suppl.Ser. **58**, 273
Sabbadin, F., Gratton, R.G., Bianchini, A., Ortolani, S.: 1984, Astron.Astrophys. **136**, 181
Schmidt-Voigt, M., Köppen, J.: 1987, Astron.Astrophys. **174**, 211
Swings, J.P., Andrillat, Y.: 1979, Astron.Astrophys. **74**, 85
Volk, K., Kwok, S.: 1985, Astron.Astrophys. **153**, 79
Walsh, J.R.: 1981, Mon.Not.R.Astron.Soc. **194**, 903
Walsh, J.R., Meaburn, J.: 1987, Mon.Not.R.Astron.Soc. **224**, 885
Webster, B.L.: 1978, Mon.Not.R.Astron.Soc. **185**, 45p
Weedman, D.W.: 1968, Astrophys.J. **153**, 49
Weinberger, R.: 1987, in prep.
Welty, D.E.: 1983, Publ.Astron.Soc.Pac. **95**, 217
Wilson, O.C.: 1950, Astrophys.J. **111**, 279

Ronald Kirkpatrick, Patrick Harrington and Mario Perinotto.

MULTIPLE SHELL PLANETARY NEBULAE

You-Hua Chu
Astronomy Department, University of Illinois,
1011 W. Springfield Avenue, Urbana, IL 61801
U. S. A.

ABSTRACT. It has been shown in several independent investigations that the multiple-shell phenomenon is prevalent in planetary nebulae. Despite the common classification, the multiple shell planetary nebulae are a heterogeneous group of objects, as testified by the wide variety of their morphologies and physical structures. There are two types of double-shell structures that are seen frequently: one has an inner shell expanding supersonically into a faint, subsonically expanding halo, and the other has a bright attached envelope co-expanding with the inner shell. The physical structures and relative elemental abundances in the shells are reviewed, and their possible formation mechanisms are discussed.

I. INTRODUCTION

Many planetary nebulae (PNe) have been reported to contain "two concentric rings," "faint envelopes," or "extended halos" by Curtis (1918), Duncan (1937), Minkowski and Osterbrock (1960), and Millikan (1974). A catalog of such nebulae, called "multiple shell planetary nebulae" (MSPNe), has been compiled by Kaler (1974). All of Kaler's MSPNe can be described by: 1) the outer structure is a regular ring that encircles at least half the extent of the inner nebula, and 2) there is a break in the surface brightness profile at the interface of the inner and outer shells, giving the double-shell appearance. In a recent compilation of MSPNe (Chu, Jacoby, and Arendt 1987), 41 nebulae are included, and the number is rapidly growing with the addition of IC 1297, NGC 6751, and many southern MSPNe reported in this symposium (Aller, Keyes, and Feibelman 1986; Gieseking and Solf 1986; Ruiz, Heathcote, and Weller 1987; Lutz, Lame, and Balick 1987).

   A30 and A78 have not been included in the list of MSPNe although they have different abundances and expansion velocities in their inner and outer regions, indicating multiple ejections (Jacoby and Ford 1983; Reay, Atherton, and Taylor 1983a; Jacoby and Chu 1987; Pismis and Moreno 1987; Manchado, Pottasch, and Mampaso 1987a). The exclusion is somewhat arbitrary, and is intended only to narrow down the scope of MSPN studies, since A30 and A78 are clearly at a very different evolutionary stage from the conventional MSPNe (Iben et al. 1983).

There are several candidates for a new kind of MSPN; they have been reported as PNe with extended halos (at a level of $10^{-3} - 10^{-4}$ times the peak surface brightness). For example, faint emission lines are still detected at twice the known radius of NGC 1535 (Bässgen et al. 1986; Weller and Heathcote 1987a) and other nebulae (Bässgen et al. 1987). The most convincing evidence for the existence of faint halos comes from the high-dispersion spectroscopic observations: unresolved lines are detected outside the boundary of the fast expanding main nebula of NGC 6210 and NGC 6309 (Chu and Jacoby 1987). It is premature to include such halos in this review; however, they deserve future consideration.

II. FREQUENCY OF OCCURRENCE

It is perhaps fortuitous that <u>ALL</u> PNe with well-known names have double shells, for example, the Dumbbell, the Eskimo, the Helix, the Owl, the Ring, and the Saturn nebulae; nevertheless, it is a good indication that the multiple-shell phenomenon is common in PNe.

In the Catalogue of Galactic Planetary Nebulae (Perek and Kohoutek 1967), about 25% of the resolved PNe have "outer envelopes." In a recent deep CCD image survey of 44 PNe, Jewitt, Danielson, and Kupferman (1986) find that 2/3 of them have extended halos, defined as material beyond the 10% isophotal boundaries. Bässgen et al. (1987) find spectroscopically that 15 out of the 20 PNe they studied have halos. Although some of these outer envelopes and extended halos are the lobes of bipolar nebulae (e.g., NGC 650-651 and NGC 6781) and some consist of only irregular wisps and filaments (e.g., NGC 6309 and NGC 6772), most of them do have shell structures. These surveys imply unanimously that a significant fraction of PNe have multiple shells.

A rigorous derivation of the multiple-shell occurrence rate is given by Chu, Jacoby, and Arendt (1987). They have examined the 126 PNe in the New General Catalogue (NGC) and Index Catalogue (IC), and carefully analyzed the selection biases in this sample. They find that in the 56 nearby (< 3 kpc) optically thin PNe in NGC/IC, at least 28 have multiple shells. They conclude that more than 50% of PNe go through the visible multiple-shell phase during their lifetime.

It should not be surprising that the multiple-shell phenomenon is so prevalent, since the multiple-shell appearance is indicative of abrupt changes in stellar mass loss during the PN formation, and the mass loss from the central star of a PN does change drastically in both rate and velocity from the progenitor AGB phase to the current PNN phase (cf. Kwok 1987; Cerruti-Sola and Perinitto 1985).

III. MORPHOLOGY OF MULTIPLE SHELLS

In an earlier attempt to classify multiple-shell morphology, Kaler (1974) defined bright double, weak double, giant halo, and triple shell nebulae, according to the relative surface brightness, size, and number of shells in a nebula. As more high-quality images become available, it is clear that some of his weak double shell nebulae (He 1-5, IC 289, and NGC 2610) are just fainter counterparts of the bright doubles (e.g., NGC 1535 and NGC 7354), and the weak double nebula NGC 2438 is morphologically similar to the giant halo nebulae (e.g., NGC 6826 and NGC 6891) except that its outer shell is less than 5 times as large as its inner shell. This problem is remedied in the new classification scheme (Chu, Jacoby, and Arendt 1987) described in the remainder of this section. In the new scheme, the class of weak double nebula is disassembled. Some of the weak doubles are combined with the giant halo nebulae and called Type I; the others are combined with the bright doubles into Type II.

For convenience, only the double-shell morphology is classified below, since nebulae with more than two shells can be decomposed into a small number of double-shell units. The triple shell nebulae may consist of different types of outer and inner double shells. Images of MSPNe can be found in papers by Chu, Jacoby, and Arendt (1987) and Balick (1987a). Some representative examples are given in Figure 1.

Type I is characterized by faint outer shells with noticeable limb-brightening, often called faint halos. The detailed outer shell structure may vary from the smooth and regular cases of NGC 2438, NGC 6826, NGC 6891, and NGC 7662 to the irregular filamentary cases of He 2-111, NGC 6543, and NGC 6720. About 30% of the known MSPNe have type I double shells.

Type II is characterized by bright outer shells that appear attached to their inner shells. There are different degrees of symmetry in geometry and surface-brightness variation in the type II double-shell morphology. Some nebulae have outer shells with similar thickness in all directions. Among these some have extremely smooth, featureless outer shells (type IIa), for example, IC 3568 and NGC 1535; while the others have significant brightness variation in the outer shells (type IIb). The variation is often amplified in the [N II] line images, for example, M2-2 and NGC 7662. Some nebulae have a marked elongation in the inner shell which seems to be breaching out of the outer shell (type IIc), for example, IC 289, NGC 6058, NGC 6804, NGC 7009, and NGC 7354. Note that there are still differences in surface brightness variation, parallel to the distinction between type IIa and type IIb, among the type IIc nebulae; perhaps further classification is warranted. There are also some peculiar nebulae with such irregular structures in their outer shells that they do not fit in any of the three subclasses above, for example, NGC 1545, NGC 2392, NGC 4361, and NGC 6026. About 70% of the known MSPNe have type II double shell structures.

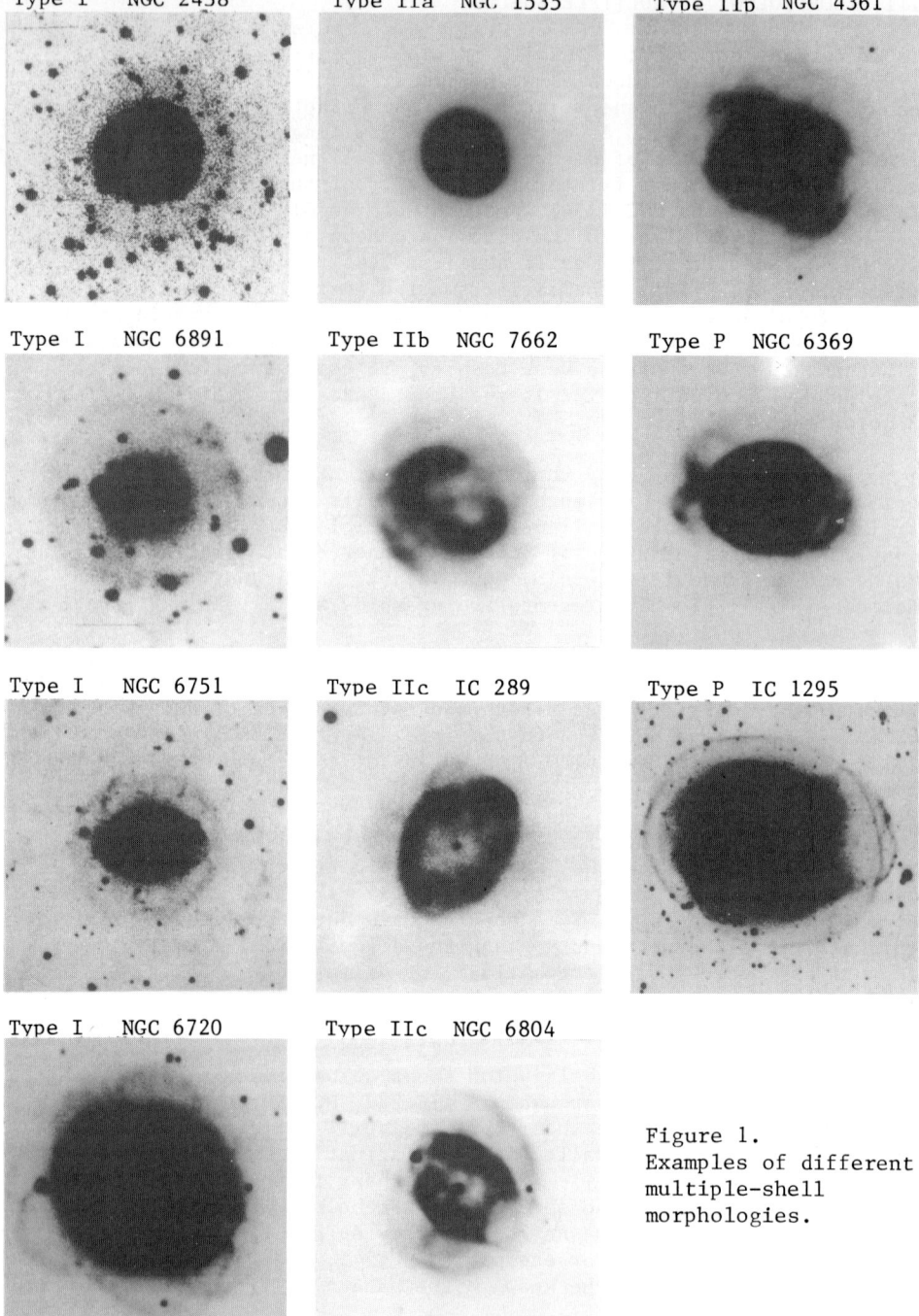

Figure 1.
Examples of different
multiple-shell
morphologies.

Finally, there are a small number of nebulae classified "peculiar" for one of the two reasons: 1) The double-shell structure seems to be a hybrid of types I and II. Since only A2 and IC 1295 are in this category, a separate classification is not justified for the time being. 2) The outer shell is so irregular that neither of the two types can describe its structure, for example, NGC 3132, NGC 6369, and the outer shells of NGC 6804. Some of these outer shells may be illuminated interstellar medium, as they seem to be part of the more extended nebulosity in the vicinity, for example, NGC 6857 and NGC 6894.

The two aforementioned morphological classes have very little overlap in properties such as size of the outer shell, relative sizes of the shells, and the relative surface brightness of the shells (Chu, Jacoby, and Arendt 1987). The type I MSPNe usually have outer diameter larger than 0.5 pc, while those of the type II MSPNe are smaller than 0.5 pc. The type I MSPNe mostly have the ratio of outer to inner radii greater than 2, while for the type II MSPNe it is smaller than 2. The surface brightness of the outer shell is usually less than 5×10^{-3} times that of the inner shell for the type I MSPNe, while the type II MSPNe have outer surface brightness about 0.1-0.5 times the inner surface brightness.

There are several triple shell nebulae. Four of them, NGC 2022, NGC 6826, NGC 6891, and NGC 7662, have type II inner double-shells and type I outer double-shells. NGC 3132 has peculiar inner double-shells and type I outer double-shells. NGC 6720 and NGC 6751 contain two type I halos (Moreno and López 1987; Chu and Jacoby 1987). The triple-shell identification may be debatable for NGC 3242, NGC 6804, and NGC 6894; these three nebulae have type II inner double-shells and peculiar outermost structure, of which a circumstellar origin is uncertain (Rosado 1986; Chu, Jacoby, and Arendt 1987).

IV. PHYSICAL STRUCTURES DERIVED FROM INTERNAL MOTIONS

The most direct way to probe the physical structure of MSPNe is through the study of internal motions. The relative motion of the shells allows us to determine the dynamical evolution and to test critically the PN-formation models. In the high dispersion spectroscopic observations it is essential to resolve the shells spatially, as well as spectrally. Welty's (1983) kinematic study of MSPNe was hampered by his lack of spatial resolution; only integrated Fabry-Perot spectra were obtained, and no definitive conclusion could be reached. Sabbadin (1984, and references therein) obtained long-slit echelle observations for a large number of MSPNe; however, most of the outer shells were undetected. The Eskimo nebula has been extensively studied by Reay, Atherton, and Taylor (1983b), O'Dell and Ball (1985), Gieseking, Becker, and Solf (1985), and Balick, Preston, and Icke (1988); the internal motion of the Eskimo is so complex that no consistent model to explain the kinematic features in both [O III]- and [N II]-lines is available yet.

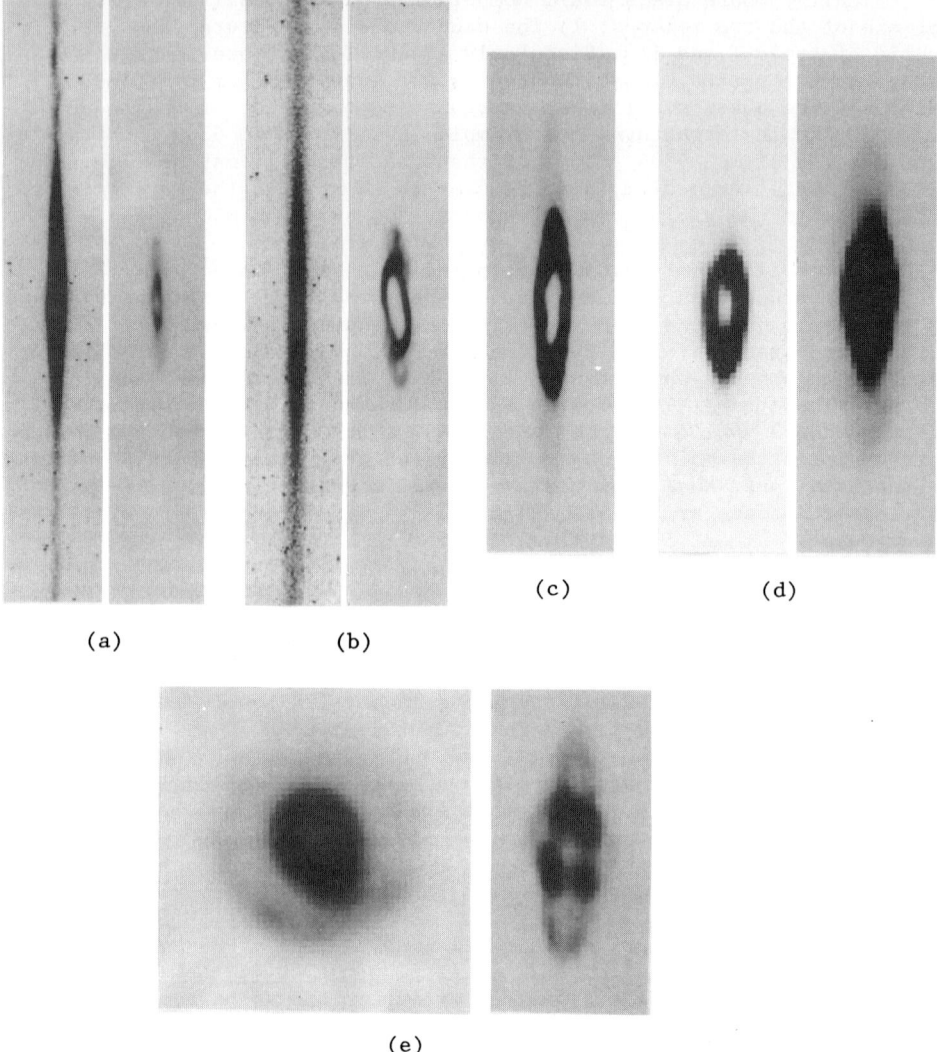

Figure 2. [O III] echelle line images of MSPNe. a) NGC 6826. The spectrum of the halo is on the left and that of the inner shells along the minor axis is on the right. b) NGC 7662. The spectrum of the halo is on the left, and that of the inner shells along the major axis is on the right. Each panel in a) and b) has 300 km/s along the X-axis and 130" along the Y-axis. c) NGC 1535. d) IC 3568. Both panels are from the same exposure, windowed differently to show the features at higher or lower brightness levels. e) M2-2. An [O III] image is on the left, and an echelle line image is on the right. The slit was NS oriented, passing through the center.

Chu and Jacoby (1988) have obtained long-slit echelle/CCD observations of 28 MSPNe, with special effort to detect the faint outer shells. The discussion in this section is based mainly on this set of homogeneous data. Some echelle line images are reproduced in Figure 2. The patterns of internal motion, archetypical examples, and their associated morphologies are summarized below:

i) The inner shell expands supersonically into a faint halo which expands subsonically or mildly supersonically.
   Example - IC 4593, NGC 6751, NGC 6826, NGC 6891, and NGC 7662.
   Morphology - type I outer shells.

ii) The outer shell is an attached envelope, co-expanding supersonically with the inner shell. For any nebulae with an ellipsoidal inner shell, the expansion velocity of the outer shell is similar to that of the inner shell along the minor axis, but slower than that along the major axis.
   Example - (round) NGC 1535 and NGC 2610.
            (ellip.) NGC 2022, NGC 6058, NGC 7009, and NGC 7354.
   Morphology - (round) type IIa, and (ellip.) type IIc.

iii) The outer shell expands independently from the inner shell.
   Example - He 2-111*, M2-2, NGC 6720, and the Helix*.
   Morphology - type I and type IIb.
      (*: He 2-111, Webster 1978; the Helix, Walsh and Meaburn 1987)

iv) The outer shell expands faster than the inner shell!
   Example - IC 3568, NGC 6826*, and NGC 6891*.
   Morphology - type IIa and type IIc.
   (*: referring to the inner two shells)

v) Peculiar motion.
   Example - NGC 2392, NGC 4361, and NGC 6369.
   Morphology - type IIp and type P.

Three important inferences can be drawn from these results. First, the nebular morphology is not a sufficient condition for the physical structure, hence should not be cited as supporting evidence for particular formation models. For example, IC 3568 and NGC 1535 both have perfectly smooth and symmetric outer shells and have been used to demonstrate the snowplow phase of the two-wind model (Balick 1987a, 1987b). However, as the line images in Figure 2 show, the outer shell of NGC 1535 expands at a velocity similar to that of the inner shell, while the outer shell of IC 3568 expands faster than its inner shell; neither case can be explained as snowplow action.

Second, there is direct evidence from the echelle line images that most of the outer shells in MSPNe are filled with material, instead of being hollow shells. The strongest evidence comes from the lack of linesplitting in the profiles which are several times the instrumental resolution. A hollow shell would have given rise to a linesplitting of

twice the expansion velocity. While the lack of linesplit can be due
to either large turbulence in the shell (Hippelein, Bässgen, and
Grewing 1985) or the presence of material moving tangentially to the
line-of-sight, the filled-envelope geometry is inevitable. A
supersonic turbulence will disperse the shell and diffuse back to fill
the gap between the shells. For a filled-envelope geometry, there is
always material moving tangential to the line-of-sight. This situation
is seen in both type I and type II outer shells, as illustrated in the
halo of NGC 7662 and the outer shell of NGC 1535 (Figure 2). The
observed [O III] velocity FWHM in the halo of NGC 7662 is 47 km/s,
while the instrumental profile is only 9.5 km/s. This result is
significantly different from what Hippelein, Bässgen, and Grewing
(1985) derived from their large aperture Fabry-Perot observations;
their results are probably spurious due to the enormous contamination
of the stray light and the lack of spatial resolution.

Third, the internal motion in MSPNs can be used to determine
whether they are bipolar nebulae viewed along the poles. Based on the
detection of molecular H_2 in NGC 6720 (Zuckerman and Gatley 1988) and
the morphology of NGC 2438, it has been suggested that these two
nebulae may be bipolar nebulae (Balick 1987a). However, the radial
expansion velocity of both halos are much smaller than their inner main
nebulae (Chu and Jacoby 1987), contrary to what is expected in bipolar
nebulae (Weller and Heathcote 1987b). While never being suspected, the
peculiar MSPNe NGC 4361 and NGC 6369 do have kinematic characteristics
of bipolar structure. The Helix nebulae, NGC 7293 might be bipolar,
too (Walsh and Meaburn 1987).

V. RELATIVE ABUNDANCES OF THE SHELLS

Very few spectroscopic investigations of MSPNe have emphasized the
detection of both shells. Future observations should be planned to
differentiate the shells.

Type I halos are faint, hence only the brightest ones have been
observed - NGC 6543 and NGC 6826. These two MSPNe have been studied by
Manchado, Pottasch, and Mampaso (1987b) and Middlemass, Clegg, and
Walsh (1987). Both conclude that the halo of NGC 6826 has higher
electron temperature and lower He and N abundances than the inner
shell. Nevertheless, they disagree on the abundance variation in NGC
6543 - one group conclude lower abundances in the halo, while the other
group conclude uniform abundances throughout the nebula. The origin of
the disagreement is in the electron temperature they derive for the
inner shell. Given the complicated ionization structure revealed in
the images of NGC 6543 (Balick 1987a), it is probably easy to get
different temperatures if different regions are sampled.

Four type II double shells have been studied with spatial
resolution - the inner two shells of NGC 3242, NGC 7009, and NGC 7662
(Barker 1983, 1985, 1986), and NGC 6826 (Jacoby, Quigley, and Africano

1987). None of the nebulae, except maybe NGC 7662, show any significant abundance differences between the inner and outer shells. In NGC 7662, there are bright [N II] knots (Balick 1987a) which may have higher nitrogen abundance, although this has not been emphasized by Barker (1986).

VI. FORMATION MECHANISMS OF MULTIPLE SHELL PLANETARY NEBULAE

Given the various morphological, kinematic, and chemical structures in the shells, the various MSPNe must have been formed differently from one another. Although theoretical calculations of PN evolution have begun to include three stages of mass loss processes (Schmidt-Voigt and Köppen 1987a, b), there is still a long way before the multiple shells can be satisfactorily modeled. Below are tentative suggestions of the formation mechanism for each type of multiple-shell structure.

The faint, subsonically expanding halos undoubtedly correspond to the slow wind from the AGB phase. The double-shell nebulae with a simple inner shell and a halo are good candidates for Kwok's (1983) two-wind formation, since both the morphology and kinematics of the shells are qualitatively consistent with the model predictions. On the other hand, the inner shell could have been created by an enhanced mass loss rate in a "superwind," too.

The "co-expanding envelope"-type outer shells are difficult to explain, since the outer shell seems to be coasting along rather than actively interacting with the inner shell. It could be that the break between the shells was formed by an abrupt change in the mass loss rate of a "superwind." Nevertheless, it remains to be explained how the shells manage to fight diffusion and maintain a steep density gradient at the interface.

The MSPNe that contain two or three independently expanding systems clearly need multiple ejections. In the case of M2-2 (Figure 2), the outer and inner shells have different expansion velocities and ellipticities. Such a rapid change (within a few x 10^3 yr) in the ejection process might be due to a binary core (Iben and Tutukov 1987).

Three known nebulae have outer shells expanding faster than the inner shells. The dynamical timescale (≡ radius/expansion velocity) is longer for the inner shell than for the outer shell. It is unlikely that a fast expanding shell sieves through a slower shell; the outer shell must have been preferentially accelerated. The acceleration could be due to simply the diffusion into vacuum (Schmidt-Voigt and Köppen 1987a) or the radiation pressure on the dust (Mathews 1978). At least one of the nebulae, IC 3568, is well-known for its high dust content (Cohen, Harrington, and Hess 1984).

The peculiar MSPNe are hard to understand because of the insufficient data available. Some of them, particularly NGC 4361 and

NGC 6369, may be bipolar nebulae projected along certain angles.

VII. FUTURE WORK

Clearly, there is a lot to be done in the future, both observationally and theoretically, in order to understand the MSPNe better. Several investigators have recently obtained a large number of data on morphologies and internal motions. In the future, it is important to model the 3-D structure of individual nebulae and extract quantitative information such as the radial density profile and the velocity law within the shells. More low-dispersion spectroscopic observations are needed to determine the elemental abundances in the shells of MSPNe. Theoretically, models of PN formation with variable wind or binary cores should be pursued further.

REFERENCES

Aller, L. H., Keyes, C. D., and Feibelman, W. A. 1986, Ap. J., 311, 930.
Balick, B. 1987a, A. J., 94, 671.
───────. 1987b, this volume.
Balick, B., Preston, H., and Vicke, V. 1988, submitted to A. J.
Barker, T. 1983, Ap. J., 267, 630.
───────. 1985, Ap. J., 294, 193.
───────. 1986, Ap. J., 308, 314.
Bässgen, M., Bässgen, G., Barnstedt, J., Grewing, M., and Bianchi, L. 1986, Mitt. Astr. Ges., Nr. 67, p. 342.
Bässgen, M., Bässgen, G., Grewing, M., Cerrato, S., and Bianchi, L. 1987, in this volume.
Cerruti-Sola, M., and Perinotto, M. 1985, Ap. J., 291, 237.
Chu, Y.-H., and Jacoby, G. H. 1987, in this volume.
───────. 1988, in preparation.
Chu, Y.-H., Jacoby, G. H., and Arendt, R. 1987, Ap. J. Suppl., 64, 529.
Cohen, M., Harrington, J. P., and Hess, R. 1984, Ap. J., 283, 687.
Curtis, H. D. 1918, Pub. Lick Obs., 13, 57.
Duncan, J. C. 1987, Ap. J., 86, 496.
Gieseking, F., Becker, I., and Solf, J. 1985, Ap. J. (Letters), 295, L17.
Gieseking, F., and Solf, J. 1986, Astr. Ap., 163, 174.
Hippelein, H. H., Bässgen, M., and Grewing, M. 1985, Astr. Ap., 152, 213.
Iben, I., Jr., Kaler, J. B., Truran, J. W., and Renzini, A. 1983, Ap. J., 264, 605.
Iben, I., Jr., and Tutukov, A. V. 1987, in this volume.
Jacoby, G. H., and Chu, Y.-H. 1987, in this volume.
Jacoby, G. H., and Ford, H. C. 1983, AP. J., 266, 298.
Jacoby, G. H., Quigley, R. J., and Africano, J. L. 1987, Pub. A. S. P., 99, 672.
Jewitt, D. C., Danielson, G. E., and Kupferman, P. N. 1986, Ap. J.,

302, 727.
Kaler, J. B. 1974, A. J., 79, 594.
Kwok, S. 1983, in IAU Symp. No. 103, "Planetary Nebulae", ed. D. R. Flower, (Reidel: Dordrecht), p. 293.
Kwok, S. 1987, in "Late Stages of Stellar Evolution", eds. S. Kwok and S. Pottasch, (Reidel: Dordrecht), p. 321.
Lutz, J., Lame, N. J., and Balick, B. 1987, in this volume.
Manchado, A., Pottasch, S. R., and Mampaso, A. 1987a, in this volume.
_____. 1987b, in this volume.
Mathews, W. G. 1978, in IAU Symp. No. 76, "Planetary Nebulae", ed. Y. Terzian (Reidel: Dordrecht), p. 251.
Middlemass, D., Clegg, R. E. S., and Walsh, J. R. 1987, in this volume.
Millikan, A. G. 1974, A. J., 79, 1259.
Minkowski, R., and Osterbrock, D. 1960, Ap. J., 131, 537.
Moreno, M. A., and López, J. A. 1987, Astr. Ap., 178, 319.
O'Dell, C. R., and Ball, M. E. 1985, Ap. J., 289, 526.
Perek, L., and Kohoutek, L. 1987, Catalogue of Galactic Planetary Nebulae (Prague: Czechoslovaak Academy of Science)
Pismis, P., and Moreno, M. A. 1987, in this volume.
Reay, N. K., Atherton, P. D., and Taylor, K. 1983a, M. N. R. A. S., 203, 1079.
_____. 1983b, M. N. R. A. S., 203, 1087.
Rosado, M. 1986, Rev. Mexicana. Astr. Ap., 13, 49.
Ruiz, M. T., Heathcote, S. R., and Weller, W. G. 1987, in this volume.
Sabbadin, F. 1984, Astr. Ap. Suppl., 58, 273.
Schmidt-Voigt, M., and Köppen, J. 1987a, Astr. Ap., 174, 211.
_____. 1987b, Astr. Ap., 174, 223.
Walsh, J. R., and Meaburn, J. 1987, M. N. R. A. S., 224, 885.
Webster, B. L. 1978, M. N. R. A. S., 185, 45p.
Weller, W. G., and Heathcote, S. R. 1987a, in "Late Stages of Stellar Evolution", eds. S. Kwok and S. R. Pottasch, (Reidel: Dordrecht), p. 409.
_____. 1987b, in this volume.
Welty, D. E. 1983, Pub. A. S. P., 95, 217.

Manuel Peimbert and Luis F. Rodríguez.

DUST IN PLANETARY NEBULAE

Patrick F. Roche,
Royal Observatory,
Blackford Hill,
Edinburgh EH9 3HJ.   Scotland.

The presence of dust in planetary nebulae can be deduced in several ways - from the observed depletions of condensable elements, internal extinction and, most directly, through the detection of infrared emission from the dust grains. We know that there is a substantial amount of dust in planetary nebulae, and that a significant fraction of the total luminosity emerges in the infrared through thermal emission in most objects. However, a number of questions still largely remain unsolved, and perhaps the most pressing of these are that we do not yet have a satisfactory understanding of the ultraviolet, optical or infrared properties of the dust grains and we also do not yet know exactly where the emitting grains are located within the nebulae; for example, are they mixed with the ionized gas, or in neutral inclusions or perhaps in a disk around the central star?

In this report, the starting point will, appropriately enough, be the review by Barlow (1983) in the last IAU symposium devoted to planetary nebulae. Since then, a number of important developments that bear on the nature, effects and location of dust in planetaries have occurred, but the most dramatic has been the successful IRAS mission which has measured the infrared fluxes of hundreds of planetaries; this is discussed in detail by Preite-Martinez in this volume.

1. Evidence of dust at short wavelengths

The effects of dust on the visible and ultraviolet emission from planetary nebulae can be seen through absorption and scattering of light. However, in almost all objects studied, these effects are small and it is difficult to measure them with sufficiently accuracy to put useful limits on the nature of the grains responsible.

Firstly, the interstellar extinction must be properly accounted for, and in many objects this can be substantial so that the shape of the adopted extinction curve is crucial; any deviations caused by internal extinction will be difficult to establish. Attempts have been made to find differences in the intensities of emission lines arising in the approaching and receding edges of pn which could be attributed to internal extinction, using high resolution spectroscopy. In this case, possible asymmetries in the nebular shells render the results uncertain, but an approach using emission lines spanning the wavelength range 3100 - 6600 A has been used by Doughty & Kaler (1982) to minimise this problem. Ultraviolet resonance lines have very high scattering optical depths, with a correspondingly high probability of being absorbed by dust. The ratio of CIV $\lambda 1549$ /CIII $\lambda 2297$ can be used as a diagnostic of the dust absorption optical depth (see Seaton 1983) for objects in which a large fraction of carbon is in the form of C^{3+}.

Light scattered off dust grains may be visible in some objects. NGC 7027 has been known to have a faint reflection nebula extending over a diameter of 50 arcsec for some years (Atherton et al 1979), probably produced by dust scatttering in a massive neutral shell around the ionized zone. An extended optical halo with a diameter of up to 2 arcminutes has been detected around IC 418 by Monk, Barlow and Clegg (this meeting). The fact that the [OII]/Hγ line intensity ratio is the same in the halo as in the ionized nebula suggests that the halo results from dust grains reflecting the nebular emission lines. The picture for other objects is not as clear, for example conflicting reports on the extent of optical emission from Vy 2-2 were presented at this meeting by Clegg, Hoare and Walsh and by Falomo and Sabbadin. Scattering by dust grains within the ionized nebula can lead to measurable amounts of polarization. A recent study by Leroy et al (1986) has detected polarization of up to 1% in the [OIII] lines in 5 pn, which they attribute to scattering by dust particles.

Ionization structure models have now been developed that include dust grains heated by both trapped UV lines and direct heating by the central star and are able to account for the observed infrared emission (e.g. for NGC 7662 and NGC 3918, Harrington et al 1982, 1987). Further developments such as including the contributions from dust in neutral regions are underway (Hoare, this meeting), and synthesis of the nebular spectrum, taking into account absorption by dust grains, together with the infrared emission promises to give a clearer understanding of the dust in pn. Overall however, it appears that dust has a relatively small effect on the ionisation structure of the nebulae, although in order to account for the observations of UV resonance lines, there must be some dust present inside the ionized regions.

2. The nature of dust in planetaries

2.1 Precursors

We believe that the precursors of planetary nebulae are the asymptotic giant branch stars. These objects are typically cool stars immersed in thick circumstellar envelopes produced by heavy mass loss. From condensation theory (see e.g. Larimer 1979 and references therein), we expect these stars to condense copious amounts of dust in their circumstellar shells and indeed, this is what is observed. Different species of grain are produced according to the details of the stellar outflow, and in particular depending upon the chemical abundances of the circumstellar material. We therefore expect the grains in planetary nebulae to reflect the conditions present in the precursor wind, though the ionizing flux from the planetary nebula will affect the dust particles, which may in turn influence the development of the ionized nebula.

2.2 Depletions

It has been shown that elements, such as Fe, Mg, Si, are underabundant in the gas phase compared to the expected values (Shields 1983); it is just these elements that would be expected to condense out as refractory grain species in planetary nebulae. However, the depletions of other species, notably carbon and oxygen, whose total abundances in the nebulae may be altered by nuclear processing, are much less certain. It is clear that carbon is not depleted by the factors of 10-20 found for iron, but some depletion probably occurs, and in carbon-rich objects, some form of carbon may well account for most of the dust.

2.3 Infrared observations of dust

Twenty years ago, Gillett, Low & Stein (1967) detected strong emission at 10 μm from NGC 7027 and interpreted this as emission from warm dust. This has of course been confirmed, and the photometric surveys of Cohen & Barlow (1974, 1980) showed that many pn display strong infrared emission from dust. Recently, near-infrared (1-4 μm) photometry has been obtained by several groups (Whitelock 1985, Persi et al (1987) and Pena & Torres-Peimbert at this meeting) pinpointing those nebulae that contain hot dust with colour temperatures up to 1000 K. In addition to the hundreds of planetaries that have been measured photometrically by IRAS, mid-infrared spectra of about 70 objects have been collected from either ground-based observations (Aitken & Roche 1982, Roche & Aitken 1986, and references therein) or the low resolution spectrometer on IRAS (Pottasch et al 1986). As well as several bright ionic fine-structure lines, these observations show a number of resolved features attributed to emission from different dust species. A few objects have been measured spectroscopically near 30 μm using the Kuiper airborne observatory (Forrest, Houck & McCarthy 1981) and here, too, there is evidence of spectral structure due to solid state features in grains in some of the objects. At shorter wavelengths spectra of a few nebulae near 3 μm have been published by Allen et al (1982) and Martin (1987), while Cohen et al (1986) and Wooden et al (1987) have presented spectra of 8 objects between 4-8 μm.

Measurement of spectral features due to dust grains can, in principle, lead to the identification of the species producing the emission features and permit investigation of the chemistry of the nebular ejecta.

Principally through spectroscopy near 10 μm, it has been established that the infrared spectral signatures and therefore the composition of the dust grains vary from nebula to nebula. Broad features peaking at 9.7 μm and 11.2 μm, were first seen in spectra of circumstellar dust shells around late type oxygen-rich and carbon-rich stars and were identified as emission from silicate and silicon carbide grains respectively on the basis of wavelength coincidence and the expected condensates in the different environments (e.g. Treffers & Cohen 1974). It was therefore likely that the dust seen in planetary nebulae could also be designated as having formed from O-rich or C-rich environments where the silicate or SiC signatures were detected; this has been verified by abundance determinations made by careful modelling of optical and ultraviolet emission lines. All the objects that show strong silicate emission bands that have been modelled have been found to have $C/O < 1$; conversely those planetaries with SiC emission have proved to have $C/O > 1$ (see Seaton 1983).

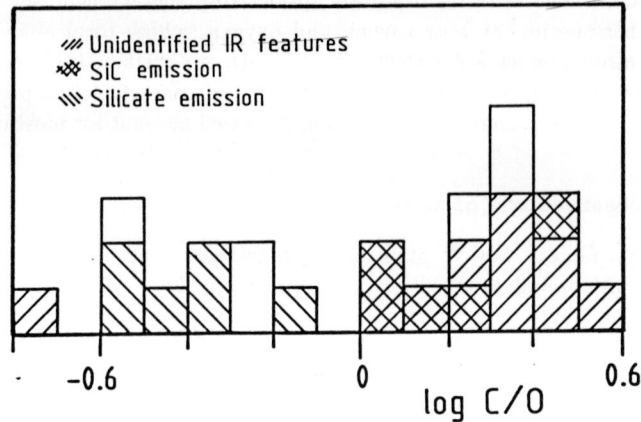

Fig 1. A histogram of those pn in the sample of Zuckerman & Aller (1986) that have been observed at 8-13 µm. Objects whose spectra are dominated by the different grain species are indicated by shading; the blank areas correspond to pn with weak continua at 10 µm.

However, of the objects for which good quality spectra near 10 µm are available only about 40% show unambiguous evidence of either silicate or SiC emission. About half of the remainder show clear emission in the family of emission bands at 11 3, 8.7 and 7.7 µm (and presumably also those at 6.2 and 3.3 µm) which were first detected in NGC 7027 (Gillett, Forrest & Merrill 1973) whilst the rest have very weak continuum emission near 10 µm and we are unable to identify clear spectral structure that could be due to dust. The family of emission bands between 3 and 13 µm has come under intensive scrutiny in the last couple of years (see Leger, d'Hendecourt & Boccara 1987 for comprehensive discussions) and it appears likely that they are produced by excitation of small (.001 µm) carbon-rich grains, possibly the polycyclic aromatic hydrocarbons proposed by Duley & Williams (1981) and Leger & Puget (1984). Cohen et al (1986) found a good correlation between the strength of the 7.7 µm band, the strongest of the family, and the C/O ratio for a small sample of planetaries, such that the emission bands are more prominent where C/O is higher. A similar conclusion has been reported by Martin (1987) from observations of the feature at 3.3 µm. This suggests that the grains giving rise to the family of emission bands are indeed carbon rich and would vindicate the suggestion of Barlow (1983) that those planetaries that are dominated by these features are very carbon rich with C/O > 2.

It appears that, through the identification of spectral features and correlations with other observable quantities, we can classify the dust from planetaries as O-rich, C-rich or very C-rich. To illustrate this, figure 1 shows a histogram of those pn in the sample of Zuckerman & Aller (1986) whose spectra have been measured at 10 µm. The different dust signatures seen in the infrared are indicated and it is clear that the qualitative estimates of C/O obtained from the dust emission features are confirmed by the detailed gas-phase abundance determinations from the sources quoted by Zuckerman & Aller. The only discrepant object is NGC 6302, a peculiar object with a very hot central star which shows the unidentified infrared bands, but also has a weak OH maser (Payne, Phillips & Terzian in these proceedings) and has been reported to be oxygen-rich. The pn which have 10 μm continua that are too weak to allow classification are scattered across the histogram, implying that they include both carbon- and oxygen- rich objects.

From the excellent agreement between the gas-phase C/O ratios and those predicted from the 10 μm spectra, it is clear that our basic view of dust species in pn is correct. Furthermore, there are some planetaries which suffer so much interstellar extinction that accurate values of the C/O ratio cannot be derived from UV and optical data. In these cases, the spectral features seen at 8-13 μm can be used to give rather more qualitative results; initial results of such a study of pn near the Galactic centre have been reported by Roche (1987).

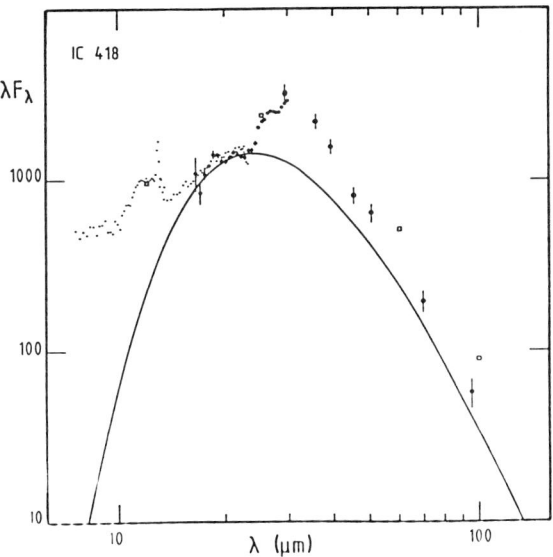

Fig 2. The infrared energy output from IC 418. The small dots shortwards of 22 μm are from the IRAS LRS, the data between 17 and 30 μm are from Forrest et al (1981), the data between 30 and 100 μm are from Moseley & Silverberg (1986) and the open squares are from the IRAS survey. The solid line represents isothermal emission from grains with $Q = 1/\lambda^2$ at T= 100 K.

Although the 10 μm spectral observations yield much valuable information about the warm (300 K) dust in pn, most of the dust is at considerably lower temperatures, and has not been studied in great detail. Spectra beyond 20 μm are available for only four pn, NGC 7027, BD +30 3639, NGC 6572 and IC 418 (Forrest, Houck & McCarthy 1981, Moseley & Silverberg 1986), but the implications of the data are far reaching. In the first two objects, no clear spectral structure is apparent, but in IC 418 (figure 2) and NGC 6572, a strong broad emission feature, peaking near 30 μm is present. This feature has also been measured in several carbon stars (Forrest et al 1981) and has been attributed to MgS (Goebel & Moseley, 1985) possibly as a mantle on grains of some other composition. Moseley and

Silverberg estimate that almost 25% of the total far-infrared flux is emitted in the "MgS" feature in IC 418, which is a very large fraction for what is possibly a relatively minor dust component. Very little information is available on the spectral structure at wavelengths beyond 20 μm, and it is quite conceivable that other emission features could lie in the far-infrared region, so that estimates of quantities such as dust mass and temperatures derived from broad band photometry alone could be subject to considerable uncertainties.

3. Location of dust in planetary nebulae.

The precise location of dust grains within pn is still uncertain. It is important to determine where the dust is located as this gives information on questions such as whether it plays an important role in softening the radiation field in the ionized region by selective absorption of high energy photons, whether emission from dust grains is important as a coolant of the ionized gas and whether the mass of dust calculated from the IR emission can be compared with the mass of ionized gas in order to derive the dust to gas ratio. From the absorption of trapped UV resonance lines, we know that at least some dust is mixed with the ionized gas but it is possible that these are the warm grains emitting at wavelengths below 20 μm rather than the bulk of the dust emitting in the far-infrared. A limited amount of spatial work has been carried out using both IRAS data and ground-based observations and these data generally show that the extent of the dust emitting region is similar, but not necessarily identical, to that of the ionized region.

Although the spatial resolution of IRAS was rather low for looking at compact objects, "super-resolution" deconvolution techniques can give information down to scales of the order of 15 arcsec at 25 μm and 30 arcsec at 60 μm and results for about 20 pn have been presented by Hawkins and Zuckerman at this meeting. Lester, Harvey & Joy (1986) have used maximum entropy deconvolution techniques on data from the Kuiper Airborne Observatory to look at the spatial structure of NGC 7027 in the far-infrared and conclude that most of the 50 and 100 μm radiation comes from a region of similar size to the ionized zone and at most 50% larger. Unfortunately, the resolution attained is insufficient to clearly show whether the emitting dust lies inside, or just outside the ionized region.

Spatial distributions at 10 and 20 μm can be measured with ground-based telescopes where 1-arcsec resolution is reachable. NGC 7027 was mapped by Becklin et al (1973) who found that the distribution of 10 μm emission was very similar to that seeen in high resolution radio maps demonstrating that the warm dust is well mixed with the gas in that object. A more complex picture emerged when Aitken & Roche (1983) scanned across the minor axis of NGC 7027 with a grating spectrometer, simultaneously measuring the spatial structure in the 10.5 μm [SIV] emission line, the 11.3 μm narrow emission feature and the continuum emission at 10, 11 12 and 13 μm. They showed that all the continua shared the same profile, ruling out a steep temperature gradient in the emitting dust, whilst the [SIV] profile was slightly narrower. In contrast, the scans in the 11.3 μm emission feature had a profile whose FWHM is greater than that of the continuum by 1.5 arcsec, clearly showing that the grains emitting the 11.3 μm feature lie outside the ionized zone and are probably contained in a neutral shell. The most natural explanation for these observations is that the grains that produce the narrow emission bands are destroyed in the ionized zone of NGC 7027; a similar conclusion has been reached in studies of the Orion ionization front where again, the dust emission features peak just outside the ionized region (Aitken et al 1979, Sellgren 1981). A consequence of this is that the dust in a planetary nebula is likely to vary in both the composition of the species present and in the mean grain size between the ionized and neutral regions.

A different spatial distribution has been found in Abell 30 where Cohen & Barlow (1974) found the 10 μm emission to be centrally condensed. Observations by Dinerstein & Lester (1984) revealed that the infrared emission from warm dust has a disk-like morphology, extended over about 20 arcsec, much less than the optical diameter. Lester & Dinerstein (1984) also found an infrared disk at the centre of NGC 6302, coincident with a dark lane that crosses the nebula. The disk lies between the optical lobes with its long axis perpendicular to the long axis of the bipolar structure and probably plays some role in shaping the outflow. The compact nebula BD +30 3639 has been observed at 10 and 20 μm by Bentley et al (1984) who conclude that the dust emission is coextensive with emission from ionized hydrogen; higher spectral resolution scans across BD +30 3639 at 10 μm were presented at this meeting by Smith et al who also conclude that the dust and gas emission come from the same volume, although the distributions of the two components may be different in detail.

The spatial distribution of near-infrared emission has been mapped out for several nebulae by scanning single detector photometers across the source. In most objects, the flux at wavelengths below 2 μm is due to free-free emission and nebular emission lines, so that measurements at J and H trace out the distribution of ionized gas, although the central star may also contribute a significant signal in some sources. At longer wavelengths, emission from dust may become more prominent and observations near 4 μm can delineate the emission from hot dust, while measurements through narrow-band filters at 3.3 μm can follow the distribution of emission in the 3.3 μm emission feature. Phillips et al (1984) found that in many cases, the near-infrared emission follows the ionized gas distribution, but in NGC 7027, after allowing for the the contribution from free-free emission, the L flux is more extended than the J,H,K and M scans. The interpretation is that the grains giving rise to the 3.3 μm emission feature lie outside the ionized region, reflecting the results cited above at 11.3 μm. In the case of IC 418, Phillips et al (1986) found that a substantial fraction of the 2 - 4 μm flux arises from a region considerably larger than the ionized nebula. In order to explain the high colour temperature of emission from dust in the neutral zone, they invoked a population of very small grains for which heating by absorption of single UV photons can give rise to high temperatures. It is just this mechanism that was proposed by Sellgren (1984) to account for the near-infrared properties of reflection nebulae, and extended by Leger and Puget (1984) to include the narrow emission features, and it is probable that the 3.3 μm emission band in IC 418 measured by Russell, Soifer & Merrill (1976) is produced in this way.

It is clear that there is no single answer to the question of where dust in planetary is located. In some objects, disk-like structures are evident where there is a substantial amount of dust girdling the central star. In others, the dust appears to be well mixed with the ionized gas, but it seems certain that the emission from some grain species is seen only, or at the least predominately, from neutral shells.

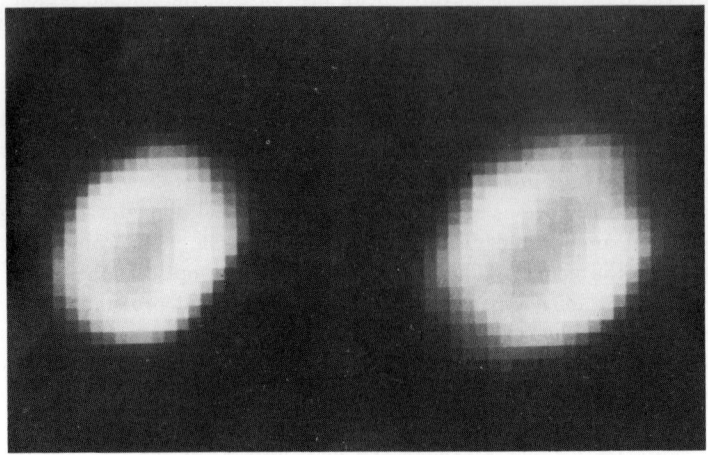

Fig 3. Images of NGC 7027 taken through filters at 2.17 (left) and 3.3 µm (right) with the infrared camera at UKIRT. The pixel size is 0.6 x 0.6"

4. Infrared Imaging.

With the release of sensitive infrared arrays, true imaging instruments are now available in the 1-5 µm spectral region. The first results on the spatial distribution of the 3.3 µm dust emission and HII in NGC 7027 made with array detectors have appeared (Woodward et al 1987). Figure 3 shows grey scale representations of the emission from NGC 7027 taken with the UKIRT common-user infrared camera (IRCAM) at 2.17 µm (Brackett γ) and 3.3 µm. It is immediately apparent that the 3.3 µm image (where most of the flux is from the 3.3 µm dust feature with small contributions from Pfund δ, free-free emission and continuum dust emission) is much more extended than the ionized gas traced by the 2.17 µm image, consistent with the data of Woodward et al, and confirming the earlier work of Aitken & Roche (1983) and Phillips (1984).

In BD +30 3639, shown in fig 4 as a contour map, the IRCAM images clearly separate the central star from the ionized gas and show a somewhat different result. It appears that the Br γ and 3.3 µm emission come from a region of similar size, but that there are differences in the detailed distribution. In particular, the 3.3 µm emission is strongest in the north-eastern edge, and shows local minima at the positions of the most intense Br γ emission. It therefore appears that in this object, which has a relatively cool central star and low flux of hard photons, the grains that emit the narrow features are located within the same volume as the ionized gas, but are located in regions away from the maximum photon flux. Clearly the spatial distribution of the grains emitting at 3.3 µm is not symmetric and it would be of interest to look at the nebular conditions in detail and at high spatial resolution to determine the conditions in the region where the 3.3 µm emission peaks.

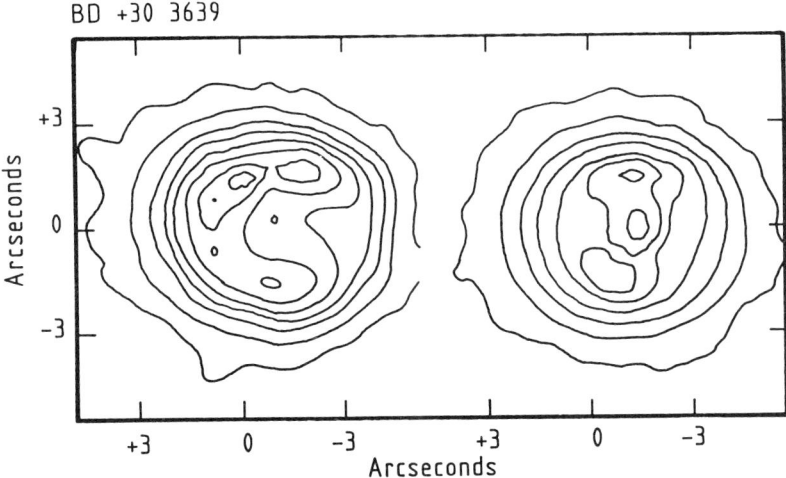

Fig 4. Contour maps of images of BD +30 3639 taken through filters at 3.28 (left) and 2.17 µm (right).

5. Cold dust

Dust which is warmer than about 30 K will have been seen by IRAS, but it is possible that there could be a substantial amount of dust outside the ionized regions which may be cooler than this; dust at 10 K emits very weakly in the far-infrared, but strongly in the submillimetre spectral region. In the last few months, two sub-mm dishes, the 15-m JCMT and the 10-m Caltech Dish, have been opened on Mauna Kea in Hawaii allowing sensitive observations in the 300 µm -2 mm region. Once these dishes are working to their design specifications, many planetary nebulae will come within their sensitivity limits for continuum observations, but the first detections of pn at 800 and 1100 µm have already been made with the JCMT (Clegg et al private communication). Using UKIRT, Gee et al (1984) measured NGC 7027 at 350, 800 and 1100 µm and found that the sub-mm flux was considerably higher than a simple extrapolation of the far-infrared data would predict. They interpreted this result as evidence of emission from cold dust in the massive neutral halo surrounding the ionized nebula. Observations of a number of pn at sub-mm wavelengths are under way (by Clegg et al and Kwok) and promise to pinpoint those objects surrounded by cold dust shells, and allow determinations of the mass of cold dust. A particular advantage of observations in the sub-mm where the grains are emitting in the Rayleigh-Jeans limit, is that the determination of the mass of dust depends linearly upon the black body function at the adopted temperature and the emissivity of the grains (e.g. Hildebrand 1983).

6. Summary

An understanding of the composition and spatial distribution of dust is vital in order for us to be able to interpret the infrared emission from pn. In particular, care is required when looking at large samples of nebulae as estimates of quantities such as absorption cross section, emissivity and grain size and hence the mass of dust, depend upon the adopted grain species whilst the derived gas to dust ratio requires knowledge of whether or not the bulk of the dust lies within the ionized region. Whilst our present understanding of these questions is inadequate, much progress can be made in answering these questions using the new infrared and sub-mm instrumentation currently under development at various observatories, although we will probably have to wait until the launch of ISO to investigate the far-infrared spectral properties of dust grains in detail. Within the next few years, we may expect sensitive infrared cameras operating in the 10 and 20 μm atmospheric windows to become available, and we will then be in a position to trace the spatial distribution of dust at temperatures down to 100 K, which is approaching the temperature of the bulk of the dust in many pn. Finally, there is no doubt that planetary nebulae offer some of the best opportunities for unravelling the properties of cosmic dust as many pn are bright objects whose history and composition are relatively well understood.

References:

Aitken, D.K. & Roche, P.F., 1982. MNRAS., 200, 217.
Aitken, D.K. & Roche, P.F., 1983. MNRAS., 202, 1233.
Aitken, D.K., Roche, P.F., Spenser, P.M. & Jones, B, 1979. Astr. Ap., 76, 60.
Allen, D.A., Baines, D.W.T., Blades, J.C. & Whittet, D.C.B., 1982. MNRAS, 199, 1017.
Atherton, P.D., Hicks, T.R., Reay, N.K., Robinson, G.J., Worswick, S.P. & Phillips, J.P., 1979. Ap.J., 232, 786.
Barlow, M.J., 1983. IAU Symp 103, "Planetary Nebulae", ed D.R. Flower, p105.
Becklin, E.E., Neugebauer, G. & Wynn-Williams, C.G., 1973. Ap.Lett., 15, 87.
Bentley, A.F., Hackwell, J.A., Grasdalen, G.L. & Gehrz, R.D., 1984. Ap.J., 278, 665.
Cohen, M., Allamandola, L., Tielens, A.G.G.M., Bregman, J., Simpson, J.P., Witterborn, F.C., Wooden, D. & Rank, D., 1986. Ap.J., 302, 737.
Cohen, M. & Barlow, M.J., 1974. Ap.J., 193, 401.
Cohen, M. & Barlow, M.J., 1980. Ap.J., 238, 585.
Dinerstein, H.L. & Lester, D.F., 1984, Ap.J., 281, 702.
Doughty, J.R. & Kaler, J.B., 1982. PASP, 94, 43.
Duley, W.W. & Williams, D.A., 1981. MNRAS, 196, 269.
Gee, G., Emerson, J.P., Ade, P.A.R., Robson, E.I. & Nolt, I.G., 1984. MNRAS, 208, 517.
Forrest, W.J., Houck, J.R. & McCarthy, J.F., 1981. Ap.J. 248, 195.
Gillett, F.C., Forrest, W.J. & Merrill, K.M., 1973. Ap.J., 183, 87.
Gillett, F.C., Low, F.G. & Stein, W.A., 1967. Ap.J., 149, L97.
Goebel, J.H. & Moseley, H., 1985. Ap.J. 290, L35.
Harrington, J.P., Seaton, M.J., Adams, S., Lutz, J.H., 1982. MNRAS, 199, 517.
Harrington, J.P., Monk, D.J. & Clegg, R.E.S., 1987. MNRAS in press.
Hildebrand, R.H., 1983. Q.Jl. R.A.S., 24, 267.
Larimer, G.W., 1979. Astr. Sp. Sci., 65, 351.

Leger, A., d'Hendecourt, L. & Boccara, N., 1987. "Polcyclic Aromatic Hydrocarbons and Astrophysics", Reidel, Dordrecht.
Leger, A. & Puget, J.L., 1984. Astr. Ap., 137, L5.
Leroy, J.L., Le Borgne, J.F. & Arnaud, J., 1986. Astr. Ap., 160, 171.
Lester, D.F., Harvey, P.M. & Joy, M., 1986 Ap.J., 304, 623.
Lester, D.F. & Dinerstein, H.L., 1984. Ap.J., 281, L67.
Martin, W., 1987. Astr. Ap., 182, 290.
Moseley, H. & Silverberg, R.F., 1985. "Interrelationships among circumstellar, Interstellar and Interplanetary Dust." ed J.A. Nuth & R.E. Stencel. p 233. NASA CP-2403.
Persi, P., Preite-Martinez,A., Ferrari-Toniolo, M., & Spinoglio, L., 1987. "Planetary and protoplanetary nebulae:From IRAS to ISO." p 221.
Phillips, J.P., Mampaso, A., Vilchez, J.M. & Gomez, P., 1986. Astr. Sp. Sci., 122, 81.
Phillips, J.P., Sanchez Magra, C. & Martinez Roger, C., 1984. Astr. Ap., 133, 395.
Pottasch, S.R., Preite-Martinez, A., Olnon, F.M., Jing-Er, Mo & Kingma, S., 1986, Astr. Ap., 161, 363.
Roche, P.F. 1987. Proc. Vulcano Workshop "Planetary and proto-planetary nebulae: From IRAS to ISO." p 45. ed A. Preite-Martinez. Reidel.
Roche, P.F. & Aitken, D.K., 1986. MNRAS, 221, 63.
Russell, R.W., Soifer, B.T. & Merrill, K.M., 1977. Ap.J., 213, 66.
Seaton, M.J., 1983. IAU Symp 103, "Planetary Nebulae", ed D.R. Flower, p129.
Sellgren, K., 1981. Ap.J., 245, 138.
Sellgren, K., 1984. Ap.J., 277, 623.
Shields, G.A., 1983. IAU Symp 103, "Planetary Nebulae", ed D.R. Flower, p259.
Treffers, R.R. & Cohen, M., 1974. Ap.J., 188, 545.
Whitelock, P.A., 1985. MNRAS, 213, 59.
Wooden, D.H., Cohen, M., Bregman, J.D., Witterborn, F.C., Rank, D.M., Allamandola, L.J. & Tielens, A.G.G.M., 1986. "Proc. Summer School on Inter-stellar Processes." p 59. ed. D.J. Hollenbach & H.A. Thronson, NASA TM-88342.
Woodward, C.E., Pipher, J.L., Shure, M.A., Forrest, W.J., Sellgren, K. & Nagata, T., 1987. "Infrared Astronomy with Arrays", Proc. Hilo Detector Workshop, p 299. ed C.G. Wynn-Williams & E.E. Becklin. Univ. Hawaii.
Zuckerman, B. & Aller, L.R., 1986. Ap.J., 301, 772.

Sun Kwok and Bob O'Dell.

MOLECULES AND NEUTRAL HYDROGEN IN PLANETARY NEBULAE

Luis F. Rodríguez
Instituto de Astronomía, UNAM
Apdo. Postal 70-264
04510 México, D.F., México

ABSTRACT. Molecules and/or neutral hydrogen have been detected in a modest number of planetary nebulae. However, when detected, these indicators of a significant neutral component can provide fundamental information on the total mass of the envelope, its chemistry and kinematics, and on the morphology and evolutionary status of the planetary nebula. A review of recent results is presented, giving emphasis to CO (carbon monoxide), H_2 (molecular hydrogen), OH (hydroxil), and HI (neutral hydrogen). A major development of the last few years has been the capability to map with considerable angular resolution these species in planetary nebulae. In the best studied cases, the neutral component appears to be located surrounding the "waist" of a bipolar planetary nebulae. These is evidence suggesting that molecules and neutral hydrogen are not as uncommon in planetary nebula as the present statistics suggest. Indeed, the outer parts of a considerable fraction of the known planetary nebulae could be neutral. It is possible that a combination of good selection criteria and long integrations with the best telescopes could increase largely the number of known cases.

1. INTRODUCTION

Until the early seventies the conception of a planetary nebula was that of an expanding, fully ionized envelope around a very hot star. There was no room in this picture for molecules or, in general, neutral gas. However, in 1975, Mufson et al. detected the $J = 1-0$ rotational transition of CO in NGC 7027. The angular extent of the CO emission is about 40" (Knapp et al. 1982), several times larger than the size of the optical object. At least in this case, we had a planetary nebula with an important molecular component. Furthermore, by then it was becoming clear that the progenitors of planetary nebula are stars at the tip of the asymptotic giant branch, of which the OH/IR stars and the carbon stars (Habing 1988) are well established examples. These stars have massive molecular winds and it was expected that remnants of these molecular winds could be found in young planetary nebulae. This expectation has been fulfilled in part; several planetary nebulae have shown associated molecules or neutral hydrogen. However, the numbers are still very small and from the present statistics, it would appear that significant neutral components are unusual in planetary nebulae. One possible explanation for these low numbers is that full ionization of the planetary nebula occurs rapidly. Assuming that the envelope was formed by a constant mass-loss-rate, constant velocity, spherically symmetric wind, one can show that the time for full ionization of the envelope since the wind stopped is given by:

$$\left[\frac{t}{\text{yr}}\right] \simeq 100 \left[\frac{\dot{M}_*}{10^{-5} M_\odot \text{yr}^{-1}}\right]^2 \left[\frac{v}{10 \text{kms}^{-1}}\right]^{-3} \left[\frac{N_i}{10^{48} \text{s}^{-1}}\right]^{-1}, \qquad (1)$$

where $\dot{M}_*$ is the mass loss rate, v the expansion velocity, and N_i the ionizing photon rate of the central star, assumed to have appeared and stayed constant since the wind stopped. For the values given in equation (1), one finds that full ionization of the envelope takes place in 10^2 years, two orders of magnitude less than the observable lifetime of a planetary nebula. However, one could also argue that other sets of parameters will give full-ionization times comparable to the observable lifetime of a planetary nebula and that in these cases, an outer molecular envelope could be present even in "evolved" planetary nebula.

Another argument that can be used to account for the small number of detections of molecules in planetary nebulae is that, even in the favorable case that, let us say, one half of the envelope is still molecular, detection is not easy. Assuming that the CO emission of the envelope is optically thin, thermalized at about 50K, and that CO/H_2 has a ratio similar to that found in molecular clouds, one expects peak line temperatures for the $J = 1 - 0$ rotational transition of

$$\left[\frac{T_L}{K}\right] \simeq 0.1 \left[\frac{D}{\text{kpc}}\right]^{-2} \left[\frac{M_{H_2}}{0.1 M_\odot}\right] \left[\frac{\Delta v}{20 \text{kms}^{-1}}\right]^{-1} \left[\frac{\theta_B}{\text{arcmin}}\right]^{-2}, \qquad (2)$$

where D is the distance of the planetary nebula, M_{H_2} its mass in molecular hydrogen, Δv is the line width (about twice the expansion velocity), and θ_B is the half-power beam width of the radio telescope. For the values in equation (2) we expect line temperature of about 0.1K. These line temperature values are still hard to detect (see Knapp 1985 for typical upper limits in CO searches toward planetary nebulae).

In this review we emphasize recent observational results regarding CO, H_2, OH, and HI. Our discussion will be restricted to objects where a significant part of the envelope is ionized and can be called planetary nebulae. Transition objects have been reviewed in Kwok (1987) and Rodríguez (1987). An excellent review of results prior to 1983 was given by Black (1983).

2. RECENT RESULTS

2.1 Carbon Monoxide

The millimeter rotational transitions of CO are expected to emit close to thermal conditions. Therefore, their measurement allows (especially when combined with observations of the optically thin ^{13}CO isotopic species) estimates of masses. CO has been detected in NGC 7027 (Mufson et al. 1975), NGC 2346 (Knapp 1986; Huggins and Healy 1986b), NGC 7293 (Huggins and Healy 1986a); NGC 6720 (Huggins and Healy 1986b), NGC 6302 (Zuckerman and Dyck 1986), and Vy 2-2 (Knapp and Morris 1985).

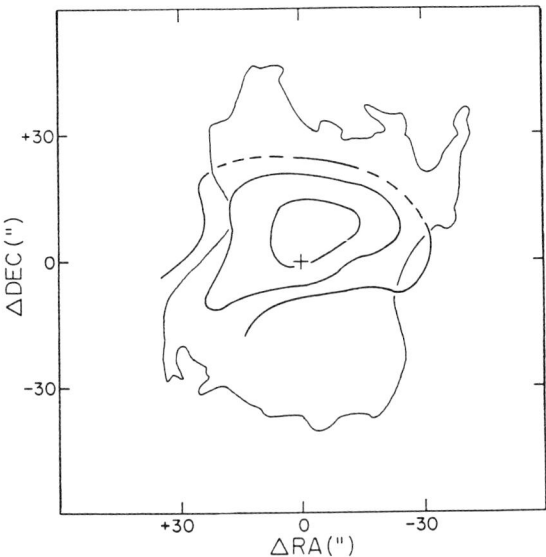

Figure 1. The integrated CO emission (thick line) down to the half-power contour is shown superposed on a sketch of NGC 2346 (thin line). The cross marks the position of the central star (actually a binary). Data from Healy and Huggins (1987) and Walsh (1983).

In these planetary nebulae it is estimated that the envelope is in good part ($> 10\%$) molecular. A surprising result, not yet fully understood, is that evolved planetary nebulae like NGC 7293 and NGC 6720 (Huggins and Healy 1986a; 1986b) still retain a molecular component.

The CO distribution has been mapped in a few cases. The best studied examples are NGC 7027 (Masson et al. 1985) and NGC 2346 (Healy and Huggins 1987). In both cases the molecular gas seems to be in an oblate structure perpendicular to the major axis of the ionized gas. In Figure 1 we show the integrated CO emission associated with NGC 2346 superposed on a sketch of the optical nebula (Healy and Huggins 1987). These data can be interpreted as indicating that the neutral gas surrounds the waist of the optical nebula. A natural explanation of the bipolar structure is then provided.

2.2 Molecular Hydrogen

The search for the infrared ($2\mu m$) vibration-rotation transitions of H_2 in planetary nebulae has been the most successful in the number of objects detected. Zuckerman and Gatley (1988) list 16 cases. The H_2 lines require excitation temperatures above 1000K and it is generally believed that they are shock-excited. Probably the shock is driven by the pressure of the expansion of the ionized gas. Under this interpretation, the $2\mu m$ lines trace, at a given moment, only a small, highly excited (that recently shocked) fraction of the molecular gas. Then, they can not be used as mass indicators. However, Black and van Dishoeck (1987) have analyzed the possibility of excitation by ultraviolet fluorescence. In this case, one would be detecting emission from a much larger fraction of the molecular component. Observations of higher transitions are required to assess the importance of

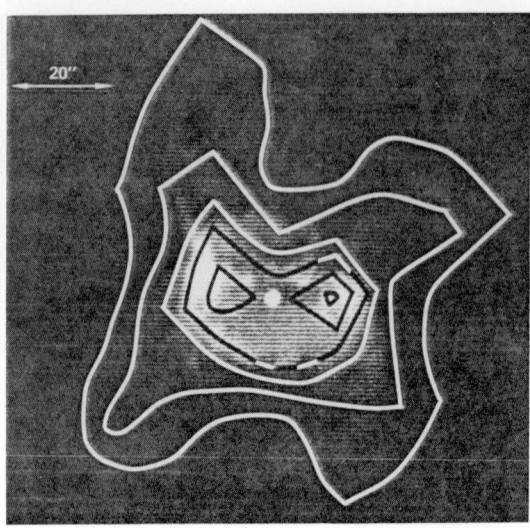

Figure 2. Contour map of the $S(1)$ $v = 1 \rightarrow 0$ molecular hydrogen emission superposed on an [NII] $\lambda 6584$ image of NGC 2346. Data from Zuckerman and Gatley (1988) and Balick (1987).

fluorescent excitation.

In this meeting we have seen striking examples of images of H_2 emission associated with planetary nebulae (Greenhouse, Hayward and Thronson 1988; Smith et al. 1988; Payne et al. 1988; Dinerstein et al. 1988). Payne et al. (1988) have proposed that there is a correlation between Type I planetary nebulae (Peimbert and Torres-Peimbert 1983) and strong molecular hydrogen emission. In general, the H_2 emission lies outside and closely traces the distribution of ionized gas.

As in the case of CO, there is evidence from the H_2 that in bipolar planetary nebulae the neutral gas is located around the waist of the object. In Figure 2 we show the $S(1)$ $v = 1 \rightarrow 0$ contour map of Zuckerman and Gatley (1988) superposed on an [NII] $\lambda 6584$ photograph (Balick 1987) of NGC 2346. As the CO, the H_2 can be interpreted to be in a torus around the waist of the nebula.

2.3 Hydroxil

The OH/IR stars, one of the most likely precursors of planetary nebulae, are characterized by the double peaked 1612 MHz maser line emission of OH. As the stellar nucleus evolved, it was expected that these objects could go through a stage where the central parts of the envelope are ionized (and could be detected via its radio continuum), while the outer parts of the envelope keep their OH maser emission. In this phase, the ionized core is optically thick at 1612 MHz and does not allow the detection of the redshifted peak of the characteristic double-peaked profile of OH/IR stars. Despite sensitive searches for radio continuum toward OH/IR-like objects (Herman, Baud and Habing 1985; Rodríguez, Gómez and García-Barreto 1985), only Vy 2-2 appeared to have blueshifted OH maser emission and radio continuum (Seaquist and Davis 1983). However, very recently, several other objects have been reported. Payne, Phillips and Terzian (1987) detected the 1612 MHz line and rotationally excited lines of OH at 5cm in NGC 6302. Using better defined selection criteria than in earlier surveys, Pottasch, Bignell, and Zijlstra (1987) found radio continuum in two OH/IR objects, G349.2-0.2 and G0.9+1.3. Two additional sources have been reported by Zijlstra et al. (1988). Obviously, OH maser emission in planetary nebulae is much more common than thought just a couple of years ago.

The best studied cases of planetary nebulae with OH are Vy 2-2 (Seaquist and Davis 1983) and NGC 6302 (Payne, Phillips and Terzian 1987). Using the VLA, these last authors have mapped the 1612 MHz emission associated with NGC 6302. The molecular gas has a considerable velocity gradient (Figure 3) and lies along the dark dust lane visible in optical photographs (Rodríguez et al. 1985). Once again, this dust lane is aligned approximately perpendicular to the axis of the ionized nebula.

Given the maser nature of the OH emission, this molecule can not be used for mass estimates.

2.4 Neutral Hydrogen

The 21 − cm hyperfine transition of HI appears as a good possibility to search for neutral components in planetary nebulae. However, confusion from line of sight interstellar gas makes the identification difficult and the measurement can be made reliably only with an interferometer. HI absorption against the radio continuum of the ionized core of NGC 6302 was observed by Rodríguez and Moran (1982) using the VLA. Other planetary nebulae with associated HI in absorption are NGC 6790 (Gathier, Pottasch and Goss 1986), IC 4997 (Altschuler et al. 1986), and IC 418 (Taylor and Pottasch 1987). Even when the amount of HI in the envelopes is subject to uncertainties as a result of the unknown excitation temperature, it seems that roughly comparable amounts of neutral and ionized hydrogen are present.

HI is, as CO and the other molecules, expected to be present as long as the envelope is not fully ionized. There is, however, an additional restriction in the case of HI; the neutral hydrogen present could well be in molecular, as opposed to atomic, form. Following the theoretical considerations of Glassgold and Huggins (1983), Rodríguez and García-Barreto (1984) have argued that detectable HI may exist only for planetary nebulae with relatively hot ($T_* > 2500K$) progenitors. For stellar temperatures above this value hydrogen is mainly atomic in the wind, while for values below, it is mainly molecular.

The HI in absorption associated with NGC 6302 has been mapped by Rodríguez, García-Barreto and Gómez (1985). The location of the HI (Figure 4) coincides with the dark dust lane and with the OH emission (Payne, Phillips and Terzian 1987; see Figure 3). Once more, the interpretation implies a neutral torus around the waist of the optical nebula.

2.5 Other Observational Results

CN has been detected in NGC 7027 (Thronson and Bally 1986), while HCN has been detected in the same object (Sopka et al. 1988) and in NGC 2346 (Walsh, Clegg and Ukita 1988). These results confirm the carbon- rich chemistry of both planetary nebulae.

An optical astronomer may wonder why lines as [O I] $\lambda 6300$ or [N I] $\lambda 5200$ are not included in this discussion of neutral gas in planetary nebulae. The usual argument is that

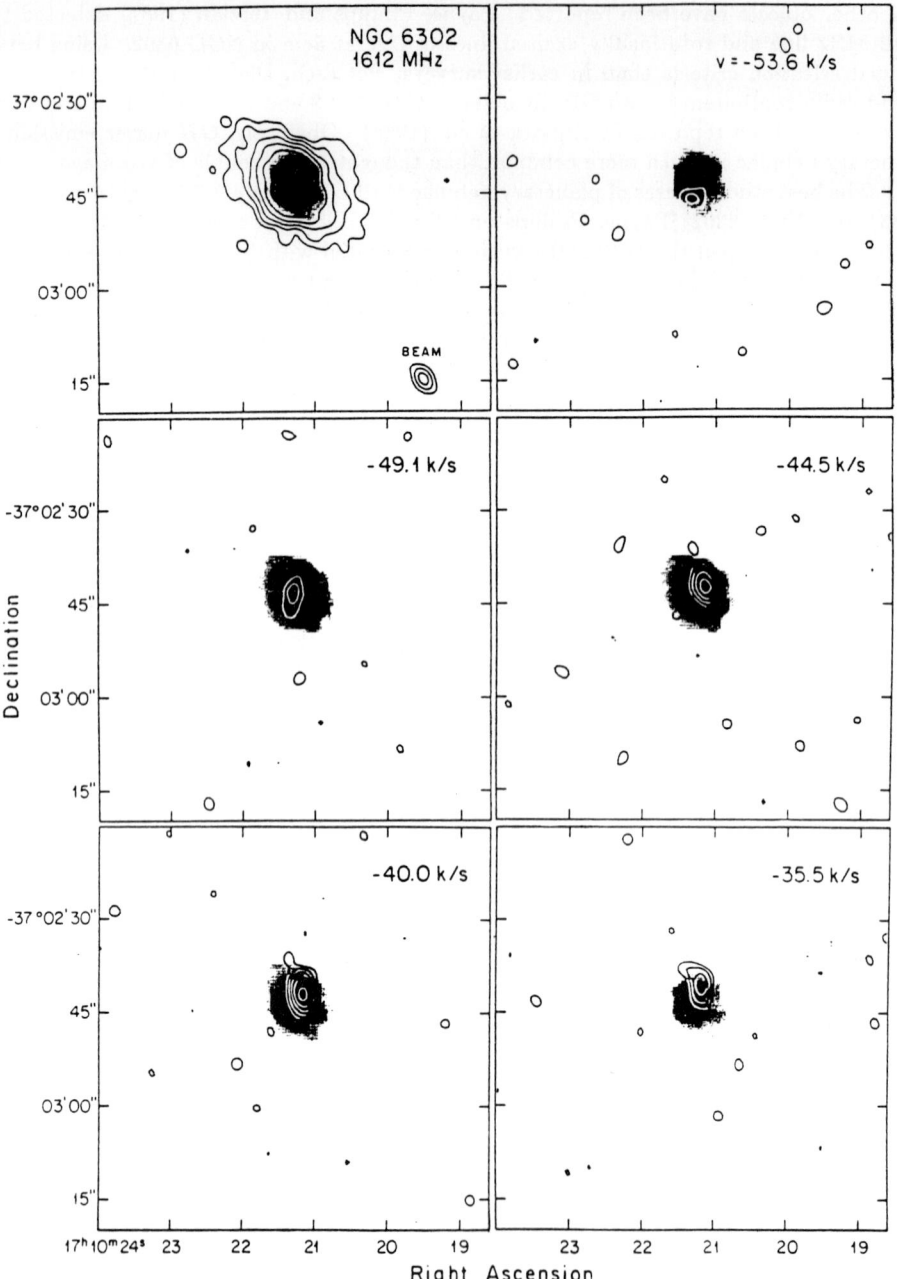

Figure 3. VLA 1612 MHz and continuum maps from Payne, Phillips and Terzian (1987). The continuum map is shown in the upper left panel. The 1612 MHz emission at different radial velocities is shown in the other panels. All panels include contour and grey scale representations.

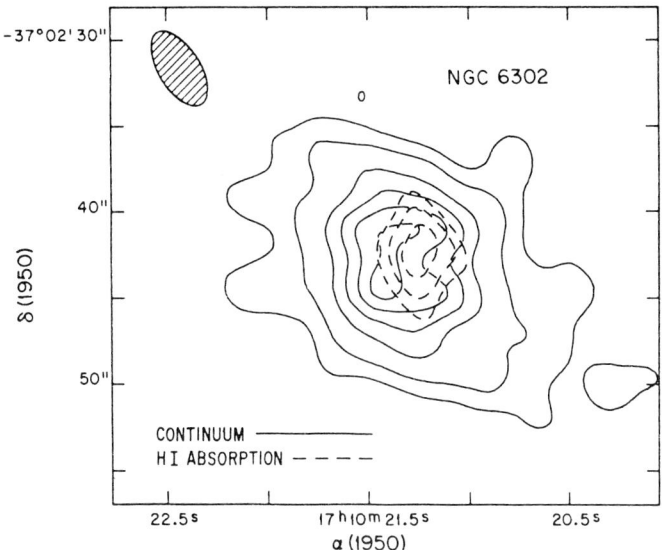

Figure 4. HI absorption (dashed lines) superposed on the continuum (solid lines) of NGC 6302, taken from Rodríguez, García-Barreto and Gómez (1985). The HI absorption coincides spatially with the optical dark dust lane and the OH emission (Payne, Phillips and Terzian 1987).

these lines originate in the transition zone where hydrogen goes from ionized to neutral and, consequently, only small amounts of gas are being traced. However, several of the planetary nebulae discussed here also have these optical lines and it is obvious that some kind of correlation could exist. Unfortunately, very little has been made to explore it in detail.

3. DISCUSSION

The field of neutral gas in planetary nebulae has had important advances in the last few years. Nevertheless, the number of objects detected in the tracers discussed here is very small. Does this mean that most planetary nebulae are density bounded (that is, fully ionized)? This conclusion is hard to accept because other results suggest the opposite. The well known mass-radius correlation (Maciel and Pottasch 1980) can be taken to imply that most planetary nebulae are ionization bounded (see Gathier 1987 for a recent discussion). In brief, it is found that $M \propto R^{1.5}$, as expected for massive envelopes that remain ionization bounded over most of the planetary nebula lifetime.

To complicate matters further, in this meeting we have seen that ionized halos are common around planetary nebulae (Chu 1988). This result is consistent with the rarity of molecules, since apparently a density bounded planetary nebula is needed to ionize a halo. However, the mass-radius correlation also seems well established and we have a controversy over such a seemingly simple issue as whether most planetary nebulae are density or ionization bounded.

Some of the results discussed here may offer an explanation. It is known that about 50% of the planetary nebulae are bipolar (Zuckerman and Aller 1986). At least for the bipolar nebulae discussed here it makes no sense to try to classify them as either ionization bounded or density bounded. They are both, ionization bounded in the waist, and density bounded in the poles. Is this situation common and thus all our classification of planetary nebulae as either ionization or density bounded dubious? I do not think the answer is known.

Given the present state of uncertainty, one could argue that most planetary nebulae have sizable neutral components yet undetected. In what refers to CO, a major search program with the new millimeter and submillimeter radio telescopes in Nobeyama, Pico Veleta and Mauna Kea appears worthwhile. Along the same lines the VLA and Arecibo could possibly yield additional OH and HI detections if the selection criteria were improved. Finally, the fast mapping that the new infrared arrays are capable of achieving should reveal H_2 in a considerable number of planetary nebulae.

4. CONCLUSIONS

I reviewed the recent observational developments regarding molecules and neutral hydrogen in planetary nebulae. When the neutral mass can be determined it turns out to be comparable with that of the ionized gas. In the best studied cases the neutral gas is located in a torus around the waist of a bipolar planetary nebulae. The number of planetary nebulae with a sizable neutral component remains small. At present it is unclear if this implies that most nebulae are density bounded or if we are still sensitivity limited. A major observational effort with the new millimeter and submillimeter radio telescopes could solve the issue.

The theoretical aspect is, in general, poorly developed. There could be more objects like NGC 6302 where hydrogen is present in ionized, atomic, and molecular form. Detailed models for the structure of these nebulae are not available.

Another important result that emerges from the study of planetary nebulae with molecules or neutral hydrogen is that, at least for the bipolars, one can not classify them as either ionization or density bounded, since they are both: ionization bounded in the waist and density bounded in the poles. Since about half of the planetary nebulae are bipolar, this ambiguity could be a major problem for methods (*i.e.* distance scales) that segregate nebulae into ionization bounded and density bounded.

5. REFERENCES

Altschuler, D.R., Schneider, S.E., Giovanardi, C. and Silverglate, P.R. 1986, *Ap. J. (Letters)*, 305, L85.

Balick, B. 1987, *Astr. J.*, 94, 671.

Black, J.H. 1983, in *Planetary Nebulae*, IAU Symp. 103, ed. W.D. Flower (Dordrecht: Reidel), p. 91

Black, J.H. and van Dishoeck, E.F. 1987, *Ap. J.*, 322, 412.

Chu, Y.H. 1988, these proceedings.

Dinerstein, H.L., Coleman, H.H., Carr, J.S., and Lester, D.F. 1988, these proceedings.

Gathier, R., Pottasch, S.R. and Goss, W.M. 1987, *Astr. Ap.*, 157, 191.

Gathier, R. 1987, in *Late Stages of Stellar Evolution*, ed. S. Kwok and S.R. Pottasch, *Ap. and Spac. Sci. Library*, 132, 371.

Glassgold, A.E. and Huggins, P.J. 1983, *Mon. Not. R. Astr. Soc.*, 203, 517.

Greenhouse, M.A., Hayward, T.L. and Thronson, H.A. 1988, these proceedings.
Habing, H.J. 1988, these proceedings.
Healy, A.P. and Huggins, P.J. 1987, submitted to *Astr. J.*
Herman, J., Baud, B. and Habing, H.J. 1985, *Astr. Ap.*, 144, 514.
Huggins, P.J. and Healy, A.P. 1986a, *Ap. J. (Letters)*, 305, L29.
Huggins, P.J. and Healy, A.P. 1986b, *Mon. Not. R. Astr. Soc.*, 220, 33p.
Knapp, G.R., Phillips, T.G., Leighton, R.B., Lo, K.Y., Wannier, P.G., Wootten, H.A., and Huggins, P.J. 1982, *Ap. J.*, 252, 616.
Knapp, G.R. and Morris, M. 1985, *Ap. J.*, 292, 640.
Knapp, G.R. 1985, in *Mass Loss from Red Giants*, ed. M. Morris and B. Zuckerman, Ap. and Spa. Sci. Library, 117, 171.
Knapp, G.R. 1986, *Ap. J.*, 311, 731.
Kwok, S. 1987, in *Late Stages of Stellar Evolution*, ed. S. Kwok and S.R. Pottasch, Ap. and Spa. Sci. Library, 132, 321.
Maciel, W.J. and Pottasch, S.R. 1980, *Astr. Ap.*, 88, 1.
Masson, C.R. *et al.* 1985, *Ap. J.*, 292, 464.
Mufson, S.R., Lyon, J., and Marionni, P.A. 1975, *Ap. J. (Letters)*, 201, L85.
Payne, H.E., Phillips, J.A. and Terzian, Y. 1987, *Ap. J.*, in press.
Payne, P.W., Storey, J.W.V., Webster, B.L., Dopita, M.A. and Meatheringham, S.J. 1988, these proceedings.
Peimbert, M. and Torres-Peimbert, S. 1983, in *Planetary Nebulae*, IAU Symposium 103, ed. W.D. Flower (Dordrecht: Reidel), p. 233.
Pottasch, S.R., Bignell, C. and Zijlstra, A. 1987, *Astr. Ap.*, 177, L49.
Rodríguez, L.F. and Moran, J.M. 1982, *Nature*, 299, 323.
Rodríguez, L.F. and García-Barreto, J.A. 1984, *Rev. Mexicana Astron. Astrof.*, 9, 153.
Rodríguez, L.F. *et al.* 1985, *Mon. Not. R. Astr. Soc.*, 215, 353.
Rodríguez, L.F., García-Barreto, J.A. and Gómez, Y. 1985, *Rev. Mexicana Astron. Astrof.*, 11, 109.
Rodríguez, L.F., Gómez, Y. and García-Barreto, J.A. 1985, *Rev. Mexicana Astron. Astrof.*, 11, 139.
Rodríguez, L.F. 1987, in *Planetary and Protoplanetary Nebulae: From IRAS to ISO*, ed. A. Preite-Martínez, Ap. and Spa. Sci. Library, 135, 55.
Seaquist, E.R. and Davis, L.E. 1983, *Ap. J.*, 274, 659.
Smith, M.G., Geballe, T.R., Aspin, C.A., McLean, I.S. and Roach, P.F. 1988, these proceedings.
Sopka, R. *et al.* 1988, in preparation.
Taylor, A.R. and Pottasch, S.R. 1987, *Astr. Ap.*, 176 L5.
Thronson, H.A. and Bally, J. 1986, *Ap. J.*, 300, 749.
Walsh, J.R. 1983, *Mon. Not. R. Astr. Soc.*, 202, 303.
Walsh, J.R., Clegg, R.E.S. and Ukita, N. 1988, these proceedings.
Zijlstra, A., Pottasch, S.R., te Lintel, P. and Bignell, C. 1988, these proceedings.
Zuckerman, B. and Aller, L.H. 1986, *Ap. J.*, 301, 772.
Zuckerman, B. and Dyck, H.M. 1986, *Ap. J.*, 304, 394.
Zuckerman, B. and Gatley, I. 1988, *Ap. J.*, 324, 501.

Sally Heap, Luciana Bianchi and Elsa Recillas.

ABUNDANCES IN PLANETARY NEBULAE

R.E.S. Clegg
Department of Physics & Astronomy,
University College London, Gower Street,
LONDON WC1E 6BT, U.K.

1. Introduction

   Good progress has been made since the last Symposium on the determination of planetary nebulae (PN) abundances. Notable features for modern abundance determinations include the availability of good collision strengths and transition probabilities (reviewed by C. Mendoza (1983) at the last meeting - IAU Symp.103), the use of the IUE & IRAS satellites to obtain UV & IR line fluxes, and the availability of modern sensitive detectors enabling measurements both of faint extra galactic nebulae and of very weak abundance-indicator lines in nearby bright nebulae. The impact of IUE on PN studies was described by Köppen & Aller (1987).
   PN abundance results tell us about mixing processes in the envelopes of red giant stars. HII regions provide much better indicators of galactic abundance gradients; the PN are a relatively old population. Discussions on PN abundances were given by Kaler (1983a), in the books by Pottasch (1984) and Aller (1984), and in the review by Kaler (1985). A related review, on nucleosynthesis and mixing processes in PN progenitor stars was given by Iben & Renzini (1983).
   For low and intermediate mass stars, Iben & Renzini describe three 'dredge-up' events. In the first, material processed in the CN cycle during main-sequence evolution is mixed out to the surface. The N abundance rises while the C and the $^{12}C/^{13}C$ ratio both drop. The second dredge-up is predicted to occur only for higher masses, $\geq 2.5 - 8M_\odot$, and the products of intensive CNO-cycle burning are mixed out. Surface He and N abundances increase, while those of C and O fall (O by less than C). The third event occurs on the Asymptotic Giant Branch (AGB): products of He-burning are dredged-up to the surface during stellar thermal relaxation oscillations. The products - mostly ^{12}C and heavy elements made in the s-process - are subject to limited CN-cycle burning on the way out, so that some conversion from C to N and production of ^{13}C are both facilitated. In general, the O abundance is predicted to be the least affected of the CNO group, and it is a prediction of the current theory that planetary oxygen abundances should be very close to the initial stellar abundances. The dredge-up episodes are subject to modification when rapid rotation produces significant internal circulation currents (Sweigert & Mengel 1979).

Peimbert (1978) classified PN in four groups: Type I are He-and N-rich, and are bi-polar with filamentary structures. Peimbert & Torres-Peimbert (1983) listed 29 such objects with defining abundance characteristics He/H > 0.125 or N/O > 0.5. They are Population I objects close to the Galactic plane and have evidently suffered the 2nd dredge-up. From the binary nuclei of NGC 2346 & 3132, and from NGC 2818 which is probably connected with the open clusters of the same name, Peimbert & Serrano (1980) estimated for these Type I nebulae that a lower limit on the initial mass which produces the He - N rich PN is about 2.4 $M_\odot$. Peimbert (1985) has reviewed additional evidence that the Type I progenitor stars have masses $\geq$ 3 $M_\odot$. This fits in with current ideas about the 2nd dredge-up, although the lower mass limit for this is very dependent on modelling parameters such as convective mixing length.

Type II and III are disk planetaries with low and high dispersions, respectively. Type IV refers to objects in the Galactic Halo, of which only 4 definite members are known (see Sec.5). Faundez-Albans & Maciel (1987) have suggested that it would be useful to divide the Type II group into subtypes (a) and (b) according to whether the nitrogen abundance log(N/H) + 12 is more or less than 8.0. Subtype (b) objects have a smaller velocity dispersion than (a), and the suggested subdivision could be useful in studies of Galactic abundance gradients and the relative ages of different PN samples.

2. Helium Abundances

He/H ratios in planetary nebulae tell us about the dredge-up of helium (eg in Type I objects), the pregalactic helium abundance (from Type IV objects) and give a calibration for $\Delta Y/\Delta Z$, the relative increase by mass of He and heavy elements during galactic evolution (eg Maciel, these proceedings). The standard method of abundance analysis is to use the effective recombination coefficients of Brocklehurst (1971, 72) for H and He together with observed fluxes of He recombination lines such as $\lambda\lambda 4471, 4686, 5876$ and 6678Å. But the interpretation of HeI lines is complicated by the meta-stability of the lowest triplet levels 2s ^{3}S. Collisional excitations from this level can excite all the observed HeI "recombination" lines. This subject has a long history and a large literature (see Peimbert & Torres-Peimbert 1987 and Clegg 1987a; respectively PTP87 and C87).

Analysis of HeI collision strengths of Berrington et al. (1985) by Ferland (1986) resulted in Ferland's correction formulae to allow for collision excitations of HeI lines. But improved calculations (Berrington & Kingston 1987) recently yielded much lower collision strengths, and updated correction formulae are available from PTP87 and C87. The two sets of correction formulae are not the same. PTP87 derived empirically the ratio, γ, of the 2 ^{3}S population to that predicted theoretically (from the balance between recombinations (arrivals) and collisional transfers to singlet levels (departures). From new line fluxes for Type I nebulae and published $\lambda 10830$ Å fluxes, they found $\gamma = 0.5 \pm 0.2$. It was suggested that additional physical processes were de-populating the 2 ^{3}S level in nebulae. In contrast C87 included collisional ionisation of He(2 ^{3}S), took only 50% of the theoretical

collision rates to the n = 4 levels, assumed γ = 1 and tried to see if
the resulting corrections were "plausible". It was shown that the
hypothesis, that He abundances should not depend on nebular temperature
and density, is better satisfied for 3 of 4 data samples when corrections
are applied. C87 found that the mean He/H ratio for non-Type I PN is
9% lower after corrections, with He/H = 0.099 ± 0.007 (He/H = 0.105 if
Type I objects are not excluded from typical data samples).

What is the cause of these discrepant findings? Clegg & Harrington
(these proceedings & in preparation) found that photo-ionisation of He
$2\ ^3S$ (I.P = 4.77 eV) can alter the metastable's population by anything
from zero to 20%. A recommended prescription will be given by them.
This, together with collisional ionisation, goes some way towards an
agreed final value for γ. Fortunately, total He/H ratios are not much
different whichever current prescription for collisional effects are used.
For example, from new line data (PTP87) for 13 Type I PN, the mean He/H
ratio is 0.131 with C87's prescription and 0.135 with PTP87's, a difference of only 3%. My current recommendation is to work with all three
lines λλ4471,5876 & 6678Å, use either C87 or PTP87 together with the
recombination coefficients of Brocklehurst (1972), and check for the
presence of neutral helium using the criteria either of Kaler (1978) or
Torres-Peimbert & Peimbert (1977) for low-excitation nebulae. If He^o
is present, a lower limit for He/H is obtained. (For Type I or very
high excitation PN, photoionisation models show that a larger fraction
of H^o than He^o exists - this also affects He/H ratios, but this time in
the opposite sense - observed ratios need to be decreased).

Although the drop in PN helium abundances is only ~ 9%, these
results indicate that the self-enrichment in He of non-Type I PN is
small, since the mean value now equals that of HII regions (eg Dufour
1984). The same result is found for Large & Small Magellanic Cloud
PN & HII regions (Monk et al. 1988): data for HII regions and non-Type I
PN are all consistent with He/H = 0.085 ± 0.004 for both Clouds. This
raises the question: are Type I objects really a separate class?
Although there is a continuous sequence of observed He/H and N/O ratios
in planetaries (eg Kaler 1983a), on theoretical grounds Type I nebulae
can be viewed as a separate group if they all arise from progenitor stars
having $M \geq 2.5_\odot$, where the 2nd dredge-up has operated. Peimbert &
Torres-Peimbert (1983) discuss additional ways in which these objects
form a coherent group.

Three interesting consequences of these collisional effects are
noted. Firstly, abundances of other elements may change because several
ionisation correction factors (icf's) depend on the derived ratio of
He^+/He^{++}, which is now altered. Secondly, the Galactic enrichment of
He by planetaries is lowered somewhat. Peimbert (1987) showed that most
of the He production is by Type I nebulae (which give 3-6 times the
contribution of Types II & III), and the helium enhancements in these
objects is now lower by ~ 25%. Thirdly, a revised estimate for the pregalactic He/H ratio can be obtained. I have re-calculated this using
the method of Peimbert (1983) based on abundances in the four known
PN in the Galactic Halo (see also Sec.5). It is assumed that the current
He/H ratios in these objects reflects the pre-galactic value plus
enhancements only from the progenitor's 1st and 3rd dredge-ups.

The results are shown in Table 1. y denotes the He/H ratio by number and Y the mass-fraction of He; 'new' refers to results including allowance for collisional effects, and ΔY values are calculated enhancements in the 1st and 3rd dredge-ups, so that $Y = Y_p + \Delta Y^1 + \Delta Y^3$.

Table 1   Pre-galactic Helium Abundance Y_p

Nebula	y^{old}	y^{new}	Y	ΔY^1	ΔY^3	Y_p
K648	0.096	0.083	0.263	0.014	0.014	0.235
H4-1	0.103	0.095	0.282	0.014	0.054	0.214
BB-1	0.101	0.090	0.276	0.014	0.036	0.226
DDDM-1	0.106	0.086	0.276	0.019	-	0.257

The mean value is Y_p = 0.23 ± 0.03 (or He/H = 0.076 ± 0.010). There are two uncertainties in the method used: the prescriptions for the ΔY values are from Renzini & Voli (1981) and are really only valid for stars of higher initial mass, and secondly for DDDM-1 which has a very low carbon abundance I assumed no 3rd dredge-up has occurred. In fact in the initial shell flashes of AGB evolution He-rich material might be picked up without the convective region reaching down to carbon-polluted layers (Schonberner 1979); but on the other hand for these low-mass Pop II stars the very first shell flash may mix out sufficient carbon to cause the surface C/O ratio to exceed unity and even drive envelope ejection (A. Renzini, these proceedings). No correction has been made for neutral helium in the nebulae, for all of which the relation O^+/O^{total} < 0.4 holds (a value above 0.4 suggests the significant presence of He^o, Torres-Peimbert & Peimbert 1977).

Although this value of Y_p is less precise than that obtainable from extragalactic HII regions, this method does give an independent check on this important quantity. Maciel (these proceedings) has used Type II PN together with extragalactic HII regions to find a value Y_p in the range 0.23 to 0.24, in agreement with and more accurate than the value from halo PN alone.

3.   Mean Abundance in PN Samples

3.1. Nebulae in the Galactic Disk

In the last ten years abundances for large samples of nebulae have been given by Torres-Peimbert & Peimbert (1977), Barker (1978), Aller & Czyzak (1983;AC) and Aller & Keyes (1987;AK). I restrict discussion to studies since the 1982 PN Symposium. AC & AK use a combination of empirical analysis of observed optical and UV ionic lines, plus results from photo-ionisation models to correct for un-observed ionic stages. AC analysed 41 nebulae, and AK give data for 51 additional PN; together with published material, abundances of some elements in up to 104 PN are listed.

Abundances can be presented relative to H or relative to O. It is well-known that on average PN have O/H roughly half the Solar value (eg <O/H> = 4.3×10^{-4} (AK) while the Solar photospheric value is 8.3×10^{-4}, Lambert 1978). Three possible reasons why O/H is low include: (i) the PN constitute an old disk population similar to the Mira variables, and this is on average metal-poor (ii) there are temperature fluctuations within the nebulae - allowance for these could raise the oxygen abundance to the solar value (eg Dinerstein et al. 1985), although such fluctuations are not understood and are arbitrary at the moment (iii) the reduction in the O abundance is larger than predicted by stellar evolution theory. In this context abundance results for 30 cool carbon stars by Lambert et al. (1986) are relevant: they found a mean O/H ratio of $4.6 \ (\pm 1.4) \times 10^{-4}$, a value very similar to that for planetaries. From Na,Ca & Fe lines it was found that the (uncertain) mean "metal" abundance (M/H) lay between $0.5 - 1.0 \times$ Solar. Thus we do not have strong evidence yet on the relative variation of oxygen and "metals" from the main sequence to PN phase.

I assume for now that these O-deficiencies are not due to mixing processes or temperature fluctuations, and present in Table 2 below mean elemental abundances, relative to oxygen, in the AC and AK samples and in a merged mean. In the last column the dredge-up process which may be responsible for a non-solar abundance ratio X/O is identified. Note that both these samples contain Type I nebulae. Incidentally, AK provide a large number of references to individual nebular abundance determinations published since 1982.

The C/O and N/O are unquestionably enhanced in PN; this is surely due to mixing processes in the progenitor stars. Similar results obtain in Magellanic Cloud nebulae: for example, Monk et al. (1988 and these proceedings) derive a very similar nitrogen enhancement for 71 Cloud PN to that listed here, and Aller et al. (1987) describe C and N enhancements in 12 nebulae in the Large & Small Clouds.

Table 2   Mean Abundances in Disk Planetaries*

Element	AC	AK	log(X/O)	[X/O]	Dredge-up
C†	8.78	8.74	0.09	+ 0.29	3
N	8.26	8.39	- 0.26	+ 0.65	1,2,3
O	8.64	8.65	0.00	-	None
F	4.6:		- 4.0:	0.3 ± 0.4?	None
Ne	8.03	8.01	- 0.64	+ 0.18	3?
Na	6.18		- 2.46	$+ 0.13 \pm 0.15$?	None
S	7.00	7.04	- 1.61	$+ 0.03 \pm 0.1$?	"
Cl	5.22	5.32	- 3.33	0 ± 0.1?	"
Ar	6.43	6.46	- 2.19	0.11 ± 0.15	"
K	4.95		- 3.69	0.04 ± 0.15	"
Ca	5.03		- 3.61	-1.1 ± 0.2	(grains)

* In the form $\log(N/H) + 12$. $[x]$ denotes $\log(x/x_\odot)$.
† From UV lines only. Mean is higher if $\lambda 4267$ is included.

Unfortunately there is an uncertainty in the carbon abundances, due to the fact that C^{++}/H^+ ratios derived from the CII $\lambda 4267$Å recombination line are often (but not always) much higher than those found from the collisionally-excited CIII 1908Å line. The cause of this is unknown, and C abundances listed here are from UV lines.

Zuckerman & Aller (1986) analysed the CNO abundances in the AC sample together with some results from the literature. They found that of 68 PN, 62% had C/O > 1. The same fraction was obtained for a subset which were clearly bipolar in shape (thus bipolarity is not a characteristic of C-rich evolved stars only). The ratio (C+N+O)/H lay between 0.34 and 1.7 times the solar value for 42 of 44 objects (here, the effects of the 1st & 2nd dredge-ups are removed, since the sum of C+N+O does not change in the CN + ON-cycle burning). It then seems unlikely, incidentally, that any significant fraction of PN progenitors are hydrogen-deficient; Goebel & Johnson (1984) suggested that cool carbon (N-type) stars might be H-deficient as an explanation for the weakness or absence of H_2 vibration-rotation lines therein, but Lambert et al. (1986) found this discrepancy vanished when model atmospheres including polyatomic molecules were used. To my knowledge there are no H-deficient planetary nebulae, except for the central cores of A30, A58 and A78 (see Sec.6).

In the above discussion it was assumed that the oxygen abundances are unaltered by mixing processes. In fact small reductions can occur in the 1st & 2nd dredge-ups. But even for massive progenitors, $M \sim 7 M_\odot$, the models (Renzini & Voli 1981, Renzini 1984) give a maximum reduction in the surface O abundance of 30% even after the thermal pulsing phase (3rd dredge-up). I draw attention to results for Type I nebulae (Peimbert & Torres-Peimbert 1983, 1987) and luminous Population I stars (Dufour, these proceedings), a few of which show large oxygen deficiencies and an anicorrelation between N and O abundances, such that O+N is approximately constant. The most likely explanation is processing of a substantial fraction of the envelope mass through the ON-cycle. In this case the O and He abundances should also be anti-correlated: when the temperature is high enough for the ON-cycle to operate, a significant fraction of H will be converted to He. Note that these objects with reduced O are sufficiently rare (thanks to their high masses and the steepness of the initial mass function) that they do not make a significant effect on the mean O abundance given in Table 2. An example of a Type I nebula with low oxygen is Hu 1-2, which has $O/H = 1.6 \times 10^{-4}$ and N/O = 1.4 (PTP87).

While there is clear evidence of alteration of carbon & nitrogen abundances in PN, the situation for neon is more marginal. Ne is predicted to be produced from fresh ^{14}N in the intershell region during AGB thermal pulsing, at least for the more massive progenitor stars, via the reactions $^{14}N(\alpha,\gamma) \, ^{18}F(\beta^+,\nu) \, ^{18}O(\alpha,\gamma) \, ^{22}Ne$. If the surface enhancement is large enough, then the Ne & C abundances should be correlated in PN. (Note that the fresh neon has mass 22, while the original surface material would be more than 90% ^{20}Ne. Unfortunately there is no known way yet of measuring neon isotope ratios in nebulae). Table 2 shows that while the mean Ne/H ratio in bright disk PN roughly equals the "Solar-System" value, the mean Ne/O of 0.24(± 0.01) for 102 nebulae (AK) is nominally higher than the recommended "Solar" value of 0.16(± 0.05) (Anders & Ebihara 1982). The latter authors show that the ratio obtained from measurements of the Ne/Ar ratio in the Solar wind (Ne/O = 0.21) is much higher than the value obtained from hot stars & HII regions (Ne/O = 0.13, Meyer 1979). Since there is evidence that elements in the Solar wind are fractionated by ionisation potential, it is better to compare PN abundances with those in hot stars and HII regions.

The mean value of Ne/O in disk PN is certainly higher than the value 0.13 from Meyer's careful review, and this suggests that PN may indeed have enhanced neon abundances. Correlations with carbon abundances should be looked for. The production of ^{22}Ne is likely to be more efficient in higher-mass stars, and indeed in a new analysis of 12 Type I nebulae (PTP87), it was found that the mean Ne/O ratio was 0.28 ± 0.05, a value even higher than that of 102 disk nebulae. Note that more work is needed on icf's for neon, for cases where UV and infrared data are unavailable (then, the NeIV and NeII may not be observed). In particular, the icf scheme Ne/H = $(Ne^{++}/O^{++})(O/H)$ will not be accurate when there is a significant fraction of H^o present, because charge-exchange reactions affect Ne and O quite differently. Since the Solar-System Ne/O ratio is not well-known, a careful comparison of PN and HII region ratios is now needed.

3.2. Galactic Bulge Nebulae

Planetaries here are of interest because they constitute a sample at a known distance - that of the Galactic Centre - and are in a region with a different chemical history from the Solar neighbourhood. For example, the ratio of carbon stars to M giants is much lower in the Bulge. Price (1981) found one out of four nebulae there to have a high oxygen abundance, so is there a metal-rich PN population in the Bulge?

The answer seems to be "generally, no". Webster (1988) derived He, N & O abundances for 49 Bulge nebulae, and found most had oxygen abundances similar to those of disk PN. One candidate metal-poor nebula was found, and ten possible O-rich objects were identified for further study. 10-20% of the sample are Type I objects, giving evidence for a young component. Mean abundances for all 49 objects were O/H = 5.4×10^{-4}, N/H = 8.1×10^{-4}. Remarkable here is the high N/O ratio - more than twice the value for disk PN - which may be due to high N abundances in the progenitor stars: Lester et al. (1987) found from measurement of the IR fine-structure lines of NIII and OIII in HII regions that the N/O ratio was about a factor three higher near the Galactic centre than near the Sun (but see Rubin et al. 1988 for a critical discussion of this question).

Pottasch & Dennefeld (1985) presented some first results from a survey of Bulge nebulae, with most nebulae again having O/H ratios typical of disk PN. Webster (1988) gives discussions and references to this question generally. C/O ratios for some of this sample would be of interest, since there are so few carbon stars in this part of the Galaxy. Kinman et al. (1988) presented lists of newly-discovered PN in the Bulge, so the size of analysable samples is increased.

3.3. Abundance Gradients in the Disk

Kaler (1983a,1985) reviewed the evidence for apparent radial abundance gradients for PN across the Galactic disk. He concluded that the apparent horizontal gradients are partly if not wholly due to vertical (ie perpendicular to the plane) gradients. There are two other problems with the interpretation of apparent gradients: PN are self-enriched in some elements (especially, He in Type I nebulae and C and N in all types of nebulae), and secondly PN belong to an old disk population with a high-velocity dispersion - so they can have non-circular orbits round the Galaxy. HII regions are better monitors of actual abundance gradients.

New results on apparent gradients were given by Faundez-Albans & Maciel (1986,1987) who find measurable gradients in Type II PN for many elements. Earlier Maciel & Faundez-Albans (1985) observed an apparent gradient in electron temperature, which is most convincing for Type II nebulae and which was interpreted as due to a general abundance gradient (cooling by heavy elements is reduced for lower abundance nebulae at large Galactocentric distances). It is in this connection that the proposed subdivision of nebulae into Types IIa and IIb may be most useful, since class IIb has the smaller velocity and should represent a 'thinner' disk population than a general sample of disk planetaries.

3.4. Correlation with position in the HR Diagram

Both Kaler (1983b) and Gathier & Pottasch (1985) found important correlations between nebular abundance and the location of the central star on the (log L, log T) theoretical HR diagram. Kaler studied 57 large nebulae, and used HI & HeII Zanstra temperatures and luminosities. In his Fig. 11 he shows that nebular N/O ratios correlate with stellar properties in the sense that higher N/O is associated with higher core mass which result from stars of higher initial mass. Gathier & Pottasch studied a sample of nebulae thought to have reasonably well-known distances, and found that He and N abundances were on average higher for stars located near higher core mass tracks.

These results give another nice confirmation that N and He are being mixed out in PN progenitor star envelopes (especially in the 2nd dredge-up). Peimbert (1985) has reviewed further evidence that the He-and N-rich Type I nebulae must originate from stars of higher initial mass ($M_i \gtrsim 2.5\ M_\odot$).

4. Depletion of Elements in Grains

The elements Aℓ, Mg, Si, Ca & Fe have abundances in nebulae which can be much lower than Solar, and this is very likely due to depletion in grains, since these refractory elements are depleted also in interstellar clouds. Shields (1983) reviewed these depletions at the last symposium.

Harrington & Marionni (1981) measured Si and Mg abundances in 6 nebulae. Si/O ratios were 0.005-0.006 in 3 nebulae (cf the Solar value of 0.05) from the Si III] λ1883 line, a depletion of an order of magnitude. For magnesium they found that, for IC2165 & NGC2440, no single abundance could match the [Mg V] λ2783 and Mg II λ2800 line fluxes simultaneously, with the predicted Mg II flux much too strong. The same result for NGC 7027 led Péquignot & Stasinska (1980) earlier to suggest that some PN had a 'Mg-gradient': in the inner, highly-ionised regions Mg was mostly gaseous, but in the outer regions (Mg^+ zone) it had largely condensed in grains. Péquignot & Stasinska alsed used [Mg IV] and [Mg V] lines in NGC 7027 to support this argument.

Mendoza & Zeippen (1987) computed new collision strengths for Mg IV & Mg V (see Sec.8), which reduce but do not remove the need for a gradient in NGC7027. In a very detailed study of NGC3918, Clegg et al. (1987a; CHBW) showed that the Mg II λ2800 line was obliterated by interstellar absorption by Mg^+ ions. This line cannot be used for abundance work unless it can be shown that the interstellar lines are shifted completely clear of the nebular emission (eg, thanks to very high PN radial velocity). Middlemass (1988; and these proceedings) gives an atlas of high-resolution Mg II line profiles; the attenuation of the nebular line flux correlates strongly with the LSR radial velocity. Models were used to derive Mg/H ratios of 4, 1 and 0.3×10^{-5} for IC418, IC4997 and NGC2440 respectively. CHBW derived Mg/H = 1.4×10^{-5} from [Mg V] and Mg I (λ4570) lines in NGC3918; the Mg I] can be used in models for abundances now that a charge-transfer rate for the reaction $Mg + H^+$ is available (Allen et al. 1988). The Solar Mg/H is 4×10^{-5}.

I conclude that (a) Mg II λ2800 must usually be corrected for interstellar absorption (this applies to C II λ1335 as well) (b) Mg depletion in grains is less than previously thought - typically a factor 3 and up to a factor 10 (c) the need for a 'gradient' in NGC7027 and NGC2440 is reduced but still not removed. It is not impossible that, with more accurate atomic data for al Mg ions and more refined PN models, the need for a gradient might vanish; the question is still open.

Depletions of iron can apparently vary greatly: Shields (1978) found for 6 nebulae a range of depletions from a factor 3 to 100. CHBW found that He was deficient by a factor 100 in NGC3918, from [Fe VI] and [Fe VII] lines. Some of the earlier work should be re-examined to (a) allow for new collision strengths for some Fe ions (b) investigate what nebular properties the depletion correlates with.

Are much of C, N & O in grains? Probably not. Clegg (1985) concluded that not more than 30% of C or O is likely to be condensed. Recently, Harrington et al. (1988; and these proceedings) modelled the thermal IR emission from NGC3918 (which has C/O = 1.6) with a size distribution of carbon grains. The implied depletions of gas phase carbon were only 11% and 4% for graphite or amorphous carbon grains, respectively. But it is not firmly known that the grains here are carbon at all - the measured depletions of Mg, Si & Fe (factors of 3, 4 & 100) are already quite sufficient to provide the required volume of dust. Pottasch et al. (1984) found a correlation between gas-to-dust ratio and nebular radius: compact PN such as BD + 30° 3639 had a ratio $\sim 10^2:1$ while for large nebulae the ratio is nearer $10^4:1$ by mass. Knapp (1985) found, from models of the dust and CO emission from the envelopes of AGB stars losing mass, that the average ratio was 400:1 for C-rich and O-rich objects. All these ratios are probably uncertain by a factor ~ 3. Ratios below about 500 are 'dangerous' in the sense that the depletion of C or O could approach 30%. We need more detailed model analyses of samples of C-rich and O-rich compact PN to see if ratios $\sim 100:1$ really can occur.

Ca is strongly depleted in many nebulae (see AC and Table 2), but S suffers either zero or rather small depletion (less than a factor two according to AC and AK). Al was found to be heavily depleted in NGC6543 and BD + 30° 3639 (Pwa et al. 1984, 1986). In an interesting new method these authors used nebular absorption lines seen in high-resolution IUE spectra, as well as emission lines, to derive abundances (the method only works if the nebular absorptions are well-shifted in velocity away from the interstellar lines). Although depletion factors are variable in PN, the pattern does resemble roughly that seen in the diffuse interstellar clouds.

5. Galactic Halo Nebulae

Four nebulae - K648, H4-1, BB-1 and DDDM-1 - are located at large distances out in the Galactic Halo. They are of special interest because (a) they have low abundances (except for carbon) and an element mixture different from Solar (b) they give information on the evolution of extreme Pop.II stars (c) they have low content of refractory elements so the dust-to-gas ratio is worthy of study (d) they can be used to estimate the pregalactic helium abundance (Sec.2). Their abundances were reviewed by Clegg et al. (1987b) and Clegg (1987b), so discussion here is highly abbreviated.

Table 3   Halo PN Abundances*

Object	C	N	O	Ne	S	Ar
K648	8.7	6.5	7.7	6.7	5.2	4.3
H4-1	9.3	8.5	8.3	6.7	5.2	5.3
BB-1	9.1	8.5	7.8	7.7	5.7	4.6
DDDM-1	<7.1	7.4	8.1	7.3	6.5	5.8
<Disk PN>	8.8	8.3	8.6	8.0	7.0	6.4

*$\log(N/H) + 12$. References in Clegg et al. (1987b).

Abundances, in the form $\log(N/H) + 12.0$ are given in Table 3. The objects fall in two groups. For K648, H4-1 & BB-1: S and Ar are much more deficient than N, O & Ne; C abundances are solar or greater so that C/O ratios are large. For DDDM-1: all elements measured have roughly 1/6 the Solar abundance, except for carbon (which is not detected at all) and Mg (which is depleted by a further factor of 3). The first group suggest that the early Galactic halo had a very different ratio of 'light' (A<11) to 'heavy' (A>15) elements from the present-day disk. K648 is in the Globular cluster M15 which has [Fe/H] = -2.1 (Fe/H 120 × below Solar), while the mean (Ar + S) deficiency for K648 is 150 × below Solar. It would appear that S & Ar have varied in step with Fe during Galactic evolution. To test this idea further I have plotted in Figure 1 (taken from Sneden 1985) all 4 Halo objects by assuming that [O/(AR+S)] represents "[O/Fe]" and [(Ar+S)/H] represents "[Fe/H]". References for data on disk & halo stars are given by Sneden.

We see that if (S & Ar) are taken as surrogates for Fe, the halo PN and old stars show the same trend with metallicity - oxygen behaves the same relative to (S & Ar) as it does to Fe. One (but not the only) interpretation of this O/"Fe" relation is that Fe is produced mainly in Type I supernovae (SNI) while O is produced in massive stars. The latter evolve so much faster than the former that O/H rises faster than Fe/H

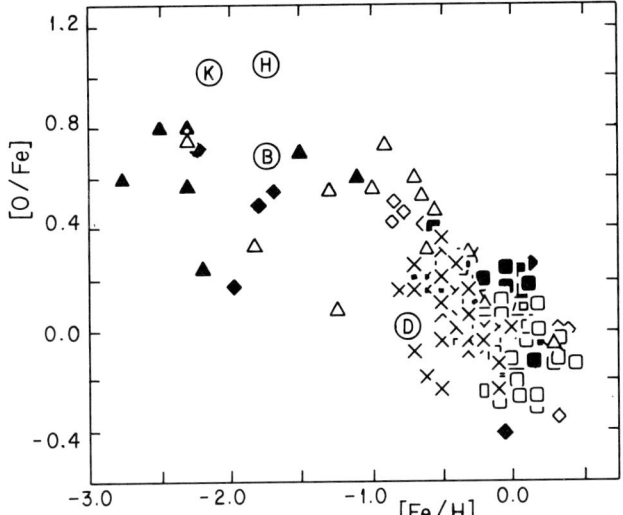

Fig.1 - Comparison of Halo PN & Halo Stars, when (S+Ar) is taken as surrogate for Fe. Halo PN symbols: H = H4-1, K = K648, B = BB-1 and D = DDDM-1. Figure from Sneden (1985).

during Galactic evolution (eg Matteucci 1987). For this scenario, we cannot have S & Ar made in massive stars or else O/(S+Ar) would be constant). The implication - that S & Ar are made in SNI-is of great interest for nucleosynthesis in supernovae: Nomoto (1985) has suggested that SNI contribute significantly to the production of elements in the Si-Ni range as well as Fe. Specifically, he has proposed a carbon-deflagration model and calculated yields of these elements. A competing, carbon-detonation model for SNI produces only Fe-peak elements. The halo PN results suggest that S & Ar are produced in a site similar to that of Fe production. Then for SNI the deflagration model is better. Analysis of old stars also shows that S,Ca and Ti vary in a manner intermediate between O and Fe (eg Francois 1987, Gustafsson 1987). Fuller discussion on nucleosynthesis in SNI & SNII is given by Nomoto (1985) and Thielemann & Arnett (1985) respectively.

The high-carbon abundances in three of the halo nebulae is presumed to be due to the 3rd dredge-up. However this presents a theoretical difficulty: that dredge-up should not have occurred at the low luminosities (a few × 10^3 $L_\odot$) corresponding to the tips of red giant branches in globular clusters (Renzini, these proceedings). Either the 3rd dredge-up is occurring at luminosities lower than predicted when the metallicity is ery low (c.f. the discovery of low-luminosity carbon stars in the Magellanic Clouds, which also disagreed with theory), or another process is at work. These three nebulae may be related to the giant CH stars (which may all be binaries). We need more accurate abundance analyses of the halo PN, more information on the central stars, and more discoveries of new halo PN to increase the sample size (currently DDDM-1 is in a class of one!).

6. CNO Abundances from Recombination Lines

Although these lines are weak in PN, they provide (a) abundance indicators for highly-obscured nebulae where no UV data are obtainable (b) possible diagnostics for electron temperature and (c) measurement of the attenuation of resonance lines by dust (eg from CIII $\lambda 2297$, NIV $\lambda 1718$. French (1983) obtained carbon abundances in 6 PN from optical recombination lines without recourse to UV data. Advantages of such lines are their insensitivity to electron temperature and their availability even in obscured objects. On the other hand they are weak, are sometimes blended, and require high-dispersion spectra. Moreover there is a serious outstanding problem for C^{++} abundances, where the CII $\lambda 4267$ recombination line can give a factor 3-10 higher abundance than the CIII $\lambda 1908$ collisionally-excited line (eg Kaler 1986; Barker 1987 and references therein).

In Table 4 I list some CNO lines that can be used. The OIV and NIV n = 6-5 lines were recently discovered in NGC3918 by Clegg et al. (1987a;CHBW); they are weak but provide useful diagnostics in bright, high-excitation nebulae.

Table 4   Some CNO Recombination Lines

Spectrum	λ(Å)	Transition	Type(a)	Rate(b)	Blends
CII	4267	3d ^{2}D – 4f ^{2}F^O	R	1	–(?)
CIII	1577-78	3d ^{3}D – 3d ^{3}F^O	D	2	[Ne V]
	2297	2p ^{1}P^O – 2p^2 ^{1}D	D	2	He II
	4070	4f ^{3}F^O – 5g ^{3}G	R	1	S II
	4187	4f ^{1}F^O – 5g ^{1}G	R	1	O II?
	4647-51	3s ^{3}S – 3p ^{3}P^O	R,D	1,2	O II
CIV	4568	5g^2 G – 6h ^{2}H^O	R	1	[Fe III]
	5801-12	3s ^{2}S – 3p ^{2}P^O	R	3	–
NII	5666-88	3s ^{3}P^O – 3p ^{3}D	R	4	–
NIV	1718	2p ^{1}P^O – 2p^2 ^{1}D	D	2	–
	4606	5g ^{3}G – 6h ^{3}H^O	D,R	2,3	–
OIV	1342	2p^2 ^{2}P – 2p^3 ^{2}D^O	D	2	C II
	4632	5g ^{2}G – 6h ^{2}H^O	R	3	–

(a) R = radiative, D = dielectronic recombination
(b) References for rates: 1 – Seaton 1977   2 – Nussbaumer & Storey 1984
    3 – see CHBW page 561   4 – Wilkes et al. 1981.

The problem about C^{++} abundances mentioned above has also been with us for some years. CII $\lambda 4267$ appears to be too strong in the integrated spectra of some nebulae, relative to CIII] $\lambda 1908$; Barker (op.cit.) has shown that the discrepancy is often worst near the inner regions of extended nebulae. The problem can also be stated another way: as derived from the flux ratio $\lambda 4267/\lambda 1908$, $T_e(C^{++})$ appears to be very often lower than T_e measured from other ionic line ratios (eg Kaler 1986). Kaler suggested that a likely solution would be to increase the effective recombination coefficient for 4267 by a factor about four. However, this would not explain the spatial variation noted above. Here are 8 possible solutions to the problem, several discussed by Barker (1982) in Paper II of his series: (a) observational errors (b) charge-transfer reaction 'feeds' $\lambda 4267$ (c) dielectronic recombination 'feeds' it (d) a blend with another line at 4267Å (e) $\lambda 4267$ excited by fluorescence (f) nebulae temperature fluctuations (g) density fluctuations (h) C-rich knots occur in the inner regions of many PN.

Space permits only brief notes on these possibilities: (a) very unlikely now several independent groups note the problem. (b) the rate would have to be 10^4 times the theoretical rate (Butler et al. 1980). (c) found to be insignificant (Storey 1981). (d) should be checked with high-dispersion spectroscopy. (e) possible. Detailed quantitative models are needed, plus observations of other CII lines which would have to be excited too. For example, one possible mechanism - absorption of starlight exciting the CII 6d ^{2}D level - involves emission at 6622Å which is not seen, with upper limits for NGC7009 (U. Carsenty, private communication) and IC418 (Monk et al., in preparation 1988) ruling out this as a significant 'route' for excitation. (f) would work. But we see no other convincing evidence for the required level of fluctuations and they are not yet explained theoretically. (g) makes the problem worse (Mihalszki & Ferland 1983), (h) not predicted by stellar evolution theory to be common at all (though see Sec.7) but worth a critical observational & theoretical look. The cool knots would contribute only to the 4267Å emission in this scheme. The minimum mass of carbon required could be calculated - is it plausible?

It is worth noting that the line CII $\lambda 4267$ gives a problem in absorption in B stars too (see Lennon 1983). LTE & NLTE analyses produce carbon abundances much lower than Solar - the line is weaker than expected. The B star and PN problems are in the same sense - the upper level of the transition (CII 4f $^2F^o$) is overpopulated relative to the lower level.

7. Abundance Variations within Nebulae

When PN show multiple shells, haloes or knots, are the abundances uniform across the whole nebula? To my knowledge the only convincing evidence for variations come from three objects - A30 and A78 (see Jacoby & Ford 1983, Iben et al. 1983, Manchado et al. 1988) and A58 (Seitter 1985, Pottasch et al. 1986) - which have H-rich outer regions and inner knots rich in He and heavy elements. Iben et al. interpreted such results for A30 & A78 in terms of a PN nuclei which had undergone a very late shell-flash, returned to the AGB and evolved off it again but now He-burning instead of H-burning. The products of helium-burning are ejected and seen in the inner knots (He,C). The central star of A58 had a 'nova-like' outburst in 1917, and in this case the central knot is only mildly He-rich. These objects are important, and Pottasch et al. note that A30 & A58 show extremely-large infrared excesses (A78 was not observed by IRAS), which might be used as a criterion for detection of further such objects.

PN with multiple-shells & haloes appear to have abundances constant within the errors. One element for which an internal gradient is of great interest is carbon, but the outstanding problem about C^{++} abundances (CII λ4267) make observational studies difficult. Jacoby et al. (1987) developed an imaging method to analyse NGC40 & NGC6826 at many positions, from calibrated CCD images taken in the light of Hβ, [OII], [OIII], [NII] & [SII]; no significant abundance variations were found. In a series of Papers, Barker (1987 and references therein) has analysed several positions in large nebulae, but again no convincing variations were seen. Space Telescope observations of the CII, CIII and CIV UV lines would aid a search for gradients and for C-rich inner knots.

8. New Collision Strength Calculations

I have included below a list of some new collision strengths published since Mendoza's (1983) compilation, and which may affect abundance determinations. There is no guarantee the list is complete; it comprises results that I or colleagues at University College know about. A reference to unpublished work implies that the data can be obtained from the authors listed. Note that the collisional strengths for the fine-structure lines of NeV (Aggarwal 1983) are not recommended here, for the reasons given by CHBW.

Some highlights are noted. The availability of good data for Ar IV is very useful. Derived Ar^{3+} abundances are affected, by up to a factor 2, and we can now use the [Ar IV] line ratio λ4711/λ4740Å more reliably as a diagnostic for electron density measurements. Secondly, an increasing number of infra-red fine structure lines can now be used for abundance work. Particular attention is drawn to observations of such lines as Ne VI at 7.6μm (Pottasch et al. 1986) and Si VI (1.96μm) & Si VII (2.48μm) by Ashley et al. (1987); these have been detected so far only in Type I nebulae. The new data permit derivation of Ne^{5+}, Si^{5+} and Si^{6+} abundances. The I.P. of Si^{5+} is 205.1 eV, so detection of Si VII attests to the very high temperatures of the central stars of Type I PN. Moreover, it suggests that such nebulae are so highly-ionized that ionisation correction factos (icf's) for heavy elements need to be evaluated very carefully in the study of these systems.

Two disadvantages of far-infra-red lines for abundance work in PN are that absolute flux calibration is difficult, and that the lines are density-sensitive (critical densities are often in the region of $10^3 - 10^4$ cm^{-3}). The lines areparticularly important for the study of highly-obscured HII regions.

Table 5   New Collision Strength Data

Spectrum

C II	10-state calculation.	Lennon et al. (1985), Hayes & Nussbaumer (1984)
N III	Fine-structure lines.	K. Butler & P.J. Storey (in prep.)
O IV		Hayes & Nussbaumer (1983)
Ne II	Fine-structure line.	Bayes et al. (1985)
Ne III	Fine-structure lines.	K. Butler (1984)
Ne IV	Fine-structure lines.	K. Butler & P.J. Storey (in prep.)
Mg IV	5-state calculation (incl. fine-str.).	Mendoza & Zeippen (1987)
Mg V	" " " " "	" " " "
Mg IV) Aℓ V)	Fine-structure lines.	K. Butler (in prep.)
Si III	Ultraviolet lines	Dufton et al. (1984)
Si VI) Si VII)	Fine-structure lines (near IR).	K. Butler (in prep.)
Ar VI	7-state calculation.	Zeippen et al. (1987)
Cℓ III	" "	K. Butler et al. (in prep.)
K V	" "	K. Butler et al. (1988)
Mg I	5-state calculation.	H.E. Saraph (In prep.)
Fe VII	Many transitions.	Keenan & Norrington (1987)

9. Concluding Remarks

New atomic data, new detectors and newly-opened spectral regions seen from space have improved our knowledge of abundances in planetary nebulae. The mixture of elements in most PN is non-Solar, with C and N the main culprits. There are some agreements with the nucleosynthesis & mixing processes described by current theory. An interesting figure is provided by Lambert et al. (1986) which shows the 'evolution' of mean C/H, N/H and O/H ratios from M/S stars via C stars to PN. The PN data arre from Aller & Czyzak (1983).

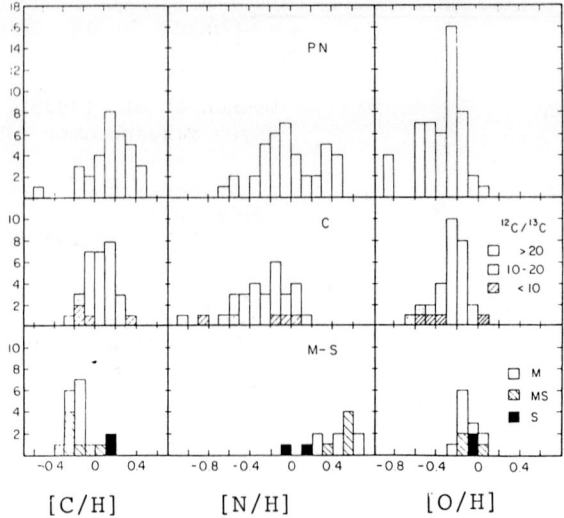

Fig.2 - Histograms of C/H, N/H and O/H for M & S stars (bottom), C stars (middle) and PN (top panel). Figure and references from Lambert et al. (1986)

Such plots are broadly encouraging. But there are plenty of problems. Examples include the large C abundances of 3/4 Galactic Halo nebulae, the low O abundances found in some Type I PN and shells around massive stars, a few high He/H ratios ($\gtrsim 0.18$) in some Type I nebulae, the C^{++}(4267/1908Å) problem, the depletion of carbon in dust. Future studies may usefully be directed at PN in specific locations (Galactic Bulge, Magellanic Clouds, other galaxies), accurate studies of Type I nebulae, search for new Halo PN and A30-type objects, and the systematics of dust in nebulae of different sizes, as this affects abundances.

I am grateful to many colleagues who sent preprints and reprints of their work, and to Dr. Butler for advice on new atomic data.

References

Allen, R.J., Clegg, R.E.S., Flower, D.R. & Dickinson, A.S., 1988. In prep.
Aller, L.H., 1984. "Physics of Thermal Gaseous Nebulae". (Reidel).
Aller, L.H. & Czyzak, S.J., 1983. Ap.J.Supp., 51, 211 (AC).
Aller, L.H. & Keyes, C.D., 1987. Ap.J.Supp., 65, 405 (AK).
Aller, L.H., Keyes, C.D., Moran, S.P., Gull, T.R., Mitchalitsianos, G. & Stecher, T.P., 1987. Ap.J. 320, 177.
Anders, E. & Ebihara, M., 1982. Geochim. et Cosmochim. Acta, 46, 2363.
Aggarwal, K.M., 1983. J. Phys.B, 16, 2405.
Ashley, M.C., Dopita, M.A. & Wood, P.R., 1987. Preprint.
Barker, T. 1978. Ap.J. 220, 193.
Barker, T., 1982. Ap.J. 253, 167.
Barker, T., 1987. Ap.J. 322, 922.

Bayes, F.A., Saraph, H.E. & Seaton, M.J., 1985. MNRAS, 215, 85p.
Berrington, K. & Kingston, A.E., 1987. J.Phys.B, 20, 6631.
Berrington, K., Burke, P.G., Freitas, L.C.G. & Kingston, A.E., 1985. J.Phys.B, 18, 4135.
Brocklehurst, M., 1971. MNRAS, 153, 471.
Brocklehurst, M., 1972. MNRAS, 157, 211.
Butler, K. & Mendoza, C., 1984. MNRAS, 208, 17p.
Butler, S.E., Heil, T.G. & Dalgarno, A., 1980. Ap.J.241, 442.
Clegg, R.E.S., 1985. ESO Workshop 'Production & Distribution of CNO Elements', (eds. I.J. Danziger et al.), p261.
Clegg, R.E.S., 1987a, MNRAS, 229, 31p. (C87).
Clegg, R.E.S., 1987b. ESO Workshop 'Stellar Evolution & Dynamics in the Outer Halo', p543.
Clegg, R.E.S., Peimbert, M. & Torres-Peimbert, S., 1987. MNRAS, 224, 761.
Clegg, R.E.S., Harrington, J.P., Barlow, M.J. & Walsh, J.R., 1987. Ap.J. 314, 551 (CHBW).
Dinerstein, H.L., Lester, D.F. & Werner, M.W., 1985. Ap.J. 291, 561.
Dufour, R.J., 1984. IAU Symp. 108, p353.
Dufton, P.L., Keenan, F.P. & Kingston, A.E., 1984. MNRAS, 209, 1P.
Faundez-Abans, M. & Maciel, W.J., 1986.Astr. Astroph., 158, 228.
Faundez-Abans, M. & Maciel, W.J., 1987. Astr. Astrophys., 183, 324.
Ferland, G.J., 1986. Ap.J. 310, L67.
Francois, P., 1987. Astr. Astrophys., 176, 294.
French, H.B., 1983. Ap.J., 273, 214.
Gathier, R. & Pottasch, S.R., 1985. See Clegg 1985, page 307.
Goebel, J.H. & Johnson, H.R., 1984. Ap.J. 284, L39.
Gustafsson, B., 1987. See Clegg 1987b, p33.
Harrington, J.P. & Marionni, P.A., 1981. 'The Universe at Ultraviolet Wavelengths', NASA Conf. Publ. 2171.
Harrington, J.P., Monk, D.J. & Clegg, R.E.S., 1988. MNRAS, in press.
Hayes, M.A. & Nussbaumer, H., 1984. Astr. Astrophys., 134, 193.
Iben, I. & Renzini, A., 1983. Ann. Rev. Astr. Astrophys., 21, 271.
Iben, I., Kaler, J.B., Truran, J.W. & Renzini, A., 1983. Ap.J. 264, 605.
Jacoby, G.H. & Ford, H.C., 1983. Ap.J., 266, 298.
Jacoby, G.H., Quigley, R.J. & Africano, J.L., 1987. Preprint.
Kaler, J.B., 1978. Ap.J. 226, 947.
Kaler, J.B., 1983a. Proc. IAU Symp. 103 (ed. D.R. Flower), p245.
Kaler, J.B., 1983b. Ap.J. 271, 188.
Kaler, J.B., 1985. Ann. Rev. Astr.Astrophys., 23, 89.
Kaler, J.B., 1986. Ap.J. 308, 322.
Keenan, F.P. & Norrington, P.H., 1987. Astr. Astrophys. 181, 370.
Kinman, T.D., Feast, M.W., 1987. Preprint.
Knapp, G.R., 1985. Ap.J. 293, 273.
Köppen, J. & Aller, L.H., 1987. In 'Scientific Accomplishments of the IUE' (ed. Y. Kondo), p589.
Lambert, D.L., Gustafsson, B., Eriksson, K. & Hinkle, K.H., 1986. Ap.J.Supp., 62, 373.
Lennon, D.J., 1983. MNRAS, 205, 829.
Lester, D.F., Dinerstein, H.L., Werner, M.W., Watson, D.M.,Genzel, R. & Storey, J.W.V., 1987. Ap.J. 320, 573.
Maciel, W.J. & Faundez-Abans, M., 1985. Astr. Astrophys., 149, 365.

Manchado, A., Pottasch, S.R. & Mampaso, A., 1988. Preprint.
Matteucci, F., 1985. See Clegg 1987b, page 609.
Meyer, J-P., 1979. 'Les Elements et leurs Isotopes', Liège Astrophys. Symposium, p489.
Mendoza, C., 1983. IAU Symp. 103 (ed. D.R. Flower), p.143.
Mendoza, C. & Zeippen, C.J., 1987. MNRAS, 224, 7P.
Middlemass, D., 1988. MNRAS, in press.
Mihalszki, J.S. & Ferland, G.J., 1983. PASP, 95, 284.
Monk, D.J., Barlow, M.J. & Clegg, R.E.S., 1988. MNRAS, submitted.
Nomoto, K., 1985. In 'Nucleosynthesis' eds. Arnett & Truran (U.Chicago Press), p202.
Nussbaumer, H. & Storey, P.J., 1984. Astr. Astrophys., Supp. 56, 293.
Péquignot, D. & Stasinska, G., 1980. Astr. Astrophys., 81, 121.
Peimbert, M., 1978. Proc. IAU Symp. 76 (ed. Y. Terzian), p215.
Peimbert, M., 1983. ESO Workshop 'Primordial Helium' (eds. P.A. Shaver et al.) p267.
Peimbert, M., 1985. Rev. Mex. Astr. Astrof., 10.
Peimbert, M., 1987. In 'Planetary & Proto-Planetary Nebulae from IRAS to ISO' (ed. A. Preite-Martinez), page 91.
Peimbert, M. & Serrano, A., 1980. Rev. Mex. Astr. Astrof., 5, 9.
Peimbert, M. & Torres-Peimbert, S., 1983. Proc. IAU Symp. 103 (ed. D.R. Flower), p233.
Peimbert, M. & Torres-Peimbert, S., 1987. Rev. Mex. Astr. Astrof., 14, 540 (PTP87).
Pottasch, S.R. & Dennefeld, M., 1985. See ref. to Clegg 1985, p303.
Pottasch, S.R., 1984. "Planetary Nebulae" (Reidel).
Pottasch, S.R., Mampaso, A., Manchado, A. & Menzies, J., 1986. Proc. 'Hydrogen-Deficient Star & Related Objexts' (eds. K. Hunger et al.) p359.
Pottasch, S.T. et al., 1984. Astr. Astrophys., 138, 10.
Price, C.M., 1981. Ap.J. 247, 540.
Pwa, T.H. & Pottasch, S.R., 1986. Astr. Astrophys., 164, 184.
Pwa, T.H., Mo, J.E. & Pottasch, S.R., 1984. Astr. Astrophys., 139, L1.
Renzini, A., 1984. In 'Stellar Nucleosynthesis', eds. C. Chiosi & A. Renzini.
Renzini, A. & Voli, M., 1981. Astr. Astrophys., 94, 175.
Rubin, R.H., Simpson, J.P., Erickson, E.F. & Haas, M.R., 1988. Ap.J. in press.
Schönberner, D., 1979. Astr. Astrophys., 79, 108.
Seaton, M.J., 1977. IAU Symp. 76, p131.
Seitter, W.C., 1985. See Clegg 1985, page 253.
Shields, G., 1983. IAU Symp. 103, p259.
Sneden, C., 1985. See Clegg 1985, page 1.
Storey, P.J., 1981. MNRAS, 195, 27p.
Sweigart, A.V. & Mengel, J.G., 1979. Ap.J. 229, 624.
Thielemann, F-K. & Arnett, W.D., 1985. Ap.J. 295, 604.
Torres-Peimbert, S. & Peimbert, M., 1977. Rev. Mex. Astr. Astrof., 2, 181.
Webster, B.L., 1988. MNRAS, in press.
Wilkes, B.J., Ferland, G.J., Hares, D., & Truran, J.W., 1981. MNRAS, 197, 1.
Zeippen, C.J., Butler, K. & Le Bourlot, J., 1987. Astr. Astrophys., 188, 251.
Zuckerman, B. & Aller, L.H., 1986. Ap.J. 301, 772.

PHOTOIONIZATION MODELS

J. Patrick Harrington
Astronomy Program
University of Maryland
College Park, Maryland 20742

ABSTRACT. A recent comparison of the photoionization models generated by five independent codes run with the same density and stellar radiation shows substantial agreement. Problems are more likely to arise with the defining parameters: the density distribution should be based on observed images, and the ionizing radiation should be from model atmosphere calculations, which, however, are inadequate for stars with winds. Models can be improved by including dust and by incorporating, self-consistently, radiative transfer in optically thick lines. Future work may extend modeling to axially-symmetric objects, to the interface with the hot, shocked, stellar wind, and to the molecular component present around many nebulae.

1. HISTORICAL INTRODUCTION

This talk could well be subtitled "Twenty Years of Nebular Modeling". For it was just 20 years ago that the first nebular models to calculate the ionization of the heavy elements and to incorporate their cooling in the determination of the run of electron temperature were constructed. The first such models published were those of Goodson (1967). Later in 1967, at the first in this series of IAU Symposia on planetary nebulae, similar ionization studies were described by Flower and by Williams, and I published some models shortly thereafter (Flower 1968, Harrington 1968, Williams 1968).

Soon during the first decade of modeling (1967-1977), models were compared with specific objects, and a problem emerged which dominated these studies during this period. It was found that the predicted emission lines from the low ionization stages were too weak when compared to observation. This problem has by now been resolved: the effects of charge-exchange with neutral hydrogen and of di-electronic recombination decrease the ionization level of many of the heavy elements, so that even high excitation nebulae can show strong lines of [N II], [O II], [O I], etc. The solution to this problem was emerging at the 1977 IAU Symposium: it was realized from the few rates then available that charge transfer could have a significant effect (Harrington 1978). Pequignot (1978) went so far as to compute what charge transfer rates would be needed to solve the problems presented by the spectrum of NGC 7027, and his predictions have proven to be in substantial agreement with the subsequent atomic physics calculations.

The first half of the second decade (1977-1982) saw the impact of ultraviolet observations of planetary nebulae on models. The results obtained with the IUE satellite were of especial importance. The ultraviolet data simultaneously made nebular modeling less arbitrary and more useful. The strongest UV lines arise in the He^{++} zone of high-excitation PNe. A lack of temperature-sensitive line ratios from this zone encourages the use of models, which provide temperatures based on the energy balance of the gas. The models were in turn constrained by the observed spectra, which were especially useful in establishing the elemental abundance of carbon, the major coolant in the He^{++} zone. The use of models for the analysis of nebular spectra helped illustrate the fact that such models provide a rational basis for the ionization correction factors (i.c.f.'s) needed to account for unobserved stages of ionization when deriving elemental abundances.

Looking over the literature since the last PN Symposium in London (1982), it seems clear that during these last 5 years modeling has finally "arrived" as a standard technique. It is a technique that is now used by a large number of groups and applied to an increasing variety of problems.

2. CURRENT STATE OF MODELING CODES

The task of evaluating the current state of nebular computational techniques was made much easier by the "Workshop on Model Nebulae", which was held at Meudon in July of 1985. This workshop included modelers of active galaxies and H II regions as well as planetary nebulae, and shock codes as well as photoionization codes. A key feature of the conference was the case studies, a set of well defined models which all the participants constructed and subsequently intercompared. Not all codes in use were represented, of course, but the 12 groups which constructed the planetary nebula case represents a substantial fraction of the active workers in this field.

The planetary nebula model consisted of a sphere of gas with a uniform density of 3000 hydrogen atoms cm^{-3} and an inner radius of 10^{17} cm, ionized by a central star with a radius of 10^{10} cm, radiating as a 150,000K black body. The nebula was to be ionization bounded, with the computation carried until the gas became neutral (at about $3.9\ 10^{17}$ cm from the star). The abundances, relative to 10^4 H atoms, of He, C, N, O, Ne, Mg, Si, and S, were set at 1000, 3, 1, 6, 1.5, 0.3, 0.3, and 0.15, respectively.

As such models go, this one is fairly demanding of the ionization program, for the high temperature star produces significant radiation at high frequencies where the absorption coefficients of H and He are small. As a result, the model must be continued to optical depths in the Lyman continuum of several hundred before the gas is predominately neutral. If the computation is terminated too soon, some of the Hβ radiation is missed and the normalization of all the lines is affected.

Table 1 is a partial list of the line intensities (on a scale of Hβ = 100) from 5 independent model codes. The agreement is overall rather good for the lines from major ionization stages and for temperature sensitive line ratios, with some problems apparent for the highest ionization stages. Differences can arise from a wide variety of sources: atomic data, numerical techniques, radiative transfer effects, ect. In fact, it emerged that the greatest problem was the treatment of the diffuse ionizing radiation which arises in the He^{++} zone. Codes used for clouds illuminated from the outside (as in the modeling of active galaxies) normally let this diffuse radiation escape from the inner face, while the planetary nebula codes assume such radiation crosses the interior and reenters the opposite side. This made a surprising difference, and it was necessary to re-run some codes assuming this radiation did not escape. But since many real nebulae (as opposed to this ideal case study) may allow leakage of such radiation, it is clear that there is an uncertainty in modeling real objects introduced by the extent of leakage. A further problem affecting the strengths of the high excitation lines from the He^{++} zone is different methods used to treat the line radiation which constitutes part of the diffuse field in the form of the He^+ Lα line and its degradation products, the Bowen fluorescence lines.

The complete list for this and other case studies (as well as the identities of the modelers!) can be found in the Workshop proceedings (Pequignot, 1986). I would urge anyone running photoionization codes to examine these proceedings and compare the results of their code with those presented there. While the workshop did not attempt to define a "standard" model, nevertheless, as Pequignot concluded, "..significant departure from these tables should reasonably call for a comment".

TABLE 1. Comparison of Independent Photoionization Codes.

Line		(1)	(2)	Model Program (3)	(4)	(5)
Hβ	(ergs/sec)	2.64E+35	2.51E+35	2.67E+35	2.29E+35	2.01E+35
He I	5876 Å	9.7	11.	11.	9.6	8.7
He II	4686	33.6	35.0	31.6	35.0	43.4
C II	2326	49.8	31.0	32.3	44.0	32.3
C III	1909	167.	173.	170.	226.	291.
C IV	1549	170.	119.	202.	212.	207.
N I	5200	2.0	1.0	1.7	0.9	1.8
N II	6548,84	149.	146.	137.	153.	145.
N III	17492	11.8	9.0	11.2	14.5	13.7
N IV	1487	11.6	9.0	16.0	15.3	23.7
N V	1240	7.9	4.0	14.8	12.0	13.8
O I	6300,63	15.5	15.0	13.4	15.2	17.6
O II	3726,29	228.	224.	218.	252.	256.
O III	5007,4959	2015.	2105.	2160.	2470.	1990.
O III	4363	15.9	15.0	15.6	21.0	17.9
O III	52 mμ	144.	142.	143.	149.	136.
O IV	1403	14.9	7.0	17.9	15.0	21.7
O V	1218	7.3	3.0	17.8	13.2	23.5
Ne II	12.8 mμ	2.35	3.0	3.2	2.5	5.2
Ne III	15.5 mμ	256.	244.	254.	256.	238.
Ne III	3869,3968	268.	252.	259.	344.	278.
Ne IV	2423	61.9	43.0	62.0	52.6	83.9
Ne V	3426,3346	65.9	60.	94.3	94.6	106.

It is perhaps worth noting that by present-day standards, the construction of photoionization models does not require "big" computers. This came to my attention recently when the Astronomy Program at Maryland acquired several SUN 3/50 workstations. These are desktop computers. I find that my code, running on this machine, can generate a full 36 zone model, iterating four times to converge the diffuse radiation field, in less than 20 minutes. The implication is that on a big computer, a full photoionization calculation can become a subroutine in a more general problem.

3. SOME REPRESENTATIVE STUDIES

As a sample of the problems which have been considered in the last few years using photoionization codes, in this section I will review a few papers from two categories: studies which apply to PNe in general, and studies of specific nebulae. This is not intended as a comprehensive survey, rather, the citations are used to illustrate particular ideas.

3.1. General Studies

Several papers applied models to evaluate the Zanstra method for determining the temperatures of central stars. Henry and Shipman (1986) constructed a large grid of models to investigate the photon-counting characteristics of optically- thick nebulae. They considered stellar radiation with a variety of dilution factors, and with flux distributions from model atmospheres computed with various He/H ratios. They found that the traditional approach is sound but that, for thick models, the hydrogen Zanstra temperature may be more useful than the He II temperature because the latter is extremely sensitive to the He/H ratio of the atmosphere.

A more unusual situation was investigated by Stasinska and Tylenda (1986), who asked whether the traditional analysis would reveal the true temperature of extremely hot central stars, i.e., T_{eff} > 200,000K. They found that the methods break down in this regime - the central star is so faint that the only viable method is to use the ratio of He II lines to H lines. However, the He^{++} recombinations produce H ionizing radiation that swamps the direct ionization of H by the stellar radiation in the 13.6 - 54.4 eV interval, so that the intensity of He II lines relative to Hβ levels off: the He II 4686/Hβ ratio never rises above about 0.6, even for a 500,000K central star. Furthermore, the exact value is sensitive to the treatment of the diffuse radiation from the central He^{++} zone and to the possible escape of this radiation as discussed above.

Models may also be used investigate the sort of errors that are introduced by the assumptions of uniform temperature and density which are implicit in the use of diagnostic line ratios (at least when the fluctuation parameter $<t^2>$ introduced by Peimbert (1967) is not used). Mihalszki and Ferland (1983), for example, investigated the effects of large fluctuations in density (and the corresponding temperature fluctuations) on the abundance ratios that would be obtained by a naive analysis. They found that the fluctuations cause the carbon abundance as determined from the collisionally excited UV lines to be higher than those inferred from recombination lines, just the opposite of what is frequently observed.

Models may be used to investigate specific physical processes. For example, Clegg and Walsh (1985) used models of nebulae to predict the strength of the O III 5592Å line in several PNe. This faint line is of interest because it arises mainly by charge transfer from H^0 to O^{+3}, and comparison with the observed value provides a check on the calculated rate of charge transfer into the ^1P channel. They found agreement to within a factor of 2, which, in view of the errors inherent in this approach, constitutes a verification of the theoretical rate.

3.2. Studies of Specific Objects

Studies of specific nebulae reveal some of the problems that have to be confronted for successful modeling. Adam and Koppen (1985) constructed two models, one of NGC 4361 and another of NGC 1535. Using the central star temperatures and gravities determined by Mendez et al. (1981) from profiles of the stellar absorption lines, they found that a consistent model of NGC 4361 could be constructed, but that NGC 1535 could not be modeled with the atmosphere flux implied by the stellar line analysis - the flux beyond 54.4 eV had to be raised by a factor of $> 10^3$.

Aller et al. (1987) used modeling techniques to derive physical parameters for 12 PNe in the Magellanic Clouds. The model atmosphere fluxes were taken from the NLTE calculations of Husfeld et al. (1984). This study found lower luminosities for the central stars of the brightest of these nebulae than an earlier study of Stecher et al. (1982), a revision due in large part to the choice of stellar fluxes.

While the previous studies encountered problems with the selection of atmosphere fluxes, the main obstacle that had to be overcome in the study of NGC 3918 by Clegg et al. (1987) concerned the density distribution. It was found that this nebula - which was selected partly because of its symmetrical appearance - could not be represented by a spherical nebula of any optical depth. Instead, it was necessary to adopt "dumbbell" density distribution with the nebula optically thick along the axis but optically thin, even in the He II continuum, in the equatorial plane. This structure allows a leakage of stellar photons in the equatorial plane and, consequently, the adoption of a stellar temperature which exceeds the He II Zanstra temperature.

Another instance where the density distribution was of critical importance was in the study of the halo planetary DDDM-1 by Clegg, Peimbert, and Torres-Peimbert (1987). In this case, the density based upon the surface brightness in $H\beta$ was higher than that based upon the density sensitive [S II] 6716,31Å line ratio. Fitting these observations required the adoption of a density distribution with a high density core and a lower density envelope: the envelope is of lower excitation and hence is the site of most of the [S II] line emission. This object has extremely low abundances of the heavy elements, in common with other halo objects, but also has a uniquely low carbon abundance. As a consequence of the low abundances, collisional excitation from the metastable He I $2\ ^3S$ level, which would normally be insignificant compared to collisional excitation of the elements such as C, N, and O, makes some contribution to the cooling of the nebular gas.

4. CURRENT CONCERNS

From the preceding discussion of recent work, we can see some of the issues that worry the modeler. While most of the computational aspects of the problem seem well enough in hand, the definition of the model often presents a challenge.

4.1. Good Models Demand Good Density Distributions

The distribution of nebular material in space is extremely important. While a simple model is better than no model at all, we should, if at all possible, use a density distribution that is consistent with the appearance of the object in the sky. The beautiful CCD images we have seen in the talks by B. Balick and Y.-H. Chu show the quality of data that is now available to the modeler. If we fail to make use of such information, then the dilution factor of the nebular material may take on unrealistic values. For many nebulae, it will not be possible to construct satisfactory models until departures from spherical symmetry are taken into account.

4.2. What Should We Use for the Stellar Flux ?

The choice of the stellar atmosphere flux is basic to any model. The Zanstra temperatures provide powerful constraints. The Stoy temperature, computed from the ratio of collisionally excited to recombination lines, is also helpful. But the detailed shape of the flux distribution should come from a model atmosphere computation. The recent NLTE models of Husfeld et al. (1984) and of Clegg and Middlemass (1987) are a welcome addition to the classic grid of LTE models by Hummer and Mihalas (1970). Even these models, however, often fail to represent central stars in the intermediate temperature range of 50,000K - 90,000K, where large deviations from predicted flux distributions seem to occur (e.g., Harrington and Feibelman 1983, and Adam and Koppen 1985, as mentioned in § 3.2 above). These stars may have strong stellar winds, which can drastically alter the emergent flux beyond 54.4 eV. The discussion in this Symposium by R. P. Kudritzki indicates that a satisfactory resolution of this long-standing problem may finally emerge as the result of self-consistent NLTE computations which explicitly include the line transfer in the accelerating wind.

Observations of x-rays from some central stars may help discriminate between different model atmosphere predictions of the far-UV fluxes. For example, NGC 1360 has been detected by EXOSAT. The implications of these data were discussed by de Korte et al. (1985). They concluded that no blackbody distribution was consistent with the x-ray data but that fluxes from LTE atmosphere models, with O V and Ne V absorption edges, were consistent. Unfortunately the analysis is complicated by the breadth of the EXOSAT filters. A further concern is that, although the nebula must be optically thin beyond 54.4 eV to see the x-ray emission at all, a small amount of absorption by He^+ could greatly affect the interpretation.

4.3. Other Concerns

Some of the other problems which make modeling less satisfactory than we might hope include incomplete atomic data and deficiencies in the treatment of optically thick lines.

The atomic data are generally good for C, N, O, and Ne, but when we turn to the next row of the periodic table, the data are less complete. Models often have difficulties with sulfur lines, for example. Without reliable charge-transfer and dielectronic recombination rates, we will not be able to model correctly the relative abundances of the different ions. While these elements have relatively little effect on the nebular conditions, we often would like to derive their abundances relative to the CNO group to help trace the nucleosynthetic processes in the progenitor star. Models could provide valuable ionization correction factors if we had more complete atomic data.

We also need to incorporate improved treatments of the transfer of optically thick line radiation in our PNe models. The prototype of such lines is the $L\alpha$ line of hydrogen. It was originally thought that trapped radiation in this line could become so intense that the radiation pressure would dominate the dynamics. It is now recognized that the universal presence of dust in PNe suppresses the line intensity in those objects with higher optical depths. This diminishes the dynamical importance of $L\alpha$, but the transfer of this line radiation is still important as an integral part of the problem of dust heating.

While we cannot observe H I $L\alpha$ directly, there are optically thick lines which can be observed. The C IV 1549Å line is very prominent in the UV spectra of high-excitation PNe, and the trapping of these photons will greatly enhance their absorption by dust. To model the degree of this attenuation requires both a solution of the line transfer problem and a model for the UV dust extinction. The solution of the problem should provide us with the expected emission line profile and the distribution of the emission over the face of the nebula - which in the case of optically thick lines is not just a projection of the emitting region. Unfortunately, IUE observations do not have quite the spatial or frequency resolution to determine much more than the total flux in this line, but the Hubble Space Telescope will be able obtain spatial maps and line profiles.

Another optically thick line which is important for the emission line spectra of PNe is the He+ $L\alpha$ line. This line, and the O III lines coupled to it by the Bowen fluorescence process, is an important contributor to the photoionization heating in high-excitation PNe. The question is not only one of what fraction of the line radiation is absorbed by photoionization, and hence heats the gas, but also of how far this radiation diffuses from its point of origin. The distribution of absorption will affect the temperature structure in the inner region of the nebula, and the collisionally excited UV lines are quite sensitive to this structure. The transfer problem depends upon the velocity field in the nebula because the O III resonance lines, with only half the Doppler width of the He II Lα line, constitute a major escape channel. But information on the velocity fields in many PNe is becoming available (e.g., H.-Y. Chu, this Symposium) and there is no reason why the models cannot incorporate this data. A by-product of the solution of this problem would be the prediction of the intensities of the O III Bowen fluorescence lines for the specific object.

The methods for solution of such transfer problems exist, but have been applied to general models rather than specific objects. Thus, we have long had solutions to the Bowen fluorescence problem (Weyman and Williams 1969). But now, with advances in computing power and algorithms, we should be able to incorporate such line transfer calculations into the photoionization codes and solve the ionization and line-transfer problems self- consistently.

5. FUTURE DIRECTIONS

Looking to future work in this field, one can predict with little risk that models will continue to be used, with varying degrees of elaboration, to obtain i.c.f.'s for chemical abundance determinations in PNe.

More interesting perhaps is the likelyhood of extending photoionization calculations into situations where other physical processes must be considered. I deliberately do not touch upon the dynamical models where the photoionization computation is carried out in conjunction with the hydrodynamic expansion of the nebula - this topic will be discussed by J. Koppen. Nor I will discuss the time-dependent ionization effects to be expected as a result of the rapid evolution of the most massive central stars, as R. Tylenda will mention this topic. In this section I will just briefly indicate a few other extensions which should be explored.

5.1 Dust in the Ionized Gas

One aspect of PNe which I feel has been somewhat neglected by modelers is the presence of the dust. Not only is the dust responsible for the IR emission, but it also causes attenuation of some observable UV resonance lines, notably C IV 1549Å as discussed above. The introduction of dust into photoionization models places a physical constraint on the dust model: the same radiation field that explains the ionization structure must be able to heat the dust to the temperature required by the observed far IR emission. The IRAS satellite observed a large number of PNe, so observational data is available for most of the bright objects. An attempt to explore this type of modeling for NGC 3918 has been presented by Harrington, Monk, and Clegg (1988). Also, see the contributed paper in this Symposium by M.G. Hoare, who has used similar techniques to model IC 418 and other objects. It is not hard to see why modelers have avoided routine inclusion of the dust. The difficulty, of course, is our poor knowledge of the optical characteristics of the dust particles.

Dust can potentially affect any of the optically thick lines in nebulae, and so should be included in treatments of the Bowen fluorescence lines and the He I 10830Å line.

The line transfer of H I Lα, C IV 1549, etc. in the paper by Harrington, Monk and Clegg (1988) is solved - or rather avoided - by the escape probability method. It is possible improve upon this treatment. If the optical depths are not too large, then partial redistribution in the line wings can be neglected, and the resulting complete redistribution problem can be solved by the powerful accelerated lambda-iteration procedure (Olson, Auer, and Buchler 1986). With this method, taking proper account of the spherical geometry is not difficult. I have obtained solutions by this method which confirm the global accuracy of the escape probability approximation, while at the same time indicating that in objects of non-uniform density structure, there are interesting effects which can only be seen when the transfer problem is solved properly. For example, the absorption by dust depends upon the dust/H^0 ratio, which is highest in low density regions, while the generation of Lα photons is highest in high density regions. Thus there is a transport of energy from the higher density to lower density regions.

5.2 The He I Spectrum

A paper by Ferland (1986), suggesting that collisional excitation of helium atoms in the metastable $2\,^3S$ state was more important than previously supposed, and that as a consequence, He abundances in PNe might have been overestimated by up to 50%, has reawakened interest in the He I spectrum. Recent discussions by Clegg (1987) and by Peimbert and Torres-Peimbert (1987) have shown that the abundance revisions are not in fact large. The discussion by Peimbert and Torres-Peimnert (1987) indicates, moreover, that the population of the $2\,^3S$ state determined empirically from the He I spectrum (the He I 10830Å line is produced chiefly by collisional excitations from $2\,^3S$) is only about half what one expects based on the rates of recombination, radiative decay, and collisions. This suggests that depopulation by photoionization may be important, and in fact computations show (Clegg and Harrington, 1988) that the population can be reduced by up to 20% in compact objects. In addition to direct photoionization by stellar radiation, the H I Lα radiation field is important. This introduces a further complication since the Lα intensity is controlled by the dust content of the nebula, so that a self-consistent treatment of the He I 10830 line brings us back to the dust problems discussed above.

5.3 Axially Symmetric Nebulae

With all the new CCD images we have seen at this Symposium, there is clearly an explosive growth in information on the structure of PNe. Unfortunately, not too many of them are "theorist's planetaries"! B. Balick has presented the case for an evolutionary sequence of axially symmetric objects, with expansion driven by the pressure of the shocked stellar wind, which forms a hot, thin gas trapped inside the denser nebular shell. Whether or not this picture is universally valid, it is clear that axially symmetric objects in various orientations can explain most of the forms we see. We will have to generalize our codes to deal with such structures. We should be grateful for any symmetry we can get!

5.4 The Interface With the Shocked Stellar Wind

When a 2000 km/sec wind from the central star strikes the virtually stationary nebular material, or rather, the previously stopped wind material, the shock results in temperatures of the order of 10^7 K. We expect that there will be some transfer of this heat into the adjacent nebular gas, either through conduction across the contact discontinuity, or if this is too strongly inhibited by magnetic fields, perhaps by conduction into small parcels of material mixed into a (mass-loaded) flow of the shocked wind. Such a flow past the nebular torus would form when the wind-blown bubble bursts, and would explain the material seen to be moving at velocities of hundreds of km/sec in some nebulae. Radiation from such a region would generally be swamped by the normal nebular spectrum, but spectra of high spatial resolution might reveal some signature of the interface. To complement the observational search for interaction of the 10^7 K component and the 10^4 K component, we need models of such an interface zone, defining the volume occupied by the various ionization species and the power radiated in different spectral lines.

5.5 The Molecular Component

Finally, we will want to consider the molecular gas which is being discovered in more and more objects. How do CO (and presumably H_2) molecules survive in such an evolved object as NGC 7293? Can we balance the heating and cooling to determine the temperature and hence predict the CO emission? The simplest questions, such as the effectiveness of self-shielding and the consequent lifetime of molecules in PNe have yet to be determined. In cases where the line ratios indicate that the H_2 IR emission may be due to fluorescence rather than shock heating, we will want to model this process. Once again, planetary nebulae may emerge as the best laboratories for the investigation of a whole set of physical processes, this time processes which have hitherto been hidden in amorphous, dark clouds.

The construction of ionization models should be regarded as the meeting ground for all the varied data we are able to collect. When thus brought together, the whole should be greater than the sum of the parts.

6. REFERENCES

Adam, J., and Koppen, J. 1985. Astron. Astrophys., **142**, 461.
Aller, L.H., Keyes, C.D., Maran, S.P., Gull, T.R., Michalitsianos, A.G., and
  Stecher, T.P. 1987. Ap. J., **320**, 159.
Clegg, R.E.S. 1987. M.N.R.A.S., **229**, 31P.
Clegg, R.E.S. and Harrington, J.P. 1988, in preparation.
Clegg, R.E.S., Harrington, J.P., Barlow, M.J., and Walsh, J.R. 1987. Ap. J., **314**, 551.
Clegg, R.E.S., Harrington, J.P., and Storey, P.J. 1986. M.N.R.A.S., **221**, 61P.
Clegg, R.E.S. and Middlemass, D. 1987. M.N.R.A.S., **228**, 759.
Clegg, R.E.S., Peimbert, M., and Torres-Peimbert, S. 1987. M.N.R. A.S., **224**, 761.
Clegg, R.E.S., and Walsh, J.R., 1985. M.N.R.A.S., **215**, 323.
Ferland, G.J. 1986. Ap. J., **310**, L67.
Flower, D.R. 1968. Ap. Letters, **2**, 205.
Goodson, W.L. 1967. Z. fur Astrophys., **66**, 118.
Harrington, J.P. 1968. Ap. J., **152**, 943.
Harrington, J.P. 1978. I.A.U. Symposium No. 76 "Planetary Nebulae", ed.
  Y. Terzian (D.Reidel; Dordrecht Holland), p. 151.
Harrington, J.P. and Feibelman, W.A. 1983. Ap. J., **265**, 258.
Harrington, J.P., Monk, D.J., and Clegg, R.E.S. 1988. M.N.R.A.S., in press.
Henry, R.B.C., and Shipman, H.L. 1986. Ap. J., **311**, 774.
Hummer, D.G, and Mihalas, D. 1970. M.N.R.A.S., **147**, 339.
Husfeld, R., Kudritzki, R.P., Simon, K.P., and Clegg, R.E.S. 1984. Astron. Astrophys., **134**, 139.
Korte, P.A.J. de, Claas, J.J., Jansen, F.A., and McKechnie, S.P. 1985.
  Advances in Space Research, **5**, No. 3, 57.
Mendez, R.H., Kudritzki, R.P., Gruschinske, J., and Simon, K.P. 1981. Astron. Astrophys., **101**, 323.
Mihalszki, J.S., and Ferland, G.J. 1983. Pub. Astron. Soc. Pacific, **95**, 284.
Olson, G.L., Auer, L.H., and Buchler, J.R. 1986. J. Quant. Spectrosc. Radiat. Transfer, **35**, 431.
Peimbert, M. 1967. Ap. J., **150**, 825.
Peimbert, M. and Torres-Peimbert, S. 1987. Rev. Mexicana Astron. Astrofis., in press.
Pequignot, D. 1978. I.A.U. Symposium No. 76 "Planetary Nebulae", ed.
  Y. Terzian, (D.Reidel; Dordrecht Holland), p. 162.
Pequignot, D. 1986. "Workshop on Model Nebulae", p 363, Pub. de l'Observatoire de Paris.
Stasinska, G., and Tylenda, R. 1986. Astron. Astrophys., **155**, 137.
Stecher, T.P., Maran, S.P., Gull, T.R., Aller, L.H., and Savedoff, M. 1982.
  Ap. J. Letters, **262**, L41.

Weymann, R.J., and Williams, R.E. 1969. Ap. J., **157**, 1201.
Williams, R.E. 1968. I.A.U. Symposium No. 34 "Planetary Nebulae", ed.
   D.E. Osterbrock and C.R. O'Dell (D.Reidel; Dordrecht Holland), p. 190.

PLANETARY NEBULAE AND THE GALACTIC BULGE

M.W. Feast
South African Astronomical Observatory
T.D. Kinman
Kitt Peak National Observatory
B.S. Lasker
Space Telescope Science Institute

ABSTRACT. Fifteen new PN have been discovered in the region of Baade's Windows using an objective prism technique. Absolute spectrophotometry, excitation classes, radii and radial velocities have been obtained. Radial velocities were also measured for eight other PN in this region. After correction for solar motion and the circular velocity at the sun, the radial velocities of bulge PN (V_c) with $|b| < 5°\!.5$ show good evidence for a rotation of the bulge. If $V_c = \alpha + \beta \Delta \ell$ then,

TABLE 1

$|\Delta \ell|$	No.	α km s^{-1}	β km s^{-1}	σ km s^{-1}
< 10°	147	−13.6±8.6	12.0±1.9	103±6
< 5°	109	−16.6±10.5	15.2±4.0	109±7

Ionized masses (M_i) for the new PN range over a factor ~ 50. These results and those of Gathier *et al.* (1983) show that $M_i < \sim 0.3$ $M_\odot$ in the bulge. This is in good agreement with the predictions of Feast and Whitelock (1987). They find $M_{Bol} > -4.7$ for bulge Miras and IRAS sources which together with pulsation masses and evolutionary theory leads to predicted nebular masses < 0.3 $M_\odot$ and an evolutionary age of the most massive bulge objects of ~ 5 Gyr. The absence of high excitation planetaries in the bulge is consistent with the lack of younger (more massive) progenitors.

REFERENCES

Feast, M.W. and Whitelock, P.A. 1987, in *Late Stages of Stellar Evolution*, ed. S. Kwok and S.R. Pottasch (Dordrecht: D. Reidel), p. 33.
Gathier, R., Pottasch, S.R., Goss, W.M., and van Gorkom, J.H. 1983, *Astron. Astrophys.*, **128,** 325.

SPECTROSCOPIC DISTANCES TO CENTRAL STARS OF PLANETARY NEBULAE

R.H. Méndez, R.P. Kudritzki, A. Herrero, D. Husfeld, and
H.G. Groth
Universitäts-Sternwarte, München, F.R.G.

ABSTRACT. We present spectroscopic distances for 22 central stars of planetary nebulae. These distances have been determined using information provided by our non-LTE model atmosphere analyses of the stellar H and He absorption line profiles. In this way, no assumptions about nebular properties are necessary.

Our spectroscopic distances turn out to be larger than many other frequently cited values. We show that our distances are not in contradiction with the available information about the interstellar extinction, and we describe additional evidence supporting them.

INVESTIGATION OF TWO PLANETARY NEBULAE AND THEIR ANGULAR VICINITY IN CYGNUS

W. Saurer and R. Weinberger
Institut für Astronomie der Universität Innsbruck, Austria

ABSTRACT. We present the results of an investigation of M1-79 and K3-82. Physical parameters like expansion velocities, spatial shapes, Zanstra-hydrogen-temperatures, etc. were obtained by use of high resolution spectroscopy and CCD-images.

The two PN are medium excitation and exhibit usual expansion velocities. To find out the spatial shapes we utilized the simple model of a truncated spherical shell. A comparison between the theoretical intensity ratios from this model and the measurements is leading to the conclusion that both PN have the structure of a ring. K3-82 is seen pole-on, M1-79 is seen edge-on.

The distances of the two nebulae were examined with an accuracy of $\lesssim 30\%$ by means of the extinction-distance method. The distances are 1900 pc (K3-82) and 2000 pc (M1-79). With this method we also get information about the galactic structure in the angular vicinity of the two PN. The absorption is rising rapidly at about 0.5-2 kpc (K3-82) and 1-2.5 kpc (M1-79) to a plateau at $E_{B-V} = 1.1$. mag.

In addition we estimated the influence of wide-band photometry on this method, especially the effect on the value $R = A_V/E_{B-V}$. There is no great effect on the measurements for early type stars, but an effect for late-type stars, which should be considered.

Part of this work was supported by the Austrian "Fonds zur Förderung wissenschaftlicher Forschung" project 5708.

STRUCTURE AND MORPHOLOGY OF NGC6369

Tapan K. Chatterjee and J. Campos
Instituto Nacional de Astrofísica, Optica y Electrónica,
Tonantzintla, A. P. 51 y 216, Puebla, 72000,
México

ABSTRACT. NGC 6369 is a remarkable object, especially in the light of structure and morphology. We studied this object by taking many red plates (103aE), coupled with red filters (F29), of varying exposures using the Schmidt Telescope of the INAOE. The structure brought out by the analysis of the plates indicate that the object consists of a prolate disk sphaped ring nebula with a central hole and featuring huge plumes emanating out of the ring which engulf an outer envelope having a diameter about twice that of the ring. The striking feature of the plumes is that they emanate almost symmetrically out of the two prolate ends of the ring and curve out almost symmetrically along opposite directions.

A comparison of the structures of this object with the temporal evolution of the gas density in the numerical simulations of a purely gaseous self-gravitating polytropic ring is conducted. Features resembling the plumes of this object are found at a certain stage of the simulations. A careful comparison of the observed and computer generated features indicate a marked similarity in the sense that both, the observed and simulated ring appear to be similar to the cross section of a prolate spheroid at the ends of which emanate the plumes. This seems to confirm that the evolution of this object is consistent with its being a planetary nebula having the appearance of a prolate spheroid with a central hole, which is a basic observational feature of most planetary nebulae.

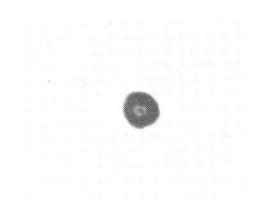 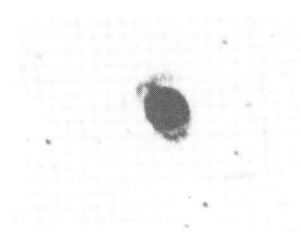

Figure 1. NGC6369 underexposed.   Figure 2. NGC6369 overexposed.

S. Torres-Peimbert (ed.), Planetary Nebulae, 169.
© 1989 by the IAU.

H_2 AND H I EMISSION LINE IMAGING OF THE RING NEBULA NGC 6720

M.A. Greenhouse, T.L. Hayward, and H.A. Thronson, Jr.
Wyoming Infrared Observatory, University of Wyoming

ABSTRACT. We present infrared emission line images of the $v = 1 \to 0\ S(1)$ transition of molecular hydrogen and Brγ recombination line of atomic hydrogen which cover the entire extent of NGC 6720, the Ring Nebula. The maps presented here are the highest angular resolution images of these transitions yet produced for this object, and have very low relative positional uncertainty. As a result, we clearly resolve the spatial stratification of the ionized and shocked molecular zones within the nebula discussed previously by Beckwith *et al.* (1978). The relative spatial distribution of molecular and ionized hydrogen we observe is typical of several planetaries which exhibit shocked H_2 emission (eg. see Zuckerman and Gatley 1987), and is similar to that predicted by the interacting-stellar-winds model of planetary nebulae formation (see Volk and Kwok 1985 and references therein).

These data, and data from the *Infrared Astronomical Satellite*, were used to determine the H_2, HI, and dust mass within the nebula. The quantitative results are summarized in Table 1. We find, using the recent model of Schonberner (1983), that our measured luminosity for the central star is consistent with evolution from a one solar mass AGB star. We also show that: 1) the Ring is optically thick in the H Lyman continuum, 2) absorption of trapped line radiation is a sufficient energy source to account for most of the observed dust luminosity, 3) the Ring displaces an insufficient volume to sweep up the observed dust mass from the interstellar medium, and 4) the measured H_2 mass is undergoing a period of net photodissociation.

TABLE 1. DERIVED PROPERTIES OF NGC 6720

Parameter	Value	
Distance	525	pc
E_m	5.9×10^4	$cm^{-6}\ pc$
IRE	1.1	
F_{IR}	3.7×10^{-9}	$erg\ s^{-1}\ cm^{-2}$
L_*	135	$L_\odot$
L_{IR}	31.7	$L_\odot$
M_{H_2}	2.7×10^{-6}	$M_\odot$
M_d	1.2×10^{-3}	$M_\odot$
M_g	3.9×10^{-2}	$M_\odot$
$N(< 912)$	9.5×10^{45}	photons s^{-1}
n_e	600	cm^{-3}
r_*	2.9×10^{-2}	$r_\odot$
T_*	1.17×10^5	K
T_d	50	K
T_e	10^4	K

REFERENCES

Beckwith, S., Persson, S.E., and Gatley, I., 1978, Ap. J., 219, L33.
Pottasch, S.R. 1984, *Planetary Nebulae*, (Dordrecht: Reidel).
Schonberner, D. 1983, Ap. J., 272, 708.
Volk, K. and Kwok, S. 1985, *Astr. Ap.*, 153, 79.
Zuckerman, B. and Gatley, I. 1987, Ap. J., in press.

S. Torres-Peimbert (ed.), Planetary Nebulae, 170.
© 1989 by the IAU.

DEEP NARROW BAND INTERFERENCE FILTER PHOTOGRAPHS OF SELECTED
EXTENDED PLANETARY NEBULAE

M. Rosado and M. Moreno
Instituto de Astronomía
Universidad Nacional Autónoma de México
México

ABSTRACT. Narrow-band interference filter photographs in the light of
Hα, [S II] (λ = 6717 and 6731 A). [N II] (λ = 6584 A) and [O III] (λ = 4363 A) of four extended PNs are shown. These photographs were obtained with a focal reducer and a single-stage image tube attached to the 2.1 m Cassegrain focus telescope of the Observatorio Astronómico Nacional at San Pedro Mártir, B.C.N. The exposure times were of one hour. For each photograph we have obtained a calibration by means of a step density wedge.

The nebulae photographed this way were: Abell 13 and Abell 24 (from the list of Abell 1966), the nebula No. 1 in the list of Weinberger and Sabbadin (1981) (hereafter, WS1) and the nebula No. 5 in the list of Dengel et al. (1980) (hereafter DHW5). The lattest two nebulae are only suspected PNs on the basis of their appearance -similar to that of a "typical" PN- in POSS plates and because they have a blue star near the center.

These photographs allow us to make a comparison between the emission of different ions. As a general trend, the morphology revealed by these photographs shows greater detail than the POSS and previous filter photographs of shorter exposures. The photographs at λ = 4363 A are interesting because they allow us the identification of blue stars, interior to the nebulae, that could be proposed as the PN nuclei.

REFERENCES

Abell, G.O. 1966, Ap. J., 144, 259.
Dengel, J., Hartl, H., and Weinberger, R. 1980, A.A., 85, 356.
Weinberger, R. and Sabbadin, F. 1981, A.A., 100, 66.

CCD IMAGES OF SELECTED PLANETARY NEBULAE

M. Turatto, E. Cappellaro, and F. Sabbadin
Asiago Astrophysical Observatory,
36012 Asiago (VI), Italy

ABSTRACT. Direct interference filter CCD frames of a number of northern planetary nebulae were obtained at the Cassegrain focus of the 182-cm telescope of Asiago Astrophysical Observatory (Italy).
   In this short communication we present preliminary results for Hu 1-2, M 2-52, M 2-55, A 2, NGC 650-1, II 2120, H 3-29, M 1-7, K 3-72, M 1-8 and M 1-18. The Type-I PN Hu 1-2 (Figure 1) appears as an irregular ring of condensations seen almost edge-on and surrounded by less dense material at high latitudes, suggesting a bi-lobed structure similar to those observed in NGC 650-1, NGC 7026 and NGC 5189. The apparent form of K 3-72 (Figure 2) closely recalls Hu 1-2 (equatorial ring + bi-lobed polar material). This morphology, along with the presence of very strong [N II] lines (spectra taken at Asiago Observatory indicate that $I(\lambda 6584$ [N II])$/I(H\alpha) = 4.5$), suggest that K 3-72 is a *bona fide* Type-I PN.
   Bi-lobed structures have been detected also in M 2-55 and M 1-8.

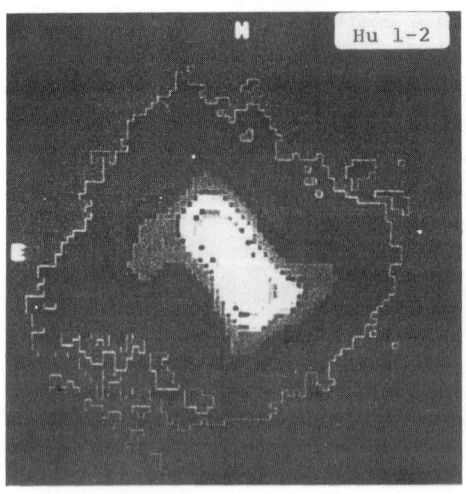
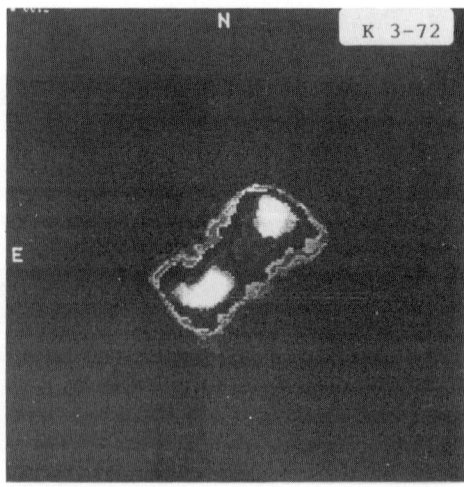

CCD IMAGES OF SOUTHERN HEMISPHERE PLANETARY NEBULAE

Julie Lutz and Nancy Jo Lame
Washington State University
Bruce Balick
University of Washington

ABSTRACT. A large, long-term survey of southern hemisphere planetary nebulae is being undertaken with several narrow-band filters (primarily [N II], H-alpha, [O III] and He II) on the 0.9-m telescope at CTIO using a TI CCD chip. The purposes of this survey are to get sizes and morphological classifications for nebulae that are little-known (indeed, some do not have size measurements at all) and to search for multiple shells and other structures that are of interest for studies of nebular formation and evolution.

CCD images have been obtained in at least two filters for approximately 70 nebulae. Some general results of the survey are:

1) Many of the nebulae have multiple structures, some of which are simple enough to characterize as multiple shells.

2) Many of the nebulae exhibit one or more types of bipolar structures. There appear to be several types of bipolar structures and it is not uncommon for different types of symmetries to show up strongly in images obtained with different filters.

3) The sizes of the nebulae can be measured by using contour maps, but specifying a single cutoff number for measuring nebular size does not appear to be appropriate. Instead, each nebula must be considered individually.

IRAS observations of extended planetary nebulae

A. Leene and S.R. Pottasch
Kapteyn Astronomical Institute,
Groningen,
The Netherlands

The pointed observations made by the IRAS satellite have been analysed. In total 67 nebulae have been observed of which 10 proved to be resolved. The majority of the observations were carried out with the CPC and with the DSD macro of the survey array (Leene & Pottasch, 1987b). Further special deep maps were made of six nebulae (Leene & Pottasch, 1987a; Leene, 1987). Typical noise levels of the deep images are 0.03–0.1 MJy ster^{-1}.

The distribution of emission around the central star confirms that the 12 and 25 μm emission are due to ionic line emission. This is very clear in the nebula NGC 7293 (Leene & Pottasch, 1987a) and in NGC 6853 (Zhang et al., 1987). The other nebulae (A7, A21, A31, A35 and NGC 1360) confirm these findings. The 12 μm emission is most likely due to the NeII (12.81 μm) or due to the SIV (10.52 μm) line. The 25 μm band shows a strong central contribution due to the OIV (25.87 μm) line and a halo of the SIII (18.68 μm) line.

The sizes of the nebulae at 50 and 100 μm of the resolved pointed observations are very similar, indicating a lack of a temperature gradient. The radial profiles of the other nebulae seem to confirm this. Thus the dust is uniformly heated, probably by trapped Lymann α photons. The infrared sizes compared to the optical sizes confirm this heating mechanism. In nearly all cases the infrared size is smaller than the Hα size. The only exception are the two "young" nebulae NGC 6543 and BD +30 3639. In these nebulae the infrared size is much larger than the Hα size.

The observations of A7, A31 and A35 show that infrared emission is still detectable for very low photographic surface brightness planetary nebulae (25–26 mag arcsec^{-2}). This allows to make images of nearly all nebulae in the Abell catalogue. No evidence could be found for infrared emission beyond the optical image.

Only in the nebulae NGC 6543 and BD +30 3639 possible evidence for the remnant of an AGB wind can be found. This confirms the scenario of Barlow (1983) in which young nebulae still have an optical thick shell. This allows a coexistence of an AGB remnant and an ionized region. However when the shell expands it becomes optical thin and the whole AGB shell becomes a HII region.

The lack of dust seen in the 12 and 25 μm band is somewhat in contrast with the observations of NGC 7027. In this nebulae there is a strong contribution of very small grains (PAH's). It is not unlikely that these small grains are destroyed by the UV radiation in the large nebulae. Only in the young nebulae with an optical thick shell the right environment is present to let these small grains survive. The lack of an infrared halo can easily be explained by the "remnant" of an AGB wind. The dust seen in planetary nebulae can be provided by an AGB wind if it reaches a maximum mass loss rate of 10^{-4} M$_\odot$ yr^{-1}. Such a mass loss rate is also observed in non-variable OH/IR stars, which are possible precursors of planetary nebulae (van der Veen et al., 1987).

References.

Barlow, M.J.: 1983, IAU Symp. **103**, p105, Reidel
Leene, A., Pottasch, S.R.: 1987a, *Astron. Astrophys.* **173**, 145
Leene, A., Pottasch, S.R.: 1987b, submitted to *Astron. Astrophys.*
Leene, A.: 1987, Ph.D. thesis, University Groningen, The Netherlands.
van der Veen, W., Habing, H.J., Geballe, T.: 1987, "Planetary and proto-planetary nebulae: from IRAS to ISO", p69, ed. A. Preite-Martinez, Reidel
Zhang, C.Y., Leene, A., Pottasch, S.R., Mo, J.E.: 1987, *Astron. Astrophys.* **178**, 247

NEBULAR DENSITY DISTRIBUTIONS; A CRITICAL LOOK

Ronald C. Kirkpatrick
Los Alamos National Laboratory

ABSTRACT. The symmetry of planetary nebulae demands an explanation. Underlying previous attempts to explain these objects is the inferred density distribution based on spatio-kinematic models such as those of Weedman (1968), Reay et al. (1983) and others. Although some observations suggest a linear velocity-radius relationship for planetaries, plausible hypothetical examples may be constructed which greatly violate a linear relationship, and hydrodynamic theory (Schmalz 1986) suggests it may not hold even when observations may be most easily interpreted as supporting a linear relationship. Therefore, an independent assessment of the nebular density distribution is needed.

For a few special cases, an independent method is readily available -that of tomographic reconstruction. The axis of symmetry for NGC 7009 and a few other objects appears to lie almost in the plane of the sky. By assuming it does and that they are truely axi-symmetric; its brightness distribution may be "peeled" to yield a density distribution. This is done by assuming the distribution consists of cylindrical layers in discs stacked along the axis of symmetry. Then one disc at a time, the outermost layer's emission measure may be deduced by dividing its brightness by the chord through the outermost layer and the contribution of that layer to brightness from all layers inside it removed mathematically, leaving a new brightness distribution for which the next to outer layer of the old brightness distribution is now the new outer layer for a partially "peeled" brightness distribution. Repeating this process until the very inside of the object is reached yields an emission measure distribution, the square root of which is a relative density distribution.

REFERENCES

Reay, N.K., Atherton, P.D., and Taylor, K. 1983, M.N.R.A.S., **203**, 1079.
Schmalz, R.F. 1986, Physics and Fluids, **29**, 1389.
Weedman, D.W. 1968, Ap. J., **153**, 49.

THE TEMPERATURE STRUCTURE OF NGC 7027

C.T. Daub
San Diego State University
J.P. Basart
Iowa State University

ABSTRACT. Radio maps of the free-free radio continuum flux (angular resolution ≅ 1.3 arcseconds) from NGC 7027 were made with the VLA operating at 20-cm, 6-cm, and 2-cm wavelengths which are near and straddle unit optical depth. Mean line-of-sight electron temperature and emission measure distributions were calculated by pairing the 2-cm and 6-cm maps, and the electron temperature distribution on the near side of the nebula was then obtained from the 20-cm map. The results suggest that the energy balance is complex in this planetary. For example, mean line-of-sight temperatures are higher than average in the direction of one of the bright lobes but not in the direction of the other. Especially noteworthy is an apparent "hot spot" on the near side of the nebula which has no apparent relation to either of the bright lobes, but it is approximately coincident with the brightest portion of the optical image.

THE DISPLAY AND MANIPULATION OF PN IMAGES ON AN IBM PC OR COMPATIBLE

M.J. Hoey
Dept. of Experimental Physics, University College Dublin
D. Whelan
Dept. of Mechanical Engineering, University College Dublin

ABSTRACT. This paper describes PC_IMega, a menu driven program that allows digital images to be displayed and manipulated on an IBM PC or compatible microcomputer, running under PC-DOS versions 2 or 3 with 128k of RAM. A sixteen colour Enhanced Graphics Adaptor card gives a screen resolution of 640×350 pixels.
    A maximum of sixteen EGA colours are assigned to a specific range of image pixel intensity values using a linear scale and image sections are scaled, panned and, if necessary, mirrored and displayed in a 512×350 window. PC_IMega can accept a viewable image of 512×700 but the program can read files with pictures of much larger widths and lengths. It can ignore file header information and allows for either one or two byte pixel intensity values. For example, the PN images presented in this paper are FITS format.
    The RAM requirements are kept to a minimum by operating on single image lines at a time. All alterations must therefore be made immediately to the stored image on disk.
    For a given screen display, the maximum and minimum intensity values can be found and a statistical distribution of intensities can be shown. A bar chart of intensities for any cross-section of the image can also be generated (slice). Individual pixel values can be verified and can be set either to a given value or to the average of the surrounding pixels. A contouring function draws contours of equal intensity.
    The images together with file and screen parameters can be stored on a floppy disk. This allows for quick subsequent regeneration of an image display and for easy transfer of images between astronomers.
    The PC_IMega will be launched as a public utility program later this year and will be available on floppy disk.

INFRARED IMAGES AND LINE PROFILES OF PLANETARY NEBULAE

M. G. Smith, T. R. Geballe, C. Aspin, and I. S. McLean
United Kingdom Infrared Telescope, Hilo, Hawaii.
P. F. Roche
Royal Observatory, Edinburgh

ABSTRACT. We present high spatial resolution infrared images of the planetary nebulae NGC 7027, M2-9, BD +30 3639, NGC 7099 and NGC 7662. These were taken through a selection of broad and narrow-band line and continuum filters (including a Fabry-Pérot interferometer) using the 2D infrared array "IRCAM" on the United Kingdom Infrared Telescope, UKIRT, in July 1987. Comparison is made with recently published high-resolution VLA radio maps (Basart and Daub 1987, Ap. J., 317, 412) and mid-IR Wyoming Infrared Telescope raster-scanning maps (Bentley et al.1984, Ap. J., 278, 665).

Significant differences are found in the fine structure of BD +30 3639 seen in the 3.3 micron emission feature when compared with the image in Brackett-gamma.

The image of M2-9 changes substantially between 1.2 and 2.2 microns. At the shorter infrared wavelengths the knots, visible in earlier optical CCD observations, are seen predominantly on the east side of the nebula. Photometry at K shows the nebula to a limiting magnitude of about 18th per square arcsecond, where it is considerably more uniform in appearance than at shorter wavelengths, J and H photometry, calibrated using earlier aperture measurements, has allowed us to form (J - K) and (H - K) colour images. These show a predominant disk-like structure 10 arcseconds in size stretching across the core region of M2-9. The reddening in the disk peaks at over 6 magnitudes in (J - K) in two locations symmetrically placed east and west of the intensity peak.

Line profiles of Brackett-gamma and $v = 1 \rightarrow 0$ S(1) molecular hydrogen emission are presented. These were obtained at UKIRT using our combination of cooled grating spectrometer and Fabry-Pérot. The molecular hydrogen line, taken through a 5 arcsecond diaphragm centred on the middle of the 6-cm VLA map of the nebula, midway between the two central molecular-hydrogen peaks, shows a splitting of about 30 km s^{-1}.

NGC 2899: AN EVOLVED BIPOLAR PLANETARY NEBULA

J.A. López[1], L.H. Falcón[1], M.T. Ruiz[2], and M. Roth[1]
1. Instituto de Astronomía, UNAM, Ensenada, B.C., México
2. Departamento de Astronomía, Universidad de Chile

ABSTRACT. NGC 2899 (PK 277-3°1, He 2-30, RCW 43) is a southern planetary nebula of fairly large angular size ($\sim$ 2!6×1!4) and moderate high surface brightness. Its morphology strongly resembles a loose bipolar structure with conspicuous bright condensations of toroidal geometry placed along the minor axis, on each side of the central object.

Results of long-slit echelle observations, low dispersion flux calibrated spectra and near infrared photometry are presented. The observations were obtained at the 3.9-m AAT and the 4-m and 1.5-m telescopes of CTIO.

The long-slit spectra obtained in the light of Hα and [N II] reveal an object with a complex kinematical field. Line splitting is present over the face of the nebula with a mean nebular expansion of $\sim$ 25 km s^{-1}. In addition, remarkably high velocity structure, up to +110 and -135 km s^{-1} is detected in the expanding equatorial toroid. Location of the slits is a critical factor that determines the spatial asymmetries found in the high velocity features. These data are interpreted in terms of the interaction of the hot stellar wind with the eroded constraining circumstellar shell/toroid that originally focused the bipolar structure.

Limited J, H and K mapping indicates that the near infrared emission arises predominantly from dust grains.

The low dispersion nebular spectrum shows a rich emission line spectrum. The [O III]/Hβ and He II/Hβ line ratios indicate an excitation class 6-7. A logarithmic extinction at Hβ c(Hβ) = 0.71 was used to deredden the line fluxes. For the electron temperature we found T_e[N II] = 10,800°K and T_e[O III] = 15,600°K. For Ne a value of 10^3 cm^{-3} is adopted from the [Cℓ III] line ratio. The derived ionic abundances are: log N$^+$/H$^+$ = -3.53, log O$^+$/H$^+$ = -3.79, log N/O = +0.26, log He^{++}/He$^+$ = -1.17, log He$^+$/H$^+$ = 0.89 and He/H = 0.196. These values place NGC 2899 among the planetary nebulae of Type I.

The spectrum of the central object denotes an early G-type star with a strong blue excess in its continuum energy distribution suggesting the presence of a central binary system.

The Structure and Kinematics of Bipolar Planetary Nebulae

W. G. Weller and S. R. Heathcote
Cerro Tololo Interamerican Observatory
National Optical Astronomy Observatories

We have obtained high-dispersion, long-slit echelle spectra at closely spaced intervals across the face of the bipolar planetary nebulae NGC 2440, NGC 6302 and Mz-3. Deep monochromatic images of these objects in lines from high (HeII, [OIII]), intermediate (HI, [OII]) and low ([NII], [SII]) excitation species have also been acquired. Taken together, these data permit us to construct self-consistent spatio-kinematic models of these nebulae and to investigate the spatial variations of excitation conditions within them.

These nebulae possess several unifying characteristics: (1) By definition they all show bi-lobate morphology; (2) All of them exhibit a disk or torus of dense material lying in the mid plane of the nebula. In NGC 2440 this disk is fully ionized, while in the other nebulae it is predominantly neutral; (3) They all show to differing but considerable extent, a remarkable degree of mirror symmetry, not only in their morphology, but also in their velocity structure; MZ-3 is the most highly symmetrical, and NGC 6302 is the least; and (4) The low surface brightness outer lobes all show expansion velocities in excess of 70 km/s, highly supersonic for a gas at a temperature of $< 10^4$ K.

Our data can be interpreted within the frame work of current models of the formation of PN through the action of a fast wind blown by the central star upon the expanding, axialy symmetric remnant of its red giant envelope. In this "colliding wind" model the dynamical evolution of the nebula is predominantly controlled by the parameters of the red giant mass loss. The observed symmetry of the nebulae studied, thus implies a consider-able degree of both axial and bi-lateral symmetry in the flow parameters of the slow wind.

COLLIMATED OUTFLOWS IN PLANETARY NEBULAE

Bruce Balick[1] and Heather L. Preston[1]
Department of Astronomy, University of Washington
Vincent Icke
University of Washington and Sterrewacht Leiden, The Netherlands.

ABSTRACT. The kinematics of the gas motions in several e-m E planetary nebulae (PNs) have been mapped with $\sim 1-2''$ resolution. All of the eE PNs (e.g., NGC 2392, 3242, 6543, 6826, 7009, and 7662) show evidence of highly focused linear flows at projected velocities of 20 - 170 km s^{-1}. The flows generally appear as linear, almost jet-like features in emission lines of low ionization. mE PNs show expanding protuberances along their polar axes.
  We propose a hydrodynamic mechanism for focussing flows in PNs with prolate elliptical symmetry whose nuclei emit an isotropic wind. The wind streamlines are bent as they pass through a prolate shock near the star, and are focussed along the polar axes of the PN. Where the gas converges and, hence, cools relatively efficiently, knots or jets can form. The subsequent evolution of the system is expected to lead to a barrel-like PN with two bubbles of rapidly expanding gas from the ends of the barrels (as observed in mE PNs), and ultimately to a large bipolar PN of low surface brightness.

1. Visiting Astronomer, Kitt Peak National Observatory, National Optical Astronomy Observatories which is operated by the Association of Universities for Research in Astronomy, Inc., under contract with the National Science Foundation.

KINEMATICAL PROPERTIES OF PLANETARY NEBULAE

Luciana Bianchi (1), Michael Grewing (2), J.Barnstedt (2),Chr.Diesch(2)

(1) Osservatorio Astronomico di Torino, Italy
(2) Astronomisches Institut Tuebingen, West Germany

We present newly determined expansion velocities for a number of Planetary Nebulae (PNe) which have been observed with the ESO 1.4m CAT and the Coude' Echelle Spectrograph (CES), operated at a r.p. of 10^5, corresponding to a resolution of about 3km/s at H_α. Two detector systems have been used: the standard ESO Reticon and a two-dimensional photon-counting imaging detector developed at the A.I.T. (AIT-MCP-Camera).
While for a number of extended objects we obtained several observations at different positions to investigate the detailed kinematical structure (see e.g. Bianchi et al. 1987) we show here only results on the PN expansion velocities obtained from data relative to the centre of each object. The expansion velocities obtained for 17 objects allowed also a statistical interpretation and comparison with current theories of nebular evolution (Bianchi and Falcetta 1987).
In the table below we list the expansion velocity determinations for the objects, obtained independently from the Reticon data and from the AIT-MCP data (given in parenthesis). We point out that for those objects where we obtained spectra with both detectors, the final results from the two independent analyses are in excellent agreement, with the exception of one object.
References.
Bianchi,L.,Grewing,M.,Falcetta,C.,Baessgen,M.,1987: in "Planetary and Proto-planetary Nebulae", Preite-Martinez ed., Reidel, p.153
Bianchi,L.,Falcetta,C.,1987: in"Mass outflows from stars and galactic nuclei", Bianchi and Gilmozzi eds., Reidel, in press

The measured expansion velocities.

	Object P.K.	Name	Exp. Velocity (km/s) H-alpha	[NII]
1	206 -40 1	NGC 1535	18.9	--
2	215-24 1	IC 418	7.4	11.9
3	234 +2 1	NGC 2440	18.7 (19.2)	17.3 (18.)
4	261 +32 1	NGC 3242	19.3 (19.4)	--
5	285 -14 1	IC 2448	11.5 (14.9)	--
6	294 +4 1	NGC 3918	19.	26.
7	315 -13 1	He 2-131	8.9 (9.1)	9.9 (10.4)
8	327 +10 1	NGC 5882	17.3	23.5
9	10 +18 2	M 2-9	32.1	--
10	334 -9 1	IC 4642	13.3	16.2
11	345 -8 1	He 2-274	8.3:	14.9 (14.7)
12	1 -6 2	SwSt 1	9.1	9.4
13	9 -51 1	NGC 6629	7.7	--
14	2 -13 1	IC 4776	10.2	17.3
15	27 -9 1	IC 4686	13.1	--
16	54 -12 1	NGC 6891	8.1:	--
17	37 -34 1	NGC 7009	18.4 (17.2)	--

KINEMATICS OF ABELL 30

George H. Jacoby
Kitt Peak National Observatory

You-Hua Chu
Astronomy Department, University of Illinois

Abell 30 is a remarkable PN with a H-depleted core. Four bright knots have been identified in the [O III] line (Jacoby 1979). The previous kinematic study has shown that J1 and J3 form a pair expanding at radial velocities of ±25 km/s with respect to the central star, and J4 has a radial velocity of -22 km/s (Reay, Atherton, and Taylor 1983).

We have obtained deep-exposure echelle spectra of the knots in the core of A30. The results, while confirming the radial velocities of J1 and J3, show previously-undetected multiple components in J2 and J4.

The average heliocentric velocity of A30 is +10 km/s. The knot J1 is dominated by one single component at +34 km/s, while J3 is dominated by a single component at -13 km/s. The knot J2 is kinematically resolved into at least four components at velocities of -13, +31, +60, and +87 km/s, with the +31 km/s component being the brightest. The knot J4 is resolved into four components at ≥+66, +36, -10, and -50 km/s; the brightest component at -10 km/s appears tilted with -2 km/s at the inner edge and -17 km/s at the outer edge.

The components in J2 and J4 form arc-like patterns in the echellograms. Such pattern can be explained by a clumpy expanding ring structure. The position angle of the ring is similar to that of the infrared disk detected by Dinerstein and Lester (1984).

HIGH RESOLUTION LONG-SLIT SPECTROSCOPY OF A78

A. Manchado[1], S.R. Pottasch[2], and A. Mampaso[1]
1. Instituto de Astrofísica de Canarias, Tenerife, Spain
2. Kapteyn Astronomical Institute, Groningen, The Netherlands

ABSTRACT. High spectral resolution bidimensional spectroscopy of A78 in the He II λ4686 A, H λ4861 A, [O III] λ4959 A and [O III] λ5007 A lines is reported, confirming the different morphology of the nebula in these lines. The resulting velocity maps suggest different episodes in the history of the nebula, with an external hydrogen-rich layer expanding at low velocity (35±10 km/s), showing little structure and extending from approximately 35 to 55 arcsec. The [O III] and the He II maps show, however, an inner shell with at least two different expansion velocities; 73 km/s and 41 km/s.
  These data, together with recent low resolution optical data for the nebula (Manchado et $al.$ 1987) allow us to calculate the mass of each shell which ranges between 0.024 $M_\odot$, 0.1 $M_\odot$ and 0.1 $M_\odot$ for the inner, intermediate and outer shells respectively and the time scales since the ejection being ∿ 1800 yrs for the inner shell, 3500 yrs for the intermediate one and ∿ 11900 yrs for the hydrogen-rich more external layer.

THE STRUCTURE AND VELOCITY FIELD OF A78.

P. Pişmiş, M.A. Moreno
Instituto de Astronomía
Universidad Nacional Autónoma de México
Apdo. Postal 70-264, 04510 México, D. F., Mexico.

We present a velocity field of the planetary nebula A78 based on three Fabry Pérot Hα (10A) interferograms taken with a focal reducer attached to the 2.1 m reflector of the Observatorio Astronómico Nacional at San Pedro Mártir, Mexico. We have used a single-stage Varo image intensifier and two different étalons with interorder separations of 283 km s^{-1} (2 interf.) and 100 km s^{-1} (1 interf.). The scale of the original photographs is 49 arcsec mm^{-1}. Our data have yielded radial velocities in the Hα line at 110 points on the face of A78; the velocity field is far from being smooth. The rings are wide around the central hole, and a few show definite splittings; from these splittings we have estimated an overall expansion velocity of 27 km s^{-1}. The average systemic velocity is found to be around -3 km s^{-1}.

Two direct images in the Hα line were obtained with the focal reducer at the 2.1 m telescope. The image in Hα has a regular oval outline with a small hole around the central star. At the edges near the "minor axis" the brightness is enhanced. There exist filamentary details within the nebula, and these coincide with the nearly circular faint filaments shown on the PSS red image. For an overall discussion of the velocity structure the Hα image was divided into two halves along the minor axis. Referred to the standard of rest of the nebula, the NW half has yielded a velocity of -6.7 km s^{-1} and the SE half, +5.9 km s^{-1}. The average velocities along the two rings in the NW half and those at the SE half are comparable with the respective average velocities in the regions where they are embedded.

Our material does not allow a unique model to be advanced for the formation of A78; however, based on the morphology of the filaments resembling a helix and the velocity field, we may state that: The outflow of gas from the progenitor has definitely not been isotropic. The outflows may have occurred from a direction oblique to the rotation axis of the central star. The helical structure of the filaments and the velocity field are consistent with this picture provided the rotation axis makes a small angle with the line of sight. Further data with higher precision will be needed before a definite mechanism can be proposed.

SPATIAL DECONVOLUTION OF IRAS OBSERVATIONS OF PLANETARIES

George Hawkins and B. Zuckerman
University of California, Los Angeles

ABSTRACT. The sizes of fifty planetaries at the four IRAS wavelengths are presented as a result of performing spatial deconvolution of survey mode data. We obtain an increase in resolving of a factor of about 2 or 3 from the normal IRAS detector sizes of 45", 45", 90", and 180" at wavelengths 12, 25, 60 and 100 microns. Most of the planetaries deconvolve at 12 and 25 microns to sizes equal to or smaller than the optical size. Some of the nebulae such as NGC 6720 and NGC 6543 show full width at half maximum at all IRAS wavelengths that are about equal to the optical size, while others give an increasing size with wavelength. The profiles of a few interesting cases are shown. The method and results should allow comparison with models for infrared emission from dust from planetary nebulae.

The one dimensional deconvolution are performed with the Richardson-Lucy algorithm. The SCANPI program is used with survey mode data to sum scans across an object from several different detectors in order to increase the sampling to a rate sufficient for deconvolving, beyond the normal survey mode sampling. The details and limitations of the method are discussed, and deconvolutions of point sources are provided for comparison.

EVIDENCE OF EXPANSION IN THE CENTRAL REGION OF NGC 2346

D.P.K. Banerjee, B.G. Anandarao, J.N. Desai
Physical Research Laboratory
Ahmedabad 380 009, India

S.K. Jain and D.C.V. Mallik
Indian Institute of Astrophysics
Bangalore 560 034, India

ABSTRACT. We present observations of the bipolar planetary nebula NGC 2346 carried out with the 1-m telescope at the Vainu Bappu Observatory in Kavalur, India using (1) a high resolution piezo-electric scanned Fabry-Pérot Spectrometer (with a velocity resolution of 10 km s^{-1}) for line studies in the 6000 A - 7000 A spectral range and (2) a pressure scanned Fabry-Pérot spectrometer (with a velocity-resolution of 5 km s^{-1}) in the green region. The nebula was observed in the H I 6563 A and [N II] 6583 A emission lines using a 15" aperture and in the [O III] 5007 A line using an 8" aperture centered on the bright central spot. A number of scans in each of these lines were co-added to improve the signal-to-noise ratio. The [O III] profile shows a well defined split between the blue and the red component, typical of an expanding shell. The [N II] profile does not show a well resolved split, although a pronounced suggestion of a split was observed in all the scans. The Hα profile was broad and asymmetric. The composite [O III] and [N II] profiles were decomposed into two individual Gaussians for obtaining the expansion velocity.

TABLE 1. EXPANSION VELOCITY

Emission Line	Expansion Velocity
[O III] 5007 A	8 1 km s^{-1}
[N II] 6583 A	11 1 km s^{-1}

A rough estimate of ion temperature was also made using the widths of the individual Gaussians in the [N II] and H I profiles. Assuming that the H I and [N II] lines originate from the same region of the nebula and that microturbulence is uniform throughout, we obtain temperatures of 7800 K and 14500 K respectively for the approaching and receding shells of the nebula. These temperatures may be compared with the electron temperature of 14200 K measured by Sabbadin (1976, *Astron. Astrophys.*, 52, 291) using the [N II] line ratio.

UNUSUAL EMISSION LINE PROFILES OF M1-1

Katsunori Shibata and Shin'ichi Tamura
Astronomical Institute, Tohoku University
Aobayama, SENDAI 980
Japan

In order to seek out intrinsically compact PN, we have observed high-excitation and angularly small PN, M1-1, with the intensified Reticon system (one-dimensional 1024 pixel-array) at the Coude focus of the 188 cm telescope. We have obtained highly resolved emission line profiles of [NII]λ6583, Hα, [OIII]λ5007, and HeIIλ4686 at several slit position angles. These lines were analyzed with the aid of Multiple Gaussian Method. From the analysis of their radial velocities and decomposed profiles, we can summarize our observational results as follows.
(a) HeII were fitted by three Gaussians, those are main blue and red-components and an extremely red-shifted sub-component. Hα and [OIII] were decomposed into blue and red components. [NII] were fitted by single Gaussian.
(b) The radial velocities at both peaks of red and blue components of HeII coincide well with those of Hα. Since it is inferable that He exist in the nearest part from the central star while Hα distribute in the whole of PN, we decided the radial velocity of M1-1 at the center between blue and red components in Hα and HeII lines, and obtained V_{LSR}= -26.0±1.1 km/sec.
(c) The radial velocities at the peak of blue components of [OIII] and [NII] shifted to blueward cosiderably than that of Hα and HeII.
(d) In the line profiles of Hα, [OIII], and HeII, red components were weaker than blue ones. In particular, the red component of [OIII] is very weak.
(e) Though we put the slit across the whole image of M1-1, the line radiation which corresponds to the components of -26.0 km/sec is hardly seen in [OIII] and [NII].
From these facts and arguments, we can get following conclusions.
(1) O^{++} and N^+ should exist in the outermost part of M1-1. The expansion velocities of these regions are 38.8±2.6, 37.8±1.8 km/sec respectively.
(2) The line profiles reveal that M1-1 is a bipolar planetary nebula and observed from a direction near the pole.

EMISSION LINE PROFILES IN THE PLANETARY NEBULAE IC 4593 AND NGC 6153

D.P.K. Banerjee and B.G. Anandarao
Physical Research Laboratory
Ahmedabad-380 009, India

ABSTRACT. The Planetary Nebulae IC 4593 and NGC 6153 are two rather compact objects not well studied. The nebula IC 4593 is about 12 arcsec in diameter and has a central star of Type O7 f; while the southern nebula NGC 6153 is about 22 arcsec in diameter and its central star is faint and of unknown spectral type. Using a high-resolution scanning Fabry-Pérot spectrometer we have made profile measurements of emission lines Hα λ6563 A, [O III] λ5007 A, and [N II] λ6584 A in the central regions of these two nebulae. We have found expansion velocities for IC 4593 of 40 km s^{-1} in [N II] and 16 km s^{-1} in [O III]. In the case of NGC 6153, we have obtained expansion velocities of 15 km s^{-1} in [N II] and 13 km s^{-1} in [O III] line. The profiles in Hα in both the nebulae dis not show a double peaked feature due to the larger thermal broadening. In the case of IC 4593, both [O III] and [N II] profiles showed complex structures. These results and their interpretation will be discussed.

A NEW STUDY OF SOME GALACTIC PLANETARY NEBULAE

Stephen J. Meatheringham, Peter R. Wood, and D.J. Faulkner
Mount Stromlo and Siding Spring Observatories
Australian National University

ABSTRACT. Expansion velocities ([O III], [O II], and He II) have been measured for a sample of 64 Southern Planetary Nebulae (PN). The ratio of [O III] to [O II] expansion velocities is used to derive a typical ionized shell thickness of order $\Delta R/R_{neb} \approx 0.12$. Nebular electronic densities have been determined from the [O II]$\lambda\lambda$3727,3729 A doublet for 23 of these objects. These data are compared with previously published values. The Dopita et al. (1987) distance scale for Magellanic Cloud PN based on a correlation between observable nebular parameters is used to derive distances to 32 Galactic nebulae. These distances are compared with published values, and lead to the conclusion that the Dopita et al., Daub (1982) and Maciel (1984) distance scales agree well, but that the Shklovsky (1956) method yields distances that are too large. Nebular ionized masses are also calculated for a subset of 30 objects.

EXPANSION VELOCITIES OF [NII] and [OIII] FROM COMPACT PLANETARY NEBULAE

Katsunori Shibata and Shin'ichi Tamura
Astronomical Institute, Tohoku University
Aobayama, SENDAI, 980
Japan

We observed the expansion velocities, Vexp[NII] and Vexp[OIII] of angularly small planetary nebulae(PNe) and examined the relations between expansion velocities and distance free parameters like relative emission line intensities of HeIIλ4686, [OIII]λ5007, and [NII]λ6583. The expansion velocities of PNe are usually obtained from the emission lines of [OIII]λ5007. But these quantities obtained from [NII]λ6583 are more suitable to investigate the evolution of PNe because O^{++} region is confined within inner part for lower and intermediate excitation PNe.

The observations were made with the one-dimensional intensified Reticon system at the Coude spectrograph of the 188-cm telescope at the Okayama Astrophysical Observatory. The dispersion was 5.3-A/mm (0.13-A/pixel). Observed line profiles are fitted by Multiple Gaussians. The expansion velocity was determined from FWHM velocity or from the velocity difference between the peaks of the Gaussians. In addition to our own samples of 18 PNe, other samples of PNe have been selected from Sabbadin et al (1984) and Ortolani and Sabbadin (1985), and analyzed on the basis of their Vexp[NII] and Vexp[OIII]. The line intensities and chemical abundances were refered from several literature.

The relations between expansion velocities and distance free parameters (line intensities, chemical abundances, and the difference of expansion velocities, Vexp[NII]-Vexp[OIII]) were examined with the samples of 40 PNe. Our conclusions are as follows.
(1) Expansion velocities obtained from O^{++} are systematically smaller than those from N^+ even in apparently compact PNe.
(2) The expansion is certainly accelerated from the center of PN even among our samples.
(3) Expansion velocities do not depend upon chemical abundances.
(4) We can show several relations between expansion velocities and line intensities which have to be explained with the expansion model of PNe. For example, the relations between Vexp[OIII] and I([NII]λ6583)/I(Hα): Vexp[NII] and I([OIII]λ5007)/I(Hβ) with the parametar of I(HeIIλ4686)/I(Hβ) seem to be the most promising ones to establish expansions of PNe.

HIGH AND LOW RESOLUTION SPECTRA OF SELECTED PLANETARY NEBULAE*

S. Cristiani[1], F. Sabbadin[2], and S. Ortolani[2]
1. European Southern Observatory, La Silla, Chile
2. Asiago Astrophysical Observatory, 36012 Asiago (VI), Italy

ABSTRACT. High (CES spectrograph + RETICON at the CAT telescope) and low (B&C spectrograph + CCD at the 2.2-m telescope) resolution spectra of selected, southern planetary nebulae allowed to obtain the $H\alpha$ and [N II] emission line profiles and the nebular emission line intensities in the spectral range $\lambda\lambda 3650-9400$ A. The $H\alpha$ and [N II] emission line parameters were derived following the procedure used by Sabbadin (*Monthly Not. Roy. Astron. Soc.*, 209, 889, 1984) and Ortolani and Sabbadin (*Astron. Astrophys. Suppl. Series*, 62, 17, 1985). Table 1 contains the relevant data for eight nebulae of the sample.

TABLE 1

P&K	Name	$V_{exp}\ H\alpha$ (km s^{-1})	V_{exp} [N II] (km s^{-1})
0+12°1	II 4634	13.6	...
37-34°1	NGC 7009	16.2	...
206-40°1	NGC 1535	19.1	...
215-24°1	I 418	6.0	12.1
309- 4°2	NGC 5315	10.7 (35.9)	37.6 (17.8-21.1)
315-13°1	He 2-131	10.6	12.0
345- 8°1	Tc 1	6.1	15.4
358-21°1	I 1297	31.0:	34.6:

Up to now the analysis of the low resolution spectra is complete only for IC 1297 (358-21°1). Main results obtained for this object are presented in Table 2.

TABLE 2. OBSERVATIONAL RESULTS FOR IC 1297

		Ionic Abundances
$He^+/H^+ = 0.093$	$O^+/H^+ = 8.7\times 10^{-5}$	$N^+/H^+ = 1.1\times 10^{-5}$
$He^{++}/H^+ = 0.024$	$O^{++}/H^+ = 4.5\times 10^{-4}$	$Ne^{++}/H^+ = 1.2\times 10^{-4}$
	$S^+/H^+ = 1.6\times 10^{-6}$	

Total Abundances
$He/H = 0.117$ $O/H = 6.7\times 10^{-4}$ $N/H = 8.0\times 10^{-5}$ $Ne/H = 1.7\times 10^{-4}$ $S/H = 1.2\times 10^{-5}$

*Based on observations obtained at the European Southern Observatory.

THE PLANETARY ESO 166 - PN21

María Teresa Ruiz
Departemento de Astronomía, Universidad de Chile
Stephen R. Heathcote and William G. Weller
Cerro Tololo Interamerican Observatory, NOAO

ABSTRACT. Long slit, high and low resolution, spectrograms of this object were obtained using the telescopes at CTIO. From the [N II] $\lambda 6584$ line in the high resolution spectrum, we obtained a radial velocity difference between the front and back expanding shells at the center of the nebula of 56 km s^{-1} corresponding to an expansion velocity of 28 km s^{-1}. From the low resolution spectra we found [S II] $\lambda\lambda 6717/6731$ line ratios indicating densities between 500 cm^{-3} and less than 200 cm^{-3}. The temperature sensitive line ratios of [O III] and [N II] were not well determined due to the weakness of the $\lambda 4363$ and $\lambda 5733$ lines. A ionization structure is clearly seen with radial distance to the central star which shows that the inner shell (A2, A4, B2, B4) is not a projection of the outer one but a separate structure.

Average abundances were determined taking $N_e = 300$ cm^{-3} and assuming a $T_e = 10^4$ K, consistent with the observed line ratios. We did not include positions A1, A5, B1 and B5 in which a considerable amount of neutral H could be present affecting the line ratios through charge transfer reactions. The abundances thus determined are He/H = 11.15, O/H = 8.66, N/H = 8.33 and Ne/H = 8.13.

The high He abundance and N/O abundance ratio are typical of planetaries with massive progenitors.

Spectrophotometry of a very blue star at the center of the nebula (indicated with an arrow in the figure) reveals a m_V = 18.1 mag star with a featureless specrum and (B-V)$_0$ = -0.38. Considering that the nebula is 160" in diameter with an expansion velocity of 28 km s^{-1} and assuming a radius of 0.6 pc we get a maximum distance of 1547 pc and an age of 2.1×10^4 years. Taking the temperature of the central star to be 10^5 then its luminosity would be L = 47 L$_\odot$, however, given that the distance estimate is an upper limit, the actual luminosity could be even lower consistent with the abundances found, both pointing to a massive progenitor star.

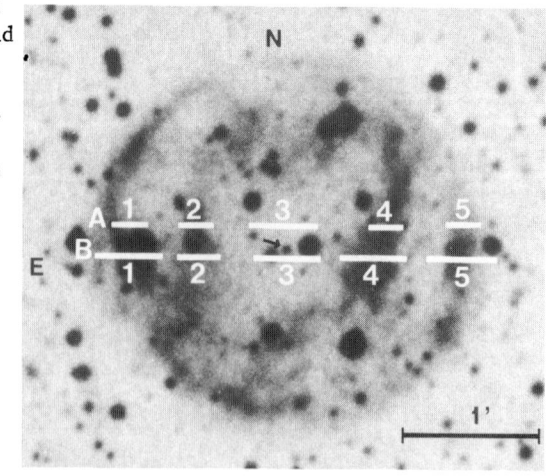

S. Torres-Peimbert (ed.), Planetary Nebulae, 192.
© 1989 by the IAU.

THE STRUCTURE OF NGC 2392

M.J. Hoey
Department of Experimental Physics
University College Dublin

ABSTRACT. This paper reports preliminary results of a recent detailed examination of the morphology of the planetary nebula, NGC 2392.
Seven monochromatic photographs were taken through narrow band filters by a CCD camera at the f/15 focus of the one metre JKT telescope at La Palma. Images of the following emission regions were taken: He II (4686 A), Hβ(4861 A), [O III] 5006 A, [O I] (6300 A), Hα(6563 A), [N II] (6583 A), [S II] 6717-32 A. The exposures were from 200s to 500s. Because the Hα filter had a halfwidth of 60 A, this frame may have some [N II] contamination but the [N II] filter was narrow with a halfwidth of 16 A. The images obtained are of very high quality and reveal interesting details of the structure of this planetary nebula. The [N II], Hα and [O III] images are shown together with sets of isophotic contours. The CCD images allow the faint outer regions of the nebula to be seen clearly. An examination of the outer edge of the nebula shows irregularities of structure and fine wispy filaments extending outwards. The resulting diameters (lower limits) are 49" for Hα, 49" for [N II], and 57" for [O III].
It is not easy to fit the observational data to a simple model of the nebula. Louise (*As. and Sp. Sci.*, 79, 1981) suggests a model of an inner toroid, surrounded by a spherical shell. Pascoli and Macron (*C.R. Acad. Sc. Paris*, t. 304 Serie II, No. 15, 1987) suggest that the double structure can result from a time-dependent magnetic field in the vicinity of the progenitor.

HIGH VELOCITY OUTFLOWS IN POST-MAIN SEQUENCE NEBULAE

J.P. Phillips
Physics Department, Queen Mary College, England
A. Mampaso
Instituto de Astrofísica de Canarias, Tenerife, Spain

ABSTRACT. We have observed a broad range of post-main-sequence, type I and irregular nebulae using the intermediate dispersion spectrograph of the Isaac Newton Telescope (La Palma, Spain). Many of these show evidence for high velocity mass outflow, and in particular we find: (i) High velocity ($\sim$ 500 km s^{-1}), and appreciable mass loss outflows within $\simeq$ 10 arcsecs of M2-9 and SH 2-71; (ii) Substantial shell expansion velocities in NGC 7026 ($\sim$ 10^2 km s^{-1}) with evidence for an appreciable driving wind at the central star, velocity $\simeq$ 10^3 km s^{-1}; (iii) Jet outflows extending over a range 260 km s^{-1} in Hb 5, with a distinctly tilted line structure suggestive of shock compression at the edges of an outflow cavity; and finally (iv) similarly strong winds (2×10^3 km s^{-1}) from each member of a WC binary in the nucleus of NGC 6905. The binary separation is approximately 3.6 arcseconds, and the core is further enveloped by two, apparently co-spatial shells, expansion velocity 90 km s^{-1}, and separation $\Delta V \simeq 130$ km s^{-1}. The [N II] emission for the shells is extraordinarily weak, although ansae located outside of the shells, and perpendicular to the binary major axis, possess I([N II]λ6584) > I(HIλ6563). We propose that the two shells were ejected at differing phases of binary evolution.

The source NGC 2392 was also re-observed. Whilst the peculiar [N II] line structure at P.A. 159° is confirmed, we do not interpret this as reflecting jet outflow. Rather, it seems likely that we are witnessing high velocity shell ejection, an unusual (asymmetric) pattern of shell excitation and/or mass distribution, and a trend for increasing velocities at larger radii.

Long-Slit 2-Dimensional Spectra of the Giant Halos Around NGC 6543 and NGC 6826.

D. Middlemass & R.E.S. Clegg.
University College London, U.K.

J.R. Walsh.
Anglo-Australian Observatory, Epping, Australia.

We have observed the large, faint halos around NGC 6543 and NGC 6826 with the Isaac Newton Telescope, using IPCS and CCD (2-dimensional) detectors. Line intensities are measured every 1.5 arcsec over a total slit length of 3 arcmin (IPCS) or 2 arcmin (CCD). Several slit positions across the halos were observed, so as to obtain average properties of these two regions. Halo spectra are compared with spectra of the bright core, so as to distinguish between reflection by dust and genuine thermal emission; the halos are not reflection nebulae.

We show that the bright edges of the NGC 6826 and NGC 6543 giant halos are not at the Strömgren radius; O^{++} does not recombine to O^+ there and the entire systems are optically thin. Average halo properties and halo masses are estimated. The [O III] electron temperature is significantly higher in the Halo than the centrally ionized nebula; we find T_e(halo) = 16 000 K and 13 000 K for NGC 6543 and NGC 6826 respectively. Upper limits of the electron density are deduced from [O II] line ratios for NGC 6543 but we measure $N_e = 550$ cm^{-3} for NGC 6826.

The mass ratios (halo/core) are ≥ 0.7 and 0.09 for NGC 6543 and NGC 6826 respectively. These are calculated from the Hβ flux measured at slit positions we judge to be representative of the Halo. Abundances of N, O and Ne in the halos are not significantly different to those in the cores.

Around NGC 6543 we present results also for the density, temperature and composition of the large bright knot due west of the PN core. Its properties appear to differ from the rest of the Halo.

We also give results on extended thermal emission from the compact nebula BD+30°3639 which indicate that its Halo also is not a reflection nebula.

Full results, including photo-ionization models, will be published in Monthly Notices of the Royal Astronomical Society.

THE HALOS OF NGC 6543 AND NGC 6826

A. Manchado[1], S.R. Pottasch[2], and A. Mampaso[1]
1. Instituto de Astrofísica de Canarias, Tenerife, Spain
2. Kapteyn Astron. Institute, Groningen, The Netherlands

ABSTRACT. Long slit low resolution (3.4 A) spectra of the planetary nebulae NGC 6543 and NGC 6826, obtained using the 2.5-m Isaac Newton Telescope (La Palma) with the Image Photon Counting System (IPCS), indicate different physical conditions in the outer halos than in the central zone, with an outward increase of electronic temperature. The estimated mass contained in these halos is considerably larger than the values of the inner nebulae.

The calculated chemical abundances seem lower in the halos than in the central parts.

From the above considerations it is clear that the emission of the halos of these nebulae is thermal and not reflection by dust.

EXPANSION VELOCITIES OF SOUTHERN PLANETARY NEBULAE

K.C. Sahu and S.R. Pottasch
Kapteyn Laboratorium, Groningen, The Netherlands

ABSTRACT. We have undertaken a programme of kinematic study of southern planetary nebulae by obtaining high-resolution (R ≥ 50,000) spectra, using the ESO 1.5-m telescope + the Coudé Echelle Spectrograph. As first results of this study, this paper presents previously unknown expansion velocities of 16 planetary nebulae. This result increases the total number of planetary nebulae for which the expansion velocities are now known by about 10%. Further, reliable distance measurements and other physical properties are available for most of these sources. Hence this sample significantly improves the previously available data for statistical analysis related to the dynamics and evolution of planetary nebulae. Some preliminary results based on such statistical analysis are presented. The details of the results will be published in *Astron. Astrophys*.

DETECTION OF AN EXTENDED OPTICAL HALO AROUND IC 418

D.J. Monk, M.J. Barlow, and R.E.S. Clegg
Dept. of Physics and Astronomy
University College London

ABSTRACT. Two-dimensional long slit AAT IPCS spectra of IC 418 (in the 3400-4400 A wavelength region) show faint extensions in the [O II] 3726, 3729 A and Hγ emission lines out to at least 60 arcsec south and 50 arcsec north of the central star, compared to the angular radius of only 6 arcsec for the bright nebula.

The [O II]:Hγ flux ratio is constant throughout the nebula and halo, although both lines show a 200-fold decrease in intensity in the halo compared to the nebula. The mean [O II] 3726:3729 A (density sensitive) ratio in the halo is 1.66±0.40, compared to 2.15±0.10 in the nebula. The errors allow the two ratios to be interpreted as equal, in which case all the data are consistent with a dust halo reflecting the nebular emission lines. However, the apparent difference in electron density between the halo and nebula could be real.

IPCS spectra in the 6400-7400 A wavelength range show a smaller extension in the Hα and [N II] 6584 A emission lines, out to radii of $\sim$ 21 and $\sim$ 28 arcsec respectively. The Hα:[N II] 6584 A ratio is constant throughout the nebula and halo, but there is a 2000-fold drop in line intensity from nebula to halo, an order of magnitude larger than that observed for the [O II] and Hγ lines.

Taking the [O II] 3726:3729 A ratio as constant, the observations are consistent with an extended reflection halo at least 110 arcsec in diameter. To explain the fact that the scattering optical depth at 6563 A is a factor of ten lower than that at $\sim$ 4000 A, we find that the scattering grains must be small, $\leq$ 0.03 µm for carbon or silicate particles. The existence of small dust grains around IC 418 is consistent with the presence of the unidentified infrared emission features in its spectrum (Willner et $al.$, 1979), since these features have often been attributed to very small carbon-rich particles. Very small dust grains have also been invoked in the past to explain an excess 2 µm continuum observed around IC 418 (Willner et $al.$, 1979, Phillips et $al.$, 1984).

Finally the flat profile of the [O II] line intensity through the halo is interpreted as being due to a second density peak in the outer nebula, and it is proposed that IC 418 is a double-shell nebula.

INTERNAL MOTIONS OF FAINT PN HALOS

You-Hua Chu
Astronomy Department, University of Illinois

George H. Jacoby
Kitt Peak National Observatory

We have obtained long-slit echelle observations of faint halos for 10 PNe - NGC 2022, 2438, 6210, 6309, 6543, 6720, 6751, 6826, 6891, and 7662. Only NGC 2022 and 6720 have linesplits indicative of a hollow expanding-shell structure. The others have observed FWHM ranging from 13 to 47 km/s in the [O III]λ5007 line, with an instrumental FWHM of 9-10 km/s. We have subtracted the instrumental and thermal widths and made geometric corrections to derive the expansion velocities.

Assuming no turbulence, we may derive upper limits of expansion velocities from the widths of velocity profiles. NGC 6309 and 6751 have the most quiescent halos, of which the expansion velocities are < 5 km/s. NGC 2438, 6210, 6543, 6826, and 6891 have kinematic FWHM's of 17-25 km/s in the halos. High S/N ratio was obtained in the data of NGC 6826 and 6891, and the variation of the velocity width is consistent with a nearly sonic expansion velocity (~10 km/s). In the case of NGC 2438, 6210, and 6543, we do not have enough S/N ratio and spatial information to derive more precisely their expansion patterns. Their expansion velocities are probably < 15 km/s. NGC 7662 has the broadest unresolved profile in the halo, 47 km/s, implying a supersonic expansion velocity of $\leq$ 24 km/s.

We detected a linesplit of 35 km/s in the halo of NGC 2022. It is probably a shell expanding at 20 km/s. NGC 6720 has larger linesplits in the [N II] line than Hα and [O III]. The linesplits in the [N II] line are ~36 and 84 km/s in the halo and the inner shell, respectively. Such velocity structure does not support the hypothesis that NGC 6720 is a bipolar nebula projected along the poles.

The swept-up interstellar mass in the halo is about 0.01 n_o $M_\odot$, where n_o is the ambient density in cm^{-3}. This amount of mass is too small to have significantly slowed down the expansion of the PN halos. The different expansion velocities in these 10 halos must be inherent. The lack of linesplit in the broad profiles can be explained by a large amount of turbulence and/or a filled envelope geometry, as opposed to a hollow shell. In the cases of NGC 6826, 6891, and 7662, it seems that the inner nebula tapers off gradually into the halo; their halos are probably filled.

SPECTROSCOPIC INVESTIGATIONS OF HALOS OF PLANETARY NEBULAE

M.BÄSSGEN[*], G.BÄSSGEN[*], M.GREWING[*], S.CERRATO[*], L.BIANCHI[**]
[*] Astronomisches Institut Tübingen (FRG)
[**] Osservatorio Astronomico di Torino (Italy)

Different ESO-telescopes with medium resolution spectrographs and CCD-detectors were used to look for emission from extended halos around Planetary Nebulae. In three out of four cases the search was successful. The ratio R_{halo}/R_{neb} ranges from 2 to about 10. Assuming that the halos have been produced by the winds of the Red Giant progenitor stars, we determine $\dot{M}/V_{AGB}$ of these progenitors, using the measured Hα- and Hβ-fluxes. These values scatter in a relatively narrow range around $2..5 \cdot 10^{-7}$ (M$_\odot$/y)/(km/s). In some objects emission line fluxes were measured at two ore more radial distances and radial density profiles were derived from them. These are in good agreement with the expected $n \sim r^{-2}$ density law expected for a freely expanding wind.
Extrapolating the halo density profile inwards we can determine n_{neb}/n_{halo} at the location of the nebula. With one exception, this ratio is larger than 4, indicating that the compression of the gas was caused by an isothermal shock rather than by an adiabatic shock.
The derived $\dot{M}/V$ values, combined with known mass loss rates and wind velocities of the central stars, are used to test the two-wind model (Kwok et al. 1978) which describes a possible formation scenario of planetary nebulae.
The conclusion seems unavoidable that in the very late stage of the evolution of the progenitor star an increased mass loss rate is required to explain the observed nebular properties.

$\dot{M}/V$ calculated from emission line fluxes in units of 10^{-7} (M$_\odot$/y)/(km/s)

Obj.	Pos.	$\dot{M}/V$	Obj.	Pos.	$\dot{M}/V$
NGC 1535	31"	3.0	NGC 3918	25"	2.2
	40"	3.5	IC 418	15"	8.0
NGC 2452	11"	17.0		29	6.0
	17"	11.0	IC 2165	10"	3.5
NGC 2792	17"	2.5		17"	3.3
NGC 2867	17"	1.8		27"	4.0
	34"	1.8	IC 2448	17"	2.5
NGC 3132	30"	7.0		29"	2.0:
	54"	4.5	IC 2501	20"	2.0
NGC 3211	19"	2.2			

KINEMATIC STRUCTURE AND CHEMICAL COMPOSITION OF THE DOUBLE SHELL PN
NGC 3242

K.C. Sahu[1], S.R. Pottasch[1], B.G. Anandarao[2], and J.N. Desai[2]
1. Kapteyn Laboratorium, Groningen, The Netherlands
2. Physical Research Laboratory, Navrangpura, Ahmedabad
   380 009, India

ABSTRACT. Kinematic study of the multiple shell PN NGC 3242 was carried out by obtaining Hα and [O III] line profiles at 9 positions of the nebula using a high-resolution (R ≅ 50,000) Fabry-Pérot spectrometer. The positions cover both the bright inner shell and the faint outer shell. It is shown here that the two apparently continuous shells are kinematically separate: the faint outer shell was ejected ∿ 5000 years earlier and has less expansion velocity than the bright inner shell.

For the study of chemical compositions, low resolution ($\delta\lambda \cong 6$ A) spectra were obtained in the wavelength region 3600 A to 8500 A using the ESO 1.5-m telescope. The chemical abundances for the bright inner shell seem higher than for the faint outer shell. The estimated mass of the outer shell is considerably larger than that of the inner shell.

THERMAL INFRARED EMISSION BY DUST IN THE PLANETARY NEBULA NGC 3918

J.P. Harrington
University of Maryland
D.J. Monk and R.E.S. Clegg
University College London

ABSTRACT. Models of the dust grains in the planetary nebula NGC 3918 are presented. The models, which are calculated for four grain materials - graphite, amorphous carbon, silicate, and iron- have a size distribution of particles based on that found for the diffuse interstellar medium. The infrared spectrum of the nebula -described mainly by IRAS photometry corrected for line emission- can be matched either with graphite grains with a size range of 0.04 - 0.30 µm, or with amorphous carbon grains having a size range 0.0005 - 0.25 µm. The implied depletions of gas phase carbon are only 11% and 4%, respectively. It is shown that iron grains cannot be the dominant dust material.

The relative heating rates for dust by resonance lines of Lα, C IV, and N V and by the stellar continuum are studied.

Comparison is made with a simple analysis using a single dust temperature and an emissivity proportional to $\lambda^{-A} B\lambda(T)$.

The small gas-phase depletion of carbon required by our IR model seems inconsistent with the high gas-phase depletions of Fe, Si, and Mg found for this nebula. This could be indicative of a population of large grains.

THE DUST CONTENT OF PLANETARY NEBULAE WITH NEUTRAL HALOS

M.G. Hoare
University College London

ABSTRACT. Young planetary nebulae (PN) which are still optically thick in the Lyman continuum can have a large fraction of material in neutral halos surrounding the ionized zone. The cool dust in the neutral region can make a significant contribution to the far infrared flux, reducing the derived dust-to-gas ratio. This is important when attempting to understand the apparent decrease in dust-to-gas ratio with nebular radius (age) suggested by Pottasch *et al*. (1984).

Comprehensive models of dusty PN with neutral regions have been developed to investigate the dust in these objects in detail. A distribution of grain sizes is used which are heated by stellar and diffuse continuum, as well as by UV resonance line radiation. The grain temperatures are calculated at each point in the nebula. Dust absorption is fully taken into account when constructing the photo-ionization model, so the predicted emission line strengths and radio flux are self-consistent with the dust emission.

Of particular interest is IC 418, in which neutral hydrogen emission was detected by Taylor and Pottasch (1987), and a faint halo seen by Monk *et al*. (1987). Thermal emission from dust, including the neutral region, is shown to agree with IR observations, using graphite with a dust-to-gas ratio of 7.9×10^{-4}. Amorphous carbon could not match the observations.

A model is also shown for DDDM-1 which is the only PN in the galactic halo detected by IRAS. An abundance analysis by Clegg *et al*. (1987) found the object to be very carbon-poor, with Mg and Fe also depleted, probably into dust. A model incorporating silicate dust with a dust-to-gas ratio of 8.5×10^{-4} fits the IRAS observations, and is consistent with the mass available from refractory elements.

SPATIALLY RESOLVED OBSERVATIONS OF THE UNIDENTIFIED DUST FEATURES IN
BD +30°3639

C.H. Smith[1], D.K. Aitken[1], and P.F. Roche[2]
1. Dept. of Physics, University College, University of N.S.W.
   Australia
2. Royal Observatory of Edinburgh, U.K.

ABSTRACT. High resolution spatial scans through the planetary nebula BD +30°3639 have been made with the UCL cooled grating spectrometer at the IRTF. A spectral resolution of $\lambda/\Delta\lambda$ = 50 was sufficient to resolve the unidentified dust features at 8.6 and 11.3 µm and separate them from the continuum emission. The scans were made in .7 arcsec steps across the nebula with a 1.8 arcsec diameter beam.

These results show that the emission from the 8.6 and 11.3 µm features, the warm dust and ionized gas are all coextensive. In contrast the features in NGC 7027 arise from a shell around the ionized region and the region of 10 µm continuum emission (Aitken and Roche 1982). A similar situation is also observed in the Orion nebula where the features arise at, or just beyond the ionization front south-east of the Trapezium (Sellgren 1981). These differences may be understood in terms of the nature of the ionizing source. In NGC 7027 the central star is extremely hot ($\sim$ 300,000 K) and θ_1^c in Orion is of spectral type O4V; the exciting star in BD +30°3639 however is relatively cool at $\sim$ 30,000 K. Thermal spiking of small carbon grains, or polycyclic aromatic hydrocarbon molecules (PAH's), on absorption of single UV photons is thought to produce the 8.6 and 11.3 µm emission features. The small grains however, are probably destroyed by the much more energetic photons found in the ionized zone around the star central to NGC 7027 and in the Orion nebula. In these sources the small grains or molecules only survive in the softer UV field outside the ionized regions, while the lower levels of excitation in BD +30°3639 allows the existence of PAH's in concert with the ionized gas, and warm dust. It appears that photons capable of triply ionizing sulphur (> 35 eV) will destroy small grains, but the $\sim$ 20 eV photons required to singly ionize neon excite without destroying them.

OBSERVATIONS OF CO AND HCN (J = 1-0) IN NGC 2346 AND NGC 7293 WITH THE NOBEYAMA 45-m TELESCOPE

J.R. Walsh, Anglo-Australian Observatory, Epping, Australia
R.E.S. Clegg, University College London, U.K.
N. Ukita, Nobeyama Radio Observatory, University of Tokyo

ABSTRACT. Many positions in the planetary nebulae NGC 2346 and NGC 7293 have been searched for CO (J = 1-0) and HCN (J = 1-0) emission. The beam was 15 arcsec at 115 GHz.

NGC 2346 is a high excitation bipolar planetary with a binary central star of late spectral type (A5), whose unseen hot companion ionizes the nebula. Various studies have suggested occultation by a moving dust cloud (e.g., Costero et al. 1987). CO was observed at three beam positions in NGC 2346 and at the brightest position, HCN was definitely detected. CO was not detected beyond the edge of the visible nebula. The radial velocity structure is compared with that of the ionized gas from echelle measurements of the [N II] $\lambda 6583$ A line. The dominant narrow CO component at the rest frame velocity of the nebula does not correspond to any line component in the [N II] profiles, but a broader component at +25 km/s LSR corresponds to the brighter [N II] component. The -ve velocity [N II] component does not have associated CO. It is suggested that +ve velocity CO component must be smaller in extent than the nebula, but it cannot be determined if it is before or behind the central star. This component may be associated with the obscuring dust cloud.

CO was detected in NGC 7293 by Huggins and Healy (1986). We observed a small region over the outer bright shell to measure small scale variations in emission. Double CO profiles were found at six positions and their separation is similar to the expansion velocity of the nebula (Walsh and Meaburn 1987). A comparison of the CO brightness with that in Hα and [N II] shows no obvious correlation, although CO tends to be brighter over the shell. The CO emission seems to arise on the outer edges of the nebula and not within the central ionized volume. No CO was detected over the outer filament observed by Walsh and Meaburn (1987) or from a bright cometary globule at the outer edge of the central hole.

The implications of these observations for the site of molecules in PN is discussed.

A YOUNG PLANETARY NEBULA WITH OH MOLECULES: NGC 6302

H.E. Payne
National Radio Astronomy Observatory
J.A. Phillips and Yervant Terzian
NAIC, Cornell University

ABSTRACT. We report the results of a sensitive survey of planetary nebulae in all four ground state OH lines. Our results confirm that evolved planetary nebulae are not OH sources in general. However, we did detect one interesting object: an OH 1612 MHz maser in the young planetary nebula NGC 6302. This nebula may be in a brief evolutionary stage, similar to the young and compact planetary nebula Vy 2-2 where OH has already been detected.

We also report the results of further observations of NGC 6302, including VLA observations of the 1612 MHz line and continuum emission and detections of rotationally excited OH lines at $\lambda 5$ cm in absorption.

CO IN THE BIPOLAR NEBULA NGC 2346

A.P. Healy and P.J. Huggins
New York University

ABSTRACT. We report on observations of the J = 2-1 line of CO to study the distribution and kinematics of the molecular gas in the bipolar planetary nebula NGC 2346. The data were obtained with the National Radio Astronomy Observatory 12-m telescope whose beamsize (FWHM = 30") partially resolves the CO emitting region. A map of the velocity integrated emission shows a roughly rectangular distribution, approximately 53"×34", oriented along the minor axis of the optical nebula. The CO spectrum towards the central star system is strongly double peaked. The mapping data show that this results from two distinct regions which are offset south-east and north-west of the center with radial velocities which are, respectively, larger and smaller than that of the star system. Overall the CO data are consistent with an expanding and partially disrupted distribution of molecular gas around the waist of the optical nebula. Mass estimates confirm that a substantial amount of the matter ejected by the star system is still in molecular form.

FLUORESCENT H₂ EMISSION IN THE PLANETARY NEBULAE BD+30 3639 AND HB 12

H.L. Dinerstein, J.S. Carr, P.M. Harvey, and D.F. Lester
Astronomy Department and McDonald Observatory
University of Texas at Austin

ABSTRACT. We report results from a program of near-infrared spectroscopic observations of the H_2 emission from planetary nebulae, being carried out at McDonald Observatory using an InSb array-detector spectrometer. Our observations employ both high spatial resolution (3" diameter aperture) and high spectral resolution ($\lambda/\Delta\lambda$ = 200 -600), thus avoiding potential problems with line blending and spatial registration. These observations provide simultaneous measurements of H I recombination lines and H_2 emission lines, thus accurately defining the relative extent and distribution of the ionized vs. molecular material. One-dimensional cuts through the compact planetary nebulae BD+30 3639 and Hubble 12, taken along east-west and north-south axes through the nebular centers, show that the H_2 emission is concentrated in a ring or shell outside the ionized nebular core. The angular extent of the H_2 emission in Hb 12, with a characteristic diameter of about 8-10" arc seconds, is strikingly larger than the dimensions of the ionized core, which is less than 2" in diameter.

In order to use the infrared H_2 lines to deduce properties of the molecular envelopes such as the gas temperature and total mass, it is necessary that one first establish what mechanism is responsible for producing the observed line emission. The two alternatives are (1) thermal emission (for example, from shocked material), and (2) "fluorescence" or radiative cascades following absorption of UV photons and molecule formation into excited states. These mechanisms can be distinguished by differences in the expected relative line intensities, particularly of lines from high-lying vibrational-rotational levels which will be strong under radiative excitation and weak in the thermal case (e.g., Black and van Dishoeck 1987, Ap. J., 322, 412). We have obtained spectra at the positions of peak H_2 emission in both BD+30 3639 and Hb 12, and find clear evidence for the fluorescence process. In particular, a 2.0-2.3 μm spectrum offset by 4" from the nebular center of Hb 12 shows at least six lines arising from the v = 1, 2, and 3 excited vibrational levels, including a 2.247 μm 2-1 S(1) line which is nearly as strong as the 2.122 μm 1-0 S(1) line. This spectrum is inconsistent with any thermal model, but is matched well by the fluorescence models. We thus conclude that radiative processes are responsible for the H_2 line emission from these particular planetary nebulae.

MOLECULAR HYDROGEN EMISSION FROM COLD CONDENSATIONS IN NGC 2440

N.K. Reay[1], N.A. Walton[2], and P.D. Atherton[1]
1. Queensgate Instruments Ltd. Sunbury-on-Thames, U.K.
2. Kapteyn Astronomical Institute, Groningen, The Netherlands

ABSTRACT. We report observations of the $v = 1-0$ S(1) line of molecular hydrogen in the high excitation Planetary Nebula NGC 2440. The emission is particularly strong at the positions of the two bright condensations which lie well within the H II region and close to the position of the very hot $T = 350,000$ K central star. The emission is consistent with an excited molecular hydrogen mass of $2-4\times10^{-5}$ $M_\odot$ in the condensations, and we estimate the total mass of excited molecular hydrogen associated with the H II region to be 6×10^{-3} $M_\odot$. We show that the radiation pressure from the central star is insufficient to excite the S(1) line emission. We also show that a stellar wind driven shock would imply a mass loss rate of 3×10^{-7} $M_\odot$ yr^{-1} if we adopt a wind velocity of 2000 km s^{-1}.

THE SYSTEMATICS AND DISTRIBUTION OF MOLECULAR HYDROGEN IN PLANETARY
NEBULAE.

P.W. Payne, J.W.V. Storey, and B.L. Webster
University of New South Wales, Australia
M.A. Dopita and S.J. Meatheringham
Mount Stromlo and Siding Spring Observatory, Australia

ABSTRACT. The infrared S(1) line of molecular hydrogen has been searched
for in twenty-two planetary nebulae using the imaging mode of the Anglo-
Australian Telescope. The line was detected and mapped in eleven objects.
It has been demonstrated that all those with strong excited molecular
hydrogen belong to a subclass of the Type I planetary nebulae, morphologically consisting of an equatorial toroid with faint bipolar extensions. Furthermore, nearly all planetaries with these characteristics
have strong molecular hydrogen. The molecular line ratios in the 2.0 to
2.5 micron window are consistent with shock excitation. The observations
suggest that the morphology of these planetaries has been controlled by
a fast stellar wind interacting with a disc of gas concentrated in the
equatorial plane.

In bright planetaries with well defined toroids the molecular
hydrogen is located outside the ionized region, while in those with less
regular and fainter toroids the ionized and molecular components are
more closely coincident. This is interpreted in evolutionary terms.

DETECTION OF OH MASER EMISSION AT 1667 MHZ FROM IC 4997

S. Tamura[1] and I. Kazes[2]
[1] Astronomical Institue, Tohoku University
SENDAI, Japan 980
[2] Observatoire de Paris-Meudon, Section D'Astrophysique
92190 MEUDON, France

We report the first detection of OH emission at 1667 MHz from a planetary nebula, IC 4997 (Figure 1). OH emission in satellite line of 1612 MHz was detected already from planetary nebulae not only Vy 2-2 (Davis and Seaquist 1979), but also two IRAS objects which were distinguished from OH/IR stars as planetary nebulae (Pottasch et al 1987). OH observations of IC 4997 have been carried out with the large radio telescope at Nançay, France. Highly resolved optical emission lines were obtained at the Okayama Astrophysical Observatory, Japan.

Detected OH line of this work shows unexpectedly different radial velocities, $V_{LSR}(OH)=-11$ km/sec from the measured values in optical regions, $V_{LSR}(Opt)=-44\pm2$ km/sec and in the recently detected HI absorption line, $V_{LSR}(HI)=-64\pm2$ km/sec (Altschuler et al 1986). The profile of our OH line is also quite different from those of usual OH/IR stars and very sharp. The OH source may belong to the expanding materials at far side from us and to the boundary between HII gas and its surrounding materials which are responsible to observed infrared excess and pumping.

A complete paper should be submitted to Astron. Astrophys..

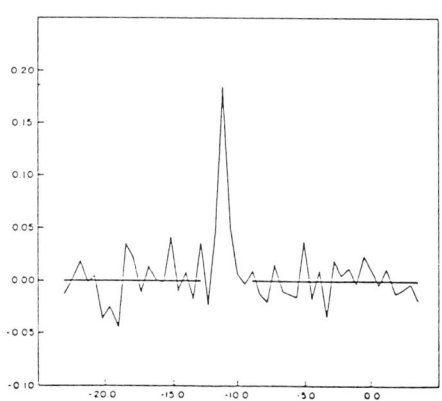

Fig. 1. OH line profile of IC 4997.

OH MASER EMISSION FROM YOUNG PLANETARY NEBULAE

A. Zijlstra[1], S.R. Pottasch[2], P. te Lintel[2], and C. Bignell[1]
1. NRAO/VLA
2. Kaypteyn Institute, Groningen

ABSTRACT. OH/IR stars are now generally accepted to be progenitors of planetary nebulae. We have carried out a project to try to find objects in the transition phase between OH/IR stars and PN. Transition objects should be characterized by having both continuum radio emission and a 1612 MHz OH maser line. The continuuum emission would indicate that the central star has become hot enough to ionize part of the envelope, while the OH maser indicates that a large part of the envelope is still neutral. Only one such object was known, namely Vy 2-2.

We have observed 70 OH/IR stars with the VLA at 2 cm. The stars were selected on the basis of low IRAS colour temperatures and a single 1612 MHz maser peak. During the transition phase the dust shell is expected to be detached from the central star, lowering the dust temperature. In the case of an ionized inner shell the red shifted peak should be unobservable due to absorption.

Three stars were found to have continuum emission at 2 cm. For those stars we obtained 6 cm measurements and measured the precise position of the OH maser. For two of the three objects, the association of the maser line with the continuum source was confirmed. In the case of the third source the line was not detected at the previously reported level.

As part of another project a number of PN were found to have a very high infrared excess, ranging from 10 to more than 100. As these sources might also be very young PN, we searched for OH maser emission using the Parkes radio telescope. Out of eleven objects seven were detected. The precise position of the OH masers still needs to be measured, but the high detection rate indicates that they probably are associated with the PN.

One of the OH/IR stars which we detected with the VLA is possibly unique, having a 2 cm continuum flux density of 7.2Jy. At 2 cm the source is still optically thick. This object would have the highest flux density of any known PN.

The results of the project are twofold. First of all, we now have a sample of PN which are known to be very young and which are known to originate from OH/IR stars. Secondly, OH emission from young PN appears to be much more common than previously thought.

HELIUM ABUNDANCES IN GASEOUS NEBULAE

R.E.S. Clegg, University College London
J.P. Harrington, University of Maryland

ABSTRACT. New collision strengths, from a 19-state quantum calculation for He I, are used to derive revised He/H ratios in planetary nebulae (PN). Empirical formulae are given, for the correction of He I recombination line fluxes for collisional effects, and for the calculation of the population of metastable helium (He I 2^3S) in gaseous nebulae. The revised He abundances for PN, for four samples of published line fluxes, show a mean ratio He/H = 0.100 ±0.007 if nebulae with neutral He and Type I PN are excluded. The mean reduction due to collisional effects is only 10% for Galactic PN. It is shown that the hypothesis, that He/H should be independent of nebular temperature and density, is better satisfied when collisional effects are allowed for. The new He abundances indicate that there is very little He enrichment in Galactic PN of Types II, III, and IV, and that the enhancement of Type I PN in He over H II regions is reduced from earlier values by one third.

Photo-ionization models are used to study the destruction of metastable helium by H I Lyman α and stellar continuum photons. It is shown that photo-ionization can reduce collisional effects on He I line strengths by up to 30%. Collisional excitation of metastable helium can provide a small but significant amount of cooling in nebulae with low heavy-element abundances. We also find that collisional ionization of metastable helium becomes significant at high temperatures, reaching 20% of all collisions out of 2^3S at 20 000 K.

We show that the line used as an indicator of self-absorption effects in the He I spectrum, λ7065 A, is mostly excited by collisions in PN. Derived optical depths for transitions from He I 2^3S are much reduced over previous values. Such depths calculated from model nebulae now appear to agree approximately with depths deduced from λ7065 A for a few objects, if velocity fields are included. Results are given for the formation of the He I λ10830 A resonance line and its attenuation by dust in PN and H II regions.

COLLISIONAL EXCITATION OF THE 10830 HE I LINE AND THE POPULATION OF
THE 2^3S HE I STATE IN GASEOUS NEBULAE

M. Peimbert and S. Torres-Peimbert
Instituto de Astronomía
Universidad Nacional Autónoma de México

ABSTRACT. From the study of the $\lambda\lambda 5876$, 7065 and 10830 He I line intensities in NGC 6572, NGC 6803, NGC 7009, NGC 7027, NGC 7662 and IC 418, it is found that the $I(10830)/I(5876)$ ratio is weaker than expected. By considering estimates of the optical depth at $\lambda 10830$ due to dust absorption and by determining the optical depth at $\lambda 10830$ due to atomic absorption, it is argued that dust absorption of $\lambda(10830)$ photons is not the cause for the low $I(10830)/I(5876)$ ratios. By assuming that the 2^3S He^0 state is depopulated only by radiative transitions to the 1^1S state and by triplet-singlet exchange collisions, it is found that its population is about a factor of two smaller than expected. This result is in agreement with a previous study of the $\lambda\lambda 3889$, 4472, 5876, 6678 and 7065 line intensities in a group of thirteen Type I planetary nebulae. One of the main implications of the underpopulation of the 2^3S level is that the collisional effects in the $N(He)/N(H)$ abundance ratios of planetary nebulae and O-poor extragalactic H II regions are smaller than previously thought.

PLANETARY NEBULAE AND THE PREGALACTIC HELIUM ABUNDANCE

Walter J. Maciel
Instituto Astronômico e Geofísico da USP
Caixa Postal 30.627
01051 São Paulo SP, Brasil

Recent work has emphasized the determination of the pregalactic helium abundance by mass Y_p and the slope $\Delta Y/\Delta Z$ based on the chemical composition of both galactic and extragalactic H II regions (Pagel, 1987; Pagel et al., 1986).

On the other hand, planetary nebulae (PN) can also be used to estimate these quantities (Peimbert, 1983; 1986), when account is taken of the different types that comprise the galactic PN. In fact, it has been recently shown that type II PN have chemical and kinematical properties closely resembling those of galactic H II regions (Maciel and Faúndez-Abans, 1985; Faúndez-Abans and Maciel, 1986; Maciel, 1987). Moreover, on the basis of nitrogen abundance and gradients of heavy-elements, it has been suggested that type II PN include two different subtypes, namely type IIa and IIb (Faúndez-Abans and Maciel, 1987), the latter showing the smallest heavy-element enrichment of all PN.

In the present work, type IIb PN are used to determine Y_p and the slope $\Delta Y/\Delta Z$. It is shown that the planetary nebulae alone imply an upper limit to the pregalactic abundance. The combination of the PN with galactic and extragalactic H II regions produces a good correlation between the helium abundance and the metallicity, from which it can be deduced that $Y_p = 0.234 \pm 0.004$, and $\Delta Y/\Delta Z = 3.5 \pm 0.3$, in good agreement with recent determinations based on H II regions and blue compact galaxies. (Work supported by CNPq-Brasil).

References

Faúndez-Abans, M., Maciel, W. J. 1986, Astron. Astrophys. 158, 228
Faúndez-Abans, M., Maciel, W. J. 1987, Astron. Astrophys. (in press)
Maciel, W. J. 1987, Late Stages of Stellar Evolution, eds. S. Kwok, S. R. Pottasch, Reidel, p. 391
Maciel, W. J., Faúndez-Abans, M. 1985, Astron. Astrophys. 149, 365
Pagel, B. E. J. 1987, preprint
Pagel, B. E. J., Terlevich, R. J., Melnick, J. 1986, PASP 98, 1005
Peimbert, M. 1983, Primordial Helium, eds. P. A. Shaver, D. Kunth, K. Kjar, ESO, p. 267
Peimbert, M. 1986, PASP 98, 1057

N/O ABUNDANCES IN PLANETARY NEBULAE FROM FAR-INFRARED LINE OBSERVATIONS

Harriet L. Dinerstein
University of Texas at Austin
Michael W. Werner
NASA Ames Research Center

ABSTRACT. Measurements of the [O III] 52, 88 μm and [N III] 57 μm fine-structure emission lines have been obtained for nine planetary nebulae, using the facility far-infrared array spectrometer on NASA's Kuiper Airborne Observatory. The N^{++}/O^{++} ratios determined from these observations range by more than an order of magnitude among the sample. Using recent improved values for the atomic parameters, we find that the N^{++}/O^{++} ratios agree fairly well with values of N^+/O^+ determined from optical lines in the same objects. The highest N^{++}/O^{++} values, found for the extreme "Type I" nebulae NGC 2440 and NGC 6302, are approximately unity. These results imply that the synthesis and mixing of nitrogen must be extremely efficient in the progenitor stars of some planetary nebulae, and that these nebulae are significant sources of nitrogen to the interstellar medium. The local electron densities derived from the intensity ratios of the two [O III] lines are generally lower than values in the literature determined from small-beam optical observations of other ions, such as [O II]. This effect can be understood in terms of the presence of clumpy structure in the nebula, since the far-infrared lines have fairly low critical densities for collisional de-excitation and therefore are preferentially emitted from low-density gas.

INVESTIGATIONS OF DDDM 1; THE FOURTH HALO PLANETARY NEBULA

A. Yu. Shchelkanova
Sternberg State University, Moscow, USSR

ABSTRACT. Spectral ($\lambda\lambda$ 4100 - 7300 A) and photoelectric observations of DDDM 1 (61 + 41°1; α_{1950} = 16^h38^m8; δ_{1950} = +38°48') were made in 1985-86 on Crimean Station of Sternberg State Institute. Photographic absolute Hβ flux is found to be F(Hβ) = (3.0±0.57) 10^{-12} erg cm^{-2} s^{-1}, angular radius ϕ = 1".0±0".5. Interstellar reddening E(B-V) for the object is found to be less than $0^m.02$. Integrated flux from central star and gaseous nebula was measured in UBV-filters V = $14^m.71$, B-V = $0^m.12$, U-B = $-0^m.9$.

Distance to DDDM1 in O'Dell's scale (optically thin case) equals 15.0 kpc, in Cudworth's scale (thick case) -9.2 kpc. Nebular parameters T_e[O III] = 9000 K, N_e[S II] = 7000 cm^{-3}. The abundance of He (11.23±0.11), O(7.7±0.4), Ne(6.7±0.2), Ar(4.3) (the scale log N(H) = 12.0) of DDDM1 is similar to that of three other halo planetary nebulae K 648, 49 + 88°1, 108 - 76°1. Zanstra temperatures of central T_H = (3.3±0.4) 10^4 K, $T_{He\ I}$ = (4.1±0.4) 10^4 K, $T_{He\ II}$ = 5.8×10^4 K. Spectrophotometric temperature (4100 - 6700 A) T_{sp} = 3.5×10^4 K. Using d = 15 kpc we obtain for the central star M_V = $-1^m.01$ and log L/L$_\odot$ = 3.28 (if T_{eff} = 58000K).

THE PECULIAR PLANETARY NEBULA M1-78

S.R. Pottasch[1], A.A. Zijlstra[1,2], N. Ukita[3], A. Manchado[4], and M. Ratag[1]
1. Kaypteyn Astron. Institute, Groningen, The Netherlands
2. NRAO (VLA) Socorro, N.M.
3. Nobeyama Radio Observatory, Japan
4. Inst. Astrofísica de Canarias, Tenerife, Spain

ABSTRACT. The question of whether M1-78 is a PN or a compact H II region is discussed. We have obtained new high resolution radio continuum maps, optical spectra and CO maps. Arguments for it being a PN include spectral information, far infrared continuum emission, and radio morphology. It is the strongest CO emitting PN known. Its abundances are peculiar: high helium and very low oxygen and nitrogen abundances. If it is a PN it must be within 4 kpc, but 21-cm absorption measurements indicate that it may be further away.

OXYGEN DEPLETION VARIATIONS IN PLANETARY NEBULAE AND SHELLS EJECTED FROM LUMINOUS POPULATION I STARS

Reginald J. Dufour
Rice University

ABSTRACT. Recent studies by the Peimberts have noted an anticorrelation between O and N abundances in the Type I He- and N-rich PN, such that N^+) is approximately constant. We report observations of the spectra and composition of several "planetary nebula-like" shells surrounding more luminous population I O- and WR-stars, which indicate that this O-N anticorrelation extends upwards in the HR diagram to among the most luminous stars known, with O/H values in the shells ranging down to -2 dex below Solar. We report optical and UV spectrophotometry of the shell nebulae NGC 2359, NGC 6164-5, NGC 6888, NGC 7635, AG Carinae, and the condensations around Eta Carinae, which generally support this anticorrelation trend. We also discuss variations in other elements such as He, C, S, and Ar in these shell nebulae, and compare the compositional variations to the expectations from stellar evolution and nucleosynthesis models of intermediate and massive stars.

MAGNESIUM ABUNDANCES IN PLANETARY NEBULAE AND INTERSTELLAR ABSORPTION
OF Mg II λ2800 A

D. Middlemass
Department of Physics and Astronomy
University College London, U.K.

SUMMARY: High-resolution IUE images have provided line profiles of emission cut by interstellar absorption in the magnesium II lines λλ2795, 2802 A in nine objects. The underlying emission has been reconstructed in seven of these by fitting Gaussian profiles to the remaining line wings. Estimates of the ratio of intrinsic to observed emission ranging from 1.5 to 23 have been obtained. Photo-ionization models of IC 418, IC 4997 and NGC 2440 have been used to obtain Mg/H ratios from the corrected Mg II line strengths. The logarithmic magnesium abundances obtained were 7.6±0.12, 6.96±0.3 and 6.5±0.3 respectively. These results indicate that depletion of Mg into grains is less than previously thought and reduce the observed gradient in the magnesium abundance. The Mg depletion may be related to the C/O ratio within the nebula; carbon-rich planetary nebulae may have less magnesium depletion than those with C/O < 1.

ABUNDANCES OF C, N AND O IN PLANETARY NEBULAE

A.A. Nikitin, A.F. Kholtygin
Leningrad State University, Astron. Observatory, USSR
A.A. Sapar, T.H. Feklistova
W. Struve Astrophys. Observatory, Toravere, Tartu, USSR

ABSTRACT. The abundances of C, N and O in planetary nebulae must correspond to the evolutionary status of their progenitor red giant stars. The best spectral features for abundance determination of these elements are the recombination lines, which depend weakly on the variations of T_e and n_e. The abundance ratio of the ions A^+ and H^+ can be given by [1-3].

$$\frac{N(A^+)}{N(H^+)} = \left\{\frac{A^+}{H^+}\right\} = \frac{\lambda_{ki}}{\lambda(H\beta)} \frac{\alpha^{eff}(H\beta)}{\alpha^{eff}_{ki}} \frac{F_{ki}}{F(H\beta)} = X(T_e) \frac{F_{ki}}{F(H\beta)}, \quad (1)$$

where F_{xi} is the observed flux in a spectral line ki of ion A, corrected for interstellar extinction and $F(H\beta)$ is the same quantity for a $H\beta$ line, the quantities α^{eff}_{xi} and $\alpha^{eff}(H\beta)$ are the corresponding effective recombination coefficients. we present some values of approximation coefficients for the quantity $X(T_e) = X_0(T_e/10^4 K)^n$ as found in papers [1-3]. Considering the spectral line C III $\lambda 4650$, the high-temperature dielectronic recombination (cf. [5]) in its modification described in [1] must be taken into account.

We estimated the total abundances of C, N and O using the following formulae obtained by L. Aller and S. Czyzak [9] for the models of highly excited planetary nebulae:

$$\{C/H\} = \{(C\,III + C\,IV + C\,V)/H^+\},$$
$$\{N/H\} = 1.7\,\{(N\,IV + N\,V)/H^+\} = 2.0\,\{N\,IV/H^+\}, \quad (2)$$
$$\{O/H\} = K\{O\,IV + O\,V\}/H^+\}, \quad K = 3.6,\ 2.4,\ 2.0$$
$$\text{for } E_w = 8,\ 9,\ 10.$$

The correction factors for nitrogen and oxygen show that the contribution of lower ionization stages even for high excitation planetary nebulae is essential, and thus (2) underestimates for low excited nebulae the abundances.

Using the observed fluxes in the spectral lines of C, N and O ions [7-9], we estimated the abundances of these elements in 40 planetary nebulae of high excitation.

The mean values of C, N and O in the high excitation planetary nebulae exceed the solar values ($\{C/H\} = 4.7$-$4, \{N/H\} = 9.8$-$5, \{O/H\} = 8.3$-4) by 2 - 3 or more times. This accords with the "two-wind" hypothesis of planetary nebulae formation. However, from Table 2 it appears that for N and O the abundances found from the fluxes in recombination lines, are essentially higher than the values found from the fluxes in ultraviolet or forbidden lines. The reason for that has been discussed [1,9], but it has remained rather obscure up to now.

SOME STATISTICS OF NEBULAR CHEMICAL COMPOSITIONS

L.H. Aller and C.D. Keyes
University of California, Los Angeles

ABSTRACT. Data from a survey of 51 planetary nebulae (PN) observed with the image tube scanner on Shane Telescope at Lick Observatory are combined with those for a comparable number of objects previously reported. For nearly all of the PN included in the later program, it was possible to obtain adequately accurate plasma diagnostics and line intensities to derive ionic concentrations for He, N, O, Ne, S, Cℓ, and Ar. To get ionization correction factors we calculated theoretical nebular models to fit the excitation level and the intensities of individual important lines. Final model parameters include the stellar radius, emergent flux, $F_\nu(*)$, from Husfeld *et al.* (1984, *Astron. Astrophys.*, 134, 139), nebular size, the optical depth at the hydrogenic Lyman limit, and chemical abundances. Many PN do not appear to be optically very thick in the Lyman continuum.

Before statistics of PN chemical compositions can be discussed, we must assess likely sources of error. Uncertainties in the electron temperature, together with possible fluctuations thereof can adversely affect derived ionic concentrations, e.g., for certain ions of neon whose lines come from metastable levels at 3eV or more. We must reinterpret lines of [Ne IV], [S II], [Cℓ III], and [Ar IV], with the aid of newly available, improved atomic parameters, and best available electron densities.

With these caveats in mind, we examine the correlation between chemical composition and distance from the center of the galaxy for PN of population Type II. The analysis differs from that by Faundez-Abans and Maciel (1986, *Astron. Astrophys.*, 158, 228), not only because of differing input data, but also because of different criteria for Pop. Type II membership. We used the lists by Kaler (1970, *Ap. J.*, 160, 887), Barker (1978, *Ap. J.*, 219, 914), and by Heap and Augensen (1987, *Ap. J.*, 313, 268), and also adopted a galactic center distance of 8500 pcs. Uncertainties in distances of individual PN is the largest source of error. We find:

$$\log N(O)/N(H) + 12 = 8.69 - 0.0156\, R(kpc)$$

ELECTRON DENSITIES IN PLANETARY NEBULAE

Letizia Stanghellini and James B. Kaler
Dept. of Astronomy, University of Illinois

ABSTRACT. A large sample of forbidden lines —[O II], [S II], [Ar IV], and [N I]— have been analyzed to obtain electron densities for 134 planetary nebulae.
Inhomogeneities in the nebulae make comparisons among different data sets difficult. Although the values show considerable scatter, we still see that the [Cl III] densities are generally higher than those derived from either [O II] or [S II] by an average factor of 2.6.
Densities from [N I] are always the lowest. In the case that [Cl III] and [S II] densities are simultaneously obtained for a given position in the nebula, the two densities appear to be better correlated, and the average factor decreases to 2.01. [O II] and [S II] densities are, in both cases, well correlated.

LOW RESOLUTION SPECTROSCOPY OF 13 LOW SURFACE BRIGHTNESS PN's

A. Manchado[1], S.R. Pottasch[2], and A. Mampaso[1]
1. Instituto de Astrofísica de Canarias, Tenerife, Spain
2. Kapteyn Astron. Institute, Groningen, The Netherlands

ABSTRACT. We have obtained long-slit low resolution spectra (7.5 A resolution) of a sample of 13 low surface brightness planetary nebulae using the 2.5-m Isaac Newton Telescope (La Palma) with the Image Photon Counting System (IPCS) covering a spectral range from 3300 A to 7300 A. From those spectra we calculated the ionic and total abundances of O, N, Ne and Ar. Variations in the ionization structure between the inner and the outer part are found in some nebulae although the total abundances appear not to change significantly along the nebulae.

EFFECT OF DENSITY VARIATIONS ON ELEMENTAL ABUNDANCE DETERMINATIONS IN GASEOUS NEBULAE

Robert H. Rubin
NASA Ames Research Center

ABSTRACT. When there are changes in gas density within a nebula, various methods of determining the electron density N_e can give different results. Irrespective of differences in ionization structure, there will be deviations in derived values of N_e due to the physics of populating the energy levels. To focus on N_e variations, the electron temperature is held constant. For two cases presented, the values of N_e inferred range over a factor of ten from nine species (line pairs); in order of increasing N_e, they are N^+(122/204 μm), O^{++}(52/88 μm), S^+(6716/6731 A), S^{++}(18.7/33.5 μm), O^+(3726/3729 A), Ne^{++}(15/36 μm), Ar^{+3}(4711/4740 A), Ar^{++}(8.99/21.8 μm), and C^{++}(1906/1909 A). This is basically a progression from lower to higher critical densities, N_c, for the lines involved, although other factors are involved. The above order can change somewhat for different mixes of densities.

Together with observations of a third line from another species, an elemental abundance ratio may be derived by standard empirical techniques. When N_c(3rd line of species X) is an extreme value relative to N_c(2 lines for obtaining N_e from species Y), the BIAS > 1, where $[N(X)/N(Y)]_{inferred}$ = BIAS $[N(X)/N(Y)]_{true}$. However when N_c(3rd line) is intermediate in value, the BIAS is closer to unity and may be < 1. This implies that when there are N_e fluctuations, chemical abundance ratios obtained with 3 lines that satisfy the latter condition should be more reliable than those satisfying the former. The degree of potential bias in the average N_e value and elemental abundance ratio inferred depends on the extent of density variations. For the cases considered, BIAS can be greater than 2 and much larger when using lines with very different N_c's.

The fact the N_e values from $C\ell^{++}$(5518/5538 A) are higher than those from S^+ and O^+, and that those from N^0(5199/5202 A) are lowest of all (Stanghellini and Kaler this volume) is consistent with what is presented here. Again, this is predominantly a progression from higher to lower N_c values.

A POSSIBLE SUBDIVISION OF TYPE II PLANETARY NEBULAE

M. Faúndez-Abans *
Laboratório Nacional de Astrofísica/ON/CNPq/MCT
Caixa Postal 21, 37500 Itajubá MG, Brasil

W. J. Maciel
Instituto Astronômico e Geofísico da USP
Caixa Postal 30.627, 01051 São Paulo SP, Brasil

* on leave from Universidad de Santiago de Chile

A revision is made of the classification scheme of planetary nebulae proposed by Peimbert (1978), taking into account the observed heavy-element abundances and radial abundance gradients (Faúndez-Abans and Maciel, 1986; 1987a). A subdivision is proposed of type II PN into the subtypes a and b, according to their nitrogen abundances relative to hydrogen (Faúndez-Abans and Maciel, 1987b). Type IIa PN show a higher enrichment in nitrogen than type IIb, which present a nitrogen abundance close to the value of population I objects and a radial N/H gradient as well.
This work was supported by CNPq and FAPESP (Brasil) and DICYT-USACH (Chile).

References

Faúndez-Abans, M., Maciel, W. J. 1986, Astron. Astrophys. 158, 228
Faúndez-Abans, M., Maciel, W. J. 1987a, Astrophys. Space Sci. 129, 353
Faúndez-Abans, M., Maciel, W. J. 1987b, Astron. Astrophys. (in press)
Peimbert, M. 1978, IAU Symp. 76, ed. Y. Terzian, Reidel, p. 215

OBSERVATIONS AND MODELS OF THE 'HELIX' NEBULA NGC 7293

R.E.S. Clegg
University College London, U.K.
J.R. Walsh
Anglo-Australian Observatory, Epping, Australia

ABSTRACT. Long-slit IPCS 2-dimensional spectra in radial directions across the inner and outer shells and across some of the brightest cometary globules are presented. The spectra show the ionization structure of the nebula quantitatively, with He II λ4686 A strong in the inner regions and [N II] λ6584 A prominent in the outer shell, almost due east of the central star. The "[Ne III] anomaly", previously reported for NGC 6720 and 7293 by Hawley and Miller, is clearly seen. It, together with the [O I] λ6300 A flux, provide constraints on our photo-ionization models, as both depend on the concentration of neutral H in the background gas.
  We find low electron temperatures, T_e = 7300-8300 K, and N_e < 300 cm^{-3}. The oxygen and nitrogen abundances are constant across the shells, with mean values O/H = 1.0×10^{-3}, N/H = 4.0×10^{-4}. The object is He-rich, with He/H = 0.13 so this is a marginal 'Type I' planetary. The average spectrum of the largest group of 'knots' (cometary globules) east of the central star is presented.
  Photo-ionization models of the smooth background gas are given. In our initial models, the central star is represented by a non-LTE H-He model atmosphere with T_{eff} = 120 000K and log g = 8.0 (Bohlin et al. 1982), and either He/H = 0.10 or 0.01. The high gravity results in a large He$^+$ absorption edge at 228 A and a rather flat spectrum in the He$^+$ continuum. The resulting, rather hard, ionizing spectrum produces a model with a high concentration of neutral H (and thus neutral O) in the outer 2/3 of the nebular volume. The model reproduces the low electron temperature, 8000K, found from [O III] and [N II] line ratios. It suggests that [O I] λ6300 A emission should be strong in the background gas as well as in the knots. A correction of approximately 10% in the derived He/H ratio is needed to allow for neutral H in this nebula; corrections may also be needed in other Type I PN.
  The implications of H^0, H$^+$ and electrons co-existing in a large volume are discussed, with special reference to the H$_2$ vibration-rotation line emission seen at 2 μm from many positions in this nebula (Storey 1984).

A PHOTOIONIZATION MODEL STUDY OF NGC 7027

Ruth B. Gruenwald
Instituto Astronômico e Geofísico, São Paulo, Brasil
Daniel Pequignot
Observatoire de Paris-Meudon, Meudon, France

The physical characteristics of NGC7027 and the nature of the observations available make it an exceptional object to check the ability of steady photoionization modelling to correctly predict the emission spectrum of nebulae. After a decade of atomic physics and observation improvements, a complete updating of the analysis performed by Péquignot et al (1978) (Astron. Astrophys. 63, 313) is in order.

We obtain a radiation bounded spherically symmetrical model, in which the central star temperature is about 2×10^5K. The star must emit less high energy radiation ($h\nu \geqslant 120$ eV) than the stellar atmosphere models of Hummer and Mihalas (1970). The gas pressure is approximately constant throughout the nebula, the mean electron density being $\sim 7 \times 10^4$ cm^{-3}. The helium, carbon and oxygen abundances by number are 0.10, 9.5×10^{-4} and 4.6×10^{-4} resp. The mean electron temperatures associated to H$^+$, He$^+$ and He^{++} are 14700 K, 12500 K and 16700 K resp. The model successfully accounts for most electron density and temperature indicators, suggesting that most collision strengths and radiative transition probabilities used are now essentially correct and that the representation of the nebula is faithful.

By contrast the ionization equilibria are not all perfectly reproduced. The discrepancies do not exceed a factor 2 in most cases, which is more satisfactory than the factors 10 found a decade ago, suggesting that all important physical processes are now taken into account. However the atomic data may not always be of adequate accuracy because most discrepancies cannot be eliminated by considering, e.g., more complex density distributions on either small or large scale. Examples of discrepancies are given in the table.

Ratio	Theo/Obs	Suggested explanation
OII/OIII	/1.4	rate O^{+2} + H too weak
OIV/OIII	x1.8	blend OIV] with Si IV
CI/CII	/4.6	[CI] from neutral shocked gas
NII/NIII	x1.4	rate N^{+2} + H too strong
NeII/NeIII	/(3 to 6)	[NeII] observation wrong
NeIV/NeIII	/2	Ne^{+3} + H and low-T diel. too strong
MgV/MgIV	x2.5	low-T dielectronic recombination
SIV/SIII	x2.1	low-T dielectronic recombination
AIII/AIV	/2.9	low-T dielectronic recombination

EMISSION LINES OF CI AND N II IN PLANETARY NEBULAE

V. Escalante
Harvard-Smithsonian Center for Astrophysics

ABSTRACT. A model potential method (Caves and Dalgarno, 1972, $J.\ Quant.\ Spect.\ Rad.\ Transf.$, 12, 1539) was used to calculate accurate non-hydrogenic radiative recombination rates and transition probabilities of singly excited states of CI and N II. The results can be used to determine the excitation mechanism of emission lines and to estimate N III concentrations in nebulae with CI and N II emission lines. In most nebulae, observed permitted lines of N II are produced by radiative recombination, but sometimes stronger recombination lines are missing in their spectra. The [CI] lines observed in NGC 7027 cannot be explained by simple radiative and dielectronic recombination. The low [CI] $\lambda\lambda 9850 + 23/\ \lambda 8727$ value may indicate that the emission is produced in high density ($N_e \gtrsim 10^5$ cm^{-3}) condensations where partial collisional deexcitation of metastable levels, takes place. N III concentrations were determined using published data of NGC 3242, NGC 3918, and NGC 6572. The procedure outlined by Wilkes $et\ al.$ (1981, $M.N.R.A.S.$, 197, 1) to determine N abundances from ($N^+ + N^{++}$)/He$^+$ ratios does not always give consistent results with UV or [N II] data. The problem may be due to errors in the calculation of transition probabilities involving the doubly excited levels 2s2p^3 $^3P^0$ and $^3D^0$ of N II that affect the branching and effective recombination rate of the multiplet N II $\lambda 5680$.

THE CONTINUUM EMISSION FROM PLANETARY NEBULAE

Sueli M. Viegas-Aldrovandi
Department of Physics, Ohio State University

ABSTRACT. The study of nebular continuum emission is important for several reasons (Pottasch 1984, *Planetary Nebulae*, Dordrecht: Reidel). First of all, it can provide information about the temperature and the density of the nebula, when the object is large enough, or when the central star is weak enough, so that the nebular continuum is easily observed without interference from the stellar continuum. On the other hand, for small planetary nebulae, both the central star and the nebula contribute to the observed continuum. In this latter case, in order to obtain the stellar continuum the theoretical nebular emission must be used. Thus, studies of the evolution of planetary nebula nuclei through the HR diagram rely on a good calculation of the theoretical nebular continuum.
　　In general, the theoretical nebular continuum is obtained following Brown and Mathews (1970, *Astrophys. J.*, 160, 939), using average values for the physical conditions of the emitting gas obtained from the observed emission lines (Shaw and Kaler 1985, *Astrophys. J.*, 295, 537). This paper aims to present results for the theoretical continuum emission of planetary nebulae, based on photoionization models. The computer code described by Gruenwald and Viegas-Aldrovandi (1987, *Astron. Astrophys.*, in press) has been used. Optically thick models have been constructed considering the central star emitting a black body spectrum. The ionizing radiation is characterized by the stellar temperature in the range $2. \leq T_*/10^4 \leq 15.$ K and ionization parameter $3. \times 10^{-4} \leq 3. \times 10^{-2}$. Three values for the hydrogen density, $n_H = 10^3$, 10^4 and 10^5 cm^{-3}, have been considered. The continuum spectrum emitted by the gas, in the range 2600 to 10,800 A, is calculated in each slab and integrated over the entire nebulae. In order to provide useful theoretical results, several diagrams have been plotted, including ratios of the nebular to the stellar continua. In particular, the B and V fluxes, and also the contribution of the emission lines to these filters, are given as a function of T. A comparison between the theoretical continuum emission, in B and V filters, calculated by the photoionization models and that obtained from the average physical conditions (also obtained from the models) shows that the latter overestimates the gas emission. The overestimate increases with increasing stellar temperature.

COMPARISON OF DIFFERENT TREATMENTS OF THE RADIATION TRANSFER IN MODEL CALCULATIONS OF PLANETARY NEBULAE

G. BÄSSGEN, M. BÄSSGEN, M. GREWING
Astronomisches Institut Tübingen (FRG)

A method of Radiation Transfer treatment (Global Shell Model (GSM)) has been developed, different from the On the Spot Approximation (OSA), taking into account all diffuse reemission from radiative recombination of hydrogen and helium in the nebula itself. For optically thick and optically thin nebulae the method automatically reproduces the On the Spot Approximation and the models without any Radiation Transfer, respectively.

In a model nebula consisting of N concentric spherical shells the diffuse radiation field at any one point is calculated along several characteristic rays through the nebula, including the outer parts. Diffuse reemission from these outer regions becomes important, when the density increases with distance from the central star.

The calculation of the ionisation and thermal balance is an iterative process.

A grid of model nebulae has been calculated in order to compare the performance of the OSA, where the diffuse reemission is calculated by the source function $\varepsilon_\lambda / \varkappa_\lambda$, and our Global Shell Model.

The grid covers a wide range of input parameters such as the temperature and luminosity of the nuclei and also the density distribution in the nebulae themselves.

In overall optically thin nebulae the On the Spot Approximation overestimates the number of reemitted H-Ly-photons with an energy close to 13.6 eV and thereby lowers the temperature in the whole nebula. Most emission line strengths turn out weaker than in the GSM.

For nebulae with moderate optical thickness the situation is different. In the innermost optically thin regions, the OSA again overestimates the diffuse radiation and lowers the temperature, emission lines turn out weaker than in the GSM. As the optical thickness increases with increasing radius, the temperature curves cross each other and the situation changes. The temperature curves cross at the same distance from the nucleus where the overestimation of the diffuse radiation becomes an underestimation in the OSA. The extent to which the diffuse radiation is underestimated depends very much on the density. It is larger for objects with a large density gradient. The resulting integrated emission line strengths depend critically on such temperature differences.

For real nebulae one has to check on a case by case basis whether a theoretical modelling without any radiation transfer treatment or with the 'On the Spot Approximation' are justified or whether the more elaborate treatment described here should be applied.

THE SPATIAL DISTRIBUTION OF LINE RATIO O III/O II IN HIGH EXCITATION
PLANETARY NEBULAE

A. Noriega-Crespo
Canadian Institute for Theoretical Astrophysics
M. McCall
DDO and University of Toronto

ABSTRACT. Previous attempts to model the integrated line ratio [O III] 5007 A/[O II] 3727 A for high excitation planetary nebulae (Che and Köppen 1983) have suggested that the charge exchange coefficient (k) for the reaction

$$O^{+2} + H^0 \rightarrow O^+ + H^+$$

could be 10 times smaller than the predicted theoretical value. The influence of other factors that may contribute to differences in the O III/ O II ratio, such as the ionizing spectrum and the nebular gas density distribution, seem to be relatively small in the high excitation nebulae.

We are undertaking a more detailed analysis of high and low excitation nebulae using "standard" ionization bounded models (Péquignot 1985), in particular to study the sensitivity of the spatial profile of [O III]/ [O II] as a function of the charge exchange rate. We report here some of our preliminary results on the high excitation cases.

Changes in k not only affect the ratio of the integrated spectra (as expected) but also the entire distribution of the emission accross the nebular gas. While the overall shape of the O III/O II spatial distribution for different k's is fairly similar near the edge of the nebula, at inner and intermediate radii the distribution changes enough that, in principle, observations could constrain the charge exchange rate to within a factor 3 or 4.

MAGNITUDES, SPECTRA, AND TEMPERATURES OF PLANETARY NUCLEI

James B. Kaler*
Department of Astronomy
University of Illinois

The purpose of this review is to examine the fundamental observational parameters of the central stars of planetary nebulae, namely their apparent magnitudes and gross spectral characteristics, and how these relate to the derivation or estimation of effective temperatures.

I. APPARENT MAGNITUDES

Probably the easiest parameter to measure for stars in general, the simple magnitude is one of the more elusive for planetary nuclei. The problem, of course, is the bright nebular background, which can in the extreme make the star impossible even to detect let alone analyze. My intention here is not to provide a listing of magnitudes — that is done in excellent fashion by Acker et al. (1982) and their supplements — but to explore and critique the various procedures used to derive these critical numbers.

The methods in use go all the way back to eye estimates from photographic plates. Remarkably, I (Kaler 1983) needed to use one of these (from Curtis 1918) only a few years ago. Fortunately, with all the recent activity in the subject, that era has mercifully ended. Nevertheless, there is yet a vast body of photographic <u>measurements</u> that are still eminently usable, starting with Hubble and van Maanen, up through the extensive work by Kohoutek and Abell: Perek and Kohoutek (1967) provide values and references. These, however, seem to suffer from a systematic trend: the nebular background apparently affects the sensitivity of the plate (in effect pre-flashes it) and makes the stars appear too bright by, crudely, $0^\text{m}.75$ (Shaw and Kaler 1985). With correction, and adoption of perhaps $\pm 0^\text{m}.4$ error (quite

*Supported by United States National Science Foundation grant AST 84-19355.

satisfactory for much work), these magnitudes are reliable to about m = 20.

The next obvious step is to employ photoelectric methods. There are a variety of interlocking ways to approach the photoelectric problem and the elimination of the nebular background, and we will examine them to evaluate the kinds of objects for which each is most suited. The simplest approach is just to choose nebulae of such low surface brightness that the background is inconsequential. Then, like Abell (1966) in his fundamental study of large nebulae, we can simply use standard methods. His UBV values are probably correct to within a few hundredths of a division. This procedure limits us to large old nebulae, and if there is any nebular contamination will cause the magnitudes to be underestimated.

Kostjakova et al. (1968) applied UBV photometry to brighter, more compact objects. They attempted to eliminate the nebular radiation, which for these broad filters consists largely of lines, by subtracting measurements made away from the star at several positions in the nebula proper. The problem here is that planetaries are highly irregular and the surface brightness at the star may not correspond to readings made elsewhere. In addition, as they point out themselves, the nebulae must be large enough to allow such measurements. Their determinations seem to be systematically too bright by about one-half magnitude (Shaw and Kaler 1985).

Shao and Liller (1972: see Liller and Shao 1968, Liller 1978, Acker et al. 1982) improved the methodology by employing small apertures and filters that avoided nebular lines, with subsequent transformation to the UBV system. However, the problem of the nebular continuum remained underappreciated. A planetary like NGC 7027 has a total continuum flux at V equivalent to an eleventh magnitude star (Kaler 1976a), so it is in fact a crucial consideration for small objects. The next logical step, therefore, is to calculate it on the basis of observed nebular parameters, with procedures developed by Webster (1969) and, more elaborately, by Kaler (1976b, 1978a) and Martin (1981), who used the detailed theoretical results of Brown and Mathews (1970). This technique culminated in extensive studies by Shaw and Kaler (1985, 1988). They extracted the stellar flux from the whole by using measured or adopted electron temperatures and densities and helium ion abundances, and attached realistic errors to these quantities and propagated them through the calcula-tions. Their work showed the Shao and Liller data to be reliable for nebulae larger than about 40 arc seconds in diameter; smaller objects again yielded under-estimates.

However, this technique has its limitations too. As the stars get fainter relative to the nebulae in which they are embedded the magnitudes become very susceptible to errors in the input parameters, and become unreliable. The problem is especially severe for small high density nebulae for which the hydrogen $2p/2s$ population ratio, critical to the calculation of the two quantum continuum, is essentially unknown. Eventually, the stars become lost in the continuum flux. Comparison with other magnitude derivations suggests that the Shaw and Kaler (1985, 1988) results are reasonably reliable to roughly V = 14.3 and 15 respectively.

A natural extension of this technique would employ spectra rather than filters. Ideally, the stellar spectrum and the appropriate nebular diagnostics for the subtraction of the nebular continuum would be observed simultaneously. Méndez, Kudritzki, and Simon (1985), for example, converted their spectra into B magnitudes, but without such correction since it was deemed to be small. As a variation on this theme, we can use IUE spectra, which have the advantage that because the stellar flux rises so steeply into the UV, the contrast between the star and the nebular continuum is notably enhanced (Heap 1983). We can either use the UV fluxes directly in our analyses or convert them to visual magnitudes via an assumption of flux distribution and a measurement of interstellar extinction. The method, however, is quite sensitive to errors in extinction, as we must calculate the nebular continuum from the Hβ flux, or extrapolate the stellar flux over an equally long wavelength baseline to the visual. In addition, if the nebula is larger than the IUE (or HST!) aperture we do not even know the convolved Hβ flux, and we may well be uncertain as to our assumed stellar model. The UV has been extensively employed by Kaler and Feibelman (1985) and by Heap and Augensen (1987). The former authors studied only large nebulae so that no correction for the nebula was needed, and they converted their UV fluxes into V magnitudes. The method is especially useful for unresolved nebulae with faint stars, and for planetaries whose nuclei are confused with nearby visual companions (e.g. K1-14); it is the only method that can be used for close binaries such as LoTr 5 (Feibelman and Kaler 1983).

These magnitudes are most commonly employed to calculate Zanstra temperatures and luminosities. The limitations on the methods discussed above conspire to produce a limit on the detectability of stars in compact nebulae, which in turn places a lower limit of roughly 125,000 K on the temperatures (Kaler 1986, Shaw and Kaler 1988), which is below the turnaround point on the log L-log T plane (where the nuclei begin their descents to the white dwarf zone) for cores of 0.6 $M_\odot$ and up (Paczynski 1971; Schönberner 1979, 1983). The highest mass cores, however, of the order of 1.4 $M_\odot$, are predicted to reach up to 10^6 K. In order to study the hotter variety of stars we must apply sophisticated imaging techniques, in hopes of detecting a faint point source against the continuum background. Reay et al. (1984) so determined the magnitudes of 8 stars by examining the nebulae in narrow line-free wavelength bands, and Walton et al. (1986) present results on 21. This type of work culminated in the detection of the elusive nucleus of NGC 2440 by Atherton, Reay, and Pottasch (1986), at a visual magnitude of 18.9 and a record temperature of 350,000 K.

This method too is not without its limitations and problems. The above authors showed it to be very sensitive to seeing. And Heap (1987), who detected the NGC 2440 nucleus at the extreme short-wave capability of the IUE, finds it to be a magnitude brighter, and consequently about 150,000 K cooler. Obviously the imaging procedure is susceptible to inhomogeneities in a nebula that could affect the appearance of the assumed stellar source. To avoid ambiguities we would have to map the object in the continuum and in the diagnostic lines, then compute a map of the true nebular continuum and subtract it from that observed to isolate the pure star: a process not yet attempted.

In summary, the magnitudes of the planetary nuclei are distinctly improving, but are still subject to serious uncertainties at the faint end, above about 15th magnitude. We need to approach the problem with the variety of techniques outlined above, adapted to the kind of nebula being observed. There are extremes for which the choice is clear. Faint stars in extended bright objects can be detected only by careful imaging. For compact, or distant unresolved nebulae, we must resort to continuum subtraction by calculation. Both methods are aided by working in the UV, which is a necessity for a nucleus unresolved from a binary companion.

II. SPECTRA

The planetary nuclei occupy an enormous portion — roughly one quarter — of the extended log L-log T plane (e.g. see Pottasch 1984 Fig. IX-2), and we see a concomitant large variety of spectral morphologies that range from pure emission through emission-absorption mixtures and near-continuous to pure strong absorption. Organization of the spectra is discussed in detail by Aller (1968, 1976, 1977), who classifies a large number of stars and presents many illustrative examples, by Smith and Aller (1969), and by Lutz (1978). We currently recognize Wolf-Rayet type spectra, O VI emission, Of, WR-Of, continuous (which do not really exist if we look closely enough), and absorption-O.

The appearance of the spectrum depends upon temperature, luminosity, and chemical composition, or more fundamentally, upon core mass and state of evolution (Heap 1982). The nuclei that are on the descending portions of their evolutionary tracks lack winds because of their lowered luminosities (Heap 1982, Kaler and Feibelman 1985) and therefore have absorption spectra. Méndez, Kudritzki, and Simon (1985) and Méndez et al. (1987, hereafter MKHHG), demonstrate that some show large composition anomalies that presage those found among the white dwarfs.

On the horizontal evolutionary tracks we find more variety, as winds can develop toward higher luminosity. MKHHG for example demonstrate that for log T > 4.45, He II $\lambda 4686$ passes from absorption to emission, producing an Of type star, at a core mass of about 0.7, or log L ≈ 4.1. Stars with powerful carbon emissions are arrayed from the lowest detectable temperatures (M4-18, below 25,000 K: Goodrich and Dahari 1985), where C II and C III dominate, to over 10^5 K (Smith and Aller 1969, Kaler and Shaw 1984) where C IV reigns and even C V is detected (Méndez and Niemela 1982).

These WC stars also exhibit a strong mixture of oxygen lines in their spectra, as well as those of silicon. O II is important at the cool limit, and in the 30,000 K range we find powerful O III and even the beginning of O V (Aller 1968). Oxygen really makes its presence felt above 80,000 K (Kaler and Shaw 1984) where O VI develops, creating the well-known O VI stars, classified by Méndez and Niemela (1982) as WC2-WC4. At the extreme these latter authors even identify strong O VII. The principal O VI feature is the $3s^2S-3p^2P$ doublet at λ3811-λ3834 Å. Wind speed is a strong function of effective temperature (Kaler, Mo, and Pottasch 1985), and at these high values these lines are usually blended into one with an effective wavelength of 3820 Å.

The critical luminosities needed for the development of the WC and O VI phenomena are unknown. Although we are able to derive Zanstra temperatures for them, the distances are too insecure to enable the absolute bolometric magnitudes to be found. The constant-mass (Shklovsky) distance method is not a fine enough discriminator to start with, the difficulty compounded by the optically thick natures of the nebulae with cool WC nuclei. And the appealing method developed in MKHHG does not work as these stars have no analyzable photospheric absorptions. Consequently we are ignorant about even fundamental matters concerning these remarkable windy stars.

Real progress will require considerable quantitative measurement of emission line fluxes with broad wavelength coverage. A strong start in this direction has been made by Aller (1977) and more recently by Aller and Keyes (1985), who treat a wide variety of objects. Extreme low and high excitation stars have been so examined by Goodrich and Dahari (1985) and Kaler and Shaw (1984) respectively. The winds that create these bizarre spectra are apparently quite significant in the evolution of planetary nuclei (Iben 1984), and it is important that we understand them better than we now do.

III. TEMPERATURES

The real controversies within the broad subject of stellar properties are reserved for temperatures. There are a variety of ways of approaching the problem and of deriving this critical parameter that are well-described in the literature: the classic Zanstra method, in which we compare the nebular recombination flux with the stellar magnitude (e.g. Harman and Seaton 1964); the energy-balance (Stoy) method, which avoids the star by comparing nebular forbidden and recombination line fluxes (Kaler 1976c, Preite-Martinez and Pottasch 1983); analysis or modeling of the nebular ionic distribution (Natta, Pottasch, and Preite-Martinez 1980; Harrington and Feibelman 1983); UV energy distribution (Pottasch et al. 1978, Lutz and Carnochan 1979; Harrington et al. 1982; Clegg and Seaton 1983; Kaler and Feibelman 1985; Grewing and Bianchi 1987); diameter correlations developed from the latter (Amnuel et al. 1985); and the modeling of stellar absorption lines (Mendez, Kudritzki, and Simon 1985; MKHHG). Extensive lists of temperatures are given in several of the references of this section, as well as by Pottasch (1984) and Khromov (1985).

The methods do not yield good agreement with one another, carrying on an argument that has existed for nearly 50 years. The fundamental problem is that blackbody Zanstra temperatures based on He II, which derive from the stellar spectrum shortward of $\lambda 228$Å, are frequently higher (and often much higher) than those based on H, which use the integrated spectrum shortward of $\lambda 912$Å. Zanstra (1961) believed this discrepancy to be caused by deviations from a blackbody, i.e. a UV excess shortward of $\lambda 228$; Minkowski (1942) and Wurm (1951) thought it due to optical depth, wherein the true effective temperature should be set equal to T_z(He II), the H value being a lower limit caused by the escape of Lyman continuum radiation. The latter view has generally prevailed in the construction of evolutionary diagrams (Seaton 1966, Kaler 1983). However, it is clear that model atmospheres (e.g. Hummer and Mihalas 1970; Henry and Shipman 1986, hereafter HS; see Harrington and Feibelman 1983, and Pottasch 1984) really do show severe departures from the blackbody, seriously compromising many extant studies.

MKHGG's recent derivation of stellar properties by line-profile fitting provides an independent way of checking the blackbody Zanstra (T_z) and Stoy (T_s) temperatures. The comparison, in Figure 1, shows T_z and T_s plotted against those found from the line fits. The T_z above 60,000 K (open symbols) are all derived from He II according to common practice. The classic Zanstra discrepancy shows clearly in that the T_z(He II) are systematically too high, consistent with HS's flux distributions for low He abundances, which are indeed found for these stars by MKHHG. In fact the average amount of the discrepancy is 19,000 K, nicely within the range of that anticipated by HS.

The T_z(H) for these hot stars (filled symbols) all fall below T(MKHHG), which can still be explained by optical depth effects; these nebulae either exhibit little in the way of low excitation ions, or they possess outer shells that demonstrate the leakage of ionizing radiation. General support for the optical depth explanation comes from the correlation between the T_z(He II)/T_z(H) ratios and the strengths of low excitation ions, which disappear when the ratios approaches high values (Kaler 1983). A particular argument against the lower MKHHG temperatures is NGC 7293 (the top point in Fig. 1), for which the H and He II Zanstra values agree.

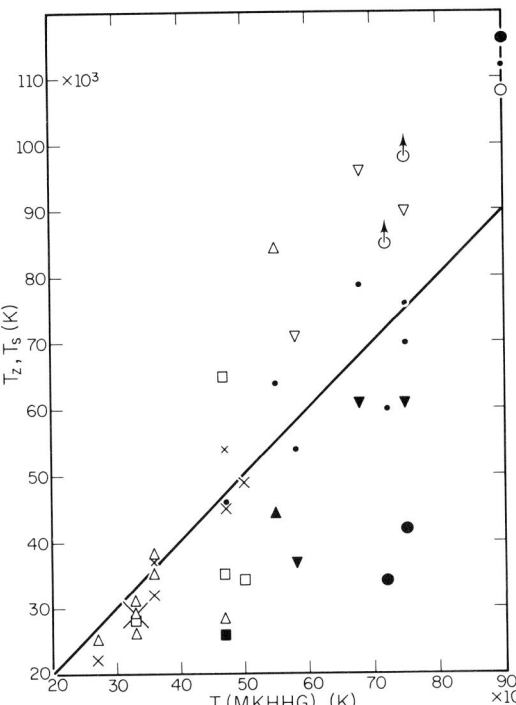

Fig. 1. T_z and T_s plotted against T(MKHHG). Above T_z = 60,000 K the open symbols represent He II temperatures; below it they indicate values derived from H for nebulae that do not exhibit He II. The filled symbols denote T_z(H) for stars for which there are T_z(He II) values. The small solid dots are the means of T_z(He II) and T_z(H). Sources are ⊙: Kaler (1983); □, △: Shaw and Kaler (1985, 1988); ▽: this paper (NGC 1535, 3242, 7009 with corrected photographic magnitudes). The X symbols represent Stoy temperatures: small: estimated from λ5007 and Kaler (1978b); middle size: computed values from Kaler (1976c,1978b); large: four points falling together.

Curiously, the average of T_z(He II) and T_z(H) (small dots) fits (probably fortuitously) with T(MKHHG) quite well. There may be no single answer to the Zanstra discrepancy: it may be caused by some combination of <u>both</u> effects, their relative importance depending upon such parameters as nebular size, and stellar temperature, luminosity, and composition.

Below 60,000 K, where we must use T_z(H) (which should now be correct since these nebulae ought to be optically thick), the Zanstra temperatures <u>still</u> fall systematically below the MKHHG values, a result <u>not</u> predicted by HS. Systematic magnitude errors (Section I) could play a role. The Stoy temperatures, however, actually agree rather well with MKHHG's results (although still a bit below them), so that we might be tempted to think that we can actually derive realistic values for these cooler stars.

A problem with the line profile temperatures arises from Kaler and Feibelman's (1985) finding that planetary nuclei can have Rayleigh-Jeans (infinite temperature) energy distributions in the accessible UV, far steeper than implied by the Zanstra temperatures. This problem has yet to be addressed: until the models used for stellar analysis can reproduce this odd characteristic (or the IUE observations are found to be flawed), we must be cautious about using the results derived from them.

The Kaler and Feibelman (1985) results have a more significant bearing on temperatures derived directly from UV energy distributions. It is hard to have confidence in any of the results so found when some of them are so obviously wrong. In order to interpret the slopes of the stellar continua that are flatter than Rayleigh-Jeans in terms of temperature, we must first be able to understand how the anomalously steep slopes can arise. Thus the 90,000 K temperature found for NGC 40 (whose Stoy temperature from Kaler 1976 is 32,000 K) by Grewing and Bianchi (1987) is probably too high, and like most of the values derived by Kaler and Feibelman (1985) should at least for now be considered an upper limit. The use of the continuum for cool WC stars is also compromised by the difficulty in locating it because of all the line features (Heap 1983; Kaler et al. 1988).

Finally, let us tie these three sections together by considering a temperature calibration of spectral class much as is traditionally done for the main sequence. Most of the planetary spectral classes (Of, WR-Of) are insufficiently described and subdivided, but the WC system given to us by Méndez and Niemela (1983) provides a good

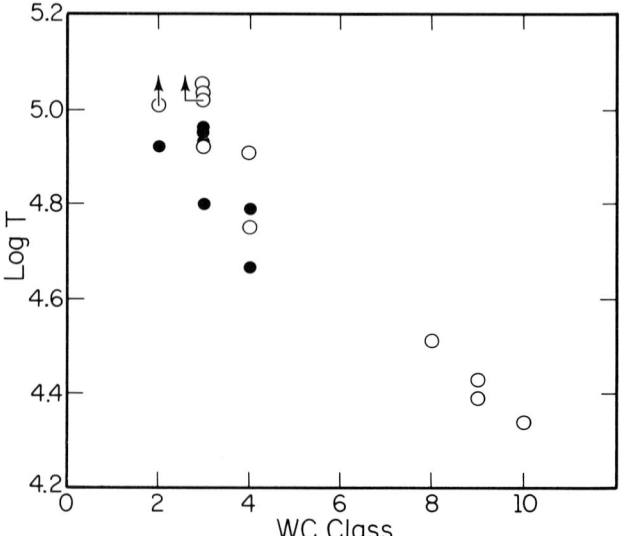

Fig. 2. Temperature calibration of the WC classification of Méndez and Niemela (1983). For early types (WC2-4) the open symbols represent T_z(He II) from Kaler and Shaw (1984) and Shaw and Kaler (1988), and the closed symbols represent equivalent MKHHG temperatures, T_z(He II) lowered by 19,000 K. Log T for NGC 5315 is an average of these with T_s derived from $\lambda 5007$. For the later types (WC8-10) T_s is used for NGC 40 and BD+30 (Kaler 1976c, 1978b) and M4-18 (Goodrich and Dahari 1985), and T_z(H) for He2-99 from Shaw and Kaler (1988).

opportunity. These are plotted in Figure 2 with T_z(He II) and the equivalent T(MKHHG) for the hot stars (WC2-4) and, with one exception, T_s for the cool ones. The result is quite similar to that previously obtained by Méndez et al. (1986). We see a nice linear relationship, which means that as the temperatures improve and we are able to resolve the discrepancies among the methods, we will be able to infer T_{eff} directly from spectral class. The next step is to subdivide the other kinds of stars properly. One disconcerting aspect of this figure is the curious gap between WC4 and WC8, which implies that the hot WC class may not be directly related to the cool one. That does not compromise their calibration, however.

In summary, the past few years have seen some significant advances that include improved magnitudes and the refinement of temperature methods. There remain important unsolved questions such as the origin of the Zanstra discrepancy and the relative effects of optical depth, deviation of stellar energy distributions from blackbodies, and the effect of model-errors on the derived temperatures. In all, however, progress has been impressive.

REFERENCES

Abell, G. O. 1966, Ap. J., **144**, 259.
Acker, A., Gleizes, F., Chopinet, M., Marcont, J., Ochsenbein, F., Roques, J. M. 1982, Catalogue of the Central Stars of True and Possible Planetary Nebulae, Publ. Spec. du CDS, No. 3.
Aller, L. H. 1968, in Planetary Nebulae, IAU Symp. No. 34, ed. D. E. Osterbrock and C. R. O'Dell (Dordrecht: Reidel), p. 339.
_____. 1976, Mém. Soc. Roy. Sci. Liège, Ser. 6, **9**, 271.
_____. 1977, J. Roy. Astron. Soc. Canada, **71**, 67.
Aller, L. H., and Keyes, C. D. 1985, Pub.A.S.P., **97**, 1142.
Amnuel, P. R., Guseinov, O. H., Novruzova, H. I., and Rustamov, Yu. S. 1985, Ap. Sp. Sci., **113**, 59.
Atherton, P. D., Reay, N. K., and Pottasch, S. R. 1986, Nature, **320**, 423.
Brown, R. L., and Mathews, W. G. 1970, Ap. J., **160**, 939.
Clegg, R. E. S., and Seaton, M. J. 1983, in Planetary Nebulae, IAU Symp. No.103, ed. D. R. Flower (Dordrecht: Reidel), p. 536.
Curtis, H. D. 1918, Lick Obs. Publ., **13**, 57.
Feibelman, W. A., and Kaler, J. B. 1983, Ap. J., **269**, 592.
Goodrich, R. W., and Dahari, O. 1985, Ap. J., **289**, 342.
Grewing, M., and Bianchi, L. 1987, Astr. Ap., in press.
Harrington, J. P., and Feibelman, W. A. 1983, Ap. J., **265**, 258.
Harrington, J. P., Seaton, M. J., Adams, S., and Lutz, J. H. 1982, M.N.R.A.S., **199**, 517.
Harman, R. J., and Seaton, M. J. 1964, Ap. J., **140**, 824.
Heap, S. R. 1982, in Wolf-Rayet Stars: Observations, Physics, and Evolution, IAU Symp. No. 99, ed. C. de Loore and A. Willis (Dordrecht: Reidel), p. 423.

———. 1983, in <u>Planetary Nebulae, IAU Symp. No. 103</u>, ed. D. R. Flower (Dordrecht: Reidel), p. 375.
———. 1987, <u>Nature</u>, **326**, 571.
Heap, S. R., and Augensen, H. J. 1987, <u>Ap. J.</u>, **313**, 268.
Henry, R. B. C., and Shipman, H. L. 1986, <u>Ap. J.</u>, **311**, 774 (HS).
Hummer, D. G., and Mihalas, D. 1970, in <u>Surface Fluxes for Model Atmospheres for the Central Stars of Planetary Nebulae</u> (JILA Rpt. No. 101).
Iben, I. Jr. 1984, <u>Ap. J.</u>, **377**, 233.
Kaler, J. B. 1976a, <u>Ap. Lett.</u>, **17**, 163.
———. 1976b, <u>Ap. J.</u>, **210**, 113.
———. 1976c, <u>Ap. J.</u>, **210**, 843.
———. 1978a, <u>Ap. J.</u>, **226**, 947.
———. 1978b, <u>Ap. J.</u>, **220**, 887.
———. 1983, <u>Ap. J.</u>, **271**, 188.
———. 1986, <u>Nature</u>, **320**, 394.
Kaler, J. B., and Feibelman, W. A. 1985, <u>Ap. J.</u>, **297**, 724.
Kaler, J. B., and Shaw, R. A. 1984, <u>Ap. J.</u>, **278**, 195.
Kaler, J. B., Shaw, R. A., Feibelman, W. A., and Lutz, J. H. 1988, in preparation.
Kaler, J. B., Mo, J.-E., and Pottasch, S. R. 1985, <u>Ap. J.</u>, **288**, 305.
Khromov, G. S. 1985, <u>Planetary Nebulae</u>.
Kostjakova, E. B., Savel'eva, M. V., Dokuchaeva, O. D., and Noskova, R. I. 1968, in <u>Planetary Nebulae, IAU Symp. No. 38</u>, ed. D. E. Osterbrock and C. R. O'Dell (Dordrecht: Reidel), p. 317.
Liller, W. 1978, in <u>Planetary Nebulae, IAU Symp. No. 76</u>, ed. Y. Terzian (Dordrecht: Reidel), p. 35.
Liller, W., and Shao, C.-Y. 1968, in <u>Planetary Nebulae, IAU Symp. No. 34</u>, ed. D. E. Osterbrock and C. R. O'Dell (Dordrecht: Reidel), p. 320.
Lutz, J. H. 1978, in <u>Planetary Nebulae, IAU Symp. No. 76</u>, ed. Y. Terzian (Dordrecht: Reidel), p. 185.
Lutz, J. H., and Carnochan, D. J. 1979, <u>M.N.R.A.S.</u>, **189**, 701.
Martin, W. 1981, <u>Astr. Ap.</u>, **98**, 328.
Méndez, R. H., Kudritzki, R. P., Herrero, A., Husfeld, D., and Groth, H. G. 1987, <u>Astr. Ap.</u>, in press (MKHHG).
Méndez, R. H., Miguel, C. H., Heber, U., and Kudritzki, R. P. 1986, <u>IAU Colloquium No. 87, Hydrogen Deficient Stars and Related Objects</u>, ed. K. Hunger, D. Schönberner, and N. Kameswara (Dordrecht: Reidel), p. 323.
Méndez, R. H., Kudritzki, R. P., and Simon, K. P. 1985, <u>Astr. Ap.</u>, **142**, 289.
Méndez, R. H., and Niemela, V. S. 1982, in <u>Wolf-Rayet Stars: Observations, Physics, Evolution, IAU Symp. 99</u>, ed. C. W. H. de Loore and A. J. Willis, (Dordrecht: Reidel), p. 457.
Minkowski, R. 1942, <u>Ap. J.</u>, **95**, 243.
Natta, A., Pottasch, S. R., and Preite-Martinez, A. 1980, <u>Astr. Ap.</u>, **84**, 284.
Paczynski, B. 1971, <u>Acta. Astr.</u>, **21**, 417.

Perek, L., and Kohoutek, L. 1967, Catalogue of Galactic
    Planetary Nebulae, Czechoslovak Acad. of Sci.
Pottasch, S. R. 1984, Planetary Nebulae, Ap. Sp. Sci. Lib., 107,
    (Dordrecht: Reidel).
Pottasch, S. R., Wesselius, P. R., Wu, C.-C., Fieten, H., and van
    Duinen, R. J. 1978, Astr. Ap., 62, 95.
Preite-Martinez, A., and Pottasch, S. R. 1983, Astr. Ap., 126, 31.
Reay, N. K., Pottasch, S. R., Atherton, P. D., and Taylor, K.
    1984, Astr. Ap., 137, 113.
Schönberner, D. 1979, Astr. Ap., 79, 108.
_____. 1983, private communication.
Seaton, M. J. 1966, M.N.R.A.S., 132, 113.
Shao, C.-C., and Liller, W. 1972, private communication.
Shaw, R. A., and Kaler, J. B. 1985, Ap. J., 295, 537.
_____. 1988, in preparation.
Smith, L. F., and Aller, L. H. 1969, Ap. J., 157, 1245.
Walton, N. A., Reay, N. K., Pottasch, S. R., and Atherton, P. D.
    1986, New Insights in Astrophysics, Proc. Joint NASA/ESA/SRC
    Conf., ESA SP-263.
Webster, B. L. 1969, M.N.R.A.S., 143, 113.
Wurm, K. 1951, Die Planetarischen Nebel (Berlin: Akademie-Verlag).
Zanstra, H. 1961, in Gaseous Nebulae and the Interstellar Medium,
    Univ. Mich. Ann Arbor, unpublished text.

Mariko Kato and Izumi Hachisu.

WIND FEATURES AND WIND VELOCITIES

Michael Grewing
Astronomisches Institut der Universität
Waldhäuserstrasse 64, D-7400 Tübingen, Germany, Fed.Rep.

ABSTRACT.

Fast winds have been detected in the spectra of many nuclei of planetary nebulae (PNNi). The wind velocities range from about 600 km/s to roughly 4000 km/s. While these winds add little to the mass of the nebular shells they may significantly effect their internal kinematics.
By studying the emission from the faint outer envelopes of PNe one can infer also the wind properties of the progenitors of the current nuclei. This will in the end allow to test quantitatively current models of the origin of PNe.

I. INTRODUCTION

The idea that stellar winds could play a significant role in planetary nebulae (PNe) arose apparently first from theoretical considerations in order to explain the shell-type structure of some of the objects. Mathews (1966) calculated the first models assuming that such winds would have velocities of several 10^2 km/s, and that the mass loss rates would be in in the order of $dM/dt \sim 10^{-5}$ M_o/yr. It was later shown that the morphology of PNe can also be explained by an initial pressure gradient (Sofia and Hunter 1968), and radiation pressure acting on dust particles (Ferch and Salpeter 1975). Still, the idea that stellar winds play a decisive role in the evolution of PNe has gained new momentum since such winds have actually been detected.

II. WINDS FROM PLANETARY NUCLEI

Among the first measurements carried out with the International Ultraviolet Explorer (IUE) were observations of planetary nebulae and their nuclei. Immediately, evidence was found for the presence of stellar winds : the UV lines of N V, O V, C IV, Si IV and others were seen to show the characteristic P Cygni profiles (Heap et al. 1978. A large number of such spectra has been compiled by Heap (1983). In Fig. 1 we show a few examples.

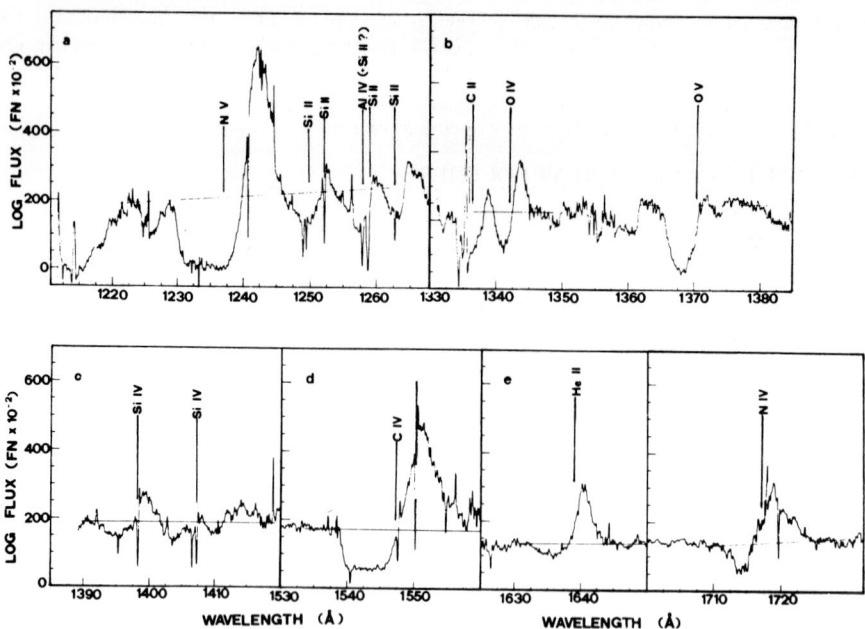

Fig. 1 : Evidence for a fast stellar wind in the high-resolution IUE spectrum of NGC 6543 (from Bianchi, Cerrato, and Grewing 1986).

While Fig.1 taken from Bianchi et al.(1986) is based on high-resolution (0.2 Å)spectra obtained with the IUE, most of the initial studies of winds in PNe relied on the much easier to obtain low-resolution spectra (6 Å). Perinotto (1983) and Cerruti-Sola and Perinotto (1985) collected such spectra for 60 PN nuclei (PNNi), 42 of which showed a measurable continuum. In 22 out of these cases P Cygni lines appear, giving clear evidence for the presence of a wind. Based on the location of these objects in the Hertzsprung-Russell diagram, Cerruti-Sola and Perinotto suggested that the occurence of a wind may actually be correlated with the temperature and radius of the stars. Indeed, by studying separately nuclei with WR, Of-Ofp, WR-Of, C-N, O-sdO, and continuum-type spectra, evidence for a wind was found in 100%, 90%, 100%, 70%, 40%, and 30% of the cases, respectively, where one must note, however, that the numbers quoted are highly uncertain due to the small numbers of objects available in each subclass.

In many cases, the IUE low-resolution spectra have been used to determine wind velocities. A list is given in the paper by Cerruti-Sola and Perinotto (loc.cit.). Already from these studies it is clear that the original conjecture of wind velocities of only a few 10^2 km/s was too pessimistic by almost an order of magnitude. This is substantiated by the much more precise measurements of wind terminal velocities from high-resolution spectra as performed by Heap (1986). She has analysed such 30 times better resolved spectra from 22 objects, and we have studied some additional PNNi at high resolution at Tübingen. Based on these results, we have computed the histogram of the frequency of occurence of certain wind velocities as shown in Fig. 2. The fact that this analysis is based on only 20 objects (because the spectra of some PNNi show no evidence for P Cygni lines) limits, of course, the statistical significance of the result. Still, I think, the important message is that the observed wind velocities range from about 600 km/s up to about 4000 km/s, i.e. they extend over a very large range.

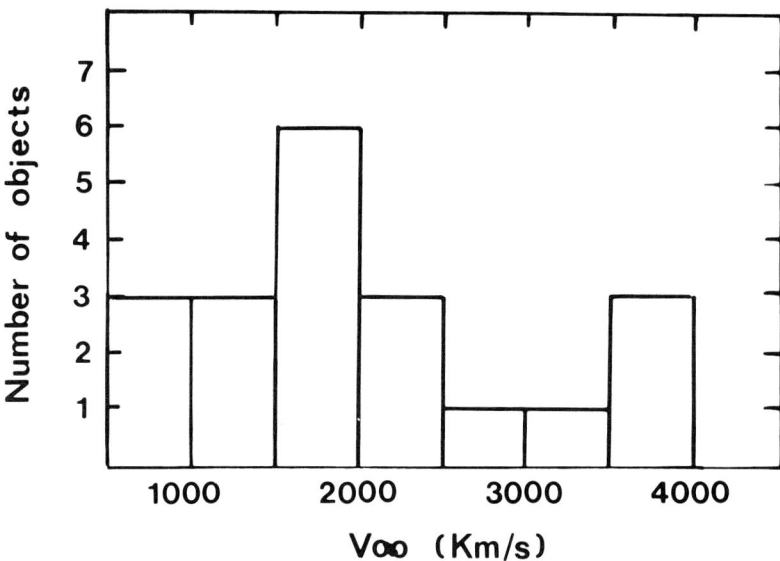

Fig. 2 : Histogram of the observed terminal wind velocities based on data from a total of 20 objects.

This result immediately leads to the question : what is driving the winds ? If we assume as a hypothesis that it is the radiation pressure from the central star, then we can use the relation (see Abbott 1982)

$$v_\infty = (\alpha/1-\alpha)^{1/2} \, v_{esc} = 774 \, [(M/R)(1-L/L_{Edd.})]^{1/2} \, [km/s].$$

Using $L = 4\pi R^2 \partial T_{eff}^4$ and $L = 6 \cdot 10^4 \, (M-0.50)$, with M, L and R in solar units, we obtain

$$v_\infty = 6.85 \cdot 10^{-3} \, T_{eff} \, [M(1-L/L_{Edd.})]^{1/2} \, [M-0.50]^{-1/4} \, [km/s].$$

Such a relation can be tested observationally as shown e.g. by Heap (1986). In Fig. 3a,b we have taken a similar approach by plotting the observed terminal velocities as a function of the PNNi's temperature and as a function of the excitation class of the corresponding PNe, respectively. In producing Fig. 3a we have made a special attempt to use the most accurate temperature determinations available rather than the Zanstra temperatures used in the original study by Heap (loc.cit.). However, irrespective of such detail, all these diagrams show a correlation as expected from the theory of radiation pressure driven winds. Furthermore, as pointed out by Heap, they show that the range over which the masses of the nuclei scatter seems to be very limited. The fit through the data points as given in Heap (1986) suggests a mean value of 0.60 M_o with very little scatter indeed.

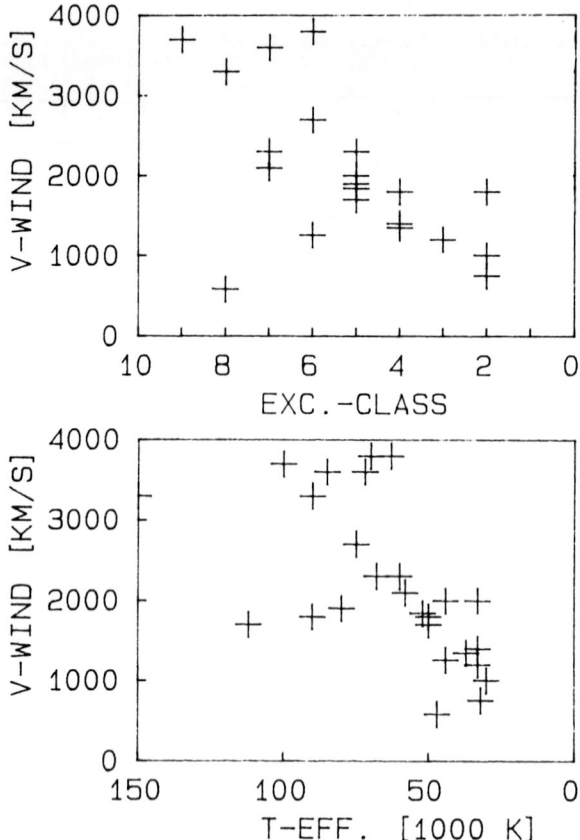

Fig. 3 : Observed terminal wind velocities plotted as a function of the temperature of the central stars (upper plot), and the excitation class of the surrounding nebulae (lower plot).

I want to point out in passing that the greatly improved atmosphere calculations carried out by Mendez et al. (1987) will probably give even better agreement between observations and theoretical predictions for radiatively driven winds.

Turning now to the physical state of these fast winds, i.e. their density, temperature, and ionization, it is somewhat surprising to see how little is known about these parameters. Obviously, such studies would need IUE high-resolution spectra, and those have until now been analysed in detail only for a very small number of objects. I would like to draw attention in particular to the work of Bombeck et al.(1986) in which the authors analyse the column densities of various wind ions (normalized with respect to N(O V)) and seem to find some weak correlation with the excitation class of the nebulae . This parameter is taken as a measure of the temperature of the central stars. Bombeck et al. find the winds to be more highly ionized for the high excitation class objects, suggesting photoionization as the main ionization mechanism in the winds. Instead of using column densities, we have collected from the literature total optical depth measurements from the P Cygni lines of several wind ions. In Fig. 4 these are plotted as a function of the excitation class of the surrounding nebulae, and we find the same kind of trend as suggested in the paper quoted, i.e. the degree of ionization seems to increase with increasing excitation class but we have far too few data to arrive at any final conclusion. Obviously, it would be extremely worthwhile to extend such studies to many more objects by analysing in a systematic way the high-resolution spectra available from the IUE archive.

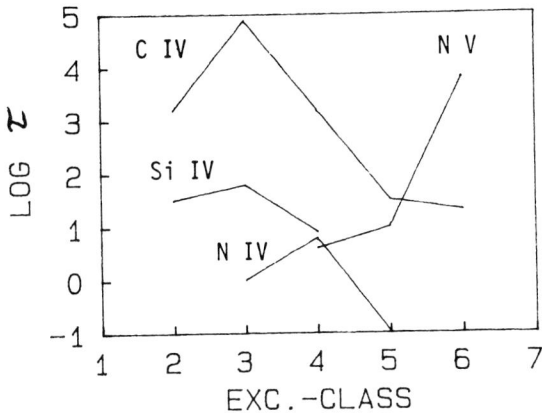

Fig. 4 : Measured total optical depths for several wind ions plotted as a function of the excitation class of the surrounding nebulae. The data pertain to NGC 6572, IC 418, IC 2149, and IC 3568.

Let me now briefly turn to the issue of how the presence of a fast wind from the central star may affect the surrounding nebular shells. I would first like to refer to the work of Adam and Köppen (1985) who computed models for the two objects NGC 4361 and NGC 1535. The interesting point of this comparative study is that the IUE spectra show clear evidence for a fast wind in the case of NGC 1535 but no such evidence for NGC 4361. The authors had chosen these particular objects because Mendez et al. (1981) had analysed the spectra of the central stars, thus providing a detailed model for the photon flux from the central objects. Using this as input in the calculations of the nebular ionisation, Adam and Köppen found consistent results for NGC 4361 but no consistency for NGC 1535. In the latter case there was a shortage of He^+ ionizing photons which they were able to overcome, however, by taking into account photons emitted from the fast wind which they assumed to be hot ($\log T_e = 5.5$).

Given the rapid improvement of self-consistent model atmosphere calculations that will in the future also include the extended wind spheres, we will probably soon know if the fast wind plasma is as hot as assumed by Adam and Köppen or if it is considerably cooler. The few observations at x-ray wavelengths that have been published so far seem to suggest that the winds may indeed not be very hot. While the ionizing effect of the fast winds remains an open issue for the the time being, there is accumulating evidence that the winds cause dramatic effects by dynamically interacting with the material that exists around the central stars from previous wind episodes. There is an increasing number of papers in the literature and also a large number of poster contributions at this conference which deal with peculiar outflow phenomena in PNe which occur at velocities of several hundred km/s, i.e. velocities which are an order of magnitude larger than typical nebular expansion velocities.

The Eskimo nebula NGC 2392 is one of the objects studied in this context. O'Dell and Ball (1985) were the first to report very high outflow velocities which they found after carefully subtracting the spectrum of the central star. Gieseking et al. (1985) found emission from material moving at about 200 km/s at a position angle of roughly 70°. In the poster contribution by Phillips et al.(1987), this result is basically confirmed but their interpretation differs somewhat from that of Gieseking et al..

Since these early findings, Gieseking and Solf (1986) reported similar results for NGC 6751. In Table 1 I have tried to summarize the current situation, making heavy use of a recent compilation of kinematical data by Weinberger (1987).

Whereas the fast flow phenomena seem to occur in places where the fast inner wind is breaking through the nebular shells and carrying away individual pockets of gas, recently some very interesting results have been reported which may relate to cases where the fast winds get stopped and conpacted at the inner edge of the nebular shells. In analysing the IUE high-resolution spectrum of NGC 40, Bianchi and Grewing (1987) saw narrow C II absorption which was clearly not of interstellar origin and which they suggested could arise at the interface between the fast wind coming from the central star and the nebula proper. This argument was based both on the velocity and of the column density of the absorbing material. Still much more intriguing are the findings reported recently by Kaler et al. (1987) for Abell 78. These authors see a number of narrow absorption lines superimposed on the broad P Cygni profiles and the continuum. They can clearly be distinguished from the interstellar lines which are also present, and they may -as the authors suggest- originate at the interface of the fast wind with the nebula.

While I have so far addressed winds ejected by the current nuclei, I would now like to briefly turn to winds from their progenitors. I think that there is a general consensus that the extended halos found in about 2/3 of all PNe searched for faint outer shells (see e.g. Chu et al. 1987) must have originated from the AGB winds of the progenitors.

As the halo gas has been decelerated to velocities of less than 10 km/s, high spectral resolution is required to study the kinematics. In view of the low surface brightness this is a very demanding requirement that has so far limited the studies to only a few objects. Fig. 5 shows the results of Hippelein et al. (1985) who have studied the three objects NGC 6543, NGC 6826, and NGC 7662, finding velocities which are indeed consistent with the AGB wind origin of the material.

Table 1

Planetary Nebulae with Large Internal Motions

PK	Name	References
10+18.2	M 2-9	AS72, Wa81, CS83, Ca83, PMR87
10+ 0.1	NGC 6537	BeS83, PMR87
36- 1.1	Sh 2-71	PMR87
36-57.1	NGC 7293	MeW80, WM87
81-14.1	A 78	MPM87, KFSH87
83+12.1	NGC 6826	BIP87
86- 8.1	Hu 1-2	PMR87
89+ 0.1	NGC 7026	PMR87
96+29.1	NGC 6543	Mu68, BIP87
106-17.1	NGC 7662	BIP87
107+ 2.1	NGC 7354	BIP87
130-11.1	M 1-1	ST87
197+17.1	NGC 2392	RAT83, ODB85, GBS85, BIP87, PMR87
208+33.1	A 30	JCh87
234+ 2.1	NGC 2440	KA74, WH87
261+32.1	NGC 3242	BIP87
277- 3.1	NGC 2899	LFRR87
315- 0.1	He 2-111	We78
331- 1.1	Mz 3	LM83, WH87
349+ 1.1	NGC 6302	MJ67, MW80, DBABT83, WH87
359- 0.1	Hb 5	PMR87

AS72	Allen, D.A., and Swings, J.P.: 1972, Astrophys.J. 174, 583
BeS83	Becker, I., and Solf, J.: 1983, Mitt.Astron.Ges. 60, 319
BIP87	Balick, B., Icke, V., Preston, H.: 1987, IAU Symp. 131, A23
Ca83	Carsenty, U.: 1983, Thesis, University of Heidelberg
CS83	Carsenty, U., and Solf, J.: 1983, IAU Symp. 103, p. 510
DBABT83	Danziger, I.J., Baade, D., Atherton, P.D.: 1983, IAU Symp. 103, p. 509
GBS85	Gieseking, F., Becker, I., and Solf, J.:1985, Astrophys. J. 295, L17
JCh87	Jacoby, G.H., and Chu, Y.-H.: 1987, IAU Symp. 131, A28
KA74	Kaler, J.B., and Aller, L.H.: 1974, PASP 86, 635
KFSH87	Kaler, J.B., Feibelman, W.A., Shaw, R.A., and Henrichs, H.: 1987, IAU Symp. 131, D13
LFRR87	Lopez, J.A., Falcon, L.H., Ruiz, M.T., and Roth, M.: 1987, IAU Symp. 131, A27
LM83	Lopez, J.A., and Meaburn, J.: 1983, MNRAS 204, 203
MeW80	Meaburn, J., and Walsh, J.R.: 1980, Astrophys.Lett. 21, 53
MJ67	Minkowski, R., and Johnson, H.M.: 1967, Astrophys.J. 148, 659
MPM87	Manchado, A., Pottasch, S.R., and Mampaso, A.: 1987, IAU Symp. 131, A29
Mu68	Muench, G.: IAU Symp. 34, p. 259
MW80	Meaburn, J., and Walsh, J.R.: 1980, MNRAS 193, 631
ODB85	O'Dell, C.R., and Ball, M.E.: 1985, Astrophys.J. 289, 526
PMR87	Phillips, J.P., Mampaso, A., and Riera, A.: 1987, IAU Symp. 131, B32
RAT83	Reay, N.K., Atherton, P.D., Taylor, K.: 1983, MNRAS 203, 1087
ST87	Shibata, K., and Tamura, S.: 1987, IAU Symp. 131, B3
Wa81	Walsh, J.R.: 1981, MNRAS 194, 903
We78	Webster, B.L.: 1978, MNRAS 185, 45p
WH87	Weller, W.G., and Heathcote, S.R.: 1987, IAU Symp. 131, A25
WM87	Walsh, J.R., and Meaburn, J.: MNRAS 224, 885

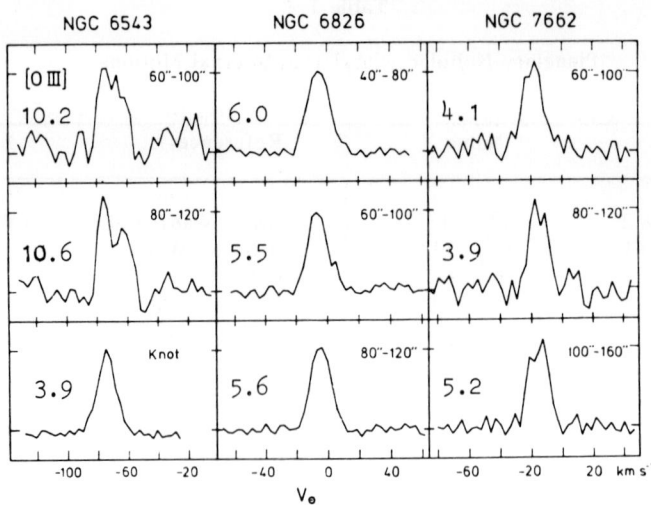

Fig. 5 : [O III] emission line profiles from the halo of NGC 6543, NGC 6826, and NGC 7662, respectively. The numbers in each plot give the size of the annular diaphragms used to observe the halo emission. Still, a correction had to be made for straylight from the bright nebular shells (see Hippelein, Baessgen and Grewing 1985 for details). Also, in each plot the expansion velocity of the halo gas is given in km/s.

Since then we have obtained a number of absolutely calibrated optical spectra which reveal the ionization structure of the nebulae, and, more importantly in the present context, yield densities in the outer wind spheres as a function of radius. This allows several tests of the AGB wind model. Firstly, we can check the r^{-2} dependence which is expected for a wind expanding at constant velocity v. Secondly, we can extrapolate the halo densities inward to those radii where the bright nebular shells are located and determine the ratio of the current nebular density to the extrapolated halo density. This should be a measure of the compression factor if we adopt e.g. the two-wind model of Kwok et al. (1978). It is interesting to note that except for two objects, the density ratio significantly exceeds the adiabatic ratio, suggesting instead isothermal shocks.

Finally, we can derive the ratio (dM/dt)/v. Results have been given by Bässgen et al.(1987) and are repeated in Table 2.

To derive dM/dt, we would need to know the expansion velocity v but the resolution of our spectra is not sufficient to actually measure v. Still, a reasonabe estimate of the mass loss rate of the progenitors can be obtained by assuming v<10 km/s. Typical values thus derived are in the range $2\ 10^{-6}$ to $2\ 10^{-5}$ M_o/yr, values that are not unreasonable for AGB stars. Still, they seem to be lower than the values found recently by Taylor et al. (1987) who analysed the radio continuum spectra of compact PNe. By fitting the observations with theoretical spectra derived from free-free emission models for photon-limited, ionized shells corresponding to stellar-wind type envelopes, Taylor et al. come to the conclusion that for the objects of their sample (dM/dt)$\geqslant 10^{-5}$ M_o/yr.

Table 2

Halo density as a function of angular radius

Object	adopted distance	Pos.	density cm^{-3}	corresp.M/V (Mo/y)(km/s)
NGC 3132	670 pc	30"	190	7.0
		54"	40	4.5
NGC 1535	1500 pc	31"	16	3.0
		40"	11	3.5
IC 418	1400 pc	15"	200	8.0
		29"	40	6.0
NGC 2452	4000 pc	11"	100	17.0
		17"	30	11.0
IC 2448	2900 pc	17"	11	2.5
		29"	3:	2.0:
NGC 2867	1600 pc	17"	25	1.8
		34"	6	1.8
IC 2165	2400 pc	10"	66	3.5
		17"	23	3.3
		27"	9	4.0

3. CONCLUSIONS

Stellar winds from the nuclei of planetary nebulae are now a well established phenomenon that occurs in a large fraction of the objects. The diagnostically most valuable information about such winds comes from high-resolution UV spectra obtained with the IUE satellite. The measured wind velocities range from about 600 km/s to more than 4000 km/s. The can be accounted for by models of radiation pressure driven winds.

The mass carried away by these winds is far too little to significantly add to the mass of the nebular shells. Still, the fast winds could have an ionizing effect, and they almost certainly cause dramatic kinematical effects as they interact with the ambient medium. They may indeed be responsible for the peculiar motions at several 10^2 km/s observed now in a number of objects, often in the form of bipolar flows.

From detailed studies of the gas in the faint outer shells of PNe, a very common phenomenon as recent studies with modern electronic detectors have shown, it seems now also possible to determine the properties of the AGB winds from the progenitors of the current nuclei thereby allowing to test models for the origin of planetary nebulae. While still at an early stage, these studies may in the end enable us to determine quantitatively the role that winds play during the various phases in the late stages of stellar evolution to accumulate the amount of material that we now find present in the different nebular shells.

Acknowledgement: I would like to thank M.Bässgen and S.Cerrato for their help in preparing the final version of this manuscript.

REFERENCES

Adam, J., and Köppen, J.: 1985, Astron.Astrophys. 142, 461
Abbott, D.C.: 1982, Astrophys.J. 259, 282
Bässgen, M., Bässgen, G., Barnstedt, J., Grewing, M., Bianchi,L.
Mitt. Astron.Ges. 67,342
Bianchi, L., Cerrato, S., and Grewing, M.: 1986, Astron. Astrophys. 169, 227
Bianchi, L., and Grewing, M.: 1987, Astron.Astrophys. 181,85
Bombeck, G., Köppen, J., and Bastian, U.: 1986, in 'New Insights in Astrophysics',
ESA SP-263, p.287
Cerruti-Sola, M., and Perinotto, M.: 1985, Astrophys.J. 291, 237
Chu, Y.-H., Jacoby, G.H., and Arendt, R.: 1987, Astrophys.J.Suppl.Ser. 64, 529
Ferch, R.L., and Salpeter, E.E.: 1975, Astrophys.J. 202, 195
Gieseking, F., Becker,I., and Solf, J.: 1986, Astrophys.J. 295, L17
Gieseking, F., and Solf, J.: 1986, Astron.Astrophys. 163, 174
Heap, S.R., Bogess, A., Holm, A., Klinglesmith, D.A., Sparks, W., West, D., Wu, C.C., Boksenberg, A., Willis, A., Wilson, R., Macchetto, F., Selvelli, P.L., Stickland, D., Greenstein, J.L., Hutchings, J.B., Underhill, A.B., Viotti, R., and Whelan, J.A.J.: 1978, Nature 275, 385
Heap, S.: 1983, IAU Symp. 103, p. 375
Heap, S.: 1986, in 'New Insights in Astrophysics', ESA SP-263, p. 291
Hippelein, H.H., Bässgen, M., and Grewing, M.: 1985, Astron.Astrophys. 152, 213
Kaler, J.B., Feibelman, W.A., and Henrichs, H.F.: 1987, IAP 87-15 (Preprint)
Kwok, S., Fitzgerald, P.M., and Purton, C.R.: 1978, Astrophys.J. 219, L125
Mathews, W.G.: 1966, Astrophys.J. 143, 173
Mendez , R.H. , Kudritzki, R.P., Gruschinkske, J., Simon, K.P.: 1981,
Astron. Astrophys. 101, 323
Mendez , R.H., Kudritzki, R.P., Herrero, A., Husfeld, D., Groth, H.G.: 1987,
Astron. Astrophys. (in press)
O'Dell, C.R., and Ball, M.E.: 1985, Astrophys.J. 289, 526
Perinotto, M.:1983, IAU Symp. 103, 343
Phillips, J.P., Mampaso, A., Riera, A.: 1987,IAU Symp. 131, B32
Sofia, S., and Hunter, J.H.: 1968, Astrophys.J. 152, 405
Taylor, A.R., Pottasch, S.R., and Zhang, C.Y.: 1987, Astron.Astrophys. 171,178
Weinberger, R.: 1987, Preprint

CLOSE-BINARY AND PULSATING CENTRAL STARS

Howard E. Bond
Space Telescope Science Institute
3700 San Martin Drive
Baltimore, Maryland 21218 USA

ABSTRACT. As a result of photometric-monitoring studies, 7 planetary-nebula nuclei are now known to be binaries with orbital periods less than one day. These systems were probably produced via a common-envelope interaction, during which a wide pair was converted to a close binary surrounded by an ejected red-giant envelope. The frequency of occurrence of such close binaries among PNNs is about 10–15%, showing that binary-star interactions are a significant production mechanism for planetary nebulae. The descendants of close-binary PNNs are probably the cataclysmic variables. Two CVs surrounded by nebulae resembling old planetaries, 0623+71 and GK Per, may provide the most direct evidence for the origin of CVs through PN ejection. The observed birth rate for close-binary PNNs is more than an order of magnitude higher than for CVs, possibly indicating that our census of the CV population is very incomplete. The nucleus of K 1-16 is a member of the GW Vir class of extremely hot pulsating pre-white dwarfs, and the only one known to be surrounded by a PN. These objects offer exciting opportunities for direct measurement of evolutionary timescales and for seismological investigations of the interiors of PNNs and their immediate descendants.

1. LIGHT VARIABILITY AMONG PNNs

This paper will review recent work on two classes of variable planetary-nebula nuclei (PNNs): (1) close binaries, in which the observed brightness of the system varies because of heating effects or actual eclipses; and (2) pulsating variables, in which a pulsational instability modulates the luminosity of the (single) central star. Over the past several years, several close-binary PNNs and one pulsating PNN have been discovered. Both classes of objects potentially offer a wealth of new information on the origin of planetary nebulae and on the evolution of their central stars.

2. SEARCHES FOR CLOSE-BINARY PNNs

The first close-binary central star, that of Abell 63, was actually discovered more than half a century ago by Hoffleit (1932). This suspected eclipsing binary received the variable-star designation UU Sge, but its faint surrounding PN was only discovered much later by Abell (1966); the coincidence of the objects was not noticed until another decade had passed (Bond 1976). Photoelectric photometry (Miller, Krzeminski, and Priedhorsky 1976; Bond, Liller, and Mannery 1978) established a remarkably short orbital period of $11^h 10^m$, showed that the eclipses are total, and demonstrated a strong "reflection effect" (heating of one hemisphere of the dK companion due to irradiation from the hot primary).

The existence of close-binary PNNs was predicted by Paczynski (1976), who proposed a scenario for the origin of compact binaries (such as the cataclysmic variables; see also Ritter 1976). The scenario involves an engulfed main-sequence companion spiralling down inside a red-giant envelope and ultimately spinning the common envelope up to breakup. This interaction would produce a close binary (the red-giant core plus the main-sequence secondary), lying inside a nebular shell that would be ionized by UV radiation from the hot core. The subsequent evolution of such a system would involve dissipation of the nebula, contraction of the hot core to white-dwarf dimensions, and, ultimately, contraction of the orbit and initiation of mass transfer due to magnetic braking (Verbunt and Zwaan 1981) or other mechanisms for loss of angular momentum.

The relative frequency of occurrence of binary vs. single PNNs would indicate the relative importance of binary-star vs. single-star PN ejection mechanisms. Moreover, the discovery of additional close-binary PNNs would provide unique opportunities for determination of fundamental properties of the central stars.

This review will concentrate on binary PNNs with $P < 1$ day; these objects are the easiest to discover during photometric runs taken during one night, and it is these objects that are the probable progenitors of the cataclysmic variables (CVs). An accompanying paper in these proceedings by Méndez covers additional aspects of PNN binarity. At present, seven PNNs are known to show periodic photometric variations due to orbital motion in binaries with $P < 1$ day. These objects are listed in Table I.

The table indicates whether the binaries show actual eclipses (UU Sge, V477 Lyr), or are variable only because of the reflection effect on the secondary stars. (Actually, our photometry of the southern object VW Pyx is not yet sufficient to rule out eclipses.) Six of the stars were described by Bond and Grauer (1987a), so few details will be given here; the references to Table I may be consulted for further information.

A seventh object, the central star of Abell 35, is included in Table I. Short-term, low-amplitude variability was first discovered by Bond and Grauer (see Bond 1985). This variability has been found to be periodic by Jasniewicz (1987). His data are not yet sufficient to distinguish between the two possible periods listed in the table, but it seems plausible that the variations are due to a heated hemisphere on the cooler star in the system. The nucleus of Abell 35 is the only binary in Table I whose optical spectrum is dominated by the late-type component, which was classified G8 III-IV by Jacoby (1981). *IUE* observations, however, have revealed the extremely hot companion that is responsible for ionizing the nebula (Jacoby, private communication).

Space limitations do not permit detailed discussion here of several PNNs that are known to be longer-period binaries. Two remarkable examples are (1) V651 Mon, the central star of NGC 2346, which is a 16-day spectroscopic binary with a highly variable light curve (Méndez and Niemela 1981; Jasniewicz and Acker 1986), and (2) IN Com (HD 112313), the central star of LT-5. The

TABLE I: Photometrically Confirmed Binary PNNs ($P < 1^d$)

Nebula	Central Star	Period	Type	Ref.
Abell 41....	MT Ser	2^h43^m	Reflection	(1)
DS 1.......	KV Vel	8^h34^m	Reflection	(2)
Abell 35....	$-22°3467$	10^h23^m or 18^h23^m	Reflection?	(3)
Abell 63....	UU Sge	11^h10^m	Eclipsing	(4)
Abell 46....	V477 Lyr	11^h19^m	Eclipsing	(5)
HFG 1.....		13^h57^m	Reflection	(6)
K 1-2......	VW Pyx	16^h05^m	Reflection	(7)

REFERENCES—(1) Grauer and Bond 1983; (2) Drilling 1985; (3) Jasniewicz 1987; (4) Bond, Liller, and Mannery 1978; (5) Bond 1985; (6) Grauer, Bond, Ciardullo, and Fleming 1987a; (7) Bond and Grauer 1987a.

optical spectrum of IN Com shows only a G-type star with rotationally broadened absorption lines and Hα emission (Bond 1985). Jasniewicz, Duquennoy, and Acker (1987) have found that the G star is itself a 2-day, double-lined spectroscopic pair, while a third body (presumably the hot component detected with *IUE* by Feibelman and Kaler 1983) orbits with a period suspected to be 41, 140, or 539 days.

Several additional PNNs have been reported to be short-period spectroscopic binaries by various authors (see references in Bond 1985 and Ritter 1986). Again, space does not permit a detailed discussion, but confirmation of these results, either through photometry or from additional spectroscopic observations, would be very important.

3. LIGHT-CURVE ANALYSES

Close-binary PNNs will provide astrophysical information about the component stars through analyses of their light curves and radial velocities. The most thorough analysis so far has been carried out by Drilling (1985) for KV Vel (LSS 2018), the central star of DS 1. The primary star was found to have $T_{\text{eff}} = 77,000$ K; the radii of the primary and secondary star are 0.2–0.3 and 0.3–0.4 $R_\odot$, respectively, and the masses are 0.4–0.7 and 0.2–0.3 $M_\odot$. These results are completely consistent with a primary star that is in a pre–white-dwarf stage, and a lower-main-sequence secondary.

J. Kaluzny (in preparation) has analyzed the light curve of V477 Lyr, the eclipsing nucleus of Abell 46. The primary star is required to have $T_{\text{eff}} = 60,000$ K to produce the observed reflection effect, and the radii of the primary and secondary are 0.2 and 0.6 $R_\odot$. Except for an apparently somewhat more massive main-sequence secondary star, V477 Lyr appears to be quite similar to

KV Vel. Both systems will become CVs when mass transfer from the secondary to the primary begins (several Gyr in the future).

4. INCIDENCE OF CLOSE BINARIES AMONG PNNs

Sufficient data are now available to permit an estimate of the frequency of occurrence of close-binary PNNs. Bond and Grauer (1987a) have listed 32 PNNs that did not show photometric variability when continuously monitored over intervals of several hours, and hence are unlikely to be close binaries. (The conspicuous heating effects described above imply that a close-binary PNN will show variations even if insufficiently inclined to show actual eclipses.) In a separate survey, Drummond (1980) monitored 17 PNNs (of which 6 are in common with the Bond-Grauer objects), and similarly found no short-term variability. Finally, Drilling (1985) mentions that KV Vel was the one variable found out of 3 objects that were monitored.

The observed incidence of close binaries among PNNs that have been monitored with photoelectric photometry is thus 7 out of 52, or 14%. Drummond's candidates emphasized the brighter PNs, while the Bond-Grauer objects have tended to be nuclei of low-surface-brightness PNs. Drilling observed 3 bright sdO stars surrounded by rather faint PNs. Therefore, the combined sample should be quite representative of the general population of PNNs.

The Bond-Grauer survey initially had a pre-selection for reported variability; for example, the nuclei of Abell 41, 46, and 63 were all suspected of variability by Abell (1966), and the variability of K 1-2 was discovered on overlapping Palomar Sky Survey prints by Kohoutek (1964). However, the list of non-variables is now large enough to begin to approach a magnitude-limited sample, and it should be noted that the remaining three binaries in Table I were discovered without any premonition of variability. The best estimate that can be made now is that *about 10-15% of PNNs are binaries with P <1 day*. The binary fraction is, of course, even higher if longer-period systems, such as those mentioned above, are included, but these statistics are still very incomplete. Nevertheless, it is clear that binary-star processes are a significant production mechanism for planetary nebulae.

5. BIRTH RATES

Since we have argued that close-binary PNNs are the progenitors of CVs, it is important to ask whether the birth rates of the two types of objects are comparable.

The birth rate of PNs in the solar neighborhood has been the subject of a number of studies, of which a few recent ones are listed in Table II. The estimates are uncertain, principally because of lingering problems with the PN distance scale, and cover the range $1-8 \times 10^{-3}$ kpc^{-3} yr^{-1}. The largest birth rate was obtained by Ishida and Weinberger (1987), who included several recently discovered, nearby, low-surface-brightness nebulae, and argue that the census of PNs is still incomplete even in the solar neighborhood.

Since it was concluded above that 10–15% of PNNs are close binaries, it follows that the birth rate for close-binary PNNs is $0.1-1.2 \times 10^{-3}$ kpc^{-3} yr^{-1}. This may be compared with Patterson's (1984) estimated birth rate for CVs of all types of 0.007×10^{-3} kpc^{-3} yr^{-1}. This is likewise a rather uncertain estimate

TABLE II: Local PN Birth Rate

Reference	Birth Rate (10^{-3} kpc^{-3} yr^{-1})
Cahn & Wyatt 1976	5
Maciel 1981	2
Daub 1982	5
Mallik 1982	2.4
Amnuel et al. 1984	4.6
Pottasch 1984	1.5
Drilling & Schönberner 1985	1
Ishida & Weinberger 1987	8

because of the uncertain lifetimes of CVs, but the discrepancy of a factor of 15 or more between the close-binary PNN and CV birth rates is still uncomfortably large. However, a post-PNN origin for CVs could still be supported if one or both of the following were true:

1. Most of the immediate descendants of binary PNNs are accumulating as inconspicuous detached binaries that, within the Hubble time, have not yet initiated the mass transfer required to produce cataclysmic activity. Several of these "pre-cataclysmic" (or V471 Tau-type) binaries are known (see Bond 1985; Ritter 1986; and references therein). As discussed by Bond (1985), the majority of them have orbital periods so long that, in fact, several Gyr will be required for magnetic braking (Verbunt and Zwaan 1981) to bring the secondary stars into contact with their Roche lobes. Thus the active CVs observed at the present epoch could represent only the short-period tail of a considerably larger group of post-PNN binaries.

2. Surveys for CVs may still be quite incomplete, so that Patterson's birth rates could be seriously underestimated. Most of the nova-like CVs, for example, are not conspicuous objects; a ninth-magnitude nova-like variable was found just 3 years ago (Garrison et al. 1984). Moreover, it has recently been argued that classical novae spend large parts of their lives in very inconspicuous states of "hibernation" (Shara et al. 1986).

We conclude that the observed birth rates for close-binary PNNs and CVs are in sufficient agreement, given the uncertainties in both rates. It may be noted that Fleming, Liebert, and Green (1986) find a local white-dwarf birth rate of 0.5–0.75×10^{-3} kpc^{-3}yr^{-1}, a factor of 2–10 below the PN birth rate. They suggest that the census of white dwarfs in the solar neighborhood may be incomplete, because many white dwarfs are concealed in binary systems with main-sequence primaries. Such systems could include not only the pre-cataclysmic descendants of close-binary PNNs, but also much wider systems (like Sirius or Procyon).

6. CATACLYSMIC PNNs?

The discussion above presents a strong, but circumstantial, argument that CVs are descended from close-binary PNNs. The most direct evidence in support of this scenario would be the existence of PNs whose central stars are already mass-transferring CVs. This might be thought rather unlikely because of the fine tuning required to eject the common envelope at a time after the secondary's Roche lobe has shrunk down to the size of the secondary, but before actual coalescence of the double core. Astonishingly, however, recent work indicates that two such PNs with cataclysmic nuclei may actually exist:

1. The object 0623+71 is a faint "bow-shock" nebula surrounding a 12th-mag nova-like CV (Ellis, Grayson, and Bond 1984). We suggested that the nebula is an ejected shell resulting from an unrecorded nova outburst. Recently, however, Krautter, Klaas, and Radons (1987) have shown that the nebula does not have the large expansion velocity expected for a nova shell (V_{exp} < 100–150 km s^{-1}), and have suggested that the shell is in fact a PN.

2. *IRAS* images of the region of the classical nova GK Per have recently been examined by Bode *et al.* (1987). GK Per was found to be surrounded by a large (radius ~0.7 pc) infrared dust shell. The surface brightness is highest in the 100μ band, and a grain temperature of 22 K was derived. The H I mass was also derived from 21-cm observations, and was found to be ~0.6$M_\odot$, which is far too large to attribute to material ejected in prehistoric nova outbursts. The authors therefore argue that the shell is in fact an extremely old planetary nebula.

It may also be worth mentioning a remarkable series of visual observations of the nucleus of NGC 7662 made by Barnard (1908), who claimed variability over a range of more than 3 magnitudes. The behavior reported by Barnard (usually faint, occasionally bright) is reminiscent of the outbursts of dwarf novae. Modern CCD photometry (which would allow accurate subtraction of the nebular background) would be of interest.

7. MORPHOLOGIES OF PNs WITH BINARY NUCLEI

The morphologies of PNs ejected from close binaries through common-envelope interactions will be the subject of a forthcoming study by M. Livio and the writer. Theoretical considerations (*cf.* Livio and Soker 1987) suggest that the initial shape of the PN will depend primarily upon the evolutionary stage of the primary star at the time it encounters its companion:

1. If the interaction occurs before the primary attains a very centrally condensed configuration, the timescale for orbital decay will drop to a value comparable to or less than the orbital period, and three-dimensional (non-spherical) effects will become important. The common envelope will be ejected preferentially in the orbital plane.

2. However, if the secondary star is not engulfed until the primary has reached an advanced (AGB) evolutionary stage, its highly centrally condensed configuration will make non-spherical effects less important. The ejected material will be less concentrated to the orbital plane.

During the subsequent evolution of a PN, a wind from the central star may lead to further modifications of the morphology, depending on the degree of density contrast (between the equatorial and polar regions of the nebula) that was set up during the envelope ejection (*cf.* Balick 1987). When a density contrast exists, the stellar wind will penetrate the nebula in the polar directions first.

In every case, the actual PNs known to contain close-binary nuclei have a non-spherical appearance. This is as expected for ejection from a binary system, but is of doubtful significance since ~80% of *all* PNs have a non-spherical shape anyway (*cf.* Zuckerman and Aller 1986). The individual PNs with binary nuclei display a wide variety of morphologies. Abell 41 (illustrated by Grauer and Bond 1983) and Abell 63 have elliptical shapes, with the highest densities at both ends of the minor axes. The lower-density major axes probably represent an intermediate to late stage of "bubble penetration." NGC 2346 has a "butterfly" shape, which in Balick's scheme requires that a large density contrast was established in the ejection from the binary nucleus.

Several of the PNNs show evidence of interactions with the surrounding interstellar medium. Most of the nebular material in Abell 46 and HFG 1 lies on one side of the nucleus, suggesting ablation of the nebula due to its motion relative to the ISM. More extreme cases are Abell 35 and 0623+71, which actually show a "bow-shock" morphology.

8. THE PULSATING PNN K 1-16 AND RELATED OBJECTS

While searching for binary PNNs, Grauer and Bond (1984) serendipitously discovered the only known *pulsating* central star, that of K 1-16 (DS Dra). Intensive photometric monitoring has shown K 1-16 to be a multiperiodic, non-radial g-mode pulsator; numerous pulsation modes are simultaneously present and are concentrated near 1500 and 1700 sec (Grauer and Bond 1987).

The optical spectrum of K 1-16 shows He II, C IV, and O VI features; no Balmer lines are present (Grauer and Bond 1984; Sion, Liebert, and Starrfield 1985). The spectroscopic and photometric properties of K 1-16 show that it is closely related to the four known pulsating GW Vir (PG 1159 − 035) white dwarfs. The GW Vir variables are extremely hot ($\sim 10^5$ K), hydrogen-deficient white dwarfs (Wesemael, Green, and Liebert 1985), whose pulsation periods generally lie in the range 400–600 sec (McGraw *et al.* 1979; Bond *et al.* 1984; Bond and Grauer 1987*b*). It is clear that they are more highly evolved than K 1-16, because of their shorter pulsation periods and because they are not surrounded by observable PNs. (The strongest upper limit for a PN around a GW Vir pulsator is at a level 60 times fainter than that of the Palomar Sky Survey, obtained for the pulsator PG 0122+200 by Reynolds 1987). K 1-16 may, in turn, be descended from the "O VI" class of hydrogen-deficient PNNs (Sion *et al.* 1985).

The pulsation mechanisms may be the same in both K 1-16 and the GW Vir variables. Starrfield *et al.* (1984, 1985) have proposed that the mechanism is cyclical ionization of C and/or O in the hydrogen-deficient envelope of the star. This suggestion is in accord with the spectroscopic appearance of the pulsators, although the detailed chemical compositions are still very uncertain (Liebert 1987). On the other hand, Iben (1984) and Wood and Faulkner (1986) have shown that hydrogen-deficient PNNs and their immediate descendants contain helium-burning shells. Kawaler *et al.* (1986) find that these shells are pulsationally unstable, suggesting that the ϵ-mechanism is responsible for the K 1-16 and GW Vir pulsations.

In a recent attempt to refine the location of the pulsational instability region in the HR diagram, Grauer *et al.* (1987*b*) showed that 14 additional hot, hydrogen-deficient PNNs and pre-white dwarfs do not pulsate. The 5 known pulsators have effective temperatures of 80,000–160,000 K and luminosities of 30–1000 $L/L_\odot$ ($\log g$ = 6–8). However, three non-pulsating members of the

PG 1159 − 035 spectroscopic class defined by Wesemael et al. (1985) also lie in this region of the HR diagram, suggesting that the instability region is narrower than just indicated, or that the instability is critically dependent on some other stellar property, such as small changes in chemical composition. The null results would appear to cast doubt either on the presence of helium-burning shells in these objects, or on the ϵ-mechanism for the pulsations, since for this mechanism all objects in this region of the HR diagram below $2000L_\odot$ would be expected to pulsate (Kawaler et al. 1987). A recent photometric survey of PNNs for short-period pulsations by Hine and Nather (1987), as well as the Bond-Grauer binary PNN search described above, also failed to reveal any pulsators other than K 1-16 itself.

Perhaps the greatest potential importance of K 1-16 is that it may provide the evolutionary timescale of a PNN through measurement of its pulsational period change, dP/dt. For GW Vir itself, Winget et al. (1985) analyzed extensive photometric data to detect a period change on a timescale of 10^6 yr. K 1-16, which lies higher in the HR diagram than the GW Vir white dwarfs, should have a timescale more than an order of magnitude shorter. Unfortunately, the light variations of K 1-16 are so complex, and the number of pulsation cycles per night so low, that it has not yet yielded to a similar analysis (Grauer and Bond 1987). It will probably be necessary to arrange continuous photometric monitoring from sites at different longitudes in order to make this important measurement.

A further important development in the study of these hot pulsators is the discovery that the modes in GW Vir appear to be spaced (nearly) uniformly in period (Kawaler 1987a,b). Since such uniform period spacings are expected for g-modes with the same degree (l) and consecutive radial wave numbers (n), it becomes possible to identify the pulsation modes. The observed periods of the modes then provide physical information about the star. Application of this technique to GW Vir provides an extraordinarily accurate (if the various assumptions are correct) mass determination of $0.60 \pm 0.02 M_\odot$. Moreover, slight departures from strictly uniform period spacings can provide information on the compositional stratification of the star (Kawaler 1987b). It would of course be of great interest to apply these techniques to the central star of K 1-16, but photometry from a single site may not be adequate for this formidable task.

REFERENCES

Abell, G.O. 1966, *Ap. J.*, **144**, 259.
Amnuel, P.R., Guseinov, O.H., Novruzova, H.I., and Rustamov, Yu.S. 1984, *Ap. Space Sci.*, **107**, 19.
Balick, B. 1987, *Sky and Telescope*, **73**, 125.
Barnard, E.E. 1908, *M.N.R.A.S.*, **68**, 465.
Bode, M.F., Seaquist, E.R., Frail, D.A., Roberts, J.A., Whittet, D.C.B., Evans, A., and Albinson, J.S. 1987, *Nature*, in press.
Bond, H.E. 1976, *Pub. A.S.P.*, **88**, 192.
_____. 1985, in *Cataclysmic Variables and Low-Mass X-Ray Binaries*, eds. D.Q. Lamb and J. Patterson (Dordrecht: Reidel), p. 15.
Bond, H.E., and Grauer, A.D. 1987a, in *IAU Colloq. No. 95, Second Conference on Faint Blue Stars*, eds. A.G.D. Philip, D.S. Hayes, and J. Liebert (Schenectady: L. Davis Press), in press.
_____. 1987b, *Ap. J. (Letters)*, in press.

Bond, H.E., Grauer, A.D., Green, R.F., and Liebert, J.W. 1984, *Ap. J.*, **279**, 751.
Bond, H.E., Liller, W., and Mannery, E.J. 1978, *Ap. J.*, **223**, 252.
Cahn, J.H., and Wyatt, S.P. 1976, *Ap. J.*, **210**, 508.
Daub, C.T. 1982, *Ap. J.*, **260**, 612.
Drilling, J.S. 1985, *Ap. J. (Letters)*, **294**, L107.
Drilling, J.S., and Schönberner, D. 1985, *Astr. Ap.*, **146**, L23.
Drummond, J.D. 1980, Ph.D. dissertation, New Mexico State University.
Ellis, G.L., Grayson, E.T., and Bond, H.E. 1984, *Pub. A.S.P.*, **96**, 283.
Feibelman, W.A., and Kaler, J.B. 1983, *Ap. J.*, **269**, 592.
Fleming, T.A., Liebert, J., and Green, R.F. 1986, *Ap. J.*, **308**, 176.
Garrison, R.F., Schild, R.E., Hiltner, W.A., and Krzeminski, W. 1984, *Ap. J. (Letters)*, **276**, L13.
Grauer, A.D., and Bond, H.E. 1983, *Ap. J.*, **271**, 259.
——————————————. 1984, *Ap. J.*, **277**, 211.
——————————————. 1987, in *IAU Colloq. No. 95, Second Conference on Faint Blue Stars*, eds. A.G.D. Philip, D.S. Hayes, and J. Liebert (Schenectady: L. Davis Press), in press.
Grauer, A.D., Bond, H.E., Ciardullo, R., and Fleming, T.A. 1987a, *Bull. A.A.S.*, **19**, 643.
Grauer, A.D., Bond, H.E., Liebert, J., Fleming, T.A., and Green, R.F. 1987b, *Ap. J.*, in press.
Hine, B.P., and Nather, R.E. 1987, in *IAU Colloq. No. 95, Second Conference on Faint Blue Stars*, eds. A.G.D. Philip, D.S. Hayes, and J. Liebert (Schenectady: L. Davis Press), in press.
Hoffleit, D. 1932, *Harvard Bull.*, No. 887.
Iben, I. 1984, *Ap. J.*, **277**, 333.
Ishida, K., and Weinberger, R. 1987, *Astr. Ap.*, **178**, 227.
Jacoby, G.H. 1981, *Ap. J.*, **244**, 903.
Jasniewicz, G. 1987, in *IAU Colloq. No. 95, Second Conference on Faint Blue Stars*, eds. A.G.D. Philip, D.S. Hayes, and J. Liebert (Schenectady: L. Davis Press), in press.
Jasniewicz, G., and Acker, A. 1986, *Astr. Ap.*, **160**, L1.
Jasniewicz, G., Duquennoy, A., and Acker, A. 1987, *Astr. Ap.*, **180**, 145.
Kawaler, S.D. 1987a, in *IAU Symp. No. 123, Advances in Helio- and Asteroseismology*, eds. S. Christensen-Dalsgaard and S. Frandsen (Dordrecht: Reidel), in press.
Kawaler, S.D. 1987b, in *IAU Colloq. No. 95, Second Conference on Faint Blue Stars*, eds. A.G.D. Philip, D.S. Hayes, and J. Liebert (Schenectady: L. Davis Press), in press.
Kawaler, S.D., Winget, D.E., Hansen, C.J., and Iben, I. 1986, *Ap. J. (Letters)*, **306**, L41.
——————————————————————. 1987, in *Late Stages of Stellar Evolution*, eds. S. Kwok and S. Pottasch (Dordrecht: Reidel), p. 403.
Kohoutek, L. 1964, *Bull. Astron. Inst. Czech.*, **15**, 161.
Krautter, J., Klaas, U., and Radons, G. 1987, *Astr. Ap.*, **181**, 373.
Liebert, J. 1987, in *Stellar Pulsation*, eds. A.N. Cox, W.M. Sparks, and S.G. Starrfield (Berlin: Springer-Verlag), p. 342.
Livio, M., and Soker, N. 1987, submitted to *Ap. J.*
Maciel, W.J. 1981, *Astr. Ap.*, **98**, 406.

Mallik, D.C.V. 1982, in *IAU Symp. No. 103, Planetary Nebulae*, ed. D.R. Flower (Dordrecht: Reidel), p. 424.
McGraw, J.T., Starrfield, S.G., Liebert, J., and Green, R. 1979, in *IAU Colloq. No. 53, White Dwarfs and Variable Degenerate Stars*, eds. H. Van Horn and V. Weidemann (Rochester: University of Rochester), p. 377.
Méndez, R.H., and Niemela, V.S. 1981, *Ap. J.*, **250**, 240.
Miller, J.S., Krzeminski, W., and Priedhorsky, W. 1976, *IAU Circ.*, No. 2974.
Paczynski, B. 1976, in *IAU Symposium No. 73, Structure and Evolution of Close-Binary Systems*, eds. P. Eggleton, S. Mitton, and J. Whelan (Dordrecht: Reidel), p. 75.
Patterson, J. 1984, *Ap. J. Suppl.*, **54**, 443.
Pottasch, S.R. 1984, *Planetary Nebulae* (Dordrecht: Reidel).
Reynolds, R.J. 1987, *Ap. J.*, **315**, 234.
Ritter, H. 1976, *M.N.R.A.S.*, **175**, 279.
─────── . 1986, *Astr. Ap.*, **169**, 139.
Shara, M.M., Livio, M., Moffat, A.F.J., and Orio, M. 1986, *Ap. J.*, **311**, 163.
Sion, E.M., Liebert, J., and Starrfield, S.G. 1985, *Ap. J.*, **292**, 471.
Starrfield, S., Cox, A.N., Kidman, R.B., and Pesnell, W.D. 1984, *Ap. J.*, **281**, 800.
─────── . 1985, *Ap. J. (Letters)*, **293**, L23.
Verbunt, F., and Zwaan, C. 1981, *Astr. Ap.*, **100**, L7.
Wesemael, F., Green, R.F., and Liebert, J. 1985, *Ap. J. Suppl.*, **58**, 379.
Winget, D.E., Kepler, S.O., Robinson, E.L., Nather, R.E., and O'Donoghue, D. 1985, *Ap. J.*, **292**, 606.
Wood, P.R., and Faulkner, D.J. 1986, *Ap. J.*, **307**, 659.
Zuckerman, B., and Aller, L.H. 1986, *Ap. J.*, **301**, 772.

BINARITY AND INTRINSIC VARIABILITY IN CENTRAL STARS OF PN

R. H. Méndez

Instituto de Astronomia y Fisica del Espacio, Buenos Aires
and Institut für Astronomie und Astrophysik, Univ. of Munich

ABSTRACT: 1. Introduction. 2. A list of binary and multiple CSPN. 3. A radial velocity study of CSPN at high spectral resolution. 4. Spectroscopic binaries or intrinsic variables? 5. Concluding remarks.

1. INTRODUCTION

We have good reasons to believe that the majority of the stars in the sky are binary or multiple systems (Abt 1983, Poveda et al. 1982). For unevolved binaries (both components at or near the main sequence) the number of binaries per logarithmic interval in P appears to be roughly constant from log P (days) = 0 to 7, with an ill-defined maximum at about 10 years (log P (days) = 3.6); see e.g. Figure 2 of Abt (1983).
 Let us briefly consider what is the effect of stellar evolution on this "family" of unevolved binaries. We shall restrict our attention to "intermediate mass stars", i.e. those that become white dwarfs in less than 10^{10} years. If we assume that the maximum possible stellar radius is of the order of 1000 solar radii (or about 5 AU), then the separation between binary components that is required to ensure their independent evolution is of about 3000 solar radii (unless the orbit is very eccentric). This limit corresponds roughly to log P (days) = 3.5 - 4.0 for a wide variety of total masses and mass ratios. We can call "wide" and "close" binaries those with separations respectively above and below that limit.
 Now we focus our attention on one given star and ask if it is a member of a binary system. If the answer is no, then at the end of its evolution we will have an envelope ejection from a single star, and subsequent transformation into a single white dwarf. If the answer is yes, we ask if the binary is "wide". If yes, we will again have an envelope ejection from a "single" (non-interactive) star. If no, we ask if the binary is "close" enough for coalescence. If yes, we will again have an envelope ejection from a single star. If no, then we will have a case of envelope ejection from a "close" binary system.
 Notice that up to now I have avoided the words "planetary nebula". Now we can state our problem with the following two

questions: (1) do all envelope ejections give rise to detectable planetary nebulae? (2) what is the relative frequency of the two cases of envelope ejection (single vs. close binary)?

I think it is fair to say that we do not have clear answers to these two questions from a theoretical point of view. As a consequence almost any number is conceivable for the percentage of close binaries among central stars of planetary nebulae (CSPN): from a few percent to 100%. Recently Paczynski (1985) presented the most extreme suggestion: perhaps all detectable PN are ejected by close binaries... (this would require a larger birthrate for white dwarfs than for PN).

The purpose of this review is to present the observational evidences about binarity of CSPN. Section 2 gives a list of binary or multiple CSPN (excluding those listed by Bond, see his review in this volume), and also provides several additional comments. Section 3 describes some preliminary results of a radial velocity study of CSPN using high spectral resolution. Several cases of radial velocity variations detected in this survey are discussed in Section 4. Finally, the review is closed with a few inconclusive but optimistic remarks.

2. A LIST OF BINARY AND MULTIPLE CSPN

The list in Table 1 is arranged by method of discovery. In the following subsections some additional comments and informations are given.

2.1. "Cool" central stars

To notice that the central star is not hot enough to ionize the nebula remains the most effective method of discovering binaries. A cautionary remark is necessary: as we go to later spectral types, the probabilities of misclassification and chance superposition increase. Notice that several of the "cool" CSPN listed by Lutz (1977) have been later reclassified as "not PN" (Acker et al. 1987). In other cases, more detailed studies have not confirmed the presence of a cool central star, or have suggested that it probably is a foreground object (Lutz and Kaler 1983). Another cool object not included in Table 1 is Abell 14 (see Abell 1966). A careful study of this CSPN appears to be lacking.

Unfortunately, to know that a given CSPN is binary is not enough; we would also like to know if it is "wide" or "close". In the case of "cool" CSPN, the very presence of the cool star complicates the investigations. To find that the cool star is a spectroscopic binary is not enough, because the hot star that has ejected and ionized the PN might be a "wide" companion of the spectroscopic binary. On the other hand, consider the visual binary CSPN of NGC 3132: the very faint, hot visual companion of the A-type star might be a close binary... An interesting example of the complications that may arise is given by LT5 (Jasniewicz et al. 1987). The G star appears to be a short period, double-lined spectroscopic binary, implying that a third object is present in the system. But it also seems that the gamma velocity of the double-lined binary is variable, implying that perhaps the hot star is not so far from the short-period binary. To this we may add the light variations, which are not yet well understood. It may require several years of careful work to understand what is happening in the central star of LT5.

Another cool CSPN that deserves additional comments is NGC 2346: it will be mentioned in subsection 2.4.

Cool CSPN have been suggested as a valuable source of reliable distances. Although this is true in a few cases, one has to be careful. It is not a good idea to take the spectral type from the literature, go to Allen's Astrophysical Quantities and extract an absolute magnitude. First we need a reliable determination of Teff and log g for the cool CSPN, using good spectrograms or spectrophotometry and good model atmospheres. This information gives the ratio of luminosity to mass of the cool star. Second, it may be necessary to check if the observed Teff and log g can be obtained using theoretical evolutionary tracks for different masses. If that is the case, there will be more than one possible distance, and it may be impossible to decide which is the correct one. It is good to remember that, spectroscopically, "giant" and "supergiant" mean "low gravity", not necessarily "massive and luminous".

2.2. Visual companion of the hot CSPN

The prototype of this method of discovery is the central star of NGC 246. Since the probability of chance superposition is not negligible, we need additional information: for example, proper motions or radial velocities. The paper by Cudworth (1973) gives proper motions for NGC 246 and for some of the pairs he found. It seems that no further work has been made on these objects.

The spectroscopic distance of the 14th magnitude G8 V - K0 V companion in NGC 246 (420 ± 40 pc, Minkowski and Baum 1960) has been traditionally considered one of the best PN distances. It was derived assuming an absolute visual magnitude Mv = +6.1 ± 0.2 for the cool star. Recently, Husfeld (1986, 1987) has obtained a spectroscopic distance for the hot companion: 960 ± 300 pc. In view of this discrepancy, it would be a good idea to study the cool star again. Even if we do not change the spectral classification, according to Allen (1973) a G8 V star can have Mv = +5.5, which would give a distance of almost 600 pc. I apologize for using the Allen tables after my remark in section 2.1.

2.3. Photometric variations

We have seen that "cool" CSPN are not a very promising source of close binaries. The search for photometric variations has been much more successful. The review by Bond in this volume brings information about 6 close binary CSPN + Abell 35 (which still needs confirmation as a close binary, see Jasniewicz 1987) + 2 cataclysmic variables surrounded by old PN (Krautter et al. 1987, Bode et al. 1987). Bond estimates that about 10 - 15% of all CSPN are binaries with P < 1 day.

TABLE 1. A LIST OF BINARY CSPN

1. "Cool" central stars

OBJECT NAME	SPECTRAL TYPE OF CSPN	HOT STAR?	RAD VEL VARIAB?	PHOTOM VARIAB?	REFERENCES
IRAS 1912+172P09	B9 V	not detected			1
NGC 1514	A	detected	no	no	2,3,4
NGC 3132	A2 V	resolved	no		5,6,7
He 2-36	A2 III	not detected	no		5,7
NGC 2346	A5 V	detected	yes	yes	5,7,8
Cn 1-1	F5 III-IV	not detected	no		9,10
M 1-2	G2 Ib	not detected	no	no	11,12,13
LT 5	G5 III	detected	yes	yes	14,15
Abell 35	G8 III-IV	detected		yes	16,17

2. Visual companion of the hot CSPN

OBJECT	REFERENCE
NGC 246	18
NGC 650-1	19
Abell 24	19
Abell 30	19
Abell 33	19
NGC 6853	19

3. Photometric variations

6 close binary CSPN
+ Abell 35
+ 2 cataclysmic variables surrounded by old PN (see review by Bond in this volume)

4. Spectroscopic binaries

OBJECT	P(days)	REF.
NGC 2346	15.99	7,20
NGC 6826	0.2377	21,22
M 1-67	2.4?	23

5. Composite spectrum

The central star of Sp 1 (PK 329 +2 1) (see text and reference 24)

1. Whitelock and Menzies 1986
2. Greenstein 1972
3. Seaton 1980
4. Bond and Grauer 1987
5. Mendez 1978
6. Kohoutek and Laustsen 1977
7. Mendez and Niemela 1981
8. Costero et al. 1986
9. Lutz 1984
10. Bhatt and Mallik 1986
11. O'Dell 1966
12. Feibelman 1983
13. Grauer and Bond 1981
14. Feibelman and Kaler 1983
15. Jasniewicz et al. 1987
16. Jacoby 1981
17. Jasniewicz 1987
18. Minkowski and Baum 1960
19. Cudworth 1973
20. Mendez et al. 1982
21. Noskova 1980
22. Acker et al. 1982
23. Moffat et al. 1982
24. Mendez et al. 1987

2.4. Spectroscopic binaries

The list of spectroscopic binaries may look disappointingly short; but notice that a few cases reported earlier have turned out to be false alarms. An outstanding example of false alarm is NGC 1360 (Mendez and Niemela 1977). When I could not confirm the velocity variations on subsequent spectrograms, I thought that perhaps the orbit was very eccentric (Mendez 1980). After several additional and unsuccessful attempts, now I believe that for some unknown reason the old stellar velocities were wrong.

Two comments are necessary about NGC 2346. First, since I have seen its central star described as an eclipsing binary, and this might be misleading, I would like to emphasize that the spectacular light variations discovered by Kohoutek (1982) were not produced by the eclipse of one star by the other, but instead by the slow passage of a dense dust cloud in front of the binary system (Mendez et al. 1982, Costero et al. 1986). If you look now (1987) at the A-type central star (it has received the name V651 Mon) you will find that it has again the constant brightness it showed before the passage of the dust cloud.

The second comment is that it has not yet been possible to check if the companion of the A-type star is really the hot star. As mentioned in 2.1, the system might be multiple, with the hot star as a wide companion of the spectroscopic binary. A few high-resolution spectrograms in the far ultraviolet, where the hot star is detectable, would probably solve the problem.

Concerning NGC 6826, it is obvious that it should be observed photometrically.

M1-67, with its WN8 central star, has been going in and out of the catalogues of PN. The last (and probably definitive) argument to consider it as a PN is by van der Hucht et al. (1985), based on their detection of IR emission from a circumstellar dust shell, with a temperature falling within the range of dust temperatures found to be common in PN (thermal emission by heated dust associated with Pop. I WR stars is quite different). If we accept M1-67 as a PN, then its central star must be included in Table 1, because according to Moffat et al. (1982) it is a spectroscopic binary and it also shows light variations. Of course, since Moffat et al. took it as a Pop. I star, several details in their paper need revision. Besides, given the small amplitudes of their light- and radial velocity curves, additional observations would be very useful.

2.5. Composite spectrum

I have added this subsection because of the central star of Sp 1 (PK 329+2 1). Observed at high spectral resolution, its spectrum is a curious mixture of low (30000 K) and high (100000 K) temperature features. Because of space limitations, I cannot give here a detailed description. We (Mendez et al. 1987) believe that Sp 1 is probably a close binary system composed of a very hot star and a cool companion, one of whose hemispheres is heated by radiation from the hot star. Further comments in Section 3.

3. A RADIAL VELOCITY STUDY OF CSPN AT HIGH SPECTRAL RESOLUTION

If we want a reliable observational determination of the percentage of close binary CSPN and of their period distribution, then a search for spectroscopic binaries is necessary, because the photometric method is not sensitive to periods longer than a few days. Unfortunately, the search for spectroscopic binaries requires a high spectral resolution. Consider the situation as it was 5 years ago, working at spectral resolutions of a few Å, when it was difficult to detect semiamplitudes of less than 30 Km/s for typical hot CSPNs. We can estimate the maximum detectable orbital period for $K1 \geq 30$ Km/s and circular orbits, as a function of the masses M1 and M2 (M1 is the visible star), using

$$P \text{ (days)} = 9.65 \cdot 10^6 \, K1^{-3} \, \sin^3 i \, M2^3 \, (M1+M2)^{-2} \qquad (1)$$

where all masses are in solar masses. The results are in Table 2, for $i = 45°$ and typical combinations of M1 and M2. Clearly, we need more accuracy if we want to extend the search to significantly longer periods.

TABLE 2

Maximum detectable period MDP (days)
for $i=45°$ and a minimum detectable $K1 = 30$ Km/s

M1	M2	MDP	log MDP
(solar masses)			
0.6	0.6	19 d	+1.27
0.6	0.3	4.2 d	+0.62
0.6	0.15	0.75 d	-0.12

The situation is much better now. In what follows I would like to present some preliminary results of a search for radial velocity variations in CSPN at a spectral resolution of 0.3 Å. The spectrograms were taken with CASPEC, the Cassegrain echelle spectrograph of the ESO 3.6 m telescope at La Silla, Chile. The selected spectral coverage is from 4000 to 5000 Å, and up to now we have extracted useful information from 62 spectrograms of 28 CSPN with apparent visual magnitudes in the range 10 - 14. A more detailed description of the results is in preparation. Some spectral descriptions can be found in Mendez et al. (1987). Typical exposure times were between 30 and 60 minutes.

A great advantage of these CASPEC spectrograms is that in many cases we can measure the radial velocities of narrow stellar absorptions and emissions of C, N, O and Si, which are much more reliable than the broad H and He lines, and are not contaminated with nebular emissions.

Table 3 shows the nebular velocities, compared with those listed by Schneider et al. (1983), and the differences between stellar and nebular velocities, for 22 CSPN. In several cases we obtained 2 consecutive spectrograms of each CSPN. Since no significant differences were found, in Table 3 the corresponding velocities have been combined (this is indicated with asterisks). The central star of EGB 5 (see Mendez et al. 1987) was not included in Table 3 because there is no information about the nebular velocity of this object. The stellar velocity is +65 and +68 Km/s on two spectrograms taken on consecutive nights. The 5 remaining CSPN will be mentioned in Section 4.

Some details in Table 3 need comment: (1) the redshift shown by the central star of NGC 7293 can be interpreted as gravitational. A more careful determination (the number we give is derived from rather uncertain measurements of the wings of the He II 4686 absorption) would give valuable independent information about the surface gravity and the distance of this CSPN (see Mendez et al. 1987). (2) the radial velocities of the central star of IC 2448 are uncertain, because the spectrograms are noisy and only one stellar line is measurable (C IV 4658 in emission). (3) the stellar He II 4686 often gives discrepant results. We interpret these discrepancies as wind effects. Sometimes we find a redshifted emission (sometimes accompanied by a clearly seen blueshifted absorption). But in some other cases we find a blueshifted emission (e.g. M1-26 and Tc 1). In the cases of H2-1 and He 2-151 we find a redshifted absorption, which may indicate the presence of an incipient blueshifted emission (more details in Mendez et al. 1987). Another case of blueshifted He II 4686 emission has been found by Heber et al. (1987) in the spectrum of LSS 1362. (4) The central star of Sp 1 shows the same radial velocity on two consecutive spectrograms, and there is no difference in radial velocity between the low- and high-temperature features. No nebular lines are present in our spectrograms of this star, and thus the difference Vstar - Vneb is uncertain. For the moment we find no support to our suggestion that the central star of Sp 1 is a close binary. But we still think it probably is, and additional observations are planned.

From Table 3 we conclude that now, given just a few spectrograms, it is quite possible to detect semiamplitudes below 6 Km/s; probably even less when the stellar spectrum shows many sharp lines. Looking at formula (1) and Table 2, we find that now the MDPs are at least 100 times longer. In such conditions, a negative result of a wide search for velocity variations would be almost as informative as a positive result.

4. SPECTROSCOPIC BINARIES OR INTRINSIC VARIABLES?

Table 4 gives some information about 5 CSPN that have shown radial velocity variations. The central star of He 2-131 was known to have a

TABLE 3. HELIOCENTRIC RV (Km/s) OF PN AND THEIR CS ON CASPEC SPECTRA

OBJECT NAME	Vneb (a)	Vneb (b)	NUMBER OF STELLAR LINES USED (c)	Vstar-Vneb (Km/s) (d)	NOTES
NGC 246	-46		4	+ 1	
NGC 246	-46		4	- 1	
NGC 246	-46		4	+ 5	
NGC 7293	-28	-29	2*	+16	Grav. redshift
LSE 125		- 6	21*	+ 1	
NGC 7009	-47	-48	10*	+ 4	
NGC 4361	+10	+12	5	+ 2	
NGC 4361	+10	+11	6	+ 1	
NGC 1360	+42	+47	5	+ 1	
NGC 3242	+ 5	+ 6	14*	- 2	
NGC 1535	- 3	- 2	4	+ 2	
IC 2448	-24	-27	1	-12	
IC 2448	-24	-26	2*	+ 1	
NGC 6891	+42	+42	16*	- 1	4686 em redshifted
NGC 5882	+10	+15	4*	- 1	4686 em redshifted
NGC 6629	+15	+13	16*	- 1	4686 em redshifted
IC 4637	+11	-10	3	+ 9	
PHL 932 (e)	+15		8*	+ 3	
He 2-182	-91	-87	10*	- 1	
M1-26	- 5	-24	4	+ 5	4686 em blueshifted
M1-26	- 5	-24	8*	- 1	4686 em blueshifted
Tc 1	-83	-96	15*	+ 8	4686 em blueshifted
He 2-108	- 8	- 8	20*	+ 3	4686 em redshifted
H2-1	-20	-21	10*	+ 6	4686 abs redshifted
He 2-162	+33	+27	29*	+ 2	
He 2-151	-128	-136	39*	- 5	4686 abs redshifted
Sp 1	-33		33*	+ 7	

(a) Schneider et al. 1983. (b) This work.
(c) Asterisks indicate that velocities from 2 consecutive spectrograms were combined to obtain the nebular and stellar values.
(d) We used our determination of Vneb whenever possible.
(e) Vneb = +15 ± 20 Km/s, taken from Arp and Scargle 1967.

---

variable spectrum (Mendez and Niemela 1979, Surdej et al. 1982) and the central star of IC 418 was known to show photometric and radial velocity variations (Mendez et al. 1986).

The reality of the variations is out of question, but a reliable interpretation is not yet possible. More spectroscopic and photometric information is necessary. It is possible to interpret these variations both as due to binary motion and to fluctuations in the photospheric outflow velocity and mass loss rate.

In the case of IC 418, which up to now has been the one most ca-

TABLE 4. VARIABLE CSPN

OBJECT NAME	HELIOC. JD (2440000+)	Vneb (a)	Vneb (b)	Vabs-Vneb (Km/s)(c)	BEHAVIOUR OF THE STELLAR EMISSIONS
PB 8	6210.480		+22	-92	
PB 8	6210.533		+24	-89	CONSTANT
PB 8	6456.792		+23	-145	
NGC 2392	6454.679	+75		-21	CONSTANT
NGC 2392	6455.589	+75		+ 4	
He 2-138	5876.639	-47	-46	- 8	
He 2-138	5876.684	-47	-40	- 8	CONSTANT
He 2-138	6455.866	-47	-36	+24	
He 2-131	6207.656	- 1	-12	+23	ANTIPHASE
He 2-131	6207.681	- 1	-12	+27	(with a few
He 2-131	6454.863	- 1	-11	-43	exceptions)
IC 418	6454.589	+62	+63	+12	
IC 418	6455.547	+62	+63	-20	
IC 418	6455.559	+62	+63	-24	ANTIPHASE
IC 418	6456.613	+62	+63	+ 6	
IC 418	6457.619	+62	+63	-12	

(a) and (b) as in Table 3. (c) The differences between the radial velocities of stellar absorption lines and the nebular velocities. In the case of He 2-138 I have not used lines that show P Cygni profiles.

refully studied (Mendez et al. 1986), we are sure that the orbital motion alone (if present) would not be enough to explain the observed variations: the velocity field near the photosphere must be variable.

An explanation in terms of variable outflow velocity appears to be most likely for PB 8, NGC 2392 and He 2-138, because the stellar emissions do not move. The central star of He 2-138 shows variable P Cygni profiles (Mendez et al. 1987). Binary motion would be more probable for IC 418 and He 2-131, because the stellar emissions move in antiphase with the stellar absorptions -the more positive the absorption velocity, the more negative the stellar emission velocity. However, a variable velocity field might conceivably produce such an antiphase effect; consider e.g. the possible behaviour of the redshifted He II 4686 emission when the outflow velocity changes.

Probably our best case for binarity is He 2-131, because we (Mendez et al. 1987) could not fit the observed H and He stellar absorption profiles with theoretical profiles, implying that perhaps the spectrum is composite. But here we might also think that we were trying to force our model atmosphere method beyond its limit of validity.

In summary, I would not claim that any of the variable objects in Table 4 is a close binary until a well defined and confirmed period is found.

5. CONCLUDING REMARKS

The existence of intrinsic variations appears to be well confirmed, at least in a few cases, and this will complicate the search for spectroscopic binaries. At the present time it is too early to suggest a number for the percentage of close binary CSPN. If only one or two of the 28 objects in our CASPEC sample are close binaries, and if their periods turn out to be less than one day, then it will be reasonable to conclude that the period distribution of close binary CSPN shows a precipitous drop at P = 1 day, and that not more than 15% of all CSPN are close binaries, because we are able to probe a much larger range of periods than with the photometric method.

However, we cannot yet rule out a much higher frequency of binaries in our sample. If these additional binaries exist, if some of them have periods > 1 day, and if frequently M2/M1 is small, then the period distribution can be flatter, and the percentage of close binaries can be substantially higher.

That the mass ratio can be frequently small is suggested by the available information about the already known close binary CSPN (see Ritter 1987). Besides, a recent paper by Halbwachs (1987) hints that perhaps many unevolved binaries have very small mass ratios.

In spite of the present uncertainties, it seems clear that we have the tools to make important progress. A careful study of the radial velocities of CSPN at high spectral resolution is likely to produce valuable information about both binarity and the almost unexplored subject of intrinsic variability.

Acknowledgements: Most of the CASPEC spectrograms mentioned in this review were taken by myself and R.P. Kudritzki, with contributions by S. D'Odorico, D. Husfeld and H.G. Groth. The reductions of the spectrograms and radial velocity measurements were made with software developed by T. Gehren. I had useful conversations with H.E. Bond, U. Heber, I. Iben, R.P. Kudritzki, J.H. Lutz, H. Ritter and A.V. Tutukov. It is a pleasure to acknowledge the hospitality of the University Observatory and the Max Planck Institute for Astrophysics, both in Munich. I am grateful to the following institutions for financial support: the Alexander von Humboldt Foundation, the Max Planck Institute for Astrophysics and the Local Organizing Committee of this Symposium.

REFERENCES

Abell, G.O. 1966, Astrophys.J. 144, 259.
Abt, H.A. 1983, Ann.Rev.Astron.Astrophys. 21, 343.
Acker, A., Gleizes, F., Chopinet, M., Marcout, J., Ochsenbein, F. and Roques, J.M. 1982, Catalogue of the CS of true and possible PN, Publ. Speciale du CDS (Obs. de Strasbourg), No. 3.

Acker, A., Chopinet, M., Pottasch, S.R. and Stenholm, B. 1987, Astron. Astrophys.Suppl. 71, 163.
Allen, C.W. 1973, Astrophys. Quantities, London, The Athlone Press.
Arp, H. and Scargle, J.D. 1967, Astrophys.J. 150, 707.
Bhatt, H.C. and Mallik, D.C.V. 1986, Astron.Astrophys. 168, 248.
Bode, M.F., Seaquist, E.R., Frail, D.A., Roberts, J.A., Whittet, D.C. B., Evans, A. and Albinson, J.S. 1987, Nature, 329, 519.
Bond, H.E. and Grauer, A.D. 1987, in IAU Coll. 95, "The Second Conference on Faint Blue Stars", Eds. Davis Philip, Hayes and Liebert, L. Davis Press, in press.
Costero, R., Tapia, M., Mendez, R.H., Echevarria, J., Roth, M., Quintero, A. and Barral, J.F. 1986, Rev.Mex.Astron.Astrofis. 13, 149.
Cudworth, K.M. 1973, PASP 85, 401.
Feibelman, W.A. 1983, Astrophys.J. 275, 628.
Feibelman, W.A. and Kaler, J.B. 1983, Astrophys.J. 269, 592.
Grauer, A.D. and Bond, H.E. 1981, PASP, 93, 630.
Greenstein, J.L. 1972, Astrophys.J. 173, 367.
Halbwachs, J.L. 1987, Astron.Astrophys. 183, 234.
Heber, U., Werner, K. and Drilling, J.S. 1987, Astron.Astrophys., in press.
van der Hucht, K.A., Jurriens, T.A., Olnon, F.M., The, P.S., Wesselius, P.R. and Williams, P.M. 1985, Astron.Astrophys. 145, L13.
Husfeld, D. 1986, Ph.D.Thesis, Univ. of Munich.
Husfeld, D. 1987, in IAU Coll. 95 (see above), in press.
Jacoby, G.H. 1981, Astrophys.J. 244, 903.
Jasniewicz, G. 1987, in IAU Coll. 95 (see above), in press.
Jasniewicz, G., Duquennoy, A. and Acker, A. 1987, Astron.Astrophys. 180, 145.
Kohoutek, L. 1982, IAU Circular 3667.
Kohoutek, L. and Laustsen, S. 1977, Astron.Astrophys. 61, 761.
Krautter, J., Klaas, U. and Radons, G. 1987, Astron.Astrophys. 181, 373.
Lutz, J.H. 1977, Astron.Astrophys. 60, 93.
Lutz, J.H. 1984, Astrophys.J. 279, 714.
Lutz, J.H. and Kaler, J.B. 1983, PASP 95, 739.
Mendez, R.H. 1978, MNRAS 185, 647.
Mendez, R.H. 1980, in IAU Symposium 88, p.567.
Mendez, R.H., Forte, J.C. and Lopez, R.H. 1986, Rev.Mex.Astron.Astrofis. 13, 119.
Mendez, R.H., Gathier, R. and Niemela, V.S. 1982, Astron.Astrophys. 116, L5.
Mendez, R.H., Kudritzki, R.P., Herrero, A., Husfeld, D. and Groth, H.G. 1987, Astron.Astrophys., in press.
Mendez, R.H. and Niemela, V.S. 1977, MNRAS 178, 409.
Mendez, R.H. and Niemela, V.S. 1979, Astrophys.J. 232, 496.
Mendez, R.H. and Niemela, V.S. 1981, Astrophys.J. 250, 240.
Minkowski, R. and Baum, W.A. 1960, Mt. Wilson and Palomar Obs. Annual Report 1959-60, p. 18.
Moffat, A.F.J., Lamontagne, R. and Seggewiss, W. 1982, Astron.Astrophys. 114, 135.
Noskova, R.I. 1980, Astron.Tsirk. 1128.

O'Dell, C.R. 1966, Astrophys.J. 145, 487.
Paczynski, B. 1985, in "Cataclysmic variables and low-mass X-ray binaries", Eds. Lamb and Patterson, Reidel, p. 1.
Poveda, A., Allen, C. and Parrao, L. 1982, Astrophys.J. 258, 589.
Ritter, H. 1987, Astron.Astrophys.Suppl. 70, 335.
Schneider, S.E., Terzian, Y., Purgathofer, A. and Perinotto, M. 1983, Astrophys.J.Suppl. 52, 399.
Seaton, M.J. 1980, Quart.J.R.A.S. 21, 229.
Surdej, A., Surdej, J. and Swings, J.P. 1982, Astron.Astrophys. 105, 242.
Whitelock, P.A. and Menzies, J.W. 1986, MNRAS 223, 497.

MODEL ATMOSPHERES AND QUANTITATIVE SPECTROSCOPY OF CENTRAL STARS
OF PLANETARY NEBULAE*

R.P. Kudritzki[1] and R.H. Méndez[1,2]

[1]Institut für Astronomie und Astrophysik
der Universität München
Scheinerstr. 1, D-8000 München 80, Germany

[2]Instituto de Astronomia y Fisica del Espacio
Buenos Aires, Argentina

1. INTRODUCTION

It is a good tradition in IAU Symposia about PN to have a paper on model atmospheres. However, this is always a difficult task for the authors, because the majority of the PN researchers still believe that the best model atmosphere for a Central Star is a black body. Of course, this puts a theorist in stellar atmospheres into a somewhat desperate position. However, Central Stars of Planetary Nebulae (hereafter CSPN) - as all other stars - show spectral lines. And we will try to use the opportunity of this paper to convince that - as for all other stars - the quantitative analysis of these lines on basis of model atmospheres yields extremely valuable information about the physical nature of the stars.
    Modern quantitative spectroscopy of hot stars has two aspects: the analysis of spectral lines formed (i) in the hydrostatic photospheres and (ii) in the supersonically expanding winds. Both aspects will be covered by this paper.

2. SPECTROSCOPY OF PHOTOSPHERIC LINES

2.1. Analysis Method

The basis for the quantitative spectroscopic analysis of photospheric lines is given by NLTE model atmospheres (Kudritzki, 1973, 1976; Husfeld et al., 1984; Husfeld, 1986; Groth, 1986) and subsequent extensive NLTE multi-level line formation calculations for H, HeI and HeII (Kudritzki and Simon, 1978; Husfeld, 1986; Husfeld et al., 1987; Herrero, 1987a and b).

*Based on observations collected at the European Southern Observatory, La Silla, Chile

The principle of the analysis is as follows: For "cool" objects (T_{eff} < 50000K) the HeI/HeII ionization ratio determines the effective temperature, the wings of the Balmer lines yield the gravity and the absolute strengths of the HeII lines gives the helium abundance defined as the number fraction $y = N_{He}/(N_H+N_{He})$. For hotter objects (T_{eff} > 50000K) the HeI lines are too weak so that no helium ionization equilibrium can be used for the temperature determination. Instead, an alternative method is applied which makes use of the fact that at these hot temperatures the profile shapes of H and HeII lines contain information about both: T_{eff} and log g. (A detailed description of this method is given by Méndez et al., 1981 and Méndez et al., 1983). In this way T_{eff} and log g can be derived simultaneously from the shapes of the H and HeII profiles, whereas y follows again directly from the absolute strength of the HeII lines.

It is of course clear that these methods require excellent spectra, i.e. well defined profile shapes of high S/N (>50) taken with (almost) linear detectors. A particular problem of CSPN is the contamination of photospheric lines with nebular emission lines. This requires in addition a rather high spectral resolution of $\lambda/\Delta\lambda > 10^4$. Since CSPN are generally faint, this is not an easy goal. However, the present day Cassegrain-Echelle spectrographs with CCD detectors (as the ESO CASPEC) are ideally suited for this purpose (see also paper by McCarthy, this meeting). These instruments form the basis for modern quantitative spectroscopy of CSPN.

In addition to excellent spectra a high quality NLTE line-formation theory is needed to extract the information about the stellar parameters. Since the pioneering developments by Auer and Mihalas (1972) and the later slight improvements by Kudritzki (1976) and Kudritzki and Simon (1978), no substantial progress was made for some time. However, recently, the consequent use of the Auer-Heasley-method (Auer and Heasley, 1976) in the NLTE code DETAIL (developed by J. Giddings, 1980) allowed a significant improvement in the detailed treatment of the transition schemes of the model atoms in question.

One of the major disadvantages of the old NLTE line formation calculations was the neglect of Stark broadening in the line profiles when solving rate equations and radiative transfer simultaneously. Stark broadening was only taken into account in the final formal solution after the iteration cycle for the occupation numbers was finished. As pointed out by Méndez et al. (1983) this approximation might affect the cores of the computed hydrogen and helium lines. Herrero (1987 a and b) using the new numerical technique of "Accelerated Lambda Iteration" by Werner and Husfeld (1985) has now overcome this approximation and treated this problem correctly. He included Stark broadening in the rate equations, added a large number of radiative transitions and took also into account the line overlap of H and HeII by treating both atoms simultaneously. As a result he obtained significantly deeper line cores, which led to much better agreement with the observations. (Examples are given in Fig. 1 and 2). As an important consequence of this result some CSPN effective temperatures obtained spectroscopically by Méndez et al. (1981, 1983, 1985) had to be revised towards 10 to 20 percent higher values.

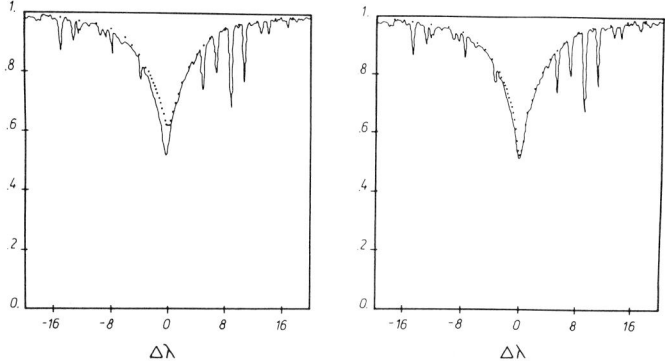

Fig. 1: Observed high S/N Hγ-profile of the O9.5V star τ Sco compared with standard NLTE calculations (left) and the recent improvements by Herrero, 1987a (right).

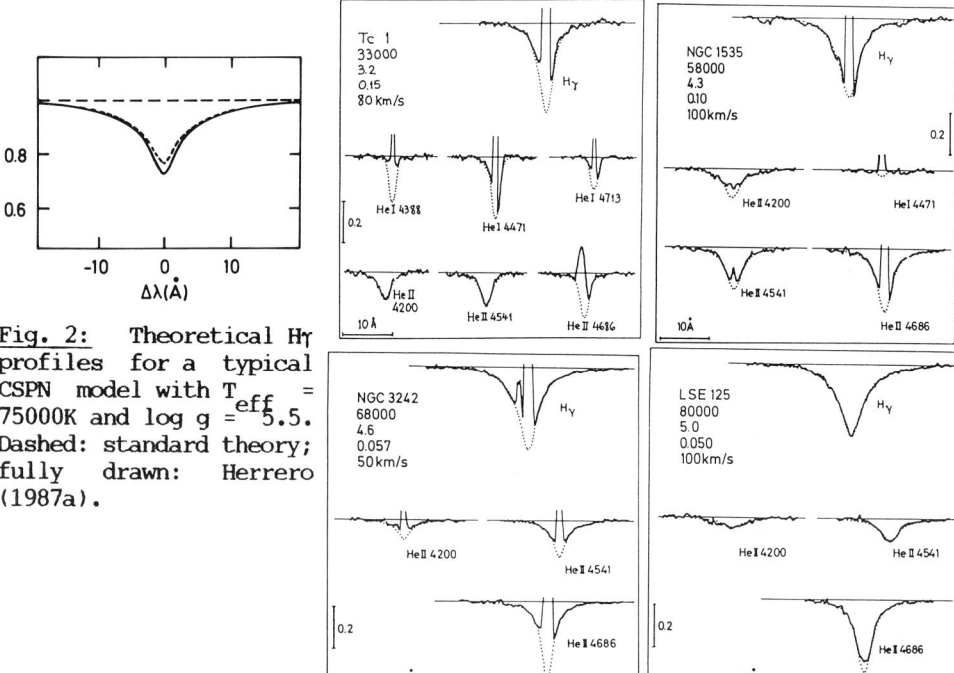

Fig. 2: Theoretical Hγ profiles for a typical CSPN model with T_{eff} = 75000K and log g = 5.5. Dashed: standard theory; fully drawn: Herrero (1987a).

Fig. 3: 4 examples of the profile fits obtained by Méndez et al.(1987). For every object T_{eff}, log g, y and the rotational (or macroturbulent) velocity (obtained from metal lines) are given.

2.2. Application to high resolution, high S/N optical spectra of CSPN

The analysis methods outlined in the preceding section have now very recently been applied on a large sample of CSPN observed with high resolution and high S/N (Méndez et al., 1987). The observational material consists of ESO 3.6 m CASPEC plus CCD spectra. The covered wavelength range is 4000 to 5000 Å, the resolution 0.2 to 0.3 Å. A S/N of 50 to 100 was achieved in most cases. The limiting magnitude is m_v = 14. 26 CSPN were observed in this way including for the first time objects embedded in nebulae with high surface brightness, which were omitted in the previous studies (Méndez et al., 1981, 1983, 1985).

The estimated uncertainties are ±10% in T_{eff}, ±0.2 in log g and ±20% in y. These estimates include probable systematic errors, which will be discussed below.

Table 1: Spectroscopic parameters of CSPN

	$T_{eff}(10^3 K)$	log g	y	$M(M_\odot)$	d(kpc)
NGC 7293	90±10	6.9±.2	.009±.005	.55±.02	0.30
LSE 125	78±5	5.0±.2	.05±.02	.60±.03	1.3
NGC 7009	75±10	4.7±.2	.05±.03	.70±.05	2.5
NGC 4361	75±5	5.4±.2	.05±.02	.55±.01	1.3
NGC 1360	72±5	5.3±.2	.07±.02	.55±.01	0.67
NGC 3242	68±5	4.6±.2	.05±.02	.65±.04	2.0
NGC 1535	58±5	4.3±.2	.09±.02	.66±.04	2.7
IC 2448	55±5	4.5±.2	.11±.02	.57±.02	4.5
NGC 6891	50±8	3.9±.2	.07±.02	.75±.07	3.8
NGC 2392	47±7	3.6±.2	.35±.10	.90±.13	2.7
NGC 6629	47±5	3.8±.2	.08±.02	.73±.06	2.4
IC 4637	47±5	3.9±.2	.09±.02	.67±.04	1.6
EGB 5	42±5	5.8±.2	.003±.002	?	?
IC 418	36±4	3.3±.2	.15±.04	.77±.07	2.0
He 2-182	36±2	3.4±.15	.09±.02	.70±.05	7.4
He 2-108	33±2	3.1±.2	.15±.03	.81±.09	8.3
Tc 1	33±2	3.2±.15	.14±.02	.72±.06	3.8
M 1-26	33±2	3.2±.15	.12±.02	.72±.06	1.9
H 2-1	33±2	3.3±.15	.08±.02	.67±.04	4.6
He 2-138	27±2	2.7±.2	.30±.10	.87±.12	5.0
He 2-162	27±2	2.9±.15	.18±.03	.68±.04	4.0
He 2-151	25±2	2.7±.15	.12±.02	.73±.06	8.0

Fig. 3 displays typical profile fits, which give an impression about the quality of the spectra and the theory applied: Tc1 is a typical example of the cooler objects, where nebular emission affects only H and HeI. (The emission of HeII 4686 comes from the stellar wind, see below). It is evident that without high resolution the HeI absorptions would be undetectable. Typical nebular contamination is also important for the hotter objects NGC 1535 and 3242. Note that NGC 3242 is the prototype of the "continuous spectral type", which according to Kudritzki et al. (1981 a and b) has no real physical meaning. It reflects simply a resolution problem in the case of narrow photospheric lines and very strong nebular lines. LSE 125 is a good example of an object within a low surface brightness nebula, which allows to fit also the very cores of the lines.

The spectroscopic parameters obtained by Méndez et al. (1987) are summarized in Table 1, which also contains already stellar masses and distances. How these quantities are derived is discussed in the two following sections.

2.3. Evolution of CSPN

The model atmosphere approach yields (besides the helium abundance) T_{eff} and log g of the CSPN. This enables us to test the predictions of stellar evolution theory in a completely alternative observational way, namely the log g, log T_{eff}-diagram. This approach has a fundamental advantage: It is independent on any assumption about nebular distances and therefore allows us to constrain the evolution of CSPN in an observationally independent way.

Fig. 4 shows the log g, log T_{eff}-diagram of 22 CSPN. From the transformation of post AGB evolutionary tracks into this diagram the evolutionary status of the CSPN is evident: They are clearly post-AGB objects with masses between 0.55 and 0.9 $M_\odot$, which nicely agrees with the masses of DA White Dwarfs (Weidemann and Koester, 1984). Since the error box arising from the fit of the observed hydrogen and helium lines just by chance has the same inclination as the tracks, a rather precise determination of individual masses is possible from Fig. 4. These masses are given in Table 1.

A few comments are necessary. First, the sample is clearly biased by the selection of more luminous (i.e. more massive) CSPN in this first pioneering step. Future observations using the ESO EFOSC spectrograph will allow to complete the sample in view of mass distribution statistics. Second, the masses of the objects close to the Eddington limit have larger uncertainties for two reasons: In the log g, log T_{eff}-plane the constant luminosity tracks of higher masses lie closer, which for the same Δlog g ≈ ±0.2 yields ΔM ≈ ±0.13 $M_\odot$ for an object like NGC 2392, whereas ΔM ≈ ±0.03 $M_\odot$ for the objects of $M/M_\odot$ ≈ 0.6. Moreover, as will be demonstrated below, photospheric geometrical extension and contamination of photospheric profiles by stellar wind emission become a problem close to the Eddington limit. This might additionally affect the analysis of an extreme object like NGC 2392.

Fig. 4 contains one CSPN, which is obviously not a post AGB-object: EGB 5. This object is located in that part of the diagram, which is normally restricted to subluminous O-stars not surrounded by a nebula (for a recent review, see Kudritzki, 1987). The evolutionary status of EGB 5 is not clear. An attractive possibility might be close binary evolution. The helium poor sdO-star LB 3459 = AA Dor (see Kudritzki et al., 1982) might be an example of such a case. Fig. 5 demonstrates that the Balmer lines of EGB 5 are significantly broader

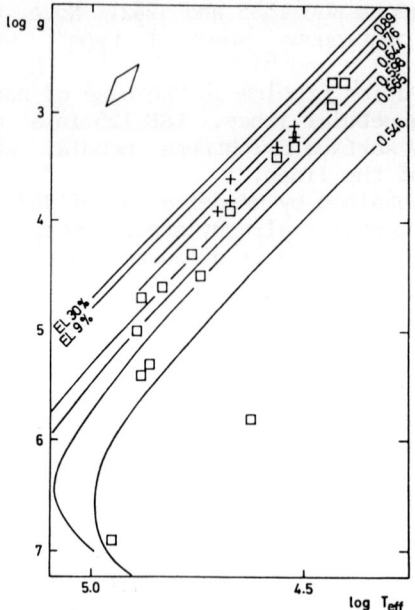

Fig. 4: The log g, log T_{eff} diagram of 22 CSPN. Crosses refer to Of-type objects, whereas squares hold for spectral type O. A typical error box is given in the upper left. Post AGB evolutionary tracks (Wood and Faulkner, 1986; Schönberner, 1983) have been transformed into this diagram and are labelled by their mass in solar units. The Eddington limits for y=0.09 and 0.3 are also shown (from Méndez et al., 1987).

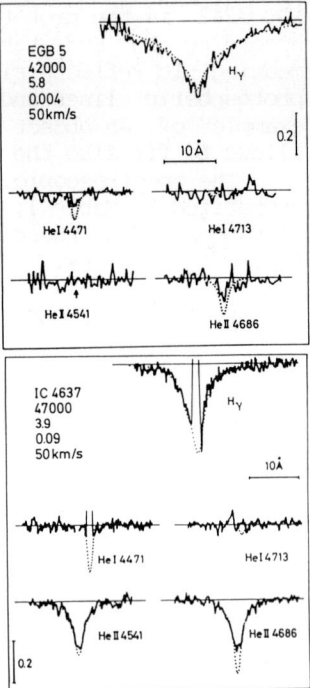

Fig. 5: Profiles of EGB 5 and IC 4637 demonstrating the high gravity of EGB 5.

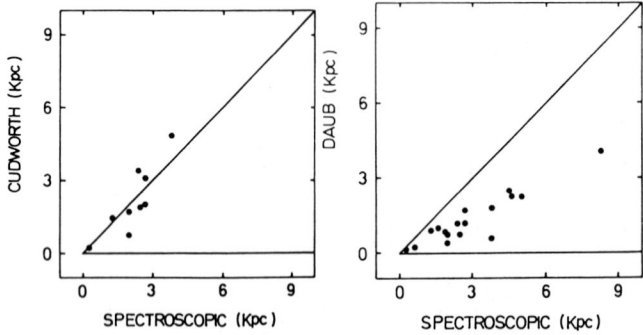

Fig. 6: PN distances by Daub (1982) and Cudworth (1974) vs. spectroscopic distances.

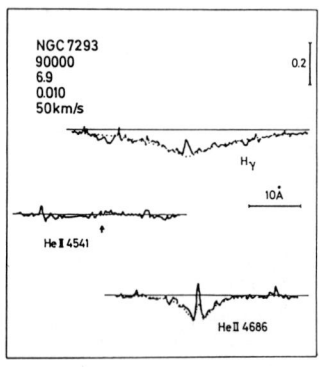

Fig. 10: Profile fit of NGC 7293.

than for IC 4637, a typical post-AGB CSPN of similar T_{eff}. Thus, the high gravity of EGB 5 provides some uncertainty for the evolutionary scenario of CSPN. Obviously not every CSPN went through the AGB channel. The question is, how many of this type do exist. This will be subject to future spectroscopic programs.

2.4. Spectroscopic distances

The model atmosphere approach allows also to obtain distances. From the log g-log T_{eff} diagram and post AGB tracks, masses can be derived. Using the gravity this yields stellar radii. On the other hand, the comparison of observed (dereddened) flux with the stellar surface flux predicted by the final model atmosphere for every individual object yields angular diameters. These combined with the radii give the "spectroscopic distances". The typical individual uncertainty of this procedure is 25%.

The spectroscopic distances are also given in Table 1. A comparison of these values with other frequently cited statistical distances (Cahn and Kaler, 1971; Daub, 1982) reveals that they are much larger. On the other hand, we find agreement with Seaton (1966) and Cudworth (1974) (see Fig. 6). We note that by the increased distances the most luminous PN in the sample of Méndez et al. (1987) become as luminous as the most luminous PN in the Magellanic Clouds, removing in this way a pronounced lack of bright "nearby" galactic PN (see Jacoby, 1980, 1983). The spectroscopic distances are also confirmed by Barlow (this meeting) in his study of LMC and SMC PN.

2.5. A test of the reliability of the model atmosphere approach

An ideal test of the reliability of the model atmosphere approach is its application on faint blue stars in globular clusters. Heber and Kudritzki (1986) have recently performed such a test for the sdO-star ROB 162 in the globular cluster NGC 6397. ROB 162 is the only known hot blue sdO star in this metal poor (Fe/H = -2) cluster. Fig. 7 shows the result of the high resolution (again ESO CASPEC spectra with S/N ≈ 50) spectroscopy of this rather faint (m_v = 13.3) object. As stellar parameters T_{eff} = 51000±2000K, log g = 4.5±0.2 and y = 0.09±0.02 were obtained.

It is now possible to determine M, R and L of ROB 162 in two alternative ways. First the spectroscopic method using the log g, log T_{eff}-diagram and second using the known cluster distance of d = 2400 pc (Alcaino and Liller, 1980), which combined with T_{eff} gives the radius directly and which combined with log g yields also the mass. The results of the two methods are compared in table 2, which shows remarkable agreement.

If we use the spectroscopically determined radius together with T_{eff}, then the comparison of observed and model calculated flux yields d = 2560 pc, which agrees well with the value mentioned above, which was obtained by main sequence and horizontal branch fitting.

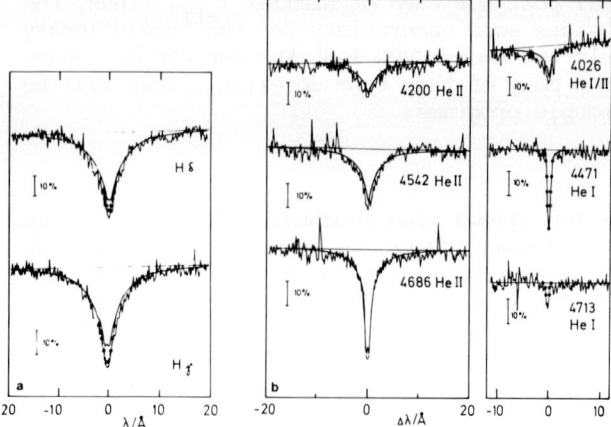

Fig. 7: Profile fits obtained in the NLTE analysis of ROB 162 (from Kudritzki and Heber, 1986).

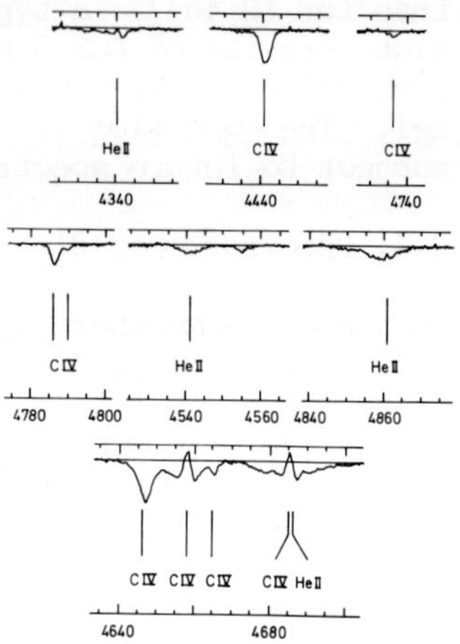

Fig. 11: ESO 3.6 m CASPEC line profiles of NGC 246. Note the enormous strength of the CIV lines and the absence of hydrogen.

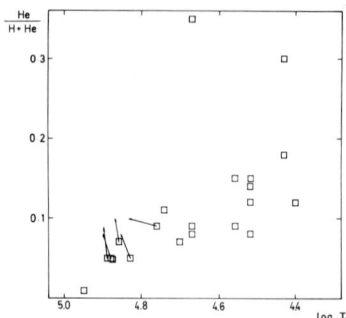

Fig. 8: Helium number fraction of CSPN as function of T_{eff}. The arrows indicate corrections applied to the results by Méndez et al. (1987), if the new broadening theory by Schöning and Butler (1988, in prep. for A&A) for HeII 4686 is used.

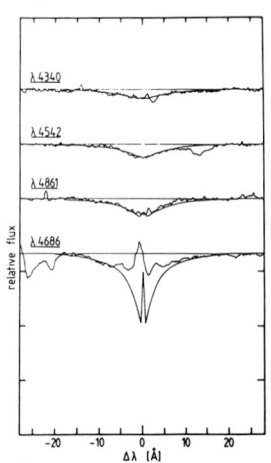

Fig. 12: Profile fits of NGC 246 (from Husfeld, 1986).

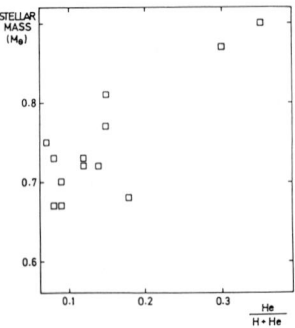

Fig. 9: Stellar mass versus helium number fraction. The diagram is restricted to objects with log T_{eff} < 4.8.

Table 2: M,R,L of ROB 162 by two alternative methods

	spectroscopic method	cluster distance method
$R/R_\odot$	$0.70^{+0.18}_{-0.14}$	0.66 ± 0.04
$M/M_\odot$	$0.56^{+0.04}_{-0.02}$	$0.50^{+0.3}_{-0.2}$
$\log L/L_\odot$	3.5 ± 0.2	3.4 ± 0.1

2.6. Photospheric abundances

2.6.1. Helium. In Figs. 8 and 9 the helium abundances of Table 1 are plotted as function of T_{eff} or $M/M_\odot$ respectively. The arrows in Fig. 8 indicate shifts, which have to be applied to Table 1, if the new broadening theory for HeII 4686 by Schöning and Butler (1988, in prep. for A&A) is used. Note that all stars have been checked by us for this effect. However, only those marked by the arrows are significantly affected, since they exhibit strong absorption wings of HeII 4686. Since $M/M_\odot$ and distance d are also slightly affected we give a correction table to Table 1.

Table 3: New parameters due to new HeII 4686 broadening theory

object	T_{eff}	log g	y	$M/M_\odot$	d(kpc)
LSE 125	85000	5.1	0.09	0.62	1.2
NGC 7009	82000	4.8	0.08	0.72	2.4
4361	82000	5.5	0.09	0.55	1.2
1360	80000	5.4	0.09	0.55	0.63
3242	75000	4.7	0.09	0.68	2.0
1535	70000	4.6	0.11	0.67	2.1
IC 2448	65000	4.8	0.13	0.58	3.5
4637	50000	4.0	0.09	0.68	1.5

From Fig. 8 it is evident that, as long as the CSPN evolution proceeds with constant luminosity, the helium abundances are larger or equal than the solar value of y=0.09. That means that in this stage the abundances obviously reflect AGB-abundances.

Fig. 9 on the other hand indicates for these objects a trend of higher helium abundances being correlated with higher post-AGB masses of our CSPN. This could be the result of stronger mass-loss or dredge up at the AGB.

The situation changes completely after the CSPN have turned down on the WD cooling sequence: The low helium abundance object at highest $\log T_{eff}$ in Fig. 8 is NGC 7293, our only high gravity object (log g ≈

7) in Fig. 4. Fig. 10 shows the profile fit for this object demonstrating the reliability of the stellar parameters, in particular the value of y=0.01 ≈ 0.1 $y_\odot$. Obviously, the process of gravitational settling, which leads to the cooling sequence of helium poor DA stars, sets in immediately after the CSPN have curved around from constant luminosity to the cooling track.

2.6.2. Photospheric metallicity. The determination of photospheric metal abundances in hot stars by means of non-LTE methods is now possible if high quality optical and UV spectra can be used for the analysis. Very detailed and improved non-LTE multi-level calculations are now available for the following ions: C II,III,IV; N II,III,IV,V; O II,IV; Mg II, Ca II; Al III, Si II,III,IV. This has been done very recently at the Munich observatory by Keith Butler, Dirk Husfeld, Sylvia Becker and Franziska Eber. This work has already been successfully applied on Population I OB-stars (Kudritzki et al., 1987a; Becker and Butler, 1987; Schönberner et al., 1988) and OB subdwarfs (Husfeld et al., 1987; Kudritzki, 1987). At this meeting a first attempt for two cooler CSPN of the Méndez et al. (1987) sample is presented by Roth et al. In our opinion, the detailed investigation of photospheric metal abundances will be one of the very interesting directions of future quantitative CSPN spectroscopy.

2.7. The extreme helium rich CSPN of NGC 246

It is well known since the work of Aller (1948) and Heap (1975) that NGC 246 is a peculiar CSPN with evidently no trace of hydrogen in its photosphere. Interestingly, the nebular abundances appear to be quite normal (Heap, 1975). Fig. 11 shows part of the ESO 3.6 m CASPEC spectra (S/N ≈ 100) of this star. Dirk Husfeld (1986) has analysed this object as part of his thesis at Munich observatory. The results of his work are given in Table 4. Fig. 12 displays the corresponding profile fits.

Table 4: Parameters of the central star of NGC 246

T_{eff}	130000±15000 K	$\dfrac{n_{He}}{n_{He}+n_H+n_C}$	: 50 to 90%
log g	$5.7^{+0.4}_{-0.2}$		
$M/M_\odot$	0.7	$\dfrac{n_C}{n_{He}+n_H+n_C}$	: 10 to 50%
$\dfrac{n_H}{n_{He}+n_H+n_C}$	< 10%		

NGC 246 is obviously the hottest CSPN studied by quantitative spectroscopy so far. Besides the absence of hydrogen it is characterized by an enormous amount of carbon in its photosphere. According to Husfeld (1986) we are looking at an extreme case of AGB mass-loss, which left the AGB intershell matter as the present photosphere.

At this point we want to stress that extreme helium rich CSPN are not rare. According to Méndez et al. (1986) 35% of all spectroscopically well studied CSPN belong to this class!

2.8. Problems

2.8.1. Systematic errors due to present NLTE-models for CSPN. The NLTE models used by Méndez et al. are still far from being physically perfect. They make use of a variety of approximations of which the most important ones are now discussed here.
- metal line blanketing is neglected:
  The inclusion of NLTE metal line blanketing was for a long time impossible for simple numerical reasons. On the other hand, LTE blanketed models are completely unreliable at these high temperatures. However, after the work by Anderson (1985), Werner and Husfeld (1985) and Werner (1986) methods are known which will allow the computation of realistic NLTE line blanketing very soon. In fact Werner (1987) in his thesis has already calculated CSPN models, which include the blanketing of more than 100 lines of H, He and C in NLTE. The effect on the temperature structure is surprisingly low in the region of formation of optical H and He lines. Thus, we expect corrections of T_{eff} only by up to 5 to 10% by this effect.
- wind blanketing is neglected:
  Wind blanketing is induced by the backscattering of photospheric photons due to the metal lines formed in the surrounding stellar wind envelope. Abbott and Hummer (1985) and Bohannan et al. (1986) have investigated this effect in the case of the O4f-star ζ Puppis. A $\Delta T_{eff} \approx 4000K$ was found. For CSPN a similar estimate is not so easy as it requires accurate knowledge of the mass-loss rates, which is observationally not available at the moment. However, following the results of radiation driven wind theory as presented in the second part of this paper, we would expect significant effects with respect to T_{eff} for objects with $M/M_\odot > 0.75$. On the other hand, since the correction in T_{eff} would proceed along the inclined error bars of Fig. 4, conclusions with respect to stellar mass would be less affected.
- atmospheric extension is neglected:
  The NLTE models are planeparallel, which means that spherical extension is neglected. This approximation was investigated by Gruschinske and Kudritzki (1979) for extended hydrostatic NLTE models for sdO and CSPN. Only small effects were found. However, for the extreme objects like NGC 2392 and some other cooler low gravity objects close to the Eddington limit systematic changes in the line profiles might be important. In these cases also the effects of stellar winds causing probably deviations from the hydrostatic equilibrium might be important. This effect will also be discussed below.

2.8.2. The Zanstra discrepancy. One of the longstanding problems of CSPN is the discrepancy between the Zanstra temperatures derived from nebular hydrogen and ionized helium lines and the spectroscopic effective temperature. Husfeld et al. (1984) indicated a way out for hot CSPN close to the Eddington limit. Henry and Shipman (1986) suggested a solution for CSPN which have a lower helium abundance due to gravitational settling. With the well defined stellar parameters of Table 1 it is possible to reinvestigate this problem. For this purpose we defined "Zanstra ratios" ZR in the following way:

$$ZR = \log \frac{\text{Number of ionizing photons (cm}^{-2}\text{s}^{-1}\text{)}}{\text{stellar continuum flux at 5480Å (erg cm}^{-2}\text{s}^{-1}\text{ Hz}^{-1}\text{)}}$$

In the next step we investigated whether our NLTE models for the final T_{eff}, log g, y are able to reproduce the observed Zanstra ratios. This is done in Fig. 13 for black bodies as well as NLTE models.

It is obvious that for both blackbodies and NLTE models a deficiency of hydrogen Lyman photons is observed. However, this is not an enormous factor, which could easily be explained by the neglect of metal line blanketing or by the fact that the nebulae are optically thin. For the HeII photons we observe a clearly pronounced excess relative to the NLTE models. Using blackbodies the effect is smaller but still present.

From Fig. 13 we conclude that the hydrostatic planeparallel NLTE models generally fail to produce the observed stellar flux shortward of the HeII edge at 228Å at least for the sample of Table 1. This casts some doubts on the reliability of the models. However, we will suggest a solution at the end of the paper.

3. SPECTROSCOPY OF STELLAR WIND LINES

3.1. Observation of winds in CSPN spectra

After the advent of the IUE satellite it became undoubtedly clear that stellar winds are present in many CPN (Heap, 1978; Perinotto, 1982). P-Cygni profiles in the UV have been detected for a variety of objects and many attempts have been made to determine terminal velocities v_∞ as well as mass-loss rates $\dot{M}$. A typical example is the Central star of NGC 3242 (see Fig. 14). Here the detailed fit of the observed NV P-Cygni profile yielded v_∞ = 2200±100 km/s and log $\dot{M}$ = -9.0±1.0 ($\dot{M}$ in $M_\odot$/yr). The large error in $\dot{M}$ reflect mainly the uncertainty of the ionization calculations, which to our eyes is typical for these objects. The general situation is best described in the paper by Cerruti-Sola and Perinotto (1985) and by Perinotto (invited paper, this meeting).

For the purpose of stellar atmosphere theory including stellar winds it is important to have a reliable hypothesis for the wind driving mechanism. Thus the question is, whether these winds are radiation driven as in the case of massive OB-stars. We present two observational arguments:

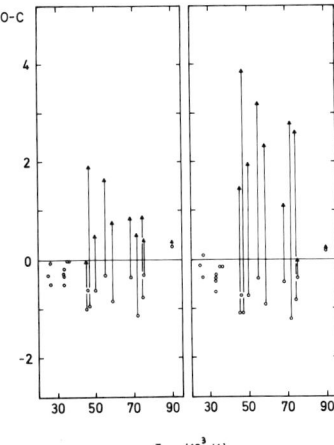

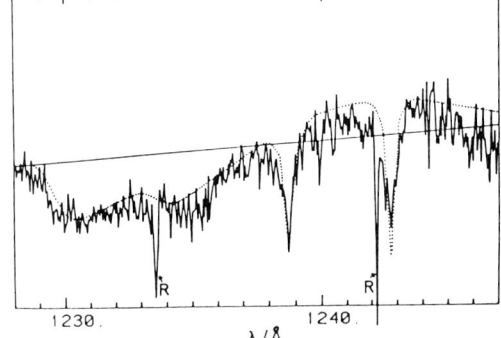

Fig. 13: Observed minus computed Zanstra ratios for H (open circles) and HeII (triangles) as function of stellar effective temperature of Table 1. The left part uses black bodies for the stellar flux, the right part the final NLTE models of Table 1.

Fig. 14: IUE high resolution profile of the NV P-Cygni profile of NGC 3242. The dotted curve shows the theoretical profile fit obtained by detailed comoving frame calculations (from Hamann, Kudritzki, Méndez, Pottasch, 1984).

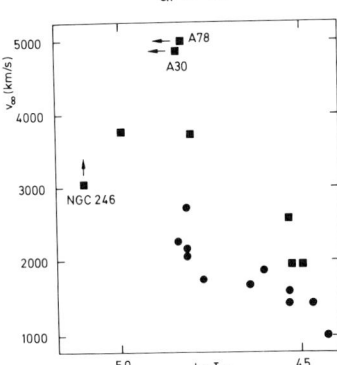

Fig. 16: Terminal velocity versus T_{eff} for CSPN. Squares are H-deficient objects; circles, normal Hydrogen abundance.

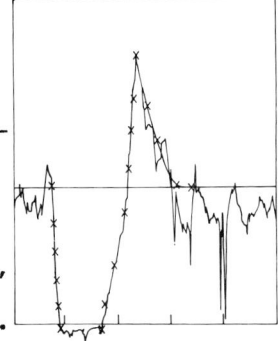

Fig. 17: NV λ 1240 IUE high resolution profile of ζ Puppis compared with prediction of a selfconsistent radiation driven wind model (crosses), which has only the stellar parameters T_{eff}, log g, R as free parameters (from J. Puls, 1987).

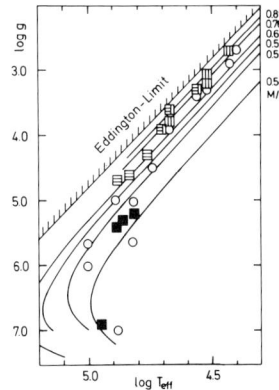

Fig. 15: The (log g, log T_{eff})-diagram of CPN according to Méndez et al. (1987, 1985). Evolutionary tracks (labelled by M/M☉) are also shown. IUE wind detections are indicated by ☰. Wind detections by optical spectra (HeII 4686 in emission <=> Of-spectral type) are given by ▥. Definitely no winds (IUE plus optical) is coded by ■. Objects with no winds in the optical but no IUE high resolution information yet available are described by o. The Eddington-limit is also indicated.

- The stellar wind features observable in the UV and/or optical become stronger for CSPN closer to the Eddington limit. This is demonstrated clearly by Fig. 15.
- The terminal velocity v_∞ of CSPN winds increases with T_{eff} (see Fig. 16). This strongly points to radiation driven winds being present in the outer layers of CSPN, since in this case the terminal velocity increases with the surface escape velocity. CSPN evolve at constant luminosity towards the blue, so that the escape velocity increases. Consequently we expect the terminal velocity to increase with T_{eff}.

3.2. Radiation driven wind models for CSPN

Motivated by the results of Fig. 15 and 16 we have calculated radiation driven wind models along the evolutionary tracks of post-AGB objects (Schönberner, 1983; Wood and Faulkner, 1986) of different masses (see Fig. 4). These calculations include the following significant improvements relative to the original wind theory by Castor, Abbott and Klein (1975):
- improved dynamics by dropping the radial streaming approximation (see Pauldrach et al., 1986; Kudritzki et al., 1987b). These improvements allowed to match the observed values of terminal velocity and mass loss rate for a variety of massive OB stars, including P Cygni.
- detailed multi-level NLTE calculations for ionization and excitation (Pauldrach, 1987). They include, simultaneously with hydrodynamics, the self-consistent treatment for 10000 line transitions of 133 ions of 26 elements, electron collisions and continuum radiative transfer. For the calculation of the line force, 100000 lines in NLTE were considered.

These calculations solved for massive O-stars the "superionization" problem: they produce NV and OVI for cool winds with a wind temperature similar to T_{eff} (see Fig. 17). The self-consistent treatment of multiple scattering, as developed by Puls (1987), has not yet been included. This will be done in a refined step of the calculations.

Fig. 18 shows the calculated relation between terminal velocity and T_{eff} along the evolutionary tracks, including the observed values for CSPN. The result is extremely convincing. It allows to read off stellar masses directly, and suggests that the masses of CSPN are in a rather narrow range between 0.5 and 0.8 solar masses (Schönberner, 1981; Méndez et al., 1985, 1987). This reveals the power of stellar wind models for the determination of stellar masses. (Note, however that Fig. 18 contains different CSPN as Table 1. Thus, the difference in mass distribution is not necessarily alarming).

Fig. 19 shows the development of theoretical mass-loss rates along the tracks. The theory predicts very strong mass-loss rates for $M > 0.65 M_\odot$, when the objects come closer to the Eddington limit. This might have consequences for evolutionary time scales and nebular dynamics. In the latter case it is interesting to note that wind momentum ($\dot{M} v_\infty$) and energy ($\dot{M} v_\infty^2$) input into the nebula increase during the CSPN evolution!

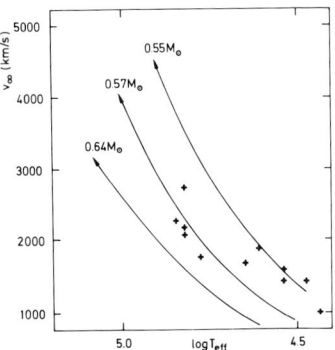

Fig. 18: Same as Figure 16 for CSPN with normal H abundance, but including wind calculations along evolutionary tracks.

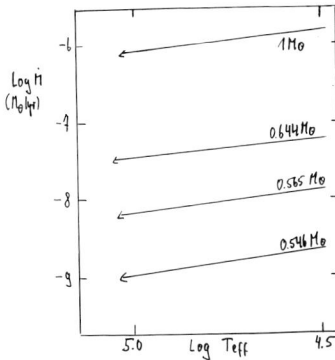

Fig. 19: Logarithm of mass-loss rate predicted by radiation driven wind theory along the evolutionary tracks as function of log T_{eff}.

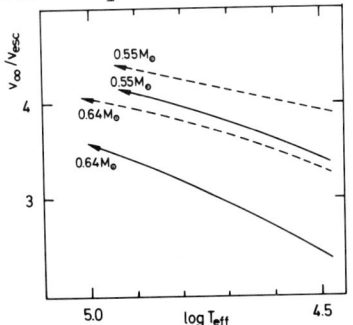

Fig. 20: The ratio v_∞/v_{esc} vs. T_{eff} for (--) hydrogen deficient objects and (-) objects containing mainly hydrogen.

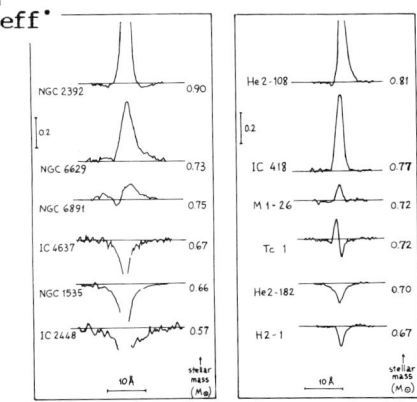

Fig. 21: The behaviour of the stellar HeII 4686 as a function of the stellar mass. The left and the right panels are for objects with T_{eff} around 50000K and 35000K, respectively.

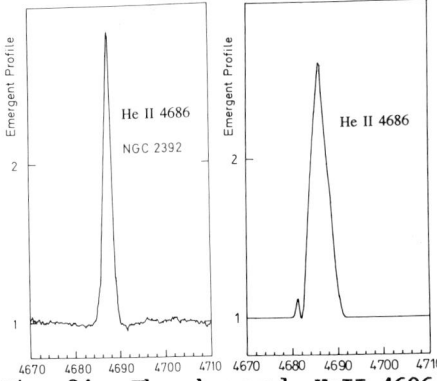

Fig. 24: The observed HeII 4686 emission of NGC 2392 (left) compared with a corresponding model calculation (right).

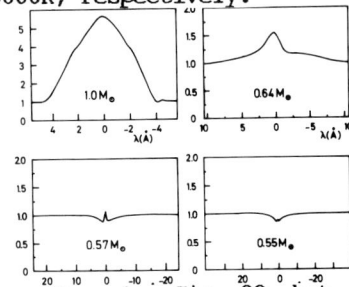

Fig. 23: As Fig. 22 but HeII 4686. Note the enormous emission at one solar mass.

3.3. Wind distances

For massive OB-stars the relation $v_\infty = 3 v_{esc}$ is observed, as long as $T_{eff} > 30000K$ (Abbott, 1978, 1982). Kaler et al. (1985) adopted this relation for CSPN also and combined it with the core-mass luminosity relation $R_\star^2 T_{eff}^4 = f(M/M_\odot)$. Since $v_{esc}^2 = 2 G M(1-\Gamma)/R_\star$ (with $\Gamma = L_E/L$), one can determine in this way the CSPN mass and radius and therefore the distance, as soon as T_{eff} is known. Kaler et al. used this method to derive "wind distances" for CPN. The crucial question is, however, whether $v_\infty = 3 v_{esc}$ holds in general for CPN. Fig. 20 shows that this is not the case. Our calculations predict that the ratio v_∞/v_{esc} depends on stellar mass, evolutionary status, photospheric helium abundance, etc. Consequently, the method of wind distances - although intrinsically very powerful - will need some refinements to become finally quantitatively reliable.

3.4. Stellar winds in the optical spectra of CSPN

As already discussed, the high resolution optical spectroscopy of CSPN by Méndez et al. shows clearly that stellar wind emission line features are systematically more strongly pronounced for objects which are located closer to the Eddington limit. A convincing example is given by the behaviour of HeII 4686 as displayed in Fig. 21. It is important to test, whether radiation driven wind models can reproduce this behaviour.

For this purpose, a new type of "unified model atmospheres" has been developed at the Munich Observatory by R. Gabler (1986) and A. Wagner (1986) in cooperation with J. Puls, A. Pauldrach and R.P. Kudritzki. These NLTE model atmospheres are spherically extended, in radiative equilibrium, and include the density and velocity distribution of radiation driven winds. The spectra of H and He lines are then calculated for these models by detailed NLTE multi-level calculations in the whole atmosphere, thus treating the contribution of subsonic deeper and supersonic outer layers to the emergent line profile in the correct self-consistent unified way, including Stark-effect broadening and velocity fields. We have calculated a sequence of such models for $T_{eff} = 50000K$ and stellar masses equal to 0.55, 0.57, 0.64, 0.75, 0.85 and 1 solar masses. The corresponding luminosities for the CSPN (or the gravities or the radii) were obtained from the evolutionary tracks mentioned already in the previous section.

By these calculations it was intended to investigate four important questions:
- Are the strategic lines used by Méndez et al. for photospheric analysis contaminated by wind emission or absorption?
Fig. 22 demonstrates that this is not the case as long as $M < 0.8 M_\odot$.
- Do typical optical wind lines increase, when approaching the Eddington-limit as demonstrated by Fig. 21?
Fig. 23 shows that in fact the new models can reproduce this behaviour at least qualitatively. Fig. 24 displays the most extreme example: NGC 2392.
- Does spherical extension together with wind outflow affect the shape

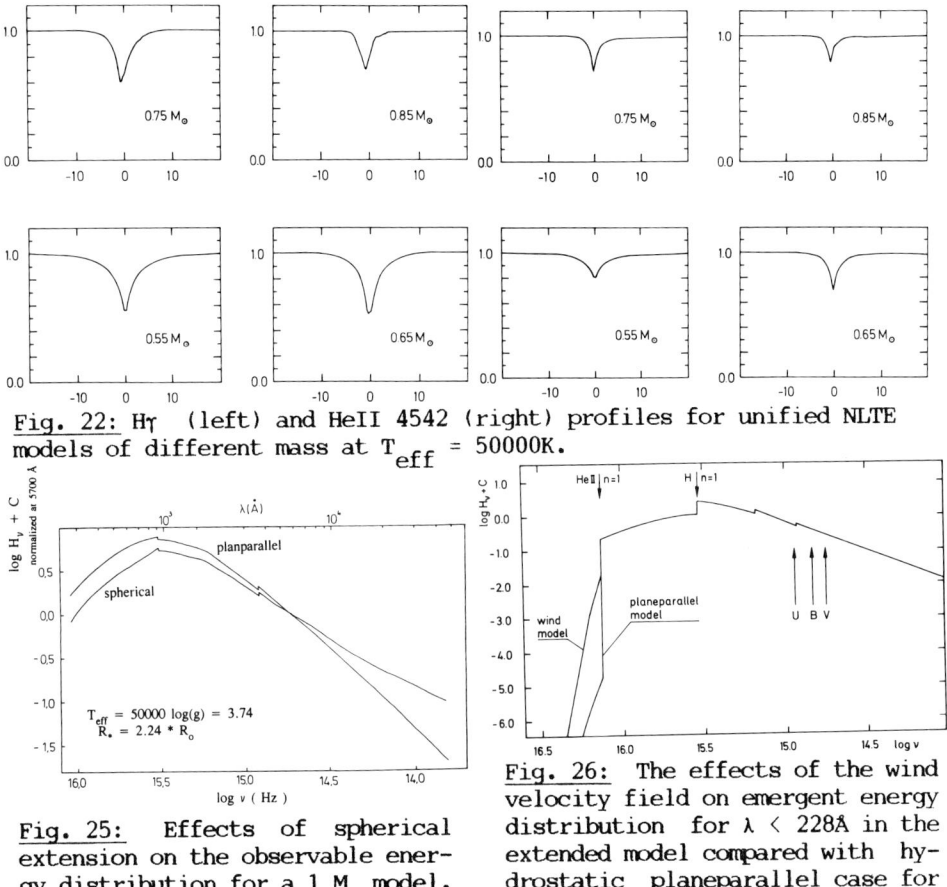

Fig. 22: Hγ (left) and HeII 4542 (right) profiles for unified NLTE models of different mass at T_{eff} = 50000K.

Fig. 25: Effects of spherical extension on the observable energy distribution for a 1 $M_\odot$ model.

Fig. 26: The effects of the wind velocity field on emergent energy distribution for $\lambda < 228$Å in the extended model compared with hydrostatic planeparallel case for M = 0.55 $M_\odot$.

Fig. 27: The departure coefficient of the HeII ground state as function of logarithm of Rosseland optical depth for a hydrostatic planeparallel model (dashed-dotted) and the extended model including the wind.

of the observable energy distribution?
No significant differences between the planeparallel and new unified models were found for 1000Å ≤ λ ≤ 10000Å as long as M ≤ 0.8 $M_\odot$. Fig. 25, however, demonstrates a significant effect for M = 1 $M_\odot$. This flattening of the energy distribution is observed for NGC 2392 (see paper by Heap and Torres, this meeting).
- Do stellar winds affect the continuum formation of HeII photons (λ ≤ 228Å)?
The answer is: yes, dramatically. Fig. 26 shows the comparison for 0.55 $M_\odot$, i.e. a model away from the Eddington-limit with a weak wind. The extended model including winds yields about 10^3 HeII-photons more than the classical hydrostatic, planeparallel model! The physical reason is as follows: Because of its enormous optical thickness the HeII-continuum is formed in the outermost layers, where the mass-ouflow is clearly supersonic. Thus hydrostatic models are in this case invalid for the formation of the HeII-continuum! In the model including the wind the very high optical thickness of the HeII resonance lines is reduced by the presence of the supersonic velocity field. This enables the strong continuum longward of 228Å to pump, by the HeII resonance transition, electrons into the n=2 stage, from which further ionization takes place. This leads to a depopulation of the ground state just in the region, where the continuum shortward of 228Å becomes optically thin (Fig. 27). As a result the usual strong HeII-absorption edge is significantly decreased, as shown in Fig. 26.
This effect will need further quantitative investigation. It looks, however, very promising to overcome the problem displayed in Fig. 13!

4. CONCLUSIONS

The modern quantitative spectroscopy of CSPN has reached a stage, where stellar parameters can be determined with high precision. Not only effective temperature, gravity and helium abundance but also mass and distance are obtained from the photospheric spectroscopy. Stellar wind lines quantitatively analysed using the improved theory of radiation driven winds allow - at least in principle - an independent determination of mass, radius and distance. Abundance determinations in CSPN photospheres and winds are at hand. A precise determination of mass-loss rates from UV lines (NV, CIV, SiIV etc.) and optical lines (Hα, HeII 4686) will be possible by means of strongly improved multi-level NLTE calculations in the winds of CSPN. Thus in total an immense potential for quantitative spectral analysis has been developed during the recent years. The future work will be characterized by the use of this potential.

ACKNOWLEDGEMENTS

We wish to thank our colleagues R. Gabler, H.G. Groth, T. Gehren, A. Herrero, D. Husfeld, A. Pauldrach, J. Puls, A. Wagner for continuous support and collaboration within this project. DFG-grants under Ku 474/11-2 and Ku 474/13-1 are gratefully acknowledged. RHM would like to thank support by the Alexander von Humboldt-Foundation, by the Max-Planck-Institute for Astrophysics, Munich, and by the Local Organizing Committee of this Symposium.

REFERENCES

Abbott, D.C.: 1978, Astrophys. J. 225, 893
Abbott, D.C.: 1982, Astrophys. J. 259, 282
Abbott, D.C., Hummer, D.G.: 1985, Astrophys. J. 294, 286
Alcaino, G., Liller, W.: 1980, Astron. J. 85, 680
Aller, L.H.: 1948, Astrophys. J. 108, 462
Anderson, L.S.: 1985, Astrophys. J. 298, 848
Auer, L.H., Mihalas, D.: 1972, Astrophys. J. Suppl. 24, 193
Auer, L.H., Heasley, J.N.: 1976, Astrophys. J. 205, 165
Becker, S., Butler, K.: 1987, Astron. Astrophys. submitted
Bohannan, B., Abbott, D.C., Voels, S.A., Hummer, D.G.: 1986, Astrophys. J. 308, 728
Cahn, J.H., Kaler, J.B.: 1971, Astrophys. J. Suppl. 22, 319
Castor, J., Abbott, D.C., Klein, R.: 1975, Astrophys. J. 195, 157
Cerruti-Sola, M., Perinotto, M.: 1985, Astrophys. J. 291, 237
Cudworth, K.M.: 1974, Astron. J. 79, 1384
Daub, C.T.: 1982, Astrophys. J. 260, 612
Gabler, R.: 1986, Diplomarbeit, Universität München
Giddings, J.: 1980, thesis, University College London
Groth, H.G.: 1986, unpublished work
Gruschinske, J., Kudritzki, R.P.: 1979, Astron. Astrophys. 77, 341
Hamann, W.R., Kudritzki, R.P., Méndez, R.H., Pottasch, S.R.: 1984, Astron. Astrophys. 139, 459
Heap, S.R.: 1975, Astrophys. J. 196, 195
Heap, S.R.: 1978, IAU Symp. 83, p. 99
Heber, U., Kudritzki, R.P.: 1986, Astron. Astrophys. 169, 244
Henry, R.B.C., Shipman, H.L: 1986, Astrophys. J. 311, 774
Herrero, A.: 1987a, Astron. Astrophys. 171, 189
Herrero, A.: 1987b, Astron. Astrophys., in press
Husfeld, D.: 1986, Ph.D. Thesis, Universität München
Husfeld, D., Butler, K., Heber, U., Drilling, J.: 1987, Astron. Astrophys., submitted
Husfeld, D., Kudritzki, R.P., Simon, K.P., Clegg, R.: 1984, Astron. Astrophys. 134, 139
Jacoby, G.H.: 1980, Astrophys. J. Suppl. 42, 1
Jacoby, G.H.: 1983, IAU Symp. 103, p. 427
Kaler, J.B., Mo, J.E., Pottasch, S.R.: 1985, Astrophys. J. 288, 305
Kudritzki, R.P.: 1973, Astron. Astrophys. 28, 108
Kudritzki, R.P.: 1976, Astron. Astrophys. 52, 11
Kudritzki, R.P.: 1987, "Spectroscopic Constraints on the Evolution of Subluminous O-stars and Central Stars of PN", Proc. of IAU Coll. No. 95 on Faint Blue Stars, ed. Davis Philip, in press
Kudritzki, R.P., Groth, H.G., Butler, K., Husfeld, D., Becker, S., Eber, F., Fitzpatrick, E.: 1987a, Proc. ESO Workshop on SN 1987A", ed. I.J. Danziger, p. 39

Kudritzki, R.P., Méndez, R.H., Simon, K.P.: 1981a, Astron. Astrophys. 99, L15
Kudritzki, R.P., Méndez, R.H., Simon, K.P.: 1981b, ESO Messenger 26, 7
Kudritzki, R.P., Pauldrach, A., Puls, J.: 1987b, Astron. Astrophys. 173, 293
Kudritzki, R.P., Simon, K.P.: 1978, Astron. Astrophys. 70, 653
Kudritzki, R.P., Simon, K.P., Lynas-Gray, A.E., Kilkenny, D., Hill, P.W.: 1982, Astron. Astrophys. 106, 254
Méndez, R.H., Kudritzki, R.P., Gruschinske, J., Simon, K.P.: 1981, Astron. Astrophys. 101, 323
Méndez, R.H., Kudritzki, R.P., Herrero, A., Husfeld, D., Groth, H.G.: 1987, Astron. Astrophys., in press
Méndez, R.H., Kudritzki, R.P., Simon, K.P.: 1983, IAU Symp. 103, p. 343
Méndez, R.H., Kudritzki, R.P., Simon, K.P.: 1985, Astron. Astrophys. 142, 289
Méndez, R.H., Miguel, C.H., Heber, U., Kudritzki, R.P.: 1986, in IAU Colloquium 87 "Hydrogen deficient stars and related objects", ed. K. Hunger et al., Reidel, Astrophys. Sp. Sci. Library 128, p. 323
Pauldrach, A.: 1987, Astron. Astrophys. 183, 295
Pauldrach, A., Puls, J., Kudritzki, R.P.: 1986, Astron. Astrophys. 164, 86
Perinotto, M.: 1982, IAU Symp. 103, p. 323
Puls, J.: 1987, Astron. Astrophys. 184, 227
Schönberner, D.: 1981, Astron. Astrophys. 103, 119
Schönberner, D.: 1983, Astrophys. J. 272, 708
Schönberner, D., Herrero, A., Becker, S., Butler, K., Kudritzki, R.P., Simon, K.P.: 1988, Astron. Astrophys., in press
Seaton, M.: 1966, MNRAS 132, 113
Wagner, A.: 1986, Diplomarbeit, Universität München
Weidemann, V., Koester, D.: 1984, Astron. Astrophys. 132, 195
Werner, K.: 1986, Astron. Astrophys. 161, 177
Werner, K.: 1987, thesis, University of Kiel
Werner, K., Husfeld, D.: 1985, Astron. Astrophys. 148, 417
Wood, P.R., Faulkner, D.J.: 1986, Astrophys. J. 307, 659

MASS LOSS RATES IN CENTRAL STARS OF PLANETARY NEBULAE

M. Perinotto
Istituto di Astronomia, Università di Firenze, Italy

ABSTRACT

Central stars of planetary nebulae (PNCS) frequently exhibit fast winds (cf. Cerruti-Sola and Perinotto, 1985; C.P. and Grewing, this volume). They may be important for the structure of the whole nebula as well as for the evolution of the central star. Their speed is typically two orders of magnitude higher than that of the classical optically visible nebula, which in turn expands a few times faster than the most external winds detected in few cases in the radio domain.

I review here the status of art in the determination of the mass loss rates ($\dot{M}$) associated with these fast winds. I restrict myself to the 'observational' determinations. Only at the end I will say something about the predictions of multi-scattering line radiation wind driven theory in connection with one best studied object: N6C 6543. This allows one to conclude that this theory may be the right explanation also for these winds.

1. INTRODUCTION

The mass loss rates associated with a stellar wind might be obtained by using:
i) P Cygni profiles of lines of heavy ions, ii) P Cygni-like profiles of hydrogen or helium lines, iii) free-free radiation in the infrared or iv) free-free radiation in the radio domain. Use of iii) and iv) has not yet successfully performed in PNCS because of the faintness of the radiation to be measured and of difficulties of taking into account the nebular contamination. Use of ii) has been done so far only in one or two objects. Thus the available $\dot{M}$ in PNCS almost all come from i).

Since various strong lines of heavy ions fall in the UV range observed by the IUE satellite (λ1200-3200 A), the observational basis for the existing values of $\dot{M}$ in PNCS rests almost exclusively in IUE data. These consist of low (6 A) and high (0.15 A) spectral resolution spectra, taken with a large aperture (oval $\approx$10"x20") or a small aperture (circular, 3" in diameter). The low resolution spectra are adequate to reveal P Cygni-type profiles, but fail if the profile is essentially in emission, as in Wolf-Rayet stars, or is dominated by a blue-shifted absorption (cf. CP). They are also useful for measuring the edge velocity associated with the wind and for first approximation determination of $\dot{M}$. The high resolution spectra are very appropriate to study in detail the whole profile, but they can be used only in objects bright enough to give a measurable stellar continuum.

The lines in which P Cygni profiles have been seen in PNCS are the resonance lines NV λ 1238.82, 1242.82; CIV 1548.20, 1550.77; SiIV 1393.73, 1402.73, the subordinate lines OIV 1338.60, 1342.98, 1343.51; OV 1371.29; NIV 1718.55 and the recombination line He II 1640.5 A.

2. METHODS TO INTERPRET THE P CYGNI PROFILES OF HEAVY IONS LINES

The methods developed to interpret these lines are: i) the escape probability method (EP), ii) the first moment of flux distribution method (FM), iii) the comoving frame method (CF), and iv) the SEI method (Sobolev plus exact integration).

In method i), the transfer equation of pure scattering line is solved using the escape probability formalism in the frame of the observer (Castor, 1970), under the Sobolev approximation. The last requires that in each point of the wind the macroscopic velocity is much larger than the stochastic velocity (local turbolence plus thermal motions). That is largely true across most of the winds in PNCS. Based on this method, Castor and Lamers (1979) have produced an atlas of theoretical profiles which has been used even in PNCS. The method has been reconsidered for the case of the excited lines and for an easier handling of the doublets by Olson (1981, 1982) respectively.

Method ii) has been developed by Castor, Lutz and Seaton (1981; CLS) just for use with the low resolution IUE data. It requires measuring the first moment W_1 of the flux distribution across the profile. The assumptions are: a) Sobolev theory of pure scattering line formation for a two-level atom in an expanding atmosphere and b) simple expressions (found successful in early type OB stars) for the velocity and opacity laws in the wind. Surdej (1982) has further investigated the CLS method showing that W_1 is actually independent of the velocity and opacity laws and of other costraints. However this is true only for very optically thin lines. Otherwise with the CLS method one underestimates M. Surdej (1983) and Hutsemékers and Surdej (1987; HS) have proposed a modification of the CLS method, for resonance and subordinate lines respectively. These improvements are useful when $W_1 > 0.24$ or $W_1 > 0.01$ respectively. We call this modified FM method FMm. The FM method has been used extensively in PNCS.

In iii) the radiative transfer equation of the line is solved in the frame of the moving fluid, under the Sobolev approximation (Lucy, 1971) or free from this assumption (Mihalas, et al. 1975). The last approach has been worked out by Hamann (1980, 1981). Based on this work, Schonberg (1985 a) has published an atlas of theoretical profiles of single resonance lines. Only one PNCS (NGC 3242) has been studied so far with the Hamann's procedure. Recently various authors have started to drop the 2-level atom approximation, used in the so far mentioned methods, for the line formation in the expanding atmosphere, introducing multi-level atoms. That may result particularly relevant for the subordinate lines observed with IUE. These lines have been treated in multi-level atom by Wessolowski and Hamann (1986). Applications to specific PNCS have not yet appeared.

Method iv), the SEI method, has been worked out by Lamers et al (1987). The source function is calculated in the Sobolev approximation, while the formal solution of the transfer equation is calculated exactly. Collisional effects are taken into account as well as the presence of a turbulent velocity across the wind. In addition the possible existence of a photospheric absorption at the base of the wind is allowed for. This method, as just illustrated, or in a less complete version (Bombek et al, 1986; BKB) has been used so far in a few PNCS. I call the method of BKB the SEIs method (SEI simplified).

3. MERITS OF THE METHODS

Method i). The EP method was found to reproduce satisfactorily the observed P Cygni profiles of thin lines in population I OB stars. At increasing optical depths, large deviations exist across the whole profile. In particular in the intermediate and inner parts of the wind, the theoretical profiles are unable to reach the observed depth of the absorption. Indeed by comparing the theoretical profiles of the EP method with those obtained with the more powerful CF method, it has been shown (Hamann, 1981;

Schonberg,1985 b) that the EP method works well for optically thin lines formed under very large velocity gradients in the wind. The profiles of the EP method differ instead substantially from those of the CF method in other cases.

Method ii). The main limitation lies in the assumed velocity and opacity laws, which may not be the same in all objects.

Method iii). The CF method is the best so far developed, because it avoids the Sobolev approximation. As a consequence, it is accurate also for low velocity winds. However Hamann has used a pure scattering line source function, therefore with occupation numbers governed only by radiation. This is true of all the studies of winds in PNCS made with the methods so far mentioned. This limitation may be not of a minor importance in the applications to PNCS, as we will see later on.

Method iv). The principle of the SEI method has been suggested by Hamann (1981). Lamers et al. (1987) have worked it out in a practical way, including various physical effects, in particular the collisions. Schonberg (1985b) has discussed the validity of the basic assumptions of the SEI method finding that it becomes inaccurate at low wind velocities or very high optical depths. On the other hand, Lamers et al. have proved that the SEI's profiles agree very chsely with the corresponding ones of the CF method in all cases published by Hamann (1981), except for a small zone close to the center of the profile. These are cases of high ratio of the wind to the stochastic velocity, appropriate to the situation in PNCS.

4. PRACTICAL CALCULATION OF $\dot{M}$

With methods EP, CF and SEI one compares the observed profiles with theoretical ones computed for assumed parametrizations of the velocity and opacity laws. Usually one parameter, β, describes the velocity law and two parameters, γ and T (the total optical depth) describe the opacity law. In the SEI method other parameters are present. In particular the role of the collisions is specified by the parameter ε. When the best fit is obtained for a given set of the parameters, the mass loss rate follows from:

$$q_i \dot{M} \propto R_* v_\infty^2 f(w)_{\beta,\gamma,T,\varepsilon} \exp(E_{gl}/kT_{rad}) / A_{el} \qquad (1)$$

where R_* is the stellar radius, $q_i(w)$ is the fractional abundance of the relevant ion, A_{el} the elemental abundance, $f(w)$ a function of the normalized velocity $w = v(r)/v_\infty$, which is known when the parameters are specified, E_{gl} the energy of the transition between the ground level and the lower level of the observed transition and T_{rad} the radiation temperature at the frequency corresponding to E_{gl}. Clearly $E_{gl} = 0$ for resonance lines. In the case of the FM method, the quantities $q_i(w)$ and $f(w)$ in equation (1) are to be substituted with $<q_i>$ and W_1, the average ionization fraction of the element and the observed first moment of the flux distribution, respectively. This in the CLS formulation. In the HS formulation one has to use instead of W_1 a function of W_1, which, for W_1 greater than the above mentioned values, depends on the velocity and opacity laws at work in the wind.

5. DETERMINATION OF MASS LOSS RATES IN PNCS

I summarize in Table 1 the available determinations of $\dot{M}$ in PNCS. The values range from $5 \cdot 10^{-11}$ to $2 \cdot 10^{-6}$ $M_\odot$ yr^{-1}, with a large scatter among the determinations of different authors in the same objects. One should bear in mind that among different authors there are not only differences in the methods but even in the way of computing q_i and in the adopted fundamental stellar quantities appearing in equation (1). I will come later on to this point. From the study of HS (see before) one infers that the determinations by CP are likely to be underestimated by factors 2-3. The values obtained from high resolution data are often larger than those obtained with low resolution data by factors much larger than the above. We cannot however assert that the first ones are automatically the more correct. All the individual determinations of $\dot{M}$ in PNCS are presented in a log $\dot{M}$ ($M_\odot yr^{-1}$), log ($L/L_\odot$) diagram in Fig. 1, together with the same numbers for population I OB stars (Garmany and Conti, 1984). The determinations belonging to the same central star have been joined by straight lines. The scatter among individual determinations in the same object is very large not only in $\dot{M}$ but even in the luminosity.

TABLE 1- Mass loss rates in PNCS

Object	Log ($\dot{M}$)	Object	Log.($\dot{M}$)
NGC 40	-7.52: (1)	NGC 6891	-9.22 (1)
1535	-9.00 (1); -7.00 (2); -7.2 (3)	7009	-9.52 (1); -8.0 (3)
2371	-7.30: (1); <-7.00 (4)	IC 418	-8.40: (1);-7.2 (3);-6.52 (11)
3242	-10.10÷-8.10 (5); -10.3 (3)	2149	-7.00 (1); -8.00 (12); -6.7 (3)
5189	-8.22 (1)	3568	-8.30 (1); -8.40 (13)
6210	-9.22 (1); -6.8 (3); -7.92 (6)	4593	-8.00 (1)
6543	-7.05 (7); -6.15 (8); -6.49 (9); -5.77 (6); -7.49 (10)	A 30	-9.52 (1)
		A 78	-9.05: (1)
6572	-9.00 (1)	Hu 2-1	-8.05: (1)
6826	-7.70 (1); -5.8 (3); -6.82 (6); -7.66 (10)	Sw St 1	-7.30 (1); -7.19 (14)
		K 648	-9.70 (15)

Sources (authors, method): 1 (CP, FM); 2 (AK, EP); 3 (BKB, SEIs); 4 (PGGW, EP); 5 (HKMP, CF); 6 (HS, FMm); 7 (CLS, FM); 8 (He, EP); 9 (BCG, EP); 10 (PCL, SEI); 11 (CBBG, No details); 12 (PBC, EP); 13 (H, EP); 14 (FV, optical); 15 (ASHAW, EP). Low resolution: 1, 6, 7, 15. High resolution: the others except 14 (optical recombination line).

6. COMPARISON WITH THEORETICAL PREDICTIONS OF $\dot{M}$

In Fig. 1 the theoretical predictions of the single-scattering line radiation wind driven theory for population I OB stars by Friend and Abbott (1986) and for PNCS by Pauldrach et al. (1987) are also shown. We see that the predictions of the theory are not far from the area where the determinations fall. However, before a close comparison with the predictions of the radiation driven theory can be made in PNCS, one must first discuss the reliability of the various experimental determinations.

7. RELIABILITY OF EXPERIMENTAL $\dot{M}$ IN PNCS

In addition to errors due to inaccuracy of the methods used to interpret the lines, errors in $\dot{M}$ come from the evaluation of q_i and of the stellar parameters entering into eq. (1). To have q_i one needs a theory of the ionization equilibrium in stellar winds complete with all the relevant physical processes: radiation, collisions and Auger ionization in multi-level atoms. Such a theory is not yet available.

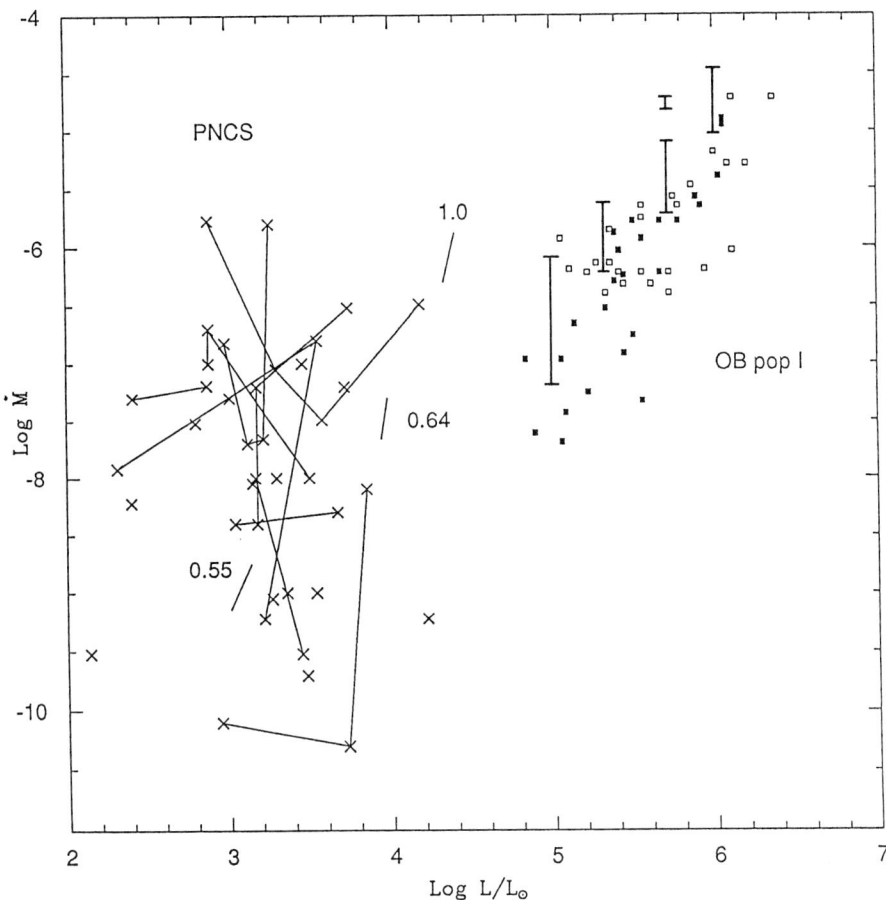

Fig 1. Log ($\dot{M}$) [M_o / yr] vs. log (L / L_o) for central stars of planetary nebulae (crosses) and for population I OB stars (squares). The last from Garmany and Conti (1984). Values belonging to the same object are joined by a straight line. Vertical bars are theoretical predictions by Friend and Abbott (1986). Tilted lines, labelled by stellar mass in M_o, are predictions for PNCS by Pauldrach et al. (1987).

A satisfactory theory, lacking only of the Auger ionization, has been worked out by Kudritzki and collaborators. Then we expect that the errors in q_i will be reduced in the future. In some of the available determinations the authors have calculated q_i using only radiative processes in atoms with the ground level and the continuum only. That is clearly inaccurate, as has been dicussed by CLS and CP. Most of the authors have however used the suggestion of CLS: $q(OIV)+q(OV) = 1$ or have searched for the maximum of $q\dot{M}$ across the wind. These last assumptions should not be very imprecise. By adopting the abundances measured in the nebula as representative of those in the winds (as done by most authors) one should not introduce a serious error. This if the lines of oxygen are used. The evolutionary theory in fact does not predict significant changes in the abundance of this element during the evolution of the central star. More severe is the problem of the stellar radius, where an error of factor of 2 can be introduced mostly by the uncertainty in the distance. Finally a very serious uncertainty may occur due to the assumed T_{rad}. For instance in NGC 6543 the different choices of T_{rad} introduce a difference in the determinations of $\dot{M}$ by CLS and by PCL of a factor of 10. To illustrate how much of the scatter of individual determinations of M may be attributed to the choice of the fundamental stellar parameters and how much may be intrinsic to the method, I consider the case of one of the best studied object: NGC 6543. In this object there are 5 values of $\dot{M}$ equal to $9 \; 10^{-8}$, $7 \; 10^{-7}$, $3.2 \; 10^{-7}$, $1.7 \; 10^{-6}$, $3.2 \; 10^{-8}$ $M_\odot yr^{-1}$ by the authors CLS, He, BCG, HS and PCL respectively. The authors have used the FM, EP, EP, FMm, and the SEI method, respectively. The first and the fourth work are made with low resolution, the others with high resolution IUE data. If, by exercise, we use in all these studies the same fundamental stellar parameter adopted by PCL, the $\dot{M}$ of CLS and HS becomes $1.2 \; 10^{-8}$ and $9.1 \; 10^{-8}$, respectively. It was not simple to reduce M from BCG to the parameters of PCL because of the particular treatment of the ionization by BCG and because of other complications. The same for the value by He because of lack of data in her study.

The difference between $9.1 \; 10^{-8}$ obtained by HS with low resolution (after reduction to the parameters of PCL) and $3.2 \; 10^{-8}$ by PCL, with high resolution IUE data and the SEI method, illustrates the residual scatter, due to the different methods, in this object. Part of this difference is due to the use of "average" velocity and opacity laws by HS with respect to the more precise ones used by PCL. Part may be due to the fact that the theory of line formation used by HS is entirely radiative while PCL have considered also collisional effects, which in this object have resulted not negligible.

How accurate is the $3.2 \; 10^{-8}$ value of PCL remains an open question, largely linked to the choice of the fundamental stellar parameters. I come to this point in the next section.

8. PREDICTION OF MULTI-SCATTERING LINE RADIATION WIND DRIVEN THEORY

It has been noticed by CLS that the observed $\dot{M}$ in NGC 6543 exceeds the limit $\dot{M}_{max}= L/(v_\infty c)$ implied by the single scattering line radiation wind driven theory. Lucy and Perinotto (1987; LP) have investigated whether the problem might be alleviated by the use of multi-scattering theory.

The theoretical treatment uses a Monte Carlo technique similar to that sucessfully used by Abbott and Lucy (1985) to study the wind of ζ Puppis. The analysis is simplified e.g. in the ionization structure across the wind, considered to occur only from the ground levels. However there are compenstion effects, because the wind is driven by some 20000 lines of several ions. Therefore, an overestimate of the abundance of a particular ion resulting in an increasing efficiency of the driving due to its lines is compensated by the contemporary underestimate of the abundance of other ions of the same element.

Input parameters are the stellar mass, the luminosity, T_{eff}, T_{rad}, the elemental abundances, the velocity law of the wind and v_∞. T_{rad} refers to the far UV part of the spectrum. Outputs are $\dot{M}$ and the synthesized spectrum of the wind. $\dot{M}$ is little sensitive to the velocity law.

When the fundamental stellar parameters of CLS are used, the theoretical predicted $\dot{M}$ is $\approx 10^{-9}$ $M_\odot$ yr^{-1} which is much smaller than observed by CLS (and by all the other authors). Moreover the computed spectrum does not match the observed one. By variating the parameters of the central star of NGC 6543, LP finds in the $\log (L/L_\odot)$, $\log T_{eff}$ plane a trajectory, which is approximately a straight line. Each point of it represents a solution which is both dynamically and spectroscopically consistent, i.e. the predicted spectrum resembles the observed one. The reproduction of the observed spectrum is better for points close to $\log T_{eff} = 4.7$, $\log (L/L_\odot) = 4.0$. Along this trajectory the corresponding $\dot{M}$ increases from $2.2\ 10^{-9}$ at $\log (L/L_\odot) = 3.1$ to 10^{-7} when $\log (L/L_\odot) = 4.7$.

A further constraint can be obtained along the trajectory. At each point one has L and T_{eff}. Thus the bolometric correction can be evaluated and then the distance of the object. From the measured expansion velocity one deduces an expansion age which can be compared with the evolutionary age in the Schonberner tracks.

A final solution is thus obtained for the central star of NGC 6543 :
$\log (L/L_\odot) = 3.8$, $\log T_{eff} = 4.7$, $M = 0.62\ M_\odot$, $\dot{M} = 1.2\ 10^{-8}$.

The solution found, totally independent, by PCL is much closer to the above than those of CLS, HS and BCG. I recall that the observational $\dot{M}$ of PCL is $3.2\ 10^{-8}$ $M_\odot\ yr^{-1}$.

9. CONCLUSIONS

There are determinations of mass loss rates from fast winds in about 20 central stars of planetary nebulae, ranging from $5\ 10^{-11}$ to $2\ 10^{-6}\ M_\odot\ yr^{-1}$. The scatter among values of different authors in the same object may amount to factors 10 - 100.

Evidence is given that the multi-scattering line radiation wind driven theory is able to predict the observed $\dot{M}$ in a well studied object to within a factor of three. This fact may also suggest that with a careful use of the best existing methods, the observational determination of $\dot{M}$ may reach a similar accuracy.

Work needs to be done to provide values of $\dot{M}$ which may be believed to be accurate to this level in all the central stars of planetary nebulae with good quality high resolution IUE spectra, both for a close comparison with the theoretical predictions and for clarifying the role of fast winds in the life of individual nebulae and of their central stars.

REFERENCES

Abbott, D.C. and Lucy, L.B. 1985, Ap.J. 288, 679

Adam, J. and Koppen, S. 1985, Astron. Astrophys. 142, 461 (AK)
Adams, S., Seaton, M.J., Howarth, I.D., Aurière, M. and Walsh, J.R. 1984,
    M.N.R.A.S. 207, 471 (ASHAW)
Bianchi, L., Cerrato, S. and Grewing, M. 1986, Astron. Astrophys. 169, 227 (BCG)
Bombeck, G., Koppen, J. and Bastian, U. 1986, in ESA SP-263, p. 287 (BKB)
Castor, J.I. 1970, M.N.R.A.S. 149, 111
Castor, J.I. and Lamers, H.J.G.L.M., 1979, Ap. J. Suppl. 39, 481
Castor, J.I., Lutz, J.H. and Seaton, M.J. 1981, M.N.R.A.S. 194, 547 (CLS)
Cerrato, S., Baessgen, M., Bianchi, L. and Grewing, M. 1987, preprint (CBBG)
Cerruti-Sola, M. and Perinotto, M. 1985, Ap.J. 291, 237 (CP)
De Freitas Pacheco, J.A. and Veliz, J.G. 1987, M.N.R.A.S. 227, 773 (FV)
Friend, D.B., and Abbott, D.C. 1986, Ap.J. 311, 701
Garmany, C.D., Conti, P.S. 1984, Ap.J. 284, 705
Hamann, W.-R.   1980, Astron. Astrophys. 84, 342
                1981, Astron. Astrophys. 93, 353
Hamann, W.-R., Kudritzki, R.-P., Mendéz, R.H. and Pottasch, S.R. 1984, Astron.
    Astrophys. 139, 459 (HKMP)
Harrington, J.P. 1982, in 'Advances in UV Astronomy: Four Years of IUE Research'
    (Nasa CP-2238), p.610 (H)
Heap, S. 1981 in 'The Universe at Ultraviolet Wavelengths' (NASA CP-2171), p.415
    (He)
Hutsemékers, D. and Surdej, J. 1987, Astr. Astrophys. 173, 101 (HS)
Klein, R.I., and Castor, J.I. 1978, Ap.J. 220, 902
Lamers, H.J.G.L.M., Cerruti-Sola, M. and Perinotto, M. 1987, Ap.J. 314, 726
Lucy, L.B. 1971, Ap.J. 163, 95
Lucy, L.B. and Perinotto, M. 1987, Astron. Astrophys., in press (LP)
Mihalas, D., Kunasz P.B. and Hummer D.G. 1975, Ap.J. 202, 465
Olson, G.L. 1981, Ap.J. 248, 1021
            1982, Ap.J. 255, 267
Pauldrach, A., Kudritzki, R.P., Gabler, R. and Wagner, A. 1987, preprint
Perinotto, M., Benvenuti, P. and Cerruti-Sola, M. 1982, Astron. Astrophys. 108, 314
    (PBC)
Perinotto, M., Cerruti-Sola M. and Lamers, H.J.G.L.M. 1987, in preparation (PCL)
Pottasch, S.R., Gathier, R., Gilra, D.P., and Wesselins, P.R. 1981, Astron. Astrophys.
    102, 237 (PGGW)
Schonberg, K. 1985 a, Astron. Astrophys. Suppl. 62, 339
             1985 b, Astron. Astrophys. 148, 405
Surdej, J. 1982, Astrophys. Space Sci. 88, 31
           1983, Astron. Astrophys. 127, 304
           1985, Astron. Astrophys. 152, 361
           1987, private communication
Wessolowski, U. and Hamann, W.-R. 1986, Astron. Astrophys. 167, 106

The Central Star of NGC 7027

N.A. Walton and S.R. Pottasch
Kapteyn Laboratorium
Groningen
The Netherlands

N.K. Reay
Queensgate Instruments Ltd.
Sunbury
Great Britain

T. Spoelstra
Radiosterrenwacht Dwingeloo
The Netherlands

Abstract

We have detected the central star of NGC 7027 by imaging the nebula through a narrow band 'continuum' filter onto the IPCS detector at the 2.5m Isaac Newton Telscope. We obtain an apparent visual magnitude for the central star of $m_v = 17.7 \pm 0.5$ mags.

Assuming that the central star radiates approximately as a blackbody, which is reasonable for the case of a hot star, then Zanstra temperatures for the central star can be calculated. We find $T_Z(H) = 3.9 \times 10^5 K$ and $T_Z(HeII) = 2.6 \times 10^5 K$. Using the correction due to Stasinska & Tylenda (1986) we estimate the central star of NGC 7027 to have a temperature, $T_{eff} = 3.1 \times 10^5 K$.

The luminosity and radii are found assuming a distance of d=1.2 kpc., giving $L = 12,600$ $L_\odot$ and $R = 0.039$ $R_\odot$. Placing the central star on the Log L - Log T diagram and comparing with evolutionary tracks for central stars with various masses from Wood & Faulkner (1986), indicates that the central star of NGC 7027 must have a mass, $M \geq 0.8$ $M_\odot$.

Radio observations of NGC 7027 have been taken using the Westerbork Radio Synthesis Telesocpe at 21cm. Self calibration techniques have been employed to give a radio continuum map of high dynamic range. These observations are being compared with a deep optical Hβ map to study the nature of the faint halo seen around NGC 7027 (Atherton et al. 1979)

References

Atherton, P.D., Hicks, T.R., Reay, N.K., Robinson, G.J., Worswick, S.P., Phillips, J.P.: 1979, *Astron. Astrophys.* **232**, 786
Stasinska, G., Tylenda, R.: 1986, *Astron. Astrophys.* **155**, 137
Wood, P.R., Faulkner, D.J.: 1986, *Astrophys. J.* **307**, 659

Note Added In Proof

Prime focus CCD observations of NGC 7027 have been obtained in Oct. 1987. These observations clearly show the central star. The results are presented in a forthcoming paper, Walton et al. 1988, *Astron. Astrophys. Letts.*, accepted.

MAGNITUDE MEASUREMENTS OF CENTRAL STARS OF PLANETARY NEBULAE

R. Gathier[1] and S.R. Pottasch[2]
1. European Southern Observatory, München
2. Kapteyn Astronomical Institute, Groningen

ABSTRACT. Magnitudes have been measured for 44 faint central stars of planetary nebulae by imaging the star and nebula of a CCD detector. The nebula is suppressed by using a continuum filter. The remaining nebular continuum is then subtracted as background as long as the star can be clearly seen. This is true in 41 of the 44 cases observed. Zanstra temperatures are calculated from the observed magnitudes, and discussed.

STROMGREN PHOTOMETRY OF THE CENTRAL STARS OF PLANETARY NEBULAE

R. Costero and J. Echevarría
Instituto de Astronomía
Universidad Nacional Autónoma de México

ABSTRACT. An attempt is made to deconvolve the contribution of the central star from that of nebular emission, in small and large Planetary Nebulae. The use of medium-band Strömgren photometry is possibly a more powerful tool than that of broad-band photometries to achieve meaningful stellar parameters, specially by means of a four-channel photometer. The various methods used to subtract the nebular contribution are analysed and the results of the first ubvy and $H\beta$ observations are presented.

NEW IDENTIFICATIONS OF FAINT CENTRAL STARS IN EXTENDED PN

K.B. Kwitter (Williams College), and G.H. Jacoby (KPNO/NOAO)

ABSTRACT. As the first step toward assessing the high-mass end of the PNN (and hence, WD) mass distributions, we report here on new PNN found in extended, low surface brightness PN. Our ultimate goal is to obtain luminosities and temperatures for these stars and to compare them with evolutionary calculations to derive stellar masses. Based on studies of a local sample of PN, Schönberner (1981, 1983) and Schönberner and Weidmann (1983) infer that the PNN mass distribution is sharply peaked near 0.6 $M_\odot$, and that 80% of PNN have masses < 0.61 $M_\odot$. Their sample was necessarily chosen for the existence of photometric data for both the nebulae and their central stars, criteria that excluded small bright PN whose central stars are masked by nebular emission, and large faint PN for which there is little photometric data for the nebulae and/or the central stars. Acknowledging these selection effects, those authors have urged further investigation, especially of low luminosity PNN.

We believe that at least some PNN currently at low luminosity are high mass cores that have undergone rapid post-AGB evolution, and have faded. Precisely because of the rapid fading time associated with high mass remnants, (e.g., Iben and Renzini 1983), they are expected to be visually quite faint. We have obtained deep (m > 21) UBV images with a CCD on the KPNO 2.1-m. Of the 26 PN on our original list, we have identified central stars candidates in 17, with certainties ranging from "extremely likely" (9) to "possible" (5) (see Table 1). Characteristics strengthening a candidate's classification are: 1) central location (symmetry); 2) blue color (when known, nebular extinction was used to calculate unreddened colors, which should be blue (B-V < 0); 3) apparent magnitude at least roughly consistent with crude estimates of nebular distance.

Of course, only follow-up spectroscopy can confirm our identifications unequivocally. However, the apparent faintness of these stars (the brightest is V = 17) will require long integration times.

TABLE 1. Nebulae with Newly Identified Central Stars

A5*	A59	K2-5	Pu-1	We-5
A9*	A80*	M1-28	Pu-2	
A18	He1-4	M4-11	S188	
A45*	K2-2	NGC 6852	We-2	

*Abell (1966) lists a "possible" candidate.

REFERENCES

Abell, G.O. 1966, Ap. J., 144, 259.
Iben, I., Jr., and Renzini, A. 1983, Ann. Rev., 21, 217.
Schönberner, D. 1981, Astron. Astrophys., 103, 119.
Schönberner, D. 1983, Ap. J., 272, 708.
Schönberner, D., and Weidemann, V. 1983, in Planetary Nebulae, IAU Symposium No. 103, ed. D. Flower (Dordrecht: D. Reidel), p. 359.

EINSTEIN X-RAY OBSERVATIONS OF PLANETARY NEBULAE AND THEIR IMPLICATIONS

S.P. Tarafdar and K.M. V. Apparao
Tata Institute of Fundamental Research
Homi Bhabha Road, Bombay 400 005, India

ABSTRACT. Central stars of nineteen planetary nebulae were observed for X-ray emission using the Einstein Observatory and four of them were detected. High resolution observations with the Einstein Observatory indicates that the X-ray source in NGC 246 is a point source. These planetary nebulae with positive observations turn out to be the nearest, have the least extinction and also have the largest size of the nebulae around them. It is possible that X-ray emission is observed from these planetary nebulae with larger ages because of the smaller extinction by the nebulae and also due to the settling of heavy elements in the central star which otherwise prevents escape of X-rays by providing opacity.

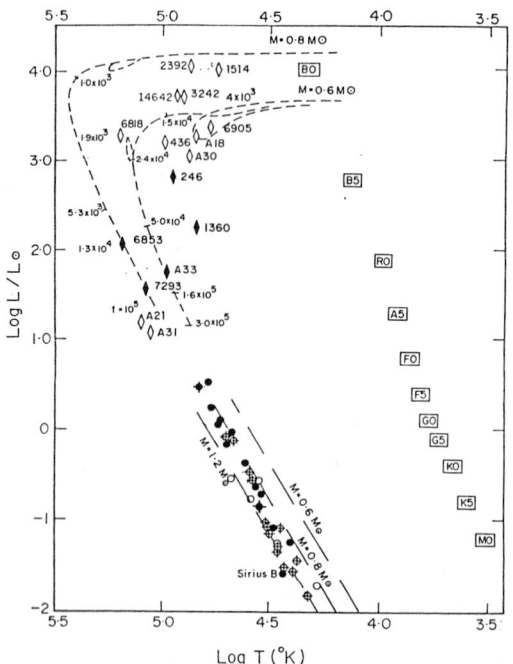

Fig. 1. Positions of planetary nebulae with X-ray observations (diamond symbols) are shown in the HR-diagram. Filled symbols for positive observations and open symbols for upper limits. Positions of main sequence stars are marked with squares. White dwarfs with X-ray observations are marked with circles; crosses superimposed on circles indicate white dwarfs other than DA type. Dashed curves are evolutionary tracks of 0.6 $M_\odot$ and 0.8 $M_\odot$ degenerate carbon-oxygen core with hydrogen and helium shell burning. Ages are on each track. Curves through the positions of white dwarfs are cooling tracks of 0.6 $M_\odot$, 0.8 $M_\odot$ and 1.2 $M_\odot$ stars given by Paczynski (1971).

THE ORIGIN OF THE ZANSTRA DISCREPANCY: UV EXCESS IN THE CENTRAL STAR CONTINUUM?

R.B.C. Henry
University of Oklahoma
H.L. Shipman
University of Delaware

ABSTRACT. The temperature of a planetary nebula central star (CPN) may be determined by observing the nebular flux in an H I or He II recombination line and the stellar flux in a continuum band. The former measures the integrated UV stellar continuum blueward of the ionization edge of the recombined ion. By assuming a continuum shape (usually a blackbody), the ratio of these two fluxes yields an effective temperature for the CPN. This particular method, first introduced by Zanstra (1931), has an advantage over others in that the observables are relatively straightforward to obtain. However, this method also carries a troublesome ambiguity with it: CPN temperatures determined using He II recombination lines. This *Zanstra discrepancy* is reviewed by Kaler (1985) and Henry and Shipman (1986). Examples of He II and H I temperatures for numerous CPNs are given in Pottasch (1984), where it is shown that the He II Zanstra temperature often exceeds the H I temperature by several times 10^4 K.

Possible explanations for the Zanstra discrepancy include arguments related to nebular optical depth, effects of dust, and a UV excess in the stellar continuum relative to a blackbody. We have employed photoionization calculations and a broad assortment of model stellar atmospheres to study the effects of UV excess.

Our results show that an atmosphere composed of pure H, such as those associated with DA white dwarfs, has an adequate UV excess to explain the size of most observed Zanstra discrepancies. We also find that the UV excess associated with the non-LTE atmospheres near the Eddington limit calculated by Husfeld *et al.* (1984) and representing an AGB star with a strong wind may also be capable of explaining many of the observed discrepancies. Finally, in each model the calculated emission line ratio,[O III]λ5007/[O II]λ3727, which is an indicator of the level of nebular excitation, is reasonably consistent with observed ratios in nebulae which exhibit Zanstra discrepancies.

REFERENCES
Henry, R.B.C. and Shipman, H.L. 1986, *Ap. J.*, 311, 774.
Husfeld, D., Kudritzki, R.P., Simon, K.P., and Clegg, R.E.S. 1984, *Astron. Astrophys.*, 134, 139.
Kaler, J. 1985, *Ann. Rev. A. Ap.*, 23, 89.
Pottasch, S.R. 1984, *Planetary Nebulae* (Dordrecht: D. Reidel).
Zanstra, H. 1931, *Publ. Dom. Ap. Obs.*, 4, 209.

ARE ZANSTRA TEMPERATURES ALWAYS REAL?

L.H. Aller
Astronomy Department, University of California, Los Angeles

ABSTRACT. The classical Zanstra method compares a H Balmer line or He II λ4686 with flux from a limited region of the planetary nuclear (PNN) spectrum to obtain the central star temperature T(PNN). Long ago it was found that generally T(He II) > T(H I), a result attributed to differing optical depths at the Lyman limits of He II and H I; T(He II) often was regarded as the "real" temperature. Much attention has been paid to discordances between Zanstra temperatures and those found by other means, such as the energy-balance (EB) method. (see e.g., Preite-Martinez and Pottasch 1984).

Now T(PNN) may also be found from the excitation level of the planetary nebula (PN). We reproduce the spectrum of a given PN by a theoretical model; adjustable parameters include the energy flux of the central star, the truncation radius, the density and size of the shell, and its chemical composition. The body of line intensity and other observable data usually suffice to fix results within a range of solution, T(PNN)±ΔT. When this program is carried out, we sometimes find large discrepancies between Zanstra temperatures and those indicated by allowable stellar fluxes.

Yet another and very powerful method (when it can be applied) is to analyse the accurately observed PNN spectrum by non-LTE model atmosphere methods. This approach, due to Kudritzki and his associates (1987) gives a T-value which is independent of the structure of the surrounding nebula.

For some objects such as the PNN of NGC 2867, 6644, and 6741, T's by Zanstra, EB, or nebular model methods all appear to be reasonably accordant. For NGC 6891, the hydrogenic Zanstra method, the EB method, and nebular model method all agree with the effective temperature found by the Kudritzki *et al.* method.

The faintness of the PNN in the bright PN, NGC 2440 and NGC 7027 require Zanstra-method temperatures ~ 350,000 K and 310,000 K, (Walton *et al.* 1987), respectively, vs. T(PNN) = 180,000 K and 190,000 K which seem to be required by the theoretical nebular models by Shields *et al.* (1981) and by Pequinot and Gruenwald (1987). A rapid fading of these PNN cannot account for the discordance as spectroscopic effects on these relatively dense nebulae would have been seen. Evidently, the far UV energy distribution must be modified in such a way as to cut down the UV flux in the region far shortward of 228 A. Otherwise the predicted intensities of [Ne V] and other highly excited ions would be too high.

REFERENCES

Preite-Martinez, A. and Pottasch, S.R. 1983, *Astron. Astrophys.*, 126,31.
Shields, G.A. *et al.* 1981, Ap. J., 248, 569.
[The 1987 references are to papers in this conference]

TEMPERATURES AND LUMINOSITIES OF PLANETARY NEBULAE NUCLEI.

Luciana Bianchi(1), Elsa Recillas(2), Michael Grewing(3)

(1)Osservatorio Astronomico di Torino, Italy
(2)Instituto de Astronomia ,U.N.A.M., Mexico
(3)Astronomisches Institut Tuebingen, West Germany

ABSTRACT. The location of Planetary Nebulae Nuclei (PNNi) on the H-R diagram is a very important clue to understand the evolution of these objects.
Several methods exist to determine the temperature of PNNi, e.g. the Zanstra method, the energy-balance method and the comparison of the stellar continuum flux with black-body or model atmospheres. For a large sample of PNNi we determined the Teff value with the last method, using IUE low resolution spectra and data from the literature at other wavelengths, and model atmospheres by Kurucz(1979), Wesemael (1981), Wesemael et al.(1980) and by van der Hucht(1987).
We have so far collected, from the IUE archive and with several observing runs, data on 44 of the 54 PNNi classified as WR-type in the catalog of central stars of Acker et al.(1982).We are also collecting data of the O-type nuclei which we analyze in the same way.
To determine the luminosity, distances were taken from the literature or re-derived by us, and the value of the stellar radius was obtained from the model fit to the absolute UV flux.
In the Table below we give results for some bright WR-type PN nuclei, where the observed spectrum is purely of stellar origin, or the nebular continuum can be unambigously subtracted from the stellar flux due to the extension of the objects. For NGC40, NGC6543, IC418 we already published detailed studies (Bianchi and Grewing,1987; Bianchi et al., 1986; Cerrato et al.1987). For the compact nebulae, the nebular flux is computed by taking into account the atomic processes, and deriving the emitting volume from the H-β flux, and it is then subtracted from the observed flux to obtain the pure stellar emission. Results for these objects, and for the O-type nuclei, will be published elsewhere.

References
Acker, A., et al., 1982: Publ. Spec. du CDS N.3 (1982)
Bianchi,L. Grewing,M., 1987: A.A., 169,227
Bianchi, L., Cerrato, S., Grewing,M., 1987: 169, 227
Cerrato,S., Bianchi,L.,Grewing,M., Baessgen,M., 1987: in"Mass Outflows from stars and galactic nuclei", Bianchi and Gilmozzi eds., Reidel, in press
van der Hucht, K., 1987, private communication
Wesemael, F., 1981: Ap.J. Suppl., 54, 2
Wesemael,F., et al., 1980: Ap.J. Suppl., 43, 2

Name	PK number	Sp.type	E(B-V)	Teff(K)	R*/D (*)	D(Kpc)	logL/Lo
NGC40	120 +9 1	WC8	0.50	90000	0.67	0.98	4.4
IC418	215 -24 1	WC7,Of7	0.25	32500	2.8	1.40	3.7
NGC2371	189 +19 1	WC8,OVI	0.05	70000	0.145	0.5-1.2	2.4-3.1
IC3568	123 +34 1	WNC,O5	0.25	40000:	0.292	.48-2.1	2.0-3.3
NGC5189	307 -3 1	WC7-8,WC2	0.25	50000	0.275	.49-1.6	2.4-3.4
NGC6543	96 +29 1	WCN,Of	0.08	80000	0.460	1.39	4.2
NGC6826	83 +12 1	WN6,OVI	0.05	35000	1.66	.69-1.1	3.6-4.0
NGC6905	61 -9 1	OW8,WC,OVI	0.15	50000	0.23	1.8	3.3

(*)R in solar radii, D in Kpc

BROAD BASELINE FLUX DISTRIBUTION OF PLANETARY NUCLEI

Sara Heap (NASA/GSFC) and Ana V. Torres (NRC Research Associate, NASA/GSFC)

ABSTRACT. We have analyzed the flux distributions of 13 planetary nuclei (NPN) spanning the full range of spectral classes known among central stars, except white dwarfs (see Table 1). We combined low-dispersion spectra from the IUE archives with absolute spectrophotometric scans taken by Dr. P. Massey at Kitt Peak with the Intensified Reticon Scanner (IRS) to obtain flux distributions covering the wavelength range $\lambda\lambda 1150$-7200 at ~ 7 A resolution. In order to get the intrinsic stellar energy distributions, we first corrected the observed fluxes for interstellar extinction, and then subtracted out the nebular fluxes, which we estimated from the Hβ flux on the IRS observations and from the nebular temperature, density and helium ionic abundances. Finally, we fitted blackbody curves to the stellar continua.

Most of the continua of the central stars are well fit by blackbody curves. However, there is no one-to-one correspondence between the blackbody temperature (T_{bb}, hereafter given in units of 10^3 °K) and the effective temperature (T_{eff}), because the stellar flux distribution depends on gravity (g) as well as on T_{eff}. Blackbody temperatures always overestimate T_{eff}, with the oberestimate increasing with g. The observations show this dual dependence on T_{eff} and g very clearly. As we go from low temperature, low-gravity stars ($T_{eff} < 50$, $\log(g) < 4$, Of-type nuclei), to high temperature, high-gravity stars ($T_{eff} \sim 70$, $\log(g) \sim 5$, sdO types), the T_{bb}'s go from ~ 40 to ~ 200.

We compared the flux distributions of the O and WR-type NPN to those of their spectral counterparts among Pop I stars. The three Of central stars in our sample have flat continua ($T_{bb} \sim 40$), also characteristic of young massive Of stars, which have $T_{bb} \leq 50$, independent of spectral type. NGC 2392 has the lowest color temperature of the three ($T_{bb} = 35$), which is even lower than its effective temperature ($T_{eff} = 47$, Méndez, Kudritzki et $al.$ 1987, preprint). We interpret its low color temperature as a consequence of atmospheric extension. In WR-type NPN, T_{bb} goes from 21 at WC11 to > 100 at WC3. This large range in T_{bb} is totally unknown for massive WC stars, which have rather flat continua ($T_{bb} \leq 45$).

TABLE 1

NPN	Sp. Type	E(B-V)	T_{bb} (10^3°K)	NPN	Sp. Type	E(B-V)	T_{bb} (10^3°K)
NGC 1535	O3	0.04	50:	IC 3568	O5f	0.18	40
NGC 6210	O3	0.05	65	NGC 6543	Of/WR	0.03	60
NGC 6058	sdO	0.05	100	M4-18	WC11	0.40	21
Abell 36	sdO7	0.05	150	NGC 40	WC8	0.40	40
NGC 4361	sdO	0.03	200	NGC 40	WC8	0.50	60-90
IC 4593	O7f	0.05	40	Sand 3	WC3/OVI	0.45	100
NGC 2392	O6f	0.05	35	NGC 2371-2	WC3/OVI	0.05	200

PHOTOMETRIC AND SPECTROSCOPY OBSERVATIONS OF PECULIAR NUCLEI OF PLANETARY NEBULAE

G. Jasniewicz and A. Acker
Observatoire de Strasbourg, France

ABSTRACT. Photometric and spectroscopic observations of some bright central stars of planetary nebulae (PN) have been conducted between 1984 and 1987 with the following tools: differential photometer P7 (70-cm swiss telescope, La Silla c/o ESO); radial velocity scanner CORAVEL (1-m swiss telescope, Observatoire de Haute-Provence = OHP); spectrograph CARELEC with CCD detector (193-cm telescope, OHP). Within the first kind of selected objects: PN with late-type central stars, LoTr 5 and Abell 35 deserve special attention. The nucleus of LoTr 5 presents some similarities with the FK Comae stars; the photometric variations are periodic with a variable amplitude, and should be caused by migrating star-spots. A triple system in the nucleus has been tentatively suggested from the shape and the behaviour with time of the CORAVEL correlation-dip (Jasniewicz et al., 1987).
  The nucleus of A35 also presents and Hα line in emission, and photometric variations with a period of a few hours and a variable amplitude, but their origin remains still uncertain (brightness inhomogeneities or binary effects?).
  Both objects share in common a very large nebula; we have undertaken the study of other PN of this type (in particular, the possible cold nucleus of DHW2).
  The second kind of selected PN are those with an O-type nucleus which is suspected to be a binary. Photometric observations are done for NGC 1360, Abell 36 and IC 418 (Jasniewicz, 1987), with the following results: no eclipses nor reflection effects were found for NGC 1360; the V-magnitude of A 36 remains constant; we have observed very important variations for IC 418, but we failed to obtain any period.

REFERENCES

Jasniewicz, G. 1987, Ph. D. Université de Strasbourg I.
Jasniewicz, G., Duquennoy, A., and Acker, A. 1987, Astron. Astrophys., in press.

HFG1: A PLANETARY NEBULA WITH A CLOSE-BINARY NUCLEUS

Howard E. Bond
Space Telescope Science Institute
Robin Ciardullo
Dept. of Terrestrial Magnetism, Carnegie Institution of Wash.
Thomas A. Fleming
Steward Observatory, University of Arizona
Albert D. Grauer
University of Arkansas, Little Rock

ABSTRACT. HFG1 (136+5°1) is a large, low-surface-brightness planetary nebula that was discovered by Heckathorn, Fesen, and Gull (*Astron. Astrophys.*, 114, 414, 1982). In the autumn of 1986, photoelectric photometry by A.D.G. and H.E.B. showed that the 14th-mag central star of HFG1 is a large-amplitude variable. Subsequent CCD photometry by R.C. and H.E.B. reveals a sinusoidal variation with a period of 13.96 hr and an amplitude of 1.1 mag in the B band.

We interpret the nucleus of HFG1 as a close binary and attribute its light variations to heating of one hemisphere of a main-sequence companion by an extremely hot primary; no true eclipses occur. Spectroscopic observations of the central star obtained by T.F. show high-excitation emission lines whose strengths are highly variable and in phase with the orbital period. The phase dependence of the emission lines indicates that they arise in the heated hemisphere of the cool companion star.

HFG1 is the newest example of a planetary nebula whose nucleus is an extremely close binary. The nebulae have probably been ejected through binary-star interactions, possibly during a "common-envelope" phase.

NONRADIAL PULSATIONAL ANALYSES OF THE PULSATING CENTRAL STARS OF
PLANETARY NEBULAE[1]

Sumner Starrfield[1] and Arthur N. Cox
Theoretical Division
Los Alamos National Laboratory
Los Alamos, NM

ABSTRACT. We have performed nonradial pulsation analyses of the central star of the planetary nebula K1-16. K1-16 is a very unusual nebulae which appears to have ejected material that is very rich in helium. The central star shows no evidence for hydrogen in its spectrum and the helium and carbon lines are in emission. Grauer and Bond (Ap. J., 277, 211, 1984) discovered that it is pulsating with periods around 1700 sec. Although its spectral characteristics are similar to those of the PG1159-035 variables, it is pulsating in much longer periods than they are.
 We have analyzed a series of stellar models that are hotter and more luminous than those that we recently analyzed to determine the helium abundance of PG1159-035. In no case was a pure carbon-oxygen composition capable of exciting the model at periods of 1700 seconds. We are continuing this study with a variety of compositions and will report on the results at the meeting. We will also discuss the connection between the central star of K1-16, the O VI central stars, and the PG1159-035 stars.

1. Supported in part by National Science Foundation Grant AST85-16173 to Arizona State University and by the DOE.
2. Also at Department of Physics, Arizona State University, Tempe, AZ.

A SEARCH FOR COOL COMPANIONS OF PLANETARY NEBULA NUCLEI

A. F. Bentley
Department of Physical Sciences
Eastern Montana College
Billings, Montana 59101

At present only a small number of planetary nebulae are known to possess binary nuclei. Since approximately 2/3 of main sequence stars are members of binary or multiple star systems, one might expect a large fraction of PN central stars to have gravitationally bound companions. Additionally, late-type stars are more numerous, and due to their low luminosities would be difficult to detect by visual observational methods at distances where PN are typically found (≥ 1 kpc). (Only 5 known PN are thought to be nearer than 0.5 kpc). It is thus possible, and in our view probable, that a significant number of PN nuclei possess cool companions, hitherto undetected.

Since K and M stars emit a considerable fraction of their energy in the infrared, we calculated the feasibility of detecting them with the Wyoming InSb photometer, and found that such detections might be possible out to distances of $\simeq 2$ kpc. To date 13 objects have been observed, including a sample of planetaries with known binary nuclei to test our hypothesis. The results are listed below.

Object	IR Excess	Remark	Object	IR Excess	Remark
NGC 7293	no		NGC 6905	no	
NGC 6853		no detection	NGC 6543	yes	known binary
NGC 246	yes	known binary	NGC 6572	yes	known binary
NGC 7008	?		NGC 6790	yes	
NGC 40	?		NGC 6210	yes	new binary?
NGC 6826	yes	new binary ?	A 63	yes	known binary
NGC 7009	yes	new binary ?			

The observations were made using standard infrared techniques in the J, H, K, and L bands. We used a 5 arc-sec aperture centered on the hot star. Any companion located within 2 arc-sec of the nucleus should have been within our beam. Thus, at a distance of 1 kpc, any binary system with a separation of ≤ 4000 A.U. should have been observed.

We caution that the results presented here are preliminary, and still subject to analysis. Specific models to fit the data are presently being constructed. We think that we may have detected 3 new binary nuclei, NGC 7009, NGC 6210, and NGC 6826. The interpretation for NGC 6790 is not clear. This work was supported by the U.S. Air Force Office of Scientific Research, the NSF, and University of Wyoming.

A CASE STUDY OF A WC NUCLEUS

J.B. Kaler
University of Illinois
R.A. Shaw
Lick Observatory
W.A. Feibelman
NASA-Goddard Space Flight Center

ABSTRACT. We have examined the extraordinarily rich WC spectrum of the nucleus of He 2-99, an object closely akin to BD +30°3639. In all, we log 25 lines in the UV ($\lambda 1240$–$\lambda 1950$; $\lambda 2550$–$\lambda 3150$) and 89 in the optical between $\lambda 3610$ and $\lambda 7065$ (including a small number of nebular features). We provide a fundamental atlas for this class of star, wherein we give fluxes and identifications of the emission lines, including a detailed accounting of blends. The most powerful emissions are those of C III followed by C II and C IV. There is good indication that C I and even C V are present as well. Oxygen is well represented by O III; O II, O IV, and O V all appear present, but are generally confused by blends. Si III and Si IV appear, as do He I and He II. Other than N V, little case can be made for stellar nitrogen. The most serious barrier to analysis of the spectrum is the problem of coincidences and blends: there are few pure lines. Analysis of the nebular spectrum, which is severely contaminated by stellar line emission, indicates enrichment in carbon, but none in nitrogen.

THE NATURE OF THE HOT COMPANION OF THE G8 IV NUCLEUS OF ABELL 35

Michael Grewing (1), Luciana Bianchi (2)
(1) Astronomisches Institut Tübingen, FRG
(2) Osservatorio Astronomico di Torino, Italy

The nucleus of the large, low surface brightness planetary nebula Abell 35 (Abell 1966) belongs to the small group of objects which are known to have a binary nucleus. From his photometric and spectroscopic study of the object, Jacoby (1981, Astrophys. J. 244,903) found the star SAO 181201, a G8 IV star, to be located near the apex of the parabolic region of enhanced [O III] emision which is completelely absent in Hα. The G8 IV star can clearly not be the ionising source for this nebulosity nor the larger scale nebulosity of the PN proper. Jacoby concluded that the central object must be a binary and suggested a hot subdwarf as the second component, which is masked in the optical by the bright SAO star. This interpretation is supported by the analysis of the DDO and UBVRI photometry, which shows that the observed colour indices can be fitted if one assumes a 50.000 K blackbody companion.

We have observed the nucleus of Abell 35 with the IUE-satellite on February 24th, 1987. Both the short- and the long-wavelength cameras were exposed in the low-resolution mode. While in the long wavelength range the continuum and the spectral features are dominated by the G8 IV star, at shorter wavelengths we clearly see the emission from a much hotter object.

The temperature of this hotter component can directly be determined from the observed and absolutely calibrated UV intensity distribution as Jacoby has shown the interstellar extinction to be effectively zero along the line of sight towards SAO 181201, a result confirmed by the absence of any significant depression in the spectrum near 2200 Å.

Fitting the short-wavelength part of the observed spectral distribution by a blackbody distribution, we find T > 120.000 K. A temperature in this range is independently found from the equivalent width of the He II 1640 Å absorption line (W_λ =1.99 Å). Using the theoretical results from Sion et al. (1982), who calculated equivalent widths for a number of absorption lines seen in the spectra of white dwarfs, we find for log g=6 T=50.000 K, a solution which is definitely ruled out by the slope of the UV continuum. For log g=8, we find (by extrapolation) a fit for the He II line for T=135.000 K. The possible range of temperatures is therefore within the uncertainty limits completely consistent with the above result.

We feel indeed justified to apply the Sion et al. white dwarf models, because from the blackbody fit to the absolutely calibrated IUE flux and taking into account the distance to SAO 181201 of d=360 ±80 pc as determined by Jacoby (1981), it follows that the star's radius must be R < 17.700 km for T > 120.000 K.

V605 AQUILAE - THE MOST EXTREME HYDROGEN-POOR OBJECT*

Waltraut C. Seitter
Astronomisches Institut der Universität Münster
Wilhelm-Klemm-Straße 10
D-4400 Münster, F.R.Germany

V605 Aquilae, whose novalike outburst was observed in 1919, is the central object of the old planetary nebula A58. It is the only 'nova' known with a hydrogen-deficient carbon-rich outburst spectrum and a remnant consisting of a WC-type star and a nebula which appears to be totally void of hydrogen. The star shows a red continuum of magnitude 22.3 and a prominent line of CIV 580.8 nm with a total width of 4400 km/sec. Other lines are extremely weak. The remnant nebula shows only forbidden lines of heavy elements. The strongest ones are due to the nebular transitions of [OIII],[NII],[OI]. IR [OII] is present, other ions are suspected. The nebula is an IRAS point source of temperature 170 K.

The lack or extreme weakness of most of the classical diagnostic lines permits only very preliminary estimates of the stellar and nebular parameters. The Zanstra temperature is near 70 000 K, the nebular parameters are determined through direct comparison with the properties of A58. Fig.1 shows the spectrum of the remnant nebula and the stellar CIV line. Table I lists the preliminary values for the nebular remnant.

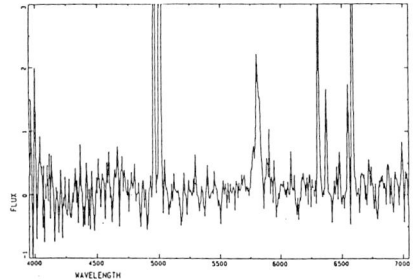

Table I. Nebular parameters

	T_e	N_e	H	He	CNO	C	N	O
A58	12 800	230	79	20	1	-	-	-
V605	14 000	$5 \cdot 10^4$	0	93	7	?	≤ 1	≤ 5

units are K, cm³, number percent.

Fig.1. Spectrum of V605 Aql in 1987 (nebular Hα is conspicuously absent).

If V605 Aql is a final helium shell flash object, then, unlike its well-known counterparts in A36 and A78, it sets a time limit of 70 years for evolution after the final flash and thus suggests helium shell burning on a dynamic time scale under total expenditure of hydrogen.

*Based on observations collected at ESO - La Silla, Chile.

ECHELLE SPECTRA OF A LARGE SAMPLE OF PLANETARY NEBULA NUCLEI

James K. McCarthy
California Institute of Technology

ABSTRACT. We have undertaken at Palomar Observatory to obtain high resolution spectra of a large sample of planetary nebula nuclei (PNN) in order to systematically investigate their spectral morphologies and then to derive temperatures and surface gravities by comparing absorption line profiles to model atmospheres. We have taken as our sample all those central stars of planetary nebulae within 1.3 kpc of the sun according to the distance determinations of Daub (Ap. J., 260, 612, 1982); of the 94 objects in this unbiased sample, 64 are in the sky visible from Palomar and 33 have central stars bright enough to be observed at a resolution of 5 000 with an "echellette" spectrograph on the 5-m Hale telescope, leaving 7 PNN (11% of the northern sample of 64 PNN) which are too faint to be observed at present.

This poster will report on work in progress, presenting representative spectra giving some idea of the quality of the data and also the range of spectral morphologies found among the central stars. Model atmosphere analysis will be performed starting in mid-October 1987 in collaboration with Kudritzki *et al.* in München, West Germany.

THE METAL-LINE SPECTRA OF CENTRAL STARS OF PLANETARY NEBULAE

M. Roth, A. Herrero, R.H. Méndez, R.P. Kudritzki, K. Butler,
H.G. Groth
Universitäts-Sternwarte, München

ABSTRACT. We present spectral descriptions based on high-resolution spectrograms of central stars of planetary nebulae, obtained with the ESO 3.6-m telescope + CASPEC (Cassegrain Echelle Spectrograph). We make preliminary determinations of stellar photospheric metal abundances, using non-LTE model atmospheres and non-LTE line formation calculations.

REVISITED MASS-LOSS RATES OF PLANETARY NEBULA NUCLEI OBSERVED WITH IUE

D. Hutsemékers and J. Surdej
Institut d'Astrophysique
Université de Liège

ABSTRACT. The first order moment of P Cygni line profiles has been computed for the case of resonance doublet and subordinate line transitions, using realistic velocity and opacity distributions. This improved method has allowed us to rederive mass-loss rates for a sample of 17 PNN observed in the low resolution mode with IUE. The average value of our mass-loss rates amounts to 10^{-7} $M_\odot$/yr.
 A detailed account of this work will be published elsewhere.

QUANTITATIVE INVESTIGATIONS ON MASS OUTFLOW FROM PLANETARY NEBULAE NUCLEI

S. Cerrato[1], L. Bianchi[2], M. Grewing[1], M. Bässgen[1], and G. Bässgen[1]
1. Astronomisches Institut, Universität Tübingen (FRG)
2. Osservatorio Astronomico di Torino (Italy)

ABSTRACT. Observed P-Cygni line profiles in the UV spectra of central stars of Planetary Nebulae (PNN) obtained with the IUE satellite has been fitted by theoretical profiles in order to determine mass loss rates.
    An attempt is made to correlate the results with the stellar parameters of both the current nuclei and their progenitors.

PLANETARY NEBULAE IN THE MAGELLANIC CLOUDS

M. J. Barlow
Dept. of Physics & Astronomy
University College London
Gower St., London WC1E 6BT

1. INTRODUCTION

The past few years have seen the study of Magellanic Cloud planetary nebulae cease to be an extragalactic sideshow and instead become one of the most quantitative branches of planetary nebula research, due largely to our ability to fully exploit the known distances to these systems since the launch of IUE and the advent of sensitive digital detectors mounted on large optical telescopes.

In this review I will concentrate on the developments that have occurred since the review given by Jacoby (1983) at the London Symposium (another relevant review is that by Peimbert 1984). Section 2 briefly describes the current situation with regard to the numbers of planetary nebulae known in each Cloud, while Section 3 describes recent work on their kinematics and internal dynamics. Section 4 reviews recent chemical abundance studies, while Section 5 describes the work that has been carried out to determine nebular masses. Finally, Section 6 surveys the work done to derive central star parameters, in particular their masses.

2. CATALOGUES

The catalogue of Magellanic Cloud planetary nebulae published by Sanduleak, McConnell and Philip (1978:SMP) lists 28 planetary nebulae (PN) in the SMC and 102 in the LMC. Jacoby (1980) found 19 faint new PN in the SMC and 35 in the LMC in his on-line/off-line filter photographic survey of selected regions. The work of Dopita et al. (1985a), Wood et al. (1987) and Meatheringham et al. (1987) has provided confirmation of the PN status of SMC J2, 4, 6, 9 and 18 and LMC J4, 5, 10, 12, 18, 20, 23, 26, 33, 38 and 41 from Jacoby's lists. Liebert & Boroson (1987) have found that SMC J10, 12, 13, 15, 16 and 22 and LMC J1, 2, 3, 6, 8, 9, 11, 13, 30 and 36 are early type stars, SMC J7, 17 and 24 are probably extended HII regions, LMC J28 and 29 are late type stars, while no object was found by them at the positions of LMC J19 and 40. The remaining Jacoby objects were confirmed by Liebert & Boroson as PN, with the caveat that a few could be peculiar emission-line objects.

Sanduleak & Pesch (1981) listed six possible SMC PN, of which four have been found not to be PN, the other two, SP 32 and 34, having been confirmed as PN (Morgan & Good 1985 and Dopita et al. 1985a). Savage, Murdin & Clark (1982) found a new high-excitation PN located outside the main body of the LMC. Finally, Morgan & Good (1985) have listed 13 new SMC PN, found on a deep UK Schmidt objective prism plate. One of these, L302, had previously been classified as a very low excitation (VLE) compact nebula by Sanduleak & Philip (1977).

Including the 32 confirmed Jacoby (1980) PN and the 2 confirmed Sanduleak & Pesch PN, a total of 54 SMC and 124 LMC PN have now been identified in papers published in the literature. At present adequate finding charts have been published for all of the PN in the SMC, but many LMC PN lack adequate charts, notably the 22 new LMC PN in the SMP catalogue for which there are no published charts (although improved positions for some of them are listed by Meatheringham et al. 1987).

Morgan (1984) has estimated excitation classes for most of the PN in the SMP catalogue, by applying to his objective prism material a development of the classification system devised by Feast (1968). This enlarged sample still shows the feature noted by Webster (1975), namely that the SMC PN have a significantly lower mean excitation class than those in the LMC.

3. KINEMATICS AND INTERNAL DYNAMICS

Dopita et al. (1985a) obtained 11.5 km s^{-1} resolution spectra of 44 SMC PN in the [OIII] 5007 Å line, increasing by a factor of three the number of PN in the SMC for which radial velocities have been determined. The PN were found to have an unstructured spheroidal distribution, with a centroid in the brightest region of the SMC Bar. Unlike the HI 21cm emission, the PN were found not to show a bimodal radial velocity distribution. Mathewson & Ford (1984) had suggested that the young gaseous component of the SMC had been disrupted by tidal interactions in the recent past, but the PN do not appear to participate in such motions, if present. Dopita et al. used the rms line-of-sight velocity dispersion of the PN to derive a mass for the SMC of 9×10^8 M$_\odot$ within a radius of 3 kpc.

Meatheringham et al. (1987; and these Proceedings) have acquired similar 11.5 km s^{-1} resolution spectra for 94 LMC PN. They re-analysed existing HI radio data in order to determine a new HI rotation solution for the LMC and found that the kinematics of the LMC PN population are in agreement with this rotation solution, the main difference being that the PN population has a much larger velocity dispersion, consistent with the PN being dynamically much older than the HI gas. Using a variety of approaches, Meatheringham et al. were able to estimate a mean age for the LMC PN population of $(2-4) \times 10^9$ years and, using the relationships of Iben and Tutukov (1984), they then predicted from this an initial main sequence mass of 1.3 − 1.6 M$_\odot$ for the precursor stars and a resulting mean central star mass of 0.61 M$_\odot$.

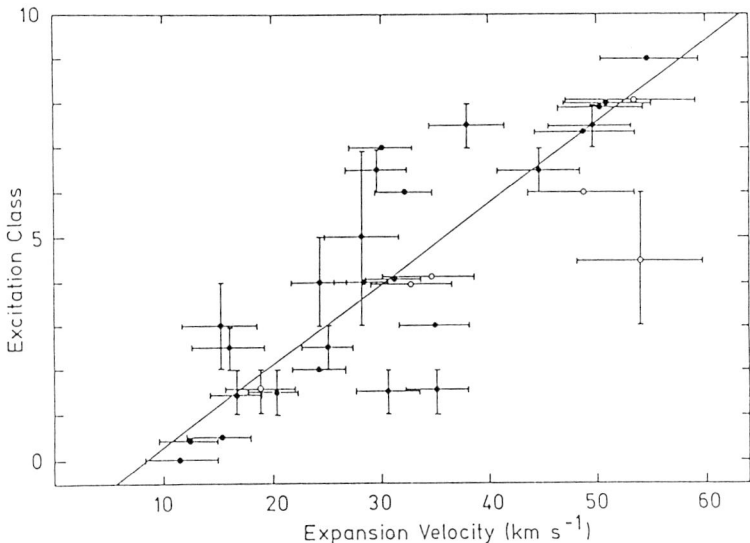

Figure 1 : OIII expansion velocities versus excitation class for SMC planetary nebulae (from Dopita et al 1985a).

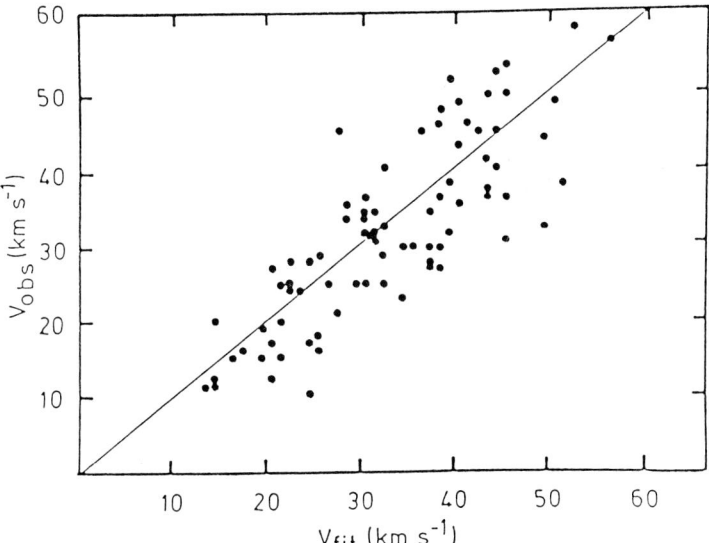

Figure 2 : Observed OIII expansion velocities for SMC and LMC PN, versus the fitted velocity from the excitation class/Hβ flux relation described in the text (from Dopita et al (1987a).

The same [OIII] spectra have also been used by Dopita et al. (1985a, 1987a,b) to study the internal dynamics of the nebulae. Figure 1, from Dopita et al. (1985a), shows a plot of nebular expansion velocity versus excitation class (on the system of Morgan 1984), for 31 SMC PN. A strong correlation is clearly evident. It should be noted that for gaussian line profiles the nebular expansion velocity is defined in the papers of Dopita et al. as equal to half the full width at 0.1 intensity, whereas for Galactic PN the expansion velocity has usually been defined to be equal to half the full width at 0.5 intensity. The expansion velocities listed by Dopita et al. are therefore 1.82 times larger than those that would be derived using the latter definition.

Dopita et al. (1985a) found that for 9 of the 44 SMC PN they observed, the [OIII] line profiles could not be fitted by a single gaussian, which was taken to indicate the existence of bipolar or multiple shell structures. Three extreme examples of this effect, in the LMC, were described by Dopita et al. (1985b) − velocity profiles covering a total range of up to 200 km s^{-1} were found.

Dopita et al. (1987b) have found that the [OII] 3727 Å expansion velocities of LMC PN are well correlated with those obtained from [OIII] 5007 Å, but systematically larger. However, the correlation between excitation class and [OIII] expansion velocity for LMC PN was found not to be as tight as for the SMC PN shown in Figure 1. Dopita et al. (1987a) and Meatheringham & Dopita (these Proceedings) find however that a good two-parameter fit can be obtained to the observed expansion velocities of both SMC and LMC PN (see Figure 2). Their best-fit to the expansion velocity, V(exp), is

$$V(\exp) = 35 + 3.1E - 14[14 + \log F(H\beta)] \text{ km s}^{-1}$$

where E is the excitation class and F(Hβ) is the observed Hβ flux (in cgs units).

Dopita et al. (1987a) also found that the dynamical age, t(dyn) = R(neb)/V(exp), could be fitted with a two-parameter formula :

$$t(\text{dyn}) = 890[M(\text{neb}).V(\exp)]^{0.6} \text{ years}$$

with M(neb) in $M_\odot$ and V(exp), the [OIII] expansion velocity, in km s^{-1}. The nebular radii and masses, R(neb) and M(neb), came from the angular diameter measurements of Wood et al. (1986,1987). The result that the nebular dynamical age depends on the 5/3 power of the momentum of the ionized gas was interpreted by Dopita et al. (1987a) in terms of an interacting-winds model, whereby PN shells are initially ejected at low velocities and are then continuously accelerated during their lifetimes.

4. ELEMENTAL ABUNDANCES

Aller (1983) presented new optical spectrophotometric data for 6 LMC PN, deriving nebular electron temperatures and densities which were then utilised for the determination of elemental abundances using the empirical ICF method. Similar data on the planetary nebula LMC N201 (P25) were presented separately by Aller & Czyzak (1983). This PN, the most luminous in [OIII] and Hβ in the Magellanic Clouds, was found to exhibit unusually strong lines of [FeV], [FeVI] and [FeVII].

AAT optical spectrophotometric data on 71 Magellanic Cloud PN have been analysed by Monk et al. (1987; and these Proceedings). Using [OIII] electron temperatures derived from their data and [OII] electron densities from Barlow (1987) and Monk et al. (these Proceedings), they determined elemental abundances using the empirical ICF method. Table I summarises the mean abundances found for nitrogen, oxygen and neon for PN in each Cloud and compares these with the mean abundances in HII regions in the same galaxy. The PN means did not include low-excitation objects, ie those with $I(5007)/H(\beta)$ < 4 (four out of the twenty-one SMC PN and six out of the fifty LMC PN) or Type I PN, the latter being defined solely by one of the Peimbert and Torres-Peimbert (1983) criteria for Galactic Type I PN status : $N/O \geq 0.5$ (this led to the exclusion of two SMC PN and ten LMC PN).

Monk et al. (1987) found that the abundances of oxygen and neon in the PN were the same, within the statistical errors, as those found in HII regions in the same galaxy (Table I). By contrast, in both galaxies the abundance of nitrogen was found to be enhanced by 0.8 - 1.0 dex in PN relative to HII regions. This clear enhancement of nitrogen was interpreted as being consistent with the exposure at the surfaces of the PN progenitor stars of the main product of the CN cycle, namely secondary nitrogen produced by the nearly complete conversion of all of the initial carbon. Such an effect (the 'first dredge-up') is predicted to occur early in the post-main sequence life of low and intermediate mass stars (Iben & Renzini 1983). The fourth column of Table I shows that the sum of the carbon and nitrogen abundances in the HII regions is equal, within the errors, to the mean nitrogen abundance found for the PN in the same galaxy. Although the CN cycle seems to have operated to near-completion in the PN progenitors, there is no evidence for any depletion of oxygen caused by the CNO cycle.

Aller et al. (1987) have carried out a detailed analysis, aided by ionization structure modelling, of IUE spectrophotometry of twelve Magellanic Cloud PN, supplementing their ultraviolet data with the optical data of Aller et al. (1981) and Aller (1983). Table II summarises the mean abundances for C, N, O and Ne found for the PN analysed by Aller et al. (1987), along with the HII region abundances which they adopted, from the work of Aller et al. (1979) and Dufour et al. (1982). Excluded from Table II are the three PN in their sample with $N/O \geq 0.5$, along with LMC N201, which Monk et al. (1987) find to have $N/O \geq 0.5$ although Aller et al. do not. The abundances derived by Aller et al. (1987) for N,O and Ne agree within the statistical errors with those found by Monk et al. Very noticeable in the results of Aller et al. (Table II) is the large enhancement, relative to the HII regions, of the carbon abundance in both the SMC and LMC PN. Aller et al. interpreted this as due to the effects of the 'third dredge-up', whereby the products of the triple-α reaction are brought to the surface late in the evolution of AGB stars, and noted that their results were consistent with the great preponderance of carbon stars amongst the AGB population of both galaxies. The Type I nebulae in their sample were found to show no enhancement of carbon at all.

Table I. N, O, and Ne abundances in the Magellanic Clouds.
(PN: Monk et al. (1987). H II regions: Pagel et al. (1978), Dufour et al. (1982).)

Mean logarithmic abundances, H = 12.

	C	N	C+N	O	Ne
13 SMC PN		7.44±0.28		8.26 ± 0.15	7.36 ± 0.22
20 SMC H II	7.16 ± 0.04	6.46 ± 0.12	7.24±0.05	8.02 ± 0.08	7.22 ± 0.12
30 LMC PN		7.81±0.30		8.49 ± 0.15	7.64 ± 0.19
18 LMC H II	7.90 ± 0.15	6.97 ± 0.10	7.95±0.15	8.43 ± 0.08	7.64 ± 0.10

Table II. Abundance results of Aller et al. (1987). (log H=12.0.)

	C	N	O	Ne
5 SMC PN	8.68±0.20	7.42±0.22	8.16±0.12	7.46±0.27
SMC H II regions	7.16	6.53	8.07	7.48
3 LMC PN	8.56±0.10	7.56±0.31	8.28±0.11	7.50±0.13
LMC H II regions	7.90	6.98	8.41	7.73
Solar	8.67	7.99	8.92	8.05

Table III. Helium abundances in the Magellanic Clouds.
(PN: Monk et al. (1987). H II regions: Dufour et al. (1982).)

Mean He/H number ratios

	SMC PN (13)	SMC H II (3)	LMC PN (32)	LMC H II (4)
He/H before correction for He I collisional excitation	0.100 ± 0.010	0.083 ± 0.004	0.105 ± 0.010	0.083 ± 0.004
He/H after correction for He I collisional excitation	0.083 ± 0.011	0.081 ± 0.003	0.087 ± 0.008	0.082 ± 0.004
Helium mass fraction, Y	0.249 ± 0.025	0.245 ± 0.007	0.258 ± 0.015	0.247 ± 0.009

Monk et al. (1987; and these Proceedings) also derived the abundance of helium in the nebulae in their large Magellanic Cloud PN sample. In doing this, account was taken of the contributions to the observed HeI line intensities resulting from collisional excitation out of the 2^3S state of HeI. They used the correction formulae of Clegg (1987; and these Proceedings), which are based upon the 19-state collisional excitation rate calculations of Berrington and Kingston (1987). The corrected HeI line fluxes, along with those of HeII 4686 Å and Hβ, were then analysed using the partial recombination rates of Brocklehurst to obtain He/H ratios. Table III summarises the mean He/H ratios found by Monk et al. (Type I and low-excitation nebulae were excluded — Zanstra analyses of the central stars of the latter indicated that helium would not be fully ionized in the surrounding nebulae). Table III shows that when no correction is made for HeI collisional excitation (row 1), the mean He/H ratio is significantly larger for PN than for HII regions, whereas once the corrections for HeI collisional excitation are made (row 2), the He/H ratios for PN and HII regions in the SMC and LMC all agree within one standard deviation. A single value of the helium mass fraction, Y = 0.247 ± 0.009, is found to be consistent with the data on PN and HII regions in both galaxies. A mean He/H number ratio of 0.085 is found for the SMC and LMC PN alone. It is interesting to note that although the CN cycle appears to have operated to near-completion in the PN progenitor stars, no significant enhancement of the surface abundance of helium has resulted.

5. NEBULAR MASSES

The well-known difficulties associated with determining the distances to Galactic PN have made planetary nebulae in the Magellanic Clouds attractive targets for studies aimed at determining nebular and central star masses. The distances to the Magellanic Clouds are known much more accurately than those to Galactic PN, the uncertainty in the distance moduli currently being discussed for each of the Clouds being typically no more than 0.3 – 0.4 magnitudes, corresponding to distance uncertainties of 15-20%. In addition, the low reddening towards the Clouds minimises one of the usual sources of error encountered when studying Galactic PN. The main problem that Magellanic Cloud PN pose is that their angular diameters are mostly less than 2 arcsec and often less than 1 arcsec, although this can be an advantage for spectrophotometric studies (see eg. 5.2 below). Ground-based methods for determining their angular diameters have now been used successfully for a significant number of PN and are discussed in 5.1 below. Once the Hubble Space Telescope and the Australia Telescope radio synthesis array become operational, high-quality direct images of many Magellanic Cloud PN can be expected.

5.1. Nebular masses from angular diameter measurements

Speckle interferometric techniques have been used on the AAT by Barlow et al. (1986) and Wood et al. (1986) to resolve Magellanic Cloud PN. The former collaboration resolved SMC N2 and the latter resolved 2 SMC and 8 LMC PN. Once the angular diameter of a nebula is known, its ionised mass can be derived straightforwardly from its dereddened absolute Hβ flux. Barlow *et al.* used the measured [OII] 3726,3729 Å electron density plus nebular ionization structure modelling to derive a filling factor of 0.45 and an ionised mass of 0.36 M$_\odot$ for SMC N2, while Wood *et al.* adopted a filling factor of 0.7 in deriving ionised nebular masses ranging from 0.02 M$_\odot$ to > 0.19 M$_\odot$ for the PN in their sample.

The signal-to-noise achievable using speckle techniques declines rapidly as the objects become fainter and more extended – the bright, compact Magellanic Cloud PN which are resolvable by speckle interferometry are usually young and optically thick. For this reason, Wood *et al.* (1987) used a direct imaging technique to determine the angular diameters of fainter, larger PN in the Magellanic Clouds. Imaging at a frame-rate of 60 Hz and using both direct and autocorrelation comparisons, they compared the gaussian fits to the azimuthally-averaged radial profile of each PN with those fitted to reference stars, in order to determine the intrinsic FWHM of the PN. The imaging was mostly done through a Hβ filter, in seeings ranging from 1.2 to 1.7 arcsec, and gave the very useful by-product of a photometric Hβ flux for each nebula. These fluxes and the derived FWHM angular diameters were then used to determine the ionised nebular masses, in the same manner as by Wood *et al.* (1986). Figure 3 reproduces the plot made by Wood *et al.* (1987) of the ionised mass versus nebular radius for their combined speckle and direct-imaging samples, together with the same quantities for the Galactic Centre PN sample of Gathier *et al.* (1983). Wood *et al.* concluded that there is a continuous increase in ionised mass up to a nebular radius of about 0.12 pc, above which radius the nebulae appear to become optically thin. For a He/H ratio of 0.10, the mean mass of the optically thin nebulae was estimated to be 0.27 M$_\odot$ by Wood *et al.* (1987). Wood *et al.* adopted distances of 52 kpc and 66 kpc for the LMC and SMC, respectively, which I shall refer to as the 'long' distance scale hereafter.

5.2. Nebular masses from forbidden-line electron densities

An alternative to angular angular diameter estimation for the derivation of nebular masses for Magellanic Cloud PN has been discussed by Barlow (1987). This makes use of the dereddened Hβ flux, the [OIII] electron temperature and the [OII] electron density, to derive the nebular mass without knowledge of the nebular radius, radial density distribution, or filling factor ϵ.

The ratio of the absolute Hβ flux to the nebular ionised hydrogen mass is given by

$$\frac{I(H\beta)}{M_H} \propto \frac{\int \epsilon n_e n_H dV}{\int \epsilon n_H dV} \equiv <n_e>$$

It can be shown (see Barlow 1987) that although O^{2+}, rather than O^+, is the dominant ionization stage of oxygen in most planetary nebulae, the density of O^+ in optically thin nebulae is directly related to that of O^{2+} (via photoionization and recombination processes), and thus to that of oxygen and of hydrogen. In addition, above and below its critical density the flux in a forbidden line is respectively proportional to the first power

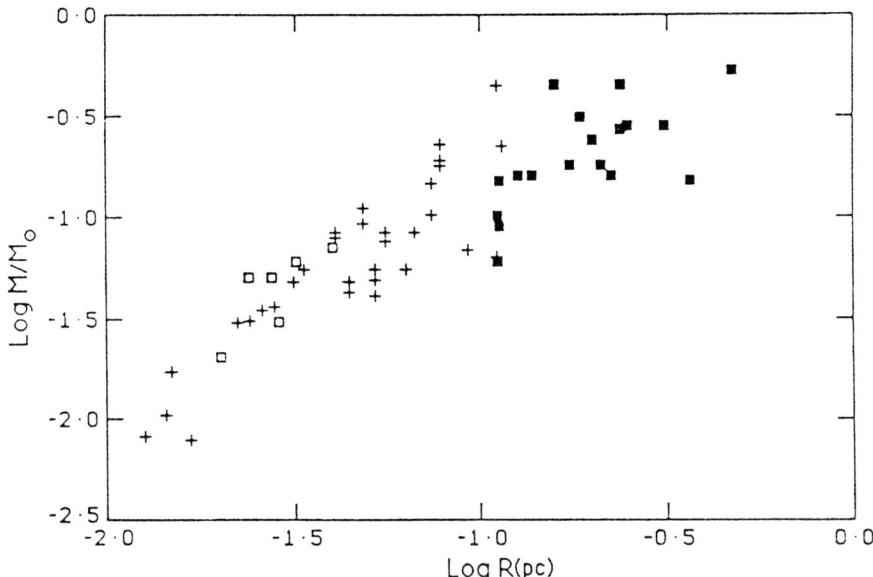

Figure 3 : Total ionised mass versus nebular radius for the Magellanic Cloud speckle and direct-imaging samples of Wood et al (1986, 1987 : open and filled squares, respectively) and the Galactic centre radio sample of Gathier et al (1983). Distances of 8.6, 52 and 66 kpc were adopted for the Galactic centre, LMC and SMC. From Wood et al (1987).

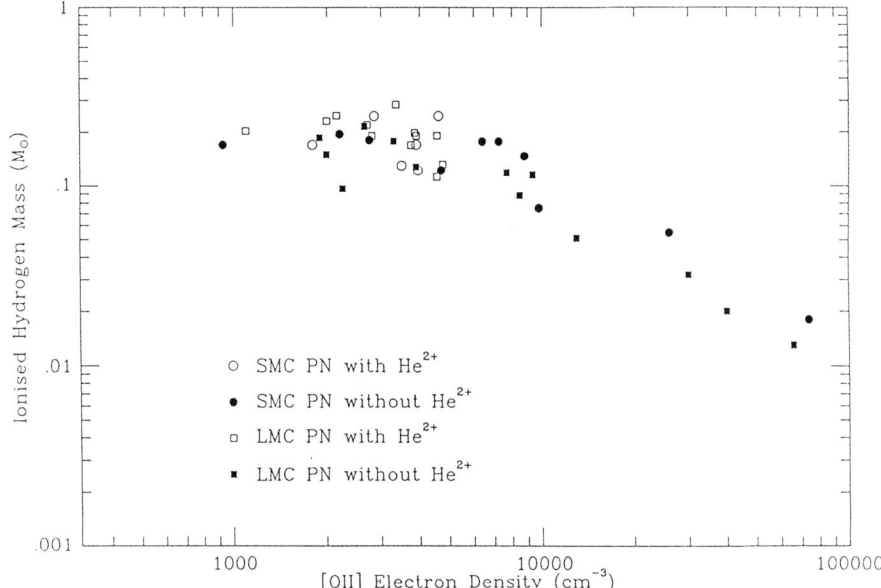

Figure 4 : Ionised hydrogen mass versus OII electron density. Distances of 47 and 57.5 kpc were adopted for the LMC and SMC. From Monk etal(1987)

of the density (Boltzmann equilibrium) and to the square of the density (collisional excitation followed by radiative decay). For typical nebular conditions, the critical densities of the [OII] 3726 Å and 3929 Å lines are respectively about 5000 cm^{-3} and 1000 cm^{-3}, so between these densities the [OII] intensity ratio will be given by

$$\frac{I(3726)}{I(3729)} \propto \frac{\int \epsilon n_e n_0 dV}{\int \epsilon n_0 dV} \equiv <n'_e>$$

It can be shown (see Barlow 1987) that $<n'_e>$ is virtually identical to $<n_e>$ for most types of nebular density distributions, so that the electron density obtained from the [OII] doublet ratio can be used to derive the ionised nebular mass via the first of the relations above. Line intensity ratios that are integrated over the entire nebula must be used, which is easy for unresolved Magellanic Cloud planetary nebulae, but more difficult for extended Galactic PN. For nebulae with electron densities in excess of 5000 cm^{-3}, the ratio of [OII] 7325 Å to 3727 Å can be used in place of 3726 Å to 3729 Å as a nebular mass tracer.

For a sample of 32 Magellanic Cloud PN with measured Hβ fluxes, electron temperatures and [OII] electron densities, Barlow (1987) found that the nebular ionised hydrogen mass increased linearly with decreasing electron density until a density of about 5000 cm^{-3} was reached. At lower densities the nebular masses remained constant, with the implication that the nebulae had become optically thin. In a continuing AAT programme, Monk et al. (these Proceedings) have already doubled this electron density sample and Figure 4 shows their plot of nebular ionised hydrogen mass versus [OII] electron density. In deriving these masses, distances of 47 kpc and 57.5 kpc were adopted for the LMC and SMC, respectively, which will be referred to as the 'short' distance scale. PN with N/O $\geq$ 0.5 (Type I) were excluded from the plot. Figure 4 confirms that the ionised hydrogen mass attains a constant value at densities less than about 5000 cm^{-3} – for the nebulae with $n_e \leq$ 4000 cm^{-3}, Monk et al. found a mean ionised hydrogen mass of $M_H = 0.186 \pm 0.044\, M_\odot$. With a mean He/H number ratio of 0.085 (Section 4), this corresponds to a total ionised nebular mass of $0.25 \pm 0.06\, M_\odot$.

Among this group of PN, no significant differences are found between the masses of SMC or LMC objects, or between those with HeII 4686 Å emission and those without (Figure 4). The implication is that the Shklovsky method, in which a constant nebular mass is adopted for optically thin PN, can be used to determine the distances to Galactic PN having $n_e \leq$ 4000 cm^{-3}; provided Type I nebulae are excluded. As discussed by Barlow (1987), the adoption of $M_H = 0.19\, M_\odot$ for optically thin PN gives a distance scale which is 40% larger than that of Cahn and Kaler (1971) and 12% smaller than that of Cudworth (1974).

Converted into the 'long' distance scale of Wood et al. (1986,1987), the turnover electron density of 5000 cm^{-3} in Figure 4 corresponds to a nebular radius of 0.1 pc at the onset of optical thinness, in fair agreement with the angular diameter method results of Wood et al. in Figure 3.

The [OII] mass tracer method will not be sensitive to emission from regions with electron densities very much less than the critical density of [OII] 3729 Å, *i.e.* densities of less than a few hundred cm^{-3}. The technique will therefore not pick up faint halos around bright nebulae. The infrared fine structure lines of [NII] at 122μm and 204μm seem best-suited to the measurement of halo masses, since their critical densities are 200 cm^{-3} and 25 cm^{-3}, respectively.

Barlow (1987) found that the 10 Magellanic Cloud PN in his sample with electron densities larger than 5000 cm^{-3} and no HeII 4686 Å emission had a narrow range of dereddened Hβ luminosities, equivalent at 1 kpc to log I(Hβ) = -8.96 ± 0.1 cgs. This calibration for optically thick PN agrees with that of O'Dell (1963) for optically thick Galactic PN, log I(Hβ) = -9.0 ± 0.1 cgs, but is brighter than several estimates subsequent to that of O'Dell. This result implies that the Zanstra/Vorontsov-Velyaminov method — the adoption of a constant absolute magnitude for optically thick PN — is viable. The above absolute Hβ flux can be used to determine the distances to Galactic PN satisfying the following criteria : n_e(OII) > 5000 cm^{-3}; no HeII 4686 Å emission; and I(5007) $\geq$ 4$\times$I(Hβ) (*i.e.* the central stars must have reached their plateau ionising luminosities, thereby excluding low-excitation PN).

6. CENTRAL STAR MASSES

Based on IUE spectrophotometry, Stecher *et al.* (1982), had derived very high central star luminosities for three carbon-rich Magellanic Cloud PN, their luminosities corresponding to stellar masses of between 0.9 and 1.2 M$_\odot$, much higher than had previously been suggested as typical for white dwarfs and PN central stars. These masses also implied very short central star evolutionary timescales, of between 20 and 300 years. Tylenda (1984) argued against the parameters derived by Stecher *et al.* on the grounds that the application of time-independent diagnostic techniques could not have yielded the correct parameters if the PN had been evolving so rapidly. Barlow *et al.* (1986) constructed ionization structure models for one of these PN, SMC N2. Using Auer-Mihalas NLTE model atmospheres (Clegg and Middlemass 1987), they were able to fit all of the observed nebular and stellar properties with a log g = 5.7, T$_{eff}$ = 110,000 K model having a luminosity eight times less than that derived by Stecher *et al.* The corresponding stellar mass was found to be 0.593 M$_\odot$.

Aller *et al.* (1987) have derived central star parameters for 12 Magellanic Cloud PN, based on nebular ionization structure fits to their IUE and optical nebular line flux data. NLTE model atmospheres from the grid of Husfeld *et al.* (1984) were used and the three PN studied by Stecher *et al.* (1982) were re-analysed. The central star effective temperature derived by Aller *et al.* for SMC N2 differed by less than 6% from that derived by Barlow *et al.* (1986), while the central star luminosities derived for SMC N2 agree within 13%, if the same distance is adopted. Excluding Type I PN, the parameters derived by Aller *et al.* (1987) yield a mean central star mass of 0.609 $\pm$ 0.021 M$_\odot$, when interpolated between the evolutionary tracks of Schoenberner (1979,1983). Aller *et al.* adopted distances of 46

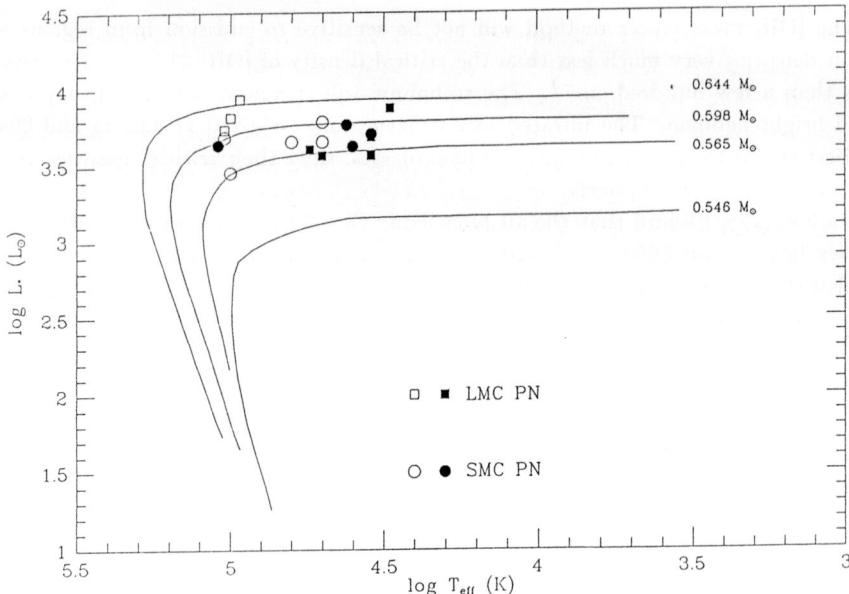

Figure 5 : Central star luminosities and effective temperatures, from Aller et al (1987 : open symbols) and Monk et al (1987 : filled symbols), are plotted along with the evolutionary tracks of Schoenberner (1979, 1983). Distances of 47 and 57.5 kpc have been adopted for the LMC and SMC, respectively.

kpc and 63 kpc for the LMC and SMC, respectively. However, to facilitate comparison with the results of Monk et al. (these Proceedings), stellar luminosities corresponding to the 'short' distance scale have been plotted in Figure 5. With this distance scale, the mean stellar mass for the Aller et al. sample in Figure 5 is 0.604 ± 0.026 $M_{\odot}$.

The sample of Aller et al. (1987) consists predominantly of high-excitation nebulae for which the central star continua are not detectable with IUE. A complementary sample of mainly low and medium-excitation PN has been analysed by Monk et al. (these Proceedings). Using IUE and AAT spectrophotometry, the Zanstra HI and Stoy energy-balance methods were applied to derive surface gravities and effective temperatures for nine Magellanic Cloud PN with detectable stellar continua, based on the NLTE model atmosphere grid of Mihalas (1972). Since model atmosphere energy distributions are almost as sensitive to the adopted surface gravity as to the effective temperature, at least two independent diagnostics are needed in order to determine both of these parameters. The model atmosphere which provides the best-fit to both the Zanstra and Stoy ratios predicts a HI Lyman continuum surface photon flux and since the absolute nebular $H\beta$ recombination line flux determines the total number of ionising photons emitted by the star, the stellar radius can therefore be derived. The resulting stellar luminosity plus the effective temperature allow the stellar mass to be derived (eg. Figure 5) and this derived mass and the stellar radius predict a new value for the surface gravity, which can be compared for consistency with the surface gravity of the model atmosphere, in an iterative manner. This extra constraint helps to increase the accuracy of the method. The stellar parameters of nine PN analysed by Monk et al. (these Proceedings) are plotted as solid symbols in Figure 5, using the 'short' distance scale, and are compared with the hydrogen-shell burning tracks of Schoenberner. The mean central star mass that they find for their sample is 0.586 ± 0.018 M$_\odot$. The combined samples of Monk et al. and Aller et al. shown in Figure 5 yield a mean central star mass of 0.595 ± 0.022 M$_\odot$ (all of the PN plotted in Figure 5 have C/O ≥ 0.5, implying carbon star predecessors). This mean central star mass agrees well with the mean mass estimated by Schoenberner (1981) for Galactic PN, using the expansion age versus stellar magnitude method, and with the mean mass of 0.58 $\pm$ 0.10 M$_\odot$ estimated for DA white dwarfs by Weidemann and Koester (1984). The latter noted that the mass dispersion of $\pm$ 0.10 M$_\odot$ that they found could be intrinsically much less, since an error of only 0.05 of a magnitude in the observed colour of a white dwarf translated into an error of 0.15 M$_\odot$ in its derived mass. In contrast, the masses derived for the PN central stars are relatively insensitive to errors in their derived luminosities, due to the very steep dependence of luminosity on core mass given by the Paczynski relation for the horizontal portion of hydrogen-shell burning evolutionary tracks :

$$L \approx 6 \times 10^4 (M/M_\odot - 0.5) L_\odot$$

As a result, a 50% error in luminosity translates into an error of only 0.025 M$_\odot$ in the derived central star mass, while an error of 20% in luminosity corresponds to an error of 0.01 M$_\odot$ error in mass. Thus the adoption of the 'long' rather than the 'short' distance scale for the nebulae in Figure 5 would lead to an increase of only 0.014 M$_\odot$ in the mean central star mass.

Four WC-type Wolf-Rayet central stars of Magellanic Cloud PN have also been analysed by Monk et al. (these Proceedings), who adopt blackbody model atmospheres, since plane-parallel, hydrostatic, NLTE models are clearly inappropriate for WR stars. In addition, since their spectra contain no lines of hydrogen, WC central stars cannot be powered by hydrogen-shell burning and so helium-shell burning tracks were used (those of Wood & Faulkner 1986, supplemented by an unpublished 0.555 M$_\odot$ track by Schoenberner). A mean central star mass of 0.59 ± 0.01 M$_\odot$ was obtained, which agrees with the mean mass of 0.55 ± 0.10 M$_\odot$ estimated for DB white dwarfs by Oke et al. (1984). The fraction of all white dwarfs that are helium-rich is estimated to be 12% for $T_{eff} > 40,000$ K, and 19% for $T_{eff} > 12,000$ K (Fleming et al. 1986). In the Magellanic Cloud PN sample of Monk et al. (1987), 15% of the medium- and low-excitation nebulae (i.e. those with detectable central star continua) had WC central stars. Although more complex evolutionary scenarios have been proposed to account for the observed fractions of DA and DB white dwarfs, these statistics are consistent with a simple scenario whereby WC central stars are the progenitors of DB white dwarfs, while the remaining central stars become DA white dwarfs.

In summary, the planetary nebulae in the Magellanic Clouds have yielded a rich body of quantitative data on both nebulae and central stars in the past few years. A mean nebular mass of 0.25 $M_\odot$ (Section 5) and a mean central star mass of 0.60 $M_\odot$ (this Section) imply a mean progenitor star mass of 0.85 $M_\odot$ at the tip of the asymptotic giant branch. The determination of halo masses is now needed in order to quantitify the amount of mass lost ascending the AGB. The extremely narrow distributions found for both nebular and central star masses are clearly related and it remains to be seen whether such narrow ranges can be explained solely by the initial mass function, or whether a steep dependence of stellar mass loss upon luminosity, prior to the PN phase, is also required.

I would like to thank Dr D J Monk for his assistance with the preparation of this review, and Drs J Liebert and T Boroson for communicating their Jacoby PN confirmations in advance of publication.

REFERENCES:

Aller, L. H., 1983, *Astrophys. J.*, **273**, 590.
Aller, L. H., & Czyzak, S. J., 1983, *Proc. Nat. Acad. Sci.*, **80**, 1764.
Aller, L. H., Keyes, C. D., & Czyzak, S. J., 1979, *Proc. Nat. Acad. Sci.*, **76**, 1525.
Aller, L. H., Keyes, C. D., Ross, J. E., & O'Mara, B. J., 1981, *Mon. Not. R. astr. Soc.*, **194**, 613.
Aller, L. H., Keyes, C. D., Maran, S. P., Gull, T. R., Michalitsianos, A. G., & Stecher, T. P., 1987, *Astrophys. J.*, **320**, 159.
Barlow, M. J., 1987, *Mon. Not. R. astr. Soc.*, **227**, 161.
Barlow, M. J., Morgan, B. L., Standley, C., & Vine, H., 1986, *Mon. Not. R. astr. Soc.*, **223**, 151.
Berrington, K. B., & Kingston, A. E., 1987, *J. Phys. B*, in press.
Cahn, J. H., & Kaler, J. B., 1971, *Astrophys. J. Suppl.*, **22**, 319.
Clegg, R. E. S., 1987, *Mon. Not. R. astr. Soc.*, **229**, 31p.
Clegg, R. E. S., & Middlemass, D., 1987, *Mon. Not. R. astr. Soc.*, **228**, 759.
Cudworth, K. M., 1974, *Astr. J.*, **79**, 1384.
Dopita, M. A., Ford, H. C., Lawrence, C. J., & Webster, B. L., 1985a, *Astrophys. J.*, **296**, 390.
Dopita, M. A., Ford, H. C., & Webster, B. L., 1985b, *Astrophys. J.*, **297**, 593.
Dopita, M. A., Meatheringham, S. J., Wood, P. R., Webster, B. L., Morgan, D. H., & Ford, H. C., 1987a, *Astrophys. J.*, **315**, L107.
Dopita, M. A., Meatheringham, S. J., Webster, B. L., & Ford, H. C., 1987b, *Astrophys. J.*, submitted.
Dufour, R. J., Shields, G. A., & Talbot, R. J. 1982, *Astrophys. J.*, **252**, 461.

Feast, M. W., 1968, *Mon. Not. R. astr. Soc.*, **140**, 345.
Fleming, T. A., Liebert, J., & Green, R. F., 1986, *Astrophys. J.*, **308**, 176.
Gathier, R., Pottasch, S. R., Goss, W. M., & van Gorkom, J. M., 1983, *Astr. Astrophys.*, **128**, 325.
Husfeld, R., Kudritzki, R. F., Simon, K. P., & Clegg, R. E. S., 1984, *Astr. Astrophys.*, **134**, 139.
Iben, I. Jr., & Renzini, A., 1983, *Ann. Rev. Astron. Astrophys.*, **21**, 271.
Iben, I. Jr., & Tutukov, A. V., 1985, *Astrophys. J. Suppl.*, **58**, 661.
Jacoby, G. H., 1980, *Astrophys. J. Suppl.*, **42**, 1.
Jacoby, G. H., 1983, *Proc. IAU Symp. 103*, p 427. ed. Flower, D.R., Reidel, Dordrecht, Holland.
Liebert, J., & Boroson, T., 1987, in preparation.
Mathewson, D. S., & Ford, V. L., 1984, *Proc. IAU Symp. 108*, p 125. ed. van den Bergh, S. & de Boer, K. S., Reidel, Dordrecht, Holland.
Meatheringham, S. J., Dopita, M. A., Ford, H. C., & Webster, B. L., 1987, *Astrophys. J.*, submitted.
Mihalas, D., 1972, *Non-LTE Model Atmospheres for B and O Stars* NCAR-TN/STR-76.
Monk, D. J., Barlow, M. J., & Clegg, R. E. S., 1987, *Mon. Not. R. astr. Soc.*, submitted.
Morgan, D. H., 1984, *Mon. Not. R. astr. Soc.*, **208**, 633.
Morgan, D. H., & Good, A. R., 1985, *Mon. Not. R. astr. Soc.*, **213**, 419.
O'Dell C. R., 1963, *Astrophys. J.*, **138**, 67.
Oke, J. B., Weidemann, V., & Koester, D., 1984, *Astrophys. J.*, **281**, 276.
Pagel, B. E. J., Edmunds, M. G., Fosbury, R. A. E., & Webster, B. L., 1978, *Mon. Not. R. astr. Soc.*, **184**, 569.
Peimbert, M., 1984, *Proc. IAU Symp. 108*, p 363. ed. van den Bergh, S. & de Boer, K. S., Reidel, Dordrecht, Holland.
Peimbert, M., & Torres-Peimbert, S., 1983, *Proc. IAU Symp. 103*, p 233. ed. Flower, D. R., Reidel, Dordrecht, Holland.
Sanduleak, N., & Philip, A. G. D., 1977, *Publs. astr. Soc. Pacif.*, **89**, 792.
Sanduleak, N., Pesch, P., 1981, *Publs. astr. Soc. Pacif.*, **93**, 431.
Sanduleak, N., MacConnell, D. J., & Philip, A. G. D., 1978, *Publ. astr. Soc. Pacific*, **90**, 621.
Savage, A., Murdin, P. G., & Clark, D. H., 1982, *Observatory*, **102**, 229.
Schönberner, D., 1979, *Astr. Astrophys.*, **79**, 108.
Schönberner, D., 1981, *Astr. Astrophys.*, **103**, 119.
Schönberner, D., 1983, *Astrophys. J.*, **272**, 708.
Stecher, T. P., Maran, S. P., Gull, T. E., Aller, L. H., & Savedoff, M. P., 1982, *Astrophys. J. Lett.*, **262**, L41.
Tylenda, R., 1984, *Astr. Astrophys.*, **138**, 317.
Webster, B. E., 1975, *Mon. Not. R. astr. Soc.*, **173**, 437.
Weidemann, V., & Koester, D., 1984, *Astr. Astrophys.*, **132**. 195.
Wood, P. R., & Faulkner, D. J., 1986, *Astrophys. J.*, **307**, 659.

Wood, P. R., Bessell, M. S., & Dopita, M. A., 1986, *Astrophys. J.*, **311**, 632.
Wood, P. R., Meatheringham, S. J., Dopita, M. A., & Morgan, D. H., 1987, *Astrophys. J.*, **320**, 178.

PLANETARY NEBULAE IN GALAXIES BEYOND THE LOCAL GROUP

H. C. Ford and R. Ciardullo,
Space Telescope Science Institute
Homewood Campus, Baltimore, MD 21218
G. H. Jacoby
Kitt Peak National Observatory, Tucson, AZ 85726
X. Hui
Boston University, Boston, MA 02215

ABSTRACT.
Planetary nebulae can be used to estimate the distances to galaxies and to measure stellar dynamics in faint halos. We discuss surveys which have netted a total of 665 candidate planetary nebulae in NGC 5128 (Cen A), NGC 5102, NGC 3031 (M81), NGC 3115, three galaxies in the Leo Group (NGC 3379, NGC 3384, NGC 3377), NGC 5866, and finally, in NGC 4486 (M87). Radial velocities of planetaries in M32 have shown that its halo velocity dispersion is most likely isotropic. Radial velocities of planetaries in M31 show that $\sim 2/3$ of the nebulae with projected radii between 15 and 30 kpc are members of a rotating thick disk with slight asymmetric drift, while $\sim 1/3$ belong to a slowly rotating halo. Velocities of 116 nebulae in NGC 5128 reveal pronounced rotation and a slowly declining velocity dispersion in the halo out to 20 kpc. The [O III] $\lambda 5007$ luminosity functions (PNLFs) in NGC 5128, M81, and the three Leo Galaxies have the same shape over the first magnitude. The highly consistent distances derived from the brightnesses of the jth nebula and the median nebula in different fields in the same galaxy and from different galaxies in the same group lend strong support to the suggestion that planetaries are an accurate standard candle in old stellar populations. Comparison of theoretical luminosity functions to the observed PNLFs shows that there is a very small dispersion in the central star masses.

1. INTRODUCTION

There are at least three reasons to search for planetary nebulae (PN) in galaxies beyond the Local Group. First, they are the only easily resolvable component in old stellar populations at large distances. The detection of PN provides a direct measure of the stellar death rate, mass return rate, and flux of ionizing radiation in an old stellar population. Second, PN are readily identified test particles whose radial velocities can be used to measure the mass distribution, mass-to-light ratios, and stellar dynamics in the innermost and outermost regions of early type galaxies. As a probe of galactic halos, they are complementary to, and, in many respects, superior to globular clusters. In galaxies which are closer than the Virgo cluster, there are far more detectable planetaries than globular clusters, and, with an on-band/off-band interference filter technique, they are much easier to find. Planetary nebulae provide one of the few, if not only, means of measuring the dynamics of the faint shells found around some peculiar galaxies. The third reason for seeking planetaries in distant galaxies is the possibility that the luminosities of the brightest PN may be a good standard candle in early type galaxies and old populations such as spiral bulges. This paper will review a continuing long term program which emphasizes the latter two points. Detailed results will be given for the galaxy NGC 5128.

2. PLANETARY NEBULAE IN GALAXIES FROM HERE TO THE VIRGO CLUSTER

Detection of faint nebulae in distant galaxies requires long integrations with a 4-m class telescope on dark nights with good seeing. The high quantum efficiency and dynamic range of a CCD camera are prerequisites for success. The nebulae are found by blinking an image taken through a narrow band interference filter which transmits a redshifted emission line against a continuum image taken through an interference filter which transmits an adjacent, emission line free region of the spectrum. Luminous high excitation planetaries are most easily detected in the light of [O III] $\lambda 5007$ for several reasons: 1) [O III] $\lambda 5007$ is usually the brightest optical line, 2) the contrast between [O III] $\lambda 5007$ and the sky plus galaxy continuum is better than any other strong optical lines, and 3) thinned CCDs have high quantum efficiency near $\lambda 5007$. A filter bandpass (FWHM) between 30 and 40 Å optimizes detection of the nebulae with a single filter. If the bandpass is much narrower than 30 Å the velocity dispersion in a giant elliptical galaxy will shift the observed wavelengths of many nebulae onto the shoulders of the filter. If the bandpass is much wider than 40 Å, the signal-to-noise ratio (SNR) is lowered. A practical consideration is the fact that the fast focal ratio (f/2.77) at the prime focus of the CTIO and KPNO 4-m telescopes will broaden the bandpass of very narrow band filters to between 20 and 30 Å and will lower the peak transmission (Eather and Reasoner, 1969). Care also must be taken when ordering narrow band filters to allow for bandpass shifts due to the ambient temperature (usually a few Å to the blue) and a fast beam (10 to 12 Å to the blue at f/2.77). The bandpass of the off-band filter can be between 200 and 300 Å wide in order to shorten the continuum exposure time. The off-band image should be at least 0.25 magnitudes deeper than the on-band image to avoid spurious detections at the noise limit.

Our total on-band exposure times range from one to six hours per field. Our identification criteria for planetary nebulae are: 1) appearance in each on-band frame and in the sum of the on-band frames, 2) absence in each off-band frame and in the sum of the off-band frames, 3) a point source profile, and 4) in the NGC 5128 and M81 fields where we took Hα images, an Hα to [O III] $\lambda 5007$ excitation ratio less than 0.9. The latter two criteria help discriminate against supernova remnants and against H II regions in later type galaxies. We are confident that the majority of our candidates are bona fide planetaries because we find them in more than one frame and derive [O III] $\lambda 5007$ fluxes which are consistent with planetaries at the estimated distance of the galaxy. In NGC 5128 we independently verified the reality of 116 planetaries by measuring their radial velocities.

The results of our recent surveys for planetary nebulae are summarized in Table 1. The last column in the table gives the total number of candidate planetary nebulae in each galaxy.

Table 1. [O III] $\lambda 5007$ On-band/Off-band Survey for Planetary Nebulae

Galaxy (NGC)	Type (RSA)	v_{obs} (km s^{-1})	Fields Surveyed	Images per Field	Exposure Time (min)	Number of Candidates
5128	S0+S$_{pec}$	526	29	1	60	222
5102	S0$_1$(5)	420	3	1	60	27
3031	Sb(r)I-II	-36	1	3	30	185
3115	S0$_1$(7)	655	2	2	90	68
5866	S0$_3$(8)	672	1	3	60	18
4486	E0	1254	1	6	60	12
Leo Group						
3379	E0	893	2	1	90	48
3384	SB0$_1$(5)	771	2	1	90	53
3377	E6	718	1	1	90	32

We have taken Hα on-band/off-band images of five galaxies in Table 1 to search for novae. The Hα filters were 65 to 75 Å wide, which maximizes the SNR for expanding nova shells but reduces the detectability of planetaries and H II regions. Nonetheless, the Hα images can be used to help separate the brightest high excitation planetary nebulae from lower excitation H II regions and supernova remnants. Details of the Hα survey are given in Table 2. The six fields in NGC 5128 were surveyed once each in 1985, 1986, and 1987.

Table 2. Hα On-band/Off-band (Nova) Survey

Galaxy (NGC)	Fields Surveyed	Number of Images per Field	Exposure Time per Image (min)
5128	6	4	45
3031	1	7	10
3115	1	9	10
5866	1	2	45
4486	1	3	60

Figure 1 shows the positions of the 2 envelope fields (107 planetary nebulae) and 27 halo fields (115 planetary nebulae) in NGC 5128. The halo fields were chosen near the major axis of the faint outer isophotes (Dufour et al. 1978) and on both sides of the galaxy in order to measure rotation in the halo.

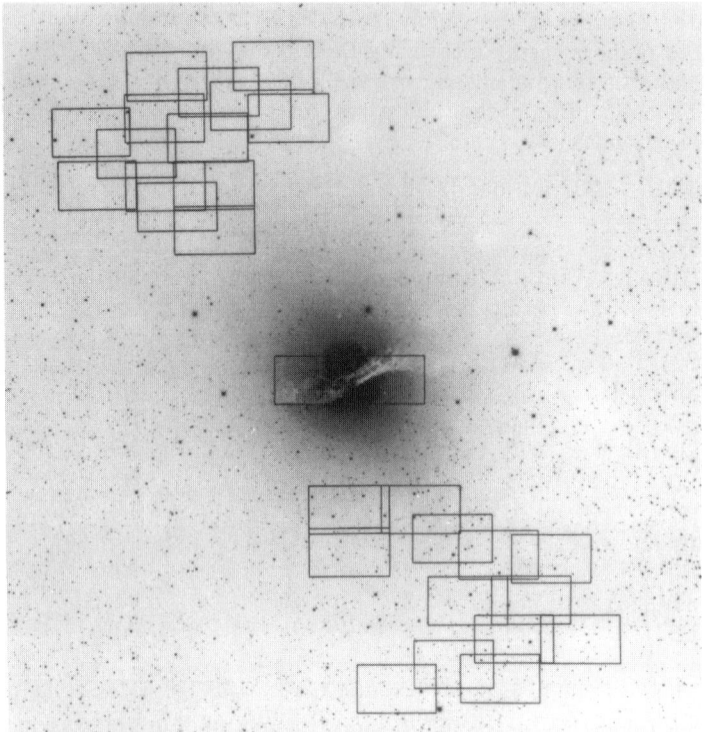

Figure 1. Each rectangle shows a CCD survey field projected onto NGC 5128. The dimensions of the CCD fields are $5' \times 3'$ (4.8 kpc $\times$ 2.9 kpc at 3.3 Mpc). The centers of the most distant fields are $\sim$ 20 kpc from the nucleus.

The number of PN per unit light in the NGC 5128 halo fields is shown in Figure 2. Because of seeing differences between the fields, the PN were counted to a constant [O III] λ5007 magnitude which was well above the detection limit in all of the images. The two envelope fields were excluded because of the continuously varying limiting magnitude and the likelihood that the western field is partially veiled by dust in the warped plane which projects into the famous dust lane. The blue light in each field was calculated by numerically integrating an analytical model which incorporates Dufour et al.'s (1979) $r^{1/4}$ law along the major axis and the observed changing ellipticity of the isophotes. The number of planetaries per unit light appears constant throughout the halo at a value of 1.14 planetaries per unit light (1 unit for $m_B = 15$). Numerical integration across the entire galaxy gives $m_B = 7.48$ for the elliptical component, a value which should be more accurate than Dufour et al.'s value of 7.14 which was derived by assuming the isophotes are spherical. The estimated total number of planetary nebulae to our flux limit (which spans 1.6 magnitudes of the luminosity function) is then 1161

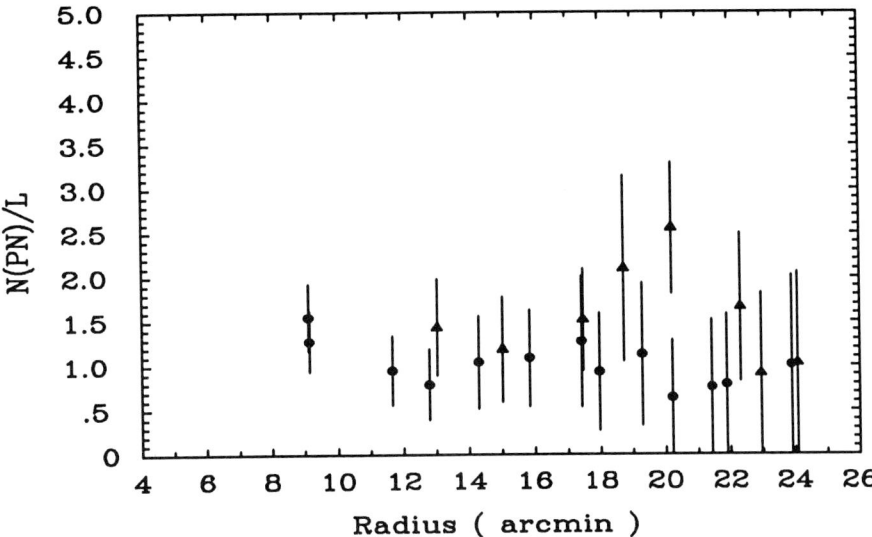

Figure 2. The number of planetary nebulae per unit light versus distance from the center of the galaxy. The filled circles are fields in the NE halo and the filled triangles are fields in the SW halo.

nebulae.

3. PLANETARY NEBULAE AND STELLAR DYNAMICS

Planetary nebulae have been used to measure the stellar dynamics in the LMC (Dopita et al., 1988), M32 (Nolthenius 1984; Nolthenius and Ford, 1986 [NF86]), M31 (Nolthenius 1984; Nolthenius and Ford 1987 [NF87]), NGC 205 (Jacoby and Ford 1988) and NGC 5128 (Ford et al. 1988). This review will not consider the LMC.

3.1 M32 and NGC 205

NF86 investigated velocity anisotropy in the halo of M32 and derived M32's mass and M/L by analyzing the radial velocities of 15 halo planetaries. Because the sample was too small for binning, Nolthenius (1984; cf NF86) developed a procedure for testing the goodness of fit between small numbers of radial velocities and trial velocity dispersions derived from models. Spherical models with constant anisotropy (Jaffe 1983) and with radially varying anisotropy (Merritt 1985) were combined with the observed nuclear velocity dispersion (Tonry 1984; Whitmore 1980) to derive the trial velocity dispersions. NF86's most important conclusion is that the stellar velocities are close to isotropic throughout M32's halo. The rms dispersion of the planetaries is 42 km s^{-1}, sufficiently lower than the nuclear dispersion of 70-80 km s^{-1} to make a strongly increasing M/L with radius seem unlikely. Finally, the most likely fits give a mass of $8.2 \pm 2 \times 10^8 M_\odot$ and a M/L_B

of 3-4 in solar units.

Jacoby and Ford (1988) have measured the radial velocities of 10 planetaries in NGC 205. The rms velocity dispersion is 27 km s^{-1} about the -219 km s^{-1} mean heliocentric velocity.

3.2 M31

Lawrie and Ford (1982) used a Velocity Modulating Camera to observe 45 planetaries within 200 pc of the center of M31. Lawrie (1983) measured the radial velocities for 33 of the nebulae to derive a systemic velocity of -309 ± 25 km s^{-1} and a velocity dispersion of 156 ± 23 km s^{-1}. His velocity dispersion is equal to the mean (156 ± 9 km s^{-1}) of the velocity dispersion estimates for M31's bulge which were derived from automated analysis techniques (see Lawrie 1983 for references). Lawrie found no evidence to support the claim that there may be systematic or subtle biases which cause the Fourier methods to overestimate the velocity dispersion in early-type galaxies.

Nolthenius (1984) and NF87 used the radial velocities of 34 planetaries with projected distances between 15 and 30 kpc to separate the nebulae into disk and halo populations. Approximately 1/3 of the planetaries are members of a slowly rotating halo whose 92 ± 43 km s^{-1} rotational velocity and 116 ± 48 km s^{-1} velocity dispersion are very similar to the 80 ± 28 km s^{-1} rotational velocity and 130 km s^{-1} velocity dispersion of the globular clusters (Huchra et al. 1982). Jacoby and Ford (1986) measured chemical abundances in two halo planetaries, one at a projected distance of 33 kpc and the other at 3.5 kpc. The distant halo nebula has a low oxygen abundance which is between the O abundance of the galactic halo planetaries 49 + 88.1 and K648, whereas the high velocity nebula at 3.5 kpc has O, N, and Ne abundances between the Orion nebula and galactic planetaries. The M31 globular clusters have a wide range of metallicity and show little, if any, correlation between kinematics and integrated line strengths (Huchra et al. 1982; van den Bergh 1969). The abundances and especially the kinematics of the halo planetaries suggest that they derive from a population similar to that of the globular clusters.

The planetaries in the outer disk ($R \geq 15$ kpc) are rotating a little more slowly (14 km s^{-1}) than the gas in the disk, a value which is comparable to the observed asymmetric drift of galactic planetaries (Cahn and Wyatt 1978; Cudworth 1974). The derived azimuthal velocity dispersion is $\sigma_\theta = 38 \pm 12$ km s^{-1}, and the inferred scale height is $z_0 = 1$ to 3 kpc, implying that the planetaries are part of a thick disk.

3.3 NGC 5128

We have begun a program in collaboration with K. Freeman and M. Dopita to study the halos of early-type galaxies, and have chosen NGC 5128 (Cen A), the nearest giant elliptical and luminous radio source, as the first galaxy for our investigation. In this review we report a preliminary analysis of the radial velocities of 116 planetary nebulae found in the envelope and remote halo of Cen A. The radial velocities and a detailed comparison of the velocities to models of Cen

A will be published elsewhere (Ford et al. 1988). The spectra of the nebulae were obtained during two nights in the spring of 1987 by using the multifiber spectrograph on the AAT. The velocities for individual PN were measured from the [O III] λ5007 line, which had between 50 to 300 photon counts. The estimated accuracy of the measurement varies from 10 km s^{-1} for nebulae with 50 to 100 counts to about 4 km s^{-1} for PN with 100 to 200 counts.

The velocities of the planetary nebulae show indisputably that the halo is rotating (NE approaching, SW receding) about the photometric minor axis. We derived the rotation curve by first subtracting the PN systemic velocity (550 km s^{-1}) from the velocities and then folding and reflecting the SW PN velocity versus minor axis distance diagram onto the NE plot. To avoid smearing due to projection, we selected the nebulae which appear near the major axis and then binned the nebulae into groups of 10. The average velocity in each bin and the dispersion about the mean yielded the rotation and velocity dispersion curves shown in Figure 3. The rotation curve rises from the center of the galaxy and then flattens to a value of ∼ 100 km s^{-1} between 10 kpc and 20 kpc. The velocity dispersion appears to decline in the halo at distances beyond 10 kpc.

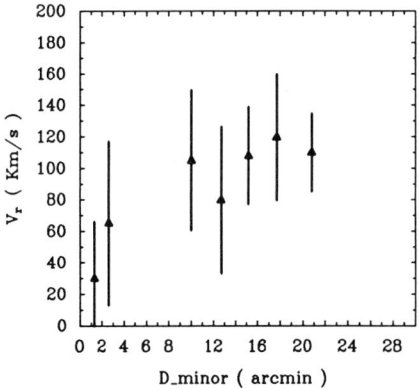
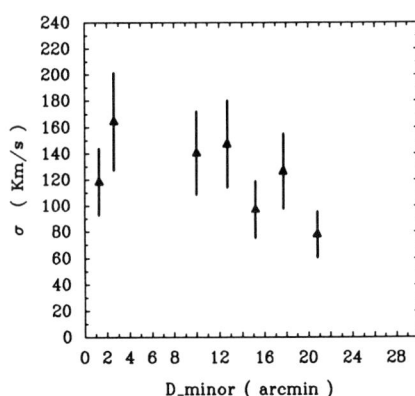

Figure 3. The left hand panel shows the observed rotation in the halo of NGC 5128 as a function of the distance from the photometric minor axis. The right hand panel shows the line-of-sight velocity dispersion in the halo. The error bars show the standard error.

We derived a *preliminary* mass distribution and M/L by approximating NGC 5128 as spherical and using the first moment of the collisionless Boltzmann equation (Hartwick and Sargent 1978). The rotational velocity was derived from a third order curve through Figure 3. The logarithmic derivative of the velocity dispersion was taken from a second order fit to the data in Figure 3, and included Wilkinson et al.'s central velocity dispersion of 145 km s^{-1}. Inclusion of this point forces a relatively constant dispersion curve in the central 10 kpc, and is consistent with an isothermal core in NGC 5128. The logarithmic gradient of the space density of PN was derived from Young's (1976) asymptotic expansion for a spherical de Vaucouleur's galaxy. Finally, the blue light was integrated assuming a spherical galaxy following Dufour et al.'s major axis light distribution and corrected for foreground extinction with Dufour's color excess. The results from this preliminary analysis are given in Table 3 for a provisional distance of 3.3 Mpc and the assumption that the velocity dispersion is isotropic. This analysis suggests that M/L increases throughout the galaxy. If the plane of rotation in NGC 5128 is appreciably inclined to the line of sight, or if the velocity dispersion is anisotropic, the masses and M/Ls in Table 3 are lower limits to the true values. These quantities will be better constrained with more data and a more sophisticated analysis.

Table 3. *Preliminary* Mass Distribution and M/L in NGC 5128

$R(kpc)$	V_{rot}	σ_V	$M(10^{11} M_\odot)$	$L_B(10^{10} L_\odot)$	M/L_B
1.3	28	146	0.2	0.7	2.4
2.6	50	147	0.4	1.1	3.1
10.0	104	137	1.9	2.4	7.8
12.7	105	128	2.4	2.6	9.0
15.2	104	116	2.7	2.8	9.8
17.7	105	102	2.9	2.9	10.2
20.8	113	81	3.0	3.0	10.1

4. PLANETARY NEBULAE AS STANDARD CANDLES

If we consider two stellar populations with similar ages, chemical compositions, and mass functions, stellar evolution should drive their stars to produce planetary nebulae with similar shell masses and central star masses. For a given age of the population there will be a well determined main sequence turnoff mass and mass of the planetary central star. For a given central star mass there will be a corresponding upper limit to the ionizing flux from that star and a maximum [O III] λ5007 flux which can be produced by the nebula. Thus, we expect two similar populations to have nearly identical [O III] λ5007 luminosity functions which are guillotined by stellar evolution at the same upper limit. Jacoby and Lesser (1981) and Ford (1983) presented data which showed that the dispersion in the reddening corrected [O III] λ5007 magnitudes of the brightest planetary nebulae in local group galaxies is $\sim 25\%$. This relative constancy of the maximum [O III] λ5007 luminosity in galaxies with a wide range in mass, chemical composition, and age is due to the fact that the strong dependence of the central star's evolutionary time on its mass results in an extremely high probability of observing central stars with a very narrow dispersion about a mean mass (Shaw 1988; however, see Pottasch 1988). These observations and arguments suggest that planetaries may be good standard candles. In particular, it is reasonable to suppose that a comparison of similar populations, such as the halo of NGC 5128 and Virgo ellipticals, will lead to even smaller dispersions, and perhaps accurate relative distances.

The best way to test distances derived from planetary nebula luminosity functions (PNLFs) would be to compare the PNLF distances to independently measured distances in the same galaxy or group of galaxies. Unfortunately, distances to even the nearest galaxies, especially early type galaxies, are so poorly determined we must at present resort to consistency checks. We first consider whether or not the PNLF appears to have the same form in different galaxies. In Figure 4 we show N(m) versus m([O III]) for M81 and NGC 5128. The two histograms are nearly identical, both rising steeply to a maximum $\sim$ 1 mag below the brightest nebula. Because our estimated completeness limit is only another 0.6 mags below the apparent peak, we think more observations and additional analysis of the completeness limit is needed before we can be certain that the peak is an intrinsic characteristic of the luminosity function. Jacoby's (1980) PNLF for the Magellanic Clouds does not show a turnover; however, there are too few planetaries in the first two magnitudes of the Clouds' PNLF to resolve such a feature.

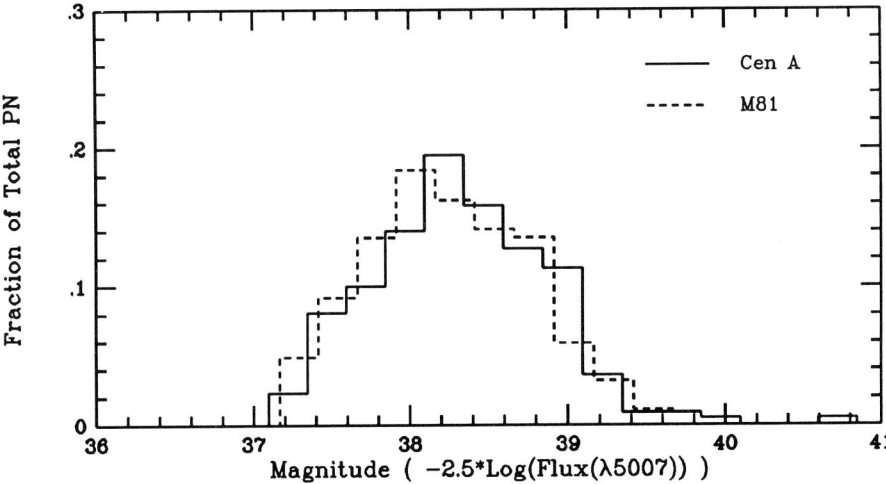

Figure 4. The planetary nebula magnitude density N(m) versus m([O III] $\lambda 5007$) for M81 and Cen A. The number of PN in each magnitude bin has been divided by the total number of nebulae.

Figure 5 shows log(N(m)) versus m([O III]) for M81, NGC 5128, and the three galaxies in the Leo Group. All of the logarithmic PNLFs exhibit the same shape over the first magnitude, a range where we think incompleteness is not a problem. Note that the shape is *not* a straight line (ie, the fall-off in the distribution is not a power law). This is very important since it means the data can be fit to its unique shape. If the PNLF were a power law, the results could always be scaled by assuming a larger galaxy; however, the observed curvature in the PNLF allows it to be used directly as a standard candle.

A second consistency test can be made by comparing the relative distances of different fields within galaxies and the relative distances of galaxies within the same group. The former shows the effects of photometric errors, variable internal extinction, and statistical fluctuations in the finite manifestation of the luminosity function. The latter provides a powerful test of the constancy of the

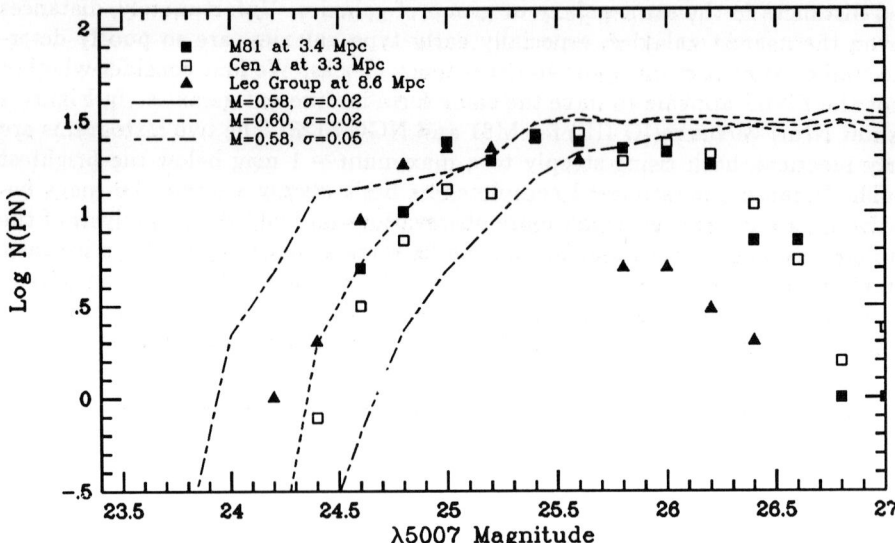

Figure 5. Logarithm of the planetary nebula number density N(m) versus [O III] λ5007 magnitude for M81, Cen A, and the sum of three galaxies in the Leo Group galaxies, corrected by the distances shown in the figure. The lines are theoretical luminosity functions for PN distributed uniformly in age, but with the Gaussian mass distribution described in the text (see §5).

PNLF in galaxies with different luminosities and Hubble types. A final goal of the consistency tests is to determine the most accurate and stable way to use the PNLF to derive distances.

There are at least three ways the PNLF can be used to derive distances: i) from the jth brightest nebula, ii) from a statistic such as the median derived from some fraction of the PNLF, and iii) fitting to a feature in the PNLF such as a peak or the curvature seen in Figure 5. The brightest nebula always must be treated with caution because of the possibility of confusion with faint supernova remnants or faint H II regions (in galaxies with star formation). The relative statistical merits of m_j and $m(med)$ can be illustrated by supposing that the PNLF $n(m)$ versus m is linear between the brightest nebula m_1 and m_2 (approximately 0.75 mag fainter than m_1), as is the case in NGC 5128 and M81. If

$$n(m) = a \times (m - m_1) \qquad m_1 < m < m_2$$

the total number of nebulae between m_1 and m_2 is

$$N(total) = (a/2) \times (m_2 - m_1)^2.$$

The magnitudes of the jth nebula and the median nebula are

$$m_j = m_1 + (m_2 - m_1) \times \sqrt{j/N(total)}$$
$$m(med) = m_1 + (m_2 - m_1)/\sqrt{2}.$$

For a fixed magnitude range $m_2 - m_1$ in either the program or reference galaxy, m_j depends on $N(total)$ whereas the median $m(med)$ does not. Simply put, as the number of nebulae in $m_2 - m_1$ increases, m_j is more likely to be found close to the bright limit. The error in $m(med)$ is 1/3 the error in m_1. Consequently, if we can confidently set m_2 above the completeness limit, $m(med)$ should be a better estimator than m_1 or m_j.

The (possible) peak and the curvature in the PNLF are potentially excellent standard candles. However, the magnitude of the peak (and $m(med)$) will artificially brighten and the curvature will change if the survey includes regions where a bright background or internal extinction moves the completeness limit above the true peak. Pushing the completeness limit deep enough to use either a peak or the shape of the PNLF will be very difficult for galaxies in the Virgo cluster.

Preliminary reddening corrected [O III] λ5007 fluxes for the 1st, 3rd, 5th, and median nebula in the first 0.75 mag of the luminosity function are given in Table 4. The last column gives the number of nebulae in the first 0.75 mag of the luminosity function. Distances for each galaxy or field are given relative to a provisionally adopted distance of 3.3 Mpc to NGC 5128. The distances will be revised when the final zero point is set from the PNLF in M31.

The relative distances in Table 4 show that the derived distances have a high degree of internal consistency and are relatively insensitive to the distance estimator. The median [O III] λ5007 flux appears to give somewhat more consistent results than the jth brightest nebula. The high degree of internal consistency for the three galaxies in the Leo Group (8.63 Mpc, $\sigma = 0.14$ Mpc) is particularly encouraging.

The distance estimates to NGC 4486 were made after eliminating the brightest nebula, which was 1.48 times brighter than the second brightest nebula. Neither any of the other galaxies nor any of the fields within the galaxies had such a large difference between the two brightest candidates. The Hα exposures of NGC 4486 (cf Table 1) did not show any nebulae at the positions of the PN candidates. Although there is little basis for guessing what the spectrum of a Type I SNR in the halo of NGC 4486 might look like, in most known SNRs Hα is stronger than [O III] λ5007 (eg. D'Odorico and Dopita 1983). The exceptions are oxygen rich Type IIs (Tuohy and Dopita 1983) in which the large expansion velocities would overfill our 30 Å λ5007 filter. Thus, even in these unlikely SNRs Hα probably would be as bright or brighter than [O III] λ5007. Although we conclude that the majority of our nebulae are not SNRs, we think the PNLF-based distance of M87 relative to the other galaxies should be viewed as highly provisional until we are able to confirm that the faint candidates in M87 are indeed planetary nebulae.

Possible sources of errors such as confusion, bandpass-redshift alignment and filter calibration, and internal reddening will be discussed in more detail in subsequent papers. We note here that the uncertainty in the magnitudes due to these difficulties can be eliminated by obtaining moderate dispersion spectra of the brightest nebulae in each galaxy. We plan to undertake this important and difficult task.

In conclusion, we think the observations presented in this review strongly support our contention the the PNLF will prove to be an accurate and reliable

Table 4. [O III] λ5007 Fluxes for the Brightest Planetary Nebulae in the Survey Galaxies

Galaxy	[O III] λ5007 Fluxes (ergs cm^{-2} s^{-1} × 10^{-15}) and Provisional Distances (Mpc) Relative to NGC 5128				
	1st	3rd	5th	median	N(median) (First 0.75 mag)
5128 (All)	1.45	1.26	1.23	0.918	46
	3.3	3.3	3.3	3.3	
5128 (SW)	1.03	0.897	0.773	0.827	7
	3.92	3.91	4.16	3.48	
5128 (NE)	1.26	1.15	1.04	0.945	18
	3.54	3.45	3.59	3.25	
5128 (Envl)	1.45	1.25	1.04	0.929	18
	3.30	3.31	3.59	3.28	
M81 (All)	1.37	1.29	1.21	0.862	52
	3.39	3.26	3.32	3.40	
3115 (All)	0.276	0.254	0.225	0.175	21
	7.56	7.35	7.72	7.56	
3115 (E)	0.254	0.179	0.159	0.141	14
	7.88	8.76	9.18	8.42	
3115 (W)	0.276	0.247	0.186	0.175	15
	7.56	7.45	8.49	7.56	
Leo Group					
3377	0.200	0.169	0.156	0.129	13
	8.89	8.98	9.27	8.80	
3384 (All)	0.202	0.187	0.174	0.137	25
	8.84	8.58	8.77	8.54	
3384 (E)	0.197	0.174	0.155	0.137	15
	8.95	8.88	9.30	8.54	
3384 (W)	0.202	0.154	0.151	0.146	16
	8.84	9.44	9.42	8.27	
3379 (All)	0.196	0.187	0.175	0.136	24
	8.98	8.57	8.75	8.57	
3379 (E)	0.196	0.175	0.170	0.137	14
	8.98	8.85	8.88	8.54	
3379 (W)	0.187	0.146	0.139	0.135	10
	9.19	9.69	9.82	8.61	
5866 (All)	0.177	0.123	0.113	0.113	10
	9.45	10.56	10.89	9.43	
4486 (All)	0.100	0.081	0.065	0.063	10
	12.57	13.06	14.35	12.59	

way to estimate the distances of early type galaxies out to the distance of the Virgo cluster.

5. IMPLICATIONS FOR PLANETARY NEBULA CENTRAL STARS

The [O III] flux in planetary nebulae is driven by the ionizing UV radiation emerging from the central star. If the central star passes through its maximum luminosity and temperature while the nebula is still optically thick to ionizing radiation, the nebula can attain a high luminosity. If the nebula expands more rapidly than the central star can evolve across to the blue side of the HR diagram (low mass central stars), only a small fraction of the nebula is ionized and its luminosity will be low. If the central star evolves very quickly (high mass central stars), the lifetime of the high luminosity period may be so short that very few, if any, will be seen.

To generate a theoretical planetary nebula luminosity function we began with Wood and Faulkner's (1986) models for the evolution of central stars. We ionized model nebulae with a series of energy distributions corresponding to ages along the evolutionary track of the central star. The $0.6 M_\odot$ nebulae were evolved by allowing them to expand uniformly at a constant rate. The models show that the precise value of the nebular mass is unimportant because only a small fraction of the total mass is ionized in the brightest nebulae.

The following simplifying assumptions were made to expedite computation of the models: 1) the nebula is a uniform density sphere, 2) the nebula expands at a constant velocity of 20 km s^{-1}, 3) the nebula is always in ionization equilibrium with the UV flux from the central star, 4) no additional energy input is available (eg. there is no stellar wind), 5) the central star flux is a black body energy distribution, 6) the central star masses have a Gaussian distribution, and 7) the chemical composition is typical of solar neighborhood nebulae (Aller and Czyzak 1983).

A series of masses of 0.6, 0.7, and $0.76 M_\odot$ were chosen from the evolutionary tracks presented in Figure 1b of Wood and Faulkner (1986), and a mass of 0.65 was interpolated from this figure. A mass track at 0.565 was also taken from Figure 2 of Schönberner (1983). This track is not entirely consistent with those of Wood and Faulkner, but is adequate for the fainter nebulae, and has little bearing on our principle interest—the high luminosity nebulae.

Wood and Faulkner present four sequences of tracks corresponding to two mass loss scenarios and two epochs for PN shell ejection. We chose the mass loss model in which PNN lose mass at a rate of $3 \times 10^{-5} M_\odot$ yr^{-1} until $\log T_{eff} = 3.8$, after which time $\dot{M}$ is zero. We selected the shell ejection epoch at the rising surface luminosity of a flash rather than midway between helium shell flashes. These choices were invoked to provide a set of models having the lowest UV fluxes. A set having higher fluxes requires smaller central star masses.

For each central star mass, the temperature and luminosity were interpolated into a grid spaced at 1000 year intervals from the moment the central star reaches $\log T_{eff} = 4.0$. Black body energy distributions for central stars at these ages were used as ionizing sources in the model nebula code described by Aller and

Keyes (1980), based on the formulation by Balick (1975).

The same nebula was used in each mass sequence, and the age of the nebula determined the size and density. Note that the age includes an offset on the order of several hundred years relative to the central star age zero point, since there is a delay from planetary nebula ejection until the central star reaches $\log T_{eff} = 4.0$.

The [O III] $\lambda 5007$ intensities predicted by the models were used to generate a grid of fluxes as a function of age and central star mass. We then randomly selected planetaries in a Monte Carlo simulation of observing many extragalactic PN according to the following prescriptions: 1) PN are distributed uniformly in age from 0 to 30,000 years, and 2) PN central stars have a Gaussian distribution such that the number of stars of mass M is proportional to $\exp\left(-0.5 \times (M/\sigma)^2\right)$. We then randomly selected 100,000 PN ages and masses, interpolated in the [O III] $\lambda 5007$ luminosity grid, and histogramed the results.

Figure 5 illustrates three such models for mass distributions selected with peaks at 0.58, 0.60, and $0.58 M_\odot$ and $\sigma(M)$ of 0.02, 0.02, and $0.05 M_\odot$. The curves in this figure have been scaled vertically to match the observed number of PN in M81. The horizontal magnitude scale is simply $-2.5 \times \log F([O\ III]\lambda 5007) - 10$ for nebulae at a distance of 1 Mpc. The principle features in the curves are: 1) a sharp cutoff at the bright end, 2) a flat distribution beginning at ~ 1 mag below the brightest, 3) a bright end shift of about 0.3 mag for a mass shift of $0.02 M_\odot$, and 4) a bright end tail of about 0.7 mag when the mass function is broadened to $0.05 M_\odot$. A comparison with the data for M81, Cen A (NGC 5128), and the three galaxies in Leo shows that the cutoff at the high mass end of the central star mass distribution must be quite abrupt.

It is important to note that the bright end cutoff is completely controlled by the high mass tail of the central star mass distribution. A model which has a long tail of low mass central stars does not change the shape of the cutoff.

We conclude that, within the limits of the models, the masses of central stars peak near $0.6 M_\odot$, and have a high mass tail characterized by a Gaussian distribution with $\sigma(M) < 0.05 M_\odot$, and probably closer to $0.02 M_\odot$. For comparison, the observationally selected distribution derived by Schönberner (1981) and Schönberner and Weidemann (1983) suggests a central star mass peak near $0.58 M_\odot$, and an equivalent $\sigma(M) \approx 0.015 M_\odot$. We caution the reader that the assumptions in the models may have a significant impact on the upper limits to the [O III] $\lambda 5007$ luminosities, but the shape of the luminosity distributions will not be altered much since these are dominated by the central star mass distribution.

This work was supported by NASA contracts NAS 5-29293 and NAGW-421. Don Neill provided excellent help in analysis of the NGC 5128 observations. We are grateful to C.D. Keyes for providing the nebula model code and instructions on how to use it.

REFERENCES

Aller, L.H. and Czyzak, S.J. 1983, *Ap. J. Suppl.*, **51**, 211.
Aller, L.H. and Keyes, C.D. 1980, *Ap. Sp. Sci.*, **72**, 203.
Balick, B. 1975, *Ap. J.*, **203**, 705.
Cahn, J. H. and Wyatt, S. P. 1978, in *IAU Symposium 76, Planetary Nebulae Observations and Theory*, ed. Y. Terzian (Dordrecht:Reidel), p. 3.
Cudworth, K. M. 1974, *A. J.*, **79**, 1384.
Dopita, M. A., Meatheringham, S. J., Wood, P. R., Ford, H. C., Webster, B. L., and Morgan, D. H. 1988, *Paper presented at the ESO Workshop on "The Evolution and Dynamics of the Outer Halo of the Galaxy"* April 7-9, 1987.
D'Odorico, S. and Dopita, M. 1983, in *IAU Symposium 101, Supernova Remnants and Their X-ray Emission*, ed. J. Danziger and P. Gorenstein (Dordrecht : Reidel), p. 517.
Dufour, R. J., van den Bergh, S., Harvel, C. A., Martins, D. H., Schiffer, III, F. H., Talbot, Jr., R. J., Talent, D. L., and Wells, D. N. 1979, *A. J.*, **84**, 284.
Eather, R. H. and Reasoner, D. L. 1969, *Appl. Optics*, **8**, 227.
Ford, H.C. 1983, in *IAU Symposium 103, Planetary Nebulae*, ed. D.R. Flower (Dordrecht : Reidel), p. 443.
Ford, H. C., Freeman, K. C., Hui, X., Ciardullo, R., Dopita, M. A., Jacoby, G., and Meatheringham, S. 1988, *in preparation*.
Keyes, C.D. and Aller, L.H. 1980, *Ap. Sp. Sci.*, **72**, 203.
Hartwick, F. D. A. and Sargent, W. L. W. 1978, *Ap. J.*, **221**, 512.
Huchra, J., Stauffer, J., and van Speybroeck, L. 1982, *Ap. J. (Letters)*, **259**, L57.
Jacoby, G. H. 1980, *Ap. J. Suppl.*, **42**, 1.
Jacoby, G. and Ford, H. C. 1986, *Ap. J.*, **304**, 490.
Jacoby, G. and Ford, H. C. 1988, *in preparation*.
Jacoby, G. H. and Lesser, M. P. 1981, *A. J.*, **86**, 185.
Jaffe, W. 1983, *M.N.R.A.S.*, **202**, 995.
Lawrie, D. G. 1983, *Ap. J.*, **273**, 562.
Lawrie, D. G. and Ford, H. C. 1982, *Ap. J.*, **256**, 120.
Merritt, D. 1985, *A. J.*, **90**, 1027.
Nolthenius, R. 1984, Ph.D. thesis, University of California, Los Angeles.
Nolthenius, R. and Ford, H. C. 1986, *Ap. J.*, **305**, 600.
Nolthenius, R. and Ford, H. C. 1987, *Ap. J.*, **317**, 62.
Pottasch, S. R. 1988, *IAU Symposium 131*, ed. S. Torres-Peimbert (Dordrecht : Reidel), *this volume*.
Schönberner, D. 1981, *Astr. Ap*, **103**, 119.
Schönberner, D. 1983, *Ap. J.*, **272**, 708.
Schönberner, D. and Weidemann, V. 1983, *IAU Symposium 103, Planetary Nebulae*, ed. D.R. Flower (Dordrecht : Reidel), p. 359.
Shaw, R. A. 1988, *IAU Symposium 131*, ed. S. Torres-Peimbert (Dordrecht : Reidel), *this volume*.
Tonry, J. L. 1984, *Ap. J. (Letters)*, **283**, L27.

Tuohy, I. R. and Dopita, M. A. 1983, in *IAU Symposium 101, Supernova Remnants and Their X-ray Emission,* ed. J. Danziger and P. Gorenstein (Dordrecht : Reidel) p. 165.

van den Bergh, S. 1969, *Ap. J. Suppl.,* **19**, 145.

Whitmore, B. C. 1980, *Ap. J.,* **242**, 53.

Wilkinson, A., Sharples, R. M., Fosbury, R. A. E., and Wallace, P. T. 1986, *M.N.R.A.S.,* **218**, 297.

Wood, P.R. and Faulkner, D.J. 1986, *Ap. J.,* **307**, 659.

Young, P. J. 1976, *A. J.,* **81**, 807.

A SURVEY OF PLANETARY NEBULAE IN THE SMC AND M31

N. Meyssonnier, M. Azzopardi and J. Lequeux
Observatoire de Marseille, France
R. Gathier
European Southern Observatory, F.R.G.

ABSTRACT. Our general method for finding planetary nebulae (PN) is to make wide field objective-prism or objective-grating low-dispersion spectra on photographic plates, PN stand up amongst other emission-line objects either as Hα + [N II] 6548-6583 A emitters or as [O III] 50007 A emitters with faint or no continuum, higher-resolution spectroscopy is used for confirming a selection of candidates.

For the SMC we used the Curtis Schmidt Telescope at Cerro Tololo equipped with the 10° prism (420 A mm^{-1} at Hα) and a 110 A band width interference filter transmitting Hα + N II; the field is 3°.4 × 3°.4 (Azzopardi and Meyssonnier 1986, IAU Symposium No. 116, p. 225), spectroscopy of about 20 new PN candidates has been partly secured at ESO using the 3.60-m telescope and both slit (Boller and Chivens) and multi-aperture spectroscopy (Optopus). See below the spectrogrammes of two new discovered planetary nebulae.

For M31 we used the CFH telescope with the prime-focus green grens (dispersion 2000 A mm^{-1}), a schott GG 435 filter and hypersensitized IIIa-J plates; the field is 1° × 1° (Lequeux, Meyssonnier and Azzopardi 1987, Astron. Astrophys. Suppl., 67, 169). We very recently completed the coverage of the galaxy (10 fields partially overlapping). The present inventory covering 2/3 of the galaxy contains about 1200 objects showing emission lines between 4350 and 5300 A, most of which have no continuum and are believed to be PN's; there is a good agreement with the surveys by Ford and collaborators in the areas in common.

THE KINEMATICS OF THE PLANETARY NEBULAE IN THE LARGE MAGELLANIC CLOUD

Stephen J. Meatheringham and Michael A. Dopita
Mount Stromlo and Siding Spring Observatories
Australian National University
Holland. C. Ford
Space Telescope Science Institute
B. Louise Webster
School of Physics, University of New South Wales

ABSTRACT. The radial velocities of a total of 94 Planetary Nebulae (PN) in the Large Magellanic Cloud (LMC) have been determined. The kinematics of the population of planetary nebulae is compared with the H I data in the context of a re-analysis of the survey by Rohlfs et al. (1984), taking into account the transverse velocity of the LMC. We find that the best solution for this transverse velocity is 275±65 km s^{-1}, and that the LMC is near perigalacticon. This is consistent with a maximum Galactic mass of order 4.5×10^{11} $M_\odot$ out to 51 kpc. The rotation curve obtained after correction for this velocity implies a mass of $(4.6 \times 0.3) \times 10^9$ $M_\odot$ within a radius of 3 degrees, or about 6×10^9 $M_\odot$, total. The rotation solution for the PN population is essentially identical with that of the H I, but the vertical velocity dispersion of 19.1 km s^{-1} is much greater than the value of 5.4 km s^{-1} found for the H I. This increase in velocity dispersion is consistent with it being the result of orbital heating and diffusion operating in the LMC in a manner essentially identical with that found for the solar neighbourhood.

N 66: A HIGH EXCITATION N RICH PLANETARY NEBULA IN THE LMC

M. Peña
Instituto de Astronomía
Universidad Nacional Autónoma de México
M.T. Ruiz
Departamento de Astronomía
Universidad de Chile

ABSTRACT. Spectrophotometric data of the planetary nebula N 66 (WS 35) has been obtained with the CTIO 4-m telescope equipped with an R-C spectrograph and a 2D-Frutti detector. The spectral range between 3700 A and 6800 A was covered at 4 A resolution.

The spectral features of N 66 show that it is a very high excitation PN. Collisionally excited lines of Ne IV, Fe VII and Ar V are clearly detected. The strength of He$^+$ $\lambda 4686$, relative to Hβ, permits us to deduce that the effective temperature of the central star is greater than 125 000 K.

The physical conditions derived for this nebula are:

T_e [O III] = 15 300 K $\pm$ 300 K, T_e[N II] = 11 100 $\pm$ 500 K, N_e(FL) = 1900 cm^{-3}.

Ionic abundance have been derived for He$^+$, He^{++}, N$^+$, O$^+$, O^{++}, Ne^{+3}, S$^+$, S^{+2} and Ar^{+3}. With these abundances and the usual ionization correction factors, the total abundances calculated for N 66 and log He = 11.09, log N = 8.37, log O = 8.34 and log Ne = 7.57.

The He, O and Ne abundances are similar to other LMC planetary nebulae abundances. The N appears enriched, like in Type I planetary nebula. However, N 66 does not satisfy the enhanced He criterion and cannot be classified as a typical Type I PN.

CHEMICAL ABUNDANCES IN MAGELLANIC CLOUD PLANETARY NEBULAE

D.J. Monk, M.J. Barlow, and R.E.S. Clegg
Dept. of Physics and Astronomy
University College London

ABSTRACT. Optical spectroscopic data for 71 Planetary Nebulae (PN) in the Large and Small Magellanic Clouds have been analysed. The line fluxes have been used to determine nebular temperatures, densities, and the abundances of He, N, O, Ne and Ar, relative to H. In our sample there are 12 nebulae with N/O $\geq$ 0.5, resembling Peimbert's Type I PN; 6 low excitation (LE) objects ($1 \leq I(5007)/I(H\beta) \leq 4$); and 4 very-low-excitation (VLE) nebulae ($I(H\beta) > I(5007)$), similar to the Galactic VLE class. Mean abundances have been calculated for the nebulae not in these special groups.

After correction for collisional excitation contributions to the nebular He I lines, the abundance of helium in the PN is found to be the same as that in H II regions in the LMC and SMC. A helium mass fraction of $Y = 0.247 \pm 0.009$ is consistent with all of the data on nebulae in both Clouds.

Compared to PN in our own galaxy, the abundances of Ne and Ar, which are the elements in our sample least affected by nucleosynthesis, are lower by 0.6 and 0.3 dex for the SMC and LMC respectively. The oxygen and neon abundances in the Magellanic Cloud PN are the same as those previously found for H II regions in the LMC and SMC, but the nitrogen in PN is enhanced by 0.9 dex and 1.0 dex in each galaxy respectively. This is consistent with the processing of all of the original carbon to nitrogen by the CN cycle operating in the progenitor stars.

CENTRAL STAR AND NEBULAR MASSES FOR MAGELLANIC CLOUD PLANETARY NEBULAE

D.J. Monk, M.J. Barlow, and R.E.S. Clegg
Dept. of Physics and Astronomy
University College London

ABSTRACT. AAT and IUE spectra of thirteen medium-excitation Magellanic Cloud planetary nebulae have been used to derive H I Zanstra effective temperatures and surface gravities for the central stars.

The known distances to the SMC and LMC allow the luminosity and the effective temperature of each central star to be plotted on the H-R diagram and its mass to be derived by interpolation between the theoretical evolutionary tracks of Schönberner (1979,1983) and Wood and Faulkner (1986). This derived mass, along with the stellar radius deduced from the central star analysis, give an estimate of log g, which can be used to check for consistency with the value of log g of the model atmosphere used in the analysis, in an iterative manner.

The mean mass for nine non-Wolf-Rayet (WR) central stars, assumed to be hydrogen shell burning, is 0.59 ± 0.02 $M_\odot$, which compares well with the mean mass for DA white dwarfs of 0.58 ± 0.10 $M_\odot$ (Weidemann and Koester 1984). Similarly, the mean mass derived for four WR central stars, assumed to be helium shell burning, of 0.59 ± 0.01 $M_\odot$, is in good agreement with the mean mass estimated for DB white dwarfs by Oke *et al.* (1984), 0.55 ± 0.10 $M_\odot$. The surface abundances and relative frequency of occurrence of each ($\sim 15\%$) are consistent with the hypothesis that DB white dwarfs originate from Wolf-Rayet central stars, the non-WR central stars giving rise to DA white dwarfs.

Nebular masses have been derived for a sample of twenty three Magellanic Cloud PN with $n_e \leq 4000$ cm^{-3}, using Hβ fluxes and [O II] doublet ratio electron densities, and a mean ionized hydrogen mass of 0.19 ± 0.04 $M_\odot$ is found, corresponding to a mean ionized mass of 0.25 ± 0.05 $M_\odot$.

EVOLUTION OF MAGELLANIC CLOUD PLANETARY NEBULAE

Stephen J. Meatheringham, Michael A. Dopita, and Peter R. Wood
Mount Stromlo and Siding Spring Observatories,
Australian National University
B. Louise Webster
School of Physics, University of New South Wales
David H. Morgan
Royal Observatory, Edinburgh
Holland C. Ford
Space Telescope Science Institute

ABSTRACT. New evolutionary correlations have been discovered to apply to the population of Planetary Nebulae (PN) in the Magellanic Clouds. Firstly, the age of the nebular shell is found to follow a relationship $\tau = 890[(M_{neb}/M_\odot)(V_{exp}/\text{km s}^{-1})]^{0.6}$ yr, which is shown to be consistent with a model in which the total energy of the ionised and swept up gas drives the expansion down the density gradient in the precursor AGB wind. Secondly, a tight correlation is found between the expansion velocity and a combination of the Excitation Class and the Hβ flux. This appears to be determined by the mass of the planetary nebula nuclear star. These correlations provide strong observational support for the idea that the PN shells are ejected at low velocity during the Asymptotic Giant Branch phase of evolution, and that they are continually accelerated during their nebular lifetimes.

PLANETARY NEBULAE AS STANDARD CANDLES FOR EXTRAGALACTIC DISTANCES

G. Jacoby
NOAO/Kitt Peak National Observatory
H. Ford
Space Telescope Science Institute
R. Ciardullo
Department of Terrestrial Magnetism/Carnegie Inst. of Wash.

ABSTRACT. Although distances to galactic planetary nebulae are typically uncertain by factors near 2, there appears to be an upper limit to the integrated [O III] $\lambda 5007$ absolute flux for planetary nebulae. This flux, 1.34×10^{-8} ergs/cm^2/sec for a maximally bright planetary nebula at a distance of 1 kpc, can be used as a distance indicator when a sufficient number of planetaries have been identified in an external galaxy. Or one may compute the upper limit to the distance for galactic planetary nebulae based on this calibration.

The original work by Jacoby and Lesser (1982, A.J., 86, 185) has been improved using CCD photometry and updated distances for Local Group galaxies, reducing the calibrator dispersion from 16% to 9%. Considering that the calibrators include ellipticals (NGC 205, NGC 185, M32), a large spiral (M31), and irregulars (LMC, SMC) have a considerable range in metallicity and galaxy luminosity, this small dispersion suggests that planetary nebulae may be excellent standard candles for all Hubble types, and can be identified easily to distances exceeding 10 Mpc. Jacoby, Ford, Booth, and Ciradullo (1987, Bull. AAS, 19, 712) derive the distance to M81 using this technique.

Kaler (1978, Ap. J., 220, 887) alludes to the upper brightness limit for the [O III] line. As $\lambda 5007$ is the principal forbidden line, and the ratios of the important cooling atoms to oxygen remain generally constant, and because forbidden lines are the dominant nebular coolants, an upper limit to the central UV flux implies a more or less constant upper limit to the [O III] flux. Evolutionary paths for central stars imply this observational upper limit: extremely massive central stars pass through the maximum UV flux region too quickly to be important, yet low mass stars have a low UV flux. The balance between lifetime in the region and the frequency of higher mass central stars defines a narrow observational window.

Detlef Schönberner, Volker Weidemann and Rolf Kudritzki.

OH/IR STARS AND OTHER IRAS POINT SOURCES AS PROGENITORS OF PLANETARY
NEBULAE

H.J. Habing, P. te Lintel Hekkert, W.E.C.J. van der Veen
Sterrewacht
Postbus 9513.
2300 RA Leiden,
THE NETHERLANDS

ABSTRACT

We briefly review the history of the search for progenitors of
planetary nebulae starting with Shklovsky's (1956) paper. The inner
structure of AGB stars (the likely progenitors) is sketched. The (l,b)
distributions and the (l,V) distributions (V is the centre of mass
radial velocity) of OH/IR stars and of planetary nebulae are compared;
it is concluded that, grosso modo, both types of objects belong to the
same galactic population and that most OH/IR stars develop ultimately
into planetary nebulae. From a comparison of the properties of OH/IR
stars and of Mira variables it is concluded that both are AGB stars
with the OH/IR stars having developed from Mira variables. Most OH/IR
stars are long period variables but the few that are not are probably
transition cases -no longer AGB stars and very early planetary
nebulae. It is argued that the IRAS catalog contains a large number of
AGB stars without (detected) OH maser emission, but otherwise similar
to OH/IR stars. An evolutionary sequence is presented from Mira's to
oxygen-rich planetary nebulae. Some speculations are added on the
formation of carbon stars and carbon-rich planetary nebulae.

1. INTRODUCTION

1.1. Some history

In the early fifties the theory of stellar evolution led to the
insight that a red giant contains a very dense and compact helium core
surrounded by a huge envelope of low density hydrogen. Hoyle and
Schwarzschild (1955) calculated the ascent of such stars on the red
giant branch and concluded that roughly half of the mass
of 1.1 to 1.2$M_\odot$ is contained inside this core, the rest is outside in
an envelope of much larger diameter and much lower density. Suppose
that one takes a planetary nebula, and shrinks it until the volume of
nebular gas has a radius of that of a red giant atmosphere, would one

not obtain a red giant- or, in other words, is a planetary nebula a
blown up red giant? This question appears to have been posed for the
first time, and answered positively by Shklovskii (1956) after he had
analyzed in detail the structure of several planetary nebulae.
Apparently the point at issue had a wide appeal although it
immediately led to another question: what causes the expansion of the
stellar envelope? Ten years after Shklovskii, and not knowing the
answer to this second question Abell and Goldreich (1966) in an
influential paper reconfirmed Shklovskii's conclusion via several
other arguments. The first of those, and one of the most powerful, had
been phrased earlier by others, e.g. Osterbrock (1964) and Minkowski
(1965): the galactic distribution of planetary nebula indicates that
they belong to the old disk population and that they must be the
result of the evolution of low-mass stars of typically, $1.2 M_\odot$; most
likely the planetary nebula stage follows that of the red giant.

Here rose, for a short time, a new problem: the red giant branch
ends with a helium flash and the emerging stars become horizontal
branch stars in a short while. Where do the planetary nebulae fit in?
Adequate answers were almost immediately available: in 1967 at the
Tatranska Lomnica Symposium on Planetary Nebulae (IAU Symposium 34)
Rose (1968) and Paczynski and Ziolkowski (1968) discussed the
existence of a second giant branch, later to be called the Asymptotic
Giant Branch or AGB, during which energy is produced in two shells
around a degenerate carbon-oxygen core: planetary nebulae emerge from
AGB stars.

There thus remains the questions by what forces and in what form
the AGB star ejects its envelope. The question "why there is ejection"
is still not answered, but the answer to the question "how ejection
takes place" has gradually been formulated over the last 15 years on
the basis of, especially, infrared, millimeter and centimeter radio
observations. A convenient starting point to describe this history is
the publication of the Caltech 2.2µm survey of Neugebauer and Leighton
(1969). This survey showed first of all that red giants, and
especially long period variables, have an infrared excess indicative
of a circumstellar shell. Subsequent microwave line measurements
(especially of OH) showed that the outflow velocity was of the order
of 15km/s and that the mass loss rate often exceeded $10^{-7} M_\odot/yr$. Of
even more significance was the discovery of optically obscured stars
with still larger mass loss rates (up to $10^{-4} M_\odot/yr$) -examples are two
of the very first objects published from the 2.2µm survey: NML Cyg and
NML Tau (Neugebauer et al., 1965). In 1968 Wilson and Barrett measured
strong 1612 MHz OH maser emission from many of the "IRC objects" and
this discovery was followed by the realisation by two groups of radio
astronomers (in Australia and in Sweden) that similar maser sources
could be detected in significant numbers by blind, unbiased 1612 MHz
radio line surveys in the galactic plane. After the radio positions of
maser sources had been determined with sufficient accuracy each maser
source could be identified with an optically invisible infrared point
source, and so the concept of an "OH/IR star" was born (Schultz et
al., 1976; Evans and Beckwith, 1977). Parallel to, but independent
from this radio astronomical route is the story of the (U.S.) Air

Force rocket survey; it produced the "AFGL catalogue" (Price and Walker, 1976) that contained many new infrared stars. One of the strongest was discovered independently as AFGL 2205 and, by Andersson et al. (1974), as OH26.5+0.6 -see also figure 6. Of course, most of the older surveys have been made obsolete by the IRAS full sky survey.

Detailed studies of OH/IR stars over the last 10 years have convincingly proven that the stars are Asymptotic Giant Branch stars ejecting neutral material at a low velocity (10 to 30km s^{-1}) and at a rate between $10^{-6} M_\odot yr^{-1}$ and a few times $10^{-4} M_\odot yr^{-1}$. Hence they are the missing link between red giants and planetary nebulae. In this article we review the arguments.

1.2 The inner structure of AGB stars

We summarize briefly some features of AGB stars -see figure 1. This

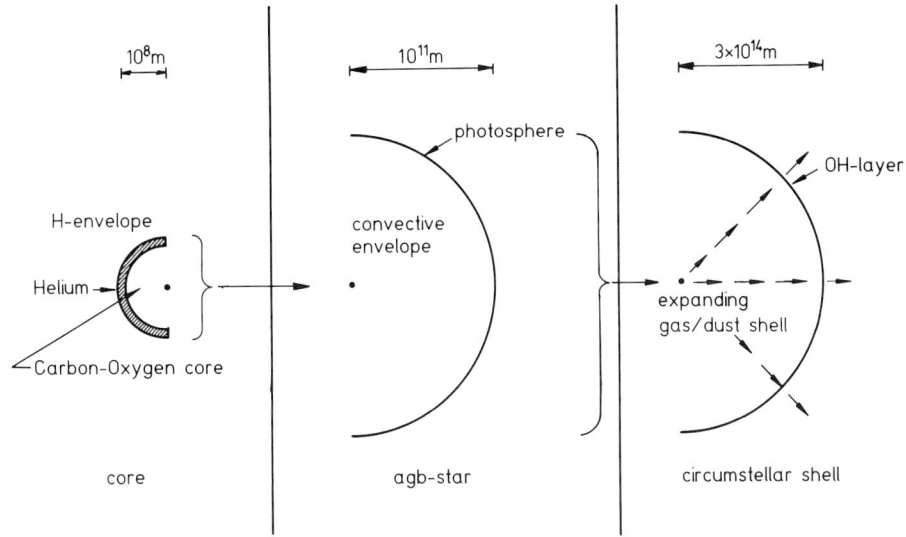

Figure 1: A schematic model for an AGB star. The core contains about 0.6 $M_\odot$, the envelope contains the rest.

figure is based partially on model calculations (Iben, 1971) and partially on observations. A core of oxygen and carbon with a degenerate electron gas is surrounded by a thin layer of helium, which in turn is surrounded by a large envelope of low density hydrogen: core and envelope contain comparable amounts of mass but their volumes have a ratio of 1 to 10^6. The envelope ends in a cool photosphere of typically 2500K. The luminosity of the star is very high (typically 6000$L_\odot$ but with a range from 2000 to 50,000 $L_\odot$); the energy is provided by the burning of hydrogen into helium at a rate dictated by the mass of the core, M_c, via the so-called Paczynski relation: $L=6\times10^4$ $(M_c-0.5)$ (L and M_c in solar units). During hydrogen burning the helium layer grows in mass until it reaches a critical limit. Then the helium will quickly burn into carbon (thermal pulse) while the hydrogen burning is shut off. For a typical core of $0.6M_\odot$ such a pulse occurs once per 10^5 yr, and it lasts about 100yr (Iben and Renzini, 1983). The upper part of the envelope pulsates with a long period between 200 and 2000 days, probably because transport of the large luminosity through the outer envelope is unstable. The pulsations cause shock waves in the atmosphere, and lift up clumps of material to such heights that solid particles can form. These absorb very efficiently the photospheric photons, and acquire an outward motion; the dust particles drag the gas along and the net result is a strong, high density ($\dot{M}>10^{-7}$ $M_\odot yr^{-1}$), slow ($\simeq$15km s^{-1}) wind. Thus a circumstellar shell is formed that extends beyond 3×10^{14}m.

2. THE STATISTICAL RELATION BETWEEN OH/IR STARS AND PLANETARY NEBULAE

2.1. Surveys, numbers and velocities of OH/IR stars

OH/IR stars are easy to pick out in a survey at 1612 MHz: they have a characteristic double peak line profile (see fig. 2). Moreover, such

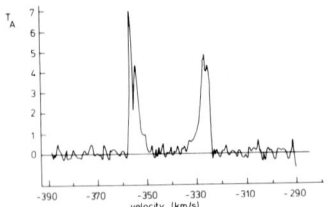

Figure 2: The 1612 MHz OH line profile of a "typical" OH/IR star. The velocity separation between the two peaks equals twice the expansion velocity of the shell; the average of the two velocities is the stellar velocity. The example line profile shown here is atypical only in that the stellar velocity is the highest known for an OH/IR star.

surveys turn up only a few (usually <5%) sources of a different kind: single peaks or complexes with several peaks; later we will come back to this point. Information derived from the double peak line profile includes the radial velocity of the star (the average of the two peak velocities) and the expansion velocity of the shell (half the difference between the two peak velocities). Several "blind" systematic OH surveys have been made: two in the southern hemisphere (Caswell and Haynes, 1975; Caswell et al., 1981) and three in the north (Johansson et al., 1977 a/b; Bowers, 1975, 1978; Baud et al. 1979a and b; 1981). In total approximately 400 sources were found. A further survey is being carried out for OH/IR stars near the galactic center; preliminary results have been published by Habing et al. (1983) using the Effelsberg 100m telescope and by Winnberg et al. (1985) using the VLA, which is a more efficient instrument so close to the galactic center; at present more than 60 OH/IR stars are known within one degree from the center. A catalogue of all OH/IR stars detected before the IRAS survey has been in preparation for a long time, but is now approaching completion (te Lintel Hekkert et al., 1988a). Considerable care has been taken in all surveys to carry them out in a uniform way; this facilitates reliable statistical analyses. One statistical conclusion is that the detected sources belong to the high-luminosity tail of a much wider maser-luminosity distribution: with better sensitivity one will find many more sources. The conclusion is supported by the results from the more sensitive searches near the galactic center and from the new, post-IRAS surveys.

Most of the energy produced by an OH/IR star is radiated in the infrared -typically between 2 and 60µm (see figure 6). A very profitable way to search for OH/IR stars is thus through an infrared survey. This is precisely what IRAS did and it is of no surprise that a few thousands of such infrared stars were found. Here we call stars "infrared" when they are surrounded by a dust/gas shell with an optical thickness exceeding 1.0 at 9.7µm. The few thousand are a significant fraction of all such stars in the Galaxy. Although IRAS was probably sensitive enough to detect the stars throughout the whole Galaxy, confusion limits their inclusion in the IRAS point source catalogue at low galactic latitudes, and at small longitudes. A hypothetical survey similar to that of IRAS, made with a larger telescope and thus a better angular resolution, but with the same sensitivity could possibly detect all infrared stars in the Galaxy.

The IRAS point source catalog is a treasure chest for new OH/IR stars. Two independent surveys search for 1612 MHz OH maser emission at the positions of suitably selected IRAS point sources. (1) Eder et al. (1987) used the Arecibo disk and discovered 184 new OH/IR stars in a sample of 474 IRAS sources in the sky visible to that telescope (declination between 0° and 37°). (2) te Lintel Hekkert et al. (1988b) used the Parkes 64m for a survey of the Southern Sky, including the galactic centre, and Le Squeren and others use the Nançay telescope to extend this survey to the north (see Sivagnanam and Le Squeren, 1986). The Parkes survey is now completed and the data are being reduced; about 900 new OH/IR stars have been detected. The Nançay survey has so far given about 200 new sources; a few hundred more are expected.

2.2. Galactic Distribution of OH/IR Stars and Planetary Nebulae

For each OH/IR star one measures at least four parameters: the galactic coordinates (l,b), the radial velocity of the star, V_* and the expansion velocity of the circumstellar shell, V_e. Figure 3a gives the (l,b) distribution for the stars from the Parkes, Arecibo and Nançay surveys together and figure 3b gives the (l,V_*) distribution. Figure 3a shows that the sources concentrate strongly to the galactic plane and that the large majority of the sources are in the inner Galaxy: there are only a few stars at |l|>90°. This is not a selection

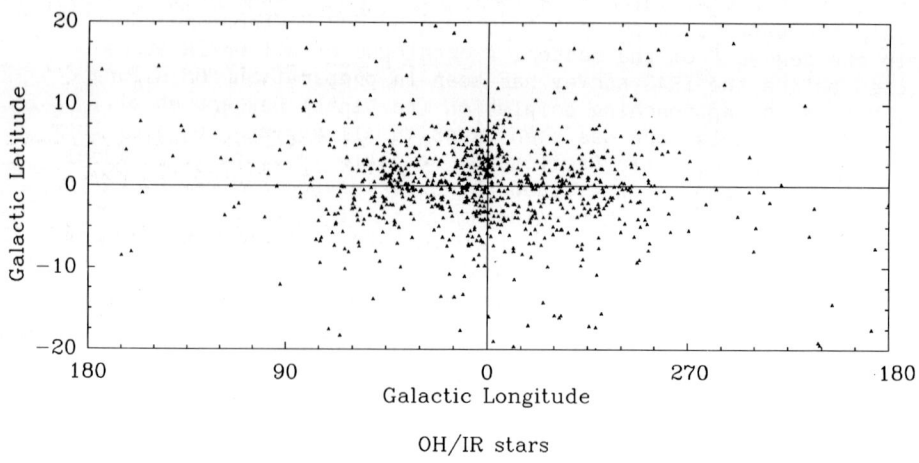

Figure 3a: (l,b) diagrams for OH/IR stars. Shown are all stars detected in the recent Arecibo and Parkes/Nançay surveys.

effect: there is a shortage of suitable IRAS sources in the Outer Galaxy. Already in the earlier, pre-IRAS surveys for OH/IR stars, especially the one by Bowers (1975, 1978), there is a shortage of stars in the anticentre. OH/IR stars are thus largely confined to galactocentric distances, $R<R_\Theta$: the Sun is close to the edge of their galactic distribution. Another feature of the pre-IRAS surveys is the strong concentration of sources toward the galactic centre; this is not so obvious in the recent IRAS based surveys. We suspect that the latter surveys have not covered completely the (confused) regions close to the galactic plane and the galactic centre. Figure 3b shows the effects of differential galactic rotation. Galactic rotation does not explain some important facts: if the stars are confined to $R<R_\Theta$ (as the l,b diagram indicates) then they will be confined to the permitted area, indicated in figure 3b, provided the stars follow closely galactic rotation. This is clearly not the case: many stars with $|l|<90°$ have velocities that deviate from galactic rotation by amounts up to 150km/s. Baud (1978; see also Baud et al., 1981) has shown that the deviations from galactic rotation are statistically correlated with the expansion velocity of the shell: larger expansion velocities imply smaller deviations from galactic rotation. The effect has been confirmed in more recent surveys (e.g. Eder et al., 1987). The explanation given by Baud also remains valid: stars with larger expansion velocities are on average more massive and thus younger;

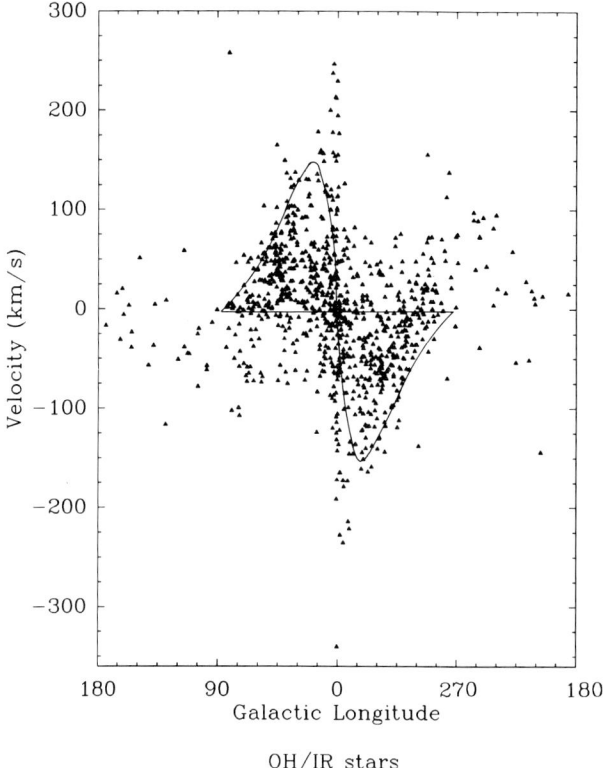

OH/IR stars

Figure 3b: (l,V_*) diagram for the OH/IR stars of figure 3a. The boundary of ruled area limits the "permitted" velocities -see the text.

thus they follow galactic rotation more closely. Comparing the average deviation from galactic rotation with that of Mira variables, Baud estimates that stars with an expansion velocity larger than the median value of 15 km/s are -on average- more massive than ≈1.2 to 1.5$M_\odot$, whereas the others are thus less massive.

As a next step compare figures 3a and 3b with similar figures for planetary nebulae (figure 4a and 4b). The shortage of planetary nebulae close to the galactic plane and to the galactic center can be explained by interstellar obscuration. The concentration of planetary nebulae toward the galactic plane seems reliable; it agrees with what is seen in figure 3a. In contrast to figure 2a, a significant number of planetary nebulae are found at $|l|>90°$. We might interpret this, however, as a relative lack of such nebulae at $|l|<90°$, because of selection effects but this interpretation we feel is not firm: are the anticentre planetary nebulae of a different kind? We conclude that figure 4a and 3a <u>may</u> represent the same distribution but real

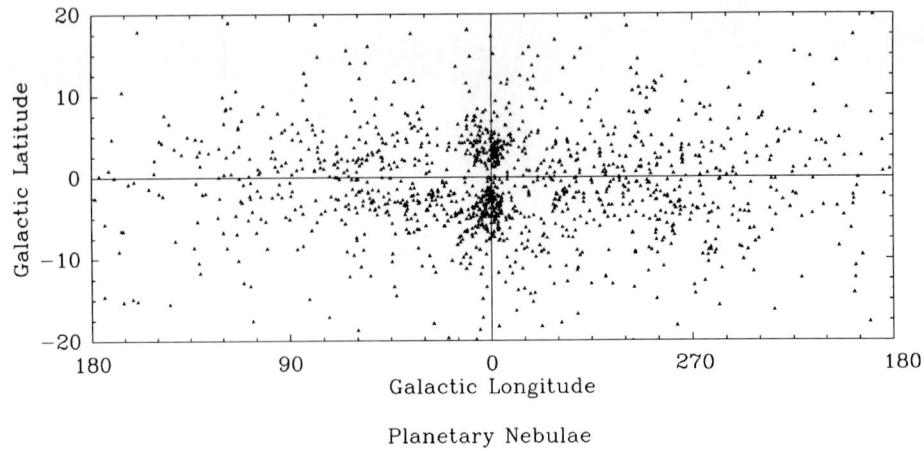

Figure 4a: (l,b) diagram for planetary nebulae. Shown are all nebulae from the list of Acker et al. 1981), plus a number of nebulae near the galactic center from data provided by A. Kalnajs (private communication).

differences may exist. A further conclusion is obtained by comparing figures 4b and 3b: these are very similar. We conclude that the agreement between the two (l,V_*) diagrams support the idea that planetary nebulae and OH/IR stars belong to the same population -and thus that the nebulae have evolved from OH/IR stars. We thus confirm Shklovskii's tentative answer (see the introduction). Selection effects in figures 4 prevent the conclusion that there is a one-to-one correspondence between OH/IR stars and planetary nebulae. It is thus possible, for example, that the most luminous OH/IR stars never become planetary nebulae or that the low-luminosity planetary nebulae discussed by Pottasch in this volume have never been OH/IR stars. However, for most objects the relation between these two kinds of objects seems well established.

Finally, there is one other argument that implies that OH/IR stars are the immediate precursors of planetary nebulae: the luminosity distribution of OH/IR stars peaks at about $6000L_\odot$ (see further down). If we translate this into a core mass using the Paczynski relation we find a peak at about $0.6M_\odot$. This agrees well with the observed peak in the white dwarf mass distribution (Koester and Weidemann, 1983).

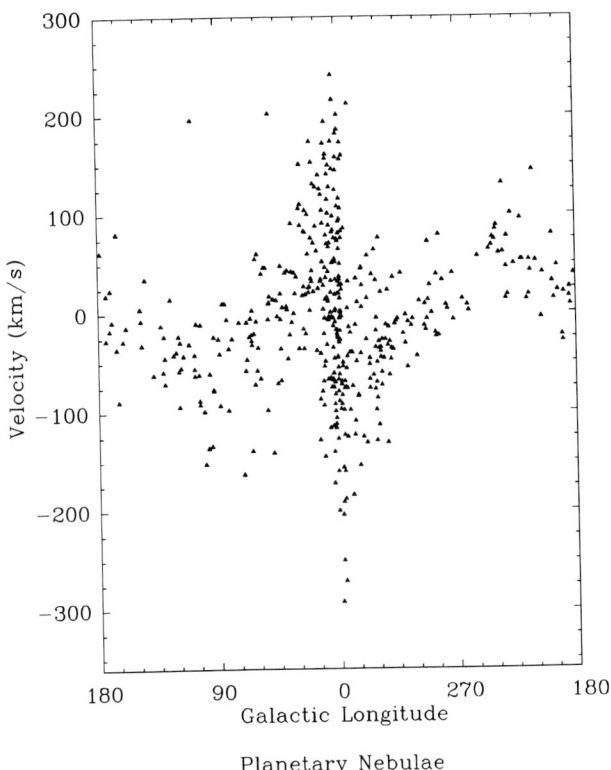

Planetary Nebulae

Figure 4b: (l, V_*) diagram for the planetary nebulae of figure 4a.

3. OH/IR AS AGB STARS

3.1. Luminosities, variation, mass loss and the relation to Mira variables

Having established a probable relation between OH/IR stars and planetary nebulae we now discuss the nature of the OH/IR stars. We will argue that OH/IR stars resemble in several ways a class of objects known already for a long time: the optically visible, oxygen-rich Mira variables. Our conclusion will be that OH/IR stars are AGB stars just as Mira's, but that they are equipped with some more extreme properties. In this section we discuss the luminosities, pulsation behaviour and mass loss of OH/IR stars, and we compare these properties with those of Mira's.

Luminosities

For a significant number of OH/IR stars the distance, d, has been measured. After measuring the total infrared flux F, and after correcting for interstellar extinction (symbolically written as a factor 10^A), we obtain $L=\pi d^2 F 10^A$. As we will see, the largest uncertainty is in the factor A, which cannot be derived from the observations, but for which a reasonable upper limit can be estimated. Because A=0 is a lower limit we obtain for each star a range in luminosity.

The distance to several OH/IR stars has been measured by geometrical means: by measuring first the angular diameter of the OH maser shell, and second its linear diameter. The ratio between the two gives the distance. The angular diameter (at most 2 arcsec but often less than 1 arcsec) is measured with radio interferometers -a remarkable achievement of radio astronomical techniques: the first such measurement was made with the British MERLIN array (Booth et al., 1981), followed immediately by a VLA measurement (Baud, 1981). At present some 30 stars have been mapped. The linear diameter is measured by a subtle light-travel time effect: in variable OH/IR stars the maser line strength varies quite regularly (see below); however, there is a small phase difference of the order of 0.01 between the "lightcurve" of the blue peak with respect to that of the red peak in the OH line profile. This phase difference, expressed in seconds of time, and multiplied by the velocity of light gives the diameter of the OH shell. The effect was predicted by Schultz et al. (1978) and first demonstrated by Jewell et al. (1980). The effect has been measured in a number of OH/IR stars by Herman (1983, see also Herman and Habing, 1985). For a total of 14 OH/IR stars Herman et al. (1986) derived thus a "geometrical" distance d; they found the total stellar flux, F, from groundbased and IRAS observations, and estimated A_v from a, we think, quite reasonable model distribution for extinction in our Galaxy (de Jong, 1986). The resulting luminosities are shown in Table 1. Notice the large spread! In our opinion the values of A_v so estimated should be considered as upper limits: the fact that precisely these OH/IR stars show up in the IRAS catalog (whereas others do not), probably means that in these directions the Galaxy is more transparant than the rather global model predicts. We thus give also luminosities for $A_v=0$. Table 1 shows that the more distant stars from Herman's bright star sample have luminosities exceeding the permitted limit of 56,000 $L_\odot$ (corresponding to a core mass of 1.4$M_\odot$, the Chandrasekhar limit); ignoring interstellar extinction brings them all except one back to below this limit. The exception is OH 127.9-0.0 with a distance of 7 kpc in the direction of the outer galaxy -a quite peculiar situation; we are convinced that in this case the distance determination has to be checked even more carefully than in the other cases. The relatively large number of very luminous sources is a selection effect; the original sample was extracted from one of the earlier surveys in the 1612MHz OH line, in which only the most luminous sources were discovered. Therefore the luminosities are not representative.

table 1

Distances and luminosities of OH/IR stars
(adapted from Herman et al., 1986)

name	distance (kpc)	luminosity[1] ($L_\odot$)	luminosity[2] ($L_\odot$)
OH16.1-0.3	0.55	530	480
OH20.7+0.1	8.3	112,000	25,000
OH21.5+0.5	11.6	151,000	52,000
OH26.6+0.6	1.0	8,500	8,200
OH30.1-0.7	1.8	5,500	4,700
OH30.1-0.2	1.1	1,500	1,500
OH30.7+0.4	7.7	99,000	44,000
OH32.0-0.5	9.3	69,000	29,000
OH32.8-0.3	8.0	128,000	47,000
OH39.7+1.5	1.2	17,000	17,000
OH44.8-2.3	2.4	26,000	26,000
OH127.9-0.0	7.0	263,000	200,000

[1] corrected for extinction
[2] without correction for extinction

A more representative luminosity distribution has been derived by Habing (1988): he made counts of IRAS point sources with colours similar to OH/IR stars. From the (l,b) distribution of the sources he derived their spatial distribution and their luminosity distribution. The only length scale that enters this derivation is the distance of the Sun to the galactic centre (8.5 kpc). The luminosity distribution peaks at $6000 L_\odot$; $16,000 L_\odot$ is an upper limit.

Comparing these results with what is known for Mira variables we find a good agreement: their average luminosity is about $5000 L_\odot$, only a small number have luminosities between 10,000 and 20,000 $L_\odot$.

Variability

Most OH/IR stars vary in time, both in the maser line and in the infrared. The infrared flux of 17 OH/IR stars has been monitored for a period of four years by Engels (1982; see also Engels et al., 1983). OH fluxes have been monitored in a program in Leiden started in 1980 by Herman and still being continued, now by Steeman and Habing. First results for 48 OH/IR stars and for 11 Mira variables have been published by Herman (1983); see also Herman and Habing (1985). There is good agreement between Herman's radio and Engels' infrared "light" curves -discrepancies occurred only for those lightcurves where the period exceeded the duration of the monitor program. The variation is periodic and the periods are very long (500 to 2000 days). The amplitude is large: up to 2 magnitudes <u>bolometric</u>. In pulsational properties the OH/IR stars are an extension of the Mira variables: amplitudes and periods are extensions toward values longer than are found for Miras, that have periods from 200 to 500 days, and bolometric amplitudes of at most 1 magnitude.

An interesting, but small group of OH/IR stars are those that do not vary at all, or vary irregularly and with a small amplitude. There are two different classes: (i) Supergiants and (ii) so-called "non-variable OH/IR stars". The first class consist of only a few stars (NML Cyg, VYC Ma and PZ Cas) with luminosities well above the AGB limit -and thus with a non-degenerate carbon-oxygen core. The second group will be discussed below in more detail.

Mass loss

The mass loss rate of OH/IR stars and of Mira's can be estimated in various ways. We will discuss the three most important methods together with their uncertainties. For a given OH/IR star these different ways lead to results that are in agreement within a factor of 2 - 5; however, in some cases discrepancies up to a factor of 100 can be found. For the Mira's a better agreement is reached usually. When mass loss rates are compared we see that the OH/IR stars are an extension to larg values larger than for Miras.
(1) Infrared mass loss rates:
One estimates the total circumstellar dust mass from the overall infrared spectrum. The weak points are the determination of the inner radius of the dust mass (or of the average density in the shell) and the gas-to-dust ratio which has to be assumed to find the total mass loss rate.
(2) OH mass loss rates:
The OH-line flux increases strongly with increasing $\dot{M}$ ($L_{OH} \sim \dot{M}^2$, Baud and Habing, 1983); this empirical fact can be used to estimate $\dot{M}$. The uncertainties involved here are the ratio $n(OH)/n(H_2)$ and the number of OH molecules necessary to get a saturated maser.
(3) CO mass loss rates:
This method is well established for Mira variables with relatively small mass loss rates, it has only very recently been used for OH/IR stars. For Mira stars the antenna temperature of the CO line is proportional to $\dot{M}$. Uncertain assumptions are in the ratio $n(CO)/n(H_2)$ and in the translation of the antenna temperature into a number of CO molecules. An unexpected problem is that the equations valid for Mira stars when used for OH/IR stars indicate mass loss rates systematically and significantly lower than those based on infrared and OH data (Omont, Forveille, Habing and Van der Veen, priv. comm.). This problem is now being studied in more detail.

Weighing in all uncertainties for the OH/IR stars the existence of mass loss rates ranging from a few times 10^{-6} to a few times 10^{-4} M_o/yr seems certain.

3.2. The IRAS-two colour diagram for OH/IR stars and Miras; a first scenario

From the discussions above it is concluded that OH/IR stars are related to Mira variables, and are probably also AGB stars. The two types are not the same: their luminosities agree, but the pulsational properties and the mass loss rates are more extreme for the OH/IR stars. This implies that OH/IR stars are "extreme Miras", and that they may have evolved from them. The suggestion finds further support in figure 5, a two-colour diagram constructed with IRAS data of OH/IR

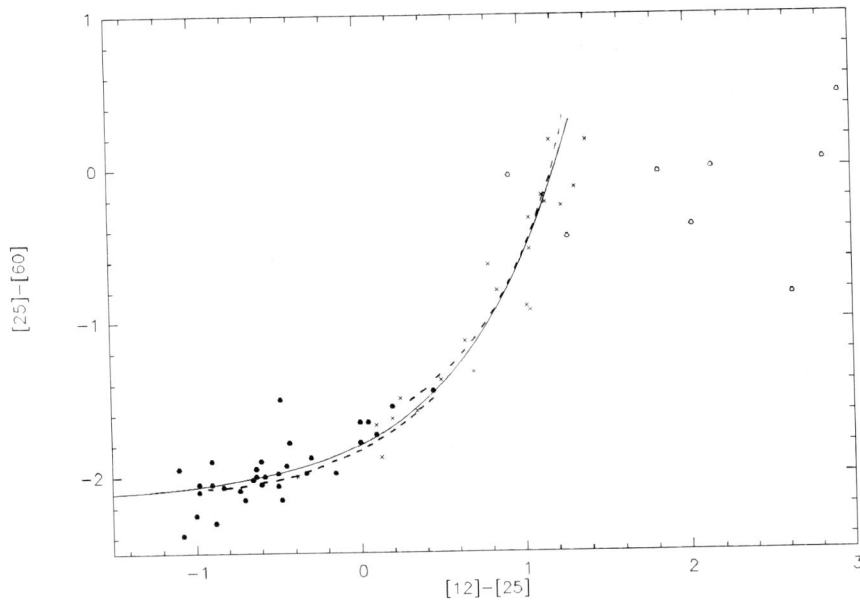

Figure 5: Two colour-diagram of (some) Mira variables and OH/IR stars constructed from IRAS 12, 25 and 60μm measurements. The Mira variables are indicated by filled circles; the variable OH/IR stars by crosses and the non-variable OH/IR stars by open circles.

stars and Miras; in its original form it was published by Olnon et al. (1983). A rather narrow sequence is defined by the Miras and the variable OH/IR stars. The non-variable OH/IR stars show deviating colours; they will be discussed later. As shown by Bedijn (1986, 1987) and by Rowan-Robinson et al. (1986) such a sequence can be interpreted as one of circumstellar dust shell models with increasing optical depth around a star of given luminosity L. The optical depth increases towards redder colours. The form and the position of the curve depend only slightly on the dust properties and not at all on the luminosity. The curve can be interpreted as one of increasing mass loss rate, where the stellar luminosity is a scaling factor at a given point on the curve.

How to interprete the curve of figure 4? A star may "start" as a Mira, stay at a fixed position in figure 4, until it switches over to become an OH/IR star, staying now at another fixed position: this viewpoint has been taken by Wood et al. (1983). Alternatively one may visualize a gradual evolution along the curve, because the mass loss rate increases continuously (Baud and Habing, 1983). Van der Veen has argued recently in favour of this latter interpretation because he observes that there is no correlation between the luminosity of a star and its position on the curve, a thing that one would expect under the first interpretation. From a count of the numbers of Miras and OH/IR stars along the curve Van der Veen concludes that the time left over on the AGB (that is: at any position the time required to reach the top of the curve) is proportional to M^α where $\alpha = \frac{4}{3}$ (van der Veen, 1987).

The non-variable OH/IR stars do not fit on the theoretical curve for increasing optical depths. This fact invalidates the suggestion given by Olnon et al. (1984) that the non-variable OH/IR stars are an extension towards larger mass loss rates. A more attractive explanation has been given by Bedijn (1986, 1987): Suppose that, when a star reaches the peak of the curve, it suddenly stops to pulsate and (at the same time) to eject mass. No longer is new, warm material fed to the inside of the shell; there is only expansion. The shell cools, and in the diagram it moves to the right. Such an OH/IR star is actually no longer an AGB star, but rather a very young planetary nebula! Bedijn predicted an interesting consequence: as the shell continues to expand, its optical depth will decrease and therefore the hot remnant of the AGB star will begin to shine through the nebula at short wavelengths: non-variable OH/IR stars will have a shoulder in the spectrum at $\lambda < 5\mu m$, in contrast to the variable OH/IR star where the spectrum drops at an exponential rate below $\lambda \approx 9\mu m$, see figure 6.

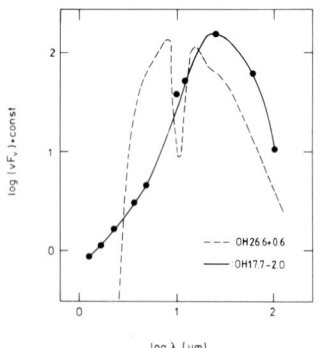

Figure 6: The infrared spectra of a variable OH/IR star (OH26.6+0.6) and of a non-variable OH/IR star (OH17.7-2.0).

This prediction has been verified in a number of cases and found to be true (van der Veen et al., 1987).

Van der Veen et al. (1987) proposed the following scenario for AGB evolution, based on the considerations given above: stars on the AGB, presumably stars with masses between 1 and 8 $M_\odot$, become pulsationally unstable and start to loose mass at an increasing rate; first they appear as Mira variables, later as OH/IR stars. The mass is lost at the "expense" of the envelope mass; when the envelope mass drops below a certain critical limit the pulsations and the mass ejection stop simultaneously and the star becomes a planetary nebula, via an intermediate stage as "non-variable OH/IR star".

In the next section we rediscuss this scenario and add (speculatively) a few new elements.

4. A LARGER SAMPLE OF DGE STARS

In a previous review we have introduced a new acronym: "DGE star" for "dust-gas envelope" stars. We will use it here again. The term was introduced to circumvent problems with two other terms: "OH/IR stars" and "CSE-stars". The term "OH/IR stars" implies that OH maser emission is detected, and this is the case in only a fraction (~1/3) of otherwise similar IRAS point sources consisting of long period variables surrounded by a dust-rich envelope. The frequently used term "CSE-star" applies also to stars with dust-free circumstellar envelopes like O and B stars -clearly a quite different category of objects.

4.1 Oxygen-rich DGE stars

In section 3 we have seen that OH/IR stars are a subsample of the point sources in the IRAS catalog that correspond to DGE stars. This allows us now to take a broader look at the evolutionary scenario developed at the end of section 3: what are the other IRAS objects, and how do they fit in? A recent analysis of this question has been made by Van der Veen and Habing (1987). Figure 7a displays all sources from the IRAS-PSC with well measured fluxes at 12,25 and 60µm ("flux quality=3" -see IRAS Explanatory Supplement). The boundaries of the diagram were chosen in such a way that less than 1% of the stellar infrared sources are not in the diagram. After consideration of (1) the association with objects from other catalogues, (2) the low resolution spectra (LRS) of IRAS and (3) the IRAS variability index, the authors divide figure 7a into several regions -see figure 7b. Most objects are in region I, II, IIIa and IV. Area I contains all objects with a blackbody spectrum with T > 2000 K longward of 6µm, as predicted by the Rayleigh-Jeans approximation. Regions II and IIIa contain stars with predominantly oxygen-rich circumstellar shells of which the ones in region IIIa show the 9.7µm feature in emission; a large number of the sources in region II and IIIa are known as Mira variables. There is a clear increase in variability of the sources going from region I via region II to region IIIa. Region IIIb contains infrared sources with the 9.7µm band in absorption, i.e. objects with a spectrum like that of OH26.5+0.6 (see figure 6); the objects are highly variable and only a few have optical counterparts. Indeed, all variable OH/IR stars are situated in region IIIb. The non-variable OH/IR stars and planetary nebulae are in region IV and V, where region V consists for about 40% of planetary nebulae. From their distribution in the sky and using a galactic model, Habing (1988) has argued that the objects in area IIIb have luminosities between 4000 and 16000 $L_\odot$

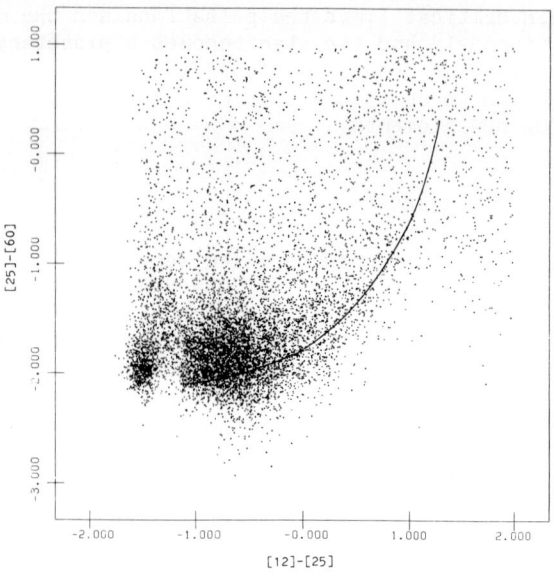

Figure 7a: Two-colour diagram of all sources in the IRAS point source catalogue, that have a "good quality" flux measurement at 12, 25 and 60µm.

with a peak at $6000L_o$; a statistical analysis of the distribution of the IRAS variability index convinced Harmount and Gilmore (1987) that more than 90% of these objects are variable with periods between 400 and 600 days. (Harmount and Gilmore analysed only stars in the galactic bulge, but we suggest that the analysis holds also outside of the bulge). In short: there are rather good indications that the objects in areas III, IV and V are similar to the OH/IR stars: the OH/IR star properties are probably representative (except for the maser emission). Thus a consideration of the point sources in II, III, IV and V lead us to the same evolutionary scenarios as proposed in section 2.

4.2 Carbon-rich DGE stars

Now turn to the other areas in figure 7b. What objects do they contain? The most interesting areas are VIa and VII: they contain large numbers of carbon stars. The objects in VIa and VII differ in some important aspects. The objects in region VIa have a very low infrared variability and the 25 to 12µm flux density ratio is that of a blackbody hotter than 1000k; however, the objects have a stronger 60µm that corresponds to such a blackbody: the indication of a cool distant circumstellar shell around a possibly very cool star. The objects in region VII have a large infrared variability and their

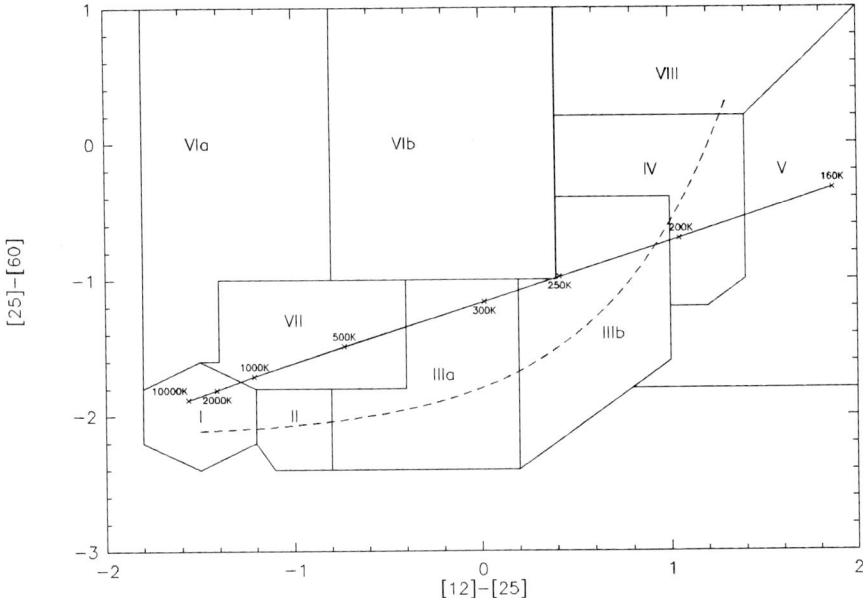

Figure 7b: The definition of the boundaries of 10 areas in the diagram of figure 6a -see text for discussion.

25μ/12μ flux density ratio is very similar to oxygen-rich Miras. The shift to larger 60μ/25μ ratio is due to the different properties of oxygen-rich and carbon-rich dust between 40 and 80μm. The circumstellar shells are carbon-rich as is evidenced by the 11.3μm emission feature in the LRS spectra (when available). Willems (1987) proposed that the stars in region VIa originate from stars with oxygen-rich circumstellar shells in region IIIa after the mass loss rate has decreased by a few orders of magnitude. Indeed in region IIIa, there is evidence for stars that have both carbon- and oxygen features. Willems (1987 see also Willems et al., 1986) and Little-Marenin (1986) found 9 <u>carbon</u> stars, well identified as such from optical spectra, to have the 9.7μm emission feature of <u>oxygen</u>-rich dust. Two of these souces were found to have H_2O maser emission as well (Benson and Little-Marenin, 1987; Nakada et al., 1987). Two others were found to have OH 1612MHz masers (te Lintel Hekkert, private comm.). Te Lintel Hekkert detected OH 1612MHz maser emission in 5 IRAS sources with the 11.3μm SiC emission feature.

This information on the simultaneous existence of oxygen-rich and carbon rich features is confusing. How to interpret this? Willems proposed, and for the moment we follow him, that we see the aftermath of the theoretically predicted third dredge-up: during a thermal pulse convection may reach in the helium/carbon burning zone and mix carbon rich material with the oxygen rich material of the envelope. If the envelope mass is small enough, which is certainly the case toward the end of the OH/IR phase, and the amount of carbon convectively mixed in is large enough, the carbon over oxygen rates will invert from (<1) to (>1). Te Lintel Hekkert (priv. comm.) suggests that the star can go through a phase of poor mixing with some parts of the stellar surface being oxygen-rich and others carbon-rich; the suggestion is based on the assumption that the surface of these stars is covered by a small number of adjacent, large convection cells.

The interpretation of a sudden transition should be considered as tentative. More detailed observations have to be made. But we point out an interesting aspect: if mass loss starts after the first thermal pulse (for some evidence that all Mira's have undergone a thermal pulse see, Little et al., 1987), then it is a critical question at what moment the superwind will set in -at what phase of the period between two thermal pulses? If the superwind starts and finishes just before a thermal pulse, the star may become an oxygen-rich planetary nebula; but if the last thermal pulse takes place during the superwind phase, then the mass of the envelope may be low and the deposition of carbon may turn the star into a carbon rich planetary nebulae. There is thus a certain randomness involved and a strict deterministic model of carbon star formation ("all stars of so much mass will eventually become a carbon star") may not work: becoming a carbon star may be an accident. The considerations above lead us to extend the proposed evolution scheme to that given in figure 8.

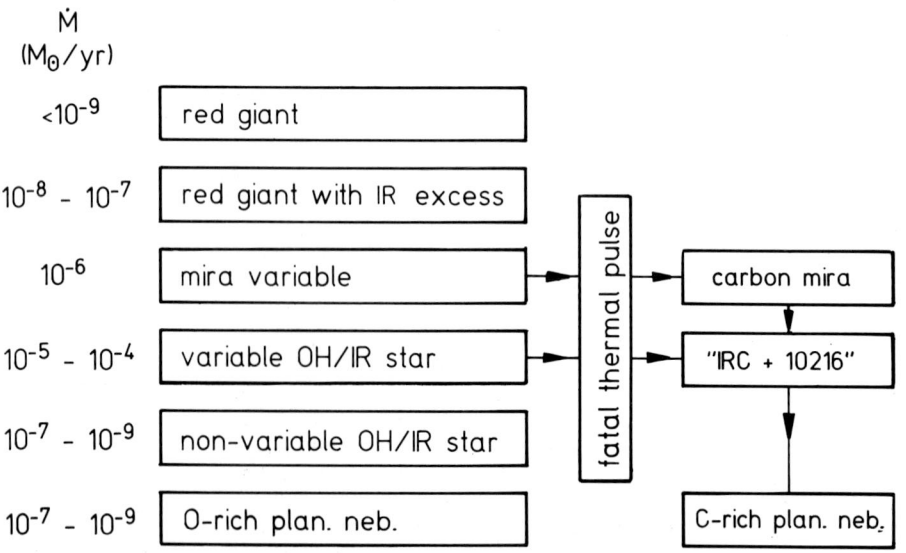

Figure 8: Evolutionary scheme as proposed in this review.

5. THE TRANSITION OF AGB TO THE PLANETARY NEBULA STAGE

5.1 Post AGB stars and proto planetary nebulae

Above we have argued that the decay of an AGB star involves two episodes, or a "two-wind model", much as Kwok et al. (1978) predicted. We have given our views of the first episode, that of the high density, low velocity wind. What about the transition to the low density, high velocity wind? We have already concluded that some objects with non-variable OH 1612MHz emission and a non-variable infrared spectrum are probably transition objects (see the posters by Volk and Kwok and by Van der Veen et al. during this symposium). The recent discovery of 1612MHz OH maser emission in several very young planetary nebulae (see the poster by Zijlstra et al. during this symposium) confirms our conclusion: the OH maser is a good tracer for a cool, distant shell (its typical distance from the star is 3×10^{14}m). The search for transition objects (or proto-planetary nebulae) has been a pastime during almost all previous planetary nebula symposia; no case survived. The IRAS data, coupled with our improved insight in the "first episode", make it likely that this symposium will have seen the first proto-planetaries of longer duration. Time (that is: time on a human scale) will show this.

5.2 The transition of a spherically symmetric wind into a bipolar flow

The observations of the 1612 MHz line emission from OH/IR stars give a result that on closer consideration is quite surprising: there are only a few objects known with mass loss in cones ("bipolar flows", thus: not spherically symmetric); such objects are rare. Examples of symmetric outflow are OH231.8+4.2 (for a recent paper with the references to earlier work see Reipurth, 1987), IRC+10420 (Diamond et al., 1983) and OH19.2-1.0 (J. Chapman, private communication). Most OH maser maps obtained by aperture synthesis arrays (VLA, MERLIN) show spherical symmetry. In all cases there are some discrete blobs present that contribute to a certain amount of chaos, but the overall impression is one of symmetry. This is supported by the fact that in 1612 MHz surveys almost all sources have exactly the two emission peaks; only rarely a single peak or three or more peaks are found. Te Lintel Hekkert and Caswell found 25 unusual objects among their 900 detections; Eder et al. report similar results.

The dominating presence of spherical symmetry contrasts with the predominance of other symmetries in the shapes of planetary nebulae (see the contribution by Balick in this volume). Why is there spherical symmetry during the first episode of mass loss and is this absent during the second episode?

6. FINAL QUESTION: WHAT ABOUT DOUBLE STARS?

A large fraction of main sequence stars are double or worse. In this review we have ignored multiplicity and discussed the objects as if they were single. Is multiplicity of importance? In our estimation multiplicity is not required to explain the observed phenomena. Again, only time wil teach us how close we are to the truth.

ACKNOWLEDGEMENT

In various ways each of the authors has been supported by Z.W.O., the Netherlands Foundation for the Advancement of Pure Research, and by the Leids Kerkhoven Bosscha Fonds. W. van der Veen holds research grant no. 782-372-020 by ASTRON, which receives its funds from ZWO.

REFERENCES

Abell, G.O., Goldreich, P. 1966, Publ. Astron. Soc. Pac. **78**, 232
Acker, A., Marcout, J., Ochsenbeid, F. 1981, Astron. Astrophys. Suppl. Ser. **43**, 265
Andersson, C., Johansson, L.E.B., Goss, W.M., Winnberg, A., Nguyen-Quang-Rieu 1974, Astron. Astrophys. **30**, 475
Baud, B. 1978, thesis, Leiden University
Baud, B., Habing, H.J., Matthews, H.E., Winnberg, A. 1979a, Astron. Astrophys. Suppl. Ser. **35**, 179
Baud, B., Habing, H.J., Matthews, H.E., Winnberg, A. 1979b, Astron. Astrophys. Suppl. Ser. **36**, 193
Baud, B., Habing, H.J., Matthews, H.E., Winnberg, A. 1981, Astron. Astrophys. **95**, 156
Baud, B., Habing, H.J. 1983, Astron. Astrophys. **127**, 73
Baud, B. 1981, Astrophys. J. Lett. **250**, L79
Bedijn, P.J. 1986, in "Light on Dark Matter", 1st IRAS conference, ed. F.P. Israel (Reidel, Dordrecht), p. 119
Bedijn, P.J. 1987, Astron. Astrophys. **186**, 136
Benson, R.J., Little-Marenin, I. 1987, Astrophys. J. (Lett.) **316**, L37
Booth, R.S., Kus, A.J., Norris, R.P., Porter, N.D. 1981, Nature **293**, 382
Bowers, P.F. 1975, Astron. Astrophys. Suppl. Ser. **31**, 127
Bowers, P.F., 1978, Astron. Astrophys. **64**, 307
Caswell, J.L., Haynes, R.F. 1975, Mon. Not. R.A.S. **173**, 649
Caswell, J.L., Haynes, R.F., Goss, W.M., Mebold, U. 1981, Astr. J. of Physics, **34**, 333
De Jong, T. 1983, Astrophys. J. **274**, 253
Diamond, P.J., Norris, R.P., Booth, R.S. 1983, Astron. Astrophys. **124**, L4
Eder, J., Lewis, B.M., Terzian, Y. 1987, Astron. J. (in press)
Engels, D., Kreysa, E., Schultz, G.V., Sherwood, W.A. 1983, Astron. Astrophys. **124**, 123
Engels, D. 1982, Veroeffentlichungen Astr. Inst. Bonn, Nr. 95
Evans, N.J., Beckwith, S. 1977, Ap. J. **217**, 726
Habing, H.J., Olnon, F.M., Winnberg, A., Matthews, H.E., Baud, B. 1983, Astron. Astrophys. **128**, 230
Habing, H.J. 1988, Astron. Astrophys. (accepted)

Herman, J. 1983, thesis, Leiden University
Herman, J., Habing, H.J. 1985, Astron. Astrophys. Suppl. Ser. **59**, 523
Herman, J., Burger, J.H., Penninx, W.H. 1986, Astron. Astrophys. **167**, 247
Hoyle, F., Schwarzschild, M. 1955, Ap.J. Suppl. **2**, 1
Iben, I. 1971, Publ. Astron. Soc. Pac. **83**, 697
Iben, I. Renzini, A. 1983, Ann. Rev. Aston. Astrophys. **21**, 271
Jewell, P.R., Webber, J.C., Snijder, L.E. 1980, Astrophys. J. Lett. **242**, L29
Johansson, L.E.B., Andersson, C., Goss, W.M., Winnberg, A. 1977a, Astron. Astrophys. Suppl. Ser. **28**, 199
Johansson, L.E.B., Anderson, C., Goss, W.M., Winnberg, A. 1977b, Astron. Astrophys. **54**, 323
Kwok, S., Purton, C.R., Fitzgerald , P.M., Ap. J. Lett. **219**, L125
Little, S.J., Little-Marenin, I.R., Hagen Bauer, W. 1987, Astron. J. **94**, 981
Little-Marenin, I.R. 1986, Ap. J. Lett. **307**, L15
Minkowski, R. 1965 in "Galactic Structure", eds. A. Blaauw and M. Schmidt (University of Chicago Press) p. 321
Nakada, Y., Izumiura, H., Onaka, T., Hashimoto, O., Ukita, N., Deguchi, S., Tanabé, T. 1987, Astrophys. J. Lett. **323**, L77
Neugebauer, G., Martz, D.E., Leighton, R.B. 1965, Astrophys. J. **142**, 399
Neugebauer, G., Leighton, R.B. 1969, "Two micron sky survey", NASA SP-3047
Olnon, F.M., Baud, B., Habing, H.J., de Jong, T., Harris, S., Pottasch, S.R. 1984, Ap. J. Lett. **278**, L37 (IRAS issue)
Osterbrock, D.E. 1964, Ann. Rev. Astron. Astrophys. 2, 95
Paczynski, B., Ziolkowski, J. 1968, "In: Planetary Nebulae", IAU Symposium 34, eds. D.E. Osterbrock and C.R. O'Dell (D. Reidel, Dordrecht), p. 396
Price, S.D., Walker, R.G. 1976, "The AFGL four colour infrared sky survey" AFGL-76-0208
Reipurth, B. 1987, Nature, **325**, 787
Rose, W.K. 1968, "Planetary Nebulae", IAU Symposium 34, eds. D.E. Osterbrock and C.R. O'Dell (D. Reidel, Dordrecht), p. 390
Rowan-Robinson, M., Lock, T.D., Walker, D.W., Harris, S. 1986, Mon. Not. R.A.S. **222**, 273
Schultz, G.V., Sherwood, W.A., Winnberg, A. 1978, Astron. Astrophys. **63**, L5
Schultz, G.V., Kreysa, E., Sherwood, W.A. 1976, Astron. Astrophys. **50**, 171
Shklovskii, I.S. 1956, Astron. J. U.S.S.R., **33**. 315
Sivagnanam, P., and Le Squeren, A.M. 1986, Astron. Astrophys. **168**, 374
Te Lintel Hekkert, P., Versteege, H., Wiertz, M., Habing, H.J., 1988a, (in preparation)
Te Lintel Hekkert, P., Caswell, J.L., Norris, R.P., Haynes, R.J., Habing, H.J., 1988b, in prep.
Van der Veen, W.E.C.J. 1987, submitted to Astron. Astrophys.
Van der Veen, W.E.C.J., Habing, H.J.: 1987, Astron. Astrophys. (in press.)

Van der Veen, W.E.C.J., Habing, H.J., Geballe, T. 1987, in "Planetary and Proto Planetary Nebulae: from IRAS to ISO", ed. A. Preite Martinez (Reidel, Dordrecht), p. 69
Weidemann, V., Koester, D. 1983, Astron. Astrophys. **121**, 77
Willems, F.J. 1987, thesis, University of Amsterdam
Willems, F.J., de Jong, T. 1986, Astrophys. J. Lett. **309**, L39
Wilson, W.J., Barrett, A.H. 1968, Science **161**, 778
Winnberg, A. Baud, B., Matthews, H.E., Habing, H.J., Olnon, F.M. 1985, Ap. J. Lett. **291**, L45
Wood, P.R., Bessell, M.S., Fox, M.W. 1983, Astrophys. J. **272**, 99

CARBON STARS AS PLANETARY NEBULA PROGENITORS

G. R. Knapp
Department of Astrophysical Sciences
Princeton University
Princeton, NJ 08544, U.S.A.

ABSTRACT. Molecular line observations show that some planetary nebulae are still only partially ionized and are surrounded by the remains of the mass loss envelope shed by the preceding AGB star. The mass loss rates and outflow velocities of these envelopes are similar to those of the cool winds from luminous AGB stars. Both the kinematics of carbon stars and observations of the molecular envelopes around young planetaries show that the carbon star progenitors have a wide range of ages and of mass loss rates. There is increasing evidence that a significant fraction of AGB stars are carbon stars and that these provide a substantial contribution to the total mass returned to the interstellar medium.

I. INTRODUCTION

The advent of infrared and millimeter-wavelength astronomy in the past twenty years has allowed the study of the cool, dusty molecular winds shed by asymptotic giant branch stars. These observations yield a wealth of information relating to many aspects of galactic and stellar evolution and of the chemistry of the interstellar medium. Such topics as the rate of mass return, the chemical composition of the returned mass, and the formation of dust are under intensive study.

Mass loss from an AGB star takes place, to zeroth order, as a spherically symmetric, steady wind whose outflow velocity becomes constant at distances of a few stellar radii. Observed values of the wind velocities range from 3 to 80 km/sec, but most objects observed so far have winds with velocities of 10-25 km/sec. The mass loss rates likewise have a large range, a few $\times 10^{-8}$ $M_\odot$ yr^{-1} to $> 10^{-4}$ $M_\odot$ yr^{-1}, but "typical" values are in the few $\times 10^{-6}$ to 10^{-5} $M_\odot$ yr^{-1} range.

These winds are observed in their infrared excess due to dust, the silicate and SiC absorption or emission features, and via a large number of molecular lines. Of these, the most useful for studies of mass loss rates etc. are the CO rotational millimeter and sub-millimeter lines, which are observed from both carbon stars and M stars (stars with $n(O) > n(C)$, hereafter "oxygen" stars), and the 1612

MHz OH maser line, observed from oxygen stars only. There can now be little doubt that it is this mass shed by the AGB star which is destined to become a planetary nebula as it becomes ionized by the hot degenerate core at the end of the mass loss phase; several planetary nebulae still have the remains of the circumstellar molecular envelope, as shown by observations of the CO lines, H_2 emission, OH maser emission, and dust.

The resulting mass-loss envelopes can reach considerable extents, > 1 pc in radius, before being physically truncated by the interstellar pressure or chemically truncated by ultraviolet radiation. The radial structure of the objects thus contains information about their time evolution, which is now being exploited as high-resolution instruments come on line. This review will first summarize the observational data for carbon stars. The local galactic kinematics and distribution will then be compared with those of planetary nebulae. The distribution of the chemical composition of planetary nebulae will be compared with that of AGB stars, and finally observations of planetary nebulae which still contain molecular material will be discussed and an evolutionary sequence constructed.

II. OBSERVATIONAL DATA - GLOBAL CO PROFILES

From the point of view of the appearance of the CO profiles, oxygen stars, S stars, carbon stars and planetary nebulae are identical. Figure 1 shows the CO(1-0) line profiles for IRC+00509 (oxygen star), CIT6 (carbon star), RY Dra (carbon star) and NGC7027 (planetary nebula), measured with the AT&T Bell Labs. 7m telescope. These profiles give the central velocity and velocity width of the outflow, and the intensity also gives the mass loss rate. A few years ago, observational sensitivity limits meant that only the stars which had the highest mass loss rates (or those few which are very nearby, <200 pc) were detected in the CO line, and the sample of AGB stars for which these data were available was unrepresentative. This is still somewhat true, but is rapidly improving. The very successful IRAS satellite mission (Neugebauer et al. 1984) provided a very large sample of target evolved stars (Olnon et al. 1984) which are being intensively observed in the CO and OH lines (Zuckerman and Dyck 1986a,b; Habing and te Lintel Hekkert 1987; Eder et al. 1987). In addition, observational sensivity has greatly improved for CO observations, making accessible measurements of a wider range of mass loss rates (e.g. Olofsson et al. 1987). At the present time, CO emission has been detected from over two hundred evolved stars, including nearly twenty planetary- and proto-planetary nebulae, and eventually it is likely that detections of several hundred stars will be made.

III. THE LOCAL KINEMATICS OF CARBON STARS

The huge data base on the mid-and far-infrared emission from carbon stars provided by the IRAS satellite (IRAS Point Source Catalog 1985) has led to the study the galactic distribution of evolved stars. Recent analyses have been undertaken by Thronson et al. (1987),

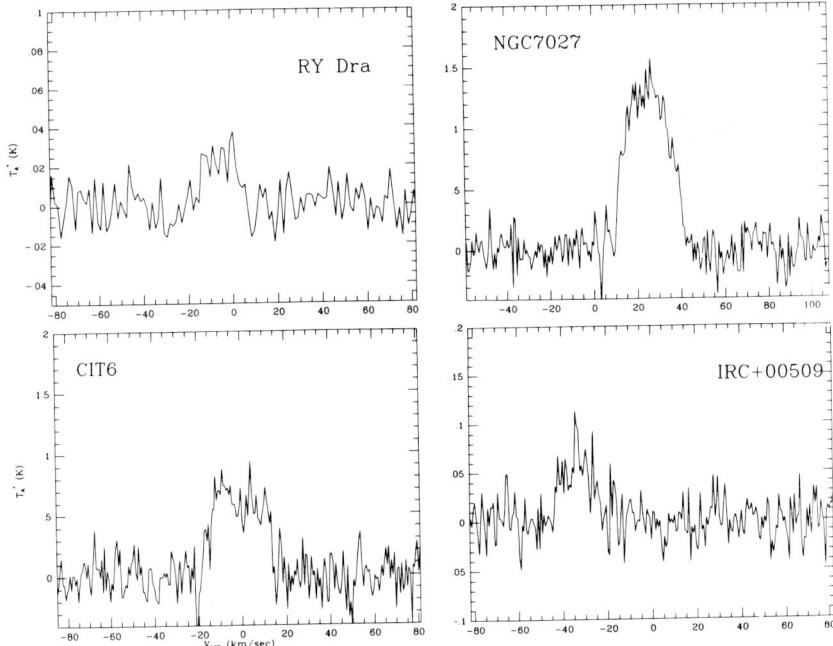

Figure 1. CO(1-0) line profiles of RY Dra (carbon star), NGC7027 (planetary nebula), CIT6 (carbon star) and IRC+00509 (oxygen star) measured with the AT&T Bell Laboratories 7m telescope.

Claussen et al. (1987) and Willems and de Jong (1987). What is needed at present is a measure of the absolute luminosity (or luminosity function) of carbon stars. The present discussion concentrates on the local distribution, i.e. the scale height and velocity dispersion of the local carbon stars.

Figure 2 shows the radial velocity histogram for carbon stars, where the radial velocities are measured using the CO(1-0) or (2-1) lines. The present set of observations has fairly good sky coverage. While objects in the northern hemisphere have been more extensively observed, the coming on line of the submillimeter telescopes on Hawaii has enabled searches for CO emission to be extended as far south as δ = -60° (Phillips et al. 1987). The data in Figure 2 are taken from

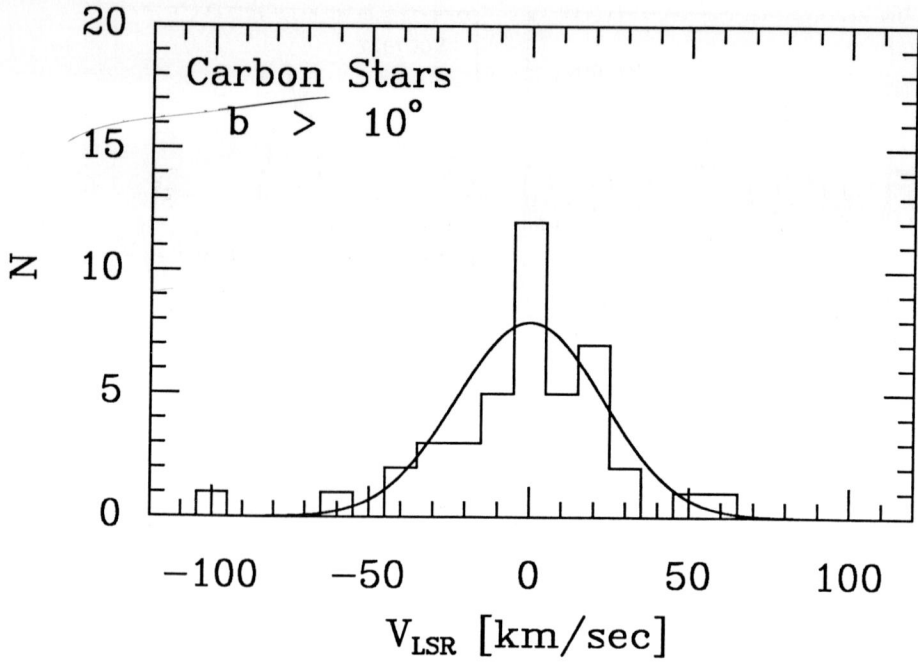

Figure 2. Histogram of radial velocities with respect to the Local Standard of Rest for carbon stars with $|b| > 10°$. The radial velocities are measured using emission in the CO J = 1-0 and 2-1 lines.

papers by Arquilla et al. (1987), Leahy et al. (1987), Rieu et al. (1987), Huggins and Healy (1986a,b), Knapp and Morris (1985), Knapp (1986, and unpublished), Olofsson et al. (1987), Phillips et al., (1987), Wannier and Sahai (1987), Zuckerman and Dyck (1986a,b, 1987) and Zuckerman, Dyck and Claussen (1986), and total 209 objects. For Figure 2, only stars with $|b| > 10°$ are plotted, to avoid galactic rotation effects. Assuming that the velocities are isotropic, σ(3-dimensional) = 39.3 ±5.9 km/sec for the carbon stars. Figure 2 also shows this gaussian plotted on top of the histogram. It is a reasonable fit, though the data marginally suggest the presence of components of different velocity dispersion. A slightly higher value, σ = 49.5 ± 6.7 km/sec, is found for the oxygen stars, suggesting that the mass-losing carbon stars are, as a group, marginally younger than the oxygen stars.

Zuckerman and Dyck (1986b) have pointed out that carbon stars with large outflow velocities tend to be found close to the galactic plane, implying that they have a small scale height and therefore arise from a

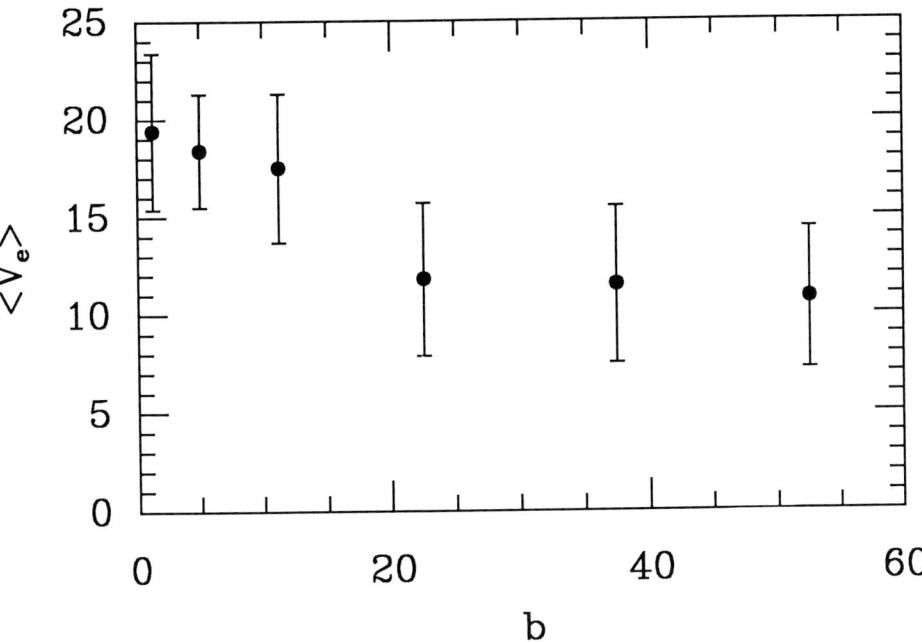

Figure 3. Mean expansion velocity for winds from carbon stars, binned by the galactic latitude of the stars.

population of larger progenitor mass. This trend is illustrated in Figure 3, where the mean carbon star outflow velocity is plotted versus latitude. These data suggest that carbon stars, like oxygen stars, have a range of progenitor masses and hence a range of ages.

The three-dimensional velocity distribution of local carbon stars has a dispersion of ~ 40 km/sec. From the compilation of Mihalas and Binney (1981) the progenitor mass of the carbon star is likely to be ~ 1.5 $M_\odot$, in agreement with the conclusions of Claussen et al (1987). This velocity dispersion is significantly less than that of optically-observed carbon stars (56 km/sec, Mihalas and Binney 1981) or of planetary nebulae and white dwarfs (~ 60 km/sec). This is further evidence that carbon stars arise from a range of progenitor masses; those losing mass copiously enough to detect in the CO line have higher progenitor masses than the average carbon star.

IV. THE FRACTION OF AGB STARS WHICH ARE CARBON STARS

Zuckerman and Aller (1986) have shown that, of the planetary nebulae with reasonably well-measured abundances, more than 50% have n(C) >

n(O). There is now growing evidence that many (most?) AGB stars may evolve from "oxygen" to carbon stars. (1) Willems and de Jong (1986) and Little-Marenin (1986) have shown that some J-type carbon stars have silicate dust in their envelopes, suggesting that the stellar atmosphere has just changed composition from oxygen to carbon-rich composition (this work is a good example of the stellar history which can be deduced from observations of the envelope properties at different radii). (2) The J-type carbon stars also have high values of $^{13}C/^{12}C$ ($\sim$ 1:10, e.g. Lambert et al. (1986). While many carbon stars have low values of $^{13}C/^{12}C$ ($\sim$ 1:50-150, Knapp and Chang 1985, Lambert et al. 1986) due to the dredge-up of 3α burning products, oxygen stars have much higher values ($\sim$ 1:4-20, references cited above) due to the production of ^{13}C during main sequence CNO burning. Again, these data suggest that J-type carbon stars may be evolving from oxygen to carbon-rich composition.

The fraction of AGB stars which are found to be carbon stars varies depending on the technique used for the observations:

Sample	Carbon Stars as Fraction of Total	Reference
Optical AGB stars	few %	Feast (1987)
IRAS sources, $S_{12\mu} > 28$ Jy	13%	Hacking et al. (1985)
IRAS stars, $|b| > 60°$	25%	Knapp & Wilcots (1987)
Stars with IRAS LRS spectra (>2Jy at 12μ)	25%	IRAS Science Team (1986)
AGB stars detected in CO lines	~50%	
Planetary nebulae with well-observed abundances	~60%	Zuckerman & Aller (1986)

Carbon grains are strongly absorbing relative to silicate grains (Draine and Lee 1984) so it is not surprising that carbon stars are poorly represented in optical surveys. On the other hand, the CO abundance is greater in carbon than in oxygen stars and so carbon stars are more easily detected by CO observations. Nevertheless, the above compilation suggests that in total carbon stars may be significant fraction (close to 50%?) of all AGB stars, contribute a significant fraction of returned mass to the interstellar medium, and form a significant fraction of the planetary nebulae.

V. MOLECULAR ENVELOPES AROUND PLANETARY NEBULAE

The link between AGB stars and planetary nebulae is most directly seen from the presence of the remnants of the AGB molecular envelopes around some planetaries, the best known case being that of NGC7027 (Mufson, Lyon and Marionni 1975). The expansion of the ionization front into the neutral circumstellar cloud has been measured by Masson (1986) for this nebula and by Kwok and Feldman (1981) for CRL618. Further evidence of the interaction between ionization fronts and circumstellar material is seen in emission from shock-excited H_2 in many nebulae (e.g. Zuckerman and Gatley 1987).

In addition to planetary nebulae with circumstellar molecular clouds, there are several circumstellar envelopes surrounding stars of much earlier type than M, and these may be evolving towards the planetary nebula phase. Examples of such objects are IRC+10420 (whose envelope is oxygen rich) and CRL2688 (carbon rich). CO observations of other possible examples of this class (Parthasarathy and Pottash 1986) have found molecular circumstellar material in the CO line (Likkel et al. 1987). A list of planetary and proto-planetary nebulae detected in the CO line is given below.

CO Observations of Planetary and Proto-Planetary Nebulae

Object	$\dot{M}$ ($M_\odot$ yr^{-1})	Type	Chemistry	Reference
CRL618	7.7×10^{-5}	PN	C	1
NGC2346	1.0×10^{-5}	PN	C	2,3,4,5
Vy2-2	2.4×10^{-6}	PN	O	1,6
IRC+10420	3.0×10^{-4}	PPN	O	1,7,8
M1-92	3.3×10^{-5}	PPN	O	6,9
NGC7027	1.1×10^{-4}	PN	C	1,2
CRL2688	1.6×10^{-4}	PPN	C	1
NGC6302	5.7×10^{-5}	PN	O	10,12
NGC7293	$\sim 10^{-5}$	PN	C	2,11
NGC6720	$\sim 2 \times 10^{-5}$	PN	O	2,4
CPD-56°8032	1.3×10^{-6}	PN	C	13
HD161796	10^{-5}	PPN		14
89 Her	6×10^{-7}	PPN		14
CRL2343	7×10^{-5}	PPN		14
SAO163075	4×10^{-6}	PPN	C?	13,14

References: 1. Knapp and Morris 1985, and references therein, 2. Pottasch 1980, 3. Walsh 1982, 4. Huggins and Healy 1986b, 5. Knapp 1986, 6. Seaquist and Davis 1983, 7. Bowers 1984, 8. Diamond et al. 1985, 9. Davis et al. 1979, 10. Zuckerman and Dyck 1987, 11. Huggins and Healy 1986a, 12. Payne, Phillips and Terzian 1987, 13. Phillips et al. 1987, 14. Likkel et al. 1987.

Here, a planetary nebula is defined as one in which ionized gas is already present, while a proto-planetary nebula is a molecular cloud surrounding an evolved star of type earlier than M (e.g. 89 Her is an F supergiant) but in which the ionization of the gas has not yet begun. There are eight planetaries currently detected in the CO lines. Of special interest is the recent CO detection of the probable low-mass planetary nebula CPD-56°8032 (Phillips et al. 1987).

The objects listed in the above table have both oxygen-rich chemistry (e.g. NGC6302, Vy2-2) and carbon-rich chemistry (e.g. NGC7027, CPD-56°8032). Further, they have a range of progenitor masses. The total mass of NGC7027 is likely to be $> 3\ M_\odot$, while

CPD-56°8032 may be a population II object. Thus carbon stars, like M stars have a range of progenitor masses, and both oxygen and carbon stars become planetary nebulae. Oxygen and carbon stars may form a parallel sequence on the AGB, and not all oxygen stars become carbon stars.

VI. EVOLUTION TO PLANETARY NEBULAE

There is increasing evidence that, whatever the fundamental cause of mass loss on the AGB, the envelope dynamics are dominated by radiation pressure on dust grains which form in the extended stellar atmospheres. A requirement for radiation pressure driven winds is $\dot{M}v_0 < kL_*/c$, where k is a number of order 1 and L_* is the stellar luminosity. Values of $\dot{M}$ can be found from CO observations; since both the observed CO line strength and the stellar bolometric flux are $\sim D^{-2}$, where D is the distance, one can take the ratio $\beta = \dot{M}v_0 c/L_*$ and require that $\beta < 1$. In this way, the distance, the largest source of uncertainty, drops out of the problem. A compilation of mass loss rates by Knapp (1986) showed that a very large fraction of AGB stars detected in molecular line emission have values of β within a factor of 3 (the observational uncertainty) of 1 while the remainder have $\beta < 1$. Thus a majority of the envelopes from which molecular emission is detected are losing mass at the radiation pressure limit.

There are, however, a few exceptions with $\beta \gg 1$. All of these have central ionized regions (i.e. are young planetary nebula) and all of the nebulae in the sample have $\beta \gg 1$ except Vy2-2 and CPD-56° 8032. This observation allows us to exploit the information about the evolutionary history of the star carried by the stellar wind. The molecular line observations tell us what the mass-loss rate used to be $\sim 10^4$ years ago, while the bolometric flux tells us what the stellar luminosity is now. The value of β thus tells us the minimum luminosity drop between the red giant branch and the present evolutionary status of the star. Figure 4 shows β versus the present spectral type of the central star. We can see stars both on Schonberner's (1987) H-burning (constant luminosity) tracks as well as stars for which the luminosity drops abruptly at the onset of planetary nebula formation, and are therefore fairly massive (cf. Paczyński 1971).

I am very grateful to Tom Phillips, Hans Olofsson, Mike Jura and Ben Zuckerman for access to their data before publication. This work is supported by NSF grant AST87-02945 to Princeton University.

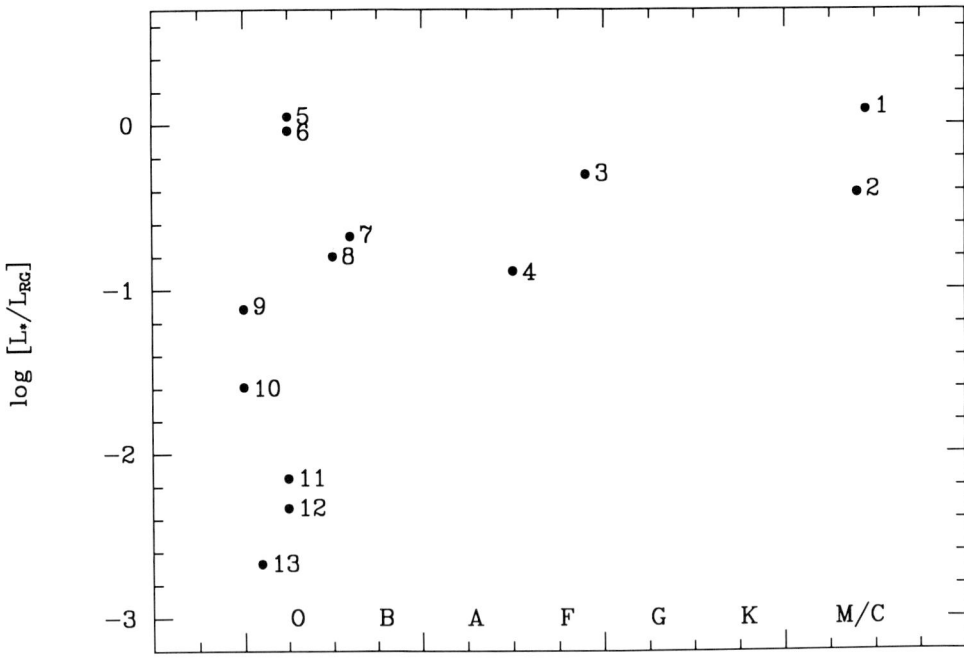

Figure 4. Luminosity decrease for PN and PPN versus approximate spectral type. The stars are (1) and (2) IRC+10011 and IRC+10216, mass losing AGB stars (3) IRC+10420 (4) CRL2688 (5) CPD-56°8032 (6) Vy2-2 (7) M1-92 (8) CRL618 (9) N7027 (10) N6302 (11) N7293 (12) N6720 (13) N2346.

References

Arquilla, R., Leahy, D.A., and Kwok, S. 1986, M.N.R.A.S. 220, 125.
Bowers, P.F. 1984, Ap.J. 279, 350.
Claussen, M.J., Kleinmann, S.G., Joyce, R.R., and Jura, M. 1987, Ap.J. (Suppl) (in press).
Davis, L.E., Seaquist, E.R., and Purton, C.R. 1979, Ap.J. 230, 434.
Diamond, P.J. et al. 1985, M.N.R.A.S. 212, 1.
Draine, B.T., and Lee, H.-M. 1984, Ap.J. 285, 89.
Eder, J., Lewis, B.M., and Terzian, Y. 1987, Ap.J. (in press).
Feast, M.W., 1987, in "The Outer Galaxy", ed. F.J. Lockman and L. Blitz, Springer-Verlag, in press.
Habing, H.J., and te Lintel Hekkert, P. 1987, in preparation.

Hacking, P., et al. 1985, P.A.S.P. 97, 616.
Huggins, P.J., and Healy, A.P. 1986a, Ap.J. (Letters) 305, L29.
                                 1986b, M.N.R.A.S. 220, 33p.
IRAS Point Source Catalog 1985, N.A.S.A., U.S. Government Printing Office.
IRAS Science Team 1986, Astron. Astrophys. Suppl. 65, 607.
Knapp, G.R. 1986, Ap.J. 311, 731.
Knapp, G.R., and Chang, K.-M. 1985, Ap.J. 293, 281.
Knapp, G.R., and Morris, M. 1985, Ap.J. 292, 640.
Knapp, G.R., and Wilcots, E.P. 1987, in preparation.
Kwok, S., and Feldman, P.A. 1981, Ap.J. (Letters) 247, L67.
Lambert, D.L., Gustafsson, B., Eriksson, K., and Hinkle, K.H. 1986, Ap.J. (Suppl.) 62, 373.
Leahy, D.A., Kwok, S., and Arquilla, R.A. 1987, Ap.J. 320, 825.
Likkel, L., Omont, A., Morris, M., and Forveille, T. 1987, Astron. Astrophys. 173, L11.
Little-Marenin, I.R. 1986, Ap.J. (Letters) 307, L15.
Masson, C.R. 1986, Ap.J. (Letters) 302, L27.
Mihalas, D., and Binney, J. 1981, Galactic Astronomy (W.H. Freeman Co.).
Mufson, S.L., Lyon, J., and Marionni, P.A. 1975, Ap.J. (Letters) 201, L85.
Neugebauer, G., et al. 1984, Ap.J. (Letters) 278, L1.
Olnon, F.M. et al. 1984, Ap.J. (Letters) 278, L41.
Olofsson, H., Eriksson, K., and Gustafsson, B. 1987, Astron. Astrophys. 183, L13.
Paczyński, B.E. 1971, Act. Astron. 21, 271.
Payne, H.E., Phillips, J.A., and Terzian, Y. 1987, Ap.J. (in press).
Parthasarathy, M., and Pottasch, S.R. 1986, Astron. Astrophys. 154, L16.
Phillips, T.G. et al. 1987, submitted to Ap.J.
Pottasch, S.R. 1980, Astron. Astrophys. 89, 336.
Rieu, N.G., Epchtein, N., Bach, T., and Cohen, M. 1987, Astron. Astrophys. (in press).
Seaquist, E.R., and Davis, L.E. 1983, Ap.J. 274, 659.
Schonberner, D. 1987, in "Late Stages of Stellar Evolution" ed. S. Kwok and S.R. Pottasch, D. Reidel Co.
Thronson, H.A., Latter, W.B., Black, J.H., Bally, J., and Hacking, P. 1987, Ap.J. (in press).
Walsh, J.R. 1982, M.N.R.A.S. 202, 303.
Wannier, P.G., and Sahai, R. 1986, Ap.J. 311, 335.
Willems, F., and de Jong, T. 1986, Ap.J. (Letters) 309, L39.
Willems, F., and de Jong, T. 1987, Astron. Astrophys. (in press).
Zuckerman, B., and Aller, L. 1986, Ap.J. 301, 772.
Zuckerman, B., and Dyck, H.M. 1986a, Ap.J. 304, 394.
                                 1986b, Ap.J. 311, 345.
                                 1987, Ap.J. (in press).
Zuckerman, B., Dyck, H.M., and Claussen M.J. 1986, Ap.J. 304, 401.
Zuckerman, B., and Gatley, I. 1987, Ap.J. (in press).

THERMAL PULSES AND THE FORMATION OF PLANETARY NEBULA SHELLS

Alvio Renzini
Dipartimento di Astronomia, Università di Bologna, Bologna, Italy

1. INTRODUCTION

Over the past decade a comprehensive, semiquantitative theoretical scenario for the final evolutionary stages of low and intermediate mass stars has been progressively elaborated and refined. It concerns the envelope ejection terminating the Asymptotic Giant Branch (AGB) phase, the AGB to Planetary Nebula (PN) transition, the fading and possible rejuvenation of PN nuclei, the formation processes of hydrogen-deficient stars, and the final production of white dwarfs (WD) of the DA and non–DA varieties (Renzini 1979, 1981a, 1981b, 1982, 1983, Iben & Renzini 1983, Iben et al. 1983, Iben 1984, 1985, 1987, Iben & Tutukov 1984, Iben & MacDonald 1985, 1986). In developing this scenario several important results of stellar evolution and hydrodynamical calculations have been incorporated, including in particular those of Paczyǹski (1971), Wood (1974), Härm & Schwarzschild (1975), Schönberner (1979, 1983), and Tuchman, Sack & Barkat (1979).

In the present communication an attempt is presented to further enrich this scenario, by explicitly considering some aspects which heretofore have remained insufficiently explored. These include the relation between thermal pulses and the envelope ejection (EE) at the tip of the AGB, with emphasis on the diversities that this relation may present depending on the initial mass of evolving stars (Section 2). The implications of EE details for the subsequent, Post–AGB evolution are then cursorily explored for the various cases. In Section 3 the question of the AGB to PN *transition time* is revisited, and the consequences of the use of a misleading definition of the *nebular age* are briefly discussed. Finally, in Section 4 some concluding considerations of general validity are recalled, together with a few elements which may be useful for the definition of the most urgent observational investigations.

2. THERMAL PULSES AND ENVELOPE EJECTION

The EE at the tip of the AGB is currently ascribed to a *short* phase of intense mass loss, conventionally called *superwind* to distinguish it from the *regular* red giant wind. This follows from the typical empirical values of the mass, radius and expansion velocity of PN shells, which require mass loss rates of $\sim 10^{-5} - 10^{-4} M_\odot \mathrm{yr}^{-1}$, having operated for $\sim 10^3 - 10^4$ yr. This, and nothing more, is what we mean by superwind, and the identification of the superwind phase with the OH/IR phase then follows very naturally (for oxygen rich stars, cf. Habing, this volume). A superwind phase can also be identified for carbon stars, i.e. with highly obscured, carbon rich objects (cf. Knapp, this volume). After the superwind has removed most of the hydrogen-rich envelope the star leaves the AGB, initiates its migration towards high temperatures, and the superwind gets quenched leaving a *residual envelope mass* M_e^R. To avoid semantic confusions it is worth recalling that two other kinds of stellar wind respectively precede and follow the superwind phase: the so-called *regular* red giant wind operating during the AGB phase, with a typical rate some 100 times smaller than the superwind, and the subsequent *fast* wind emitted during the PN stage.

Thermal pulses, also known as helium shell flashes, have often been regarded as a possible trigger for the EE (e.g. Rose & Smith 1972, Trimble & Sackman 1978, Tuchman & Barkat 1980). Indeed, following a thermal pulse the surface luminosity increases by $\sim 3/4$ of a magnitude (cf. Iben 1982), and the transition from the regular wind to the superwind obviously becomes more likely. It is well known that PNe and WDs are produced by stars in a rather wide range of initial masses, from $\sim 0.85 M_\odot$ up to perhaps $\sim 8 M_\odot$, and correspondingly we should expect that the EE may assume different quantitative and qualitative characteristics, depending on the initial mass M_i, and the related mass of the hydrogen exhausted core M_H. For example, the duration Δt_{peak} of this post-flash luminosity peak is a strong function of M_H, with a rough analytic fit giving:

$$\Delta t_{\mathrm{peak}} \simeq 60 \, M_H{}^{-8} \text{ yr.} \qquad (1)$$

The variation of the peak duration with M_H, coupled with the large variation of the available envelope mass, gives rise to a variety of situations that are explored next.

2.1. Case A: a low mass, Pupolation II star

Evolving Population II stars, such as those in galactic globular clusters, have an initial mass $M_i \simeq 0.85 M_\odot$. Their mass is then decreased by stellar winds during both the first red giant branch, and the early portion of the AGB. By the time they experience the first thermal pulse on the AGB their core mass has grown to $M_H \simeq 0.54 M_\odot$, and their envelope mass M_e has thinned to just a few $10^{-2} M_\odot$. With $M_H = 0.54 M_\odot$, Eq. (1) gives $\Delta t_{\mathrm{peak}} \simeq 10^4$ yr, and with a superwind mass loss rate $\dot{M}_{\mathrm{SW}} \simeq 10^{-5} M_\odot \mathrm{yr}^{-1}$ the mass that could be lost during one pulse peak

($\Delta M = \Delta t_{\text{peak}} \dot{M}_{\text{SW}} \simeq 0.1\, M_\odot$) largely exceeds the envelope mass M_e. This implies that in such low mass stars the envelope ejection may be possible during just one thermal pulse, and most likely during the very first one (Renzini & Fusi Pecci 1988).

The superwind envelope removal during one pulse peak has several important implications for the subsequent evolution of Population II stars. In fact, helium burning Post–AGB stars are produced, and since their luminosity is independent of M_e, the fast wind (typical of PN nuclei) is likely to complete the removal of the residual envelope during the bright Post–AGB phase. With $M_e^R \simeq 10^{-3}\, M_\odot$, and a Post–AGB fading time $\sim$ several 10^5 yr, a mass loss rate $\lesssim 10^{-8}\, M_\odot \text{yr}^{-1}$ is indeed sufficient to expose the helium/carbon intershell material, thus producing a hydrogen deficient Post–AGB star which will eventually evolve into a non-DA WD.

This scenario may offer a chance to explain the otherwise puzzling high carbon abundance observed in three out of the four known PNe in the galactic Halo (Torres-Peimbert 1984, Clegg 1985), one of which is K648, the PN in the globular cluster M15. With $C/O \simeq 10$, these PNe should have been ejected by AGB carbon stars, which neither are found in galactic globular clusters, nor are produced by current AGB models for $M_H \lesssim 0.6\, M_\odot$ (cf. Iben & Renzini 1984). There is therefore both an empirical and a theoretical embarrassment for dredge-up in the precursor having caused the carbon overabundance. In another alternative, the carbon rich wind ($X_C^{\text{wind}} \simeq 0.2$) from a bare PN nucleous could *inplant* carbon/helium rich, high density pockets into these very young PNe. In the prototype case of K648 the only photospheric spectral feature is a strong CIV asymmetric absorption (Adams et al. 1984), perhaps suggestive of a He/C atmosphere. With $M_{\text{PN}} \simeq 0.01\, M_\odot$ and an estimated nebular abundance $X_C^{\text{PN}} \simeq 0.005$, the PN would contain $\sim 5 \times 10^{-5}\, M_\odot$ of carbon: too much for having been provided within the K648 lifetime (~ 2000 yr) by the present carbon mass loss rate $X_C^{\text{wind}} \times \dot{M} \simeq 10^{-12}\, M_\odot \text{yr}^{-1}$ estimated by Adams et al. On the other hand, in the carbon *inplantation* scenario, the nebular carbon lines originate from shocked (and then cooled) high density wind material, rather than from a homogeneous nebula. A much smaller carbon mass may therefore be sufficient to account for the observed emission. In any case, carbon rich PNe of Population II would have carbon rich nuclei (e.g. Wolf-Rayet type of the WC variety), and carbon rich inhomogeneities, and both predictions are easily testable by Space Telescope (HST) observations.

HST observations of globular clusters will provide further opportunity for a decisive test of this scenario. Indeed, only non-DA WDs should be produced and detected, if in low mass stars the envelope ejection is always linked to a thermal pulse. Taking advantage of the excellent observational conditions at the CFH telescope, Richer & Fahlman (1987) may have recently anticipated the space detection of globular cluster WDs, and Ortolani & Rosino (1987) may have achieved the same result for the cluster ω Cen. From the location in the $U - (U - V)$ diagram of their six WD candidates in M71, Richer & Fahlman conclude that most likely they all belong to the non-DA variety, thus providing circumstantial support to the idea of a flash-triggered EE in globular cluster stars.

2.2. Case B: a typical old disk star

As we consider stars of larger and larger initial mass, the core mass M_H at the first thermal pulse tends to increase (although slowly), Δt_{peak} decreases according to Eq. (1), while the available envelope mass obviously increases. The fraction of the envelope ($\Delta t_{\text{peak}} \dot{M}_{\text{SW}}/M_e$) that can be lost during one post-flash peak therefore decreases, provided $\dot{M}_{\text{SW}}$ does not increase too rapidly. For a typical disk star, such as the precursors of most PNe in the solar neighborhood, we can adopt $M_i \lesssim 2 M_\odot$, $M_H \simeq 0.65 M_\odot$, $M_e \simeq 1 M_\odot$, and $\dot{M}_{\text{SW}} \simeq$ few $10^{-5} M_\odot \text{yr}^{-1}$. Correspondingly, we have $\Delta M = \Delta t_{\text{peak}} \dot{M}_{\text{SW}} \simeq$ few $10^{-2} M_\odot$, i.e. much less than the available envelope mass M_e. Envelope removal during a post-flash peak is therefore relatively infrequent, while the superwind can easily remove most of the envelope during one interpulse period (duration = several 10^4 yr) when the star is burning hydrogen in the shell. In this case the production of hydrogen burning Post–AGB stars is then more likely, as so is the formation of DA remnant WDs, in agreement with their larger frequency in the solar neighborhood, compared to non–DAs.

2.3. Case C: young disk stars

It is now several years that an embarrassingly large discrepancy persists between theoretical AGB models on one side, and the observations of AGB stars in the Magellanic Clouds on the other side (e.g. Iben & Truran 1978, Renzini & Voli 1981, Blanco et al. 1980, Reid & Mould 1985, and references therein). Theory indeed predicts much more bright AGB stars ($M_{\text{bol}} \lesssim -6.0$) than actually observed, and several possible explanations have been suggested. The latest, perhaps most promising of these suggestions moves from the finding of severe *convergence problems* in thermally pulsing AGB models brighter than this limit, or, equivalently, with core mass $M_H \gtrsim 0.85 M_\odot$ (Wood & Faulkner 1986, Mazzitelli 1987). Actually, shortly after a thermal pulse, the local luminosity may reach dangerously close to the Eddington limit, as the energy released by the flash leaks out through the base of the envelope. It is then speculated that a *radiation pressure levitation* can lead to a rapid, hydrodynamical ejection of the envelope, as first suggested by Rose & Smith (1972).

Several aspects of the idea certainly require careful examination. For example, rapid but still quasi-static envelope expansion subtracts energy from the local energy flow, thus potentially providing a self-regulating mechanism contrasting the approach of the local luminosity to the Eddington limit. The use of envelope models neglecting the so-called gravitational energy coefficient ϵ_g (as done by Wood & Faulkner) could therefore give misleading indications, especially in this particular case. Certainly, if one wants to study the onset of this radiation pressure instability, rather than dropping the ϵ_g term, it would actually be more appropriate to include the acceleration term in the equation of hydrostatic equilibrium, thus turning it into the equation of motion! This could be most efficiently obtained by using an implicit hydrodynamics stellar structure program, such as the *KEPLER* code of Weaver et al. (1978), which is able of generating a sequence of hydrostatic evolutionary models, as well as of following the onset and development of an hydrodynamical instability, if any. On the other hand, apart from the limitations of the current approach to the problem, the mere fact that models corresponding to the limits of the observed AGB present numerical and physical difficulties is highly suggestive, as it may offer an attractive solution to a long lasting discrepancy between theory and observations.

Before proceeding further, it is worth emphasizing the physical difference between the pulse triggered envelope ejection discussed in the previous two sections, and this one, based on the radiation pressure mechanism. In the former ones the superwind may be due to a *pulsational instability* of the whole envelope, like in the case of the transition from the overtone to the fundamental mode, as proposed by Wood (1974). Here, the hydrodynamical runaway seems to develop in a very restricted region, at the base of the hydrogen rich envelope. The large inflation of this region (perhaps followed by recollapse, bounce, and shock formation) would then lead to the rapid ejection of the envelope. The details of the phenomenon remain however totally unexplored, including in particular its timescale and the number of pulses required to complete the envelope removal, which may considerably increase with increasing initial mass.

Observations may greatly help the development of a satisfactory theory for this kind of ejection, which should be experienced by the more massive ($3 \lesssim M_i \lesssim 8\, M_\odot$) and brighter ($-6 \lesssim M_{bol} \lesssim -7$) AGB stars, leaving the more massive Post–AGB remnants ($0.85 \lesssim M_H \lesssim 1.06 M_\odot$). OH/IR sources in this luminosity range are therefore the natural candidates to look at, although the observational situation is complicated by the relative rarity of such massive AGB stars, compared to lower mass ones. In fact, it could be hard to extract the objects belonging to the real high-luminosity tail of the OH/IR luminosity distribution, as the apparent tail of the luminosity function may be substantially contaminated by bulk, intrinsically fainter sources having *diffused* to higher luminosities because of errors, for example in distance modulus. Moreover, the Post–AGB progeny of these massive AGB stars must be searched among the faintest PN nuclei, as the *fading time* dramatically drops with increasing M_H.

On the other hand, these observational difficulties have to be overcome in one way or another, if one wants to understand the final evolutionary stages of ~ 3 to $\sim 8\, M_\odot$ stars, and to assess their contribution to galactic nucleosynthesis. Indeed, with the prompt ejection envisaged in this radiation pressure scenario there is little space for the third dredge-up and envelope burning processes to work, and correspondingly the contribution of ~ 3 to $\sim 8\, M_\odot$ stars to galactic enrichment would be limited to the first and second dredge-ups, with a corresponding drastic reduction over the predictions of Iben & Truran (1978) and Renzini & Voli (1981).

Concerning the Post–AGB evolution, Wood & Faulkner (1986) argue that, thanks to the radiation pressure ejection process, all AGB stars which produce core masses greater than $\sim 0.85\, M_\odot$ should produce helium rich PN nuclei, and therefore non-DA WD remnants. The fact that Sirius B is a DA with $M \simeq 1\, M_\odot$ apparently militates against this suggestion, in particular if one rejects the speculative hypothesis of a subsequent hydrogen accretion onto this WD. More likely, for such large M_H values the *fading time* is too short to allow the complete envelope removal by the fast, Post–AGB wind.

3. ENVELOPE EJECTION AND POST–AGB TIMESCALES

3.1. The AGB to PN Transition

It is well known that the production of an observable PN requires a fine tuning between two physically independent timescales: the dispersion time of the nebula, and the *transition time* t_{tr} taken by the central star to evolve from the AGB to its hot configuration. For practical purposes t_{tr} is defined as the time spent by the star to evolve from the superwind quenching somewhere close to the AGB, to the effective temperature $T_{eff} = 30,000$ K, high enough for the excitation of the PN. Following this definition, the transition phase begins when the superwind phase ends, and a regular wind resumes. It is also useful to introduce the quantity M_e^N, as the envelope mass when the above effective temperature is first reached. Note that in this section one deals with the evolution of the central stars, rather than with the behavior of the ejecta through the protoplanetary phase, an aspect discussed e.g. by Kwok (1987).

In the case of hydrogen burning Post–AGB stars the transition time is primarily controlled by the residual envelope mass M_e^R, whatever the physical mechanism responsible for the transition. One can indeed define three relevant timescales, respectively nuclear, wind and thermal:

$$\text{Nuclear} \quad \tau_N = \frac{X_e \Delta M_e}{L/E_H}$$

$$\text{Wind} \quad \tau_W = \frac{\Delta M_e}{\dot{M}}$$

$$\text{Thermal} \quad \tau_{th} = \frac{G M_H M_e^R}{LR},$$

where $\Delta M_e = M_e^R - M_e^N$, L is the luminosity during the transition phase, E_H is the energy released by the nuclear burning of one gram of hydrogen, X_e is the envelope hydrogen abundance, $\dot{M}$ is the average wind mass loss rate operating during the transition, and R is the stellar radius at the beginning of the PN phase, conventionally fixed at $T_{eff} = 30,000$ K. The first two timescales are simply given by the amount of envelope consumed during the transition, over the rate of this consumption, due to either nuclear burning or mass loss. In the third case what is consumed is the gravitational energy of the envelope, rather than the envelope itself, but the corresponding timescale is defined in a formally equivalent way. The wind timescale depends on the mass loss rate, which is very poorly known for stars in the relevant temperature range, when also the regular red giant wind gets quenched, and a the *fast* wind typical of hot stars is about to start.

Between the first two timescales, nature will automatically choose the shortest for the transition time $t_{\rm tr}$:

$$t_{\rm tr} \simeq {\rm Min}\,(\tau_{\rm N}, \tau_{\rm W}), \qquad (2)$$

and would actually chose the third ($t_{\rm tr} \simeq \tau_{\rm th}$) if $M_{\rm e}^{\rm R} \lesssim M_{\rm e}^{\rm N}$, i.e. when after superwind quenching the envelope is left very far from its thermal balance. It would be astrophysically interesting to assess which one of the various possibilities is practically realized, depending on the stellar initial mass. The existence of so many PNe with arguably very small nebular age (see next section) indicates that quite often the transition is faster than the nuclear timescale. However, the point worth stressing here is that, whatever the actual timescale, this will in any case depend on the residual envelope mass $M_{\rm e}^{\rm R}$, whose value is practically unknown any better than in order of magnitude ($\sim 10^{-3} M_\odot$). In other terms, nobody knows whether the superwind will cease at, say, Log $T_{\rm eff}$ = 3.65, 3.70, 3.75, or whatever, but these small differences actually imply very large differences in $M_{\rm e}^{\rm R}$, and therefore in $t_{\rm tr}$.

The fact that $t_{\rm tr}$ remains indeterminate has important implications for our understanding of the Post–AGB evolution. Indeed, as most recently emphasized by Wood & Faulkner (1986), this implies that Post-AGB timescales remain uncertain by the additive, *unknown* term $t_{\rm tr}$, and therefore time marks on evolutionary sequences must be regarded as relative to an arbitrary *zero point*, even when this is not explicitly emphasized. Moreover, the inavoidably hydrodynamical nature of the superwind envelope ejection strongly supports the notion that sizable fluctuations of $M_{\rm e}^{\rm R}$ values are likely even among virtually identical stars, and therefore no strict determinism is possible for the Post–AGB evolution. Pretending the contrary may lead to serious misinterpretations of the observational data.

All the above considerations refer to the case of hydrogen burning Post–AGB stars, i.e. to stars completing the superwind envelope ejection while burning hydrogen during a so called *interpulse* phase. In the case of helium burning Post–AGB stars, the evolution across the HR diagram is primarily driven by the secular decrease in the luminosity released by the helium burning shell, and is still affected by the residual envelope mass $M_{\rm e}^{\rm R}$ (cf. Iben 1984, Wood & Faulkner 1986). Also in this case the transition time is therefore controlled by $M_{\rm e}^{\rm R}$, but the physics is somewhat more complicated, and less straightforward is the relation between $t_{\rm tr}$ and the basic stellar timescales.

3.2. What is the *Nebular Age*?

The quantity $R_{\rm PN}/v_{\rm exp}$ if often used as a measure of the *age* of optically thin PNe. Here $R_{\rm PN}$ is the nebular (outer) radius and $v_{\rm exp}$ is the expansion velocity, which is assumed constant through the nebular evolution. This is a rough definition indeed, good at most for order of magnitude estimates, but which should be used *cum grano salis* to avoid erroneous conclusions. In fact, the outer edge of a thin nebula in case corresponds to the transition from the regular wind to the superwind, and therefore $R_{\rm PN}/v_{\rm exp}$ represents the time elapsed since the beginning of the superwind phase, i.e.:

$$\frac{R_{\rm PN}}{v_{\rm exp}} \simeq t_{\rm SW} + t_{\rm tr} + t_{\rm PN}, \qquad (3)$$

where $t_{\rm SW}$ is the duration of the superwind phase ($10^3 - 10^4$ yr), and $t_{\rm PN}$ is the proper nebular age, defined as the time elapsed since the first shining of the nebula. As noted by Méndez et al. (1987), $t_{\rm PN} \simeq R_{\rm PN}/v_{\rm exp}$ only if $t_{\rm SW}$ and $t_{\rm tr}$ are both negligibly small. This does not appear to be the case in most instances. In principle, one could get rid of the term $t_{\rm SW}$, by using the *inner*, rather than the outer nebular radius, so that:

$$\frac{R_{\rm PN}^{\rm inner}}{v_{\rm exp}} \simeq t_{\rm tr} + t_{\rm PN}, \qquad (4)$$

but the indeterminacy of $t_{\rm tr}$ clearly remains. In this connection, browsing through an atlas of PNe (e.g. Chu et al. 1987) one can appreciate a number of interesting situations, ranging from *butterfly* PNe for which it seems difficult to define either an outer or an inner radius, to *filled* PNe (such as NGC 6894 or IC 3568) where the inner radius is vanishingly small. Certainly, in these cases the nebular age must be much shorter than $R_{\rm PN}/v_{\rm exp}$ (i.e. $t_{\rm PN} \ll t_{\rm SW}$), while only in an *overwhelming minority* of cases can $R_{\rm PN}^{\rm inner}$ be unambiguously defined.

From the above considerations, it therefore appears that in most istances it is currently impossible to define a precise kinematic age of the nebulae, which unavoidably remains uncertain by at least the first of the two additive terms $t_{\rm SW}$ and $t_{\rm tr}$. In conclusion, it appears that both the evolutionary clock and the kinematic clock are affected by *zero point* uncertainties that cannot be presently eliminated, and their synchronization is correspondingly precluded. It seems therefore fair to firmly discourage the use of diagrams involving an admixture of theoretical evolutionary times and ill-defined nebular ages, such as plots of the luminosity of central stars versus nebular radii. This would in fact be equivalent to use a pair of unsynchronized clocks, each affected by an unknown zero point bias which can even differ from one PN to the next, and from one evolutionary sequence to another. The claimed success of investigations making use of such plots is then more apparent than real, as the occasionally reasonable results (such as for example an average mass $<M_{\rm H}> \simeq 0.6 M_\odot$ for PN nuclei) actually follow from the very strong mass dependence of the fading time of Post–AGB stars, where $t_{\rm f} \propto \sim M_{\rm H}^{-8}$ (cf. Iben & Renzini 1983). For example, this dependence ensures that even a factor of 2 error in the age translates into only a $\sim 10\%$ error in the inferred mass $M_{\rm H}$. Moreover, any desired fine adjustement in the inferred value of $<M_{\rm H}>$ is easily achieved by fudging with the arbitrary choice for the zero point of the Post–AGB timescale.

4. CONCLUSIONS

As a conclusion, I would like to summarize here some of the main points so far discussed.

• Stellar evolution theory predicts the production of single star PN nuclei and WDs in the mass range from $\sim 0.54\,M_\odot$ up to at least $\sim 1.06\,M_\odot$, which corresponds to the minimum core mass for non-degenerate carbon ignition.

• Thermal pulses are likely to cause envelope ejection either for low mass, Population II stars $(M_i \lesssim 1\,M_\odot)$, or for relatively young Population I stars $(M_i \gtrsim 3\,M_\odot)$.

• In older disk stars $(1 \lesssim M_i \lesssim 3\,M_\odot)$, i.e. in most cases, superwind envelope ejection appears more likely during one interpulse phase, rather than in coincidence with a thermal pulse.

• It appears that progress in understanding the final evolutionary stages of the more massive PN/WD producers (i.e. stars with $3 \lesssim M_i \lesssim 8\,M_\odot$) can be achieved by concentrating on the study of the high luminosity tail of the luminosity function of OH/IR and IR sources, and on their likely progeny, i.e. the bolometrically faintest PN nuclei.

Acknowledgments. I am grateful to Dr.s Robin Clegg, Rolf Kudritzki, Leon Lucy, and R.H. Méndez for useful and stimulating discussions, and to ESO for its hospitality during the period in which part of the points discussed in this paper have been worked out. This research has been partly supported through grants of the Italian Ministry of Public Education (MPI) for the scientific research.

REFERENCES

Adams, S., Seaton, M.J., Howarth, I.D., Aurière, M., Walsh, J.R. 1984. M.N.R.A.S. **207**: 471
Blanco, V.M., McCarthy, M.F., Blanco, B.M. 1980. *Ap. J.* **242**: 938
Chu, Y.-H., Jacoby, G.H., Arendt, R. 1987. *Ap. J. Suppl.* **64**: 529
Clegg, R.E.S. 1985. *Production and Distribution of CNO Elements* ed. I.J. Danziger, F. Matteucci, & K. Kjär (Garching: ESO), p. 261
Härm, R., Schwarzschild, M. 1975. *Ap. J.* **200**: 324
Iben, I.Jr. 1982. *Ap. J.* **260**: 821
Iben, I.Jr. 1984. *Ap. J.* **277**: 333
Iben, I.Jr. 1985. *Quart.J.R.A.S.* **26**: 1
Iben, I.Jr. 1987. *Late Stages of Stellar Evolution*, ed. S. Kwok & S.R. Pottasch (Dordrecht: Reidel), p. 175
Iben, I.Jr., Kaler, J.B., Truran, J.W., Renzini, A. 1983. *Ap. J.* **277**: 333
Iben, I.Jr., MacDonald, J. 1985. *Ap. J.* **296**: 540
Iben, I.Jr., MacDonald, J. 1986. *Ap. J.* **301**: 164
Iben, I.Jr., Renzini, A. 1983. *Ann. Rev. Astron. Astrophys.* **21**: 271

Iben, I.Jr., Renzini, A. 1984. *Physics Reports.* **105**: 329
Iben, I.Jr., Truran, J.W. 1978. *Ap. J.* **220**: 980
Iben, I.Jr., Tutukov, A.V. 1984. *Ap. J.* **282**: 615
Kwok, S. 1987. *Late Stages of Stellar Evolution*, ed. S. Kwok & S.R. Pottasch (Dordrecht: Reidel), p. 321
Mazzitelli, I. 1987. *Mem.S.A.It.* **58**: 117
Méndez, R.H., Kudritzki, R.P., Herrero, A., Husfeld, D., Groth, H.G. 1987. preprint
Ortolani, S., Rosino, L. 1987. *Astron. Astrophys.* **185**, 102
Paczyǹski, B. 1971. *Acta Astronomica.* **21**: 417
Renzini, A. 1979. *Stars and Star Systems*: ed. B. Westerlund (Dordrecht: Reidel), p. 155
Renzini, A. 1981a. *Physical Processes in Red Giants*: ed. I. Iben Jr. & A. Renzini (Dordrecht: Reidel), p. 431
Renzini, A. 1981b. *Mass Loss and Stellar Evolution*: ed. C. Chiosi & R. Stalio (Dordrecht: Reidel), p. 319
Renzini, A. 1982. *Wolf-Rayet Stars*: ed. C.W.H. de Loore & A.J. Willis (Dordrecht: Reidel), p. 413
Renzini, A. 1983a. *Planetary Nebulae*: ed. D.R. Flower (Dordrecht: Reidel), p. 267
Renzini, A., Fusi Pecci, F. 1988. *Ann. Rev. Astron. Astrophys.* in press
Renzini, A., Voli, M. 1981. *Astron. Astrophys.* **94**, 175
Rose, W.K., Smith, R.L. 1972. *Ap. J.* **173**: 385
Schönberner, D. 1979. *Astron. Astrophys.* **79**: 108
Schönberner, D. 1983. *Ap. J.* **272**: 708
Torres-Peimbert, S. 1984. *Stellar Nucleosynthesis*: ed. C. Chiosi & A. Renzini (Dordrecht: Reidel), p. 3
Trimble, V., Sackman, I.-J. 1978. *M.N.R.A.S.* **182**, 97
Tuchman, Y., Sack, N., Barkat, Z. 1979. *Ap. J.* **234**: 217
Tuchman, Y., Barkat, Z. 1980. *Ap. J.* **242**: 199
Weaver, T.A., Zimmerman, G.B., Woosley, S.E. 1978. *Ap. J.* **225**: 1021
Wood, P.R. 1974. *Ap. J.* **190**: 609
Wood, P.R., Faulkner, D.J. 1986. *Ap. J.* **307**: 659

PROGENITORS OF PLANETARY NEBULAE

Sun Kwok
Department of Physics
University of Calgary
Calgary, Alberta
Canada T2N 1N4

ABSTRACT. Over the past decade, we have come to realize that mass loss on the asymptotic giant branch (AGB) plays a significant role in the formation of planetary nebulae (PN). Mass ejected during the AGB can now be observed in haloes of PN and we believe that the main shell of PN is formed by the interaction of this material with a later-developed central-star wind. In this review, we show that the evolution from AGB to PN can be traced in a continuous infrared sequence. This sequence predicts properties of proto-PN which allow them to be identified.

1. INTRODUCTION

Since both the nebula and the central star of a planetary nebula (PN) originate from a progenitor star on the asymptotic giant branch (AGB), the study of the formation of PN cannot be isolated from the evolutionary stages immediately preceeding the PN. Figure 1 shows the evolutionary pathway leading to the formation of PN. As a star evolves up the AGB, it loses mass at an increasing rate until it is totally obscured by its own circumstellar dust envelope. At this time, the photosphere is no longer visible and the star will appear as an infrared object. It is useful to designate this stage as the late AGB (LAGB). When mass loss reduces the mass of the hydrogen envelope (M_e) to a certain value ($M_e \sim 10^{-2}$ $M_\odot$ for a core mass [M_c] of 0.60 $M_\odot$, Schönberner 1983), the star will turn to the left and evolve towards the blue side of the H-R diagram. When M_e is down to 10^{-3} $M_\odot$ (again for M_c=0.6 $M_\odot$), the envelope is so disrupted that large-scale mass loss is no longer possible to continue. We will define this point as the end of the LAGB and the beginning of the proto-planetary nebula (PPN) phase. The effective temperature of the star will continue to increase due to the loss of envelope mass as the result of hydrogen shell burning. The PPN phase will last ~1500 yr until the central star is hot ($T_*\sim 30,000$K) enough to ionize the circumstellar nebula. Recombination line of hydrogen and forbidden lines of metals will make the nebula easily observable in the visible. In this talk, I will discuss the physical processes in the LAGB and PPN phases and their effects on the formation of PN.

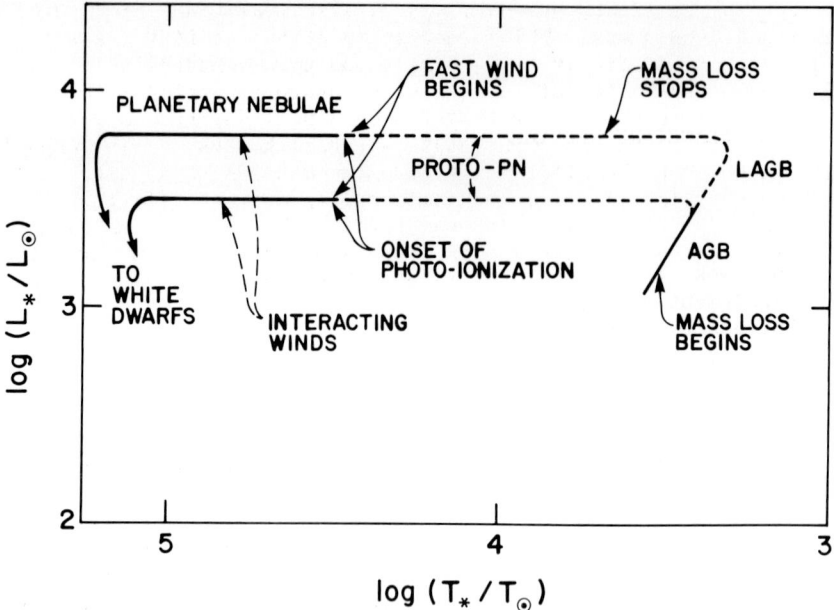

Figure 1. Evolutionary paths from AGB to PN in the H-R diagram.

2. ASCEND OF THE ASYMPTOTIC GIANT BRANCH

While conventional spectral classification schemes stop at about spectral class M10 for evolved stars, it is now recognized that there exist AGB stars that have evolved beyond this limit. The *IRC*, *AFGL*, and *IRAS* infrared sky surveys have discovered many heavily reddened stars which have luminosities higher than Mira Variables and are likely to be LAGB stars. Since the stellar photosphere of LAGB stars are obscured by dust ejected during the mass loss process, one has to rely on radio and infrared techniques as probes of their properties. For oxygen-rich stars, the circumstellar envelope (CSE) generally shows the 9.7 μm silicate feature either in emission or absorption (Merrill and Stein 1976) in the infrared and OH maser emission in the radio (Herman and Habing 1985). For carbon-rich stars, the 11.3 μm SiC feature is usually present and the structure of CSE can be studied by rotation transition of CO in the radio (Knapp and Morris 1985).

Among these four diagnostic tools, the 9.7 μm silicate feature is particularly useful. It is observed in over 2000 stars by the *IRAS* Low Resolution Spectrometer (LRS) and shows a variation in strength from strong emission to strong absorption (Volk and Kwok 1987a). The inferred optical depth in the feature ranges from 0.1 to >100, implying a change in mass loss rate of over 3 order of magnitudes among AGB stars. Since the transition from emission to absorption occurs at $\tau(9.7$ μm$)\sim 4$, almost all stars showing silicate absorption features are without optical counterparts (Kwok, Hrivnak, and Boreiko 1987), and are therefore members of LAGB stars. While precise locations of LAGB stars on the H-R diagram are difficult to determine due to uncertainties in both L_* and T_*, the distribution of the silicate absorption objects in a colour-colour diagram shows that they lie on a well-defined band (Olnon et al. 1984; Bedijn 1988; Kwok, Hrivnak, and Boreiko 1987). In comparison, stars which show the silicate feature in emission (e.g. Mira variables) occupy part of the colour-colour diagram to the left of the absorption objects (Figure 2). If we interpret this band as an evolutionary sequence, then we have a picture of AGB stars evolving from the colour temperatures of >600 K for Mira variables to ~250 K for extreme LAGB stars.

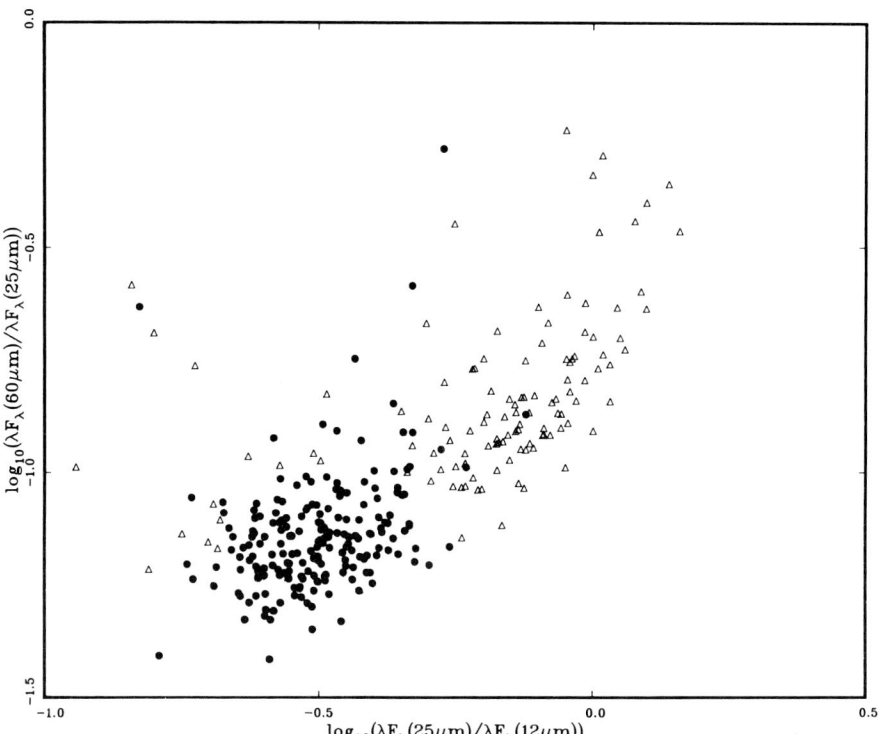

Figure 2. Colour-colour diagram of AGB and LAGB stars. The filled circles are LRS class 25-29 sources with good fluxes at all four bands and the triangles are LRS class 31-39 sources with good fluxes at 12,25 and 60 μm bands. A number of class 30 sources with possible confusion with HII regions are not plotted.

3. INFRARED LINK BETWEEN THE AGB AND PN

For a number of years, I have argued that the CSEs created by mass loss on the AGB should be observable in PN and their presence may have significant effects on the formation of PN (Kwok 1980, 1982). The recent *IRAS* sky survey has revealed that PN have cool dust components, and Pottasch et al. (1984) have found that the colour temperatures of evolved PN lie in the range of 40-100 K. An analysis of young PN by Kwok, Hrivnak, and Milone (1986) shows that the near infrared emission from PN is due to thermal free-free emission from the ionized gas whereas dust emission is responsible for the far infrared emission. They also found that the colour temperatures of young PN are higher than those of evolved PN, with typical values in the range of 100-200 K.

The fact that the colour temperatures of young PN are in between the colour temperatures of LAGB and evolved PN strongly suggests that the infrared emission in these objects all originates from the dust CSEs which are ejected during the LAGB phase but have cooled during the PN stage. Figure 3 shows a plot of the radio brightness temperature against 60 μm optical depth for 266 PN. These two parameters are used because they both are distance-independent and are gauges of the dynamical ages of the gas and dust components respectively. The strong correlation found is evidence that the gas and dust components of PN are both expanding with time and the location of individual PN in Figure 3 is a measure of its relative age.

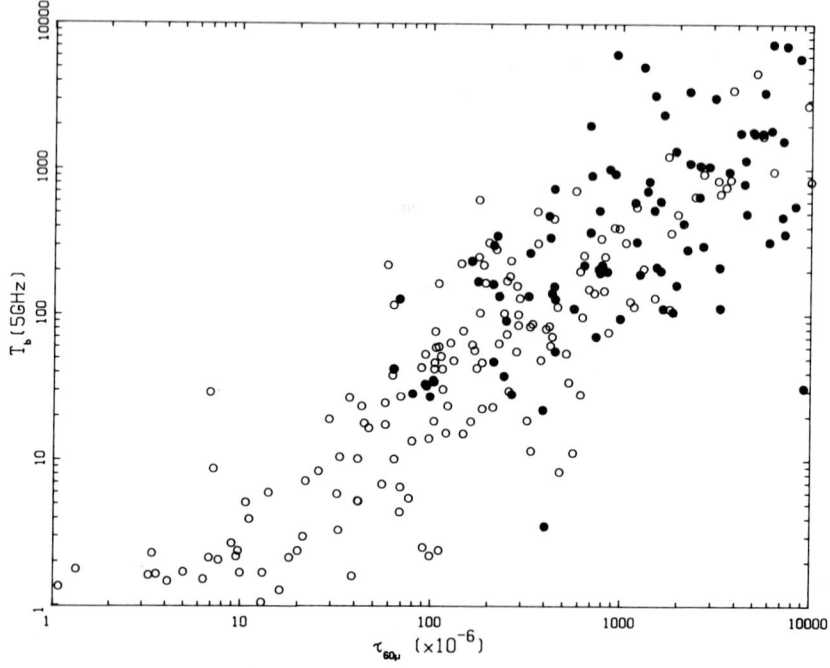

Figure 3. The 5 GHz radio brightness temperature plotted against 60 μm optical depth for 266 PN. The filled and open circles are respectively VLA and single-dish measurements.

From the above discussions, it has become obvious that there exists an continuous infrared sequence connecting the evolutionary stages of AGB and PN. The colour temperature decreases monotonically from >600 K for Mira Variables to 250-600 K for LAGB stars, to 100-200 K for young PN and then to 40-100 K for evolved PN. It is important to note that while the change in colour temperature is monotonic and reflects the evolution of the dust CSE, the physical reason for the decrease is different in the AGB and PN phases. During the AGB and LAGB when mass is continuously ejected from the star, the lowering of colour temperature is the result of increasing optical depth in the CSE. However, beyond the LAGB, optical depth of the envelope begins to decrease and the change in colour temperature is the result of geometric dilution.

4. THE PROTO-PLANETARY NEBULA PHASE

Renzini (1983) has suggested that many PN may remain undetected if the central stars evolve too slowly. The common occurrence of PN therefore implies that the PPN stage cannot be more than a few thousand years in duration, otherwise the nebula would have been dispersed before it is ionized. If we adopt the transition time of ~1500 yr calculated by Schönberner (1983) then the ratio of PPN to PN is approximately the ratio of their respective lifetimes, or 5-15%. From the number of PN in the *IRAS* Point Source Catalogue (>1000) we therefore expect that there should be 50-150 PPN detected by the *IRAS* survey.

If we accept the existence of the infrared sequence discussed in §3, then the infrared properties of PPN can be interpolated from those of LAGB and young PN. This suggests an intermediate colour temperature of 150-250 K. As a PPN has very little hydrogen atmosphere, we also do not expect it to be a strong pulsator (Habing, van der Veen, and Geballe 1987). A search of PPN using these criteria were carried out by Kwok and Hrivnak at the Canada-France-Hawaii Telescope (CFHT). A number of *IRAS* sources of low colour temperatures were found with blackbody-like energy distributions and with no optical counterparts.

As the remnant dust CSE continues to expand and the envelope optical depth continues to drop, the star will eventually become visible again. By extending the radiative transfer models for AGB stars to beyond the AGB, Volk and Kwok (1987b) find a distinct spectral shape for PPN. A search of the LRS catalog resulted in a number of sources showing agreement with the model prediction, one of them being 18095+2704. This *IRAS* source was identified at the CFHT with a relatively bright star of 11 mag. Optical spectroscopy at the Dominion Astrophysical Observatory shows that it has a spectral type of F3 Ib (Hrivnak, Kwok, and Volk 1988). A combined visual and infrared spectrum from 0.36 to 100 μm for 18095+2704 is shown in Figure 4. Assuming $L_*=10^4$ $L_\odot$ and $T_*=6000$ K, a model fit to the spectrum suggests that the inner radius of the dust shell is located at 7×10^{15} cm. Using the expansion velocity of 7 km s^{-1} derived from OH observations (Lewis, Eder, and Terzian 1985), it is estimated that the star had a mass loss rate of 2×10^{-5} $M_\odot$ yr^{-1} at the AGB and the shell was detached from the star ~325 yr ago.

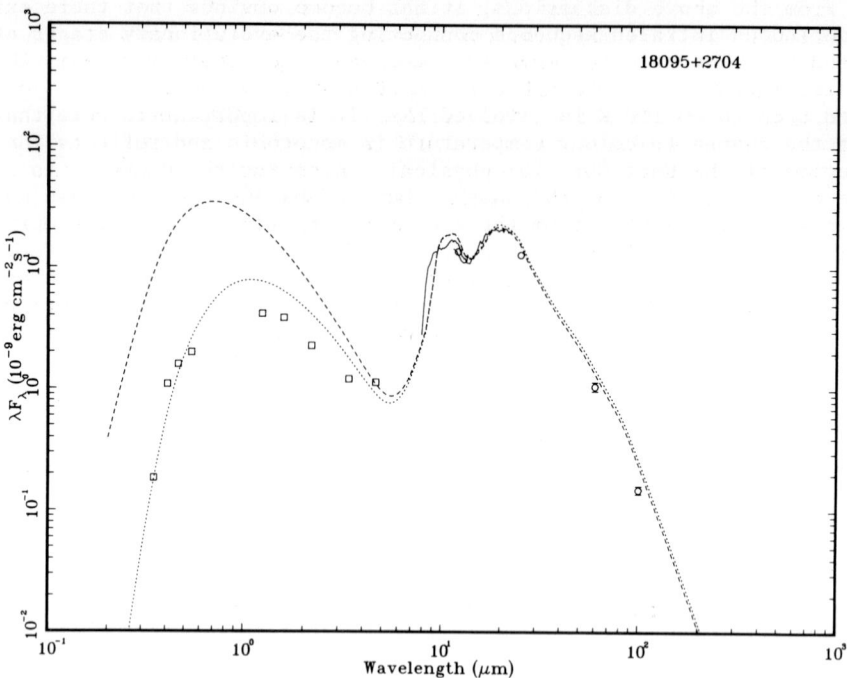

Figure 4. The IRAS LRS spectrum of 18095+2704 combined with the ground-based and IRAS photometry, and plotted together with a PPN model. The temperature and luminosity are assumed to be 6000K and 10^4 $L_\odot$ respectively. The dotted line shows the resultant spectrum after the addition of an extinction component of the form $\exp[-1.406/\lambda(\mu m)]$.

The dual-peak spectral behaviour of this PPN candidate is similar to the PPN candidates found from non-variable OH/IR stars (van der Veen, Habing, and Geballe, these proceedings). As PPN evolve into the early stages of PN, part of the nebulae will begin to be ionized. The OH sources with radio continuum emission found by Zijlstra et al. (these proceedings) could represent some of the youngest PN observed.

5. TERMINATION OF THE AGB

An important question on the origin of PN is the time scale of the ejection of the nebula. This ranges from the traditional theories of sudden ejection where the process occurs in a few years (cf Roxburgh 1978), to $\sim 10^3$ yr (Tuchman, Sack, and Barkat, 1979) and $\sim 10^4$ yr (Herman and Habing 1985) in the "superwind" models. The term "superwind" has its origin in that the mass loss rate during this phase is high in comparison to the Reimers' formula (Reimers 1975). However, infrared and radio (CO) observations show that mass loss rates of AGB stars with spectral types later than M3 are already much higher than that given by

the Reimers' formula. Since the mass loss rates of AGB stars increase
with luminosity (or time), one could refer to the very last stage ($<10^4$
yr) of mass loss as a "superwind" phase, although the distinction is
strictly semantic. In my opinion, it is not appropriate to refer to the
OH/IR phase as the "superwind" phase because the LAGB phase, which
coincides with the observation of OH/IR stars with no optical counterparts, certainly lasts more than 10^5 yr. Furthermore, a mass loss rate
of only few times 10^{-6} $M_\odot$ yr^{-1} will completely obscure the photosphere
of the star and such mass loss rates cannot in any sense be considered
"super".

If the AGB is indeed terminated by the complete removal of the
hydrogen envelope through steady mass loss, then can PN be nothing more
than ionized CSEs? This possibility was first raised by Paczynski
(1971) and has since been re-discussed by other authors (Harpaz and
Kovetz 1981), and most recently by Taylor, Pottasch, and Zhang (1987).
This scenario of PN formation was considered by Kwok (1981) to be
inadequate for three reasons:

1. Densities of PN shell are higher than the observed densities in AGB CSEs.
2. Expansion velocities of PN shell ($\sim$20-50 km s^{-1}) are higher than expansion velocities of AGB star winds ($\sim$3-20 km s^{-1}).
3. Many PN have well-defined shell-like morphologies whereas AGB winds have smooth density distributions.

All these three points are still relevant today. Direct comparisons of the densities of shell and halo of PN also suggest a difference
in density of a factor $\sim$10 (Jewitt, Danielson, and Kupferman 1986;
Bässgen et al., these proceedings). Expansion of AGB stellar winds are
well determined by CO and OH measurements and the discrepancy between
these values and PN velocities are still significant. Brightness
distributions of CO in IRC+10°216 shows an inverse square density law
with no definite shell structure (Kwan and Link 1982). These problems
indicate that another physical process is needed to adequately explain
the formation of PN.

6. THE INTERACTING WINDS MODEL

The Interacting Stellar Winds model of Kwok, Purton, and FitzGerald
(1978, hereafter KPF) was an attempt to address the problems mentioned
in §5. This model postulates the presence of a fast wind which shapes,
accelerates, and compresses the remnant AGB wind material into PN. If
this model represents the universal formation mechanism of PN, then
three conditions must be satisfied:

1. Remnants of CSE of AGB progenitor are still present in PN;
2. high velocity wind from the central star is a common phenomen-on;

and 3. masses of PN increase with age as the remnant CSE is swept up by the central-star wind.

While the existence of central-star winds and haloes around PN has been known for some time (Smith and Aller 1969; Duncan 1937), their common occurrence in PN was not confirmed until recently. The *IUE* satellite found P Cygni profiles in resonance lines of central stars of many PN (Heap et al. 1978) with velocities ranging from 1400 to 5000 km s^{-1}, and mass loss rates between 10^{-9} to 10^{-7} $M_\odot$ yr^{-1} (Grewing, these proceedings; Perinotto, these proceedings). Faint haloes are now believed to be present in over 50% of all PN (Jewitt, Danielson, and Kupferman, 1986; Chu, Jacoby, and Arendt 1987; Balick 1987). Maps of H_2 molecules show that the molecules are distributed outside of the main shell (Zuckerman and Gatley 1987). OH and CO molecules which are common in AGB CSEs are found in increasing number of PN (OH: Seaquist and Davis 1983; Payne, Phillips, and Terzian, these proceedings; Zijlstra et al. these proceedings; CO: Thronson 1983; Healy and Huggins, these proceedings; Walsh, Clegg, and Ukita, these proceedings). Atomic hydrogen has also been detected (Rodriguez and Moran 1982; Taylor and Pottasch 1987). The presence of dust continuum emission from the remnant CSE has already been discussed in §3.

The masses of PN remain a controversial issue. There is, however, good evidence that the ionized masses of PN range over three order of magnitude (Pottasch 1980). Ionized masses determined from the radio survey of compact PN also point to similar conclusions (Kwok 1987). It is possible that some of this variation in mass is due to ionization effects, but the interacting winds process may also contribute toward the large observed mass range.

The above summary of recent observations suggests that the qualitative predictions of KPF have been largely confirmed. It is, however, important to test the quantitative predictions of the model as well. It has been shown that at least in the energy conserving case, the observed central-star wind mass loss rates and velocities are adequate to explain the observed masses and velocities of the PN shell (Volk and Kwok 1985). With kinematic information of the halo now available through spectroscopic observations (Chu, these proceedings; Bässgen et al., these proceedings), it is possible not only to test the validity of the dynamical models but also to determine whether the wind interactions are energy or momentum conserving.

Previous dynamical calculations usually assume an *ad hoc* mass loss formula for the central star (e.g. Volk and Kwok 1985). The theoretical formula reported by Kudritzki (these proceedings) may allow the incorporation of a realistic formula in dynamical calculations. With the improved models of central star evolution (Schönberner, these proceedings) and models of nebular dynamics, calculation of the evolution of the nebular spectrum has become possible (Schmidt-Voigt and Köppen 1987a,b). Comparison of the theoretical results and observations may become one of the most fruitful area of research in this subject.

7. CONCLUSIONS

In this conference, Chu raised the question "What is the definition of planetary nebula?" This is a very valid question. My response is that we should consider the term "the PN system" which consists of a central-star wind, a shell, and a halo, in addition to the central star. The traditional meaning of PN refers only to the shell component, which may have densities, velocities, or compositions which are different from the other two components.

Significant progresses have been made in the understanding of the transition phases between AGB and PN since the last IAU symposium. Observations in the far infrared from the *IRAS* satellite have led to the identification of a continuous infrared sequence connecting the AGB, LAGB, PPN and PN phases. This proposed sequence of evolution is summarized in Table 1.

A number of candidates of PPN have now been identified. Further identification and observations of PPN in the next few years will hopefully settle the question of the origin of planetary nebulae.

TABLE 1
EVOLUTION FROM AGB TO PLANETARY NEBULAE

evolutionary phase	example	optical image	period (days)	colour temperature (K)	silicate dust	OH
AGB	Mira Variables	bright	300-600	>600K	emission	yes
LAGB	OH/IR stars	no optical counterpart	600-2000	250-600K	absorption	strong
post-AGB	19454+2920	no	non-variable	150-250	?	weak
proto-PN	18095+2704	yes	non-variable	150-250	emission	weak
young PN	Vy2-2 Hb 12	bright	non-variable	100-200	emission	single peak
PN	many	bright	non-variable	<100	no	no

The ground-based observations of *IRAS* sources and the models of PPN evolution are done in collaboration with Drs. B.J. Hrivnak and K.M. Volk respectively. This work is supported by an operating grant from the Natural Sciences and Engineering Research Council of Canada.

REFERENCES

Balick, B. 1987, *Astron. J.*, **94**, 671.
Bedijn, P. 1988, preprint.
Chu, Y.H., Jacoby, G.H., and Arendt, R. 1987, *Astrophys. J. Supp.*, **64**,

529.
Duncan, J.C. 1937, *Astrophys. J.*, **86**, 496.
Habing, H.J., van der Veen, W., and Geballe, T. 1987, in *The Late Stages of Stellar Evolution*, eds. S. Kwok and S.R. Pottasch (Reidel:Dordrecht), p. 91.
Harpaz, A., and Kovetz, A. 1981, *Astron. Astrophys.*, **93**, 200.
Herman, J., and Habing, H.J. 1985, *Physics Reports*, **124**, 225.
Heap, S.R. et al. 1978, *Nature*, **275**, 385.
Hrivnak, B.J., Kwok, S., and Volk, K.M. 1987, submitted to *Astrophys. J.*
Jewitt, D.C., Danielson, G.E., and Kupferman, P.N. 1986, *Astrophys. J.*, **302**, 727.
Kwan, J., and Linke, R.A. 1982, *Astrophys. J.*, **254**, 587.
Knapp, G.R. and Morris, M. 1985, *Astrophys. J.*, **292**, 640.
Kwok, S. 1980, *Astrophys. J.*, **236**, 592.
Kwok, S. 1981, in *Effects of Mass Loss on Stellar Evolution*, eds. C. Chiosi and R. Stalio (Reidel:Dordrecht), p. 347.
Kwok, S. 1982, *Astrophys. J.*, **258**, 280.
Kwok, S. 1987, in *The Late Stages of Stellar Evolution*, eds. S. Kwok and S.R. Pottasch (Reidel:Dordrecht), p. 321.
Kwok, S., Purton, C.R., and FitzGerald, M.P. 1978, *Astrophys. J. (Lett.)*, **219**, L125.
Kwok, S., Hrivnak, B.J., and Milone, E.F. 1986, *Astrophys. J.*, **303**, 451.
Kwok, S., Hrivnak, B.J., and Boreiko, R.T. 1987, *Astrophys. J.*, **321**, 975.
Lewis, B.M., Eder, J., and Terzian, Y. 1985, *Nature*, **313**, 451.
Merrill, K.M., and Stein, W.A. 1976, *Publ. Astron. Soc. Pac.*, **88**, 874.
Olnon, F.M., Baud, B., Habing, H.J., de Jong, T., and Pottasch, S.R. 1984, *Astrophys. J. (Lett.)*, **278**, L41.
Paczynski, B. 1971, *Astrophys. Letters*, **9**, 33.
Pottasch, S.R. 1980, *Astron. Astrophys.*, **89**, 336.
Pottasch, S.R. et al. 1984, *Astron. Astrophys.*, **138**, 10.
Reimers, D. 1975, *Mém. Soc. Roy. Sci. Liège*, Ser. 8, p. 369.
Renzini, A. 1983, in *I.A.U. Symp. 103: Planetary Nebulae*, ed. D.R. Flower (Reidel:Dordrecht), p. 267.
Rodriguez, L.F., and Moran, J.M. 1982, *Nature*, **299**, 323.
Roxburgh, I.W. 1978, in *I.A.U. Symp. 76: Planetary Nebulae*, ed. Y. Terzian (Reidel:Dordrecht), p. 295
Seaquist, E.R., and Davis, L.E. 1983, *Astrophys. J.*, **274**, 659.
Schmidt-Voigt, M. and Köppen, J. 1987a, *Astron. Astrophys.*, **174**, 211.
Schmidt-Voigt, M. and Köppen, J. 1987b, *Astron. Astrophys.*, **174**, 223.
Schönberner, D. 1983, *Astrophys. J.*, **272**, 708.
Smith, L.F., and Aller, L.H. 1969, *Astrophys. J.*, **157**, 1245.
Taylor, A.R., and Pottasch, S.R. 1987, *Astron. Astrophys.*, **176**, L5.
Taylor, A.R., Pottasch, S.R., and Zhang, C.Y. 1987, *Astron. Astrophys.*, **171**, 178.
Thronson, H.A. 1983, *Astrophys. J.*, **264**, 599.
Tuchman, Y., Sack, N., and Barkat, Z. 1979, *Astrophys. J.*, **234**, 217.
Volk, K., and Kwok, S. 1985, *Astron. Astrophys.*, **153**, 79.
Volk, K., and Kwok, S. 1987a, *Astrophys. J.*, **315**, 654.
Volk, K., and Kwok, S. 1987b, in *The Late Stages of Stellar Evolution*, eds. S. Kwok and S.R. Pottasch (Reidel:Dordrecht), p. 305.
Zuckerman, B., and Gatley, I. 1987, *Astrophys. J.*, in press.

MODELS OF PLANETARY NEBULAE: GENERALISATION OF THE MULTIPLE WINDS MODEL

F. D. Kahn
Department of Astronomy
The University
Manchester M13 9PL
England

ABSTRACT. According to the multiple winds model a planetary nebula forms as the result of the interaction of a fast wind from the central star with the superwind that had previously been emitted by the progenitor star. The basic theory which deals with the spherically symmetrical case is briefly summarised. Various improvements are then considered in turn. A better history is clearly needed of the way that the central star becomes hotter, it is unrealistic to make the assumption that the superwind is spherically symmetrical, and finally there are likely to be important instabilities at some of the interfaces in the PN, notably that between the shocked superwind and the HII layer. These changes in the theoretical description produce a better understanding of the conditions in the outer parts of a PN and of the nature of its general shape, and they should lead to an explanation for the occurrence of high speed motions, and of highly ionized species and high excitation spectral lines.

1. INTRODUCTION

The multiple winds model (Sun Kwok 1982, 1983, 1988; Kahn 1983) gives a very good description of the overall properties of planetary nebulae. In this paper it will be discussed first in its simplest form, and then various restrictions will be removed. The basic model, with spherical symmetry, has its uses and makes it possible to establish various important physical parameters, as well as the relations between them. However, it fails to allow for the fact that the shapes of most PNe are far from spherically symmetrical and it is based on idealised assumptions concerning the onset of the fast wind from the central star which constitutes the planetary nebula nucleus (PNN). Finally there is a stability problem for those PNe which are ionization bounded. The HII layers here are squeezed between the hot shocked stellar wind (HSSW) inside and the non-ionized gas outside, deriving from the superwind (SW); such configurations are in general unstable, and in a wide range of cases the growth time of disturbances is short enough for finite amplitude wave motions to develop within the lifetime of the PN. If so, there is a strong likelihood that the cool gas deriving from the SW will mix with hot gas from the HSSW,

2. THE BASIC MODEL

The superwind is assumed to carry away mass at rate $\dot{M}$ from the progenitor of the central star up to time $t = 0$. It has terminal speed u, and is spherically symmetrical. At time $t = 0$ the fully fledged PNN is assumed to appear. It has luminosity L, blows a fast wind with terminal speed $V(\gg u)$. The wind carries off mechanical energy at rate $L_W \equiv \eta L$, and the star emits Lyman continuum photons at rate $S_* \equiv jL$.

Two shocks develop as the result of the interaction between the winds. On the outside a shock is driven into the gas from the SW, which it compresses into a thin, well-cooled shell at radius r. Closer to the PNN there is another shock facing into the fast wind. The HSSW fills the region between this shock and the HII layer, which lines the inside of the cool shell at radius r.

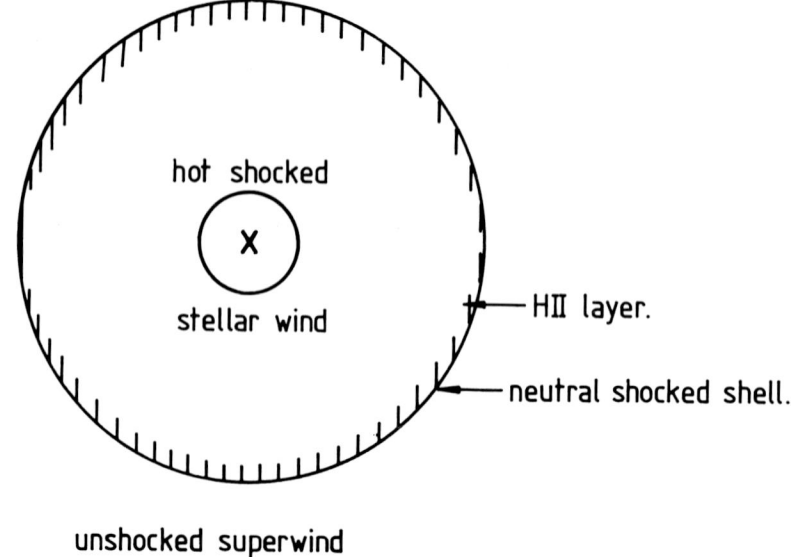

Figure 1. The various regions in the multiple winds model.

The equations to describe the motion are

$$L_W \equiv \eta L = \frac{d}{dt} 2\pi r^3 P + 4\pi r^2 P \dot{r} \quad (1)$$

$$P = \rho_0 (\dot{r} - u)^2 + \frac{M_s \ddot{r}}{4\pi r^2} \quad (2)$$

$$M_s = \frac{\dot{M}}{u}(r - ut) , \qquad (3)$$

and here

$$\rho_o = \frac{\dot{M}}{4\pi u r^2} \qquad (4)$$

is the density in the SW just ahead of the outer shock, P is the pressure in the HSSW and M_s is the mass of the shell at radius r.

Equation (1) shows how the mechanical energy supplied by the fast wind goes partly to heat the HSSW and partly to push back the shell outside; equation (2) relates the pressure behind the shell to the ram pressure at the outer shock and the force per unit area needed to accelerate the shell; equation (3) is self-evident, but is based on the assumption that the fast wind starts up at the moment that the SW stops.

The solution of the equations leads to the results

$$r = \lambda u t , \qquad \dot{r} = \lambda u , \qquad P = \frac{\dot{M}(\lambda - 1)^2}{4\pi \lambda^2 u t^2} \qquad (5)$$

where λ satisfies

$$\lambda(\lambda - 1)^2 = \frac{2}{3} \frac{\eta L}{\dot{M} u^2} \qquad (6)$$

Balancing the ram pressure of the fast wind against the pressure in the HSSW shows that the inner shock is at radial distance

$$r_i = \left(\frac{2L_W u}{\dot{M} V}\right)^{\frac{1}{2}} \frac{\lambda t}{\lambda - 1} \qquad (7)$$

and therefore

$$\frac{r_i^3}{r^3} = \left(\frac{3\lambda u}{V}\right)^{3/2} \qquad (8)$$

is the ratio of the volumes enclosed respectively by the inner and the outer shocks. Here λu is the expansion speed of the PN, typically 30 km/s, and the fast wind speed V is typically 3000 km/s, so that the ratio of volumes becomes about 1:200.

The temperature in the HSSW is found from

$$\frac{kT}{m} = \frac{P}{\rho_H} = \frac{V^2}{9} \qquad (9)$$

where

$$\rho_H = \frac{3M_H}{4\pi r^3} = \frac{2L_W t}{V^2} \qquad (10)$$

is the density, and M_H the total mass in the HSSW. The fast wind is typically raised to a temperature of 7×10^7K on passing through the inner shock. It is easy to verify that radiative cooling of the HSSW will be quite negligible under realistic conditions. Further the sound speed in the HSSW is typically 1300 km s^{-1}, and the expansion of the PNe therefore takes place very subsonically with respect to the hot gas. The pressure will consequently be (almost) uniform in the HSSW bubble.

For illustrative purposes the values of important parameters are here taken to be

$$u = 10 \text{ km s}^{-1}, \quad \dot{M} = 9 \times 10^{-5} M_\odot/\text{year},$$

$$L = 6300 \, L_\odot = 2.5 \times 10^{37} \text{ erg/s}, \quad L_w = \eta L = 10^{35} \text{ erg/s}$$

$$S_* \equiv jL = 2.5 \times 10^{47} \text{ photons/s}$$

and are (almost) consistent with a value $\lambda = 3$, and an expansion speed $\dot{r} = 30$ km/s for the shocked shell.

A PN will be optically thick as long as the mass M_s of the shell exceeds the mass M_i of ionized gas in the HII layer. The number of electrons in this layer is

$$N_e = M_i/m_a \tag{11}$$

where m_a is the mass of gas per free electron, and the electron density, at pressure P, is

$$n_e = P/m_a c_i^2, \tag{12}$$

and here c_i is the isothermal sound speed in the HII region. If b ($\doteq 2 \times 10^{-13}$ cm³/s) is the recombination coefficient, then, after some algebra, it is found that

$$M_i = m_a N_e = m_a S_*/bn_e \equiv jL \, m_a/bn_e$$

$$= \frac{6\pi j \lambda^3}{\eta b} m_a^2 c_i^2 u^3 t^2. \tag{13}$$

The ratio of masses is

$$M_i : M_s = t : t_* \tag{14}$$

where

$$t_* = \frac{\eta b \dot{M}(\lambda - 1)}{6\pi j \lambda^3 m_a^2 c_i^2 u^3} \tag{15}$$

or, with the present typical values, 15000 years. A PN that is younger will be optically thick. In many cases the PNN evolves through its hot phase in a rather shorter time than 15000 years, so that greater ages cannot occur. Evolution rates for central stars have been discussed in detail by Schönberner (1983, 1986, 1988).

The thickness of the HII layer is also physically important; it equals

$$\Delta_i = \frac{M_i}{4\pi r^2 \rho_i} = \frac{M_i c_i^2}{4\pi r^2 P}$$

and the sound travel time across the HII layer is

$$t_c = \frac{\Delta_i}{c_i} = \frac{6\pi j}{b\eta} \frac{\lambda^3}{(\lambda - 1)^2} \frac{m_a^2 c_i^3 u^2 t^2}{\dot{M}} \qquad (16)$$

At a typical age t, say 3000 years, or 10^{11} s, it is found that

$$t_c = 10^{10} \text{ s}$$

and is thus smaller by an order of magnitude than t. The ratio $t_c:t$ increases linearly with time.

3. A MORE REALISTIC START TO THE EXPANSION

One obvious shortcoming of the basic model lies in the assumption that the fast wind starts to blow at the moment that the superwind stops. Computed models of the evolution of the central star allow intervals of between 1000 and 10000 years for the transition from the AGB to a hot PNN, depending on the mass of the star concerned. There is as yet no detailed description of the history of such a transition period. At present the best that can be done is to make an estimate of how the model will change if an interval of length t_o elapses between the end of the superwind and the sudden appearance of a PNN, with a copious photon output in the Lyman continuum and a fast stellar wind.

At time t_o a cavity, with radius ut_o, will have formed in the distribution of gas from the superwind. The fast wind will fill this volume with hot shocked gas, whose pressure at time $t_o + \tau$ will be

$$P = \frac{L_w \tau}{2\pi u^3 t_o^3} = \frac{3}{4\pi} \frac{\lambda(\lambda-1)^2 \dot{M}\tau}{ut_o^3} \qquad (17)$$

The pressure builds up quite rapidly and a shock propagates into the surrounding material, forming a dense shell of cool gas. If v is the speed of the shell, then the rate of growth of its mass is given by

$$\frac{dM_s}{d\tau} = \frac{\dot{M}}{u}(v - u) \qquad (18)$$

and the equation for v is

$$P = \frac{M_s \dot{v}}{4\pi r_o^2} + \frac{\dot{M}(v-u)^2}{4\pi r_o^2 u} \quad . \tag{19}$$

These equations apply early on during the acceleration, before the radius of the shell has increased much beyond r_o. It follows that

$$v - u = \frac{3}{2} \lambda^{\frac{1}{2}}(\lambda - 1) u \left(\frac{\tau}{t_o}\right)^{\frac{1}{2}} \tag{20}$$

and

$$M_s = \lambda^{\frac{1}{2}}(\lambda - 1) \dot{M} \frac{\tau^{3/2}}{t_o^{\frac{1}{2}}} \quad . \tag{21}$$

Both the mass of the shell and its outward velocity build up quite rapidly towards the values that they would have at radius r_o according to the basic model. The mass of gas in the HII layer is again given by

$$M_i = N_e m_a = jLm_a^2 c_i^2 / bP$$

so that now

$$M_i = \frac{4\pi j L m_a^2 c_i^2 u t_o^3}{3\lambda(\lambda-1)^2 \dot{M}\tau^3} = \frac{2\pi j}{\eta} \frac{m_a^2 c_i^2 u^3 t_o^3}{b\tau^3} \quad . \tag{22}$$

The model is consistent only once M_s exceeds M_i; the HII shell is then optically thick, and this applies when

$$\frac{\tau}{t_o} > \left(\frac{2\pi j}{\eta b \dot{M}}\right)^{2/5} \frac{m_a^{4/5} c_i^{4/5} u^{6/5} t_o^{2/5}}{\lambda^{1/5}(\lambda-1)^{2/5}} \quad . \tag{23}$$

With the usual values for the parameters of the PN, and setting $t_o = 10^{11}$ s or 5×10^{10} s, the condition becomes, respectively,

$$\frac{\tau}{t_o} > 0.074 \quad \text{or} \quad 0.056 \quad , \tag{24}$$

so that there is a (relatively short) period when the PNN is turned on, but the shell is transparent to Ly c photons. On average about 71 per cent of the photons reach the SW beyond the shocked shell. Using typical parameters and taking t_o to be 10^{11} s or 5×10^{10} s again, the number of Ly c photons thus made available is 1.3×10^{57} and 5×10^{56}, respectively. They are enough to ionize 1.3 $M_\odot$ or 0.5 $M_\odot$ of atomic hydrogen in the wind, or to dissociate and then ionize 0.87 $M_\odot$ or 0.33 $M_\odot$ of molecular hydrogen. This ionized gas in the unshocked SW later recombines, but does so only slowly, so that at time $\Delta\tau$ after the flash of Ly c radiation the electron density still is approximately

$$n_e = (b \, \Delta\tau)^{-1},$$

so
$$n_e = 50 \text{ cm}^{-3},$$

typically, after 10^{11} s, or 3000 years. Of course the electron density is as large as this only where there is an adequate mass density in the superwind.

Finally to estimate the emission measure of this low ionization halo. Take standard values for the parameters again, and assume that the superwind originally contained hydrogen in molecular form, and that the transition time t_o is 5×10^{10} s, then the mass in the halo is 0.33 $M_\odot = 6 \times 10^{32}$ gm $= M_h$ and its boundary is at distance

$$r_b = \frac{M_h u}{\dot{M}} = 6 \times 10^{17} \text{ cm} = 0.2 \text{ pc}.$$

The low ionization halo then has an emission measure

$$EM = n_e^2 r_b = \frac{r_b}{b^2 \Delta\tau^2} = \frac{5 \times 10^{24}}{(\Delta\tau)^2} \text{ cm}^{-6} \text{ pc}$$

where $\Delta\tau$ is expressed in seconds. Some 3000 years after the flash the value of EM is 500, so that the low ionization halo is detectable until then, but not much longer.

4. DEPARTURES FROM SPHERICAL SYMMETRY

The multiple winds model must be generalised if it is to describe the variety of shapes of PNe. The most recent observational data are discussed by Balick (1987, 1988) who gives a classification of the different types of structure. He finds that, though there is a large variety among the images of PNe, they can all more or less be accounted for by the multiple winds model, with the simple modification that the superwind is assumed to carry away more mass per unit time in the equatorial regions of the progenitor star than in the polar regions.

An elementary treatment of the dynamics involved has been given by Kahn and West (1985). Their underlying premise is that the superwind contains cool gas, which has a low sound speed, so that any inhomogeneities are smoothed out only slowly. The HSSW, on the other hand, is very hot and has a high speed of sound. It will therefore restore isobaric conditions extremely quickly. Consequently it is much more profitable to study the effects of inhomogeneities in the SW since they are likely to produce a lasting effect on the shape of a PN. The fast wind cannot do so, unless it is very highly collimated and jet-like, and this seems most improbable.

The basic multiple winds model has the useful property that all motions proceed at a uniform rate, and this permits an easy extension of the dynamical treatment to flows which are not spherically symmetrical. Kahn and West deal with axially symmetrical cases where the terminal velocity and the mass loss rate in the superwind are constant, but the mass flux depends on the polar angle θ like $1 + \varepsilon \sin^n \theta$. They consider a variety of positive values of ε and n. Increasing ε increases the departure from spherical symmetry, increasing n sharpens the concentration of the mass flux towards the equatorial regions.

The bubble of HSSW will obviously expand more easily in the polar direction where there is least obstruction from the gas released in the superwind. The bubble is often pinched in at the equator, and sometimes the model predicts the existence of a cusp there, and (formally) accumulation of so much material that infinite surface density occurs in the shell along the equator. The reader is referred to the paper for the detailed argument. Here it will be enough to illustrate the variety of possible shapes that can be generated by this model. The important parameters are ε, n and λ; the first two have already been defined, and

$$\lambda \equiv \dot{r}(0)/u \ ,$$

where $\dot{r}(\theta)$ is the radial velocity in the shocked shell at polar angle θ, and u the speed of the superwind, as before. The polar direction is $\theta = 0$ and is the horizontal direction in the three figures below.

5. INSTABILITIES IN THE HII LAYER

So far the flow in PNe has been taken to be orderly, with each component in the structure moving smoothly. There has been no mixing between the different kinds of gas; in particular it has been assumed that there is negligible thermal exchange across the contact discontinuity between the HSSW and the HII region. One consequence of this assumption is that the HSSW loses negligible amounts of energy by radiation (and none by conduction) and therefore expands adiabatically.

However a closer investigation shows the HII layer to be subject to instabilities, with a short growth time (Kahn and Breitschwerdt 1988). The motion is driven by flows at the adjacent ionization front and in general grows better in regions where the illumination from the PNN is incident obliquely. It is therefore to be expected that the effects of the instability will be most marked in those regions of a PN shell where the distortions from spherical symmetry are largest.

In the simplest treatment of the effect, the HII layer has thickness $\mathfrak{z}$ and the contact discontinuity with the HSSW is along the plane $z = 0$. The layer of non-ionized gas lies beyond the plane $z = \mathfrak{z}$, and is assumed to be so massive that it can be regarded as immobile. The boundary condition at $z = 0$ is determined by the fact that the HSSW is so hot, and has so high a sound speed, that its pressure remains constant throughout any disturbance. Finally the Ly c radiation from the central star is incident on the layer at angle β with the z-direction, and the x-axis is coplanar with Oz and the direction to the PNN.

A sound wave passing along the HII layer propagates with the isothermal sound speed c_i. It has to satisfy the boundary conditions that there shall be no pressure changes on the plane $z = 0$, the interface with the HSSW; on the plane $z = \mathfrak{z}$ there must be (almost) no Ly c flux, because

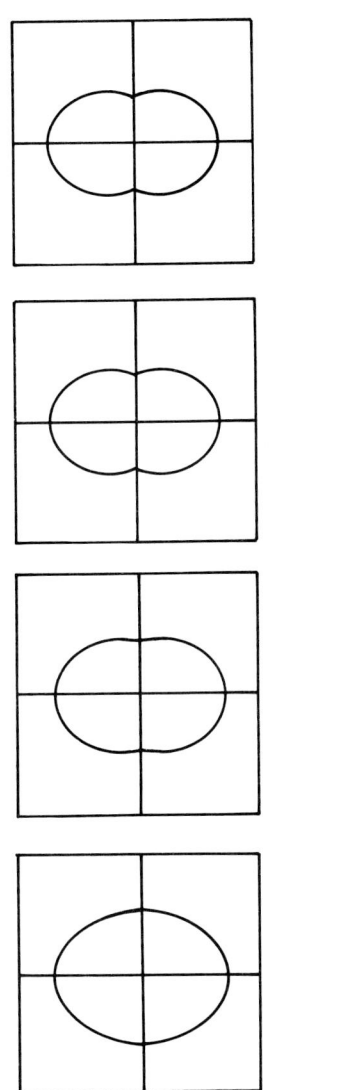

Figure 2. Shape of the nebula for $\varepsilon = 1$, $n = 2$ and $\lambda = 2$ (bottom), 3, 4 and 5, showing the effects of fast expansion.

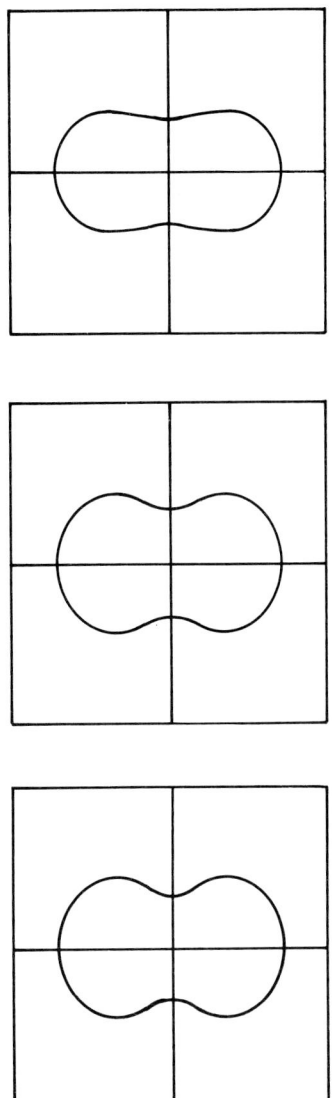

Figure 3. Shape of the nebula for $\lambda = 3$, $\varepsilon = 5$ and $N = 3$ (bottom), 4, 5, showing the effects of increased concentration to the equatorial plane.

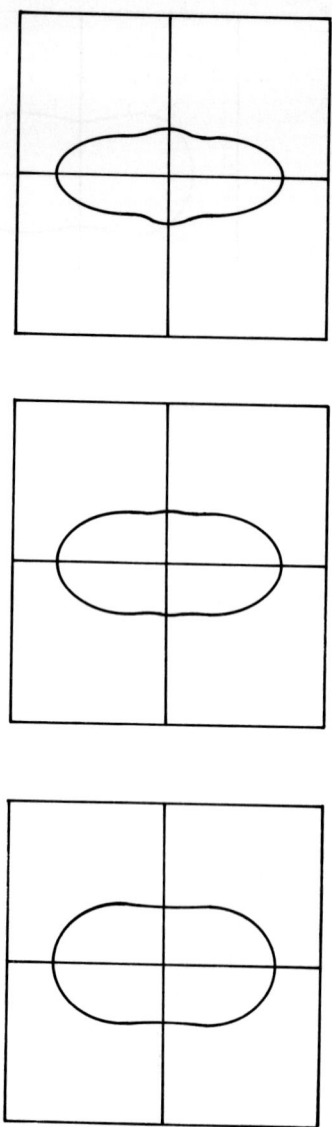

Figure 4. Shape of the nebula for $\lambda = 3$, $n = 2$ and $\varepsilon = 3$ (bottom), 6 and 18 showing the effect of increased departure from spherical symmetry.

ionization fronts have to be well shielded from the ionizing flux. This second condition applies whenever the timescale of a motion is much longer than the recombination time in the HII region. The properties of such waves are then found from the usual equations of motion, and the condition of (almost) perfect shielding; a wave propagating parallel to the x-axis has the space-time dependence

$$\cos mz \exp i (kx - \omega t), \tag{25}$$

with ω, k and m satisfying the relations

$$\frac{\omega^2}{c_i^2} = k^2 + m^2 , \tag{26}$$

and

$$\frac{m}{\bar{z}(k^2+m^2)} = \frac{1 - \exp\{i(m\bar{z}+\alpha)\}}{m\bar{z} + \alpha} \frac{1 - \exp\{-i(m\bar{z}-\alpha)\}}{m\bar{z} - \alpha} , \tag{27}$$

where $\alpha \equiv k\bar{z} \tan \beta$.

Relation (26) just relates the frequency of the waves to the wavevector; relation (27) expresses the condition of no Ly c flux at the ionization front. Possible variations of the transmitted flux arise from the changing shape of the interface of the HII layer with the HSSW, and from internal changes in electron density during the passage of the wave. Their combined effects must cancel. The angular frequency ω for such waves is in general complex, with imaginary parts (i.e. growth or decay rates) of order $c_i/\bar{z}$. Instabilities occur for all values of the wave number k, and the growth time is typically the sound crossing time of the HII layer, in the interesting cases therefore smaller by an order of magnitude, or more, than the dynamical time scale of the PN.

In favourable cases the unstable waves will grow to such large amplitudes that turbulence occurs and mixing takes place at the interface between the HSSW and the HII layer. The temperature of the hot gas is sharply reduced by this dilution with cooler gas, but no thermal energy is lost until radiative cooling becomes significant. For electron density n_e the cooling time is

$$t_c = \frac{3kT}{\Lambda(T)n_e} \tag{28}$$

where $\Lambda(T)$ is the usual cooling function (see e.g. Dalgarno and McCray 1972). The electron density can be expressed in terms of the pressure P, and for a PN, in which $\lambda = 3$, it follows that

$$t_c = \frac{54\pi(kT)^2 ut^2}{\Lambda(T) \dot{M}} \tag{29}$$

At a representative time, say 10^{11} s, and with the usual parameters, the cooling time equals the dynamical time provided that the mixing of

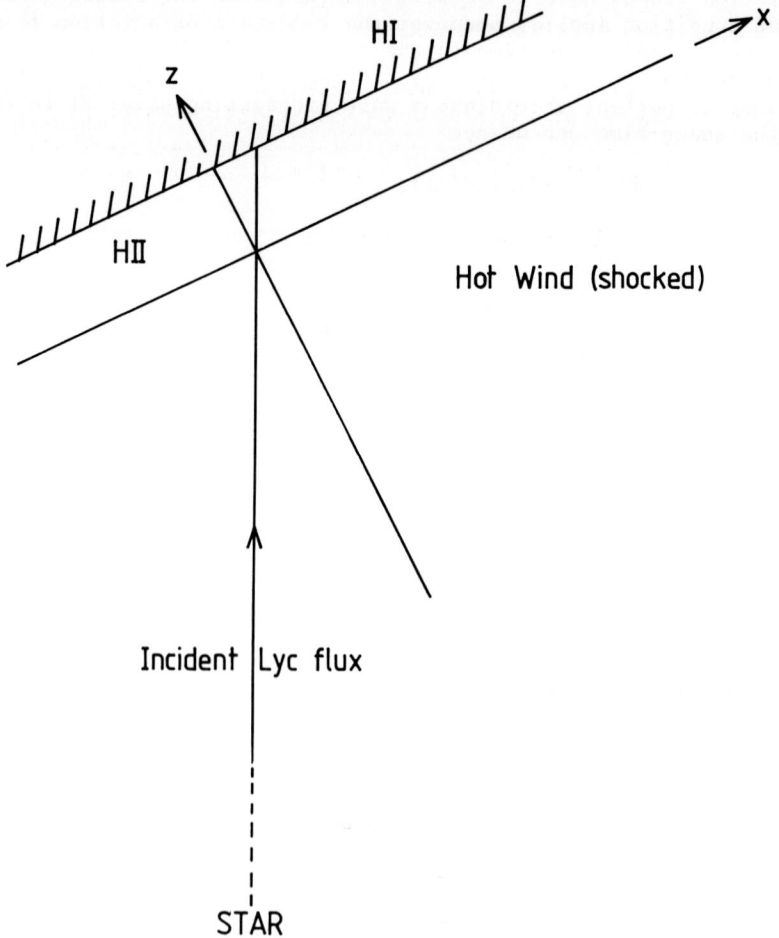

Figure 5. The setting for an unstable sound wave in the HII layer. The Ly c flux from the star is incident at angle β to the normal of the interface.

the gas has reduced the temperature to about 10^6K. The gas begins to cool well at that temperature and should emit photons with typical energies in the range of a few hundred eV.

An absolute upper limit is set to the rate at which such mixing can occur by the condition that the gas in the HII layer cannot flow at a speed faster than c_i. At radius r the maximum rate of addition of mass is therefore given by

$$\dot{M}_{add} < 4\pi r^2 P/c_i$$

$$= \frac{(\lambda-1)^2}{\lambda^2} \frac{\dot{M}r^2}{uc_i t^2} = (\lambda - 1)^2 \frac{\dot{M}u}{c_i} \qquad (30)$$

and the coefficient $(\lambda - 1)^2$ equals 4 when $\lambda = 3$. The actual rate of addition of gas from the HII layer will be $\varepsilon(\lambda - 1)^2 \dot{M}u/c_i$, where ε is an efficiency factor. The mass in the HSSW increases at rate $2L_w/V^2$ and the mixture therefore settles to a temperature given by

$$\frac{kT}{m} = \frac{2}{9} \frac{L_w c_i}{\varepsilon(\lambda-1)^2 \dot{M}u} = \frac{\lambda}{3\varepsilon} cu_i \qquad (31)$$

after some reduction. Inserting standard values for parameters here leads to the conclusion that the HSSW can be cooled to 10^6K by the mixing process, provided that the efficiency factor ε exceeds 0.01. If this does occur it will be followed by a catastrophic loss of pressure. There are two further consequences: the drop in pressure allows the HII layer to thicken and the instability rate will drop, and there will be large scale motions in the HSSW to compensate for the pressure imbalance. The second of these conclusions is interesting in connection with the observations reported by Lopez et al (1988) of high velocity components, with speeds up to 130 km/s, in NGC 2899.

6. CONCLUSIONS

The multiple winds model continues to provide valuable insight into the nature of PNe. In its simplest form it deals with objects having spherical symmetry and leads to the definition of interesting length and time scales. Real PNe will differ in various respects from the basic models. Some simple modifications have been discussed in this paper, and it has been argued that they can lead to a better understanding of the properties of PNe that are actually observed.

REFERENCES

Balick, B.: 1987, Astron. J., 94, 671.
          1988, Paper given at this Symposium.
Dalgarno, A. and McCray, R. A.: 1972, Ann. Rev. Astron. Astrophys., 10, 375.
Kahn, F. D.: 1983, Planetary Nebulae, I.A.U. Symp. No. 103, p.305, ed. Flower, D. R., Reidel, Dordrecht, Holland.
Kahn, F. D. and Breitschwerdt, D.: 1988, in preparation.
Kahn, F. D. and West, K. A.: 1985, Mon. Not. R. astr. Soc., 212, 837.
Kwok, Sun: 1982, Astrophys. J., 258, 280.
          1983, I.A.U. Symp. No. 103, p.293.
          1988, Paper given at this Symposium.
López, J. A., Falcón, L. H., Ruiz, M. T. and Roth, M.: 1988, Poster paper at this Symposium.

Schönberner, D.: 1982, Astrophys. J., 272, 708.
1986, Astron. Astrophys., 169, 189.
1988, Paper given at this Symposium.

STELLAR EVOLUTION AND THE PLANETARY NEBULA FORMATION RATE

J.P. PHILLIPS
Physics Department, Queen Mary College,
Mile End Road, London, E1 4NS,
England

1. INTRODUCTION

A knowledge of the formation rate of planetary nebulae is crucial to our understanding of a wide range of physical environments and processes, and as such has been fitfully investigated over the last forty or so years. The elemental composition of the interstellar medium, for instance, and in particular the seeding of the IS medium by CNO and s-process materials, turns out to be rather sensitively dependent upon the local rate of PN formation $\chi(PN)$ (Salpeter, 1978, Pottasch, 1984; Tinsley, 1978; and Wood et al, 1983). Similarly, and to varying degrees, the ionisation balance of the I.S. medium (cf. Pottasch, 1984; Salpeter, 1978), composition and number density of I.S. grains (Natta and Panagia, 1981), and ultimately the galactic star formation rate (cf. Miller and Scalo, 1979) can all (more or less directly) trace a dependency upon $\chi(PN)$. Finally, $\chi(PN)$ has proven historically important in establishing the evolutionary status of PN (cf. Abell and Goldreich, 1966), and is expected to be directly related to the rate of formation of white dwarfs, the death-rate of AGB-type stars (Mira variables, OH/IR sources), and the main-sequence turn-off rate in the mass range $2 \stackrel{<}{\sim} M/M_\odot \stackrel{<}{\sim} 10$, some 10^{10} years ago.

It is clear therefore that, be it ever so insecure, the value of $\chi(PN)$ has a bearing upon the broad range of important astrophysical questions. In the following, we shall briefly consider the procedures whereby $\chi(PN)$ is determined, together with the vagaries inherent in such estimates. We shall also review the full range of current estimates for $\chi(PN)$ together with $\chi_G(PN)$, the galactic formation rate, and suggest best current estimates for these parameters. Finally, we provide a brief survey of χ for various related stages of pre- and post-planetary evolution, and their relation to $\chi(PN)$.

2. DETERMINATION OF $\chi(PN)$: METHODS AND UNCERTAINTIES

The evaluation of $\chi(PN)$ is in principle straightforward, and for the optically thin regime of expansion may be determined through general expressions of form

$$\chi(PN) \simeq 1.02 \cdot 10^{-15} \left\{ \frac{\sigma_o(R)}{kpc^{-3}pc^{-1}} \right\} \left\{ \frac{V_{exp}(R)}{km \cdot sec^{-1}} \right\} pc^{-3} yr^{-1} \quad (1)$$

where $\sigma_o(R)$ is the local number density of PN per unit radius about R, and $V_{exp}(R)$ is the corresponding mean nebular expansion velocity. For a variation

$$\sigma(R, Z) = \sigma_o(R) e^{-kz} \; kpc^{-3} \; pc^{-1} \quad (2)$$

with respect to height z above the galactic plane, then the local density may in turn be estimated through

$$\sigma_o(R) = N(D, R) \left\{ 2\pi \left[\frac{D^2}{k} - \frac{2}{k^3} + e^{-kD} \left(\frac{2D}{k^2} + \frac{2}{k^3} \right) \right] \right\}$$

where N(D, R) represents the number of nebulae within unit radius of R, and within a given distance D of the sun. A summary of values χ(PN) determined through this, or similar procedures is outlined in table 1, together with optically thin PN number densities σ_o^*, galactic formation rates $\chi_G(PN)$, and total cumulative galactic number N_T.

Whilst the methodology is therefore simply stated, there are several uncertainties that require careful consideration:

(a) <u>The value of D</u>

It is clearly necessary, in determining $\sigma_o(R)$, to obtain a reasonably complete sample of nebulae for any particular radius R. This is increasingly difficult at large D, however, due to the effects of galactic IS absorption, which will preferentially obscure nebulae having large R (and low surface brightness). The consequences of this have been illustrated by Phillips (1984) and Daub (1982), from which it is apparent that N(D, R) is a very much steeper function of R when D is >1 kpc, than appears to be the case where D < 1 kpc.

Two primary methods have been outlined to overcome this difficulty. In the first case (cf. Cahn and Kaler (1971)), the variation of N(D, R) with R is determined for a variety of (decreasing) distances D, until the trend appears reasonably invariant, and consistent with current nebular expansion models (thus, Cahn and Kaler adopt the Seaton (1966) evolutionary model in explaining a deficit of sources at radii R ≳ 0.4 pc, whilst Phillips (1984) evaluates N(R) on the basis of an empirical function $V_{exp}(R)$. Both of these procedures have defects). Alternatively, $\sigma_o(R, D)$ may be evaluated for a variety of values R, and increasing D, until incomplete sampling causes $\sigma_o(R)$ to decrease. This

TABLE 1

PLANETARY NEBULAE FORMATION RATES

Reference	$\chi(PN)$ $10^{-12}pc^{-3}$ yr^{-1}	N_T $\times 10^4$	$^+\chi_G(PN)$ yr^{-1}	σ_0 kpc^{-3}	k kpc^{-1}	PN Lifetime yrs	Distance Scale
Paranego (1946)	-	.6-1.0	0.3-0.5	-	5.08	-	Own
Vorontsov and Velyaminov (1950)	-	10	5	-	-	-	Own
Minkowski (1956)	-	-	-	-	-	-	-
O'Dell (1962)	-	6	3	-	-	-	Own
Abell & Goldreich (1966)	0.4	4.8	2.4	-	-	$2.0.10^4$	Own
Seaton (1966)	-	0.5-5.0	-	-	-	$1.5.10^4$	Own
Cahn (1968)	0.1	-	-	12	-	-	Seaton (1966)
O'Dell (1968)	1.4-0.5	-	-	14	-	$3.5.10^4$	Own
Perek (1968)	.39-.46	-	-	30	-	-	Various
Seaton (1968)	-	-	-	-	-	-	Own
Gurzadyan (1969)	2.0	-	-	-	-	-	-
Cahn & Kaler (1971)	-	-	1-10	-	-	-	-
O'Dell (1971)	3.2	29-43	42	40-54	11.1	$1.6.10^4$	Seaton (1968)
Osterbrock (1973)	0.4	-	14	-	-	-	Seaton (1968)
Osterbrock (1974)	2.5	-	-	-	8.0	-	-
Alloin et al (1976)	3.1	-	-	-	6.25	-	Cudworth (1974)
	.63-3.1	1→2.25	0.5-1.1	-	-	-	/Seaton (1968)
Cahn & Wyatt (1976)	5.1±1	3.8±1.2	1.9	80±15	8.70	$1.6.10^4$	Seaton (1968)
Smith (1976)	11±3	-	-	150	10.0	3.10^4	Seaton (1968)
Weidemann (1977)	1.8-2.6	-	-	-	-	$1.6.10^4$	Own
Cahn & Wyatt (1978)	-	1.35-4.5	.3-.7	-	8.70	-	Cudworth (1974)
Acker (1978)	3.0	2.5	1.25	48	5.0	$1.6.10^4$	Own
Tinsley (1978)	1.2-4.8	-	-	-	8.0*	-	Cudworth/Seaton
Jacoby (1980)	-	1±.4	0.5	-	-	-	-
Maciel (1981)	2.0	2.1	1.0	41	6.94	2.10^4	Own
Mallik (1982)	2.4±.2	2.8	1.4	44±4	-	-	Own
Daub (1982)†¹	5±2	1.4	0.7	55	8.0	$4.8.10^3$	Own
Isaacman (1983)	2.9	2.1	1.0	-	-	-	-
Phillips (1984)	4.4±1.5	-	-	-	8.0	-	Daub (1982)
Pottasch (1984)	1.5	0.2-0.5	0.1-0.3	50	4.0	-	-
Amnuel et al (1984)	4.6±2.2	4.0	1-2	117±49	7.69	$1.6.10^4$	Own
Ishida & Weinberger (1987)	8.0	14	7.0	326	10.0	4.10^4	Various
Phillips (1987)†²	2.39±.32	2.96	1.31±.18	90±6	6.7	-	Own

* Assumed value of k $^+\chi_G$ determined from $N_T/(2.10^4$ years), unless otherwise
specified.   †¹ σ_0* determined for optically thin and thick nebulae;
†² This review.

procedure has been discussed in some detail by Daub (1982), and it is clear that D, in consequence, may become a variable function of R. As a result of the application of this procedure, however, Daub (1982) finds a severe increase in $\sigma_0(R, D)$ for radii R > .24 pc; a result which is not readily explicable in terms of most current models of nebular expansion or, indeed, of observed velocity trends (see Section 2e).

Finally, several investigators have adopted the simpler recourse of taking a low limiting distance D < 1 kpc (cf. Ishida and Weinberger (1987), who chose D = 0.5 kpc), or have attempted to correct for the effects of IS absorption using plausible models of the galactic dust distribution (cf. Smith, 1976).

(b) <u>Distribution with height z above the Galactic Plane</u>

The adoption of an exponential law for σ_0 (expression 2) is in many ways no more than an analytical convenience; it is clear that such a trend can only very approximately represent the true z-distribution of PN. Nevertheless, various investigators have shown that such an expression represents a reasonable working assumption (cf. Daub, 1982), and one that is unlikely to lead to gross errors in $\chi(PN)$. The value of the scaling factor k, on the other hand, has varied widely over the years, and such uncertainties feed directly through to χ - although most estimates appear to cluster about $\sim 7-8 \mathrm{kpc}^{-1}$.

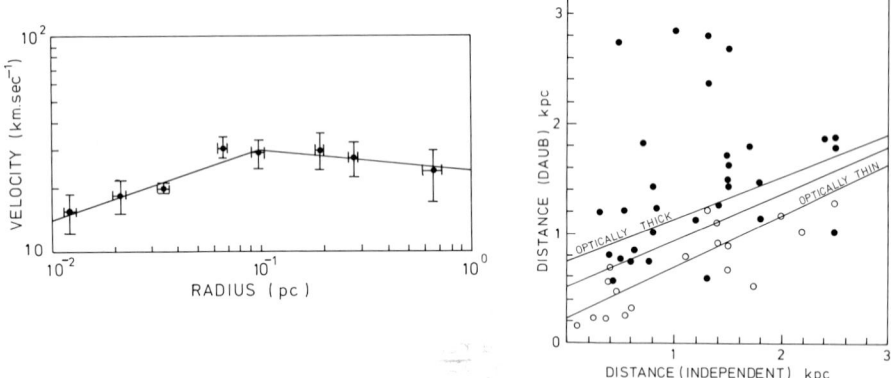

Figure 1. Variation of mean nebular expansion velocity with radius (see text for details). The solid lines indicate the trends for δ = 0.3 (R < .1 pc) and δ = -0.1.

Figure 2. Comparison of Daub (1982) distances with values based upon model independant measures, where the upper line represents a least-squares regression analysis for optically thick sources ($\bullet$; correlation coefficient $r_c \sim .37$), and the lower line corresponds to the optically thin results ($\circ$: $r_c \sim .76$). The least-squares fit to the entire data set is indicated by the central line.

In this regard, the values $K \lesssim 5 \mathrm{kpc}^{-1}$ adopted by Acker (1978) and Pottasch (1984) appear aberrant, and are unlikely to be consistent with the PN progenitor mass distribution (see later).

(c) The Distance Scale

Earlier estimates of χ depended upon a variety of distance estimates, ranging from the early Shklovsky (1956) scale through to the analyses of O'Dell (1962), Seaton (1966, 1968) and Cudworth (1974). All of these predicated upon some assumed invariant nebular property (usually ionised mass) and they have all, for this reason, been subsequently found wanting. More recently, Pottasch (1980) has noted that if such assumptions are side-stepped, and distances based on extinction trends and the like are employed, then a steeply varying trend of nebular mass with radius is deduced for $R \lesssim 0.1$ pc. This analysis has subsequently been expanded and generalised by Daub (1982) and Maciel and Pottasch (1980), and forms the basis of what must be regarded as the more reliable, recent estimates of $\chi(PN)$.

Having noted this, however, the current reliability of such scales is by no means so great as to leave us sanguine, and Ishida and Weinberger (1987) and Mallik (1985) have recently emphasized the extreme sensitivity of $\chi(PN)$ to any such uncertainties - for a distance scaling parameter κ, then in general $\chi(PN) \propto \kappa^4$.

Given the range of scales espoused by Phillips and Pottasch (1984) and Cudworth (1974), say, then it is clear that $\chi(PN)$ may potentially vary by a factor ~ 10.

The degree of uncertainty is further illustrated in figure 2, where we compare the Daub distances with a range of 'model independent' values based upon extinction trends, HI absorption measurements, kinematic parallax, and central star gravities (cf. Pottasch (1983, 1984) and Gathier et al (1986 a, b); see also the summaries of Sabbadin (1986) and Phillips (1984)). Apart from the obviously poor correlation between the respective distance scales (the correlation coefficient r_c is ~ 0.6) and the disparate trends between optically thick and thin nebulae, the scaling factor $\kappa \sim$ < d (independent) >/< d (Daub) > $\sim .84$ would imply a reduction of the Daub/Phillips formation rates by a factor ~ 2.0, to of order $\chi(PN) \sim (2.2 \rightarrow 2.5).10^{12}$ pc^{-3} yr^{-1}, together with a revised scale height $k \sim 6.7$ kpc^{-1}.

(d) Influence of Progenitor Mass

The rate of PN formation is usually calculated on the assumption that central star masses are similar for all PN - with the consequence that the rate of PN evolution is also invariant. In fact, this is far from being the case, and it is clear that whilst central stars having masses $M_{NPN} \sim .55 M_\odot$ evolve rather slowly within the HR plane, a relatively small change in M_{NPN} to $\sim .6 M_\odot$, say, is likely to result in (i) a 20-fold secular increase in evolution within the L-T_{eff} plane (cf. Schonberner, 1981, 1983); (ii) a grossly enhanced envelope mass (cf. Weidemann, 1987; Weidemann and Koester,1983); and (iii) an appreciable alteration in envelope C, He, and N abundances, reflecting the influence of convective-zone dredging of CNO-process elements (Renzini and Voli, 1981). Higher mass (population I) sources appear also to be characterised by distinctive morphologies and high expansion velocities (cf. Peimbert and Torres-Peimbert, 1982), and it is clear

that the straightforward application of expression (1) must be viewed
with caution. Similarly, it may be noted that the differing
progenitor masses would imply appreciably differing scale heights
$z_0 (\equiv k^{-1})$, and the dependency of velocity dispersion and nebula type
upon z_0 is by now reasonably well established (cf. Cudworth, 1974;
Amnuel et al, 1984). A significant dispersion of this kind is also to
be found in the progenitor OH (IRAS, Mira, and OH/IR) sources (cf.
Herman and Habing, 1985), and any accurate assessment of σ_0 should
therefore fully take account of this hierarchy of central star masses.
Having said this, however, the available information does not encourage
us to believe that such a detailed analysis is yet possible. Similarly,
the sharply peaked nature of the central star mass-distribution (cf.
Schonberner, 1981) implies that errors arising from a single mass
analysis are likely to be low.

(e) Expansion Velocities

The value of $V_{exp}(R)$ is usually taken to be constant, and of order
20 km.sec^{-1} in the optically thin regime ($R \lesssim .1$ pc). Larger values
have on occasion been employed, however, and these account for at
least some of the variation in $\chi(PN)$ to be noted in table 1.

It is clear, from this, that although $\chi(PN)$ may depend rather
sensitively upon V_{exp}, most previous investigations have been happy to
adopt a rather hand waving estimate of this parameter. Phillips (1984)
has investigated this question in a little more detail, by establishing
an empirical relation of form $V(R) = V_0 R^\delta$, although even for this case
the relation is strongly biassed towards optically thick nebulae, for
which dR/dt may take a value appreciably at variance with V_{exp}, and of
order $\sim V_I$ (the velocity of the ionisation front). This disparity may
also account for the apparent inability of such velocity trends to
replicate the sharp rise in $N(R)$ ($\propto \chi/dR/dt$)) for low values of R (see
later).

For the purposes of this review we have updated this analysis to
include a range of new and improved expansion velocities due,
primarily, to Sabbadin et al (1983, 1984, 1985 and 1986) and Sabbadin
(1984 a, b; 1986), together with earlier data referenced in Phillips
(1984). Combining these with the model independent distance estimates
referenced in Section 2c, we then obtain results which are closely
similar to those of the earlier analysis, implying $\delta \sim 0.3$ for
$R \lesssim 1$ pc (see figure 1). In the $R > 0.1$ pc optically thin regime, on
the other hand, there is evidence for a peaking of velocities close to
$V_{exp} \sim 30$ km.sec^{-1}, followed by a more-or-less gradual decline to
higher radii.

A mean for the optically thin nebulae in figure 1 yields the
expansion velocity $\bar{V}_{exp} \sim 26.1 \pm 2.9$ km.sec^{-1}, whilst a more balanced
weighting with respect to the complete sample of nebulae in figure 3
would imply $\bar{V}_{exp}$ ($0.1 < R < 0.6$ pc) 26.0 km.sec^{-1} for optically thin
sources, and $\bar{V}_{exp} \sim 24.5$ km.sec^{-1} for all nebulae having $R < 0.6$ pc.

(f) Sample Completeness

Finally, recent investigations of large ($R \gtrsim .5$ pc), previously unrecognised, and/or high galactic latitude nebulae are increasing the corpus of known PN close to the sun; examples of such surveys include the recent analysis by Ishida and Weinberger (1987), and the careful discussion of Pottasch (1984). Nevertheless, perhaps of order $\sim 25\%$ of low surface brightness nebulae possess neither independantly estimated distances, nor integral Hβ fluxes with which to evaluate Shklovsky distances, and this results in a rather sharp cut-off in $\sigma(R)$ for $R > 0.6$ pc (see later).

Finally, we may remark that question of sample completeness also arises for low values of R, where nebulae are likely to be unresolved, and possibly mis-identified or unrecognised. Where a mean value $\sigma_0(R) \approx \int_{\Delta R} \sigma_0(R) \, dR/\Delta R$ is determined for the full range of nebulae (as in, say, Daub (1982) or Ishida and Weinberger (1987)) then it is evident that $\chi(PN)$ may be correspondingly underestimated. However, the trends for local compact nebulae (cf. figure 3) suggest, on the contrary, that the values of σ_0 ($R \lesssim .1$ pc) are significantly in excess of the mean, and any further enhancement would compound an already puzzling trend.

Given all these caveats and uncertainties, it is perhaps not surprising to find that estimates of $\chi(PN)$ have varied appreciably over the years, with more recent estimates extending through the range $\sim (1\rightarrow 4).10^{-12}$ pc^{-3} yr^{-1}, implying a mean $\sim 3.4.10^{-12}$ pc^{-3} yr^{-1}. Ishida and Weinberg (1987) have recently attempted to buck this trend, however, and follow Smith (1976) in deriving a high value $\chi(PN) \sim 8.10^{-12}$ pc^{-3} yr^{-1}; an estimate which, if true, would be not a little perplexing, implying a ratio $\chi(WD)/\chi(PN) \lesssim 0.25$. A re-analysis of the Ishida and Weinberger data base, however, suggests that for $R \lesssim 0.4$ pc and $k = 10$ kpc^{-1}, then σ_0^{*} takes a value $\sim 83.$ kpc^{-3}, and $\chi(PN)$ is of order $\sim 4.3.10^{-12}$ pc^{-3} yr^{-1}. Similar values are also obtained for $R < 0.6$ pc. We have not allowed for a N-S disparity in PN numbers for this more limited range of radii, a feature which increases the $R < 0.8$ pc sample analysed by Ishida and Weinberger by a factor 1.4. Similarly, we express some surprise at the apparent disparity in volume number densities between samples having $D \lesssim 1$ kpc (see later), and $D \lesssim 0.5$ kpc; a difference which is unlikely to be entirely due to extinction.

Taken as a whole, therefore, it is apparent that whilst such high values of $\chi(PN)$ can probably be discounted, the derived range in $\chi(PN)$ remains uncomfortably broad - and somewhat at variance with the optimistic prognoses of Weidemann (1977, 1978). Nevertheless if a mean of certain of the more recent and realistic estimates of $\chi(PN)$ are selected, then we find a value $\sim (2.60 \pm .24).10^{-12}$ pc^{-3} yr^{-1} which is tolerably consistent with earlier values of $\chi(WD)$ (see Section 4a). Alternatively, model independant distances and radii (see section 2(c) and (e)) have now been determined for the larger fraction of nebulae having $D < 1$ kpc; and where these are not available (in general for the larger ($R \gtrsim .5$ pc) optically thin sources) then Shklovsky distances may be derived. The results, in the form of the trend of $\sigma_0(R)$ with R are illustrated in figure 3, where sampling for $R > .5$ pc is clearly increasingly incomplete (a full description of this analysis will be provided in Phillips (1987)). The comparative function $T_{\sigma_0}(R) \propto \chi(PN)/V_{exp}(R)$, based upon the velocity trends in figure 1 and scaled to match the observed integral of sources for $R \lesssim 0.6$ pc, appears to reproduce the observed trend for optically thin sources tolerably well. At shorter radii, on the other hand, it is clear that there is a strong peaking in $\sigma_0(R)$ for $R \lesssim 0.05$ pc, a feature which it is impossible to replicate given the observed velocity trends.

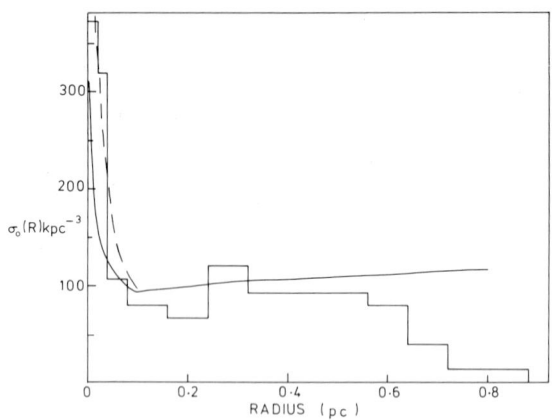

Figure 3. Variation of $\sigma_0(R)$ for sources within $D = 1$ kpc of the sun. The solid curve (fitted to the $R < 0.6$ pc results) represents the trend expected for the velocities in figure 1, whilst the dashed curve indicates the possible consequence of ionisation fronts.

In the optically thick expansion regime, on the other hand, it is likely that $\sigma_0(R) \propto \chi(PN)/(dR/dt) \propto V_I^{-1}$, where V_I is the velocity of the ionisation front. If for illustrative purposes we therefore assume the secular relations

$$V_G = V_{GO} (t/t_o)^\alpha \qquad V_I = V_{IO} (t/t_o)^\beta$$

where V_G is the gas expansion velocity ($\equiv V_{exp}$), then we determine

$$R_I = \frac{V_{IO} t_o}{\beta + 1} \left(\frac{V_G}{V_{GO}} \right)^{\frac{\beta + 1}{\alpha}}$$

$$\sigma_o(R) = \frac{\chi(PN)}{V_{IO}} \left\{ \frac{(\beta + 1) R_I}{V_{IO} t_o} \right\}^{\frac{-\beta}{\beta + 1}}$$

Fitting these functions to the observed trends in figures 1 and 3 then yields the solutions $\alpha = 2.75$, $\beta = 7.33$, and the trends in $\sigma(R)$ indicated by the dashed curve in figure 3.

It is clear, in short, that for such an hypothesis to reproduce the observed trends in $\sigma_o(R)$ requires an extremely rapid variation in V_I, presumably preceded by a more benign phase of I-front development (unless local densities are to become extraordinarily high). Alternatively, of course, it is possible that our zeal in identifying compact nebulae has resulted in a spate of mis-identifications, or that the distances to these sources have for some reason been grossly undervalued.

In any case it is apparent that by taking the optically thin results alone ($0.1 \leq R \leq 0.6$ pc), and adopting a value $\bar{V}_{exp} \sim 26$ km.sec^{-1} (Section 2e), then we find a reasonably invariant $\sigma_o(R) \stackrel{\sim}{=} 89.6 \pm 6.4$ kpc^{-3} pc^{-1}, $g_o \stackrel{\sim}{=} 44.8$ kpc^{-3}, and a formation rate $\chi(PN) \sim (2.39 \pm 0.32) 10^{-12}$ pc^{-3} yr^{-1}.

3. GALACTIC FORMATION RATE

The surface density of PN throughout the galaxy is frequently characterised by a function

$$\mu(R_G) = \mu_o e^{-k_G R_G}$$

where R_G is the galactocentric radius and μ_o the central density of PN; a trend which also appears to be representative of kinematic, density, and surface brightness trends in other galaxies (cf. de Vaucouleurs, 1959; Freeman, 1979; Toomre, 1972, and Cruz-Gonzalez, 1974). By scaling this function through comparison with $\mu(R_o)$ (the local surface density) it is then possible to obtain an estimate for N_T, the total number of nebulae in our galaxy (cf. Cahn and Kaler, 1971). Earlier attempts to estimate k_G utilised the observed galactic radial decrement in σ_o near the sun, yielding values $k_G \sim 0.9$ kpc^{-1} which would have implied phenomenal galactic centre densities, and total numbers $N_T \sim 2.9 \rightarrow 4.3 \cdot 10^5$ - and in turn suggesting, for a typical expansion period $\sim 2.10^4$ years, a formation rate $\chi(PN) \sim 20$ PN.year^{-1}. More recent analyses have resulted in a down grading in k_G to values ~ 0.3 kpc^{-1}, however, more clearly relating to the distribution of F-M type stars and PN number densities between $R_G = 5$ and 10 kpc (cf. Alloin et al, 1976; Cahn and Wyatt, 1978; and Amnuel et al, 1984), and this has resulted in a commensurate reduction in N_T to $\sim 10^4$. Other procedures employed to estimate N_T include (a) estimates of the local mass specific number density κ_m in the range $1.1 \rightarrow 2.5 \; 10^{-7} \; M_\odot^{-1}$, a value which is presumed to apply galaxy-wide. By taking an overall galactic mass $M_G \sim 1.3 - 1.5 . 10^{11} \; M_\odot$, therefore, it follows that N_T must reside in the range $(1.3 \rightarrow 3.8).10^4$. There are several problems with such an analysis, the more serious of which concerns the presumed invariance of κ_m - such a value will certainly not apply in the galactic halo, for instance. Nevertheless, values of κ_m determined for the galactic centre PN (Isaacman, 1981), and planetary nebulae in the (presumably similar) galaxy M31 (Jacoby, 1980), appear to confirm the widespread viability of such procedures.

(b) Jacoby (1980) has investigated the luminosity function of PN in the Magellanic clouds, and concludes that the trends are well represented by a simple model of nebular expansion due to Henize and Westerlund (1963). Under these circumstances, he argues that luminosity functions are likely to be similar in other galaxies - the effects of differing progenitor mass distribution, of chemical composition and nebular expansion velocities being merely to 'smear' these trends. A rather more important uncertainty may however arise from his extrapolation to fainter nebulae which, although uncertain, roughly doubles the presumed number of PN.

A specific mass number density for M31 ($\sim 1.1.10^{-7} \; M_\odot^{-1}$) is thereby found which appears closely similar to that of our own galaxy - although Jacoby prefers to use the M31 luminosity specific density $(6.1 \pm 2.2) .10^{-7}$ PN $L_\odot^{-1}$ to obtain a final estimate $N_T \sim (10 \pm 4).10^3$ PN.

Other, and in certain cases simpler procedures have also been employed; Cahn and Wyatt (1976), for instance, adopt the local column density of PN as typical for the entire galactic plane, and integrate over the disc area to obtain $N_T \sim 1.3.10^4$ - a procedure which, with all its faults, at least lacks the contrivance of certain more model-intensive estimates.

Despite the adoption of these varying procedures, and the wide range of σ_o^* upon which they are based, the values of N_T thus obtained (see table 1) are for the most part tolerably consistent - a mean of values over the past decade yields $N_T \sim (2.1 \pm 0.4).10^4$ PN. For the purposes of this review, these estimates have been subsequently divided by a typical expansion period 2.10^4 years to yield the galactic formation rate $\chi_G(PN)$, implying a characteristic value $\sim 1.0 \pm 0.2$ PN. year^{-1}. This may be compared with the values $N_T \sim 2.47.10^4$ (for $R < 0.6$ pc), and $\chi_G(PN) \sim 1.31 \pm .18$ yr^{-1} determined from a local formation rate $\chi(PN) \sim 2.39.10^{-12}$ pc^{-3} yr^{-1} (section 3), and assuming a galactic mass $M_G \sim 1.4.10^{11} \; M_\odot$, and $\mu_m \sim 75 \; M_\odot$ pc^{-2} (Schmidt, 1963).

Finally, and for the future, it is plain that the vast amount of FIR data collected through IRAS is likely to yield yet further estimates of χ_G. Specifically, it is clear that PN occupy a rather closely defined zone within FIR colour-colour plots, and the distribution of such sources within the galactic plane should allow a similar investigation to that of the GC radio sources (cf. Isaacman, 1983). Further details of such an analysis are provided in Phillips (1987).

4. THE PLANETARY NEBULA FORMATION RATE AND STELLAR EVOLUTION

The broad progress of stellar evolution, the behaviour of stars along the AGB branch, and the general development of these stars through the subsequent white dwarf and PN phases of evolution are by now tolerably well understood. Nevertheless, this understanding is far from complete, and several phases of both AGB and RGB evolution remain obscure and ill-defined. In this respect, values of χ for the Mira stars, the OH/IR sources and the white dwarfs not only act as a constraint upon our estimates of $\chi(PN)$, but also enable us to test the viability (or otherwise) of current models of stellar evolution - the proposed rates of evolution through the HR diagram, the assumed relation between PN, progenitors, and post-PN phases, and (through comparison with main-sequence turnoff) the variation of the star-formation rate over the last 10^{10} years, together with the presumed PN progenitor mass range. In the following, we briefly review our current understanding of these issues.

(a) White Dwarf Formation Rate

A summary of values for the white dwarf formation rate is provided in table 2, whence it is seen that prior to 1986, a consensus had been reached whereby $\chi(WD)$ varied between $1.4 \cdot 10^{-12}$ pc^{-3} yr^{-1} and $\sim 2.6 \cdot 10^{-12}$ pc^{-3} yr^{-1}, suggesting a plausible average $\sim 2 \cdot 10^{-12}$ pc^{-3} yr^{-1}. Given that all PN ultimately evolve into white dwarfs, it was apparent (within errors) that $\chi(PN)$ would be of order similar to $\chi(WD)$. Several recent developments in our understanding of stellar evolution, PN central star (M_{NPN}) and white dwarf (M_{WD}) mass distributions, and the value of $\chi(WD)$, however, have somewhat muddied these waters. In the first place, estimates of M_{NPN} based on the location of PN with respect to evolutionary tracks (cf. Schonberner and Weidemann, 1983) suggest a trailing-off in numbers to higher masses, but an extraordinarily sharp cut-off at $M_{NPN} \sim 0.55 M_\odot$. This compares with a distribution in M_{WD} which appears to be symmetrical, and peaked at $M_{WD} \sim 0.55 M_\odot$.

A description of the underlying reasons for this trend has been outlined by Drilling Schonberner (1985), whereby it is clear that $\sim 2-3\%$ of stars evolving off the main sequence, and ending their lives as white dwarfs, have masses $\lesssim .53 M_\odot$ which may imply direct evolution from the horizontal branch to the white dwarf phase - the intermediate AGB-PN stage is altogether avoided (cf. Schonberner and Drilling, 1984). For another and more important fraction of stars having masses $M \sim .55 M_\odot$, constituting perhaps 5-70% of white dwarf progenitors, the evolution from the AGB to the white dwarf sequence is extremely slow, and PN shells have generally dissipated before the central star temperatures have appreciably increased, and shell ionisation takes place. For these cases, therefore, we observe no associated PN shell. Finally, the remaining 30-95% of higher mass stars evolve at a reasonably rapid rate, and proceed to the observed PN.

TABLE 2

White Dwarf Formation Rates

Reference	χ_{WD} 10^{-12} pc^{-3} yr^{-1}
Weidemann (1968b)	2.0
Weidemann (1977)	1.4
Weidemann (1978)	2.0
Liebert (1980)	1.4
Wesemael (1981)	1.4
Guseinov et al (1983)	2.5
Fleming et al (1986)	.49 - .75

What does all this entail for $\chi(WD)$, and its relation to $\chi(PN)$? Superficially, at least, if most of the white dwarfs having mass > .55 $M_\odot$ are assumed to derive from a PN progenitor, then we might expect $\chi(PN) \sim 0.6\ \chi(WD)$, and given a value $\chi(WD) \sim 2.10^{-12}$ pc^{-3}yr^{-1}, then a rate $\chi(PN) \sim 1.2.10^{-12}$ pc^{-3} yr^{-1} would be anticipated. The situation has recently, however, become even more severe, with Fleming et al (1986) re-analysing the data base of Green (1980), and spectrally re-classifying a large fraction of the previously identified white dwarf population to lower gravities. The result is an apparent lowering of $\chi(WD)$ to $\sim 4.5 - 7.5.10^{-13}$ pc^{-3}yr^{-1} which, if confirmed, would imply $\chi(PN) \sim (2.7 - 4.5).10^{-13}$pc^{-3}yr^{-1}. Such a range of values would, surprisingly, be more consistent with earlier estimates of $\chi(PN)$ (e.g. O'Dell (1968); Seaton (1966), and Cahn (1968)), and disagree with practically the entire corpus of more recent values. If we take these recent estimates to be the more reliable, therefore, it is clear that we are confronted with an interesting paradox - the rate of PN production is approximately 4→7 times greater than would be predicted from $\chi(WD)$.

In this respect, we note that the procedures for evaluating $\chi(WD)$ (which depend upon a cooling rate defined from stellar theory) are almost certainly reasonably accurate - more so, indeed, than those employed in the determination of $\chi(PN)$. Under these circumstances, one might be inclined to argue that the white dwarf statistics are at fault - perhaps many white dwarfs, for instance, represent hitherto undetected binary companions (cf. Grauer and Bond, 1983; Fleming et al, 1986). Under these circumstances, however, and given an incidence of spectroscopic binaries in PN no greater than $\sim$ 10-15% (see Bond, these proceedings), we would require between 65 and 80% of PN nuclei to contain longer period binaries. Alternatively, Kudritzki and Barlow (these proceedings) have argued in favour of the Cudworth (1974) scale of PN distances, and Weidemann (these proceedings) suggests a value $\sim 1.3\ \kappa$ (Cahn and Kaler). Either of these would result in a sharp reduction in $\chi(PN)$ to of order$(0.6 \rightarrow 1.2).10^{-12}$ pc^{-3} yr^{-1} - closer, certainly, to $\chi(WD)$, although more disparate with $\chi(Mira)$ and, on a galactic scale, $\chi_G(OH/IR)$. Similarly, such re-scaling would imply appreciable errors in the M.I. scales adopted here (section 2c), a result which would be difficult to readily comprehend.

We should finally remark that the scale height z_0 of white dwarfs appears to lie somewhere in the range 250 and 300 pc (Fleming et al (1986), Downes (1984) and Ishida et al (1982)), and is similar to that of Mira variables ($\sim$ 300 - 400 pc according to Wood and Cahn (1977) and Wyatt and Cahn (1983)). The disparity with $z_0(PN) \sim 150$ pc is readily understood if only the more massive stars proceed through a PN phase of evolution, and indeed similar arguments have been used to argue for a progenitor mass range < 2.5 $M_\odot$ (Mallik, 1985). Subsequent dynamical interactions with the galactic disc may also operate to increase $z_0(WD)$ even further (cf. Wielen, 1977; Koester and Wiedemann, 1980). The more massive Miras are also expected to translate through the OH/IR phase prior to becoming PN (cf. Habing 1987; Kwok 1987), and it is therefore gratifying to note that the mean height $|z(OH/IR)|$ $\sim$ 56-194 pc is indeed similar to that of PN (Herman and Habing, 1985; the higher values corresponding to a ZAMS mass $\sim$ 1.6 $M_\odot$, and shell expansion velocities < 15 km.sec^{-1}, whilst the lowest mean height appears consistent with $M(ZAMS) \sim 8 M_\odot$, and $V_e > 28$ km.sec^{-1}). The distribution of z_0 with stellar type is therefore consistent with only a fractional post-AGB evolution through the PN phase (ignoring, of course, those that subsequently chose to become supernovae, in the progenitor range $M \gtrsim 10 M_\odot$), and confirms the expected low value of $\chi(PN)$ with respect to $\chi(WD)$.

(b) <u>Rate of Evolution of OH/IR stars</u>

The evolution of Mira variables along the AGB is ultimately believed to result in a change in pulsation mode, from first overtone to fundamental. Whilst the origins and consequences of this behaviour are far from being well understood, it is believed that the resulting, pulsationally driven mass loss leads to mass-loss rates $\sim 10^{-4} M_\odot.yr^{-1}$ over a period of between < 10^3 and (for low mass sources) 10^5 years, giving rise to an extensive, opaque dusty thermosphere. Recent analysis of the statistics of such sources by, in particular, Hermann and Habing (1985), implies a galaxy wide formation rate $\chi_G(OH/IR) \sim 0.9$ yr^{-1}. Given that the larger part of these sources are expected to become PN, it is clear that $\chi_G(OH/IR)$ is in tolerable agreement with the best current estimate of $\chi_G(PN)$ (see section 3).

(c) <u>Estimates of χ(Mira)</u>

Wood and Cahn (1977) have determined a value χ(Mira) of order 3.10^{-13} pc^{-3} yr^{-1}, using a scale height z_0(Mira) of 314 pc, and implying a local space density $\sim$ 245 kpc^{-3}. In the case of these sources, the rate of evolution is by no means so well established, and Willson (1981) for instance takes account of mass loss to evaluate a very much higher figure $\sim 3.10^{-12}$ pc^{-3} yr^{-1}. Given that only $\sim$ 60% of Miras are expected to result in PN, then very approximately we might expect $\chi(PN) \sim 0.6 \chi(Mira) \sim 1.8.10^{-12}$ pc^{-3}, in tolerable conformity with our earlier derived values. Note also that Wyatt and

Cahn (1983) find a local density ~ 200 kpc^{-3}, whilst Oort van Tulden (1942) determine a value $\sim 10^2$ kpc^{-3} - estimates which may imply somewhat lower values of χ(Mira).

(d) <u>Main-Sequence Turn-off Rate</u>

Finally, we note that several attempts have been made to relate the rate of main-sequence turn-off some $5-10.10^9$ years ago with the current density of degenerate and PN central stars. Koester and Wiedemann (1980) for instance find it possible to match degenerate star number densities and mass distributions providing the SFR varies slowly ($\propto e^{-t/5.10^9}$ yr, for instance), and either a Salpeter (1955) or Larson and Tinsley (1978) IMF is adopted, whilst Miller and Scalo (1979) also require (from a less detailed comparison) that the SFR varies only slowly, and by less than a factor ~ 2 over the last 10^{10} years. Cahn and Wyatt (1976) deduce a PN formation rate $\sim 2.3.10^{-12}$ pc^{-3} yr^{-1} which is largely independent of the assumed galactic age, or indeed

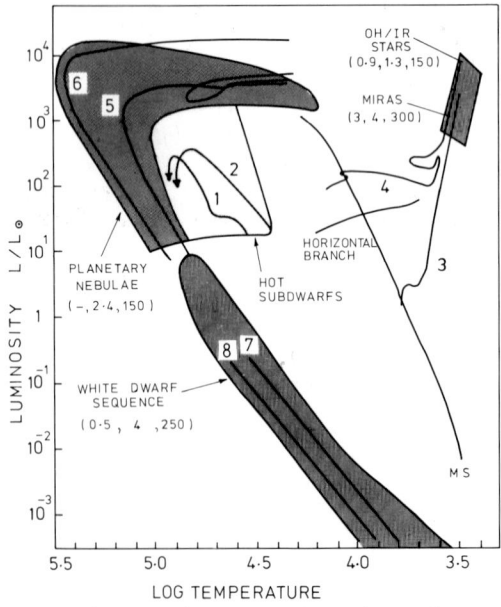

Figure 4. Schematic diagram illustrating the primary regimes occupied by planetary nebulae, hot sub-dwarfs, Miras, and OH/IR stars. For comparison, we also show evolutionary tracks for 0.5 M$_\odot$ stars away from the EHB (1, 2), 1.1 M$_\odot$ and 3 M$_\odot$ main sequence stars (3, 4), 0.6 M$_\odot$ and 0.8 M$_\odot$ central stars (5, 6), and 0.6 M$_\odot$ and 1 M$_\odot$ white dwarfs (7, 8). The codes (χ, χ_p, z_o) for Miras, PN, and white dwarfs refer respectively to the observed formation rate (x 10^{12}), the predicted formation rate for χ(PN) = $2.4.10^{-12}$ pc^{-3} yr^{-1}, and the typical mean scale height. The codes for OH/IR stars correspond to galactic formation rates χ_G(PN.yr^{-1}) and are based on a rate χ_G(PN) = 1.3 yr^{-1}.

of the fractional gas content, although uncertainties in the adopted correlation between M_{BOL} and stellar mass, for instance, may lead to factors of two uncertainty in $\chi(PN)$. If the relevant input parameters of Abell and Goldreich (1966) are adopted, for instance, then a PN formation rate $1\text{-}2.10^{-12}$ pc^{-3} yr^{-1} would be implied. Similar results are also deduced by Tinsley, who predicts $\chi(WD) \sim 2.10^{-12}$ pc^{-3} yr^{-1}.

Finally, we note that by using an improved understanding of the mass range M_{NPN}, an initial/final mass relation as given by Weidemann and Koester (1983), and assuming a secularly invariant SFR, Mallik (1985) determines a formation rate $\sim 5.10^{-13}$ pc^{-3} yr^{-1} - similar, indeed, to the values quoted by Alloin et al (1976), O'Dell (1962, 1968), and Cahn (1968) - although less than our best current estimate.

In summary, it is clear that the overall picture of PN as emerging from the higher mass AGB sequence via a high mass-loss OH/IR phase is reasonably well attested through the respective main-sequence turn-off rate, Mira death rate, and OH/IR formation rate (see figure 4 for a summary of these parameters). The errors (and uncertainties) in all of these estimates are not such as to create an overwhelming confidence, but appear nevertheless to confirm a value close to $\chi(PN) \sim 2.4.10^{-12}$ pc^{-3} yr^{-1}. The relation of $\chi(PN)$ to $\chi(WD)$ is somewhat more problematic. Whilst earlier values of $\chi(WD)$ appeared to relate quite well to $\chi(PN)$, the realisation that only $\sim 60\%$ of white dwarfs pass through an observable PN phase of evolution makes the agreement less secure. Similarly, the recent estimates of $\chi(WD)$ by Fleming et al (1986) appear consistent with only the lowest estimates of $\chi(PN)$, and may require a complete re-evaluation of the M.I. distance scale, or alternatively a large ($\sim 80\text{-}90\%$) population of PN central star binaries.

Clearly, therefore, there is room here for a considerable improvement in either $\chi(PN)$, $\chi(WD)$, or perhaps both, unless current models of stellar evolution are grossly in error.

REFERENCES

Abell, G.O., and Goldreich, P., 1966.
  Pub. Astr. Soc. Pacific 78, 232
Acker, A., 1978. *Astron. Astrophys. Suppl.* 33, 367
Alloin, D., Cruz-Gonzalez, C., and Peimbert, M., 1976
  Astrophys. J. 205, 74
Amnuel, P.R. Guseinov, O.H., Novruzova, H.I., and Rustamov, Ya.S., 1984. *Astrophys. Space Sci* 107, 19
Cahn, J.H., 1968 in *I.A.U. Symposium No.43*, eds. D.E. Osterbrock and C.R. O'Dell, D. Reidel Publishing Co., Dordrecht, Holland, p.44
Cahn, J.H., and Kaler, J.B., 1971. *Astrophys. J. Suppl.* 22, 319
Cahn, J.H., and Wyatt, S.P., 1976. *Astrophys. J.* 210, 508
Cahn, J.H., and Wyatt, S.P., 1978 in *I.A.U. Symposium No. 76*, ed. Y. Terzian, D. Reidel Publishing Co., Dordrecht, Holland, p.3
Cruz-González, C., 1974. *Mon. Not. R. Astron. Soc.* 168, 41

Weidemann, V., 1968a.   *Ann Rev. Astron. Astrophys.* 6, 351
Weidemann, V., 1968b in *IAU Symposium No. 43,* ed. D.E. Osterbrock and
   C.R. O'Dell, D. Reidel Publishing Co., Dordrecht, Holland, p.423
Weidemann, V., 1977.   *Astron. Astrophys.* 61, L27
Weidemann, V., 1978 in *IAU Symposium No. 76,* ed. Y. Terzian, D. Reidel
   Publishing Co., Dordrecht, Holland p. 353
Weidemann, V., 1987 in *Late Stages of Stellar Evolution,* ed. S. Kwok
   and S.R. Pottasch, D. Reidel Publishing Co., Dordrecht, Holland,
   p.347.
Weidemann, V., and Koester, D., 1983.   *Astron. Astrophys.* 121, 77
Wesemael, F., 1981.   *Astrophys. J.* 243, 328
Wielen, R., 1972.   *Astron. Astrophys.* 60, 263
Willson, L.A., 1981 in *Effects of Mass Loss on Stellar Evolution*
   ed. C. Chiosi and R. Stalio, D. Reidel Publishing Co., Dordrecht,
   Holland, p. 353

The Proto-Planetary Nebula Vy 2-2.

R.E.S. Clegg & M.G. Hoare
University College London, U.K.

J.R. Walsh
Anglo-Australian Observatory, Epping, Australia.

High and low-resolution optical and near-IR spectroscopy of the candidate proto-planetary (or very young PN) Vy 2-2 (P-K 45 − 2°1) is reported. This object has associated OH maser emission and an angular diameter of only 0.4 arcsec, found from VLA and optical speckle interferometry. Empirical analysis gives the values $N_e \approx 3 \times 10^5$ cm^{-3}, T_e=11000($\pm$1500)K. The electron temperature is quite uncertain because of the high density. Abundances of He, C, N, O, Ne and Ar are reported; the carbon abundance is uncertain as it relies on the C II λ4267Å line, since the object is too highly-reddened (c=1.8 $\pm$ 0.2) to be observed with IUE. We find He/H=0.10, O/H=4x10^{-4} and C/O=0.8. The HI Zanstra temperature is 38 000 K (for black-body). The spectrum shows broad stellar lines of He II λ4686, C III λ4647 and N III λ4640; the central star may be of type Of.

A photo-ionization model is presented for this young, dense object. The central star is represented by a non-LTE H-He model atmosphere with T_{eff}=38 000 K, log g=3.5. We adopt a distance of 2.5 kpc, based on a calibration for optically-thick Magellanic Cloud PN (Barlow 1987). Major constraints for the modelling include the observed angular diameter, the stellar continuous flux level, the absolute (optically-thin) 100 GHz radio flux and the [O III] 5007 & 4363Å line fluxes. The stellar luminosity is 3500 L⊙ for the adopted distance. We introduce silicate dust grains into this model together with the same grains in a neutral region surrounding the ionized zone. The dust parameters are adjusted so as to match the observed $IRAS$ 4-channel photometry and the measured silicate emission feature at 9.7 μm.

It is concluded that Vy 2-2 is a bona-fide young planetary nebula. The stellar parameters are those of a post-AGB object and the nebula abundances are typical of disk PN.

THE OPTICALLY RESOLVED PLANETARY NEBULA/OH MASER Vy 2-2[*]

R. Falomo and F. Sabbadin
Asiago Astrophysical Observatory, 36012 Asiago (VI), Italy

ABSTRACT. Vy 2-2 is, to our knowledge, the only planetary nebula exhibiting OH emission, indicating that it is a very young PN still enveloped by the neutral shell of the progenitor AGB star.

Low and mean resolution spectroscopy of Vy 2-2 was obtained at the European Southern Observatory (Chile) to study physical and evolutionary characteristics of this peculiar object, considered as the missing link between Mira variables-OH/IR sources and planetary nebulae.

Vy 2-2 appears as an extended object (apparent diameter about 20 arcsec) in a CCD spectrum obtained with the B&C spectrograph attached to the 2.2-m telescope. This optical size is much larger than that derived with the Very Large Array: at 15 GHz the ionized nebula appears as a thin shell about 0.5 arcsec in outer diameter and about 0.2 arcsec in inner diameter. For an assumed distance of 1.5 kpc, the optical radius of the nebula is 0.073 pc; for an expansion velocity of 10 to 15 km s^{-1} its dynamical age is 4500-7000 years.

All these facts indicate that Vy 2-2 is a compact PN, similar to NGC 6572, BD+30°3639, I 418, II 5117 and some other well studied objects. The exceptionality of Vy 2-2 consists in the presence of the OH maser emission.

[*] Based on observations obtained at the European Southern Observatory, La Silla, Chile.

NEW IR-OBSERVATIONS OF POST AGB STARS AND PROTO-PLANETARY NEBULAE

W.E. van der Veen, H.J. Habing, Leiden, The Netherlands
T.R. Geballe, UKIRT, Hawaii

ABSTRACT. A sample was selected from the IRAS Point Source Catalogue based on the following selection criteria: very red ("cold") IRAS-colours: roughly $F_{25}/F_{12} > 2.5$ and $F_{60}/F_{25} < 1.2$; and low IR-variability: VAR < 30. These non-variable IR-sources may be stars that have evolved beyond the AGB (Asymptotic Giant Branch); a large fraction (40%) is associated with known planetary nebulae (Van der Veen and Habing, 1987, *Astron. Astrophys.*, in press). To determine the nature of the other 60% additional observations were made mainly in the infrared: 1-13 μm, during 4 observing runs: ESO (La Silla, Chile) in July 1986 and June 1987; UKIRT (Hawaii) in August 1986 and June 1987. A total number of 58 sources was observed. A summary of the observations: --IR broad band photometry at 1.2, 1.6, 2.2, 3.8 and 4.6 μm for all 58 sources. --IR broad band photometry at 8.4, 9.7 and 12.8 μm for 19 sources. --IR small band photometry for 4 sources in the ranges 2-2.5 μm and 3-3.5 μm. --IR spectroscopy for 10 sources in the ranges 2-2.5 μm and 3-3.5 μm. --V,R, I observations (0.55, 0.7 and 0.9 μm) for 5 sources associated with a star of visual magnitude 8-9. These observations were carried out by D. de Winter (Amsterdam) with the 0.5-m ESO telescope at La Silla (Chile). --Walraven photometry (0.32, 0.36, 0.38, 0.43 and 0.54 μm) for 21 stars brighter than V = 15 and within 10" from the IRAS position. These observations were carried out by M. van Haarlem (Leiden) with the 0.9-m Dutch telescope at La Silla (Chile).

Although all 58 sources have similar IRAS colours, they show large differences at wavelengths shorter than 5 μm. In most of the sources a stellar component and one or two dust components can be distinguished, but the relative strength of the components differs from object to object. A simple model of a star surrounded by one or two optically thin dust shells, all radiating as black bodies, was used for interpretation. The observations are consistent with a central star, temperature typically between 4000 and 20,000 K, that is surrounded by a cold distant dust shell with a characteristic temperature of 80-150 K. This distant shell probably results from the mass loss at a very high rate when the star was at the top of the AGB; the mass of this shell ranges from a few times 0.1 $M_\odot$ to a few times 1 $M_\odot$. Another indication for the AGB origin of the distant dust shell is the double peaked OH 1612 MHz maser profile, which is characteristic for OH/IR stars situated at the top of the AGB, and is still present in about 30% of our sources. If we assume a typical AGB expansion velocity of 15 km/s we find that these distant dust shells are ejected between 1000 and 4000 years ago. A relation between stellar temperature and time elapsed since the ejection of the circumstellar shell is found when the expansion velocity of 15 km/s is assumed: $T_* = 2500 + 5 (t/yr)$ K. This result suggests a transition time from AGB to the planetary nebula stage ($T_* = 30,000$ K) of 5000 yr, in rough agreement with theoretical predictions of 3000-4000 yr (Schönberner, this conference). There is also evidence for a second, relatively hot wind following the cool AGB wind from a gradual increasing IR excess at 12 μm.

IRAS 17516-2525: THE BIRTH OF A PLANETARY NEBULA?

W.E. van der Veen and H.J. Habing
Leiden, The Netherlands
T.R. Geballe
UKIRT, Hawaii

ABSTRACT. IRAS 17516-2525 is a cool object at infrared wavelengths between 1 and 60μm. Spectroscopy in the wavelength ranges 2 - 2.5μm and 3 - 3.5μm shows the presence of Br_α, Br_γ, Pf_γ emission lines and of a weak C_2H_2 absorption band, all clearly associated with the IRAS source. At the infrared position a faint star with a visual magnitude of about 20 is found.

A double peaked 1612 MHz OH maser profile, which is characteristic for OH/IR stars situated at the top of the AGB, coincides with the source -the accuracy of the OH measurement is modestly good. Assuming that the OH source is indeed associated with the IRAS object, one concludes that the expanding distant dust shell must have been formed on the AGB, when the star was much cooler (typical 2500 K) and the mass loss rate much larger. The shell has an expansion velocity of 17 km/s. From the line profile of Br_α we find an expansion velocity of the hot inner region of about 50 km/s.

The ionizing region has to be small ($R < 10^{15}$ cm) because a VLA measurement of the radio continuum flux (P. Katgert, private communication) gave an upper limit of 5mJy. The fact that the redshifted OH maser peak is still visible supports the smallness of the ionized region.

A simple model (Van der Veen $et\ al.$, this conference) fitted to the observed energy distribution is consistent with a hot star ($T_* > 30000$ K) surrounded by a hot nearby dust shell (900 K) and a cold distant dust shell (100 K). The nearby dust shell contains about 10^{-4} $M_\odot$ and must have been ejected recently (5-20 yr). This result is practically independent of the assumed luminosity of the central star. If we assume a typical core mass of the central star equal to 0.58 $M_\odot$ ($L_* = 6000\ L_\odot$), the mass of the distant dust shell is of the order of one solar mass and was ejected about 2000-3000 years ago.

Finally: the observed C_2H_2 absorption at 3μm in combination with the OH maser emission suggest that the star is now being transformed from an oxygen-rich AGB star into a carbon-rich proto-planetary nebula.

THE SHOCKING TRUTH ABOUT SOME "PROTO-PN"

R.W. Goodrich[1] and Luciana Bianchi[2]
1. Lick Observatory, Santa Cruz, California, U.S.A.
2. Osservatorio Astronomico, Pino Torinese, Italia

ABSTRACT. A small number of bipolar planetaries -including M2-9, M1-91, GL 618, and M2-56- exhibit very strong emission lines of low ionization species such as [O I], [N I], [N II], and [S II]. Most previous authors have attempted to analyze the spectra of these objects assuming that they are photoionized by their central stars. Closer examination, however, suggests that a different excitation mechanism may be at work -that of shock heating.

In the relatively high excitation objects M2-9 and M1-91 the ratio of [O III] $\lambda\lambda$4959, 5007 to λ4363 may be used to define an [O III] temperature, T([O III]). Similarly, [N II] $\lambda\lambda$6548, 84/λ5755 defines a [N II] temperature, T([N II]). Under the assumptions of photoionization and low density we find that T([O III]) for the wings of M2-9 and M1-91 is typically 4 times larger than T([N II]). This is typical of shocks, where the [O III] comes from the high-T region just behind the shock and the [N II] arises from cooler gas further downstream. On the other hand, T([O III]) in the cores is undefined, indicating that there, at least, the $\lambda\lambda$4959, 5007 lines are collisionally deexcited, indicating $N_e \sim 10^6$ cm^{-3} in the [O III] region. However, this high electron density would almost completely quench the [N I], [N II], and [S II] lines, and these lines are already unusually strong in these nebulae. Hence there must exist two phases of gas at very different densities. ([N II] λ5199 has a critical density of only 2000 cm^{-3}). Previous models have not taken this into account, but in any case still would have difficulty reproducing the strengths of the low ionization species relative to Hβ.

The low-ionization objects GL 618 and M2-56 have no [O III] lines, so the excellent shock discriminant T([O III]) is not available. However, the relative line intensities of these two objects compare rather well with the spectra of HH 43C and HH 43N. It is now accepted that HH objects are shock heated, and we may infer by analogy that GL 618 and M2-56 are also shock heated. Further detailed modeling is currently under way.

Clearly care must be exercised in interpreting the spectra of these objects, and it is likely that we will require more sophisticated models than are usually used for PN.

NEW OH/IR STARS: PROTO-PLANETARY NEBULAE?

J. Eder
Yale University
B.M. Lewis
NAIC, Arecibo Observatory
Yervant Terzian
NAIC, Cornell University

ABSTRACT: The IRAS infrared colors, $(60 - 25)\mu m$ and $(25 - 12)\mu m$, allow efficient identification of Type II OH/IR stars. We present Arecibo[1] OH (1612 MHz) observations of 474 IRAS point sources chosen to define the exact regions of the two-color diagram occupied by OH/IR stars. Our observations are complete within the boundary regions of the two-color locus and within the region, 16^h < right ascension < 22^h, $0°$ < declination < $37°$. The sensitivity of the Arecibo telescope allows the identification of many weak sources that would not have been detected by previous surveys and the weak end of the masing phenomenon has been studied for the first time.

Within the OH/IR star color region defined by the 184 detections, 171 IRAS sources were not detected in OH, providing a detection rate of 52%. The spectra of most of the detections show the characteristic double-peaked profiles associated with circumstellar shells. The velocity widths, central velocities, and integrated fluxes derived from these profiles are listed along with their infrared colors. Detections with $(60 - 25)\mu m > - 0.8$ are confined to the galactic plane and are predominantly strong sources; but, those with $(60 - 25)\mu m \leq -0.8$ can be found at a large range of galactic latitude and integrated flux strength. A correlation was found between the infrared flux and the 1612 MHz flux. Despite a low OH threshold, we fail to detect any sources with a $25\mu m$ flux less than 1 Jy. Detailed analysis of the data is deferred to the completion of our survey of all appropriate, color-selected sources accessible from Arecibo.

1. The Arecibo Observatory is part of the National Astronomy and Ionosphere Center, which is operated by Cornell University under a management contract with the National Science Foundation.

INFRARED PHOTOMETRY OF OH/IR STARS

Miriam Peña and Julieta Fierro
Instituto de Astronomía
Universidad Nacional Autónoma de México

We have initiated a near infrared photometric study of OH/IR stars and proposed protoplanetary nebulae, using the 2.12 m telescope and the IR photometric system at the Observatorio Astronómico Nacional in Baja California, México.

The aim of this project is to derive near IR characteristics of suspected protoplanetary nebulae and their relation with those of young PN.

J, H, K, L and M magnitudes for 17 objects, obtained from three observing seasons, are presented. Reported variable and non-variable OH/IR stars have been observed; in the sample, we have included the objects with detected ionized gas (Pottasch and Zijlstra, 1987, A. A. Lett., in press), as well as some "photoplanetaries".

Near IR color-color diagrams have been constructed for the observed objects. From the (H-K) vs (K-L) and (J-K) vs (K-L) diagrams, the following trend is noted: the variable OH/IR star seem redder than the objects with ionized gas which are redder than the non-variable OH/IR stars.

A plot of the energy distribution of the observed objects shows that variable OH/IR's have a steeper near-IR spectra than non-variable objects. These results are similar to those reported by Habing, van der Veen and Geballe ("The Late Stages of Stellar Evolution", ed. S. Kwok and S. Pottasch). However, the objects with ionized gas have both kinds of energy distribution (flat and steep).

THE PRESENCE OF WATER MASERS IN COLOR-SELECTED IRAS SOURCES

B.M. Lewis
NAIC, Arecibo Observatory
D. Engels
Hamburger Sternwarte

ABSTRACT. Eder, Lewis, and Terzian (1987) examined $\sim$ 400 sources from the IRAS Point Source Catalogue with colors appropriate to OH/IR stars, for the presence of 1612 MHz emission. We examined a proportion of these objects at Effelsberg for the presence of water-maser emission. In sources with $|b^{II}| > 10°$ which are therefore relatively local, we find a 68% detection rate for water-masers among objects associated with 1612 MHz masers, as opposed to a 17% detection rate among sources with similar colors but without 1612 MHz emission. Those conditions in a circumstellar shell that favor the presence of water-masers also favor the presence of a 1612 MHz maser. These results are consistent with most Type II masers being associated with water-masers. Since Cooke and Elitzur (1985) show that water-masers are collisionally excited, this result excludes stirring of the envelope by a companion star with an associated loss of velocity coherence, as the primary cause for the existence of the color-analogue sources without 1612 MHz masers. We discuss an alternative scenario.

Six sources have water emission without any OH masers. These are explicable either as objects in which the 1612 MHz maser is suppressed by a companion, or as objects in which the circumstellar shell has yet to develop sufficient depth for OH masers to operate.

SOME DEPENDENCES FOR LONG-PERIOD VARIABLES AND A POSSIBLE SCHEME OF THEIR EVOLUTION

I.L. Andronov and L.S. Kudashkina
Astronomical Observatory of the Odessa State University, USSR
G.M. Rudnitskij
Sternberg Astronomical Institute of the Moscow State University, USSR

ABSTRACT. Some dependences between the parameters for approximately 150 stars, of which 81 are sources of maser emission in molecular lines, are constructed. The following parameters are considered: period P, asymmetry (M-m) of the visual light curve, visual amplitude A, color index (I-K). We use the data of the General Catalogue of Variable Stars. For the stars Z Cyg, R Tau, R Peg, RT Vir, RX Boo, PZ Cas, U Her, and R Cas, some parameters were determined by the authors.

On the A - (M-m) plot, the maser LPVs lie, on the average, higher than the non-maser ones; the SR stars lie lower. With increasing A, (M-m) varies but very weakly. From the A - P plot it can be concluded that for "short-period" LPVs (P < 280^d), A for all stars is almost the same. For the "longer-period" stars (P > 280^d), (M-m) varies strongly. It can be supposed that, with increasing P, light curve becomes very non-stationary and its asymmetry varies from one cycle to another.

We discuss the dependences found from the point of view of the character of the LPVs' pulsational instability at different stages of evolution. In our opinion, different peculiarities of the LPVS' light variations allow to construct the following evolutionary scenario for these stars:

SR("short-period) $\rightarrow$ mira ("short-period ", non-maser) $\rightarrow$

mira ("long-period" , M-maser, S-, or C-type) $\rightarrow$

$\rightarrow \begin{cases} \text{supergiant (SRAC, "long-period")} \rightarrow \text{pre-SN} \rightarrow \text{explosion} \\ \text{giant (SRB, "long-period)} \rightarrow \text{"quiet"outflow} \rightarrow \text{PN} \end{cases}$

The choice of the pathway after the "long-period"-mira stage depends on the star's mass. However, only the most massive stars (M $\gtrsim$ 10 $M_\odot$) may follow the branch ending in an SN explosion. Less massive ones (5 - 10 $M_\odot$) may lose a few solar masses in the violent mass loss stage, forming thick circumstellar shells ("black" planetary nebulae); these objects may be observable as OH/IR stars. Finally, low mass stars (1 - 5 $M_\odot$) form, through "quiet" mass outflow,"ordinary" PN.

PROTO-PLANETARY NEBULAE: MODELS AND IRAS OBSERVATIONS

K. Volk and S. Kwok
University of Calgary

ABSTRACT. A number of high-galactic latitude supergiant stars of intermediate spectral types have been suggested to be proto-planetary nebulae (PPN), such as the 89 Her and R CrB stars. A number of these stars along with some IRAS sources expected to be PPN -either sources with IRAS low Resolution Spectra (LRS) showing features which may indicate unusually cool dust shells or unusually red IRAS sources for which CO emission from a circumstellar envelope has been observed- where chosen for study. The IRAS 12/25 and 25/60 µm colours of 32 such stars from 3 groups on a colour-colour diagram. Class I show colours similar to ordinary stars; Class II have a 60 µm excess but have normal 12/25 µm colours; Class III are much redder than ordinary stars.

The Class I sources may be binaries or normal late-type stars with poor quality spectra as observed by IRAS, leading to mis-classification. 89 Her, R CrB, HR 4049 and υ Sgr are in this group. Any PPN which may be found in this area cannot be separated from ordinary stars based upon the IRAS colours. The nature of Class II is unknown, although 2 of them are A0e stars. The Class III sources have IRAS colours intermediate between normal late-type stars and planetary nebulae. Thus they are the best PPN candidates.

Searching the IRAS data for sources of similar colour to the Class III objects yielded 371 sources after eliminating 24 objects associated with H II regions, galaxies, etc. 140 of them are planetary nebulae and another 200 have no identifications. 11 of them are SAO stars with spectral types between G7 and A5.

Radiative transfer models of stars with cool stellar wind dust shells have been carried out for comparison with the IRAS PPN candidates. The implications of these models will be discussed.

HCN, THE FIRST STRONG MASER IN CARBON-RICH STARS

A. Omont, S. Guilloteau, and R. Lucas
Observatoire de Grenoble
Université de Grenoble, France

ABSTRACT. Maser emission has been observed at the frequency of the (0,2,0) J = 1 - 0 line of HCN (89088 MHz, energy 2050 K above the ground state) in 7 C-rich circumstellar envelopes: CIT 6, S Cep, IRC+ 50096, IRC+30374, FX Ser, AFGL 2513 and IRAS 17581-1744. This is the second molecule showing strong maser emission in millimeter lines, and the first strong maser ever observed in a C-rich circumstellar envelope. The emission is particularly strong (70 Jy) in CIT 6 because of its proximity (Guilloteau, Omont, and Lucas 1987). The masers in the stars have luminosities and linewidths (FWHM about 1 km/s) similar to CIT 6; they are also blueshifted by a few km/s with respect to the velocity of the star given by the central velocity of the ground-state HCN emission (Lucas, Guilloteau, and Omont 1987). These properties are somewhat similar to those of SiO masers in O-rich stars, although the intensities of HCN masers are weaker.

85 C-rich circumstellar envelopes were searched for HCN maser emission. Strong maser emission seems confined to envelopes with intermediate mass-loss rates between 10^{-6} and 10^{-5} $M_\odot$/yr which have very similar infrared properties. HCN maser emission is present in about 20% of C-stars in this particular subclass.

Time variations occur on scales of months in HCN masers, since one maser (IRC+50096) had disappeared 80 days after its detection. The maser emission is strongly linearly polarized in CIT 6 (Goldsmith *et al.* 1987). No emission was detected in CIT 6 from the J = 3-2 transitions of different vibrationally excited states of HCN.

The existence of HCN maser emission, especially in CIT 6, could be related to the particular activity and structure in the vicinity of the photosphere of these extreme AGB stars.

REFERENCES

Goldsmith, P.F., Lis, D.C., Guilloteau, S., Lucas, R., Omont, A. 1987, in preparation.
Guilloteau, S., Omont, A., Lucas, R. 1987, *Astron. Astrophys.*, 176, L24.
Lucas, R., Guilloteau, S., Omont, A. 1987, *Astron. Astrophys.*, submitted.

LOWER LIMIT FOR NPN's MASSES

Amos Harpaz
Department of Physics, The Technion, Haifa 32000, Israel

The lowest mass observed for a nucleus of a planetary nebula (NPN) is about $0.55\ M_0$ (Weidemann and Koester, 1983, Schonberner, 1983). Hence, Lower mass WD's should have been produced without going through the phase of a visible PN ejection. Recently, Harpaz et al. (1987), have shown that very low mass WD's (up to $0.45\ M_0$) can be formed by a single star evolution from red giant branch (RGB) stars, due to mass loss along the RGB. It turns out that WD's in mass range of $0.46-0.55\ M_0$ formed by a single star evolution should be formed from the AGB, without an observable PN.

We suggest that WD's with masses in this range are formed from the lower part of the AGB. During this phase the mass loss rates are about $1-5 \times 10^{-7}\ M_0/Y$. Harpaz and Kovetz (1981) have shown that a mass loss rate of $2 \times 10^{-6}\ M_0 Y$ is a lower limit for the production of an expanding nebula which might be observed as a faint PN. Lower mass loss rate will yield very dilute gas clouds, which will not produce the ionized shells observed as a PN. Hence, only stars which turn into WD's from the upper part of the AGB ($L > 5000\ L_0$, $m_c = M_{WD} > 0.55\ M_0$) will be observed as NPN's while stars which turn into WD's from the lower part of the AGB ($M_{WD} = m_c < 0.55\ M_0$) will not produce an observable PN. The gap between stars, which turns into WD's from the RGB, without any PN ($M_{WD} < 0.45\ M_0$), and those that pass through the stage of a PN ejection ($M_{WD} > 0.55\ M_0$) is bridged by stars, which expel their envelopes while still on the lower part of the AGB, with dilute gas cloud around them, which is too sparse to be observed as a PN.

Acknowledgement: This work was supported by a grant from the Eppley Foundation for Research.

Harpaz A., Kovetz A., 1981, Astron. Astrophys., 93, 200.
Harpaz A., Kovetz A., Shaviv G., 1987, Astrophys. J., in press.
Schonberner D., 1983, Astrophys. J., 272, 708.
Wei demann V., Koester D., 1983, Astron. Astrophys., 121, 77

SOME HYPOTHESIZED OBSERVATIONAL ASPECTS OF MAGNETIC FIELDS IN PROTOPLANETARY NEBULAE

G. Pascoli
Faculté des Sciences
Départment of Physique, Amiens, France

ABSTRACT. Polarization mapping of some reflection nebulae (e.g., NGC 6729) reveals parallel bands of polarization vectors across pre-main sequence stars (Ward-Thomson et al. 1985, Mon. Not. R. Astr. Soc., 215). These authors have suggested a model in which the bands would be explained by dust discs, the grains being aligned by toroidal magnetic fields $\sim 10^2$ µG.

Although the nebulae associated with pre-main sequence stars and the protoplanetary nebulae PPNS (associated with most evolved stars) be morphologically distinct objects, some likeness appeared in evidence from an inspection of the centro symmetric pattern of polarization (CPP).

Besides the substantial amount of polarization ($\sim$ 10 - 40%), produced by the reflection of the light star continuum on aligned (or not) grains throughout the nebula (Cohen, Proc. IAU Symp. No. 103, 1983); some part remains ($\sim$ 2-3%), which can be approximately represented by parallel bands of polarization vectors in an essentially perpendicular direction to the major axis of the PPN (small departure from the CCP or ellipticity of the CCP: Aspin and Mc Lean, 1984, Astron. Astrophys., 134; King et al. 1985, Mon. Not. R. Astr. Soc., 213).

Without neglecting the conclusions of these authors (presence of a central object not point-like, but elongated along the minor axis of the nebula); we propose another and new interpretation assuming a (toroidal) magnetic field within the PPN.

Some support of this can be drawn from the Reid et al. observations (1979, Ap. J., 227) of a M-type star (U Ori) embedded in a OH maser region.

Indeed, a strong circular polarization, interpreted to be a result of the Zeeman effect, indicate magnetic field strengths $\sim$ 10 milligauss in the masing region.

Theoretical results have been published (Pascoli, Astron. Astrophys.,1987, in press) showing the peculiar morphology of PPNs (and planetary nebulae) as a direct consequence of an internal toroidal magnetic field. New matter (origin and geometry of this fossil magnetic field convectively ejected from the Red Giant progenitor) has also been discussed and will be subsequently published.

STALLED WINDS: INTERACTIONS BETWEEN NEBULAE AND STELLAR WINDS

J.B. Kaler
University of Illinois
W.A. Feibelman
NASA-Goddard Space Flight Center
R.A. Shaw
Lick Observatory
H. Henrichs
University of Amsterdam

ABSTRACT. Spectra of the nuclei of two planetaries show what appear to be features caused by fast stellar winds as they encounter the surrounding nebulae. Superimposed upon the high velocity (3670 km s^{-1}) ultraviolet P Cygni profiles of Abell 78 are low velocity absorption lines that likely arise from a density enhancement in the wind as it brakes and builds up against the inner edge of the nebula. The deepest portions of these narrow absorptions fall at -78 km s^{-1} for N V and O V and -26 km s^{-1} for C IV, which implies a gradient in the decelerating wind, as does the profile of the strongest C IV line. The lower value may be related to the expansion velocity of the inner helium-rich nebulosity, which we associate with a sharp absorption feature. Another density enhancement, evidenced only by absorption lines, appears at -250 km s^{-1}, and may be caused by a rebound shock of the sort envisioned by Kahn (*IAU Symposium No. 103, Planetary Nebulae*, 1983, 305) and Okorokov *et al*. (*Astr. Ap.*, 1985, 142, 441).

In addition, the nucleus of NGC 2371 exhibits narrow O VI lines at λ3811 and λ3834 (first detected by Aller, *IAU Symposium No. 38, Planetary Nebulae*, 1968, 339) superimposed upon the broad underlying blend produced by the fast (3400 km s^{-1}) stellar wind that seem to be produced by the same phenomenon. M3-30 is an additional candidate. These observations provide at least part of the evidence needed to demonstrate the idea that the fast winds can affect and even shape the surrounding nebulae.

OPTICALLY THICK WIND FROM POST-AGB STARS AND FORMATION OF PLANETARY NEBULAE

Mariko Kato
Department of Astronomy, Keio University, Japan

ABSTRACT. Self-consistent mass-loss solutions are computed for post-AGB stars, and the evolution of the central star is followed by utilizing the sequences of the steady mass-loss and the static solutions. The mass loss is driven by the radiation pressure gradient and the matter is accelerated in the inner region to the photosphere. The optically thick wind occurs when the H-burning luminosity is larger than the Eddington limit at the surface region. Such a situation is realized in the low-temperature region of the H-R diagram. Therefore no wind solution exists when the surface temperature becomes high enough.

Both the mass-loss rate and the wind velocity are large when the surface temperature is relatively low, and both of them decrease as the central star evolves toward the high temperature region of the H-R diagram. For an 1.2 $M_\odot$ WD, for instance, the mass-loss rate $\dot{M}$ is 4.21×10^{-4} $M_\odot$/yr and the wind velocity at the photosphere is v = 38 km/s at log T_{ph} = 3.81, and $\dot{M}$ is 1.25×10^{-5} $M_\odot$/yr, v = 17 km/s at log T_{ph} = 3.97; just before the mass-loss stops. The wind ceases when the surface temperature becomes higher than log T_{ph} = 4.05.

DIRECT EVIDENCE FOR A BIPOLAR STELLAR WIND IN NGC 2392

C.R. O'Dell
Rice University

ABSTRACT. Very high spectral resolution slit spectra have been used to investigate the kinematic structure of the double shell planetary nebula NGC 2392, the Eskimo Nebula. Each shell produces a spectrum corresponding to the projected radial velocity, a pattern which varies according to the position angle of the spectrograph's entrance slit. Multiple slit observations were used to determine that the inner shell is an incomplete prolate spheroid pointed almost at the observer and missing the tip ends. The outer shell is nearly spherical and more uniform.

We present evidence that the stellar wind is the dominant force in determining the kinematic structure of the shells and is producing a continuous bipolar flow of material at velocities up to at least plus and minus 190 km/s through the ends of the prolate spheroid. We argue that the distribution of material along the equator of the inner spheroid is evidence for equatorial loss of material from the central star during its second period of high mass-loss.

TWO-DIMENSIONAL HYDRODYNAMICAL MODELS OF PLANETARY NEBULAE (PNe)

I.V. Igumenshchev, B.M. Shustov, A.V. Tutukov
The Astronomical Council of the USSR Academy of Sciences
Moscow, USSR.

To study the evolution of an expanding PNe formed by non-spherical stellar (super) wind we have computed seven evolutionary sequences for the dusty envelopes with mass 0.05 $M_\odot$ taking angular distribution of mass loss and of the outflow velocity as parameters. In the case described below the gas velocity on the inner boundary is assumed: $V_z=25$ $(1-0.4\ (3\cos^2\theta -1)12)$ km s^{-1} (where θ is the polar angle) with the density of ejected matter independent on θ. The mass loss rate on the (super) wind phase is 6 10^{-5} $M_\odot$/yr. The structure of the model at the moment of ionization breaks through PN in the polar direction is illustrated in Fig. 1. The density distribution in a meridional plane is shown in Fig. 1a together with the ionization boundary (the dashed line). Contours of equal emission measure nebula model are shown in Fig. 1b. After complete ionization of PN hydrogen distributions become smoother. We found that some of double-shell PNe may be the product of single mass loss event.

Other six computed models expand significantly the range of possible configurations. The final reason of the great variety of PNe forms can be explained by the geometry of (super) wind forming them. The latter can depend on the duplicity of nuclei.

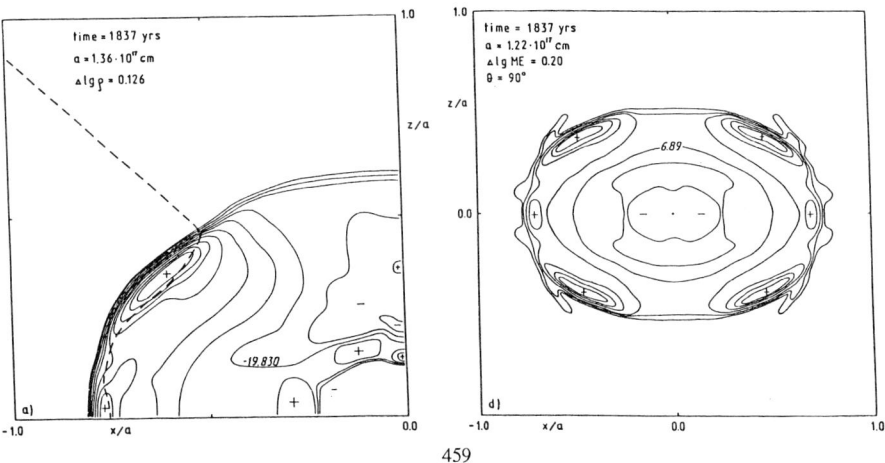

S. Torres-Peimbert (ed.), Planetary Nebulae, 459.
© 1989 by the IAU.

THE SPATIAL STRUCTURE OF PLANETARY NEBULAE WITH BINARY CENTRAL STARS

I.G. Kolesnik and L.S. Pilyugin
Main Astronomical Observatory of the Academy of Sciences of
the Ukrainian SSR, Kiev, USSR

ABSTRACT. The planetary nebulae formation in the detached binaries is considered.

We assumed that the process of the planetary nebula formation (or the envelope ejection) is the process of the unlimited expansion of the red giant envelope. The expansion velocity is invariant with distance from the star. The envelope ejection by the single star leads to the spherical planetary nebula formation.

When the envelope ejection takes place in the detached binaries then a few factors determine the spatial structure of the ejected shell. The main of them is that the shell ejected by one of the stars flow past a gravitating body (binary companion). If a gravitating body moves through the medium of the non-interacting particles then the density is maximal on the downstream axis. Conversely, if a gravitating body moves supersonically through the gas then the density on this axis is lower than the one away from the axis (R. Hunt, $M.N.$, 1971, 154, 141; 1979, 188, 83). The planetary nebulae is a gas but not a medium of the non-interacting particles. Hence, the density minimum can be expected in the orbital plane.

It is known that the star rotation results in the increase of the density of the ejected shell in the equatorial plane.

Thus, two factors (a. the flow past of the binary companion, b. the rotation of the ejecting star) mainly determine the spatial structure of envelope. If the first factor is dominant, then the main structure (the enhanced density area) has the form of an hour-glass. If the second factor is dominant then the main structure has the toroidal form.

The peripherical structure (a relative fainter shell around main structure) has the form of an oblate ellipsoide. The observed form of the peripherical structure of the ionization-bounded nebula can differ from the real one.

The results have been published in $Astron.$ $Zh.$, 1986, 63, 279; 1987, 64, 537.

MORPHOLOGIES OF PLANETARY NEBULAE WITH CLOSE-BINARY NUCLEI

Howard E. Bond
Space Telescope Science Institute
Mario Livio
Dept. of Astronomy, University of Illinois
Michael Meakes
Space Telescope Science Institute

ABSTRACT. We will present photographic and CCD images of planetary nebulae that are known, on the basis of photometric observations of the central stars, to possess close-binary nuclei. All of the orbital periods range from 2.7 to 16 hours, except for the 16-day binary nucleus of NGC 2346.

We attribute the ejection of a planetary nebula from a close-binary system to the interaction that occurs when the more-massive star expands and engulfs a main-sequence companion in a common envelope. The companion spirals in toward the core of the giant, until the envelope is spun and up and ejected.

Theoretical considerations suggest that the morphology of a planetary nebula ejected via a common-envelope interaction will depend on the evolutionary stage of the primary star at the onset of the interaction: (1) If the interaction occurs when the primary star is on its first ascent of the giant branch, or low on the AGB, the envelope is ejected preferentially in the orbital plane, creating a large density constrast between the equatorial and polar regions and leading to a "butterfly"-shaped PN (e.g., NGC 2346). (2) In the case of a primary star that attains a highly centrally condensed supergiant configuration well up on the AGB before encountering its companion, the ejected material is less concentrated toward the orbital plane, a more moderate density contrast is created, and the PN will have an elliptical morphology (e.g., Abell 41, Abell 63, K 1-2).

The initial morphologies created by common-envelope ejection can subsequently be modified by stellar winds and/or interaction with the interstellar medium. Abell 35 and possibly HFG1 and Abell 46 show evidence of the latter.

CCD IMAGES OF THREE PLANETARY NEBULAE WITH BINARY NUCLEI

Julie Lutz and Nancy Jo Lame
Washington State University

ABSTRACT. A 14, H 3-75 and K 1-2, three planetary nebulae with binary nuclei, were imaged with narrow-band [N II], Hα, [O III] and He II filters by using the TI CCD chip on the 0.9-m telescope at CTIO. The purpose of doing the observing was to see if planetaries with known binary nuclei exhibit particularly peculiar morphologies. In some cases (e.g., NCC 1514, NGC 3132), planetaries with binary nuclei have morphologies that are exhibited by a number of nebulae. In other cases (e.g., He 2-36, the nebulae with binary central stars have unique morphologies.

A 14 shows highly symmetric, complicated structures that are unique among those found in a large survey of southern hemisphere planetary nebulae. H 3-75 is a round, double-shell nebula with a marked asymmetry on one side of the inner shell. K 1-2 shows interesting and unique enhancements in the [N II] images, including one small "jet-like" structure. The unique structures exhibited by these three nebulae make them candidates for nebulae whose morphologies may have been influenced by the presence of a binary central star.

EVOLUTIONARY TRACKS FOR CENTRAL STARS OF PLANETARY NEBULAE

Detlef Schönberner
Institut für Theoretische Physik und Sternwarte
Universität Kiel
Olshausenstrasse, 2300 Kiel, Fed. Rep. Germany

1. INTRODUCTION

Our understanding of the evolution of Central Stars of Planetary Nebulae (CPN) has made considerable progress during the last years. This was possible since consistent computations through the asymptotic giant branch (AGB), with thermal pulses and (in some cases) mass loss taken into account, became available (Schönberner, 1979, 1983; Kovetz and Harpaz, 1981; Harpaz and Kovetz, 1981; Iben, 1982, 1984; Wood and Faulkner, 1986). It turned out that the evolution depends very sensitively on the inital conditions on the AGB. More precisely, the evolution of an AGB remnant is a function of the phase of the thermal-pulse cycle during which this remnant was created on the tip of the AGB by the planetary-nebula (PN) formation process (Iben, 1984, 1987). This was first shown by Schönberner (1979), and then fully explored by Iben (1984). In short, two major modes of PAGB evolution to the white dwarf stage are possible, according to the two main phases of a thermally pulsing AGB star: the hydrogen-burning or helium-burning mode. If, for instance, the PN formation, i.e. the removal of the stellar envelope by mass loss, happens during a luminosity peak that follows a thermal pulse of the helium-burning shell, the remnant leaves the AGB while still burning helium as the main energy supplier (Härm and Schwarzschild, 1975). On the other hand, PN formation may also occur during the quiescent hydrogen-burning phase on the AGB, and the remnant continues then to burn mainly hydrogen on its way to becoming a white dwarf.

In order to classify the different internal structures of an AGB star over a thermal-pulse cycle, we define that phase zero be at the surface-luminosity peak occurring shortly after the helium shell flash. The following classification is then possible:
i) Phase 0.....0.15, the star is burning helium, hydrogen is shut off ("helium burner");
ii) Phase 0.3 .... 1.0, the star is burning hydrogen, heliums burns only on a low level, $L_{He}/L_H \cong 0.01$ ("hydrogen burner");
iii) Phase 0.15... 0.3, helium and hydrogen are burning at comparative levels.

The timing of the PN formation, which is of crucial importance for

our understanding of the late phases of stellar evolution, is a priori
not known owing to our poor knowledge of the mass-loss processes on the
AGB. The high sensitivity of PAGB evolutionary tracks to the initial
phase ϕ_i at the tip of the AGB allows, however, a distinction to be
made between the helium-burning (ϕ_i = 0) and hydrogen-burning
($\phi_i \gtrsim$ 0.3) mode of evolution by observations. Using hydrogen-burning
models (with $\phi_i >$ 0.5), Schönberner (1981), Schönberner and Weidemann
(1981) and Schönberner (1984) demonstrated that the temporal evolution
of central stars can be very well explained by models with masses between 0.55 and 0.64 $M_\odot$. The conclusion then follows that obviously the
PN ejection is generally not initiated by a thermal pulse, but occurs
during the quiescent hydrogen-burning phase on the AGB. Also, the observed shape of the luminosity function of CPN could only be explained
by hydrogen-burning PAGB models (e.g. Fig. 9 of Schönberner, 1981, and
discussion in Schönberner and Weidemann, 1983). One special feature of
this luminosity function is a deficit ("gap") of CPN with $M_v \approx$ 5. This
"gap" can be explained by hydrogen-burning PAGB models of $\approx 0.6\ M_\odot$ because they experience a rapid luminosity drop of $\approx$ 1 dex within only
about 10^3 years when hydrogen burning starts to cease. Conversion into
a luminosity function leads to a pronounced dip between $M_v \approx$ 4.5 and
6.0, the exact position depending somewhat on the mass of the models.
Such a luminosity drop is not found in models that leave the AGB while
burning helium (cf. Fig. 1 in Iben, 1984), and this fact clearly indicates that at least the majority of CPN must be hydrogen burners. Additional observational support for a fast luminosity drop during the
CPN evolution comes from the variation of the nebular ionization during
the later phases of evolution. Schönberner (1986) showed that a correlation exists between the luminosities of the CPN and the degree of
nebular ionization, in that PN with a lower ionization also belong to
intrinsically faint CPN, whereas highly ionized PN also have luminous
central objects (see also Schmidt-Voigt and Köppen, 1987).

In this review, I will concentrate only on models that are evolving off the AGB in thermal equilibrium under the influence of hydrogen
shell burning and mass loss. The possibility of a final helium shell
flash, and its consequences, is extensively discussed in Iben (1984,
1987).

2. POST-AGB EVOLUTION

The structure of an AGB star is rather complicated. It has a hydrogen-
exhausted core, M_H, which contains two burning shells, namely the
hydrogen-burning shell at the core's surface and the helium-burning
shell further inwards. The helium-exhausted inner part of the core consists of carbon and oxygen and is electron degenerated. The core is
actually nothing else than a very hot white dwarf which is surrounded
by a huge, nearly fully convective envelope, M_e, containing the
unprocessed stellar matter. The stellar radius exceeds the core radius
by factors up to about 10^4! In the course of evolution along the AGB,
the hydrogen-exhausted core is growing in mass at the expense of the
envelope due to nuclear burning in the hydrogen-burning shell, while

its radius is shrinking. The core of an AGB star may contain up to more than 99% of the stellar mass! The evolutionary track of an AGB star in the H-R diagram is entirely due to the response of the envelope to the masswise growing core: expansion of the envelope along the AGB, and finally contraction to white-dwarf dimensions if the envelope mass becomes too small (PAGB evolution).

The evolution along the AGB is terminated if either M_e becomes very small by the combined effect of nuclear burning in the hydrogen-burning shell and mass loss from the surface, or M_H approaches 1.4 $M_\odot$. The second possibility leads to an SN explosion and will not be discussed here. The transition from an AGB star to a white dwarf can be split into two steps:

i) If M_e is only of the order of several percent of the stellar mass, the envelope starts to shrink, but is still able to release enough gravitational energy to maintain the burning temperatures at its base. Consequently, the luminosity stays about constant ("plateau" luminosity), and the star evolves horizontally across the HR diagram. The core evolution is still independent from that of the envelope.

ii) If $M_e/M_\odot$ becomes about 10^{-4}, the hydrogen-burning shell starts to cool, the effective temperature reaches its maximum value (turnaround point), the luminosity drops rapidly and the star enters the white-dwarf regime, living mainly from its gravitational energy (Iben and Tutukov, 1984; Koester and Schönberner, 1986).

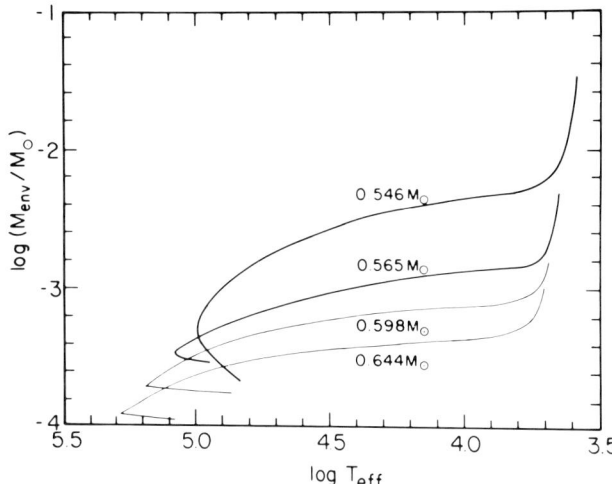

Fig. 1: Envelope mass M_e vs. T_{eff} for PAGB models of different core masses M_H according to Schönberner (1983).

Fig. 1 shows the variation of the envelope mass M_e with effective temperature for different PAGB models as given by Schönberner (1983). Note that in these evolutionary phases, M_H practically equals the total stellar mass M because of the smallness of M_e ($M = M_e + M_H$). For a given M_H, a unique relation $T_{eff}(M_e)$ exists for the horizontal evolution from the AGB till the turn-around point at $T_{eff} > 10^5$ K. The shapes of the relations $T_{eff}(M_e)$ are similar, but $\dot{M}_e$ increases with decreasing M_H. Similar relations for a larger range of M_H are given in Paczynski (1971).

The timescale for the crossing of the HR diagram with the "plateau" luminosity L is determined by the total amount of the available fuel ΔM_e and the fuel consumption $\dot{M}_e$:

$$\Delta t = \Delta M_e / \dot{M}_e .$$

Following Schönberner (1987), we define a horizontal "speed" as follows:

$$\dot{T}_{eff} = \dot{M}_e (dT_{eff}/dM_e),$$

where $\dot{M}_e$ consists of two terms, one of which being due to nuclear burning, $\dot{M}_H$, at the bottom of the envelope, the other describing mass loss from the surface by a stellar wind, $\dot{M}_W$:

$$\dot{M}_e = -(\dot{M}_H + \dot{M}_W).$$

With $\dot{M}_H = L/E_H X_e$, where E_H ($= 6 \, 10^{18}$ erg g^{-1}) is the energy release per gram of hydrogen, and X_e the hydrogen abundance (by mass) in the envelope, we have for a typical PAGB star of 0.6 $M_\odot$ with $L = 6000 \, L_\odot$: $\dot{M}_H \approx 10^{-7} \, M_\odot$ yr^{-1}. This value may be compared with typical mass-loss rates as they are found in the CPN regime which are, in most cases, well below $10^{-7} \, M_\odot$ yr^{-1} (Cerruti-Sola and Perinotto, 1985). Thus, it appears that only the nuclear term controls the horizontal speed of hydrogen-burning PAGB stars throught the CPN region.

The situation is different at the cool side of the H-R diagram. Without mass loss, all models evolve extremely slowly in the vicinity of the AGB, as can be understood from the shape of the $T_{eff}(M_e)$ relation. Observed mass-loss rates at the tip of the AGB seem to reach values of $\dot{M}_W \approx 10^{-4} \, M_\odot$ yr (e.g. Knapp, 1987), about 3 orders of magnitude larger than the nuclear term $\dot{M}_H$. Even a Reimers-like wind (Reimers, 1975) with its $\dot{M}_W \approx 10^{-6} \, M_\odot$ yr^{-1} exceeds $\dot{M}_H$ by a large amount. Thus, it is the mass loss which terminates the AGB evolution and also controls the evolutionary speed in the vicinity of the AGB. The different transition times from the AGB to the CPN region for a hydrogen-burning remnant of 0.6 $M_\odot$ are collected in Table 1 for 3 different cases. Case 1 means no mass loss at all, $\dot{M}_W = 0$. Case 2 means that mass loss is included according to the Reimers formula ($\eta = 1$), which is assumed to hold, for convenience, also for hotter stars (Schönberner, 1979, 1983). Finally, Case 3 is the model adopted by Schönberner (1983): $\dot{M}_W$, as in Case 2, with the exception that $\dot{M}_W = 10^{-4} \, M_\odot$ yr^{-1} for $T_{eff} \leq 10^{3.7}$ K ("superwind", Renzini, 1981).

Table 1 demonstrates clearly the sensitivity of the transition time from the tip of the AGB to 30000 K to the assumed mass-loss model. Especially the details of the "superwind" are important for this transition time, since if the "superwind" stops too early (i.e. at a lower

Table 1: Transition times $\Delta t = \Delta M_e/(\dot{M}_H + \dot{M}_W)$ in different parts of the H-R diagram for a PAGB model with $M_H = 0.6\ M_\odot$ and $\dot{M}_H = 9\ 10^{-8}\ M_\odot yr^{-1}$ from Schönberner (1979).

$\Delta \log T_{eff}$	$M_e/M_\odot$	t/yr		
		Case 1	Case 2	Case 3
3.55 .. 3.7	$5\ 10^{-2}$	$5.5\ 10^5$	$6\ 10^4$	$5\ 10^2$
3.7 ... 4.5	$6\ 10^{-4}$	$7\ 10^3$	$3\ 10^3$	$3\ 10^3$
4.5 ... 5.0	$3\ 10^{-4}$	$3.5\ 10^3$	$3.5\ 10^3$	$3.5\ 10^3$

T_{eff} than assumed in Case 3), the remnant will spend too much time in the vicinity of the AGB ("lazy" CPN, cf. Renzini, 1981). With reasonable assumption about $\dot{M}_W$ (cf. Case 3), it is possible to get short

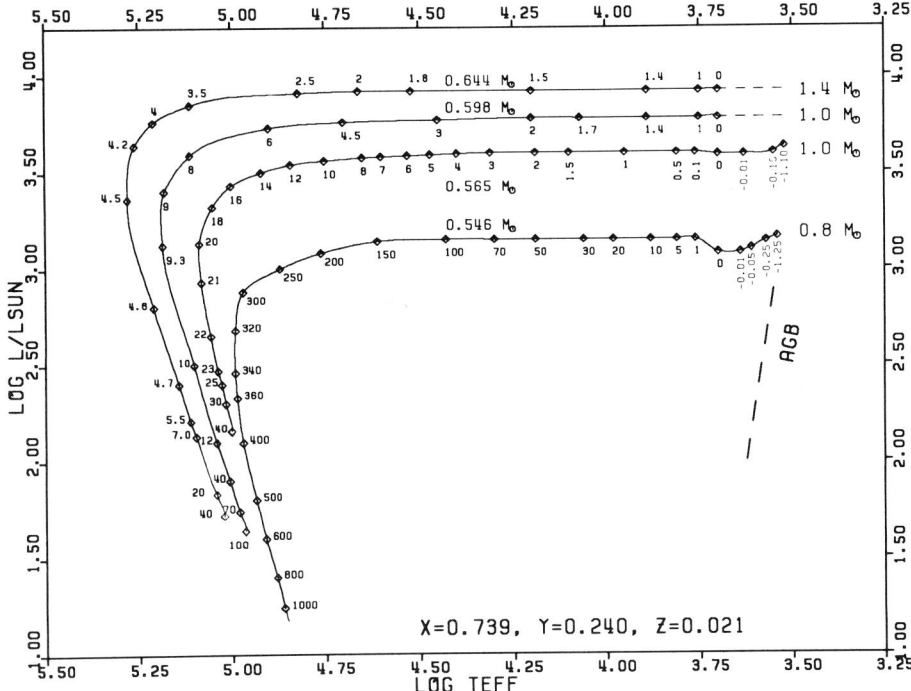

Fig. 2: Evolutionary tracks of four hydrogen-burning post-AGB models (Schönberner, 1981, 1983). The numbers give the ages in 1000 yr; age zero is assumed at $T_{eff} = 10^{3.7}$ K.

transition times which are consistent with the observations. Fortunately, the mass-loss term appears to be unimportant at hotter temperatures (see above), and this fact facilitates the modelling of CPN evolution by hydrogen-burning PAGB remnants without the details of previous mass-loss phases being known. A fuller discussion of mass loss and PAGB evolution can be found in Schönberner (1987).

Fig. 2 shows the evolutionary tracks of four hydrogen-burning post-AGB models according to the computations of Schönberner (1979, 1983), generated from Pop I stars with inital masses varying from 0.8 to 1.4 $M_\odot$. The two lower-mass remnants were generated according to Case 3 of Table 1, the remaining two according to Case 2. Except for the lowest remnant mass (0.546 $M_\odot$), the transition times are rather short and not in contradiction with the observations. This is due to the inclusion of mass loss which considerably accelerates the evolution below 10000 K. The horizontal evolution through the CPN region is highly mass-sensitive: $\Delta t \sim M_H^{-10}$. The reason is the larger luminosity (core-mass luminosity relation) and the smaller available amount of fuel (see Fig. 1) if M_H is increased. A similar mass dependence holds for the luminosity drop when hydrogen burning extinguishes.

This rapid drop of the stellar luminosity is possible because the hydrogen-burning shell is so thin (in mass, $\approx 10^{-4} M_\odot$) and the helium-burning shell so weak ($L_{He} \approx 0.01 L_H$). This luminosity drop is expected to be larger and faster the smaller the hydrogen-shell mass and the lower the helium-shell luminosity is. The former decreases with increasing core mass M_H (i.e. with increasing luminosity), and the minimum of L_{He} during a thermal-pulse cycle decreases with increasing pulse number (Gingold, 1974, Fig. 2). Thus, we expect that only post-AGB stars which went on the AGB through full-amplitude helium shell flashes experience fast luminosity drops when hydrogen burning ceases.

Indeed, the 0.546 $M_\odot$ model is still below the threshold for the occurrence of thermal pulses, and helium burning still contributes about 30% to the stellar luminosity. The 0.565 $M_\odot$ model experienced 4, the 0.598 $M_\odot$ model 10 and the 0.644 $M_\odot$ model 24 thermal pulses. The initial thermal-pulse cycle phase for the post-AGB evolution of the latter three models is about 0.7. In passing, we note that helium-burning models do not show rapid luminosity drops (Iben, 1984), obviously because the mass contained in the helium-burning shell is about 100 times larger than that of the hydrogen-burning shell.

3. A "STANDARD" 0.6 $M_\odot$ PAGB MODEL

In this section I will try to extract the properties of a typical 0.6 $M_\odot$ hydrogen-burning PAGB model, as they follow from computations of different authors. The models are the following:
1: 0.598 $M_\odot$, Z = 0.02, Schönberner (1979);
2: 0.593 $M_\odot$, Z = 0.02, Kovetz and Harpaz (1981);
3: 0.599 $M_\odot$, Z = 0.001, Iben (1984);
4: 0.6 $M_\odot$, Z = 0.02, Iben and MacDonald (1986);
5: 0.6 $M_\odot$, Z = 0.02, Wood and Faulkner (1986).
Important features of these evolutionary models are compiled in

Table 2, as there is the "plateau" luminosity L, the envelope mass ΔM_e burnt between 30000 K and 100000 K, the envelope mass at the turn-around point, $M_e(TA)$, the luminosity drop within 10^3 yr, $\Delta \log L/L_\odot$, starting at the turn-around point, the absolute magnitude, M_v, after that drop, the absolute magnitude, M_v', after 50000 yr, and the number of thermal pulses, N, on the AGB.

Table 2 shows that all the pop I models (Nos. 1,2,4,5) have practically the same "plateau" luminosity (the table entries are not corrected for the slightly different model masses). Despite its larger envelope mass, the evolution of the pop II model of Iben (No. 3), beyond the turn-around point is essentially identical with that of

Table 2: Important properties of 0.6 $M_\odot$ hydrogen-burning PAGB models

Mod.	$\log L/L_\odot$	$M_e/M_\odot$	$M_e(TA)/M_\odot$	$\log L/L_\odot$	M_v	M_v'	N
1	3.78	$3\ 10^{-4}$	$1.9\ 10^{-4}$	0.8	6.0	6.8	10
2	3.81	$3\ 10^{-4}$	$1.0\ 10^{-4}$	1.2	6.2	7.0	5
3	3.73	$7\ 10^{-4}$	$2.7\ 10^{-4}$	0.9	6.2	7.2	10
4	3.80	$2.5\ 10^{-4}$	$1.2\ 10^{-4}$	-*	-*	-*	-*
5	3.79	$6\ 10^{-4}$	$3\ 10^{-4}$	0.5	4.9	6.5	12

* No information given.

model No. 1 and 2. Of course, the horizontal evolution of model 3 is about 2 times slower because it burns more matter. The larger envelope mass for a given effective temperature, and the slightly lower luminosity, are obviously due to the lower metallicity, as the pop I model (No. 4) of Iben and McDonald demonstrates (computed with the same evolutionary code). Only model No. 5 of Wood and Faulkner disagrees in all its properties (except for the luminosity) from the other (pop I) models. A possible explanation will be given at the end of this section.

Neglecting for the moment the model of Wood and Faulkner (1986), the following properties emerge for a typical hydrogen-burning PAGB model of 0.6 $M_\odot$:

i) the "plateau" luminosity is 6200 $L_\odot$;
ii) the transition time from the AGB to 30000 K depends on the assumed mass-loss rates but may be as small as 3000 yr;
iii) the evolution from 30000 K till the turn-around point occurs in 6000 yr ($\dot{M}_w = 0$);
iv) the luminosity drops by ≈ 1 dex within 1000 yr when hydrogen burning stops;
v) the limiting CPN magnitude is predicted to be $M_v \approx 7$ (or $L \approx 80\ L_\odot$);

vi) it follows from the evolutionary rates that at least 75% of a complete sample of CPN should be fainter than $M_v \approx 6$ (or $L \approx 300\ L_\odot$).

I will close this review with a discussion on the discrepant behaviour of Wood and Faulkner's (1986) 0.6 $M_\odot$ hydrogen-burning PAGB model. It has already been shown above (cf. Table 2) that the computations of Schönberner (1979), Kovetz and Harpaz (1981) and also Iben (1984) - if opacity differences are taken into account - give practically the same results. The model of Wood and Faulkner (1986) differs considerably in that it burns more hydrogen, resulting in a reduced horizontal speed of evolution. Furthermore, the final luminosity drop is only one third as large (0.5 dex) and the limiting magnitude brighter by 0.5. Overall, the temporal evolution of Wood and Faulkner's (1986) hydrogen-burning 0.6 $M_\odot$ PAGB model mimics the corresponding 0.565 $M_\odot$ model of Schönberner (1983). Before going further into detail, it should be noted that theory predicts in fact a variation of the evolutionary speed with ϕ_i in the sense that the speed increases slightly with ϕ_i (Wood and Faulkner, 1986). This effect, however, cannot explain the discrepancies discussed here.

One might speculate that a possible explanation for these differences comes from the model histories on the AGB. In the Schönberner, Kovetz and Harpaz, and Iben calculations, mass loss was either included according to Reimers' formula (1975) or simply neglected. Wood and Faulkner, however, applied a rate as high as $\approx 1\ M_\odot\ yr^{-1}$ till the star was stripped down to $M_e = 0.015\ M_\odot$. Then a much lower rate was used ($3.10^{-5}\ M_\odot\ yr^{-1}$). A rate of $\approx 1\ M_\odot\ yr^{-1}$ certainly destroys the thermal equilibrium in the deeper layers. For instance, Schönberner (1983) found that already rates of $\dot{M}_w \approx 10^{-4}\ M_\odot\ yr^{-1}$ lead to small de-adjustments of the nuclear-burning regions (cf. also Fig. 2 above). Much larger effects are expected for even higher mass-loss rates. A not thermally adjusted PAGB model has a larger hydrogen-burning shell mass and, as a consequence, also a larger envelope mass for a given effective temperature. Also such a model should have a larger gravitational energy release. Both effects result in reduced evolutionary speed and luminosity drop. It would be desirable to make a direct comparison between the internal structures of the Wood and Faulkner models and those of the other authors.

For the time being, the following conclusions can be drawn: since the properties of PAGB models are extremely sensitive to the previous treatment on the AGB, realistic models for central stars are only expected if

i) the applied mass-loss rate does not largely exceed $\approx 10^{-4}\ M_\odot\ yr^{-1}$,
ii) all thermal pulses are taken properly into account,
iii) the initial masses are roughly consistent with an empirical initial-final mass relation.

ACKNOWLEDGMENTS. The author gratefully acknowledges a travel grant from the Deutsche Forschungsgemeinschaft.

REFERENCES

Cerruti-Sola, M., Perinotto, M.: 1985, Astrophys. J. **291**, 237.
Gingold, R.A.: 1974, Astrophys. J. **193**, 177.
Harpaz, A., Kovetz, A.: 1981, Astron. Astrophys. **93**, 200.
Härm, R., Schwarzschild, M.: 1975, Astrophys. J. **200**, 324.
Iben, I. Jr.: 1982, Astrophys. J. **260**, 821.
Iben, I. Jr.: 1984, Astrophys. J. **277**, 333.
Iben, I. Jr.: 1987, "Late Stages of Stellar Evolution", S. Kwok and S.R. Pottasch (eds.), Reidel, Dordrecht, p. 175.
Iben, I. Jr., Tutukov, A.V.: 1984, Astrophys. J. **282**, 615.
Iben, I. Jr., MacDonald, J.: 1986, Astrophys. J. **301**, 164.
Knapp, G.R.: 1987, "Late Stages of Stellar Evolution", S. Kwok and S.R. Pottasch (eds.), Reidel, Dordrecht, p. 103.
Koester, D., Schönberner, D.: 1986, Astron. Astrophys. **154**, 125.
Kovetz, A., Harpaz, A.: 1981, Astron. Astrophys. **95**, 66.
Paczynski, B.: 1971, Acta Astron. **21**, 417.
Reimers, D.: 1975, "Problems in Stellar Atmospheres and Envelopes", B. Baschek, W.H. Kegel, G. Traving (eds.), Springer, Berlin, p. 229.
Renzini, A.: 1981, "Physical Processes in Red Giants", I. Iben Jr. and A. Renzini (eds.), Reidel, Dordrecht, p. 431.
Schmidt-Voigt, M., Köppen, J.: 1987, Astron. Astrophys. **174**, 223.
Schönberner, D.: 1979, Astron. Astrophys. **79**, 108.
Schönberner, D.: 1981, Astron. Astrophys. **103**, 119.
Schönberner, D.: 1983, Astrophys. J. **272**, 708.
Schönberner, D.: 1984, IAU Symp. No. 105 "Observational Tests of the Stellar Evolution Theory", A. Maeder and A. Renzini (eds.), Reidel, Dordrecht, p. 209.
Schönberner, D.: 1986, Astron. Astrophys. **169**, 189.
Schönberner, D.: 1987, "Late Stages of Stellar Evolution", S. Kwok and S.R. Pottasch (eds.), Reidel, Dordrecht, p. 337.
Schönberner, D., Weidemann, V.: 1981, "Physical Processes in Red Giants", I. Iben and A. Renzini (eds.), Reidel, Dordrecht, p. 463.
Schönberner, D., Weidemann, V.: 1983, IAU Symp. No. 103, "Planetary Nebulae", R.D. Flower (ed.), Reidel, Dordrecht, p. 359.
Wood, P.R., Faulkner, D.J.: 1986, Astrophys. J. **307**, 659.

Harm Habing, Michael Feast and Wil van der Veen.

THE DISTRIBUTION OF PLANETARY NEBULA NUCLEI IN THE LOG L-LOG T PLANE: INFERENCES FROM THEORY

R. A. Shaw
Lick Observatory
University of California
Santa Cruz, CA 95064

ABSTRACT. The expected distribution of planetary nebula nuclei (PNNs) on the log L-log T plane is calculated based upon modern stellar evolutionary theory, the initial mass function (IMF), and various assumptions concerning mass loss during post-main sequence evolution. The distribution is found to be insensitive to the assumed range of main-sequence progenitor mass, and to reasonable variations in the age and the star forming history of the galactic disk. Rather, the distribution is determined primarily by the heavy dependence of the evolution rate upon core mass, and secondarily upon the steepness of the IMF and other factors. The distribution is rather different than any found from observations, and probably reveals strong observational selection effects.

1. INTRODUCTION

Nearly a quarter-century has passed since the first serious attempts to study the evolution of the nuclei of planetary nebulae. The observational studies of O'Dell (1962, 1963), Harman and Seaton (1964), Seaton (1966), Schönberner (1981), Kaler (1983), and Heap and Augensen (1987) have traced the evolution of cool, luminous PNNs surrounded by dense, compact planetary nebulae, to temperatures exceeding 10^5 K, followed by sharply declining luminosities as the nebular shells expand and the stars contract to white dwarf dimensions. On the theoretical side, Paczyński (1971), Schönberner (1979), Iben (1982, 1984), and Wood and Faulkner (1986) have published evolutionary post-asymptotic giant branch (post-AGB) stellar models. They revealed the importance of shell nuclear burning in providing the source of luminosity and determining the rate of evolution through the PNN phase, as well as the extraordinary dependence of the evolutionary timescales upon PNN mass. Together, the observations and theory have established the planetary nuclei unambiguously as the link between the AGB and the white dwarfs. Iben and Renzini (1983, hereafter IR83) provide a general review of this subject.

Although our understanding of post-AGB stellar evolution has improved greatly, our ability to compare observational results with predictions from theory has not kept pace. Indeed, previous observational studies, such as those of Schönberner (1981), Kaler (1983), and Heap and Augensen (1987), have dealt only obliquely with the question of, *e.g.* the distribution of PNNs on the log L-log T plane that one would expect directly from theory. This paper will derive just such a theoretically-determined distribution, and compare it with those determined from observations. The analysis will closely parallel that in Shaw, Truran, and Kaler (1984), and Shaw's (1985) doctoral dissertation.

2. CALCULATION OF THE LOG L-LOG T DISTRIBUTION

2.1. Theoretical Assumptions

The distribution expected from theory is possible to calculate in principle if the initial mass function (IMF), the relation between initial (main sequence) mass and post-AGB core mass, and the PNN evolution paths (including timescales) as a function of mass are all known. The following simplifying assumptions apply to this distribution calculation:

1. The adopted IMF is from Miller and Scalo (1979), which is well-known for stars less massive than 10 $M_\odot$. Furthermore, the slope of the IMF is assumed to be constant, although the effects of a slowly decreasing star formation rate are considered explicitly.

2. The function relating initial (main sequence) stellar mass to remnant (post-AGB) core mass is given by the nearly linear relation of Iben and Truran (1978), which is based upon Reimers' (1975) empirical formula for single, isolated stars. Here, stars up to 5 $M_\odot$ will produce PNNs, although a steeper linear function for stars up to 10 $M_\odot$ will be considered separately. In both cases it is assumed that no star less than 0.8 $M_\odot$ has evolved past the AGB in the lifetime of the galaxy, although a larger lower-limit will be adopted when a variation in galactic age is considered.

3. All stars that form degenerate cores less massive than the Chandrasekhar limit will illuminate a planetary nebula shell, or at least the fraction of those that do not does not change with PNN mass.

4. The evolutionary path through the log L-log T plane is uniquely defined by the PNN mass (see § 2.2 below), and each point in the log L-log T plane is intersected by one and only one evolution path (*i.e.* the evolution paths do not cross).

5. Only those stars that are passing through the PNN domain for the first time were included: those that suffer a post-AGB helium flash (see Iben 1984; Wood and Faulkner 1986) will be considered in § 4.1.

6. Finally, stars do not lose a significant amount of mass during the PNN phase, although the effect of the mass-loss process itself upon the rate of evolution will be discussed in § 4.1.

2.2. Adopted Stellar Models

Three specific post-AGB evolution models were adopted from Schönberner (1983): they have core masses of 0.546, 0.565, and 0.644 $M_\odot$. Two other models were taken from Paczyński (1971): they have core masses of 0.80 and 1.20 $M_\odot$. These models were chosen largely because they cover virtually the entire range of interest in mass, and because each investigator has provided a set of internally consistent evolution tracks that can be compared with the other investigator's models.

Unfortunately, this comparison reveals a major problem with the Paczyński models, namely the post-AGB helium flashes that result from an improper initial description of the structure of the AGB star (Schönberner 1981). Fortunately, as IR83 point out, a model PNN of a given, constant mass will always follow roughly the same path in the log L-log T plane, and the evolution timescales are not especially sensitive to details of the models. The discrepancy was resolved by comparing the 0.60 M$_\odot$ models and scaling those from Paczyński along the following guidelines: first, the luminosity of a remnant core that has just evolved off the AGB is approximately given by the linear relation (Schönberner 1983): $L_{PNN} = 6.0 \times 10^4 (M_{PNN} - 0.513)$; and second, the time required for the PNN to fade by 2.5 bolometric magnitudes after first ionizing the nebular shell scales roughly as $M_{PNN}^{-9.6}$ (IR83). This scaling also agrees well with the more recent models of Wood and Faulkner (1986) for higher PNN mass. In this same way, using the mass-radius relation for white dwarfs as an asymptote to the evolution path, the evolution track of an Chandrasekhar-limited PNN (*i.e.* at 1.4 M$_\odot$) was estimated.

2.3. The Domain

The calculation of the PNN distribution was done on a grid whose extent on the log L-log T plane extends over $1.6 < \log L/L_\odot < 4.8$, and $4.4 < \log T_{eff} < 6.1$. These limits were chosen partly for the availability of evolutionary models, and partly to correspond roughly to the domain over which the surrounding planetary nebula (PN) can be observed: the PNNs do not ionize the proto-planetary shell when they are cooler than $\log T_{eff} = 4.4$; the PNs are too distended to be detected easily after they expand for more than $\sim 30,000$ yr, which corresponds to about $\log L \geq 1.6$; and the other boundaries are set by the tracks of the most and least massive PNNs that pass through this domain (see § 2.1 and § 2.2).

The scheme was as follows: the PNNs were assumed to be forming at a constant rate, and the zero-point of the model timescales was adjusted such that $t = 0$ corresponds to the moment that the PNNs reach $\log T_{eff} = 4.4$. The domain was divided into grid intervals of 0.2 in log L, and 0.1 in log T. The evolution tracks were used to determine the mass track that passes through each grid point, and the time it takes a PNN on that track to evolve to that point. Each adjacent pair of grid points defined a boundary of a cell, within which the number of PNNs contained was calculated, according to the description of Fig. 1.

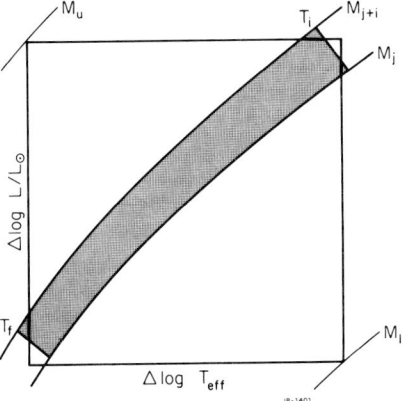

Fig. 1. Schematic representation of the calculation. Integration was performed over a strip spanning the evolution tracks M_j and M_{j+1} to yield the total number of PNNs in that mass interval. The product of that integral and the difference between times T_f and T_i (when the PNN leaves and enters the cell, respectively) yields the total number of PNNs in that strip. The total from each strip between M_l and M_u was summed and normalized such that the total from all cells in the PNN domain was unity.

3. THE DISTRIBUTIONS

The calculated distributions are listed in Fig. 2a, and shown as a contour plot in Fig. 2b, for the steady-state case. (A color representation of this distribution may be found in Kaler 1986.) This calculation, where progenitor masses between $0.8 < M_*/M_{PNN} < 5$ were included, will serve as a benchmark of comparison for the other calculations. The most striking feature in the figures is the tremendous range in relative number: about 8 orders of magnitude. This large range results from the great dependence of the evolution time-scales upon core mass and, hence, is also the greatest source of uncertainty in the calculations. The contours of constant PNN density are nearly coincident with lines of constant age (beyond $\log T_{eff} = 4.4$), and show that the vast majority of PNNs should be found along the lowest-mass evolution tracks, and at relatively low luminosity.

The next illustrative case is considered in Fig. 3, where progenitor stars up to 10 $M_\odot$ are assumed to produce PNNs. This change, though drastic in the broad context of stellar evolution, produces only a small change in the high mass portion of the log PNN distribution. And yet this subtle difference is not at all surprising in view of the steepness of the IMF—i.e. the number of stars formed in the 5 to 10 $M_\odot$ range is a very small fraction of the total number of PNN progenitors. The finite age of the galactic disk is addressed in the next calculation. Figure 4 shows the distribution where only those stars that evolve past the AGB within 1.0×10^{10} years (or $M_* > 1\ M_\odot$) are included. The disappearance of PNNs along the lowest-mass tracks is significant—reflecting the steepness of the IMF—but although the normalization for the distribution has changed, the relative numbers in each cell above the adjusted low-mass cut-off have remained constant. The next calculation includes an exponential decrease in the star formation rate, normalized to be a factor of 2 lower at present than at the birth of the galactic disk. This effect, shown in Fig. 5, is the smallest found in this study, and is naturally most pronounced in the high mass portion of the plane, due to the shorter stellar lifetimes.

4. INTERPRETATION

4.1. Lessons from Theory

Of the several conclusions that can be drawn from the distributions presented above, the surest is this: There is no hope of divining the range of PNN progenitor masses purely from considerations of an observed distribution in the log L-log T plane. Indeed, only ~1600 galactic PNs are known, which is far too few to examine a distribution which covers 8 orders of magnitude. There are other means of approaching the problem, however, which include a comparison of nebular He and CNO element abundances with those predicted by AGB dredge-up theory. Such a study by Kaler (1983), using the theory of Becker and Iben (1979, 1980), showed qualitative agreement. Another promising approach would be to use the abundances derived from stellar emission spectra for those PNNs with winds. Such an analysis may provide an interesting *ex post facto* radial probe of the nuclear burning and mixing processes in AGB stars, since the nebular shell and the subsequent winds came from different depths within the remnant AGB star.

IR83 note that the rate of evolution through the PNN domain is governed largely by the rate at which the star's fuel is depleted, be it through nuclear burning or mass-loss through a stellar wind. For typical mass-loss rates of 10^{-9}

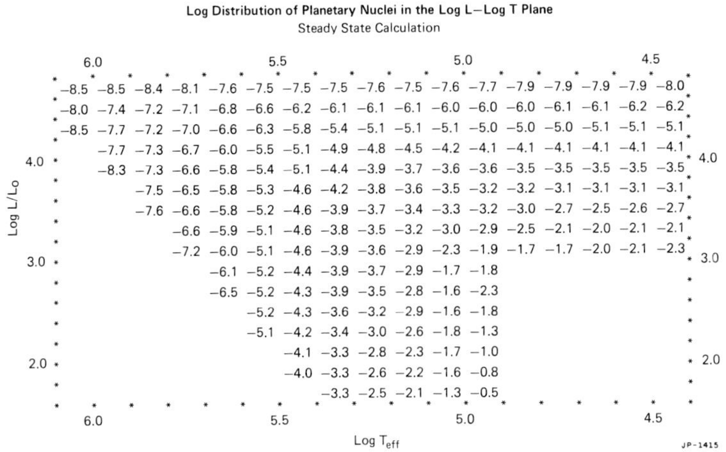

Log Distribution of Planetary Nuclei in the Log L–Log T Plane
Steady State Calculation
6.0 5.5 5.0 4.5
−8.5 −8.5 −8.4 −8.1 −7.6 −7.5 −7.5 −7.5 −7.6 −7.5 −7.6 −7.7 −7.9 −7.9 −7.9 −7.9 −8.0
−8.0 −7.4 −7.2 −7.1 −6.8 −6.6 −6.2 −6.1 −6.1 −6.1 −6.0 −6.0 −6.0 −6.1 −6.1 −6.2 −6.2
−8.5 −7.7 −7.2 −7.0 −6.6 −6.3 −5.8 −5.4 −5.1 −5.1 −5.1 −5.0 −5.0 −5.0 −5.1 −5.1 −5.1
−7.7 −7.3 −6.7 −6.0 −5.5 −5.1 −4.9 −4.8 −4.5 −4.2 −4.1 −4.1 −4.1 −4.1 −4.1 −4.1
−8.3 −7.3 −6.6 −5.8 −5.4 −5.1 −4.4 −3.9 −3.7 −3.6 −3.6 −3.5 −3.5 −3.5 −3.5 −3.5
−7.5 −6.5 −5.8 −5.3 −4.6 −4.2 −3.8 −3.6 −3.5 −3.2 −3.2 −3.1 −3.1 −3.1 −3.1
−7.6 −6.6 −5.8 −5.2 −4.6 −3.9 −3.7 −3.4 −3.3 −3.2 −3.0 −2.7 −2.5 −2.6 −2.7
−6.6 −5.9 −5.1 −4.6 −3.8 −3.5 −3.2 −3.0 −2.9 −2.5 −2.1 −2.0 −2.1 −2.1
−7.2 −6.0 −5.1 −4.6 −3.9 −3.6 −2.9 −2.3 −1.9 −1.7 −1.7 −2.0 −2.1 −2.3
−6.1 −5.2 −4.4 −3.9 −3.7 −2.9 −1.7 −1.8
−6.5 −5.2 −4.3 −3.9 −3.5 −2.8 −1.6 −2.3
−5.2 −4.3 −3.6 −3.2 −2.9 −1.6 −1.8
−5.1 −4.2 −3.4 −3.0 −2.6 −1.8 −1.3
−4.1 −3.3 −2.8 −2.3 −1.7 −1.0
−4.0 −3.3 −2.6 −2.2 −1.6 −0.8
−3.3 −2.5 −2.1 −1.3 −0.5
4.0 3.0 2.0
Log L/Lo
6.0 5.5 5.0 4.5
Log Teff
JP-1415

Fig. 2*a*. A two-dimensional histogram for the steady-state calculation showing the log of the relative number of PNNs in each cell on the log L-log T plane. This calculation includes the contribution from all progenitor stars in the mass range $0.8 < M_*/M_\odot < 5.0$.

-7.5
-6.0
-4.5
-3.0
4.0
3.0
2.0
log L/L⊙
1.40 M⊙
1.20 M⊙
0.80 M⊙
0.644 M⊙
0.598 M⊙
0.565 M⊙
0.545 M⊙
-1.5
6.0
5.5
5.0
4.5
log Teff
JP-1402

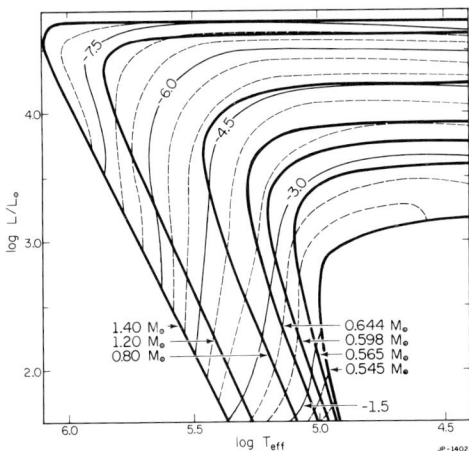

Fig. 2*b*. A contour plot of the log distribution presented in Fig. 2*a*. The intervals refer to the log of the relative number of PNNs expected to lie within each cell. Also shown are the adopted PNN evolution tracks (***heavy solid curves***): the 0.546, 0.565, 0.598 and 0.644 $M_\odot$ models are from Schönberner (1981), the 0.80, and 1.20 $M_\odot$ models are from Paczyński (1971, scaled), and the 1.4 $M_\odot$ model is from an estimate for the Chandrasekhar-limited case (see text).

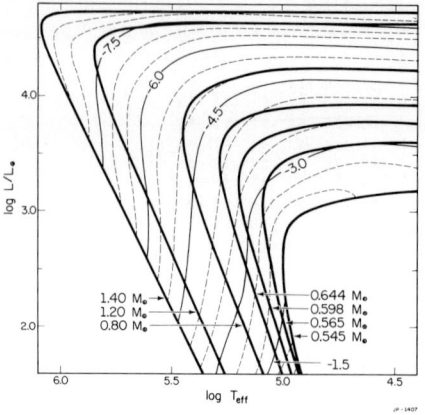

Fig. 3. Same as Fig. 2b, except that stars up to 10 $M_\odot$ are assumed to produce PNNs.

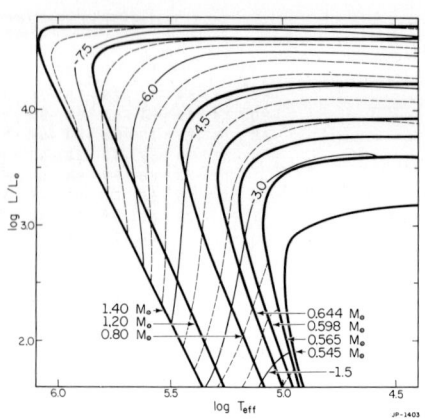

Fig. 4. Same as Fig. 2b, except that only those progenitor stars that have produced PNNs within the last 10 billion years are included.

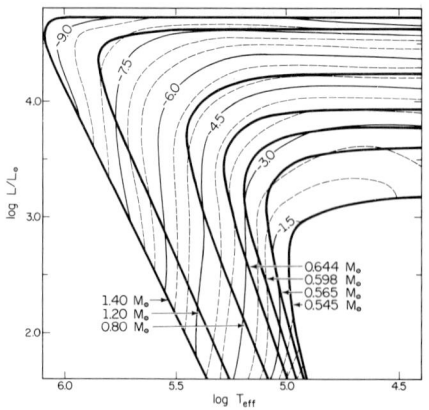

Fig. 5. Same as Fig. 2b, except that the SFR is assumed to have declined exponentially by a factor of 2 during the life of the galactic disk.

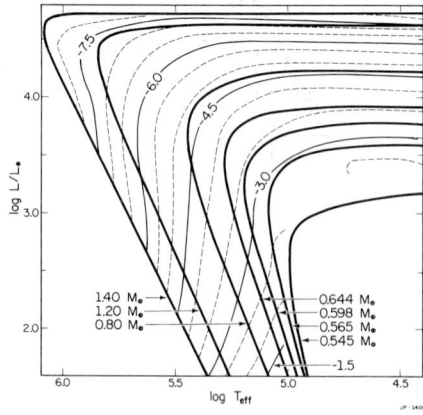

Fig. 6. Same as Fig. 2b, except that the PNNs are assumed to be undetectable after 30,000 yr.

to 10^{-7} $M_\odot$ yr^{-1}, and a typical hydrogen envelope mass of $\sim 10^{-3}$ to 10^{-4} $M_\odot$ (Iben 1984), a stellar wind could speed the PNN evolution by up to a factor of about 3. Although the explicit effect on the distribution of some fraction of PNNs (perhaps 50%) possessing winds was not calculated, it must compound the already high PNN density at low L, particularly for high mass PNNs if they preferentially engender high-velocity winds (see Wood and Faulkner 1986). On the other hand, if about 10% of all PNNs produce their luminosity through shell helium burning (Iben 1984), then somewhat more PNNs would be found at higher L (due to the slower evolution rate), near the lower-mass evolution tracks.

4.2. Observational Selection Effects

While the observed distributions cannot yield the PNN progenitor mass range directly, the theoretical distributions can reveal selection effects in the observed sample. In particular, the observed distributions of Schönberner (1981), Kaler (1983), and Heap and Augensen (1987) are not based upon a volume-complete sample, and do not match the broad features of that expected from theory. Furthermore, no such complete study has been done of those rare PNNs that are both extremely hot ($T_{eff} \geq 150,000$ K) *and* extremely luminous (L $\geq 10^3$ $L_\odot$).

Several effects may be operating to skew the observations. These include the systematic difficulties with the distance scale for young, optically thick PNs (Maciel and Pottasch 1980; Daub 1982); systematic contamination of the stellar continuum by the nebula in compact planetaries (Shaw and Kaler 1985); and the difficulty of discovering older, low surface brightness PNs due both to the drop in their emission measure as the shell expands, and to their propensity to lie in the galactic plane, where they suffer heavy interstellar extinction. The effect of aging nebular shells fading from view is shown in Fig. 6, where PNs older than 30,000 yr are assumed to be undetectable and are omitted. Although the contribution of low-mass PNNs at low L to the distribution is significantly smaller, the qualitative features are the same as in Fig. 2b, in that this distribution still predicts far more PNNs at low L than are seen observationally.

5. CONCLUSIONS AND FUTURE WORK

Clearly the theoretical distribution of PNNs on the log L-log T plane differs substantially from that determined from observations. It is probably also true that the theory for PNN evolution is far more secure than the observations, although an observationally complete, volume-limited sample of PNNs should be analyzed to understand fully the discrepancy. Those PNNs that have suffered a post-AGB helium shell flash (as is likely the case for A 30 and A 78: see Iben *et al.* 1983) must also be identified observationally, both to guide the theorists on the relative frequency of such events, and to allow for their effect on the observed distribution. It would be misleading to say that any one of the distributions calculated here on the basis of simplified stellar evolutionary theory represents the true distribution. Rather, the broad features of the distribution are presented to illustrate the nature of the discrepancy between observations and theory, and to suggest promising lines of investigation to resolve them.

Support for this research is acknowledged from NSF grants AST 80-23233 and 83-14415 to the Univ. of Illinois, and AST 86-11457 to the Univ. of California.

REFERENCES

Becker, S. A., and Iben, I. Jr. 1979, *Ap. J.*, **232**, 831.
Becker, S. A., and Iben, I. Jr. 1980, *Ap. J.*, **237**, 111.
Daub, C. T. 1982, *Ap. J.*, **260**, 612.
Harman and Seaton, M. 1964, *Ap. J.*, **140**, 824.
Heap, S. R., and Augensen, H. J. 1987, *Ap. J.*, **313**, 268.
Iben, I. Jr. 1984, *Ap. J.*, **277**, 333.
_____. 1982, *Ap. J.*, **260**, 821.
Iben, I. Jr., Kaler, J. B., Truran, J. W., and Renzini, A. 1983, *Ap. J.*, **264**, 605.
Iben, I. Jr., and Renzini, A. 1983, *Ann. Rev. Astr. Ap.*, **21**, 271 (IR83).
Iben, I. Jr., and Truran, J. W. 1978, *Ap. J.*, **220**, 980.
Kaler, J. B. 1986, *Am. Scientist.*, **74**, 244.
_____. 1985, *Ann. Rev. Astr. Ap.*, **23**, 89.
_____. 1983, *Ap. J.*, **271**, 188.
Maciel, W. J., and Pottasch, S. R. 1980, *Astr. Ap.*, **88**, 1.
Miller, G. E., and Scalo, J. M. 1979, *Ap. J. Suppl.*, **41**, 513.
O'Dell, C. R. 1962, *Ap. J.*, **135**, 371.
_____. 1963, *Ap. J.*, **138**, 67.
Paczyński, B. 1971, *Acta Astr.*, **21**, 417 (PZ).
Reimers, D. 1975, *Mem. Soc. R. Sci. Liege*, 6^e Ser. **8**, 369.
Schönberner, D. 1983, *Ap. J.*, **272**, 708.
_____. 1981, *Astr. Ap.*, **103**, 119.
_____. 1979, *Astr. Ap.*, **79**, 108.
Seaton, M. J. 1966, *M.N.R.A.S.*, **132**, 113.
Shaw, R. A. 1985, Ph. D. thesis, Univ. of Illinois.
Shaw, R. A., and Kaler, J. B. 1985, *Ap. J.*, **295**, 537.
Shaw, R. A., Truran, J. W., and Kaler, J. B. 1984, *Bull. A. A. S.*, **16**, 530.
Wood, P. R., and Faulkner, D. J. 1986, *Ap. J.*, **307**, 659.

THE POSITION OF THE CENTRAL STARS OF PN ON THE HR DIAGRAM

S.R. Pottasch
Kapteyn Astronomical Institute
P.O. Box 800
9700 AV Groningen
The Netherlands

1. INTRODUCTION

Central stars can be placed on the HR diagram if their effective temperature (T_{eff}) and radii are known. Knowledge of the radius can sometimes be replaced by another indication of the luminosity. The distance, which always plays an important, really critical role, is not well known. This is the essential reason that there is so much uncertainty about the position on the HR diagram.

The situation as it was several years ago is the following. Kaler (1983) studied the central stars of an extensive group of large, presumably evolved nebulae. He assumed the distance could be obtained from the Shklovskii method. His resultant diagram, which is shown as Fig. 5c, covers mainly the region after the nuclear burning has stopped. The central star positions are consistent with theoretical evolution calculations for core masses between 0.5 and 0.8 $M_\odot$.

I have approached the problem in a somewhat different way (Pottasch, 1983). The statistical methods of determining distance were discarded. Only those central stars were used whose distance could be determined in an independent way. The fact that they are independent of the statistical methods does not mean they are correct, because determining accurate distances is difficult. The general conclusions indicated: (1) the existence of less luminous central stars than predicted by the Schönberner (1981) 0.55 $M_\odot$ evolutionary track, and (2) the existence of very high temperature central stars.

A third approach to the problem is given in the work of Mendez et al. (1981, 1985). Here the profiles of the stellar H and He lines are measured at high resolution. With the help of model atmospheres, the profiles are fitted to derive T_{eff} and the gravity g. Using theoretical evolution tracks, g may be separated into the central star mass M_{cs}, and the stellar radius R. This leads directly to the luminosity. The distance is a by-product. The method can only work if the model atmospheres is approximately correct. The atmospheres used at present can only reproduce stars with an absorption line spectrum which limits the applicability of the method.

A related method of obtaining information from the evolutionary tracks makes use of a comparison between the predicted nebular ages and those observed. This has especially been applied by Schönberner (1981) who plots the predicted absolute magnitude M_v as a function of predicted age, and compares these with the observed values. He has obtained the result that most central stars have core masses between 0.55 $M_\odot$ and 0.6 $M_\odot$. In the comparison knowledge of the distance is necessary and these have been obtained from the Shklovskii method. Another important assumption which goes into this method is a knowledge of at what time in the theoretical evolution the nebulae is ejected. Schönberner (1981) assumes that this occurs when the star has a temperature of 5000 K, but this is apparently an arbitrary choice.

We shall report here on the most recent developments in these approaches, hopefully in a critical way.

2. DISTANCES TO PLANETARY NEBULAE

As mentioned above, the distance determination is critical for placing the central star on the HR diagram. The use of statistical distance scales is becoming more and more suspect, not only by myself (e.g. Pottasch, 1984; 1987; Gathier, 1983) but by others as well (e.g. Wood et al., 1986; Kinman et al., 1987). I feel that it is very dangerous to use for an arbitrary sample of nebulae because of the evidence that young nebulae are much more affected than older ones. This immediately biases the evolution. Kalers results using Shklovskii distances will be discussed below.

It is necessary to separate the groups of nebulae according to how the distance has been determined. This is not only because there may be inherent systematic errors which refer only to that particular group, but also because there are important selection effects present which should be known when intercomparing the results.

The four groups are:

(1) that diccussed by Gathier (1984) and Gathier and Pottasch (1987). They have used distances determined in various ways, but two methods dominated the sample. These are the extinction distance diagrams discussed in detail by Gathier et al. (1986a) and the 21 cm absorption method (Gathier et al., 1986b). The sample of the above authors has been modified to remove the expansion distances determined using photographic plates, for reasons which will become clear presently.

(2) that discussed by Mendez et al. (1987), which uses the analysis of the central star line profiles as discussed above. Only the latest results have been used, since these authors consider them more reliable. They are clearly selected as the brightest known central stars having an absorption line spectrum.

(3) that discussed by Kaler (1983). This group represents distance determination using the Shklovskii method and was chosen as a comparison with the group having independent distance. This particular study was chosen above others using the same distance method for two reasons. First, it has been widely cited and appears to be representative. Secondly, only large evolved nebulae are included in the sample, for wich the errors in the mass are probably limited to an order of magnitude and the distance is probably correct to about a factor of 2.

(4) a sample of galactic center nebulae selected for their faintness. Many are recently reported by Kinman et al. (1987). Some have been discussed by Webster (1975). Radio continuum flux densities have recently become available for all (Gathier et al., 1983; Zijlstra, unpublished).

These samples will now be compared assuming the distance given by the above authors is correct. Fig. 1 is a histogram of the intrinsic 6 cm continuum flux density which the nebulae would have if placed at the galactic center, for the four samples. The results of Jacoby (1980) in a survey of nebulae in the Magellanic Clouds is included for comparison. For many of the nebulae studied by Kaler (1983) and the Magellanic Cloud nebulae radio measurements are not available. In that case Hβ fluxes were used and converted to radio continuum assuming $T_e = 10^4$ K and $He^+/H = 0.1$. The extinction given by the authors was used.

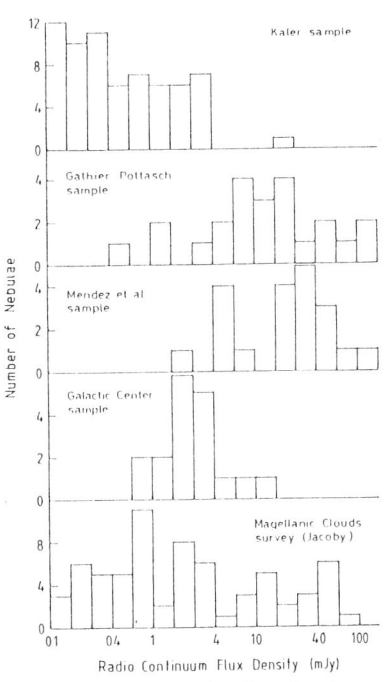

Fig. 1 - Histogram showing the intrinsic 6 cm radio continuum flux density which the nebula would have if placed at the galactic center, for the four samples of nebulae discussed in the text.

From the figure it is clear that the nebulae in both the Mendez et al. sample and (to a slightly lesser extent) the Gathier-Pottasch sample are intrinsically equally bright and are an order of magnitude brighter that the average nebula as defined by the Magellanic Cloud objects. By contrast the nebulae in Kaler's sample are generally very faint objects indeed. Some of them are fainter than the best survey's of the Magellanic Clouds could detect. By comparison, the galactic center sample of faint nebulae seem to be bright.

If the nebular brightness of the Gathier-Pottasch (GP) and the Mendez et al. (M) samples appear similar, in most other ways they are substantially different. First of all most of the GP nebulae are at low galactic latitude, because the method of distance determination is applicable only for low latitudes. The M nebulae in contrast are usually higher latitude objects. In that respect they are similar to the Kaler sample. Another important difference between the GP and M samples is the brightness of the central star. A histogram of the absolute visual magnitude of the central star for the GP and M samples is shown as Fig. 2. As can be seen from the figure there is a difference of at least 5 magnitudes (a factor 100) on the average between the samples.

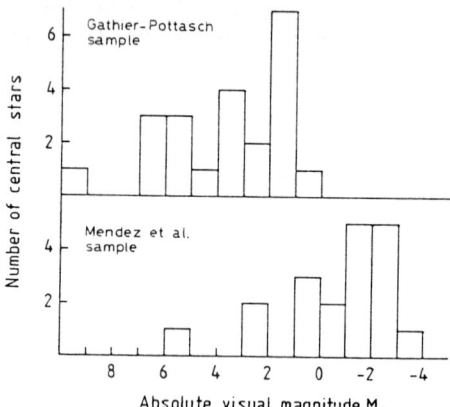

Fig. 2 - Histogram giving the absolute visual magnitude of the central star for two of the samples discussed in the text.

Thirdly there is a difference in morphology between the two samples. The M sample contains mostly symmetric nebulae which are designated as Type II in the classification of Peimbert. The GP sample contains a substantial number of Type I nebulae as well. This is also reflected in the abundances of helium and nitrogen. For example, the histogram in Fig. 3 shows the nitrogen-oxygen ratio for those nebulae in the two samples where it has been well determined (references in the two papers cited). A comparison with all galactic planetaries (e.g. Pottasch, 1984) is given in the bottom part of the diagram. It can be

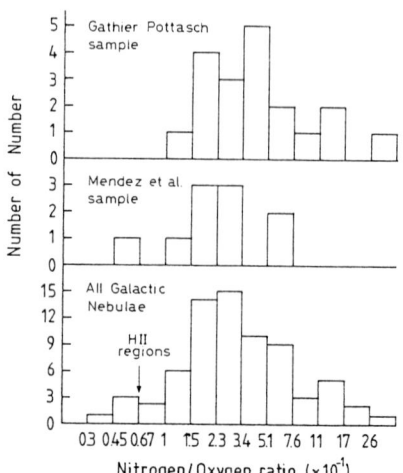

Fig. 3 - Histogram comparing the nitrogen-oxygen ratio of two of the samples with all known galactic planetary nebulae. An arrow in the lower diagram indicates the N/O ratio in HII regions.

seen that the higher N/O values are well represented in the GP sample but are hardly represented in the M sample. The Kaler sample is not plotted but high values of N/O are also present (Kaler, 1983). The consequences of this will be discusses presently.

3. TEMPERATURE OF THE CENTRAL STAR

Before presenting the HR diagrams a few words can be said about the stellar temperatures T_{eff}. Much recent work has been done. The energy balance method has been improved and applied to a large number of nebulae (Preite-Martinez and Pottasch, 1983). New magnitudes have become available for improved Zanstra temperatures (Reay et al., 1984; Shaw and Kaler, 1985; Walton et al., 1986; Gathier and Pottasch, 1987). Finally Mendez et al. (1987) have determined T_{eff} from the (absorption) line profiles. Only in the last case have model atmospheres been used in interpreting flux ratios as temperatures. In the other cases blackbody radiation has been assumed. While this may not be correct, model atmospheres only represent an improvement if they are the correct model for the particular star under consideration.

One of the problems in the temperature determination has been the interpretation of the difference between the hydrogen Zanstra temperature, $T_z(H)$ and the ionized helium Zanstra temperature $T_z(HeII)$. It has long been known that the latter is often higher than the former. Two interpretations have been given: (1) the nebula is optically thin in hydrogen ionizing radiation, while it is optically thicker in radiation which will doubly ionize helium, (2) the actual atmosphere departs from a blackbody in the sense that there is excess radiation shortward of λ 228 Å. While the first explanation probably plays a role in a few of the very large nebulae, evidence is now accumulating that departure from blackbody radiation are the important effect in most cases. Two arguments may be cited.

The first argument is the following. If a plot is made of the ratio $T_z(HeII)$ to $T_z(H)$ against $T_z(H)$, the ratio is much larger for small values of $T_z(H)$ and approaches unity at temperatures above 10^5 K. Such a plot is shown in Fig. 4. Since electron scattering becomes relatively

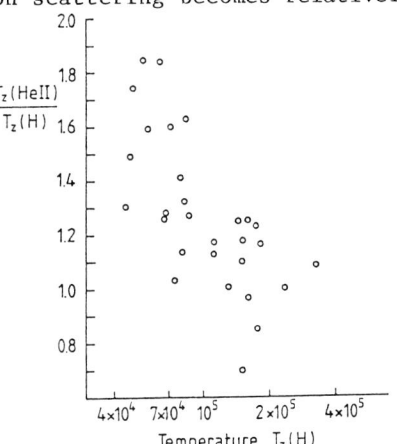

Fig. 4 - The ratio of the HeII Zanstra temperature to the H Zanstra temperature is plotted against the latter quantity (from Gathier and Pottasch, 1987).

a more important source of opacity with increasing temperature, it may be expected that the higher temperature stars emit more like blackbodies than the low temperature objects. Hence the ratio will decrease with increasing temperature as observed. The scatter in the points in the diagram is large and probably real. It may reflect different atmospheric structure at any given temperature.

The second argument derives from a comparison of the various temperature determinations for individual objects, such as shown in Table 1. Here all the stars are listed whose temperature is determined both from a study of the line profiles ($T_{PROFILE}$, Mendez et al., 1987) and from the energy balance method (T_{EB}, Preite-Martinez and Pottasch, 1983). As can be seen from the Table, $T_{PROFILE}$ always lies between $T_z(H)$ and $T_z(HeII)$, usually closer to $T_z(H)$. This is not expected if the difference in the two Zanstra temperatures is due to optical depth effects in which case $T_{PROFILE}$ should be the same as $T_z(HeII)$. It may be concluded that, except for the large, low surface brightness nebulae, a value in between $T_z(H)$ and $T_z(HeII)$ should be used when only Zanstra temperatures are available.

TABLE 1 - STELLAR TEMPERATURES OBTAINED BY VARIOUS METHODS

NEBULA	$T_z(H)$	$T_z(HeII)$	$T_{PROFILE}$	T_{EB}
NGC 1535	37	70	58	67
NGC 2392	27	66	47	78
NGC 3242	59	91	68	60
NGC 6891	34	<50	50	40
NGC 7009	68	90	75	60
IC 418	36	-	36	30
IC 2448	49	86	55	70

References: Preite-Martinez and Pottasch, 1983; Mendez et al., 1987; Shaw and Kaler, 1985; Gathier and Pottasch, 1987.

The values of T_{EB} given in the Table are computed assuming the star radiates as a blackbody. They would be reduced, especially for the higher temperatures if the model atmospheres given by Mendez et al. (1987) were used. This would sometimes make the agreement with $T_{PROFILE}$ better, but sometimes it would be worse. In conclusion, it appears that the combination of the different methods yields an effective temperature which is probably reliable to 20%. Obtaining more accurate values of T_{eff} is now very difficult because of present uncertainties in the atmospheric structure. If $T_z(H)$ and $T_z(HeII)$ are equal, the temperature may be more accurate.

4. RESULTANT HR DIAGRAMS

The HR diagrams for each of the samples are shown as Fig. 5a, b and c. On Fig. 5a the GP sample is plotted. The filled circles indicate those nebulae which have high helium/nitrogen abundance and which are known as

Type I nebulae. About half of the sample have T_{eff} greater than 10^5 K while two of the stars (NGC 2440 and 7027) have temperatures of between 3 and 4×10^5 K. There is a tendency for the stars falling near the high core mass tracks to also have high nitrogen abundance but there is one exception (NGC 6369) which should be better studied. The stars whose position falls near the low mass tracks all appear to have normal He and N abundances.

Two stars distinguish themselves in Fig. 5a. The star at log T_{eff} = 4.42, log $L/L_\odot$ = 2.2 is from the nebula He 2-131. It's distance has been measured by the extinction method both by Maciel (1985) and by Gathier

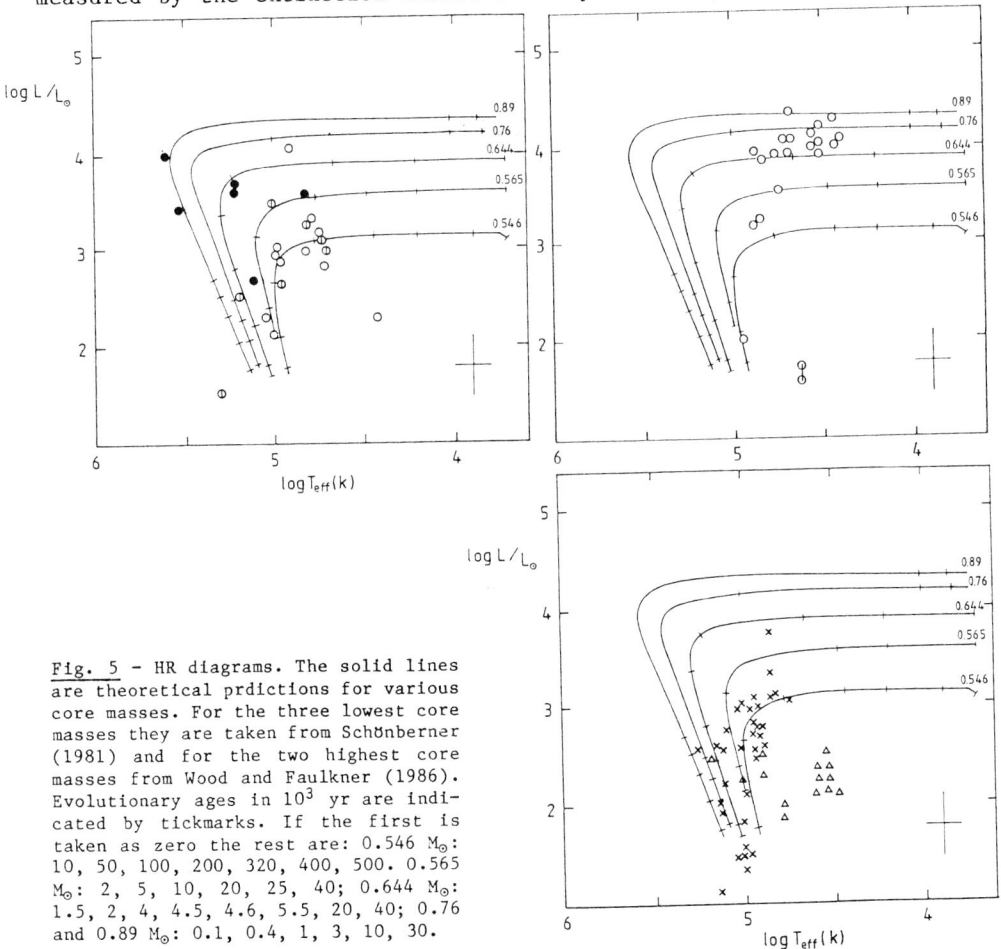

Fig. 5 - HR diagrams. The solid lines are theoretical prdictions for various core masses. For the three lowest core masses they are taken from Schönberner (1981) and for the two highest core masses from Wood and Faulkner (1986). Evolutionary ages in 10^3 yr are indicated by tickmarks. If the first is taken as zero the rest are: 0.546 $M_\odot$: 10, 50, 100, 200, 320, 400, 500. 0.565 $M_\odot$: 2, 5, 10, 20, 25, 40; 0.644 $M_\odot$: 1.5, 2, 4, 4.5, 4.6, 5.5, 20, 40; 0.76 and 0.89 $M_\odot$: 0.1, 0.4, 1, 3, 10, 30.

On these diagrams the various samples are plotted:
a. The Gathier-Pottasch sample. The filled circles are those stars whose nebula shows high nitrogen and/or helium abundance. The open circles indicate normal abundance, while for the other nebulae not enough abundance information is available.
b. The Mendez et al. sample. The two circles connected by a line indicate the same star with two different assumptions going into the distance determination.
c. Two samples are given. The crosses are the Kaler sample while the triangles are galactic center sample.

et al. (1986a). It is in a part of the diagram where central stars are not expected according to the evolution caculations. As Maciel points out, the star is somewhat below the galactic plane (b = -13°) which could make the extinction method less reliable. Mendez et al. (1987) also indicate that the distance may be greater. This point is therefore less certain than the others. On the other hand it should be considered as evidence that central stars may indeed populate that part of the diagram.

The other star which is worthy of special note is that in the lower left (log T_{eff} = 5.3, log $L/L_\odot$ = 1.4). This is the central star of the little studied nebula NGC 6565. The distance seems to be well determined. It is a small nebula and therefore quite young. One might therefore expect that the central star is intrinsically bright. Instead it is very faint (Reay et al., 1984; Gathier and Pottasch, 1987) and the Zanstra temperature is much lower; this lower temperature is confirmed by the rather low nebular excitation class (5). This strange behaviour may be caused by an extremely rapid evolution of the central star, so that the nebulae has not yet reached equilibrium with the radiation from the central star. Further study of this nebula is desireable.

Fig. 5b shows the Mendez et al. sample. The majority of these stars have a high luminosity (log $L/L_\odot \simeq 10^4$) and temperatures in the range 25 to 75000 K. There are almost no stars in this range in the GP sample. The theory predicts that very few stars should be found in this range because the evolution proceeds very rapidly. We shall return to the question of the time scale presently. It is also remarkable the none of these high luminosity stars shows a clearly higher nebular nitrogen and/or helium abundance, whereas most of the high luminosity objects in the GP sample clearly do.

A further point to note on Fig. 5b is the presence of a central star at low luminosity and low temperature. The recently discovered nebula, EGB 5, is faint and not well studied. However it falls clearly in the same region of the diagram excluded by the evolution calculations.

The last two samples are shown in Fig. 5c. The crosses are the Kaler sample. As can be seen, there is only one high luminosity central star in the sample. The stars fall mainly in that part of the HR diagram predicted by evolutionary calculations for stars which have stopped nuclear burning and are slowly cooling. The evolution is predicted to be much slower in this region and it would be expected that most central star fall there. This is all the more true since the nebulae were selected as large, low surface brightness and thus presumably old objects. The character of this sample will remain the same even if the individual distances used are in error by a factor of 2.

The sample of faint galactic central stars is shown by the triangle in Fig. 5c. For these stars the distance is better known than for any of the objects discussed so far. All the other properties are not so well known. The temperature has been taken from the excitation class of the nebulae (Kinman et al., 1987) using the calibration given in Pottasch (1987b). The luminosity has been taken as 150 times the Hβ luminosity which is approximately what is expected theoretically for an optically thick nebula and also what has been found in practice in the GP sample.

As can be seen from the figure, many of these faint nebulae fall in the same region of the HR diagram as EGB 5 and He 2-131, and which is excluded by the evolution calculations. It seems that at least some of them are real. This poses a problem for the theory.

In calculating the luminosity of the galactic center nebulae it was assumed that the nebulae are optically thick for radiation which can ionize hydrogen. The evidence for this is presented in Fig. 6, which is a plot of the nebular mass against its radius. The galactic center nebulae form a sequence in which the mass varies as the radius over the entire range of mass observed (0.01 $M_\odot$ to 0.4 $M_\odot$). The only reasonable interpretation of this is that the nebula is optically thick at every stage. The mass then increases with radius because as the density decreases the same number of ionizing photons can ionize an increasingly greater mass. Kinman et al. (1987) have reached the same conclusion for these nebulae.

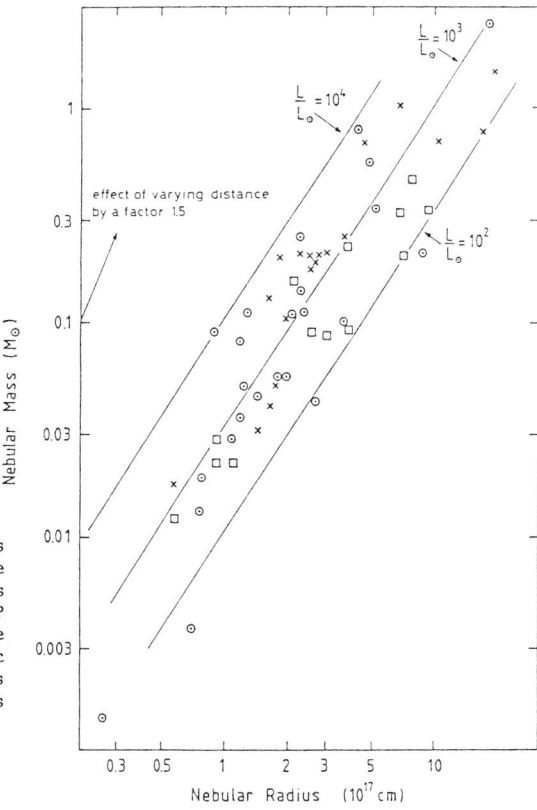

Fig. 6 - A plot of the nebular mass against the nebular radius for the nebulae studied in the various samples. The circles are from the GP sample, the crosses are the M sample while the squares are the galactic center sample. The Kaler sample falls on a line at M = 0.18 $M_\odot$ with radius between 5 and 20 × 10^{17} cm.

The other samples have also been plotted on Fig. 6. Both the GP and M samples show the same kind of mass increase as the radius increases, indicating that most must be optically thick (ionization bounded). In contrast the Kaler sample would fall on a horizontal line at M = 0.18 $M_\odot$ between a radius of 5 × 10^{17} cm and 20 × 10^{17} cm, only partly overlapping with any of the individually determined values.

5. EVOLUTIONARY TIME

On the theoretical evolution tracks in Fig. 5, the time it takes for the evolution is indicated by tik marks which are given values in the figure caption. In general the evolution becomes very rapid for high masses because the nuclear burning occurs more rapidly. The zero point is arbitrary however, if it is defined as the time since nebular ejection. The ejection time is not known in the evolution calculations as it could occur on the AGB or at any time after the star leaves the AGB. This time is known observationally however assuming that the nebular expansion velocity has been constant. The time, or nebular age, is then the ratio of the nebular radius to the expansion velocity.

Schönberner (1981) and others since have tried to make use of a comparison of the predicted time with the nebular age to derive the mass of the central star. Plots have been made of the absolute magnitude of the star against the nebular age. An example of such a plot is shown in Fig. 7. The theoretical curves make use of the assumption that the nebular ejection occurs when the star has a temperature of 5000 K. On the diagram we have plotted the individual stars from the sample of M and GP. The sample falls in the mass range lower than 0.64 $M_\odot$, with the majority less than 0.57 $M_\odot$. This is completely inconsistent with the mass found for this sample of stars from the HR diagram. If the M luminosities and distances are correct, either the theoretical times are not correct or the zero point (assumed time when ejection occurs) is not correct. Mendez et al. favor the latter conclusion. It has the consequence that all masses derived using this diagram are incorrect, because the abscissa (for the theoretical curves) must be shifted by an unknown amount, which may have a different value for each nebula.

But one of the other possibilities could also be wrong. For example, Mendez et al. give a luminosity for NCG 2392 which places it slightly above the M = 0.89 $M_\odot$ track. On this track the times for evolution from a star of 15000 K to its present 47000 K is calculated to be

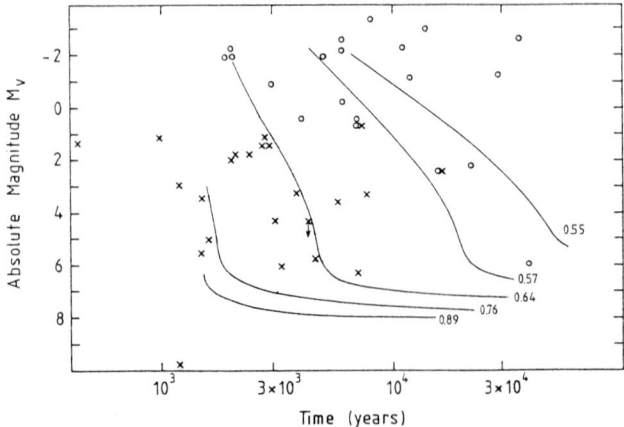

Fig. 7 - Absolute visual magnitude of the central star is plotted against nebular age. The theoretical curves assume that the nebula was ejected when the star reached a surface temperature of 5000 K. The circles from the M sample while the crosses are from the GP sample.

less than 100 years. The picture of the nebulae made by Curtis (1918) more than 70 years ago does not noticeably differ from its present morphology, suggesting that 70 years ago the amount of ionizing radiation was the same or at least very similar. Thus either the luminosity given is too high or the theoretical times are not correct.

There are also indications from the GP sample that the theoretical times are wrong. For example, NGC 2440 and 7027 are predicted to have evolved from a star of T = 10000 K in 100 to 200 years, yet they have observed ages of 8000 and 1000 years respectively. One must again conclude that the ejection must have taken place before the star reached 5000 to 7000 K, thus invalidating the use of diagrams such as Fig. 7 to determine the core mass. But again Curtis' pictures of these nebulae show that they are essentially the same 70 years ago as they are today. There was no trace of a central star 70 years ago in either of the nebulae indicating thta they had a high tepmerature even then. They do not appear to evolve as quickly as predicted.

The opposite effect is also present, especially in the GP sample. The stars clustering near the M = 0.55 $M_\odot$ track have predicted ages of the order of 10^5 years or older yet the observed nebular age is usually younger than 10^4 years. This problem may be related to that of the stars which fall in the lower righthand part of the HR diagram and which, according to theory, evolve too slowly to be seen as planetary nebula, yet which apparently are real nebulae.

6. SUMMARY

While new observations have become available in the past 5 years, they confirm only the rough outline of the theoretical evolution and leave many problems for future consideration. For example, there is a direct conflict between the distances found by Mendez et al. (1987) and those determined earlier by Liller et al. (1968) for the same nebulae from nebular expansion. It is easy to say that nebular expansion is a very difficult technique, but whether the results are wrong should be carefully investigated. The model atmosphere technique, also contains assumptions which are doubtful. While it has been tested with success on some hot stars, PN central stars may be different enough to cause important errors.

Taken at face value we must take the tentative conclusion that there are two distinct groups of central stars with core masses greater than 0.65 $M_\odot$. The one, represented in the GP sample, are those faint stars associated with Type I nebulae, having very high temperature and whose nebulae have high nitrogen and helium abundance. The other high core mass stars, found in the M sample, have more symmetric nebulae at high galactic latitudes. They have nebulae with average N and He abundances. The stars are intrinsically much brighter and evolve more slowly.

It appears from both samples that a conflict between observed and predicted nebular ages exists. Part of it can be removed by assuming that the nebula was ejected at a very early stage in the evolution from the AGB. Since this ejection time cannot be predicted, the comparison of observed and predicted ages which have appeared in the literature are very questionable. The core masses derived from this comparison and the conclusion that most PN have core masses close to 0.58 $M_\odot$ is therefore doubtful.

Further progress involves the determination of more accurate distances. It also involves the more detailed study of nebulae in the galactic bulge, whose distance is quite reliably known.

REFERENCES

Barlow, M. 1987, Mont. Not. Roy. Astron. Soc.
Curtis, H.B. 1918, Lick Obs. Publ. Vol XIII, p. 57
Gathier, R. 1984, Thesis, U. of Groningen
Gathier, R., Pottasch, S.R., Goss, W.M., van Gorkom, J.H. 1983, Astron. Astrophys. 128, 325
Gathier, R., Pottasch, S.R., Pel, J.W. 1986a, Astron. Astrophys. 157, 171
Gathier, R., Pottasch, S.R., Goss, W.M. 1986b, Astron. Astrophys. 157, 191
Gathier, R., Pottasch, S.R., 1987 (to be submitted)
Jacoby, G.H. 1980, Astrophys. J. Suppl. 42, 1
Kaler, J.B. 1983, Astrophys. J. 271, 188
Kinman, T.D., Feast, M.W., Lasker, B.S. 1987
Maciel, W. 1985, Rev. Mex. Astron. Astrofis. 10, 199
Mendez, R.H., Kudritzki, R.P., Gruschinske, J., Simon, K.P. 1981, Astron. Astrophys. 101, 323
Mendez, R.H., Kudritzki, R.P., Simon, K.P. 1985, Astron. Astrophys. 142, 289
Mendez, R.H., Kudritzki, R.P., Herrero, A., Husfeld, D., Groth, H.G. 1987 (submitted to Astron. Astrophys.)
Liller, M.H., Liller, W. 1968, IAU Symp. 34 (Reidel, Dordrecht)
Pottasch, S.R. 1984, IAU Symp. 103, ed. D. Flower (Reidel, Dordrecht)
Pottasch, S.R. 1984, Planetary Nebulae (Reidel, Dordrecht)
Pottasch, S.R. 1987a, Torino Workshop Proc. 'Mass loss in stars' (Reidel, Dordrecht)
Pottasch, S.R. 1987b, ESO Workshop Proc. 'Stellar evolution ......'
Preite-Martinez, A., Pottasch, S.R. 1983, Astron. Astrophys. 126, 31
Reay, N.K., Pottasch, S.R. 1983, Astron. Astrophys. 126, 31
Schönberner, D. 1981, Astron. Astrophys. 103, 119
Shaw, R.A., Kaler, J.B. 1985, Astrophys. J. 295, 537
Walton, N.A., Reay, N.K., Pottasch, S.R., Atherton, P.D. 1986, Proc. IUE Conf. (ESA SP-263), p. 497
Wood, P.R., Bessell, M.S., Copita, M.A. 1986, Astrophys. J. 311, 632
Wood, P.R., Faulkner, D.J. 1986, Astrophys. J. 307, 659

INITIAL MASSES

D.C.V. Mallik
Indian Institute of Astrophysics
Bangalore 560034
India

1. INTRODUCTION

Planetary nebulae represent a transitory stage in the life of the majority of stars as they proceed towards the end of their nuclear evolution and descend to the domain of white dwarfs. The immediate precursors of the central stars are probably red giants which populate a part of the HR diagram far removed from the region inhabited by the central stars of well recognised nebulae. The problem of determining the initial masses is complicated by the widespread occurrence of massloss on the red giant branch. The total amount of mass lost by a star must depend upon a number of stellar parameters including the initial mass, but the exact nature of this dependence remains to be discovered and a unique relation between the final masses and initial main sequence masses is not yet available. Thus even though the mass distribution of the nuclei of planetary nebulae (NPN) has been derived in the last few years, it has not been possible to deduce from this an unambiguous initial mass distribution of the progenitors. Further, an observed sample always suffers from selection effects and, in the particular case of NPN mass distribution, this has led to irretrievable loss of information.

Planetary nebulae are observed to have a great diversity of chemical and kinematical properties. This must, in some way, relate to the diversity of stars that they descend from. They are fairly numerous in the Galaxy and considering that their lifetimes are very short on astronomical timescales, it is obvious that their production rates are quite high, which, in turn, implies that they must originate from stars at the lower end of the mass spectrum. Since it is also firmly established that the nuclei are incipient white dwarfs, we must seek a common origin for both. The theory of stellar evolution puts rather severe constraints on the masses of stars evolving to white dwarfs and these limits should also apply to stars evolving to planetary nebulae.

2. CLASSIFICATION INTO POPULATION TYPES - BROAD HINTS ON THE INITIAL MASSES

The diversity of morphology, kinematical properties and chemical composition of the nebulae as well as their galactic distribution has led various authors to the classification of these objects into different groups indicating their Population types. Since the lifetime of the nebulae is very short, the galactic distribution and kinematics closely reflect the properties of the progenitor stars and hence carry information on the initial masses. In Table I we summarise these various schemes with emphasis on the information relevant to initial masses. For Classes B and C nebulae as well as Heap and Augensen's Population types Schmidt's (1963) calibration of the mean distance from the plane versus the main sequence mass has been used to derive the mean initial mass. Although widely used, this procedure of comparing the mean height of planetary nebulae with the mean height of main sequence stars to deduce a mean intial mass is not correct, since the mean height of stars in Schmidt's calibration samples stars of all ages for any particular mass, while by its very nature the planetary nebula population selects a particular age sample (the oldest) for each mass.

TABLE I. CLASSIFICATION SCHEMES OF PLANETARY NEBULAE

Class	Criterion	Kinematics	Mean Height above plane (pc)	Initial Mass ($M_\odot$)	Source
B	Large ansae or tubular or filamentary structure	Circular motion $\sigma_z = 15$ kms^{-1}	160	1.5	}
C	Centric increase in surface brightness	noncircular motion $\sigma_z = 26$ kms^{-1}	320	< 1.0	} 1,2
Pop I	$P^5 < 1.5$	Circular motion	144	1.5	}
Pop I$^+$	$1.5 < P < 2.0$		340		} 3
Pop II	$P > 2.0$	$\Delta V_r > 60$ kms^{-1}	774	< 1.0	}
Type I	He/H > 0.125, N/O > 0.50			3-5	}
Type II	He/H = 0.11	$\Delta V_r < 60$ kms^{-1}	< 1000	1.5-3	} 4
Type III	He/H = 0.11	$\Delta V_r > 60$ kms^{-1}	> 1000	1.0-1.5	}
Type IV		Halo		<1.0	}

Notes to Table I: 1. Grieg (1971, 1972), 2. Cudworth (1974), 3. Heap and Augensen (1987), 4. Peimbert (1985), 5. Population index $P = 1 + \sqrt{[z^2 + (\Delta V_r/60)^2]}$

A better discriminant of the initial masses is the more elaborate classification scheme proposed originally by Peimbert (1978) and rediscussed by Peimbert and Serrano (1980) and Peimbert and Torres-Peimbert (1983, hereafter PTP 83). Since the chemical composition is a sensitive function of the evolutionary history and hence the initial main sequence mass of the progenitor star, this scheme provides a better idea of the range of initial masses producing the different types of nebulae. In assigning the limits in this case use has been made of evolutionary calculations and theories of dredge-up. We should remember that the existing theories of dredge-up are not fully in agreement with observational data.

A comparison of these various classification schemes shows a general one-to-one correspondence between Types I and I-II nebulae of Peimbert and Class B nebulae studied by Cudworth and the Population I nebulae of Heap but there are some ambiguities. For example, NGC 6629, 6894 and 7008 are all classified as Type I in PTP 83 while they belong to the list of Class C nebulae according to Cudworth. Some of the Population I objects in Heap and Augensen also belong to Class C nebulae. NGC 6058 belongs to Population II category according to Heap but to Class B according to Cudworth.

3. PLANETARY NEBULAE AND WHITE DWARFS IN CLUSTERS

Perhaps the most direct of the indirect ways to determine the initial masses is to look for planetary nebulae in clusters with well defined turn-offs. Unfortunately only one globular cluster (M15) and one open cluster (NGC 2818) contain a planetary nebula each. Based on the cluster data the progenitor of K648 in M15 is deemed to be a 0.8 $M_\odot$ star and that of NGC 2818 a 2.1 ± 0.3 $M_\odot$ star (Peimbert 1981).

Much more extensive data are available on white dwarfs in open clusters. Their presence in young clusters like the Pleiades and NGC 2516 is the best evidence we have for the high initial masses of the progenitors. The upper limit on the progenitor mass is an important number predicted by the stellar evolution theory. However, the number depends upon metallicity, massloss rate and the treatment of convection during the core burning phases and a range of values (5 $M_\odot$ to 11 $M_\odot$) is obtained (Becker and Iben 1979, Iben and Renzini 1983, Castellani et al. 1985). For a set of chosen parameters, theory predicts a well defined relation between the masses of degenerate remnants and the initial masses. In general, the more massive a progenitor, the more massive is its stellar remnant. Therefore, if massive NPN's or white dwarfs are found, one would infer large initial masses for their progenitors. Observationally, such massive nuclei would be difficult to discover, since the fading time for the massive degenerate cores is very short.

The cluster white dwarf data have been used by Weidemann and Koester (1983, hereafter WK) and Weidemann (1984) to derive an empirical relationship between the white dwarf masses and the initial masses. This empirical initial-final mass relation is much flatter than the ones predicted by theory. The flatness of the relation implies that although stars over a wide range of initial masses may evolve to the PN/WD stage, the final mass distribution of these objects should be rather narrow. The theoretical curves, on the other hand, would result in a broader final mass distribution for the same range of initial masses. Alternatively, a narrow final mass distribution coupled with the theoretical relation would suggest that the observed sample descends from a restricted range of initial masses. The white dwarf mass distribution of Koester, Schulz and Weidemann (1979) is rather narrow with two thirds of all DA white dwarfs confined to about 0.58 ± 0.10 $M_\odot$. The NPN mass distribution consistency each one would require a different value of the parameter η. The situation is rather unsatisfactory unless the choice of η could be restricted by other means. However, if the empirical WK relation were used instead, only the Cudworth scale is seen to yield a consistent lower limit. Weidemann (1984) has argued in favor of the flat WK relation. If this relation proves to be the correct one in further analysis, two important results would follow: (1) among the different distance scales, the Cudworth scale is to be preferred, (2) the lower limit on the initial mass of a planetary nebula progenitor is 2.5 $M_\odot$, significantly above 1.0 $M_\odot$.

Recently, the observed birthrate of planetary nebulae has been revised upward by a factor of 2 with respect to Daub's and a factor 9 with respect to Cudworth's (Ishida and Weinberger 1987). This value is higher than the white dwarf birthrate. It is obvious that this birthrate cannot be accomodated at all with theoretical deathrate estimates given above.

5. HEIGHT DISTRIBUTION

Planetary nebulae show a fairly strong concentration to the galactic plane. The mean height is low, typically between 100 pc and 200 pc. Since they originate from stars of different lifetimes, the current population contains nebulae formed from old low mass stars and from the more recent heavier ones. The diffusion of stellar orbits increases the velocity dispersion perpendicular to the plane with age. Thus the less massive planetary nebula progenitors will be on an average farther from the plane than the more massive ones. It is possible to relate the age to the velocity dispersion and the latter to the scale height of the distribution (Wielen and Fuchs 1983). The distribution itself is of the form $n(z) = n(z=0) \text{sech}^2 z/H$ where the scale height $H = \sigma_W^2 / \pi G \mu$, where σ_W is the velocity dispersion and μ the surface density of stars. These relations have been used to obtain the height distribution displayed in Figure 2.

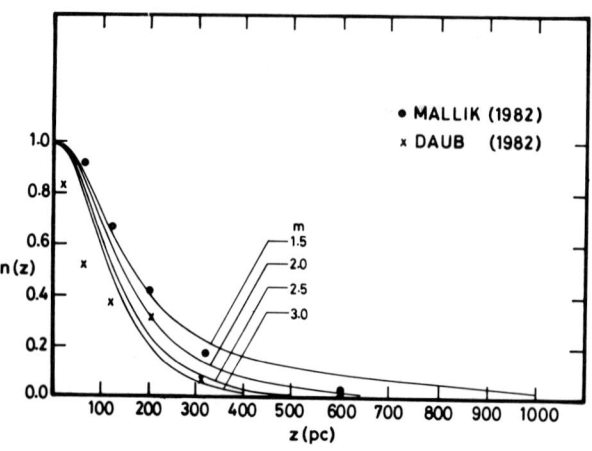

Figure 2. Normalised height distribution for different values of m.

The different curves correspond to the different lower limits on the initial mass since $n(z) = \int_m^{8} n_m(z) \, dm$. It is clear that for low values of m a larger fraction of old less massive stars is present in the sample and n(z) decreases less rapidly with increasing z. The observational points from two different samples are also plotted in the Figure. The observed height distribution shows a rapid drop which, in turn, suggests the absence of light progenitors in the observed samples. While it is not worth attempting an exact fit, we see from Figure 2 that with an m ~2.0 $M_\odot$ the observed height distribution is rather well reproduced.

6. CHEMICAL COMPOSITION OF NEBULAE AS INDICATOR OF THE INITIAL MASS

A significant fraction of planetary nebulae show definite signs of enrichment as a result of nuclear processing in the parent stars. The stellar evolution theory has been able to identify the processes leading to such enrichment and to predict the amount of enrichment as a function of the main sequence mass. Certain trends are very clear: (i) the He/H ratio increases monotonically with the initial mass, (ii) the He and N abundances show proportionate increases with initial mass during the second dredge-up phase, (iii) the C and He abundances increase monotonically with mass due to the third dredge-up. The functional relationships depend upon a host of parameters - the massloss rate, initial metallicity, the assumed dredge-up law, treatment of convection etc. These factors introduce uncertainties in any initial mass that may be inferred by matching the observed nebular abundances with model predictions. The theoretical predictions from the models of Becker and Iben (1979, 1980) and Renzini and Voli (1981) have been compared with the observed abundances of planetary nebulae with conflicting inferences (PTP 83, Kaler 1983, Pottasch 1984). Several authors have used these comparisons to arrive at the initial masses. It appears that the Type I planetary nebulae originate from stars in the mass range 3-5 $M_\odot$ and that their C/O and N/O ratios require moderately efficient envelope burning (Torres-Peimbert 1984).

The main problem in comparing the dredge-up theories with observations is the conflict with the core masses. The models show that the core masses for effective dredge-up are rather large, in the neighbourhood of 1.0 $M_\odot$ while the observed mass distribution of NPN shows very few nuclei, if any, of this mass. More specifically, Heap and Augensen (1987) have derived masses for several Type I central stars and found them to be low (~ 0.65 $M_\odot$). The dredge-up theories predict hardly any enrichment for such small core masses. Convective overshooting so far neglected in the dredge-up models will produce larger cores and enhancement at smaller initial masses and will not solve the problem of dredge-up at small core masses. PTP 83 suggests that NGC 6302 is the most massive well-documented object. However, its He and N abundances are beyond the range produced in the models. A He/H > 0.18 cannot be satisfactorily produced in the dredge-up models. Even as one requires an initial mass > 6 $M_\odot$, an inconsistency creeps in because, according to the same dredge-up models, stars more massive than 5 $M_\odot$ do not evolve to planetary nebulae but explode as supernovae. The latter event will surely modify the envelope abundances, nor shall we see a pn. Similarly, the observed N/O ratio in NGC 2440 is much too high to match any of the models.

Type II and Type III nebulae show very similar abundance trends. Comparison with dredge-up models indicates initial masses in the range 1.0-2.5 $M_\odot$ for them. However, their initial CNO values are rather diverse and this introduces an extraneous scatter whose effects are difficult to disentangle. Considering that the vast majority of planetary nebulae belong to Types I, II and III, we conclude from chemical composition alone that they originate from stars with main sequence masses 1.0-5.0 $M_\odot$ although individually some of the nebulae (e.g. NGC 6302) do not fit into the scheme at all.

7. PLANETARY NEBULAE IN BINARY SYSTEMS

Approximately 25% of the known planetary nebula population is expected to be in binary systems; the number actually discovered is somewhat less. Besides those which are members of visual binary systems and the ones which are spectroscopic and/or eclipsing binaries with well observed periods, there are many which are inferred to be binaries since optically only cool stars are seen associated with the nebulae. In some cases IUE observations have clearly revealed the presence of the hot star.

Peimbert and Serrano (1980) considered the binary nature of NGC 3132 and NGC 2346 and based on the evolutionary lifetime considerations inferred the initial masses of the central stars to be above 2.0 $M_\odot$. Recently Whitelock and Menzies (1986) have discussed the binary nature of the nuclues of the new nebula IRAS 1912 + 172 P09 where the optically visible companion is a B9V star of mass 2.9 $M_\odot$. This is by far the earliest companion discovered amongst the binary central stars. Again based on pure evolutionary timescale considerations they infer an initial mass of 3.0 $M_\odot$ for the progenitor. While NGC 3132 is a wide binary system, it is not clear if NGC 2346 and IRAS 1912 + 172 could be assumed to be so. If the central stars of these objects were members of interacting binary systems, the intial mass could be significantly higher.

The interacting binary scenario for the origin of the planetary nebulae is radically different since Roche lobe overflow and common envelope evolution introduce new aspects to the evolutionary problem. A recent review on the subject is due to Paczyński (1985). Several systems which are definitely the result of a common envelope evolution leading to pn formation are known and generally classed as precataclysmic binaries (Bond 1985, Ritter 1986). In Table III, I summarise the relevant information on them. The secondaries are all low mass stars presumably unevolved. The primaries are hot subdwarfs and exciting the nebulae. The fact that is crucial to our discussion is the shortness of the periods. During the common envelope (CE)

TABLE III. CLOSE BINARY CENTRAL STARS

Name	Spectra	$M_2(M_\odot)$	q	P(d)	Source
a) Short period systems - precataclysmic binaries					
A 63	sdO+K-MV	0.6	0.57	0.465	1
A 46		0.2-0.3	0.43	0.472	1
A 41	sdO+MV	0.1-0.3	0.2-0.5	0.113	2
DS1	sdO+MV	0.3	0.43	0.357	1
K1-2				0.671	1
LoTr5	sdO+G5III-V			0.35	3
b) Systems with P > 1.0					
NGC 2346	?+A2V			16.	4a
c) Systems with unknown P					
NGC 1514	sdO+AOIII				2
A 35	?+G8III-IV				2
He 2-36	sdO+A2III(?)				4b
Cn 1-1	sdO+F5III-IV				5
IRAS1912+172P09	?+B9V				6

Notes to Table III. 1. Ritter (1986) 2. Bond (1985) 3. Acker (1985) 4a. Mendez and Niemela (1981) 4b. Mendez (1978) 5. Lutz (1984) 6. Whitelock and Menzies (1986).

phase large orbital contraction is possible due to the loss of mass and orbital angular momentum. It is possible to make reasonable guesses about the initial mass ratios of these systems from the currently observed periods and mass ratios. Some idea of the initial period of the system is necessary. This determines, in turn, the stage at which mass exchange takes place preceding the CE formation. For Roche lobe overflow during the AGB evolution of the primary, the periods are respectively in the range 55d to 3y and 25d to 16y for a $5M_\odot$ and a $3M_\odot$ star. Conservatively one may assume a value of about a year for the initial period. Since a CE formation followed by orbital contraction needs a large mass ratio it is unlikely that the initial mass of the primary be less than $3M_\odot$. From the formula due to Tutukov and Yungelson(1979) the initial mass ratios could be obtained. For a period 11^h as in A 63 and a current mass ratio of 0.5 this yields an initial mass ratio of about 17 and hence an initial mass of about $5M_\odot$ for the primary. For the short period systems the initial mass of the pn progenitors is thus $5M_\odot$ or more. According to Iben and Tutukov (1985) stars initially more massive than $5M_\odot$ are able to shed their He-rich envelope during a second phase of Roche-lobe overflow. Thus many of these nebulae may have He-rich inner zones.

Two nebulae in Table III are associated with binary central stars having periods much longer than a day indicating that binaries can emerge from a CE phase with fairly long periods. For the rest periods are not definitely known. A remarkable fact about these systems is that the companions are all within a narrow Spectral

Type range lying close to or slightly off the main sequence suggesting that they have masses between 2.0 and 3.0 $M_\odot$. If these systems evolved out of a CE phase we may immediately conclude that the initial masses of the primaries are < 5 $M_\odot$. According to Iben and Tutukov (1985) stars more massive than 5 $M_\odot$ go through two episodes of Roche-lobe overflow and it is highly unlikely that we see a planetary nebula emerging out of a CE phase following the second such episode, since most of the donor mass would already have been extracted earlier. For primaries less massive than 5 $M_\odot$ only one episode of Roche-lobe overflow takes place following which these stars burn He for a very long time and achieve surface temperatures compatible with those found in central stars of planetary nebulae. Then, primordial mass ratios are in the neighbourhood of 2, and it seems plausible that large orbital contraction has not taken place in these systems. The periods may have been marginally shortened through CE evolution and one should search for long periods in these systems.

8. CONCLUSION

Wisdom gained from stellar evolution theory had led to the belief that planetary nebulae originate from all stars in the mass range 1-8 $M_\odot$. Recent calculations with convective overshoot have brought down the upper mass limit to the neighbourhood of 6 $M_\odot$. However, observational considerations based on birthrate, height distribution and the empirical m_f/m_i relation suggest that the lower mass limit of pn formation is near 2.0 $M_\odot$. This will also explain the paucity of planetary nebulae in globular clusters and the galactic halo. Chemical composition of the Type I nebulae suggests that they come from rather massive progenitors (3-5 M). Planetary nebulae in binary systems classed as precataclysmic variables probably evolve from massive progenitors ($M \gtrsim 5 M_\odot$). The others with cool companions and unknown periods have evolved from progenitors in the range 4-5 $M_\odot$. Further observations are needed to establish these limits firmly. Abundance analyses of nebulae with binary nuclei may provide important clues.

REFERENCES

Acker,A. 1985, Astron.Astrophys. **151**, L13.
Becker,S.A., and Iben,I., Jr. 1979, Ap.J. **232**, 831.
Becker,S.A., and Iben,I., Jr. 1980, Ap.J. **237**, 111.
Bond,H.E. 1985, Cataclysmic Varaibles and Low-Mass X-ray Binaries, ed. D.Q.Lamb and J.Patterson, D.Reidel, p.15.
Cahn,J.H., and Wyatt,S.P. 1976, Ap.J. **210**, 508.
Castellani,V., Chieffi,A., Pulone,L., and Tornambe,A. 1985, Ap.J.Lett. **294**, L31.
Cudworth,K.M. 1974, Astron.J. **79**, 1384.
Daub,C.T. 1982, Ap.J. **260**, 612.
Greig,W.E. 1971, Astron.Astrophys. **10**, 161.

Greig,W.E. 1972, Astron.Astrophys. **18**, 70.
Heap,S.R., and Augensen,H.J. 1987, Ap.J., **313**, 268.
Iben,I.,Jr. and Renzini,A. 1983, A.Rev.Astron.Astrophys. **21**, 271.
Iben,I.,Jr. and Tutukov,A.V. 1984, Ap.J.Suppl. **54**, 335.
Iben,I.,Jr. and Tutukov,A.V. 1985, Ap.J.Suppl. **58**, 661.
Ishida,K., and Weinberger,R. 1987, Astron.Astrophys. **178**, 227.
Kaler,J.B. 1983a, Ap.J. **271**, 188.
Kaler,J.B. 1983b, Planetary Nebulae, IAU Symposium 103, ed. D.R.Flower, D.Reidel, p.245.
Koester,D., Schulz,H., and Weidemann,V. 1979, Astron.Astrophys. **76**, 262.
Lutz,J.H. 1984, Ap.J. **279**, 714.
Mallik,D.C.V. 1982, Bull.astr.Soc.India **10**, 73.
Mallik,D.C.V. 1985, Astrophys.Lett. **24**, 173.
Mendez,R.H. 1978, Mon.Not.Roy.astr.Soc. **185**, 647.
Mendez,R.H. 1980, Close Binary Stars, IAU Symposium 88, ed.M.J.Plavec, D.M.Popper and R.K.Ulrich, D.Reidel, p.567.
Mendez,R.H., and Niemela,V.S. 1977, Mon.Not.Roy.astr.Soc. **178**, 409.
Mendez,R.H., and Niemela,V.S. 1981, Astron.J. **250**, 240.
Miller,G.E., and Scalo,J.M. 1979, Ap.J.Suppl. **41**, 513.
Paczyñski,B. 1985, Cataclysmic Variables and Low-Mass X-ray Binaries, ed.D.Q.Lamb and J.Patterson, D.Reidel, p.1.
Peimbert,M. 1978, Planetary Nebulae, IAU Symposium 76, ed.Y.Terzian, D.Reidel, p.215.
Peimbert,M. 1981, Physical Processes in Red Giants, ed.I.Iben,Jr. and A.Renzini, D.Reidel, p.409.
Peimbert,M. 1985, Rev.Mex.Astron.Astrofis. **10**, 125.
Peimbert,M., and Serrano,A. 1980, Rev.Mex.Astron.Astrofis. **5**, 9.
Peimbert,M., and Torres-Peimbert,S. 1983, Planetary Nebulae, IAU Symposium 103, ed.D.R.Flower, D.Reidel, p.233.
Pottasch,S.R. 1983, Planetary Nebulae,IAU Symposium 103, ed.D.R.Flower, D.Reidel, p.391.
Pottasch,S.R. 1984, Planetary Nebulae,D.Reidel, p.235.
Reay,N.K. 1983, Planetary Nebulae, IAU Symposium 103, ed.D.R.Flower, D.Reidel, p.31.
Renzini,A., and Voli,M. 1981, Astron.Astrophys. **94**, 175.
Ritter,H. 1986, Astron.Astrophys. **169**, 139.
Schmidt,M. 1963, Ap.J. **137**, 758.
Schonberner,D. 1981, Astron.Astrophys. **103**, 119.
Torres-Peimbert,S. 1984, Stellar Nucleosynthesis, ed.C.Chiosi and A.Renzini,D.Reidel, p.3.
Tutukov,A.V., and Yungelson,L. 1979, Mass Loss and Evolution of O-type Stars, ed.P.S.Conti and C.W.H. de Loore, D.Reidel,p.401.
Weidemann,V. 1984, Astron.Astrophys. **134**, L1.
Weidemann,V., and Koester,D. 1983, Astron.Astrophys. **121**, 77.
Whitelock,P.A., and Menzies,J.W. 1986, Mon.Not.Roy.astr.Soc. **223**, 497.
Wielen,R., and Fuchs,B. 1983, Kinematics, Dynamics and Structure of the Milky Way, ed.W.L.H.Shuter, D.Reidel, p.81.

Yervant Terzian and Lawrence Aller

BINARY STARS AND PLANETARY NEBULAE

Icko Iben, Jr.*
Dept. of Astronomy, University of Illinois, 1011 W.
Springfield, Urbana, IL 61801
Alexander V. Tutukov**
Astronomical Council, USSR Academy of Sciences

ABSTRACT. A non-negligible (~ 15-20%) fraction of planetary nebulae is expected to be formed in close binaries in which one component fills its Roche lobe after the exhaustion of hydrogen or helium at its center. The nebula is ejected as a consequence of a frictional interaction between the stellar cores and a common envelope; the ionizing component of the central binary star may be a relatively high luminosity contracting star with a degenerate CO core, burning hydrogen or helium in a shell, or it may be a lower luminosity shell hydrogen-burning star with a degenerate helium core or a core helium-burning star. Even more exotic ionizing central stars are possible. Once the initial primary has become a white dwarf or neutron star, the secondary, after exhausting central hydrogen, will also fill its Roche lobe and eject a nebular shell in a common envelope event. The secondary becomes the ionizing star in a tight orbit with its compact companion. In all, there are roughly twenty different possibilities for the make-up of binary central stars, with the ionizing component being a post asymptotic giant branch star with a hydrogen- or helium-burning shell, a CO dwarf, a core helium-burning star, a shell helium-burning star with a degenerate CO core, a shell hydrogen-burning star with a degenerate helium core, or a helium degenerate dwarf, while its companion is a main sequence star, a CO degenerate dwarf, a helium star, a helium degenerate dwarf, or a neutron star. We estimate the occurrence frequency of several of these types and comment on the prior evolutionary history of 4 observed binary central stars.

---

*Supported in part by the United States NSF Grant AST 84-13771.
**Supported in part by the Instituto de Astronomia, University of Mexico, and by the USSR Academy of Sciences.

1. INTRODUCTION

It is commonly assumed that most planetary nebulae (PNe) originate from single stars which eject a shell of matter after they have exhausted helium in their cores and have become asymptotic giant branch (AGB) stars. The nature of the ejection mechanism has yet to be satisfactorily identified, but it may involve a secular instability in the envelope itself (Wood 1974; Tuchman, Sack, and Barkat 1979) and/or the development of a wind driven by pulsational energy and by radiation pressure on grains (Wood 1981, Willson 1986, Draine 1981, Fadeev 1986). A low mass representative reaches the thermally pulsing (TP) AGB phase and develops a chemically evolved surface composition and may pass through a Mira phase before ejecting most of its hydrogen-rich envelope. In the case of more massive progenitors, the ejection event may be evident in the OH/IR phenomenon. Much of this evolution is described in Kwok and Pottasch (1987).

The conventional picture is, of course, an oversimplification. In a significant fraction of all cases, membership in a binary may play an important if not crucial role either in ejecting matter or in influencing the morphology and chemistry of the extended nebula, or both. Theory suggests that many close binaries (initial orbital separation less than ~ $1500 R_\odot$) should experience one or more common envelope events during which they eject from the system most of the mass of one of the stars, which thereafter evolves either directly into a luminous, hot (>100,000K) shell nuclear burning star (which can cause the ejected material to fluoresce) and then into a white dwarf, or first into a less luminous and cooler helium star (which also can, if hot enough, cause nebular fluorescence).

The characteristics of the exciting central star and of its companion can conspire against the serendipitous detection of duplicity. The central star is, as a rule, extremely bright and lines in its spectrum are rather broad, making it difficult with conventional techniques to detect a relatively dim main sequence or degenerate dwarf companion. Only the shortest period systems with a favorable orientation of orbital plane can be detected as eclipsing binaries or as variable stars in which the intrinsically dim component is illuminated by its bolometrically brighter companion.

To estimate the probabilities of different binary evolutionary scenarios we need to know the initial distribution of newly forming binary stars with respect to semimajor axis A_0 and primary mass M_1. In first approximation, this is given by (Popova et al. 1982, Iben and Tutukov 1984a, 1985)

$$d^2N \sim 0.1 \; d\log A_0 \; dM_1/M_1^{2.5} \; yr^{-1}, \tag{1}$$

where A_0 and M_1 are in solar units (which we shall use hereinafter, unless otherwise specified). This relationship is valid for $1 < \log A_0 < 6$ and $1 < M_1 < 100$. The numerical value of the coefficient in equation (1) is uncertain by at least a factor of two and the uncertain dependence on mass ratio has been suppressed. From this equation it follows that almost all planetary nebula nuclei (PNNi) may be in binaries, with the separation of components on occasion being of the order of the nebular size itself. However, only about 15-20% of all initial binaries will be close enough (A < 1500) that the primary will, after exhausting central hydrogen, fill its Roche lobe before it can reach the thermally pulsing AGB phase and eject a PN shell of its own accord. An even larger percentage may be far enough apart to avoid mass loss by Roche-lobe overflow, but close enough to exercise some shaping influence on the nebula emitted by one of the stars.

Curiously enough, approximately 15% of all central stars that have been examined carefully for duplicity are binaries with periods less than 1 day (Bond 1987, Ritter 1987, and references therein) and Bond (1987, this conference) infers that a much larger percentage of planetaries contains close binary central stars with periods larger than 1 day, with perhaps all PNe containing close binary central stars. We suspect that the apparent high frequency of very short period binary stars may be a consequence of selection.

Most early computations of close binary evolution were carried out in the conservative approximation (total mass = constant, total angular momentum = constant). Now it is known that, in binaries with initial mass ratio q_0 (= M_1/M_2) exceeding ~ 2 when the donor has a radiative envelope and ~ 0.6 when the donor has a convective envelope, Roche-lobe filling leads to common envelope formation ([Ostriker and] Paczynski 1976, Meyer and Meyer-Hofmeister 1979, Tutukov et al. 1982). Matter entering the common envelope is driven from the system because of friction between the central binary and the common envelope. Most of the matter originally in the hydrogen-rich surface layers of the primary is in this way lost from the system. Mass loss ceases when the compact remnant of the primary shrinks within its Roche lobe.

After the loss of the common envelope, the system consists of the essentially unaltered secondary and a compact remnant, the two stars being much closer together than the original pair. If it becomes hot enough quickly enough, the remnant can ionize the ejected common envelope material, and the system can be recognized as a PN. In the further course of evolution, as many as three additional major common envelope events may occur, each event leading to a decrease in orbital separation. After each ejection a PN phase may occur.

The reduction in semimajor axis during the common envelope stage can be estimated from the expression (Tutukov and Yungelson 1979)

$$M_1^2/A_0 \sim \alpha\, M_2 M_{1R}/A_f, \qquad (2)$$

where A_0 is the semimajor axis before the common envelope event, A_f is the final semimajor axis, M_1 is the initial primary mass,

M_{1R} is the mass of its remnant, M_2 is the mass of the companion, and α is the "efficiency". Equation (2) permits us to estimate the semimajor axis after each successive common envelope event which the system experiences during the evolution of its components and equations (1) and (2) together may be used to estimate birth rates of different types of binaries at centers of planetary nebulae.

Equation (2) follows from the fact that energy is required to drive matter against gravity from the "surface" of the donor, through the common envelope, and endow this matter with sufficient kinetic energy to escape from the system. A measure of the energy required to eject all of the matter M_{lost} lost by the primary may be written as $\delta E_{eject} \sim GM_{lost} \langle M_{stars} \rangle / A_0$, and a measure of the change in orbital binding energy may be written as $\delta E_{bind} \sim GM_{1R}M_2/2A_f - GM_1M_2/2A_0$. Here $\langle M_{stars} \rangle$ is the mean mass of the stellar system from which matter is escaping, and we have assumed that the secondary has not gained any mass. We may write $M_{lost} = M_1 - M_{1R}$ and $\langle M_{stars} \rangle \sim (M_1 + M_{1R})/2 + M_2$. If M_1 is large compared with both M_{1R} and M_2, setting $\delta E_{eject} \sim \delta E_{bind}$ gives $M_1^2/A_0 \sim M_2 M_{1R}/A_f$. If we assume that $M_1 \sim M_2$, but still retain the assumption of small M_{1R}/M_1, the same argument gives $M_1^2/A_0 \sim (1/4) M_2 M_{1R}/A_f$. Thus, α in equation (2) is a parameter of order 0.2-1, subject, of course, to the validity of the assumption that orbital energy is efficiently used up in driving off matter in the common envelope.

Several attempts have been made to model a common envelope event using two dimensional hydrodynamics (Bodenheimer and Taam 1984, Bond and Livio 1987, Hachisu 1987), and these studies suggest that a large fraction of the orbital energy goes into escaping radiation and into producing large terminal velocities, making α considerably smaller than unity. It must be cautioned, however, that the problem is three dimensional and important physical processes, such as the transport of angular momentum by turbulent viscosity, have been left out of the calculations. Further, with α as small as, say, 0.1, it becomes difficult to understand the formation of cataclysmic variables.

In the following sections we will (II) explore the various possibilities for PN formation in close binaries and estimate the occurrence frequencies for several types; (III) attempt to divine the nature of the progenitors of several observed PNe with binary central stars; and (IV) compare and contrast the morphology, chemical composition, brightness, and lifetimes of PNe formed by single stars with those formed by close binaries in common envelope events.

2. TYPES AND FORMATION FREQUENCIES OF PLANETARY NEBULAE IN CLOSE BINARIES

The various possibilities for PN formation may be discussed best with reference to Fig. 1, where we summarize critical orbital

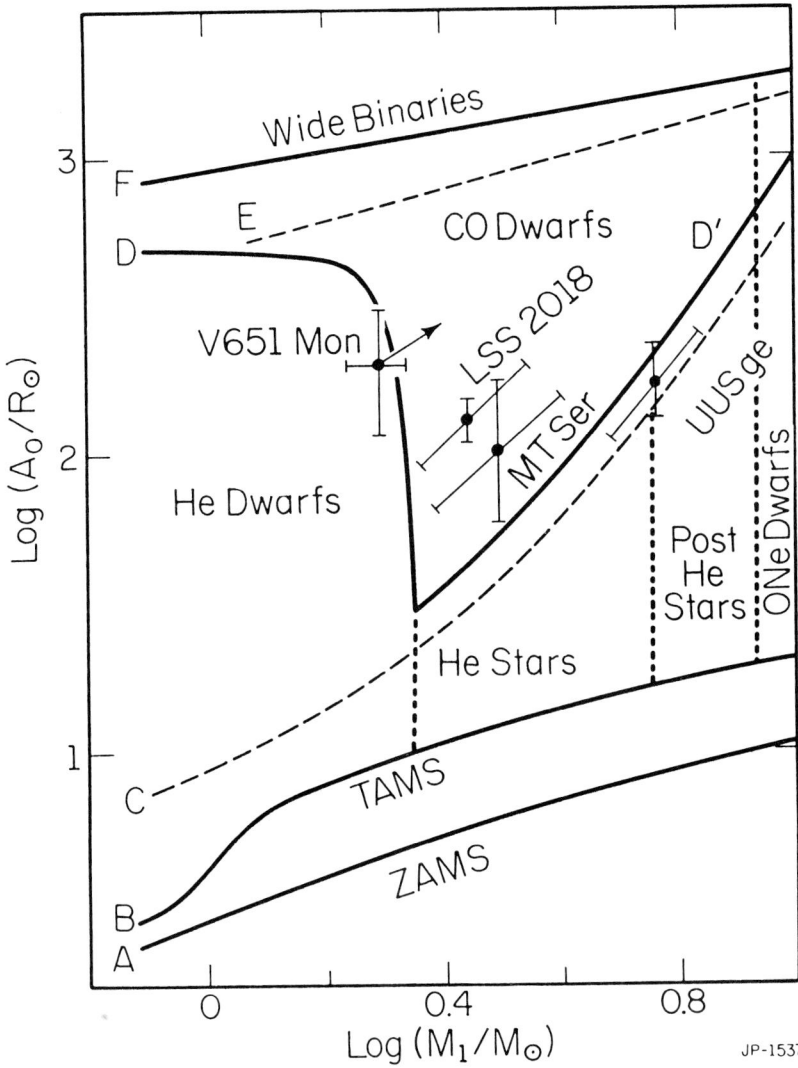

Figure 1. Critical borders influencing the nature of the mass exchange process and the nature of the remnants in close binaries. A_0 = orbital separation and M_1 = primary mass. F is the maximum separation for interaction, E is the start of the TP-AGB phase, D is where the primary ignites helium in its core, C is the border between systems in which the primary has a radiative envelope and those in which it has a deep convective envelope. Curves A and B are main sequence boundaries. The dotted lines are separate systems whose components produce different remnants.

separations as a function of the initial primary mass. Curve A gives the minimum separation A_0 that allows a zero age main sequence primary to fit within its Roche lobe, and curve B gives the separation allowing a terminal age main sequence primary to fit within its Roche lobe. For systems with A_0 below curve C, the primary has a large radiative envelope, and for those with A_0 above curve C, the primary has a deep convective envelope. For systems with A_0 larger than on curve D, the primary ignites helium in its core (which is degenerate if $M_1 < 2.3$ and non degenerate otherwise) and cannot fill its Roche lobe until after it has exhausted central helium and become an AGB star. For systems with A_0 between curves D and E, the primary fills its Roche lobe after it has developed a degenerate CO core, but before it has begun to experience thermal pulses. For systems with A_0 between curves E and F, the primary fills its Roche lobe while in the TP-AGB phase. Finally, in systems with A_0 larger than on curve F, the primary ejects its hydrogen-rich envelope of its own accord, largely uninfluenced by being in a binary.

The locations of the critical curves are to be considered only as rough qualitative guides. Not only has the dependence of critical separations on the mass ratio prior to each common envelope event been suppressed, but the critical separations are functions of the composition and input physics. More details are described in Iben and Tutukov (1985, 1987). We will concentrate first on what happens as a consequence of the initial Roche-lobe overflow event.

In systems bounded by curves B and C, if $q_0 \sim 1$, the thermal time scales of the radiative envelopes of both components are comparable when the primary first fills its Roche lobe, and it is probable that mass transfer will not lead to the establishment of a common envelope. The net result may be the transfer of much of the hydrogen-rich envelope of the primary to the secondary and an increase in the orbital separation. Evolution of this sort explains the properties of low mass Algols such as RY Gem ($M_{tot} \sim 2.6$) and XY Pup ($M_{tot} \sim 2.3$). In these systems, the initial primary was probably of mass less than 2.3 and did not fill its Roche lobe until after it had developed an electron-degenerate helium core. More massive Algols such as U Sge ($M_{tot} \sim 7.6$) and RS Vul ($M_{tot} \sim 5.9$), which derive from systems with $M_1 > 2.3$, have probably experienced conservative mass transfer as well; even though the helium core of the primary was not degenerate at the start of mass transfer, it ultimately became degenerate in the course of mass loss. This must be so, since, in both instances, the primary has evolved into a subgiant. In Algols, mass transfer continues at a rate controlled by angular momentum loss due to a magnetic stellar wind (Verbunt and Zwaan 1981). When the mass in its hydrogen-rich envelope has been reduced sufficiently, the primary will contract within its Roche lobe and, when shell hydrogen burning in it ceases, it will evolve into a helium white dwarf. The more massive secondary continues for a while to burn hydrogen as a main sequence star.

In neither case does a PN phase follow the inital mass-reversing mass-transfer phase. Not only is there very little mass ejected to be ionized, but the primary, which contains a degenerate helium core, does not become hot enough to ionize what little mass is ejected until after it has evolved for millions of years at low surface temperatures burning hydrogen in a shell.

A better chance for PN formation is presented by systems bounded by curves C and D and by $2.3 < M_1 < 10$ in Fig. 1. In such systems, the primary, with a non degenerate helium core, will have developed a deep convective envelope before filling its Roche lobe. After ejecting most of its hydrogen-rich envelope through its Roche lobe, it will evolve for a time as a compact core helium-burning star. If q_0 is large enough so that envelope shrinkage forces the remnant of the primary to have a surface temperature > 30,000K as it shrinks within its Roche lobe, the ejected hydrogen-rich material will become ionized and a PN will result. If $M_1 < 5$, the primary remnant, of mass < 0.75, will remain within its Roche lobe as it exhausts helium at its center and evolves eventually into a CO dwarf. If $M_{1R} < 0.4$, both hydrogen and helium shell flashes occur (Iben and Tutukov 1985, Iben et al. 1986) and there is the possibility of the ejection of a nebula of mass ~ $10^{-4} M_\odot$.

If M_1 is between 5 and 10, the remnant of the primary will, after it has exhausted central helium, swell again to fill its Roche lobe. In a second mass loss event, a common envelope is again formed and mass loss from the system continues until the mass of the primary remnant is reduced to ~ 0.75-1.1. The ejected material, of mass between ~ 0 and 1.4 is essentially pure helium (with a small admixture of ^{14}N). The remnant burns helium in a shell on a time scale of ~ 10^4yr at high luminosity (few x$10^4 L_\odot$) and at temperatures large enough to cause the helium-rich nebula to fluoresce (Iben and Tutukov 1985). Similar behavior is expected from systems with A_0 less than along curve C in Fig. 1, if q_0 is large enough.

Other possible sources of observable PNe are systems bounded by curves C and D and by $M_1 < 2.3$ in Fig. 1. The primary star has both an electron degenerate helium core and a deep convective envelope. Roche-lobe overfilling will therefore lead to a common envelope and loss from the system of most of the hydrogen-rich envelope of the primary. The ejected envelope can potentially be ionized by the shell hydrogen-burning remnant. From published models, however, one might infer that the remnant spends so much time contracting at temperatures less than 30000 K, that the ejected nebular material will be dispersed before the remnant becomes hot enough to excite the nebula into fluorescence (Iben and Tutukov 1986a). Configurations such as these have been dubbed "lazy" PNe by Renzini (1979, 1983). Since the remnant may eventually achieve temperatures far in excess of 30,000 K on a long time scale and at a relatively large luminosity, we have the interesting situation of a very hot and bolometrically bright star at the center of an invisible, or at least very dim, nebula.

The situation may not be as bleak as this. The published models do not explicitly take into account the shrinkage of the Roche lobe about the primary and the possible dynamical motions induced in the envelope of the mass-losing primary. When the primary first fills its Roche lobe, its luminosity and radius are related to the mass M_{He} of its degenerate helium core by

$$L_1 \sim 10^{5.5} M_{He}^{6.6}, \quad R_1 \sim 10^{3.5} M_{He}^4. \qquad (3)$$

The luminosity of the remnant as it emerges from the common envelope event is approximately $L_{PNN} \sim L_1$, its mass is $M_{PNN} \sim M_{He}$, and its radius R_{PNN} is related to its surface temperature T_{PNN} by

$$R_{PNN} \sim 10^{1.32} M_{PNN}^{3.3} (30000K/T_{PNN})^2. \qquad (4)$$

This tells us that, if the radius of the Roche lobe of the remnant as it emerges from the common envelope phase is smaller than, say, 1.5 (if $M_{PNN} \sim 0.45$), or 0.4 (if $M_{PNN} \sim 0.3$), a PN phase with $L_{PN} \sim L_1$ is ensured.

Consider a system with $M_1 = 1.5$, $M_2 = 0.5$, and $A_0 = 283$. The Roche-lobe radius of the primary is given approximately by

$$R_{1L} \sim 0.52 \, (M_1/M_{tot})^{0.44} \, A_0, \qquad (5)$$

where $M_{tot} = M_1 + M_2$. Combining equations (3) and (5), we have that the primary fills its Roche lobe when its core mass reaches 0.45. Equations (3) and (2) tell us that, following an assumed common envelope phase, $A_f \sim 15.9\alpha$ and the radius of the Roche lobe of the remnant is $R_{1L,f} \sim 6\alpha$. Since nebular excitation requires that $T_{PNN} > 30000K$, we conclude that a PN phase will not immediately follow the common envelope event, if indeed it occurs at all, unless $\alpha < 0.25$. Similarly, choosing $M_1 = 1.5$, $M_2 = 0.5$, and $A_0 = 56$, we find $M_{1R} \sim M_{He} = 0.3$, $A_f \sim 2.1\alpha$, and $R_{1L,f} \sim 0.7\alpha$ and conclude that a PN phase is ensured only if $\alpha < 0.6$.

Actually, as the common envelope event nears its close, the radius of the primary remnant may (due to dynamical effects alluded to earlier) be inflated considerably beyond that of a formal quasistatic model of the same luminosity and mass and, following Roche-lobe detachment, the radius may shrink on a time scale much smaller than given by the quasistatic model (for an elaboration of this theme, see our discusion in III of UU Sge). Hence, the upper limits on α which we have estimated are only suggestive and serve primarily to indicate that (1) whether or not a PN phase follows the common envelope phase is very sensitive to α, and (2) all other things being equal, the smaller the mass of the remnant relative to the mass of its progenitor, the more likely is the formation of a PN immediately following the first common envelope event in a low mass system.

This second statement is true only up to a point. There is a maximum surface temperature T_{max} which a remnant can achieve and this maximum is smaller, the smaller the mass of the remnant. If the remnant eventually evolves quasistatically, burning hydrogen in a shell, the maximum may be estimated quantitatively:

$$T_{max} \sim 10^{5.5} M_{PNN}^{1.8}. \tag{6}$$

To obtain this estimate, we have normalized to the $0.3 M_\odot$ model of Iben and Tutukov (1986a), assuming that the luminosity at T_{max} has the same mass dependence as does L_1 in equation (3) and that the radius at T_{max} has the same mass dependence as the radius of a low mass white dwarf, namely $R \propto M_{PNN}^{-1/3}$. For the $0.3 M_\odot$ model, the luminosity at T_{max} is approximately $L_1/2.6$ and the radius is approximately $5.7 R_{WD}$, where $R_{WD} \cong 10^{-1.9} M_{PNN}$ is the final radius of the model after it has become a white dwarf. We conclude that the mass of the remnant must be larger than $0.27 M_\odot$ if the ejected nebula is to become ionized.

In systems bounded by the curves E and F in Fig. 1, the primary will have developed a degenerate CO core before it fills its Roche lobe and will also have a very deep convective envelope. One therefore expects a common envelope to be formed when Roche-lobe overflow occurs. Since it is in the thermally pulsing stage, the primary may have developed chemical peculiarities at its surface. We note that the "phase space" for the formation of such systems (the area between curves E and F is quite small, so they arise infrequently.

Consider next what happens after the initial primary has evolved into a degenerate dwarf (or neutron star). We refer again to Fig. 1 as a qualitative guide, replacing A_0 by A_f and M_1 by M_2, and remember that, if A_0 is between curves C and F, then A_f is much smaller than A_0 as a result of common envelope action. When using equation (2), we must replace M_{1R} by M_{2R}, M_2 by M_{1R}, and M_1 by M_2. In all cases, A_f is expected to be less than given by curve D, and in many instances less than given by curve C (unless the initial mass-transfer was conservative). This time, however, the potential accretor is a compact white dwarf (or sometimes a helium star or a neutron star) and accretion at a rate larger than the Eddington accretion limit ($\sim 10^{-6} M_\odot yr^{-1}$ for a $1 M_\odot$ white dwarf) will always cause the accreted layer to swell to fill the compact object's Roche lobe, so that a common envelope is formed. Even if the donor has a radiative envelope when it fills its Roche lobe, it is expected that mass transfer rates will in general exceed the Eddington limit. Both because a common envelope is formed and because the mass ratio of donor to accretor is frequently large, considerable orbital shrinkage will occur and the remnant of the donor should quickly become hot enough to ionize the ejected material.

If A_f is between curves B and C and $M_2 < 2.3$ in Fig. 1, the secondary will become a shell hydrogen-burning star with a degenerate helium core and ultimately evolve into a degenerate helium dwarf. If $M_2 \sim 2.3-5$, the secondary will become a compact core helium burning star before evolving into a CO degenerate dwarf. If $M_2 \sim 5-10$, the secondary will swell to fill its Roche lobe and, in a common envelope event, will eject a helium envelope which it will ionize as a shell helium burning star with a degenerate CO core.

To summarize, our evolutionary scenarios suggest that the ionizing star after the first mass exchange may be (1) shell hydrogen-burning star with a degenerate helium core, (2) a helium star, (3) a post helium star object with a helium-burning shell and a degenerate CO core, (4) a post AGB star with a hydrogen or helium burning shell, and (5) a helium or CO degenerate dwarf. Possibilities (1)-(3) can occur only in binaries. In case (3), the matter of the PN is pure helium. Possibilities (4) and (5) differ from single-star PNe in the manner of ejection of nebular material. After the second mass loss event, the ionizing star may be one of possibilities (1), (2), (3), or (5), while its companion may be (1) a CO dwarf, (2) a helium dwarf, (3) a helium star, or (4) a neutron star. There are also cases in which the central stars merge in the process of ejecting a common envelope, but we decline to explore this possibility further here.

The formation frequency of the most likely variants can be estimated with the help of equation (1) and Fig. 1. Equation (1) implies that almost all stars are binaries and that the frequency of PN formation around components of all binaries (most of which begin and remain wide) is ~ $0.56 yr^{-1}$. We estimate a formation rate of $0.065 yr^{-1}$ for PNe with central stars consisting of a shell hydrogen-burning star with a degenerate CO core (or its CO white dwarf descendant) and a main sequence star and a formation rate of ~ $0.007 yr^{-1}$ for PNe consisting of a post core helium-burning star with a degenerate CO core (or its CO white dwarf descendant) and a main sequence star. About ten percent of these two types of systems evolve eventually into cataclysmics. The second type forms a helium-rich nebula. In about 1% of all cases the central star may be a core helium-burning star, but this may be a gross underestimate since, if q_0 is large, common envelope events occur even if A_0 is below curve C.

About one tenth of all post common envelope systems consisting of a hot compact star and a main sequence star will evolve by angular momentum loss into cataclysmic variables. Two conditions must be satisfied for this transformation to occur. First, $q_0 < 1.2$ if $M_2 > 0.8$ and $q_0 < 0.6$ if $M_2 < 0.8$. Second, there are restrictions on A_f. If $M_2 < 0.3$, the secondary is completely convective and the magnetic stellar wind (MSW) is weak. So, only gravitational wave radiation (GWR) can drive the components closer together; only if the period of the system after emerging from the common envelope satisfies $P < 5^h M_2^{3/8}$, will the system become a cataclysmic variable in less than $10^{10} yr$. If $0.3 < M_2 < 1.5$, the MSW can drive the components together in less than $10^{10} yr$ if $P < 32^h M_2^{1.2}$. These limits are discussed in more detail by Iben and Tutukov (1984b), for example.

After the second common envelope event, the most probable systems will be a shell hydrogen-burning star with a CO core in orbit with a CO degenerate dwarf deriving from systems with 2.3 <

$M_1 < 4$, $2 < \log A_0 < 3$, at a formation frequency of $0.015\mathrm{yr}^{-1}$, and systems of CO and He degenerate dwarfs (one of which may have a hydrogen-burning shell) deriving from systems with $2.3 < M_1 < 4$, $1 < \log A_0 < 1.5$, at a frequency of $\sim 0.005\mathrm{yr}^{-1}$. The total frequency of close binaries as PNNi is thus estimated to be about $0.10\mathrm{yr}^{-1}$ or about 20% of the total frequency of PN formation.

Systems with $A_0 < 500$ and $M_1 < 2.3$ form double degenerate helium dwarfs (Iben and Tutukov 1986b). Their shell hydrogen-burning progenitors may evolve too slowly on their first trip to the blue in the HR diagram to be able to ionize the ejected common envelope material before it merges with the interstellar medium. However, following subsequent hydrogen shell flashes during which envelopes of mass $\sim 1\text{-}10 \times 10^{-4} M_\odot$ are ejected through Roche-lobe overflow, the remnant travels to the blue rapidly enough to excite the emitted matter (Iben and Tutukov 1986a) and, if it is massive enough ($> 0.4 M_\odot$), the system may become a detectable PN. But, because the lifetime of the central star is so short during its most luminous state and because the nebular material is still very close to the central star, the probabilty of seeing such a system is not very large. In mergers of low mass degenerate dwarfs a small common envelope may be ejected and lit up be the merged product. We note that at least one case of an extremely low mass ($\sim 10^{-8} M_\odot$) PN exists (Liebert, this volume).

One intriguing final possibility which has not yet been explored numerically is that, in a common envelope event initiated by the secondary, the cold CO white dwarf remnant of the primary can accrete enough material to reignite hydrogen and be thereby resurrected to become the dominant ionizing star for the nebula ejected by the secondary.

3. OBSERVED PLANETARY NEBULAE WITH BINARY NUCLEI

Approximately 10 planetary nebulae are known to contain a central star of established duplicity (Bond 1987, Ritter 1987) and the characteristics of several provide direct evidence for the common envelope ejection hypothesis (e.g., Bond 1976, Grauer and Bond 1983, Bond 1985, Drilling 1985, Bond and Grauer 1987, Bond and Grauer 1987).

We will attempt to reconstruct the prior evolution of four of these systems. We emphasize at the outset, however, that the results of this sort of exercise are not only sensitive to the theoretical assumptions, but are exceedingly sensitive to the stellar and orbital characteristics estimated from the observations. Because the theory of binary star evolution is in such a rudimentary state, it is crucial for its development that every effort be made to establish these observational characteristics as carefully as current technology will allow.

V651 Mon (the central star of NGC 2346) consists of a hot dwarf of mass ~ 0.4 and an unevolved main sequence star of mass ~ 1.8 (Mendez and Niemala 1981). The orbital period is 16 days on a long time scale burning hydrogen in a shell at luminosities larger than and at surface temperatures smaller than given by equations (7) before it ignites helium at its center. From Fig. 1 of Iben and Tutukov (1985) we see that the rate at which the radius of a model of an 0.76 $M_\odot$ shell hydrogen-burning star changes with time as it approaches the helium-burning main sequence is of the order of $d\log R/dt \sim 0.3/10^6$ yr when $R \sim R_\odot$. The implication is that the sdO star in UU Sge may have required of the order of 2×10^6 yr to reach its present state after detaching from its Roche lobe. It may therefore not yet have ignited helium at its center and so our estimates based on equations (7) might be wholly inappropriate.

However, the lifetime of a typical planetary nebula is orders of magnitude less than 10^6 yr, which means that the sdO star cannot have required as much time as suggested by the formal quasistatic models whose evolution rate is controlled by the rate of hydrogen burning in a shell. What may have happened in the real situation is that, when Roche-lobe detachment occurred, the radius of the mass-losing precursor of the sdO star was considerably larger than the radius of a constant-mass model in quasistatic equilibrium and that the mass of hydrogen-rich matter remaining near the surface was much less than that necessary to sustain hydrogen-burning in quasistatic equilibrium at high luminosity. The fact that the spectrum of the ionizing star is sdO rather than sdB supports this interpretation. Immediately after Roche-lobe detachment, the surface layers of the precursor of the sdO star, inflated by dynamic effects, may have begun to contract on a time scale less than the thermal one. The surface layers of the sdO star may now be contracting on a thermal time scale. Helium has probably ignited at the center, thus accounting for the current luminosity.

The current masses and separation of the components of UU Sge permit us to estimate an appropriate value for the parameter α. From Iben and Tutukov (1985), $M_{1R} = 0.76$ implies $M_1 \sim 5$. Setting $M_2 = 0.7$ and $A_f = 3$ in equation (2) gives $A_0 = 141/\alpha$. But, in order for the remnant to become a helium star, we have that the radius of the Roche lobe of the primary before it fills its Roche lobe must be less than $R_* \sim 10^{0.24} M_1^{2.25}$ (used to establish curve D in Fig. 1). This means that $R_{1L} \sim 0.49 A_0 = 69/\alpha < 65$ or that $\alpha > 1$! Had we selected $M_{1R} = 0.86$, then $M_1 = 5.5$ and $\alpha > 0.86$. Since curve D is model and composition dependent, we must not take these estimates too literally. We may, however regard them as confirmation of our anticipation that, for initially widely spaced components of quite unequal masses, the parameter α is indeed of the order of unity.

As a final word of caution, however, we note that our placement of UU Sge progenitor parameters in Fig. 1, is very close to the critical curves C and D, and that any change in the parameters of the current system inferred from the observations could alter the most likely scenario considerably. This strengthens the lesson of V651 Mon.

4. COMPARISONS BETWEEN PNe OF SINGLE AND OF BINARY ORIGIN

The nature of the (super)wind that produces most PNe is still unclear. It is possible that pulsational instability of a red supergiant envelope in combination with radiation pressure on dust and molecules can produce the mass-loss rate required by the observations, but an explicit demonstration from first principles is still missing. In contrast, it is clear that common envelope action in close binaries can eject large quantities of matter on a very short time scale.

In addition, binary scenarios provide a natural way of accounting for the bipolar structure and multiple shells seen in many PNe (Balick 1987, Chu and Jakoby 1987, Chu, this volume). There are at least two ways in which a binary core can influence nebular shape. In close binaries, angular momentum from the stellar orbit can be transferred in an axially symmetric way to the material ejected in a common envelope event. In wide binaries, if the orbital velocity of the components exceeds a typical nebular expansion velocity (~ 10 kms^{-1}), the motion of the superwind material ejected by one of the components can be influenced by the second component. Thus, all binaries with $A_0 < 10^4$ are possible precursors of PNe with cylindrical symmetry. Consequently (see equation [1]), perhaps half of all PNe have experienced shaping by a central binary. Of course, the wind emitted by the ionizing star will also exert a shaping influence (Kwok, Purton, and FitzGerald 1978, Kwok 1982, 1987, Kahn 1982) and, under the proper conditions, can also produce bipolar and shell effects which might be difficult to disentangle from those due to duplicity.

One of the most important pieces of information required for a further development of our understanding of close binary evolution is a firm, empirically based estimate of the value of the parameter α describing the degree of orbital shrinkage during the common envelope stage. The example of V651 Mon shows that fairly precise estimates can be made in those situations where the ionizing star has a degenerate helium core. This is because, between the mass of such a core and the radius of the precursor red giant, there exists a tight relationship which can be used to estimate the orbital separation of the precursor system. However, a definitive estimate by this means requires that the mass of the ionizing star can be unambiguously determined to be less than ~ 0.45.

PNe formed around close binaries can have systematiclly different chemistrys which could help to distinguish the two. Since the evolution of close binary components does not proceed as far as the evolution of single stars, the chemical composition of the nebulae around close binaries might be expected to be less chemically evolved than the nebulae around single stars. Certainly, because the range in A_0 which will allow a primary component to reach the TP-AGB phase and begin to dredge up carbon and s-process isotopes (A_0 between curves E and F in Fig. 1) is so

Tutukov, A. V., Fedorova, A. B., and Yungelson, L. R. 1982, Pis'ma Astron. Zh., **8**, 365.
Verbunt, F., and Zwaan, C. 1981, Astron. Ap., **100**, L7.
Willson, L. A. 1987, in Late Stages of Stellar Evolution, ed. S. Kwok and S. R. Pottasch (Dordrecht: Reidel), p. 197.
Wood, P. R. 1974, Ap. J., **190**, 609.
_____. 1981, in The Physics of Red Giant Stars, ed. I. Iben, Jr., and A. Renzini (Dordrecht: Reidel), p. 135.

THE EVOLUTION OF THE COMMON PLANETARY NEBULA

J. Köppen
Inst. f. Theor. Astrophysik
Im Neuenheimer Feld 561
D-6900 Heidelberg
Fed. Rep. Germany

1. INTRODUCTION

When we look at evolutionary aspects of PNe, we hope to test our ideas and understanding of the origin of PNe - i.e. why and how gas is expelled from the progenitor star - and of the processes relevant in the evolution of the nebula as well as the central star.

Evolutionary aspects of the central stars are covered elsewhere in this Symposium, so let us look at what we can observe in the nebulae that tells us about their evolution. As the PN's lifetime of about 30000 yrs is much greater than the time basis for the observations, we mainly depend on comparing objects of different ages. Expansion velocities are of the order of 20 km/s, so the radius R of a nebula can be taken as a measure of its age. It depends however on the assumed distance. Also, we not only assume that all PNe have a common origin and history, but we also ignore the presence of multiple shells in many PNe.

The basic quantities of a PN, formed of gas expanding away from and photoionized by the central star, are:

$$\text{density } n \qquad \text{velocity } v \qquad \text{ionization } x$$

The state of the gas is further described by the electron temperature, which in PNe however is always near 10^4 K. There are two ways to study these nebular properties: First, from observations integrating over the nebula's face one gets average values $N(R)$, $V(R)$, $X(R)$ for a nebula of size R. Secondly, for angularly resolved objects one may try to derive the radial dependences $n(r)$, $v(r)$, $x(r)$ and compare them with distributions $n(r,t)$ etc. from theoretical models.

Let us have a first look at the density: The average electron density decreases with increasing nebula size.

The measurements can be fitted quite well with a simple law $N R^3$ = const (Fig.4 of Schmidt-Voigt and Köppen, 1987b, henceforth SKb). The very crude model of a spherical nebula whose ionized mass M_{ION}, outer radius R and thickness D, yields: $N R^3 \propto M_{ION} / (D/R)$. So the observed dependence $N(R)$ can be understood in terms of a nebula with constant ionized mass, if its shell geometry D/R remains also constant. What do **proper** models tell us?

2. ASSUMPTIONS OF NEBULAR MODELS

Numerical models which solve the equations for the flow, ionization, and energy balance of the nebular gas at various levels of sophistication have been constructed by Mathews (1966), Sofia and Hunter (1968), Ferch and Salpeter (1975), Okorokov et al.(1985), Bedogni and D'Ercole (1986), Schmidt-Voigt and Köppen (1987a). The physical assumptions used were (the degree of underlining denotes the degree of importance in PN):

Geometry: spherical symmetry
Forces: thermal pressure, radiation pressure(gas+dust),
          gravity
Ionization: by photoabsorption, electron collisions
    Treatment:
        complete ioniz. (O.K. for principal effects)
        ioniz.equilibrium (O.K. for main nebula only)
        time dependent (best)
Central Star:
    Parameters: theor.evolut.tracks
    Spectra: blackbody (!!)
Energy: photoabs.,compression, heat conduction,
        line+contin.emission, expansion, coll.ioniz.
    Treatment:
        isothermal (O.K. for most purposes)
        balance of gains=losses (better)

A model is characterized by its initial conditions, in particular the initial distribution of density and velocity. These can be described by three major periods of mass loss which the central star underwent:
(1) as a red giant it had a slow (~10 km/s) "AGB wind".
(2) at the end of its life on the AGB, it might expel matter in the form of a short, slow, massive "superwind".
(3) central stars have fast (2000 km/s) winds.
The mass loss rates (velocities, duration,...) of each component are considered to be free parameters.

The interface between the fast wind and the slowly moving gas expelled earlier was studied by Lazareff (1981), Bedogni and D'Ercole (1986): A shock develops which separates an inner region of unshocked fast wind matter and

an outer region of shocked, very hot gas. The pressure by
this "hot bubble" prevents the exterior nebular gas to flow
towards the central star, and is thus responsible for the
formation of the central cavity typical for PNe (Mathews,
1966; Okorokov et al.1986; SKa). Although the structure
and physics of this interaction between fast wind and slow
nebula do deserve more detailed studies, it is often
sufficient and numerically more convenient to take into
account the pressure as an inner boundary condition for the
nebula (cf. Mathews, 1966; SKa; analytical model of Volk
and Kwok, 1985). SKa found that the fast wind - as massive
as observed - is responsible essentially only for the
central cavity. Therefore we shall concentrate in the
following on the exterior regions where superwind and AGB
wind material form the visible PN shell.

In our discussion of the evolution of different types of
models we shall use this classification:
1 Wind Model: essentially superwind ejecta, representing
the idea of PN formation in a single, sudden event. Very
dilute AGB wind; fast wind present to make inner cavity.
Termed "non-accreting 3WM" by SKa,b).
2 Wind Model: no superwind; only AGB and fast wind
interact, as proposed by Kwok et al.(1978).
3 Wind Model: superwind and AGB wind form nebula (see
below), fast wind present.

3. PATTERNS OF THE GAS FLOW

According to theory (Schönberner, 1981) a star leaving
the AGB takes about 3000 yrs to heat up before it can
ionize hydrogen to make the surrounding nebula visible.
Little is yet known about this first phase of nebular
evolution. The gas is still neutral and probably rather
cool, so thermal pressure is expected to be less important
than radiation pressure on gas and dust (c.f. Okorokov et
al., 1986). The dust is heated by the stellar continuum
and its radiation in the infrared should help investigating
this "neutral phase".

Then the nebula becomes ionized, and soon after a nebula
of a typical mass of 0.3 $M_\odot$ is completely ionized. Now the
nebula can well be treated as a shell of ionized gas with a
temperature of about 10^4 K. The increased thermal pressure
dominates the further evolution of the shell: The outer
boundary of the nebula moves into the surrounding AGB wind
remnant with a velocity that increases with increasing
density contrast across the boundary.

Thus a 1WM expands (in the extreme case: into vacuum)
with a greater speed than a 3WM which runs into a denser
AGB wind. The velocity increases toward the outer
boundary, while the density shows a decrease (Fig.2 in
SKa).

Though the velocity profile in a 3WM (Fig.4 of SKa) also shows an increase with radius, the density profile changes over from an initially decreasing function to an increasing one. This change of slope is caused by the mass flow across the shock from the AGB wind into the nebula. The accretion rate for the whole nebula

$$\dot{M}_{acc} = \dot{M}_{AGB} (v_s - v_{AGB}) / v_{AGB}$$

depends on the shock velocity v_s and on the velocity v_{AGB} and mass loss rate $\dot{M}_{AGB}$ of the AGB wind. The time during which the nebula has swept up as much mass as its original mass (i.e. M_{sw} of the superwind ejecta):

$$t_{sweep} = M_{sw} / \dot{M}_{acc}$$

is the age when the density profile changes its slope. If this occurs during the visible life of a nebula, the object then has a structure distinctly different from a 1WM. This condition is the proper definition of a 3WM as opposed to a 1WM (SKa, SKb used the terms (non-) accreting 3WM).

In a 2WM the nebula is created solely by the accretion process, and its structure resembles that of a developed 3WM: the density profile always increases with radius; the velocity increases, and tends to be smaller than in 3WMs.

How do these types of models compare with observations? Since most PNe show rather well defined outer borders, 2WM and 3WM because of their increasing density profile seem more appropriate than 1WM which have a rather extended decline of their surface brightness.

The decline of the average density **N(R)** - Fig.7 in SKb - is quite well reproduced with almost any model. There is a tendency for 1WMs to disperse too quickly and 2WMs too slowly.

Nothing is known about the density profile **n(r)**, which should provide another test between 1WM and 2WM/3WM.

The observed expansion velocities **V(R)** - Fig.1 of SKb - are not reproduced by the rapidly expanding 1WMs. 3WM and the even slower 2WM fit the velocity pattern much better.

Weedman (1968) found that the velocity **v(r)** within the nebula increases linearly with radius. Unfortunately, all types of models show this type of velocity profile, so no distinction can be made with this parameter.

4. IONIZATION

To understand the gas flow one essentially needs to take into account the fact that the star has become hot enough to ionize tha nebula. The degree of ionization, however, depends on the stellar temperature and luminosity, both of which vary strongly with time. Therefore, the ionization reflects the actual stellar evolution more closely. The star's temperature first rises quickly, goes through a maximum, after which the stellar energy sources are exhausted and the star starts to cool off very slowly. So,

for each value of temperature there are two possible places in the HR-diagram where a star might be. To resolve this ambiguity, one can make use of the fact that the maximum temperature in Schönberner's (1981) tracks occurs at a stellar absolute visual magnitude of about 5, and thus separate observed objects into two groups:
*** the UPs: nebulae with bright central stars that can be tought of still heating up.
*** the DOWNs: nebulae with faint stars which we interpret as already cooling down.

To measure the volume averaged degree of ionization, the HeII/Hβ ratio is attractive for a first step: Recombination lines are much less sensitive to nebular temperature and density than collisionally excited lines. If the nebula is optically thick in both H and He^+ Lyman continua, these lines count the number of ionizing photons from the central star. Finally, since both elements are the main contributors to the opacity, they set up an opacity and ionization structure which the other elements have to follow.

The evolution of the line ratio up to the time of the maximum stellar temperature, as observable in the UP nebulae (Fig.2 of SKb), is dominated by the time scale of the star's heating up; differences in models are negligible. The HeII/Hβ ratio increases until the nebula becomes optically thin first in H, then also in He^+ continuum when it levels off. Since the stellar time scale is a steep function of the star's core mass, one obtains a very good leverage on this mass: SKb find Schönberner's hydrogen burning stars of 0.6 to 0.64 $M_\odot$ to give the most favourable fit.

After the temperature maximum the number of ionizing photons drops rather sharply. From now on, the stellar evolution proceeds at snail's pace, so one may consider all stellar parameters as constant for the rest of the nebula's life. Therefore, the evolution of the ionization is now strongly determined by the properties of the nebular models (Figs.3, 4 in SKb).

The observed line ratio of DOWN objects is lower than that of the late UPs, optically thin in both H and He^+. Therefore, the nebulae must not only recombine partially, but also keep this lower degree of ionization. Only nebular models that meet **both** conditions can be acceptable.

1WM that are massive enough to recombine (both in H and He^+) fail to meet the second constraint: as the nebula expands and its density decreases, the constant star is able to ionize a greater mass. The nebula eventually become optically thin in H again, the HeII/Hβ ratio increases.

2WM and 3WM of suitable and reasonable parameters can be found that have accreted enough matter to recombine when

the star's photon output drops. They are also able to meet
the second condition: their mass increases by accretion at
such a rate that the total number of recombinations remains
constant, despite the decline in density. Consequently,
the ionization structure in the shell does not change
(Fig.5 of SKa). SKb find as most suitable models either a
   3WM: $M_{BW} = 0.1\ M_\odot$, $\dot{M}_{AGB} = 6\ 10^{-6}\ M_\odot/yr$
or a
   2WM: $\dot{M}_{AGB} = 10^{-5}\ M_\odot/yr$.
The 3WM has the advantage that it explains the high Zanstra
temperatures of rather small nebulae better than 2WM (Fig.6
in SKb).

   In this way the low ionization of old PNe enables us to
measure the mass loss rate on the AGB which are of the
order of $10^{-5}\ M_\odot/yr$.

   Additional information can hopefully obtained by looking
at other line ratios. Also, the distribution of ions in
the nebula might be tried, especially since the nebula can
well be assumed to be in ionization equilibrium and its
structure can be well calculated with static models.

5. MASS RADIUS RELATION AND OPTICAL DEPTH

The various types of models can be nicely characterized by
their mass-radius relation, which for almost all models of
SKb can be well represented by:
         $M(R) = M_{BW} + 10^5\ \dot{M}_{AGB}\ R$
(in the convenient units of solar masses, years, parsec).
In the following figures we show the ionized mass-radius
relations for a number of models from SKb (labelled by
their number). Full lines show when the nebula is
optically thick in H – the adjoining numbers are the
optical depths at 911 Å – dotted lines when it is optically
thin both in H and He^+. When the nebula is optically thin
or slightly thick, the ionized mass equals the total mass,
and the above relation holds also for the curves in the
figure, exept for the time when the ionization front
transverses the nebula for the first time (R < 0.07pc).
Exept for No.4, a rather massive 1WM, the models shown then
become optically thin in H, then in He^+. When the star's
(here: Schönberner's 0.644 track) photon output drops
sharply, the nebulae recombine, and become (moderately)
optically thick in H again. In Nos.4 and 11 the mass is so
large that the optically depth gets rather large and the
ionized mass decreases appreciably. Thereafter the nebula
enters the "DOWN" phase (R > 0.15pc) and finally becomes
optically thin in H and He^+ again. As a consequence of
their high accretion rate, 2WM and 3WM are able to stay
optically thick much longer than 1WM.

   The empirical ionized mass-radius relation from Phillips

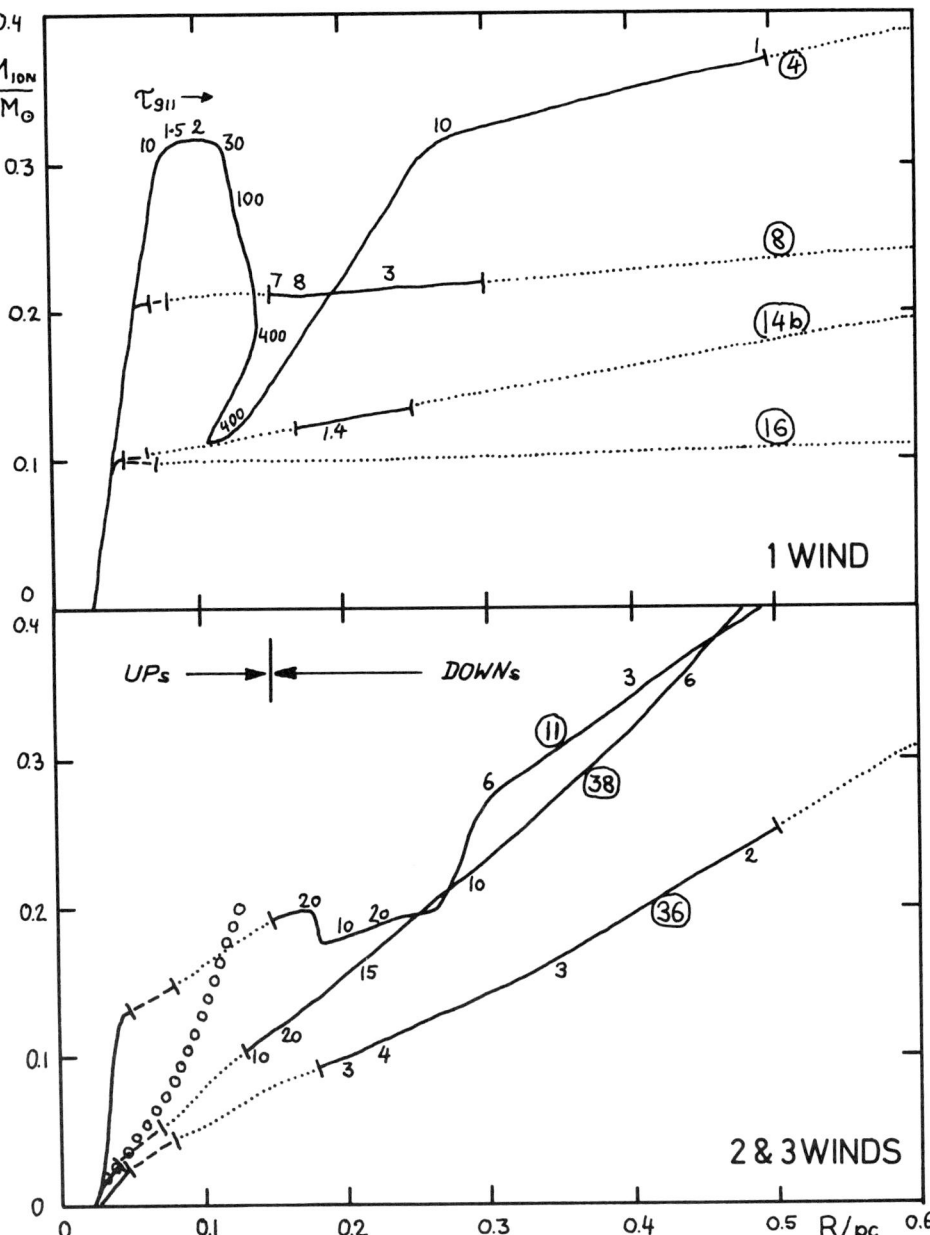

Ionized Mass-Radius Relation for models of SKb with the 0.64 $M_\odot$ star, labelled by circled numbers. The nebula is optically thick in H (full line), and thin in H and He^+ (dotted). The numbers are the optical depths at 911 Å.

and Pottasch (1984) (open circles) could be explained by a 2WM with $\dot{M}_{AGB} \approx 2\ 10^{-5}$ $M_\odot$/yr.

6. FINAL REMARKS

To summarize:
** From younger nebulae (UPs) one can determine the stellar evolutionary time scale.
** Old, extended nebulae (DOWNs) are extremely useful to probe into the initial conditions and to measure the AGB mass loss rate.
** Accretion from a rather massive AGB wind is a most important process in the formation of PN shells.
   Apart from the problems and possibilities already mentioned, here are some more items for our shopping bag of things to be done:
** Interaction of fast wind with slow nebula. Are the obnoxious filaments created by this process?
** Influence of stellar wind on central star evolution.
** Use of proper stellar flux distributions rather than blackbodies.
** Departure from spherical symmetry. What happens when the nebula is optically thin in one direction but thick in another?
** Multiple shells: constraints on the ejections.

7. ACKNOWLEDGEMENTS

Financial support by the Deutsche Forschungsgemeinschaft (SFB 132) is gratefully acknowledged.

8. REFERENCES

Bedogni, R., D'Ercole, A.: 1986, *Astron.Astrophys.* **157**, 101
Ferch, R.L., Salpeter, E.E.: 1975, *Astrophys.J.* **202**, 195
Kwok, S., Fitzgerald, P.M., Purton, C.R.: 1978, *Astrophys.J.* **219**, L125
Lazareff, B.: 1981, These, Université Paris Sud, Centre d'Orsay
Mathews, W.G.: 1966, *Astrophys.J.* **143**, 173
Okorokov, V.A., Shustov, B.M., Tutukov, A.V., Yorke, H.W.: 1985, *Astron.Astrophys.* **142**, 441
Phillips, J.P., Pottasch, S.R.: 1984, *Astron.Astrophys.* **130**, 91
Schmidt-Voigt, M., Köppen, J.: 1987, *Astron.Astrophys.* **174**, 211 (SKa) and **174**, 223 (SKb)
Schönberner, D.: 1981, *Astron.Astrophys.* **103**, 119
Sofia, S., Hunter, J.H.: 1968, *Astrophys.J.* **152**, 405
Volk, K., Kwok, S.: 1985, *Astron.Astrophys.* **153**, 79
Weedman, D.W.: 1968, *Astrophys.J.* **153**, 49

PLANETARY NEBULAE WITH MASSIVE CENTRAL STARS

R. Tylenda
Laboratory of Astrophysics
Copernicus Astronomical Center
Chopina 12/18
87-100 Toruń, Poland

ABSTRACT. The paper discusses the problem of planetary nebulae with massive nuclei from the point of view of their theoretical evolution and observational appearances. The available data suggest that NGC 2440, 6302, and 7027 have central stars with masses greater than 0.8 $M_\odot$.

1. INTRODUCTION

It is now commonly accepted that most of planetary nebula nuclei (PNNi) have masses close to 0.6 $M_\odot$. The main controversy concerns the high mass tail of the distribution (cf. Schonberner, 1981; Kaler, 1983; Renzini, 1983; Heap & Augensen, 1986). At present this problem cannot be solved definitively because of observational and theoretical uncertainties. In this review we present the main points concerning massive PNNi and their planetary nebulae (PNe). The principal evolutionary aspects of massive PNNi are outlined in the next section. Section 3 is devoted to the evolution of PNe surrounding massive PNNi. Finally, in Section 4 we discuss the methods for observational determination of masses and the results for individual PNNi. We conclude that at present we know three PNe, i.e. NGC 2440, 6302, and 7027, whose PNNi have masses above 0.8 $M_\odot$.

2. EVOLUTIONARY CHARACTERISTICS OF MASSIVE CENTRAL STARS

The luminosity of a PNN, L, with active shell sources depends only on its mass, M, and can be determined from a widely known formula (e.g. Paczyński, 1971):

$$L/L_\odot = 5.9 \times 10^4 \, (M/M_\odot - 0.52). \qquad (1)$$

The effective temperature varies greatly during the PNN evolution. It is determined by the mass of the H-rich envelope and the nuclear burning activity. However, the maximum value of the effective temperature, T_M, that can be achieved by a PNN is again a function of its mass only. This can be estimated from the formula:

where n is the nebular electron density in cm^{-3}. Condition (6) can be used as a self-consistency test when applying the Zanstra method to PNe with luminous PNNi.

After the nuclear fuel has been exhausted the PNN is decreasing in luminosity. Initially the fading is fast and down to $\log L/L_\odot \simeq 3$ it proceeds on a time scale not much longer than τ_n (Eq. 4). The reduced flux of ionizing photons can maintain the ionization only in the innermost layers of the PN. The outer regions are now recombining and cooling off. The nebula is in the recombination phase. The most spectacular aspect of this phase is the appearance of a double-envelope structure in the image of the PN, i.e. an inner, bright, high-excitation ring is surrounded by a faint, low-excitation halo (Tylenda, 1983). The halo is fading with time. However, the time scale of this process increases because of the decreasing electron density (cf. Eq. 5). Consequently, even after several thousands of years a tenuous, very low-excitation halo showing $n \simeq 10 - 100$ cm^{-3} can still be visible.

Finally, it is worth of noting that the appearance of a double-envelope structure, similar to that discussed above, is predicted for PNe with less massive PNNi ($M > 0.6$ $M_\odot$) as well, provided that the PNNi burn hydrogen quiescently (Tylenda, 1986; see also Schmidt-Voigt and Koppen, 1987). This is because the PNNi of this sort have a fast decline in luminosity after the cessation of the nuclear burning (Schonberner, 1981).

4. OBSERVED CASES OF PLANETARY NEBULAE WITH MASSIVE CENTRAL STARS

The most classical method for observational testing of the PNN evolution theory is the H-R diagram (see e.g. reviews by R.A. Shaw and by S.R. Pottasch in this volume). After having determined luminosities for PNNi lying in the horizontal part of the H-R diagram one may hope to derive their masses from Eq.(1). Unfortunately, this method cannot give reliable results for individual cases because of the well known problem of the distances to the galactic PNe. However, it can be applied to PNNi in the Magellanic Clouds. A first attempt to determine luminosities of PNNi of three bright PNe in the Magellanic Clouds made by Stecher et al. (1982) gave very high values implying masses close to 1.2 $M_\odot$. It appeared, however, soon that the Zanstra luminosities derived by Stecher et al. violated condition (6) (Tylenda, 1984). In other words, the observed PNNi evolved much slower than a 1.2 $M_\odot$ PNN should have done. Subsequent reanalyses of the observational data suggested much smaller masses, i.e. 0.6 - 0.7 $M_\odot$ (Tylenda, 1984; Heap & Augensen, 1987). Recently Aller et al. (1987) have derived masses for 12 PNNi in the Magellanic Clouds from the observed luminosities. The values range from 0.58 to 0.71 $M_\odot$ including the three controversial PNNi. In all cases condition (6) is satisfied. In conclusion, we do not see massive PNNi within the brightest PNe in the Magellanic Clouds. This is not surprising since massive PNNi evolve very quickly while being luminous. We, therefore, expect to find them rather among hot, low-luminosity PNNi.

Schonberner (1981) has elaborated a method which compares theoretical models with observations on the Abell's (1966) diagram: stellar M_v versus nebular radius. Since the evolutionary time scale is very sensitive to the stellar mass (Eq. 4) the theoretical tracks are well separated in this diagram even for small differences in mass. Hence its potential usefulness for empirical mass determination. However, with the present uncertainities both in the observations (distances) and in the theory (mechanism and time of the PN formation, subsequent dynamics of the PN, importance of residual stellar winds) conclusions drawn from the M_v-R_n diagram alone can be, sometimes, very misleading. Recently Heap & Augensen (1987) have derived individual PNN masses using the discussed method. For about 30% of the objects they have obtained values in excess of 0.65 $M_\odot$. Most of this comes from compact PNe having R_n < 0.1 pc. A closer analysis shows that precisely these stars have in majority $\log L/L_\odot$ < 3.5 and $\log T$ < 4.85. Consequently they lie below the horizontal part of the 0.6 $M_\odot$ track in the H-R diagram (e.g. Pottasch, 1983). And this suggests that these are very low mass PNNi. The situation is, therefore, not clear and requires thorough study.

Very recently Mendez et al. (1987) have derived PNN masses for 21 PNe from a model atmosphere analysis of the observed stellar H and He absorption profiles. For a half of the sample they have derived masses above 0.7 $M_\odot$ - a result really surprising in view of other studies in our Galaxy (Schonberner, 1981; Kaler, 1983) and in the Magellanic Clouds (Aller et al. 1987). Let us concentrate on two PNNi, i.e. NGC 2392 and He 2-138, for which Mendez et al. have found masses close to 0.9 $M_\odot$. It appears that with the distances proposed by Mendez et al. the Zanstra lminosities for these two PNe do not satisfy condition (6). In the case of NGC 2392 the HeII Zanstra luminosity is $\log L_Z/L_\odot$ = 4.69 (HeII λ 4686 line intensity taken from Aller & Czyzak, 1979; other data from Mendez et al. 1987) whereas the HI Zanstra luminosity for He 2-138 is $\log L_Z/L_\odot$ = 4.41. The upper limits from Eq.(6) are 4.34 ± .06 ($\log n$ = 3.5 - Aller & Czyzak, 1979; Shaw & Kaler, 1985) and 4.42 ± .05 ($\log n$ = 3.9 - Torres-Peimbert & Peimbert, 1977), respectively. Even the case of He 2-138 cannot be regarded as mariginally consistent since Eq.(6) has been obtained using Eq.(4a) which gives the time scale for the overall nuclear evolution. The two PNNi, according to Mendez et al., have $\log T$ < 4.7 and for these temperatures the stellar evolution is much faster than τ_n. The model PNNi of Schonberner (1981) evolve from $\log T$ = 4.4 to 4.7 during a time span 10 times shorter than τ_n. Thus an 0.9 $M_\odot$ PNN requires only some 10 years in order to pass this temperature interval. NGC 2392 has been seen for more than 100 years (it was discovered in 1853 - Perek & Kohoutek, 1967). The discrepancy is important and we conclude that Mendez et al. (1987) have overestimated the PNN masses, at least for NGC 2392 and He 2-138.

A lower limit to the mass of a PN can be obtained from Eq.(2) if we know its effective temperature. This method has the advantage of being independent of the distance. The main problem is, of course, the temperature determination the more so as we expect $\log T$ > 5.3 for massive PNNi. The existing methods for effective temperature determination have recently been analysed by Stasińska & Tylenda (1986) from the point of view of their usefulness in the case of very hot stars. They conclude that only the Zanstra method can give a reliable estimate of the effective temperature of a hot PNN. Other methods underestimates, sometimes seriously, the temperature in this case.

Table 1
Planetary nebulae with massive central stars

Name	log T_z	log T_s	$M/M_\odot$
NGC 2440	5.50	5.26	>0.81
NGC 6302		5.48	>0.80
NGC 6445	5.27	5.18	>0.65
NGC 6537	5.29		>0.66
NGC 6741		5.37	>0.72
NGC 7027	5.78	5.48	>1.05
IC 2165	5.29	5.05	>0.66

In Table 1 we list the PNNi which have log T > 5.2. The second column gives the Zanstra temperatures calculated from the data recently compiled by Stasińska & Tylenda (1987). In all cases T_z(HI) was greater than or comparable to T_z(HeII) so we took the mean value from the two estimates. The third column contains the Stoy temperature calculated by Preite-Martinez & Pottasch (1983). The lower limits to the PNN masses are given in the last column. They have been derived from Eq.(2) using T_z or T_s if the former was not available. All PNNi listed in Table 1 have masses well above the canonical value of 0.6 $M_\odot$. But in three cases, i.e. NGC 2440, 6302, and 7027, the PNNi are very massive. For NGC 6302 no PNN has been observed as yet. It has one of the highest Stoy temperatures. Since the Stoy method underestimates the effective temperature (Stasińska & Tylenda, 1986) it is clear that this PN has a PNN much more massive than 0.8 $M_\odot$. NGC 7027 has been extensively discussed in Tylenda (1984). This PN has an extended, very faint halo observed in Hα by Atherton et al.(1979). The nature of the halo is not clear but it may suggest that the PN is now in the recombination phase. An analysis of the observational data in the frame of this hypothesis leads to the mass estimate for the NGC 7027 nucleus of 1.0 ± 0.2 $M_\odot$ (Tylenda, 1984). Observations of the halo in other lines, in particular in [OIII], would serve as a conclusive test to this hypothesis.

Finally, as it is often argued (e.g. Renzini, 1983) and as it has been mentioned in the beginning of this section we should expect to find massive stars among hot, low-luminosity PNNi. At present an investigation of this sort is difficult and cannot give conclusive results, mostly because of the distance problem. However, we can mention 5 candidates, i.e. A 21, A 31, Jn 1, K 2-2, and PW 1. The observational data available at present indicate that these PNNi have log $L/L_\odot$ < 1.5, log T = 5.0 - 5.1, and Mv > 8.0 (Stasińska & Tylenda, 1987). Consequently, they lie in the region occupied by evolved, massive model PNNi both in the H-R diagram and in the Abell's diagram.

REFERENCES

Abell, G.O.: 1966, Astrophys. J. 144, 259.
Aller, L.H., Czyzak, S.J.: 1979, Astrophys. Space Sci. 62, 397.
Aller, L.H., Keyes, C.D., Maran, S.P., Gull, T.R., Michalitsianos, A.G., Stecher, T.P.: 1987, submitted to Astrophys. J.
Atherton, P.D., Hicks, T.R., Reay, N.K., Robinson, G., Worswick, J., Phillips, J.P.: 1979, Astrophys. J. 232, 786.
Heap, S.R., Augensen, H.J.: 1987, Astrophys. J. 313, 268.
Iben, I.: 1982, Astrophys. J. 259, 244.
Iben, I.: 1984, Astrophys. J. 277, 333.
Kaler, J.B.: 1983, Astrophys. J. 271, 188.
Mendez, R.H., Kudritzki, R.P., Herrero, A., Husfeld, D., Groth, H.G.: 1987, Astron. Astrophys. in press.
Paczyński, B.: 1971, Acta Astron. 21, 417.
Perek, L., Kohoutek, L.: 1967, Catalogue of Galactic Planetary Nebulae, Czechoslovak Academy of Sciences, Prague.
Pottasch, S.R.: 1983, in Flower, D.R. (ed.): Planetary Nebulae, IAU Symp. 103, Dordrecht, Reidel, p. 391.
Preite-Martinez, A., Pottasch, S.R.: 1983, Astron. Astrophys. 126, 31.
Renzini, A.: 1983, in Flower, D.R. (ed.): Planetary Nebulae, IAU Symp. 103, Dordrecht, Reidel, p. 267.
Schmidt-Voigt, M., Koppen, J.: 1987, Astron. Astrophys. 174, 211.
Schonberner, D.: 1981, Astron. Astrophys. 103, 119.
Schonberner, D.: 1983, Astrophys. J. 272, 708.
Schonberner, D.: 1987, in Proceedings of the Calgary Workshop "On the Late Stages of Stellar Evolution", in press.
Shaw, R.A., Kaler, J.B.: 1985, Astrophys. J. 295, 537.
Stasińska, G., Tylenda, R.: 1986, Astron. Astrophys. 155, 137.
Stasińska, G., Tylenda, R.: 1987, in preparation.
Stecher, T.P., Maran, S.P., Gull, T.R., Aller, L.H., Savedoff, M.P.: 1982, Astrophys. J. Lett. 262, L41.
Torres-Peimbert, S., Peimbert, M.: 1977, Rev. Mex. Astron. Astrophys. 2, 181.
Tuchman, Y.: 1984, Monthly Notices Roy. Astron. Soc. 208, 215.
Tylenda, R.: 1983, Astron. Astrophys. 126, 299.
Tylenda, R.: 1984, Astron. Astrophys. 138, 317.
Tylenda, R.: 1986, Astron. Astrophys. 156, 217.
Wood, P.R., Faulkner, D.J.: 1986, Astrophys. J. 307, 659.

You-Hua Chu, Karen Kwitter, Jim Kaler, and George Jacoby.

A NEW METHOD FOR OBSERVATIONAL TESTING OF THE PLANETARY NEBULAE NUCLEI EVOLUTION

R.Szczerba,
Polish Academy of Sciences,
N.Copernicus Astronomical Center,
Laboratory of Astrophysics, Toruń, Poland

Planetary nebulae (PNe) are very useful as a tool for testing the theory of stellar evolution. The most widely applied method in this respect is the Hertzsprung-Russell (H-R) diagram. However, the observed positions of planetary nebulae nuclei (PNNi) on the H-R diagram are subject to large uncertainties, mostly due to inaccurate distances to them. On the other hand, the (absolute visual magnitude, age)-diagram also is not free of this problem. Therefore, an attempt has been done to develop a new method which is distance-independent. For comparison between theory and observations we propose the $I(HeII\ \lambda 4686)/I(H\beta)$ versus $\log[I(H\beta,PN)/I_c(H\beta,PNN)]$ diagram. Both ratios reflect the evolutionary status of the central star and the surrounding nebula. Consequently, such diagram is a valuable tool for studying common evolution of the PNN-PN system.

The appropriate observational data have been collected from the literature for about 120 objects. The dereddened PNN continuum at $H\beta$ was calculated from the corrected B and/or V magnitudes in the Rayleigh-Jeans approximation. In our sample are not included these PNe the nuclei of which are close binaries and nebulae with uncertain central stars.

The evolutionary tracks of post-AGB models (Schönberner 1979, 1983; Wood and Faulkner 1986) and the simplified model of PN structure and evolution (Szczerba 1987) has been adopted for the numerical experiments. Results of comparison between theory and observation show that:
1. Our diagram yields some information on the phase of the flash cycle at which PN ejection occurs. Namely, only Schönberner H-burning models can explain a gap around 80 in the $I(HeII\ \lambda 4686)/I(H\beta)$.
2. The evolution along the brightest portion of tracks is too slow if mass loss by fast wind is not taken into account.
3. The PNNi studied seem to have masses below 0.7 M_o.

REFERENCES
Schönberner, D.: 1979, Astron. Astrophys. <u>79</u>, 108
Schönberner, D.: 1983, Astrophys. J. <u>113, 125</u>
Szczerba, R.: 1987, Astron. Astrophys. <u>181</u>, 365
Wood, P.R., Faulkner, D.J.: 1986, Astrophys. J. <u>307</u>, 659

MASS DISTRIBUTION AND BIRTH RATE OF CENTRAL STARS OF PLANETARY NEBULAE: COMPARISON WITH WHITE DWARFS, AND INFLUENCE OF SELECTION EFFECTS.

V. Weidemann
Institut für Theor. Physik und Sternwarte
Universität Kiel, F.R. Germany

The mass distribution of central stars (CPN) as derived by the Schönberner method (1981) M_v vs. age, v(exp) = const, for an enlarged local ensemble, as presented at the London Symposium, 1983, appears to be much narrower and more strongly peaked towards smaller masses than the one recently derived by Heap and Augensen (1987) (HA) using the same method, but IUE data and M_v (λ 1300) vs. age, corrected for individual v(exp). Whereas according to Schönberner 65% of all CPN have $M < 0.64\ M_\odot$, HA find only 44% below the same limit. We demonstrate that this discrepancy is entirely due to the fact, that HA use Daub and 0.9 x Cahn/Kaler distances, whereas Schönberner used 1.3 x CK. We list a number of arguments which favor the larger distances, especially the recent work by Méndez et al. (preprint, 1987) (T_{eff}/g determinations) and investigations of Magellanic Cloud PN by Aller et al. (1987), Wood et al. (1987) and Barlow (1987) which all indicate a scale ≥ 1.4 x CK. If one uses Barlow's recalibration formula for optically thick PN, the distances for those - which mainly contribute to the massive CPN in the HA analysis - are increased so much as to remove most of them from the local ensemble. We thus obtain for the revised IUE ensemble 84% CPN with $M < 0.64\ M_\odot$, in better agreement with results for white dwarfs (70%) (cf. Weidemann, 1987).
It is furthermore argued that agreement between PN and white dwarf birthrates can only be achieved if the PN distances are increased to above 1.3 x CK. We finally present CPN distributions in HR diagrams which are calculated with a galactic evolution program and demonstrate selection effects operating against high mass CPN and in favor of helium-burning CPN (details to be published elsewhere).

Aller, L.H., Keyes, C.D., Maran, S.P., Gull, T.R., Michalitsianos, A.G. and Stecher, T.P.: 1987, Astrophys. J. **320**, 159.
Barlow, M.J.: 1987, Mon.Not.Roy.Astr.Soc. **227**, 161.
Heap, S.R., Augensen, H.J.: 1987, *Astrophys*. J. **313**, 268.
Schönberner, D.: 1981, Astron. Astrophys. **103**, 119.
Weidemann, V: 1987, 2nd Conference on Faint Blue Stars, Tucson A.G. Davis Philip, D.S. Hayes, J. Liebert eds., in press.
Wood, P.R., Meatheringham, S.J., Dopita, M.A., Morgan, D.H.: 1987, Astrophys. J. **320**, 178.

COMMON ENVELOPE EVOLUTIONS OF BINARY SYSTEM AND FORMATION OF PLANETARY NEBULAE

Izumi Hachisu: Department of Physics and Astronomy, Louisiana State University and Department of Aeronautical Engineering Kyoto University
Mariko Kato: Department of Astronomy, Keio University

The more massive component star evolves faster than the less massive one. When it fills its inner critical Roche lobe at the red-giant stage or at the asymptotic-giant-branch (AGB) star stage, the mass transfer begins from the more massive to the less massive component. Since the separation decreases with the mass being transferred, the more massive component star eventually overfills its outer critical Roche lobe. The mass outside the outer critical Roche lobe flows out of the system and the outgoing matter carries away the orbital angular momentum. As a result, the separation of the binary shrinks and the size of the outer critical Roche lobe drops. This shrinkage of the Roche lobe enhances the systemic mass outflow. This process is almost dynamically unstable because the deep convective envelope responses the loss of the envelope mass in an almost dynamical time scale. This dynamical process will stop when most of the hydrogen-rich envelope of the more massive component is lost and its radius becomes less than the radius of the outer critical Roche lobe.

The matter outside the outer critical Roche lobe is accelerated by the gravitational torque and gets the outward velocity, which was estimated by Sawada, Hachisu, and Matsuda (1984) to be about one third or one fourth of the orbital velocity of binary.

Three typical models of binary planetary nebula formation are calculated. After spiral-in, a small amount of hydrogen-rich envelope remains on the white dwarf surface ($<10^{-3} M_\odot$). The time scale of the nuclear burning depends mainly on the white dwarf mass. If the white dwarf mass is $0.6 M_\odot$, its elapsed time until the extinction of the hydrogen-shell burning is about 10^4 yr.

	Model 1	Model 2	Model 3
initial masses	$1 M_\odot + (1+\alpha) M_\odot$	$1 M_\odot + (1+\alpha) M_\odot$	$1 M_\odot + 2 M_\odot$
separation	$300 R_\odot$	$1000 R_\odot$	$1200 R_\odot$
white dwarf mass	$0.45 M_\odot$ (He)	$0.6 M_\odot$ (C+O)	$0.8 M_\odot$ (C+O)
orbital period	10 day	100 day	3 day
nebula mass	$0.5 M_\odot$	$0.4 M_\odot$	$1.2 M_\odot$
separation	$30 R_\odot$	$100 R_\odot$	$10 R_\odot$
time scale	7×10^5 yr	1×10^4 yr	2×10^3 yr
outward velocity	10-30 km/s	10-20 km/s	10-40 km/s

REFERENCES
Sawada, K., Hachisu, I., and Matsuda, T. 1984, N. N. R. A. S., **206**, 673.

THEORETICAL MODELS FOR THE EVOLUTION OF PLANETARY NEBULAE NUCLEI
TESTED BY OBSERVATIONS

Romuald Tylenda[1] and Grażyna Stasińska[2]
1. Copernicus Astronomical Center, Toruń, Poland
2. DAEC, Observatoire de Paris-Meudon, France

ABSTRACT. We compare theoretical evolutionary tracks of planetary nebulae nuclei with observational data on over a hundred planetary nebulae in the ($\log L$, $\log T_{eff}$, $\log t_{exp}$) space. The simultaneous use of the three coordinates eliminates some interpretation that might be proposed when looking at the ($\log L$, $\log T_{eff}$) plane and at the (M_v, $\log t_{exp}$) only. The inconsistencies which we find between theory and observations could be partly removed by adopting a different distance scale.
 Another plot using coordinates which are both independent of distance tends to confirm this view.
 The observational data do not specially favour either of the two families of models: hydrogen-burning models or helium-burning ones.

SNAPSHOTS OF EVOLVING MODEL PLANETARY NEBULAE

Grażyna Stasińska
DAEC, Observatoire de Paris-Meudon, France

ABSTRACT. We have constructed a series of model planetary nebulae along a sequence of evolutionary models for the central star, taking into account the expansion of the nebula. The calculations have been performed using the computer code PHOTO, which calculates the intensities of the emission lines emitted by a nebula in ionization and thermal equilibrium.
 The results for the behaviour of the optical emission lines as a function of time are compared to the observations.
 Predictions are made for the infrared emission lines, with special attention to planetary nebulae with large overall heavy element abundances, such as are expected to be found in the vicinity of the galactic center.

EVOLUTION OF PLANETARY NEBULAE: A COMPARISON WITH OBSERVED CENTRAL STARS

M. Schmidt-Voigt
Max-Planck Institut für Physik und Astrophysik
Institute für Astrophysik, München (FRG)

ABSTRACT. The relation between nebular excitation E(He II $\lambda 4686$/Hβ-ratio) and absolute visual magnitude of the central star (CS) is compared with hydrodynamical models of planetary nebulae (PNe) from Schmidt-Voigt and Köppen (*Astron. Astrophys.*, 174, 211 and 223) (see figure below, data from D. Schönberner, *Astron. Astrophys.*, 169, 189). Models marked by drawn lines have a 0.644 $M_\odot$ CS following a Schönberner track, an initially expelled PN of 0.1 $M_\odot$, and different mass loss rates of the precursor star on the AGB, described by the Reimers parameter η; $\eta = 1$ corresponds to a mass loss rate of $1.55 \times 10^{-6} M_\odot$ a^{-1}. The dashed line model has a higher initially expelled mass (0.3 $M_\odot$), the dash-dotted line model a CS of 0.6 $M_\odot$ which evolves more slowly. Model numbers refer to the above cited studies. Since M_V increases with evolutionary time, the M_V axis represents a (highly) nonlinear time axis: for $M_V < 4$ the CS heats up towards its temperature maximum and the PN is optically thin. Differences for high excitation nebulae are most probably due to different helium abundances. When the rate of ionizing photons decreases as the nuclear energy sources extinguish ($M_V > 4$), the excitation may decline, depending on the density in the nebula. For the so called "accreting models" ($\dot{m} > 10^{-6} M_\odot$ a^{-1}) the mass accretion from the AGB wind determines the density hence nebular excitation. For an AGB mass loss rate $\dot{m} < 10^{-5} M_\odot$ a^{-1} the numerical results approximately fit an exponential law $E = E_0 \exp(-\dot{m}/\dot{m}_0)$ with $E_0 \simeq 1.1$ and $\dot{m}_0 \simeq 6.1 \times 10^{-6} M_\odot$ a^{-1}. From the spread of the observed $E(M_V = 4)$ we conclude a mean AGB mass loss rate of $6.7^{+3.3}_{-2.3}$ $10^{-6} M_\odot$ a^{-1} within 1σ error bars. Obviously the model 11 reproduces the data best since most of the observed objects are found in the dark shadowed regions of the histogram. This is totally consistent with our previous results (cited above). The colliding-wind models, having no initially PN, behave quite similar as model 11.

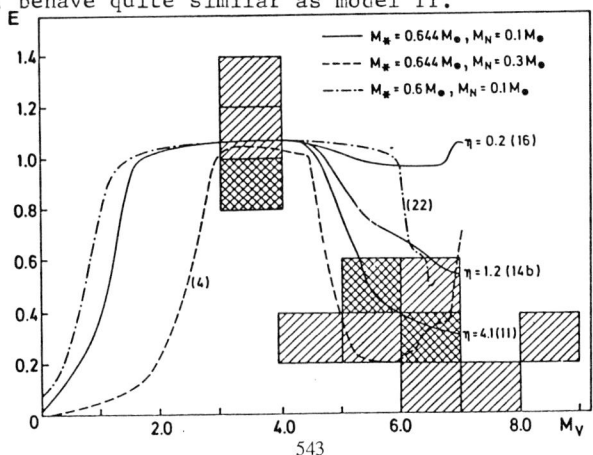

S. Torres-Peimbert (ed.), Planetary Nebulae, 543.
© 1989 by the IAU.

Escuela Nacional Preparatoria, main patio.

WHITE DWARFS AND PLANETARY CENTRAL STARS

James Liebert
Steward Observatory
University of Arizona
Tucson, Arizona 85721, USA

ABSTRACT. Studies of hot white dwarf samples constrain the properties and evolution of planetary nuclei and the nebulae. In particular, the white dwarf and planetary nebulae formation rates are compared. I discuss the overlap of the sequences of white dwarfs having hydrogen (DA) and helium-rich (DO) atmospheres with known central stars of high surface gravity. There is evidence that the hydrogen atmosphere nuclei have "thick" outer hydrogen layers ($\gtrsim 10^{-4}$ M$_\odot$), but that DA white dwarfs may have surface hydrogen layers orders of magnitude thinner. Finally, a DA planetary nucleus is discussed (0950+139) which has undergone a late nebular ejection; this object may be demonstrating that a hydrogen layer can be lost even after the star has entered the white dwarf cooling sequence.

1. STUDIES OF HOT WHITE DWARFS: CONSTRAINTS ON THE PLANETARY NEBULA EVOLUTION

It is now commonly accepted that the principal "channel" into the white dwarf sequence is post-asymptotic giant branch (post-AGB) evolution with the ejection of a planetary nebula. It follows that the studies of samples of hot white dwarfs are relevant to the masses and compositions of planetary nuclei, their evolution, and the formation rate and evolution of the nebulae. Harry Shipman has summarized nicely some of the ongoing scientific questions raised in studies of the degenerate dwarfs. We shall revisit the question of whether and how the surface abundances in white dwarfs change with cooling age in addressing how the central stars (CSPN) evolve into white dwarfs.

1.1 Comparison of the Birthrates

The formation rate of white dwarf stars should be an upper limit to the planetary nebula formation rate, but close to the actual value, provided one can sample objects from the same galactic population. Yet a recent comprehensive determination of the white dwarf birthrate (Fleming, Liebert and Green 1986) yields a number which is substantially <u>smaller</u> than most published estimates for planetary nebulae. It is approximately a factor of two lower than the value for the nebular formation rate favored by Phillips (these proceedings).

Hot white dwarfs may be found in complete, color and magnitude limited surveys, such as the Palomar Green Survey (Green, Schmidt and Liebert 1986). Because these stars may be readily assigned photometric or spectroscopic parallaxes, and because the cooling evolution for the hotter stars is based on relatively simple and well understood physics, relatively accurate space densities and birthrates for these stars should be derivable. It is arguably more difficult to construct a complete sample, and assign distances and nebular lifetimes to the planetary nebulae.

On the other hand, the white dwarf number derived from an ultraviolet excess survey does not include those degenerates in binary systems with companions more luminous than M dwarfs. We have already heard strong evidence from Bond and Mendez that the CSPN include significant fractions of binary systems. Moreover, the white dwarf number is drawn from a sample at typical distances of ~100 pc with an indicated scale height (from evidence other than the Palomar Green Survey) in the 250-350 pc range. The planetary samples generally cover distances of order 1 kpc, and Phillips has reviewed the evidence that their scale height is closer to 150 pc. A local population might be more affected by inhomogeneities in the space density of the disk population. For these reasons, it is not clear that the preferred planetary number is wrong, even though the white dwarf result should be a fairly accurate determination for the sample that it represents.

1.2. Targets of Opportunity in Searches for Residual Nebulae

Samples of hot white dwarfs and field subdwarfs have recently been the targets of sensitive searches for residual nebulae. Such surveys for faint, extended emission lines are potentially useful in comparing the evolutionary time of the CSPN with that for dispersal of the nebula. This may permit a testing of the validity of "kinematic ages" and the identification of a few rapidly-evolving CSPN of high core mass. However, it is also becoming clear (see Section 3) that nebulae can exist around white dwarfs which result from an episode of mass loss by the star long after it has left the AGB.

In the last few years, several CSPNs have been found to have white dwarf surface gravities based on spectra and model atmospheres analyses. Certainly the majority of these were studied because of the prior discovery of a nebula. A few previously-known field white dwarfs have shown evidence for residual nebulosity, though the largest imaging survey of known hot subdwarfs and white dwarfs (to my knowledge) by Kwitter and Massey (1987) and other collaborators has yet to yield a definite detection of a nebula. A smaller set observed by Reynolds (1987) has yielded one interesting candidate. Some of the successful matchups of nebula and white dwarf CSPN are highlighted in the next section.

2. MATCHING UP THE SEQUENCES OF H AND HE-RICH CSPN AND WHITE DWARFS

Planetary nebulae are relatively rare objects, due to the relatively short nebular lifetimes, and can be found in significant numbers only at distances relatively large in comparison to those for samples of white dwarfs. The degenerate stars are of course sufficiently faint that they cannot be found easily at distances beyond a few hundred parsecs. Accordingly, some low luminosity CSPN in known nebulae may be too faint to be detected or studied. The detection of very faint, residual nebulae around nearby, hot degenerate stars offers a way of bridging this luminosity gap and seeing how well the CSPN and white dwarf sequences match up, despite the differing discovery techniques and search volumes. In particular, since the relative frequencies of hydrogen and helium-dominated atmospheric compositions are approximately 80% to 20% (respectively) for both the CSPN and the hot white dwarfs, it is logical to start with the assumption that the dominant atmospheric constituent does not change as the CSPN evolve into white dwarfs. This can be tested by looking for continuity and overlap at the CSPN/white dwarf H-R Diagram boundary for both composition sequences.

2.1 The He-Rich CSPN/White Dwarf Sequence

For the minority He-rich post-AGB stars, there is ample evidence for continuous evolution across this boundary. At or just above the hot end of the helium atmosphere white dwarf sequence lies the group of pulsating stars whose prototype is PG1159-035 (GW Vir). These have effective temperatures near to or exceeding 100,000 K, surface gravities near white dwarf values ($\log g \sim 7$) and photospheric lines of He II, C IV and O VI. A nonpulsating object H1504+65 discussed by Harry Shipman appears to be even hotter (Nousek et al. 1986). Several more field stars with similar spectra apparently do not pulsate (Grauer et al. 1987). Searches for residual nebulosity by Kwitter and collaborators have apparently yielded only null results so far.

That these objects nevertheless invade the domain of He-rich CSPN is indicated by the similar pulsating nucleus of K1-16 (Grauer and Bond 1984) and possibly the nucleus of VV47 (Liebert et al. 1988a). They have strong spectroscopic similarities to the CSPN of the "O VI" type (Sion, Liebert and Starrfield 1985). Some of these show vigorous, photospheric mass loss in the form of Wolf-Rayet emission features, although this activity has abated in some of the hottest helium-rich CSPN (Mendez et al. 1986). In fact the K1-16 nucleus shows weak O VI 3811,3830Å emission; the ultraviolet spectrum also indicates that mass loss is ongoing (Kaler and Feibelman 1985).

The planetary nebula link could extend downward in the H-R Diagram into the region of the hot DO degenerates, which show primarily lines of He II. Reynolds (1987) has found evidence for very extended nebulosity around PG0108+101. Wesemael, Green and Liebert (1985) estimate that this star has $T_{eff} \sim 80,000$ K, $\log g > 7$, and $He/H \geq = 2$ by number. In

Kaler's talk, the spectrum presented for the nucleus of Jn1 appears also to be that of a hot, helium-rich star of high surface gravity.

In summary, a broad overlap is developing between the sequences of known He-rich CSPN and the samples of He-rich degenerate stars. It is logical to hypothesize that a continuous evolutionary sequence of decreasing luminosity and stellar mass loss carries the WC/O VI nuclei into the pulsational instability strip and that they then cool into DO white dwarfs.

2.2 The Hydrogen Rich Sequence

A few known H-rich CSPN have temperatures and gravities which also overlap the hot end of the sequence of DA white dwarfs. The best analyzed cases are the nuclei of Abell 7 and NGC 7293 (Mendez, Kudritzki and Simon 1983). The former has a $T_{eff} \sim 75,000$ K, $\log g \sim 7$ while estimates for the latter are approximately 90-100,000 K and $\log g \sim 6.5-7$ (see also Mendez' talk, this meeting). PG0950+139, discussed in the next section, also has a temperature near 70,000 K and a surface gravity (as indicated by broader lines) higher than than for the Abell 7 nucleus.

Yet the field DA stars do not extend continuously up to $T_{eff} \sim 100,000$ K. The hottest known field white dwarfs with H-rich compositions and $\log g \gtrsim = 7$ appear to have temperatures near 70,000 K (Holberg 1987). Below 70,000 K, the DA stars constitute a large fraction (~80%) of all hot white dwarfs. Thus it is reasonable to ask why the helium sequence includes hot DO stars near 80,000 K and even hotter objects of high gravity past 100,000 K, while there are very few if any such hot counterparts in the majority H-rich sequence.

2.3 The H Layer Masses and Evolutionary Lifetimes

The lack of DA stars with $T_{eff} > 80,000$ K may be due to the fact that the lifetimes in the H-rich CSPN phase are much shorter, provided that these objects have thick enough outer hydrogen layers that they leave the AGB as H shell-burning objects. Schönberner (1981, 1986, and this conference) has advanced two arguments that most CSPN are hydrogen burning objects:

(1) The gap in the absolute magnitude distribution of CSPN near $M_v = +5$ divides PNNs into high and low luminosity groups, and is attributed to the onset of rapid evolution as the H shell source shuts off at the end of the luminous phase of horizontal evolution in the H-R diagram. In particular, Schönberner's tracks for helium shell-burning CSPN are not compatible with the existence of such a gap;

(2) A similar argument can be made in a distance-independent way, using the nebular excitation parameter, and Kaler's sample of large (old) CSPN (see also Schmidt-Voigt and Köppen 1987).

It follows that such PNNs with thick outer hydrogen envelopes should enter the white dwarf stage as DA stars. The models of Koester and Schönberner (1986) and of Iben and Tutukov (1984) predict that the phase of rapid evolution for those objects with quenched H shells should last until the CSPN reach approximately $\log L/L_0 \sim 2$ (see Iben and Tutukov, Fig. 1). This corresponds to the approximate luminosities of the PG1159-035/H1504+65 objects and may account for the scarcity of H-rich counterparts. In fact the observed beginning of the DA sequence at $\log L/L_0 \sim 1$ (70,000 K) suggests that the phase of rapid evolution lasts to lower luminosities than predicted, although the observed samples are small. Moreover, there should be some allowance for the uncertainties in the derived stellar L, T_{eff} parameters and in the interiors models.

2.4 Do DA White Dwarfs Have Very Thin Outer Hydrogen Layers?

In any case, the distributions of the hottest H and He-rich objects forming white dwarfs are explainable only if the former retain thick hydrogen shells ($M_H \gtrsim = 10^{-4} M_\odot$), a necessary condition for the luminous H shell-burning phase. The distributions of these low luminosity stars are thus consistent with and support Schönberner's arguments concerning the majority of CSPN.

There are several arguments, however, which lead to the conclusion that DA white dwarfs have outer layer masses many orders of magnitude thinner than $10^{-4} M_\odot$. These arguments are summarized in Fontaine and Wesemael (1987), and are discussed in other review papers presented at IAU Colloquium 95. Four of the major arguments are outlined below:

1. He/H abundance ratios observed in hot DA stars (> 30,000 K) are orders of magnitude too large to be explained by radiative levitation of helium (Vennes et al. 1987). These authors favor as the origin a diffusion tail reaching upwards to the surface through very thin hydrogen layers of only 10^{-13} to $10^{-16} M_\odot$.

2. There are currently no known white dwarfs with He-rich atmospheres in the temperature range $45,000 < T_{eff} < 30,000$ K (Liebert et al. 1986). Yet further statistics indicate that some 25% of the DA turn into He-rich atmospheres below 30,000 K (Sion 1984). A possible explanation of this peculiar observational result, due primarily to G. Fontaine, is that convective overshooting in the helium envelope could mix away an H-rich surface layer, provided that the latter consisted of less than approximately $10^{-15} M_\odot$. Shipman, in the previous talk, suggests that the "DB gap" might be explainable instead by differences in the cooling times of the two sequences.

3. The pulsating ZZ Ceti stars include all or nearly all DA degenerates cooling through the 10-13,000 K instability strip. The mode-trapping model for the pulsational driving favored to produce the rather long periodicities observed in these objects requires modest hydrogen layer masses of order $10^{-8} M_\odot$ or less (Winget et al. 1982). However, this conclusion is disputed by Cox et al. (1987).

4. Finally, it is known that most of the remaining DA stars must evolve into objects with helium atmospheres $T_{eff} < 10,000$ K. In fact Winget et al. (1982) and others have argued that the mixing due to penetration of the helium layer by the hydrogen convective zone may terminate the ZZ Ceti instability strip. In Greenstein's (1986) compilation of high signal-to-noise ratio observations, it is apparent that some 70% of the stars have become helium-rich at the temperatures approaching the low extreme (~6,000 K) that hydrogen lines would be visible. The calculations indicate again (D'Antona and Mazzitelli 1979) that convective mixing will not take place unless the outer hydrogen layer is thinner than approximately 10^{-6} $M_\odot$.

Individually, these arguments for thin outer hydrogen layers depend on the assumptions that we understand accurately enough the physics of (1) envelope convection, especially the behavior in regions having large gradients in the chemical composition, (2) selective radiative acceleration processes and their competition against gravitational and thermal diffusion, and/or (3) pulsational driving theory. Individually, each argument may not be very convincing. Together, they constitute a case which must be taken seriously. On balance, the observational evidence that the fraction of DA white dwarfs decreases markedly with decreasing surface temperature is difficult to explain if the majority of these stars retain outer layer masses of hydrogen of order 10^{-4} $M_\odot$.

Fontaine and Wesemael (1987) suggest that all DA stars have very thin outer envelopes, and that the most logical explanation of the absence of DA stars near 100,000 K is that the pulsating helium-rich and the hot DO stars retain enough residual envelope hydrogen for this later to float to the surface, form a thin outer hydrogen layer, and convert the He-rich atmosphere into a DA star.

This simple hypothesis has difficulty in accommodating the arguments that the majority H-rich PN nuclei must have thick hydrogen layers in order to undergo hydrogen shell burning. At the low luminosity end, one wonders how the H-rich CSPN would lose these thick layer masses and turn into ~ 10^5 K pulsating He-rich objects just after the stars have evolved rapidly through the M_v ~ +5 region; the corresponding mass loss rates would exceed $10^{-8} M_\odot$ per year for typical masses near 0.6 $M_\odot$. In particular, low luminosity hydrogen-rich CSPN such as Abell 7 are most difficult to understand. However, the object discussed in the next section offers a clue that this process of converting a thick CSPN envelope into a thin DA white dwarf envelope may actually happen!

3. EGB/PG 0950+139: EVIDENCE FOR ONGOING/RECENT MASS LOSS FROM A HOT DA WHITE DWARF

One of the very hottest known DA white dwarfs is associated with an extended planetary nebula of low surface brightness. Ellis, Grayson and Bond (1984), EGB) identified nebulosity of roughly circular morphology some 12 arc minutes in diameter around a 15th magnitude blue star. The stellar counterpart was independently catalogued in the Palomar Green Survey as PG 0950+139. Both sets of authors were aware of a compact nebular line spectrum, essentially unresolved in a two-dimensional Palomar 5-m SIT Vidicon scan (by R. Green), and in two additional spectra with the Kitt Peak National Observatory 4-m telescope and cryogenic camera/CCD system (by H. Spinrad and R. Green). The extended nebula is so faint that it is not detected in H or [O III] lines on any of these long slit spectra. H. Bond (private communication) comments that this component has highest surface brightness near an outer rim. The analysis of both the compact nebular component and the stellar photospheric spectrum is described in a forthcoming paper (Liebert et al. 1988b).

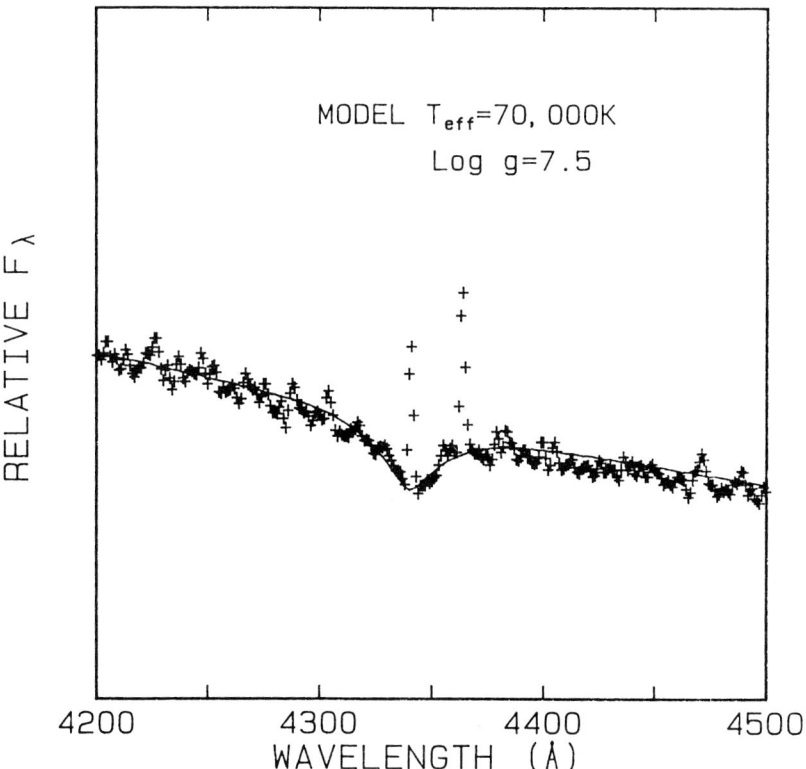

Figure 1. The spectrum of the photosphere of 0950+139 centered on the Hγ line with the synthetic spectrum of a pure hydrogen, T_{eff} = 70,000, log g = 7.5 model by F. Wesemael (courtesy of collaborators J. Holberg and K. Kidder). Nebular emission lines of Hγ [O III] 4363Å and He I 4471Å are also visible.

3.1 Properties of the DA Central Star

The nucleus of this planetary is a white dwarf at the high temperature extreme of the DA distribution. J. Holberg and K. Kidder have analyzed the photospheric H-gamma and other absorption lines. The best fit using pure hydrogen, LTE models by F. Wesemael is $T_{eff} \sim 70,000(\pm 7,000$ K), log g $\sim 7.5(\pm 0.25)$, and H/He > 1. The helium abundance is not estimated because of the problem of nebular contamination of any narrow He II 4686A absorption. The star has broader hydrogen lines (i.e., higher log g by +0.5) than the CSPN of Abell 7, which has a similar T_{eff}.

3.2 Properties of the Unresolved Nebular Component

The compact nebula is quite remarkable. Detectable forbidden lines of [O III], [Ne III] and [O I] are seen, along with permitted lines of H and possibly He. The flux ratio usually used in calculating the [O III] temperature provided the first clue to the unusual physical conditions. The ratio

$$R[O\ III] = [\ f(5007Å) + f(4959Å)\ ] / f(4363Å)$$

was found to be only 10.9, while typical planetary nebulae values are as much as an order of magnitude larger. Even for ion densities approaching $N_e = 10^6$ cm^{-3}, this value leads to very high electron temperature (T_e) estimates, in excess of 20,000 K.

Normally the density of a nebular region may be estimated from the ratios of doublets of [O II] or [S II], but these lines do not appear. Yet the very large [O III]/[O II] excitation ratio is inconsistent with a similar measure of nebular excitation using permitted lines, the ratio of the He II 4686Å to the H-β 4861Å lines. Even stranger at first glance is the appearance of [O I] 6300, 6360 auroral lines in two red CCD spectra.

Ionized gas at densities above 10^6 cm^{-3} may occur in the line-emitting regions of active galactic nuclei and radio galaxies. In this context, the critical densities at which quenching of various forbidden lines occurs are listed in Fillipenko (1985, Table 3). For example, $N_{e,crit}$ for the [O III] 4959 and 5007 transitions occurs at 7.9 x 10^5 cm^{-3}, while the value for [O III] 4363 is much higher at 3 x 10^7. Thus the small ratio of R[O III] may be understood in terms of a density higher than 10^6 cm^{-3}. The unseen [O II], [N II] and [S II] doublets are all quenched at densities below 10^5 cm^{-3}. The visible [Ne III] and [O I] transitions survive at higher densities, the former at 1.2 x 10^7 cm^{-3}, while the weak [O I] lines suggest that some material in the inner nebula may have densities as low as 1.2 x 10^6 cm^{-3}.

On balance, we argue that the mean density of the unresolved nebular component of 0950+139 is near 10^6 cm^{-3}. Using this number, R[O III] yields an estimate of $T_e \sim 19,100$ K, still quite high. Assuming a fully-ionized, uniform spherical nebula at this single density and temperature, we estimate a nebular mass of order $10^{-8} M_\odot$, and a radius of order 10 a.u.! There is no indication that the helium abundance is high, though the abundance constraints on He and heavier elements are sparse.

These results lead to the inescapable conclusion that the mass loss has been a recent or ongoing event. While the expansion rate of this component is as yet unknown, velocities of 10-50 km/s would extend the material to 10 a.u. in 1-10 years. The cooling age of a 0.6 $M_\odot$ white dwarf at 70,000 K is approximately 5×10^5 years!

3.3 0950+139: A Case of Envelope "Thinning"?

These results suggest the possibility that significant mass loss may evaporate most or all of the outer hydrogen envelope as the stars approach and enter the white dwarf stage. Since this CSPN is one of the few known DA white dwarfs near the 70,000 K limit of the known sequence, the frequency of observable nebulae ejections around such objects is not determined. However, one may estimate that an average mass loss rate of 10^{-10} $M_\odot$ per year would be sufficient to reduce the outer layer mass to nil, if the instability lasted for the approximately 10^6 years required for a white dwarf to cool to 60,000 K. Steady mass loss rates this small are difficult to exclude, even for stars observed with the International Ultraviolet Explorer Satellite at high resolution. It is also possible that such mass loss may occur in the form of sporadic ejections, so that most objects may not show readily detectable nebular emission at any given time. However, there are no detected changes between nebular spectra obtained in the late 1970s and those obtained in the mid-1980's.

This discovery should spur further attempts to detect residual nebulosity around hot DA white dwarfs, and evidence for ongoing mass loss around hydrogen-rich CSPNs with known extended components. Further identifications of the rare DA white dwarfs near the high temperature limit may be necessary in order to determine the frequency of the late mass loss phenomenon observed in 0950+139. Moreover, a helium-rich DO star of similar temperature also shows some evidence for recent mass loss (Downes et al. 1987).

I acknowledge support from the National Science Foundation (grant AST 85-14778). I thank J. Holberg and K. Kidder for providing Fig. 1.

REFERENCES

Cox, A. N., Starrfield, S. G., Kidman, R. B. and Pesnell, D.: 1987, Astrophys. J., 317, 303.
D'Antona, F. and Mazzitelli, I.: 1979, Astron. Astrophys., 74, 161.
Downes, R. A., Sion, E. M., Liebert, J. and Holberg, J.: 1987, Astrophys. J., 321, 943.

Ellis, G. L., Grayson, E. T. and Bond, H. E.: 1984, Publ. Astron. Soc. Pacific, **96**, 283.
Fillipenko, A. V.: 1985, Astrophys. J., **289**, 475.
Fleming, T. A., Liebert, J. and Green, R. F.: 1986, Astrophys. J., **308**, 176.
Fontaine, G. and Wesemael, F.: 1987, A. G. D. Philip, D. S. Hayes and J. Liebert (eds.), 'Second Conference on Faint Blue Stars,' IAU Coll. 95 (in press).
Grauer, A. D. and Bond, H. E.: 1984, Astrophys. J., **277**, 211.
Grauer, A. D., Bond, H. E., Green, R. F., Liebert, J. and Fleming, T. A.: 1987, Astrophys. J., (in press).
Green, R. F., Schmidt, M. and Liebert, J.: 1986, Astrophys. J. Suppl., **61**, 305.
Greenstein, J. L.: 1986, Astrophys. J., **304**, 334.
Holberg, J.: 1987, A. G. D. Philip, D. S. Hayes and J. Liebert (eds.), 'Second Conference on Faint Blue Stars,' IAU Coll. 95, (in press) and private communication.
Iben, I., Jr. and Tutukov, A. V. 1984, Astrophys. J., **282**, 615.
Kaler, J. B. and Feibelman, W. A.: 1985, Astrophys. J., **297**, 724.
Koester, D. and Schönberner, D.: 1986, Astron. Astrophys., **154**, 134.
Kwitter, K. and Massey, P. 1987, in preparation.
Liebert, J.,Fleming, T. A., Green, R. F. and Grauer, A. D.: 1988a, submitted to Publ. Astron. Soc. Pacific.
Liebert, J., Green, R. F., Bond, H. E., Holberg, J., Kidder, K., Fleming, T. A. and Wesemael, F.: 1988b, preprint.
Liebert, J., Wesemael, F., Hansen, C. J., Fontaine, G., Shipman, H. L., Sion, E. M., Winget, D. E. and Green, R. F.: 1986, Astrophys. J., **309**, 241.
Mendez, R. H., Kudritzki, R. P. and Simon, K. P.: 1983, D. R. Flower (ed.), 'Planetary Nebulae,' IAU Symp. **103**, 343.
Mendez, R. H., Miguel, C. H., Heber, U. and Kudritzki, R.: 1986, K. Hunger, D. Schönberner and N.K. Rao (eds.), 'Hydrogen Deficient Stars and Related Objects,' IAU Coll. **87**, 323.
Nousek, J. A., Shipman, H. L., Holberg, J. B., Liebert, J., Pravdo, S. H., White, N. E.and Giommi,P.: 1986, Astrophys. J., **309**, 230.
Reynolds, R. J.: 1987, Astrophys. J., **315**, 234.
Schmidt-Voigt, H. and Köppen, J.: 1987, Astron. Astrophys., **174**, 223.
Schönberner, D.: 1981, Astron. Astrophys., **103**, 119.
Schönberner, D.: 1986, Astron. Astrophys., **169**, 189.
Sion, E. M.: 1984, Astrophys. J., **282**, 612.
Sion, E. M., Liebert, J. and Starrfield, S. G.: 1985, Astrophys. J., **292**, 471.
Vennes, S., Pelletier, C., Fontaine, G. and Wesemael, F.: 1987, A. G. D. Philip, D. S. Hayes and J. Liebert (eds.), 'Second Conference on Faint Blue Stars,' IAU Coll. 95 (in press).
Wesemael, F., Green, R. F. and Liebert, J.: 1985, Astrophys. J. Suppl., 58,379.
Winget, D. E., Van Horn, H. M., Tassoul, M., Hansen, C. J., Fontaine, G. and Carroll, B. W.: 1982, Astrophs. J. Letters, **252**, L65.

PROPERTIES AND EVOLUTION OF WHITE DWARF STARS

Harry L. Shipman
Physics and Astronomy Department, University of Delaware

ABSTRACT

   This paper reviews the properties and evolutionary status of white dwarf stars, focusing most closely on those aspects which are likely to be of significance to understanding the ultimate fate of planetary nebulae and their central stars. White dwarf stars show a broad variety of chemical compositions. Broadly speaking, they are divided into the DA stars (with H-rich photospheres) and the non-DA stars (with He-rich photospheres), though there are a fairly large number of subtypes. The mass distribution of white dwarf stars is quite narrow, with a mean value near 0.6 and with extremes at 0.43 and 1.05. Different varieties of trace elements (such as C, N, O, Si, Ca, and Mg) are quite common. I will review several recent proposals for explaining these abundance patterns. A particularly significant question is whether processes operating while the star cools as a white dwarf can account for their variety, or whether at least part of the white dwarf phenomenology is related to events which took place when the object was a planetary nebula or even earlier.

I. INTRODUCTION: A COOK'S TOUR OF THE BOTTOM OF THE HR DIAGRAM

   In the last several years, it has become clear that the white dwarf stars are a phenomenologically very rich class of stellar objects. Broadly speaking, they are divided into a group of stars with H-rich photospheres, called the DA stars ("D"=degenerate and "A" refers to the main sequence analog) and the non-DA stars, which (in all cases but one) almost certainly have He-rich photospheres. However, a number of very peculiar subclasses have been recently identified in which substantial quantities of trace elements are introduced into the photospheres of these objects. The roster of white dwarf stars now includes a number of different types of objects illustrated in Figure 1 on the next page. Complex as Figure 1 seems, it considerably oversimplifies our understanding of white dwarf stars; a figure illustrating white dwarf evolution in its full glory is Figure 1 of Sion's (1986) review article.

T(eff)/10³ -->	150	75	50	30	15	10	8	6
Type:								
H		/////////	DA	V471*?		ZZ Ceti*		DC
Hybrid H/He		DAO			DBA		DZA	
He		PG1159* DO		///DB*		DQ		DZ
?		H1504						
DA/non-DA ratio:		0(!)	7:1	oo(!)	4:1			1:1

* denotes variable stars.

Figure 1: A simplified description of the classes of white dwarf stars. Effective temperature decreases from left to right, following the cooling sequence. Each of the major classes of white dwarf star has its place in this table; diagonal slashes denote those places along the four major cooling sequences where no stars are found.

The broadest, most obvious division of the white dwarf stars is into the DA and non-DA categories, but each of these categories includes a number of spectroscopically, chemically, and evolutionarily distinct subclasses. The spectral classification system is fully defined in Sion, Greenstein, Landstreet, Liebert, Shipman, and Wegner 1983; only a resume will be provided here. All white dwarf spectral classes have the prefix "D" for degenerate, which in the present context generally means $\log g > 7$. Thermal pressure plays a role in the structure of hot white dwarfs, so that an object with T(eff) > about 50,000 K may be fully degenerate and not lie on the Hamada-Salpeter (1961) zero temperature mass radius relation. The first letter following the "A" indicates the element with the strongest lines in the star, with "A" being hydrogen, "B" being neutral helium, "O" being ionized helium, "Q" being carbon, and "Z" referring to other elements, sometimes referred to by astrophysicists as "metals." (In the case of white dwarf stars, these usually are genuine metals like Ca, Mg, or Fe rather than substances like oxygen which no respectable chemist would call metallic, despite the astrophysical nomenclature.) A white dwarf with no clear spectral features is designated as "C" (for continuous); improvements in spectral resolution, sensitivity, and coverage of the electromagnetic spectrum have resulted in a considerable decrease in the fraction of white dwarf stars classified as "DC." If a second element is present in the spectrum, it is indicated by a second letter; thus a DBA star (discussed in some detail below) is one showing both He I and H I, with H I being weaker than He I. The numerical digit found in catalogs of white dwarf stars (e.g., McCook and Sion 1987) refers to 50400/T(eff). This catalog, incidentally, is a good reference to the rather confusing nomenclature of white dwarf stars; many investigators continue to prefer to refer to a star by its original catalog designation rather than by a coordinate-based system. This classification system is an elaboration of earlier ones by Greenstein (1960) and Luyten (1952). The "A","B", and "O" designations came from the main sequence analogs to white dwarf spectra. "P" is used to designate polarized white dwarfs, which have strong magnetic fields.

Most broadly speaking, about two thirds of the white dwarfs are DA stars (either just plain DA, or DA with a suffix like DAO,DAZ,DAB), which with only one known exception, the DAB star Gr 488, really have photospheres which are dominated by H. The highest He abundances in the DA stars are found in the very hot DAO stars, where $N(He)/N(H) = 10^{-2}$ (Wesemael, Green, and Liebert 1985). The remainder have He-dominated photospheres; if $T(eff) > 11,000$ K, they appear as DB or DO stars since the He I or He II lines in accessible spectral regions (which all arise from excited states) are only visible at sufficiently high temperatures.

2. TEMPERATURES, MAGNETIC FIELDS, AND ROTATION

Some of the most extreme stellar properties are found in the white dwarf region of the HR diagram. High magnetic fields, high temperatures, low rotational velocities, low luminosities, and extreme chemical compositions are all aspects in which particular white dwarfs can represent the extremes of directly observable stellar properties. Since the second half of this review deals extensively with chemical peculiarities, the remainder of this introductory section will deal with the other aspects in which white dwarfs are peculiar.

A few percent of them have very strong magnetic fields, ranging from a few to a few hundred megagauss. The highest field found to date is the >500 MG value recently reported for PG 1031+234 (Schmidt, West, Liebert, Green, and Stockman 1986. These fields are recognized by the distortion of the usual pattern of hydrogen Balmer lines as well as by the circular polarization of the light from the white dwarf.

The hottest white dwarf is the very peculiar object H1504+65, with $T(eff) = 160,000$ K(Nousek, Shipman, Liebert, Holberg, Pravdo, Giommi, and White, 1986). This object, along with several other hot white dwarfs of the PG 1159 class, with $T(eff)$ near 130,000 K, is hotter than all but a very few planetary nebula nuclei, but there are no visible nebulae around H1504 and the PG 1159 objects. (The nucleus of the planetary nebula K 1-16, which is pulsationally and spectroscopically similar to PG 1159, may not be a degenerate object.)

All other things being equal, one would expect white dwarf stars to rotate quite rapidly, as neutron stars do when they are born. A star of one solar mass rotating with a period of 1 month would have a rotational period of about 6 min (or an equatorial velocity of 140 km/sec) if it conserved angular momentum per unit mass and collapsed to a white dwarf of 0.6 solar masses and 0.012 solar radii. The most complete set of white dwarf rotational velocities has been determined by Pilachowski and Milkey (1987) from high resolution observations of the narrow cores of Balmer lines in DA white dwarfs. Of fifteen white dwarfs, they find that ten have negligible velocities (the lowest 2 sigma limits are 20 km/sec), while the rest have velocities detected at the 2 sigma level or better, the highest being 60 ± 10 km/s for the DA star GD 140. These values are consistent with the limit of v sin i < 30 km/s set from the C and Si lines in the hot white dwarf Feige 24 (Wesemael, Henry, and Shipman 1984).

Rotational velocities can be determined for the magnetic white dwarfs because the observed magnetic field can change as the star rotates. Changing fields can show up as changing polarization patterns, changing spectra (as the Zeeman shifts change), or both. The five magnetic white dwarfs with detected (or possibly detected) rotation periods all have periods which are appreciably longer than expected: 1.6 hr (PG1015+014), 3 hr (L795-7), 31 hr (G 195-19), 67 hr (tentative detection; BPM 25114), and 3.4 hr (PG1031+234; data from Angel, Borra, and Landstreet 1981 and references therein and Schmidt, West, Liebert, Green, and Stockman 1986). Some extreme lower limits can be set by the absence of polarization changes in three objects; the position angle of linear polarization has changed by less than 3° in several years for three magnetics (GD 229, G 240-72, and Grw+70°8247), suggesting rotation periods exceeding 100 years! While it is unlikely that all three of these stars are rotating pole-on, it would be useful to have similar data for a larger sample of magnetic white dwarf stars.

The conventional explanation for these low rotation periods is that the white dwarf progenitors rotate as solid bodies when they are red giants. The angular momentum in a slowly rotating red giant is concentrated in the outer layers, and mass loss will then carry away a disproportionate amount of angular momentum. I am not aware of any investigations which have tried to explain whether this scenario can account for century long rotation periods. It would be quite interesting if someone could observe consequences of angular momentum losses during the final red giant stages (for example, effects on planetary nebula morphology or determinations of nebular rotation), though I suspect that such a task is quite difficult. It might also be interesting to explore why some white dwarf stars do appear to retain a fraction of their angular momentum and why others do not.

The coolest white dwarfs are the lowest luminosity stars known. Their temperatures are difficult to determine, largely because their photospheres are partially degenerate in the sense that the perfect gas equation of state cannot be used in model atmosphere calculations. Kapranidis and Liebert (1985) determined T(eff) = 4500 K for the cool degenerate LP 701-29, which at the moment is probably the best analyzed of a small collection of very cool objects. ER 8, discovered as a by-product of a supernova search program at the Universidad de Chile, may be cooler still, since it is as red as LP 701-29, though it is probably not as cool as claimed in the discovery paper (Ruiz, Maza, Wischnjewsky, and Gonzalez 1986; Ruiz, private communication).

About ten years ago, Liebert and co-workers (Liebert, Dahn, Gresham, and Strittmatter 1979; see also Liebert 1980, Liebert, Dahn, and Sion 1983) showed that despite exhaustive searches, very cool white dwarfs like those mentioned in the previous paragraph represent the extreme cool end of the white dwarf distribution. This conclusion is reinforced by the failure to discover any very cool white dwarfs as astrometric companions to nearby stars (Shipman 1983). The interpretation of this cutoff in the white dwarf luminosity function is complicated. If you believe that we can correctly calculate the cooling rate of white dwarfs, then the existence of this cutoff can set limits on the age of the galactic disk. The most recent such determination by Winget et al. (1987) indicates an age for the disk of 9.3 ± 2 Gyr. However, the physics of white dwarf cooling, particularly at the very cool end of the white dwarf sequence which is crucial for using white dwarfs to determine the age of the disk, is complex. The equation of state and opacity in the partially degenerate layers, which overly the core and are the throttle that determines the rate of cooling, must be known accurately in order to calculate correct cooling times. Winget et al. take the audacious step of extrapolating from the age of the galactic disk to the age of the Universe. Whether one can reliably state that the Universe is only 1 Gyr older than the galactic disk is a matter of taste and judgment; consumers should be wary of uninformed, hasty use of cosmic ages based on white dwarf cooling times, in my view.

3. MASSES OF WHITE DWARF STARS AND WHITE DWARF PROGENITORS

Nearly a decade ago, two comprehensive investigations of the masses and radii of DA white dwarf stars were completed, one by the Kiel group (Koester, Schulz, and Weidemann 1979) and the other by Shipman (1979) and Shipman and Sass (1980). In general, the results of the two discussions were quite similar. Both found that the observed sample of DA white dwarfs had a mean mass of 0.6 solar masses. Both investigations suggested that the mass distribution of white dwarfs is quite narrow, far narrower than the range from 0.45 to 1.4 solar masses which is allowed in principle by the physics of white dwarf stars. (The lower mass limit is set by core masses of main sequence stars which can evolve in less than a Hubble time; the upper limit is the maximum mass of a C white dwarf according to the mass radius relation of Hamada and Salpeter 1961). The principal difference between the two investigations is that Shipman (1979) found a significantly higher mass spread, leading to a selection effect which would skew any observed sample of white dwarf stars towards those with larger radii (since they can be observed to greater distances). This selection effect will only apply if the cosmic scatter in the white dwarf distribution is sufficiently large.

Subsequent investigations have tended to confirm these results on the mean masses of the DA white dwarfs, and have suggested that the lower value of the mass spread is more likely to be correct. Weidemann and Koester (1984) and Greenstein (1985) find that the mass distribution of white dwarfs is sufficiently sharply peaked that the selection effect will only skew the mean mass of white dwarfs by 0.05 solar masses or less. While Guseinov, Novruzova, and Rustamov (1983, 1984) find a larger spread of white dwarf masses, they used UBV colors to define the white dwarf temperatures and thus obtained a mass distribution which may well be much broader than the real one, since UBV colors from heterogeneous sources can be subject to uncertainties which are both large and difficult to determine (Koester 1984). At the moment, my best estimate is that the mean mass of the DA white dwarf stars is between 0.58 and 0.63 solar masses.

The observed range of white dwarf masses extends, at its extreme ends, to the reasonably precise masses for white dwarfs in binary systems: 0.43 solar masses (40 Eri B) and 1.05 solar masses (Sirius B). Recent work by the Kiel group on the masses of white dwarf stars in galactic clusters confirms the existence of a high-mass component to the white dwarf mass distribution (see references below). At the moment, neither the observations nor their theoretical calibration in terms of various methods of determining white dwarf masses are sufficiently precise that the characteristics of the high-mass and low-mass components of the white dwarf mass distribution can be given in detail.

Shipman (1979) also determined the mean mass of a sample of non-DA white dwarfs, using model atmospheres and parallaxes, and found no appreciable difference between the masses of the non-DAs and the masses of the DAs. The difficulty with the non-DAs is that the model atmospheres are less certain (because of convection) and that the nice separation between white dwarfs of different gravity in the two-color (U-B vs. B-V or its analog in other color systems) diagram does not occur because non-DAs don't have a large Balmer jump. Oke, Weidemann, and Koester (1984) did use the two-color diagram to try to find the masses of a sample of DB stars, and agree that there seems to be no appreciable difference between the masses of the DA's and the DB's.

For a number of years, investigators have sought to determine whether there is any relation between the mass of a white dwarf and the mass of its progenitor. If a white dwarf is found in a star cluster, the cooling age of a particular white dwarf can be subtracted from the age of the cluster to determine the nuclear burning age (and hence the mass) of that particular white dwarf's progenitor. Koester and Weidemann (1984; see also Weidemann 1984, 1987) provide a recent summary of these efforts. Investigations of a number of young clusters (see particularly Reimers and Koester 1982, Koester and Reimers 1985) show that even comparatively massive progenitors, with masses of roughly 8 solar masses, still produce white dwarfs with relatively low masses (around 0.8 solar masses).

A second, important result from the investigations of white dwarfs in star clusters is the determination of the initial mass M_W of a star which will die as a white dwarf rather than something else. Reimers and Koester (1982) estimated M_W as being 8 (+3,-2) solar masses. There seems to be no compelling reason to challenge this estimate, though a value of M_W which was as low as solar masses would be difficult to reconcile with the data on the cluster NGC 2451.

4. CHEMICAL COMPOSITIONS: THE ORIGIN OF DA AND NON-DA WHITE DWARFS

"The uniformity of composition of stellar atmospheres appears to be an established fact," wrote one of the pioneers, Cecilia Payne, in her celebrated thesis in 1925. (Payne 1925, quoted in Bidelman 1986). In retrospect, I find it indeed remarkable that Payne could recognize the uniformity of composition of main sequence stars, despite their very dissimilar appearance. In 1925 the Saha equation, the key to understanding the many spectroscopic faces of the "Russell mixture," was scarcely five years old. It took great insight to understand that the great change of hydrogen line strength from spectral types A through O was nothing more than the results of the Saha-Boltzmann equation and radiative transfer, not of compositional differences.

However, the exceptions to Cecilia Payne-Gaposhkin's dictum are indeed quite interesting, and white dwarf stars constitute one of the most numerous exceptions to the uniformity of stellar compositions which tends to prevail elsewhere in the HR diagram. In 1925, only two or three

white dwarf stars were known to exist, but there are now over 10^3 of
them, and next to main sequence stars they are the most common type of
star. They are anything but uniform in composition. It was in the the
late 1950s and early 1960s, when spectra of a reasonable variety of white
dwarf stars became available and when model atmosphere techniques were
reasonably well developed, that the existence of two compositionally
distinct classes of white dwarf stars had become clear (Greenstein 1960).
Certainly ever since the early 1970s a number of us have struggled with
the question of why this compositional dichotomy should exist.

The existence of the non-DA stars is particularly puzzling because a
naive view of white dwarf evolution would suggest that all white dwarfs
should be DA stars. The photospheres of DA white dwarf stars are very
thin indeed. A white dwarf which makes a single passage through a
reasonably dense region of the interstellar medium will accrete enough
hydrogen to appear as a DA, if there is no fractionation during the
accretion process. Furthermore, the high gravity of white dwarf stars
means that heavier elements settle to the bottom quite rapidly, in a
process which is generally referred to as diffusion. One would thus
expect that whatever hydrogen is left or accreted in the outer layers of
a white dwarf would float to the top and make the star look like a DA.
Why do non-DA white dwarf stars even exist? There is no obvious
difference in mass or kinematics between DA and non-DA white dwarf stars,
so it is no longer possible to appeal to different accretion regimes in
order to explain the existence of the two types of stars (as was popular
several years ago, where a mass difference was reputed to be the cause).

As an aside, before considering various explanations for the
existence of both DA and non-DA white dwarfs, a definition of what
constitutes an acceptable "explanation" is in order. For a number of
years many of us in the white dwarf business have sought to delineate the
channels of white dwarf evolution, explaining whether DB's become DZ's
and how the DQ's fit in, and so forth. Fig. 1, presented some pages
back, represents one such attempt, but the true picture is considerably
more complex (see Fig. 1 of Sion 1986). Earlier I cited a classic volume
of Payne-Gaposhkin; here I go considerably further back in time (Ockham
1488). "Entia non sunt multiplicanda praeter necessitatem." (For
bibliographical sources, see Sarton 1947.) In common engineer's
parlance, "Keep it simple, stupid." I have elsewhere (Shipman 1987)
quoted E.W. Kolb's way of expressing Occam's Razor: "A theory which is
too complex to fit on a T-shirt is too complex to be correct."
Complicated pictures are not the whole story; while we seek to delineate
the channels of stellar demise, we also seek to understand why? What are
the underlying causes?

Let me set forth two extreme scenarios for explaining the
distinction between DA white dwarfs and non-DA white dwarfs, illustrated
schematically below. The first scenario, which I will call "primordial,"
postulates the distinction between DA's and non-DA's lies in the
planetary nebula stage if not before, certainly predating the white dwarf
stage itself. In this scenario, DA stars remain DA stars for most if not
all of their cooling lifetimes, possibly dredging up subsurface He to
become He-dominated DC stars at very cool temperatures. A specific
mechanism for producing approximately the right number of non-DA's (~ 25

%) was suggested by Iben and Renzini (1983; see also Iben 1984), where the phase of the final thermal pulse determined whether a star would become a DA or a non-DA. However, the basic outline of the primordial scenario as discussed here does not depend on one particular mechanism for the origin of the two types of white dwarfs; binaries might indeed play a numerically significant role, as suggested by Tutukov (1987).

Another scenario is a "mixing" hypothesis, in which the establishment of convection zones and other events such as accretion from the interstellar medium or nuclear burning transform DA's into non-DA's and vice versa. The contrast between the two scenarios is illustrated in Figure 2 below. Liebert (1987) and Liebert, Fontaine, and Wesemael (1987) have discussed this scenario in the context of contemporary theory. Its origins go back to earlier discussions by, e.g., Strittmatter and Wickramasinghe (1971) and Shipman (1972).

Convective mixing certainly seems to make sense as an explanation for the origin of carbon in the DB stars. A widely cited mechanism for changing the surface composition of white dwarf stars is the establishment of a deep convection zone which will mix interior layers with the surface. In one case, a combined theoretical and observational effort has demonstrated quite convincingly that mixing does occur. The DQ white dwarfs are non-DA stars which contain a trace abundance of carbon; model atmosphere analyses indicate that the carbon abundance peaks at about T(eff) ~ 10,000 K (Wegner and Yackovich 1984, Koester, Weidemann, and Zeidler 1984). These authors suggested the mixing explanation, which was confirmed by more detailed "ab initio" models (Fontaine, Villeneuve, Wesemael, and Wegner 1984; Pelletier, Fontaine, Wesemael, Michaud, and Wegner 1986).

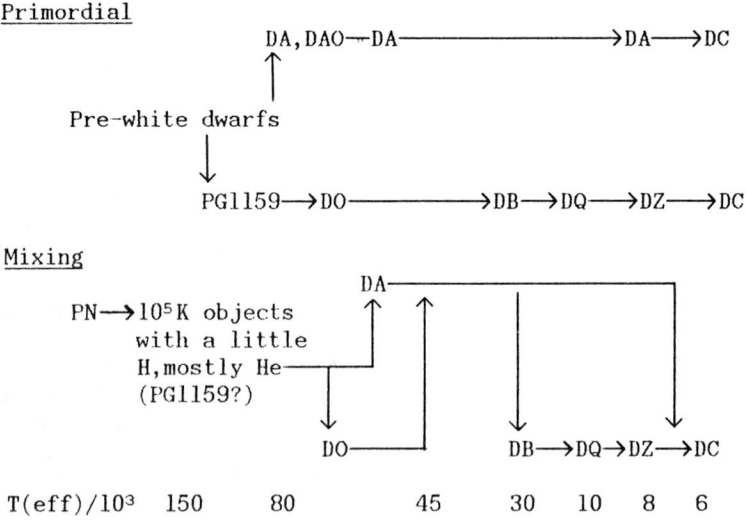

Figure 2. Two extreme explanations for the chemical evolution of white dwarf stars.

There are two important pieces of evidence which tend to favor the primordial scenario. First, as emphasized by Renzini in the discussion at this conference, about 20-30 % of central stars of planetary nebulae are extremely H-deficient (Mendez, Miguel, Heber, and Kudritzki 1987; Kudritzki, this conference). The existence of this abundant class of planetary nebula central stars strongly suggests an evolutionary connection between these central stars and the non-DA sequence, which represents (very!) approximately 20 % of the white dwarf stars. A second, less direct reason for believing in the primordial hypothesis is that white dwarf stars which do originate as DA's should have fairly thick hydrogen shells, with masses of roughly 10^{-4} solar masses, overlying a He envelope of 10^{-2} solar masses, overlying a C/O core. Such hydrogen envelopes are too thick to be mixed with the underlying He. While it is possible that diffusion-induced H burning could thin the H layer down (Michaud, Fontaine, and Charland 1984), Iben and MacDonald (1985) have questioned whether this process can work with the required effeciency to reduce the mass of the H layer by of order 10^4 or more.

The complex phenomenology of the white dwarf stars, outlined in Figure 1, has been used by several of us to argue in favor of mixing in at least some circumstances. The varying ratio of DA's to non-DA's as a function of T(eff), illustrated in Figure 1 at the beginning of this paper, has been used by many to suggest that there is considerable crossing over between the two sequences (Sion 1984, Greenstein 1986, Liebert 1986, and references therein). There are zero DA stars above about 90,000 K, where the ratio turns around and we find 7 DA stars for every non-DA star. Between 30,000 K and 45,000 K, the Palomar-Green survey of the north galactic pole contains no non-DA stars (which, in this temperature range, would be hot DB's); five would be expected from normal statistics (Liebert, Wesemael, Hansen, Fontaine, Shipman, Sion, Winget, and Green 1986) Between 10,000 K and 30,000 K, there are four DA stars for every non-DA star, a ratio which a few years ago was thought to prevail at all temperatures. Cooler than something like 10,000 K, the ratio becomes 1:1. Cooler than about 5,000 K, the ratio is unknown, because Balmer lines are no longer visible. In principle, infrared absorption by the pressure induced dipole of hydrogen could be used as a composition diagnostic, but clear conclusions have not yet appeared.

However, there are other ways to explain these data. One simple one suggested by MacDonald (1987, private communication; see also Iben and MacDonald 1985, 1986) is that the cooling rates, which underlie the conclusions about the DB "gap," might be incorrect, making the significance of the gap considerably less. MacDonald, Iben, and Jason Tillett are currently undertaking calculations to see whether this idea will work. Another suggestion appeals to changes in the star formation or evolution processes in the last few billion years. White dwarfs of differing temperatures are the end products of stars which formed at different times. If star formation in the Galaxy is patchy, then the changing numbers of DB and DA stars could simply reflect a different population of ancestors. It is especially interesting to consider this suggestion in light of the fact that all white dwarfs that we know of, and which are the basis of these statistics, lie within roughly 100 pc of the sun. It is not too unreasonable to suppose that star formation on

such a short length scale is inhomogeneous in time. Still another
possibility, relevant to the shortage of high-temperature DA's is that H
shell burning in the DA precursors will affect their evolutionary tracks
in such a way that DA degenerates will join the white dwarf sequence at
T(eff) near 80,000 K, rather than a hotter value.

The behavior of trace quantities of H in the DB stars and He in the
DA stars provides some pieces of evidence which remain to be successfully
interpreted. In particular, Shipman, Liebert, and Green (1987) discussed
the DBA stars, a phenomenon which may (or may not) suggest an
evolutionary connection between DA and non-DA stars. These stars are
predominantly He, with H abundances of order 10^{-4}; about 20 % of the non-
DA stars in the temperature range where H and He can be seen
simultaneously are DBA stars. Shipman, Liebert, and Green discuss two
scenarios for the origin of the H in these stars: convective mixing in
the star (in which a DA would mix and turn into a DBA) and accretion.
There are problems with both scenarios. A very recent discovery of Ca in
two DBA stars (Kenyon, Shipman, Sion, and Aannestad 1987) suggests that,
in these cases at least, accretion is the origin of the H seen in their
spectra. We still have to explain how objects like GD 40 accrete Ca but
no H, while these DBA stars accrete Ca and H in approximately solar
proportions.

Thus we have no simple answers which can explain why the lower part
of the HR diagram contains such a wide diversity of white dwarf stars.
At the present time, white dwarf researchers are endeavoring to sort out
what pieces of evidence really are clues to the correct fundamental
answer and which are red herrings. If the "primordial" scenario is even
partially correct, then the chemical properties of white dwarf stars will
provide an important boundary condition for understanding the way that
planetary nebulae are formed and evolve. If, on the other hand, "mixing"
(and accretion from the interstellar medium) can explain the chemical
diversity of white dwarfs, then their chemical composition will be only
loosely coupled to prior evolutionary stages.

This research has been supported by the National Science Foundation
(AST 85-15747) and by NASA.

REFERENCES

Angel, J.R.P., Borra, E.F., and Landstreet, J.D. 1981, Astrophys.J.
Suppl. 45,   457.
Bidelman, W.P. 1986, in IAU Colloquium No. 87. Hydrogen Deficient Stars
and Related Objects, K. Hunger, D. Schonberner, and N.K.Rao, eds.,
(Dordrecht: Reidel), p. 3.
Fontaine, G., Villeneuve, G., Wesemael, F., and Wegner, G. 1984,
Astrophys. J. Letters 277, L61.
Greenstein, J.L. 1960, in Stars and Stellar Systems, vol 6: Stellar
Atmospheres, J.L. Greenstein, ed., (Chicago: University of Chicago
Press), p. 676.
Greenstein, J.L. 1985, Publ. Astron. Soc. Pacific 97, 827.
Greenstein, J.L. 1986, Astrophys. J., 304, 334.
Guseinov, O.H., Novruzova, H.I., and Rustamov, Y.S. 1983, Astrophys.
Space Sci. 96, 1.

Guseinov, O.H., Novruzova, H.I., and Rustamov, Y.S. 1984, Astrophys. Space Sci. 97, 305.
Hamada, T., and Salpeter, E.E. 1961, Astrophys.J. 134, 683.
Heber, U. 1986, in Hydrogen Deficient Stars and Related Objects: Proceedings of IAU Colloquium 87, Hunger, K., Schonberner, D., and Kameswara Rao, N., eds., p. 33.
Iben, I., jr. 1984, Astrophys.J. 277, 333.
Iben, I., and MacDonald, J. 1985, Astrophys.J. 296, 540.
Iben, I., and MacDonald, J. 1986, Astrophys. J. 301, 164.
Iben, I., and Renzini, A. 1983, Ann. Rev. Astron. Astrophys. 21, 271.
Kapranidis, S., and Liebert, J. 1985, Astrophys.J. 305, 863.
Kenyon, S., Shipman, H.L., Sion, E.M., and Aannestad, P. 1988, Astrophys. J. Letters, submitted.
Koester, D., Schulz,H., and Weidemann, V. 1979, Astr.Astrophys. 76, 262.
Koester, D., Weidemann, V., and Zeidler K.-T., E.-M. 1984, Astron. Astrophys. 116, 147.
Koester, D. 1984, Astrophys. Space Sci. 100, 471.
Koester, D., and Reimers, D. 1985, Astron. Astrophys. 153, 260.
Luyten, W. J. 1952, Astrophys.J. 116, 283.
Liebert, J. 1980, Ann. Rev. Astron. Astrophys. 18, 363.
Liebert, J., Dahn, C.C., and Sion, E.M. 1983, in Proceedings of IAU Colloquium No. 76: the Nearby Stars and the Stellar Luminosity Function, ed. A.G.D. Philip and A. Upgren, (Schenectady, NY: L. Davis Press, 1983, p. 103.
Liebert, J., Dahn, C.C., Gresham, M., and Strittmatter, P.A. 1979, Astrophys.J., 233, 226.
Liebert, J. 1986, in Hydrogen Deficient Stars and Related Objects: Proceedings of IAU Colloquium 87, Hunger, K., Schonberner, D., and Kameswara Rao, N., eds., p. 367.
Liebert, J., Wesemael, F., Hansen, C.J., Fontaine, G., Shipman, H.L., Sion, E.M., Winget, D.E., and Green, R.F. 1986, Astrophys.J. 309, 241.
Liebert, J., Fontaine, G., and Wesemael, F. 1987, Memorie Societa Astronomica Italiana 58,17.
McCook, G., and Sion, E. 1987, Astrophys. J. Suppl., in press.
Mendez, R., Miguel, I., Heber, U., and Kudritzki R.P. 1987, in Hydrogen Deficient Stars and Related Objects: Proceedings of IAU Colloquium 87, Hunger, K., Schonberner, D., and Kameswara Rao, N., eds.
Michaud, G., Fontaine, G., and Charland, Y. 1984, Astrophys. J., 280, 247.
Nousek, J., Shipman, H., Liebert, J., Holberg, J., Pravdo, S., Giommi, P., and White, N. 1986, Astrophys.J., 309, 230.
Ockham, W. (1488). Summa Totius Logicae (Paris).
Oke, J.B., Weidemann,V., and Koester, D. 1984, Astrophys. J., 281, 276.
Payne, C.H. 1925, Stellar Atmospheres, (Cambridge, Mass.: the Observatory), p. 189.
Pelletier, C., Fontaine, G., Wesemael, F., Michaud, G., and Wegner, G. 1986, Astrophys.J. 307, 242.

Pilachowski, C., and Milkey, R.W. 1987, Publ. Astron. Soc. Pacific 99,

836.
Reimers, D., and Koester, D. 1982, Astron. Astrophys. 116, 341.
Ruiz, M.T., Maza, J., Wischnjewsky, M., and Gonzalez, L.E. 1986, Astrophys.J.(Letters) 304, L25.
Sarton, G. 1947. Introduction to the History of Science, vol. III. Science and Learning in the Fourteenth Century (Baltimore: Williams and Wilkins), pp. 549-557.
Schmidt, G., West, S.C., Liebert, J., Green, R.F., and Stockman, H.S. 1987, Astrophys.J. 309, 218.
Shipman, H.L. 1972, Astrophys. J., 177, 723.
Shipman, H.L. 1979, Astrophys. J., 228, 240.
Shipman, H.L., and Sass, C.A. 1980, Astrophys.J. 235, 177.
Shipman, H.L. 1983, in Proceedings of IAU Colloquium No. 76: the Nearby Stars and the Stellar Luminosity Function, ed. A.G.D. Philip and A. Upgren, (Schenectady, NY: L. Davis Press, 1983, p. 417.
Shipman, H.L. 1987, in Proceedings of IAU Colloquium 95: the Second Conference on Faint Blue Stars, A.G.D. Philip, D.S.Hayes, and J.Liebert, eds., (Schenectady, NY: L.Davis Press), in press.
Shipman, H.L., Liebert, J., and Green, R.F. 1987, Astrophys.J. 315, 239.
Sion, E.M. 1984, Astrophys. J. 282, 612.
Sion, E.M. 1986, Publ. Astron. Soc. Pacific. 98, 821.
Sion, E.M., Greenstein, J.L., Landstreet, J.D., Liebert, J., Shipman, H.L., and Wegner, G. 1983, Astrophys.J. 269, 253.
Wegner, G. and Yackovich, F. 1984, Astrophys.J. 284, 257.
Weidemann, V. 1984, Astron. Astrophys. 134, L1.
Weidemann, V. 1987, Memorie Societa Astronomica Italiana 58, 33.
Weidemann, V., and Koester, D. 1984, Astron. Astrophys. 132, 195.
Wesemael, F., Henry, R.B.C. and Shipman, H.L. 1984, Astrophys.J., 287, 868
Wesemael, F., Green, R.F., and Liebert, J. 1985, Astrophys. J. Suppl., 58, 379.
Winget, D.E., Hansen, C.J., Liebert, J., Van Horn, H.M., Fontaine, G., Nather, R.E., Kepler, S.O., and Lamb, D.Q. 1987, Astrophys.J. (Letters) 315, L77.

SUMMARIZING REMARKS ON THE STRUCTURE AND EVOLUTION OF PLANETARY NEBULAE AND PROPERTIES OF THEIR CENTRAL STARS.

L.H. Aller
Department of Astronomy
University of California
Los Angeles, CA 90024
U.S.A.

To review or attempt to summarize a conference such as this recalls the request our department office once received from a little fourth-grade schoolgirl: "I'm doing a science project on the stars. Tell me all about them on two pages." We can only try to look at the broad picture and attempt to identify problems that require special emphasis or attention.

First of all, I'd like to remark on the wide geographic distribution of planetary nebulae enthusiasts. In addition to capable, well-organized and well-led groups at larger centers such as Groningen, London, and Mexico City (to mention only three), there are outstanding investigators at many other places, for example in South America and India. One has only to glance at the bibliography of the new comprehensive planetary nebular catalogue by Acker et al., — a truly formidable undertaking — to appreciate this point. Like the whole science of astronomy, PN research is an endeavor in which the whole world can rejoice.

Impressive advances have been made both in theory and observation, the latter providing foundations for the former. New technologies now permit definitive attacks on a broad range of problems. Developments in CCD's make possible quantitative monochromatic direct imaging of PN as illustrated by the beautiful pictures shown to us by Bruce Balick, Ms. Chu, J. Jacoby and associates, by J. Lutz, and many other participants, particularly in the poster sessions. Applications to high-dispersion spectroscopy as in the outstanding survey of the infrared spectrum of NGC 7027 by D. Pequinot and R.B. Gruenwald and the remarkable echelle spectra of planetary nebulae nuclei (PNN) by J.K. McCarthy are two examples of the power of new techniques that will open inspiring vistas to PN investigators. These CCD detectors also enable the enthusiast for nebular kinematics to detect heretofore unknown speedy blobs in many nebulae.

IRAS data, described by A. Preite-Martinez constitute one of the truly great leaps forward in PN study. The survey covered 98% of the sky from 8 to 12 μm. Alas, NGC 7027 was missed! The survey provided energy distributions for 70 PN, low-resolution spectra (LRS) for 60 PN, and detailed maps for three large objects, including NGC 7293. These

IRAS data proved fundamental for determinations of temperatures, energetics, and spatial spread of dust. It also detected halos at $\lambda > 50$ μm. IRAS supplied very important abundance information for Ne, S, and Ar, whose ions have heretofore been observed with difficulty from the ground. Spatial distribution of [O IV] in NGC 7293 has been measured. Much more information remains to be extracted from the IRAS tapes. Preite-Martinez and Pottasch estimate that more than 500 objects, both PN and proto-PN, observed in the far-IR, remain to be confirmed with radio and near-IR observations.

As IR arrays are developed and applied to PN problems we can make detailed maps of the dust distribution and compare them with maps made in the radio-frequency (r.f.) domain which shows the H+ gas distribution. Roche noted that by comparing images of Brackett and of dust in nebulae such as BD + 30°3639 and NGC 7027, one finds that the dust and gas do not peak in exactly the same places. In NGC 7027 the dust peaks just outside the H+ region although some must exist therein. Nebular images in hot dust can be found in NGC 7027, but we need to know where the cool dust that radiates in the 30- to 100-μm region is located.

With the VLA, Terzian notes that it is possible to construct r.f. isophotic maps with a resolution between 0.1 and 1.0" not only in the H II (H+) continuum but also in H I (observed in NGC 6302, IC 418, and IC 4997) and in the OH 1612 Mhz maser line seen in Vy 2-2 (a very young PN) and in NGC 6302. Important progress is being made in studying the transition from OH-IR stars through the pre-PN phase. Terzian reports that first epoch observations have been secured for measurements of expansion velocities in a number of PN; these may prove useful in handling the troublesome problem of nebular distances.

Thus, we now can obtain PN images in the radiations of H, lines excited by collisions, molecules and dust which may emit over a wide range of temperatures and densities. Supplementing these data are kinematical measurements. We obtain line-of-sight velocities over the nebular surface and eventually the proper motions of knots and filaments across the line of sight.

Balick summarized the structures and morphologies of PN as revealed by CCD direct imaging. Most of the mass destined to form the PN was ejected in the slow wind, but the topology was later largely fixed by the fast wind (~ 2000 km/sec) emerging from the central core remnant. This scenario, proposed by Sun Kwok, seems well established as a basic working hypothesis. Balick emphasized that the actual appearance of the nebula depends on the age or stage of evolution of the nebular shell and the configuration of the ejected material. If the ejected shell is spherically symmetrical, we observe a structure such as IC 3568 or an annular nebula such as NGC 6894. More often the shell is not spherically symmetrical but is thicker and/or denser in the equatorial plane so the outrushing wind bursts through in the polar regions first. For a detailed summary, see Balick (1986).

The next step is to build numerical models where the history of a PN can be followed by two-dimensional hydrodynamics with radiation pressure taken into account. Kahn sketched some possibilities involved, suggesting that the superwind might be nonsymmetrical.

Detectable effects on PN shapes might appear in 10 to 400 years. The shapes of PN are fashioned by winds and radiation pressure but the ionization and heating of the detached red giant shell is due to the stellar radiant flux. In bright rims, we might observe sometimes the dissipation of energy from winds. Before detailed modeling can be undertaken, we need improved, high spatial resolution observational material from the r.f. to UV regions with supplementary kinematical data to match. Velocity patterns must be measured over the entire PN image; O.C. Wilson tried this long ago with a multi-slit; Fabry-Perot interferometers are now often employed.

Multiple shells are common phenomena in PN. Ms. Chu finds that 50% of nearby objects in the NGC and IC catalogues have multiple shells. Kaler had classified them on the basis of their surface brightnesses: I = faint detatched shells that show limb brightening and with outer diameters exceeding 0.5 pcs. The masses are low ($\sim$ 1%) compared with the inner bright ring. Type II shells have an attached outer ring that is about 25% as bright as the inner ring. These shells have radii less than 0.5 pcs, show no limb brightening, and have masses comparable with that of the main ring. Some objects such as NGC 2392 are classified as "peculiar." The motions can be complex and follow no uniform pattern. Ms. Chu notes that sometimes the attached outer ring co-expands supersoniocally with the inner shell, suggesting a possible multiple ejection. Other times a supersonically moving inner shell collides with a faint, subsonically expanding halo. In PN such as NGC 6826, a slow moving inner shell penetrates a faint outer one.

In M2-2, however, a detached outer shell expands independently of the inner shell. Different formation mechanisms are clearly involved. Curiously, elemental abundances in the outer shells usually did not differ from those found in the inner shells. Tylenda's suggestion that a faint outer halo may be a fossil ionized sphere left behind as the PNN fades certainly allows an observational test.

Weinberger's discussion highlighted the great complexity of the kinematical data. Unique, single-valued expansion velocities cannot be specified for individual objects. For example, it is not possible to define a specific expansion-velocity versus radius relationship for PN. Instead, one has to describe an entire velocity pattern for each object, a point emphasized by Chu et al. (1984). Although average expansion velocities of the order of 20 km/sec for H I and [O III] seem to be found in the majority of PN that have been measured, it must be emphasized that within some objects, such as NGC 2392, 6302, 6537, 6543, 6826, and M2-3, there occur jets moving with velocities up to 100 km/sec or more. These kinematical phenomena must be revealing essential clues to the initial, almost certainly non-uniform ejection and acceleration mechanisms.

The nature of planetary nebular evolution lies at the very center of our concerns. Before the London meeting our ideas on this subject were rather general as many essential observational details had not yet been provided. Now the scenario seems to be well-defined. Important steps are described in the reviews by Habing, Knapp, Renzini, and Kwok, while Roche and Rodriguez have described important clues provided by dust and molecular data.

The precursors of PN are Mira stars (long-period variables) or other asymptotic giant branch (AGB) stars. These evolve into OH-IR objects that are essentially invisible optically; they emit in the IR and r.f. spectral regions. Large numbers of them were "fingered" by IRAS and then verified by r.f. observations of the 1612 MHz line of the OH maser. The PN and OH-IR objects seem to show the same pattern in velocity and galactic distribution. The OH-IR stars are presumed to evolve into proto-PN and eventually into young PN such as Vy 2-2. We might expect the number of pre-PN to be the order of 10% to 25% of the number of PN, to show color temperatures of the order of 130 K to 200 K, to be non-variable (unlike their Mira or youngest OH-IR predecessors), to show silicate features, possibly also CO or OH, and to possess no optical counterparts.

Not all OH-IR objects are expected to evolve into PN and some PN (such as the strongly C-rich nebulae) may evolve by a different route. The "standard" evolutionary scenario starts with a Mira star with a period between 250 and 500 days, a variation of about one magnitude in bolometric luminosity, and an expansion velocity of about 5 to 10 km/sec. The mass loss may be triggered by pulsations and fall between a rate of 10^{-7} and 10^{-6} solar masses/year. As their outer envelopes expand further and cool, the Mira stars evolve into OH-IR variables with periods between 500 and 1000 days and amplitudes of about 2 magnitues. They attain luminosities of the order of 1000 to 10,000 that of the sun and the mass loss rate increases to 10^{-5} or even 10^{-4} solar masses/yr in the slow wind of about 10 to 15 km/sec. This stage may last 10,000 years. The outer shell is finally detached; the dying AGB star becomes now a non-variable OH-IR source with a hot core concealed under a thick, dusty, circumstellar envelope.

The next stage is the appearance of a fast wind from the hot core with a velocity of 1000 to 3000 km/sec, but a low-mass loss rate of about 10^{-7} solar masses/yr. The rich UV radiation field of the hot core now ionizes the slowly moving shell of the defunct red giant and turns on the PN. In many PN, much of the intact, cool, outer AGB envelope is still there; it is revealed by solid grains and molecules such as those of hydrogen and CO. There are found in NGC 2346, 6302, 6720, and 7027. OH is detected in the neutral torus around the waist of NGC 6302. It is also found in Vy 2-2 and IC 4997. Intensive searches have revealed molecules in only a few PN. Why? Rodriguez estimates that in a typical PN, molecules would last only about 100 years. Why are they found in an aged object such as NGC 7293? Possibly some PN contain dense blobs that shield them from radiation.

Dust occurs within the H II regions of PN as well as in cool, neutral regions, which may show a considerable temperature range. Among the IR lines is the 9.7-μm silicate feature characteristic of O-rich material. In sooty clouds are observed the 11.2-μm band of SiC and the 3.3- to 13-μm feature attributed to C-rich grains and polycyclic aromatic hydrocarbons (PAH's). Sometimes O-rich envelopes are associated with C-rich central stars, but never vice versa. As a result of He burning to C, and subsequent mixing to the atmosphere, O stars become C stars, but there is no evidence of a C star ever turning into an O star.

Studies of PN chemical compositions, reviewed by Clegg, put some contraints on evolutionary models. Of particular importance is the C/N/O ratio. With respect to the Sun, PN tend to be O-deficient, while C and N are often enhanced in PN. The N-rich objects, Peimbert's type I (1983), are believed to originate from the more massive progenitors. Large numbers of PN show C/O > 1 and are believed to come most recently from C stars. Neon is probably not affected by nuclear reactions in stars, while S, Cℓ, and Ar must relfect the abundance pattern of what are loosely called "metals." Thus, their abundances reflect that of the interstellar medium (ISM) from which PN progenitors were formed. New, improved atomic data such as collision strengths for lines of [S II], [Cℓ III], and Ar IV] will yield improved abundances of the corresponding elements. Calculations of increasingly accurate models such as that by Clegg et al. (1987) for NGC 3918 are very valuable.

Perhaps the best candidate to the missing link between OH-IR sources and PN is Vy 2-2 = 45-2°1, described in poster papers by Clegg, Hoare, and Walsh, and by Faloma and Sabbadin. VLA measurements at 15 GHz show it to have outer and inner shell radii of 0.25" and 0.1", N_ε = 300,000 cm^{-3}, and T_ε = 11,000°K. With an expansion velocity of 10 to 15 km/sec, it appears to have a dynamical age of 1200 to 2000 years. Thus, it is an object younger than NGC 6572, BD + 30°3639, or IC 4997. We hope that even younger objects will be revealed by IRAS data.

Any discussion of the evolution of PN and PNN brings us back to the distance problem. Methods now frequently employed include the use of 21-cm H I absorption, the reddening of nearby field stars, and stellar winds. Eventually, expansion velocities measured with the VLA may be useful. Although progress has been made, as Ms. Lutz has described, we cannot pretend that the distance problem is in a satisfactory state. For a particular PN, errors ranging from 30% to a factor of 2 are often quoted. This uncertainty enters in all types of discussion. As an example, let me illustrate how errors affect a dependence of N(O)/N(H) on R (distance from center of the galaxy) for type II PN. Faundez-Abans and Maciel (1986) have recently discussed this problem. C.D. Keyes and I selected objects primarily on the basis of their kinematical properties. We relied on lists by Kaler (1970), Barker (1978), and by Heap and Augensen (1987). We assume the distance of the sun, R(sun), to be 8500 pcs from the galactic center. Chemical compositions obtained via theoretical models were generally employed.

Our purpose in showing this diagram is not actually to discuss log N(O)/N(H) as a function of R, 8.69 - 0.0156 R(kpc), but rather to emphasize uncertainties that certainly accrue from distance errors. These may be less than 1 kpc for nearby PN but could be a factor of 2 for nebulae at a distance of 4 kpc or more. There is also an error of the order of 0.1 dex in the N(O)/N(H) determination, largely a consequence of uncertainties in T_ε. It is likely that much of the scatter in the diagram must arise from errors in the input data. Errors in the distance scale frustrate efforts to make reliable estimates about PN formation rates. Using Cudworth's distance scale, J.P. Phillips finds the formation rate of PN to be 6.2 to 13 pc^{-3}/yr^{-1}.

The number of objects for which accurate distances can be found, PN in binary systems, in the galactic bulge, and in the Magellanic Clouds, is inadequate for most purposes. Few PNN are binaries, while for nebulae at great distances, essential information on structure and internal kinematics is not available.

For planetaries whose central stars show absorption lines, a powerful method of distance determination has been developed by Kudritzki and his associates. Long ago it was suggested that by comparing the absorption line spectra of PNN with those of type I population O stars one could determine both their effective temperatures and surface gravities, expressed as log g (Aller 1948; Wilson and Aller 1954). The effort failed for two reasons, the inadequacy of too-low dispersion photographic spectra and use of LTE atmospheres. By using high-dispersion spectra measured with CCD's and modern non-LTE atmospheric models, Kudritzki et al. have shown that the effective temperature can be obtained to an accuracy of 10% and log g to 0.2 dex. Since stellar evolution theory for PNN gives a relationship between log g and the effective temperature for which one does not need to know the distance, one can deduce the mass, M; with M and g known, R can be found. With R and the emergent stellar flux, one can predict the absolute magnitude of the star of known apparent magnitude and thus deduce its spectroscopic parallax. A check on the method is provided by applying it to a hot subdwarf in NGC 6397. It gave the same distance to this globular cluster as did conventional, well-established methods. Furthermore, the absolute magnitudes of the brightest PN in the sample match those found in the Magellanic Clouds. These results favor the Cudworth distance scale. Where they can be determined for individual PN, the distances found by the Kudritzki method are to be preferred.

The perplexing problem of Zanstra temperatures, reviewed by Kaler and discussed in posters by Henry and many others, is not yet completely solved. Although the Kudritzki procedure seems to give a reasonable interpretation of a star such as the He-rich nucleus of NGC 246, (T_{eff} = 130,000°K, log g = 5.71), troubles appear to abound for yet hotter PNN, e.g., the nucleus of NGC 7027, for which model nebular methods give $T(*)$ = 190,000°K (Pequinot and Gruenwald) but the PNN magnitude $m(*)$ = 17.7 (Walton et al.) implies $T(*)$ = 310,000°K. For NGC 2440, $T(*)$ = 350,000°K was found from $m(*)$ by Zanstra methods, while Shields et al. (1981) found $T(*)$ = 180,000°K by model nebular methods. Similar difficulties occur for NGC 6565 where Reay et al. (1984) found T_Z(H I) = 185,000°K, T_Z(He II) = 130,000°K, while the level of the spectrum requires a $T(*)$ of about 85,000°K. One can scarcely account for these discordances by assuming that the PNN is fading rapidly; the response of the nebular spectrum easily would have been observed. Presumably, we must assume that if the star is as hot as indicated by Zanstra methods, heavy absorption must occur well shortward of 228 Å. Otherwise, [Ne V] would be predicted much too strong. These new theoretical stellar fluxes can help us in another important way. They provide excellent input data for the calculation of theoretical PN models so that we can concentrate on other matters such as geometrical factors in interpretations of nebular spectra.

Harrington emphasized the crucial importance of nebular geometry in his review of current developments in nebular modeling. Refinements in the basic theory as well as in the input atomic data are now available. Nebular modeling is becoming more and more popular; several codes are now used and intercomparisons between some of them have been made. In spite of refinements, small structural irregularities, knots, and condensations present difficulties. For example, when models are used to estimate ICF (ionization correction factors), large errors are possible as when we try to deduce the nitrogen abundance from [N II] lines. New trends in theory will include dynamical and evolutionary effects in a format applicable to individual observed objects. We hope theory will enable us to identify possible shock excited atomic transitions. Shocks are invoked to explain the shapes and forms of PN, but forecasts of observable spectroscopic effects are few.

The importance of stellar winds is clearly evident for many PN. For example, in constructing their model for NGC 1535, Adam and Köppen (1985) adopted $T(^*)$ from the work of Kudritzki et al., so they had to postulate a wind in order to explain the excitation level of the spectrum. Stellar winds can modify the far UV output flux; in some stars Kudritzki et al. found the $\lambda < 228$ Å flux to be enhanced by a factor of 1000! Grewing reviewed the general observed features of winds which blow with velocities between 500 and 4000 km/sec. As expected, all Wolf-Rayet-type stars show P Cygni profiles, as do many stars with bland spectra in the optical domain. Empirical or "observational" determinations or mass-loss rates depend on mechanisms invoked to produce and maintain these winds. Perinotto evaluated current popular theories and analyses based on P Cygni profiles observed with the IUE. He notes that the scatter in dM/dt for the same objects can extend over a factor of 100, but if we remove unnecessary discordances arising from various assumptions in fundamental parameters, this factor can be reduced to as little as 3 to 5, essentially the residual difference due to method employed. With the most recent version of a radiation wind-driven theory, it has been found that in the "best observed" PNN, NGC 6543, theoretical and "best empirical" dM/dt values differ by a small factor. Ultimately, we can hope to achieve an accuracy in dM/dt of a factor of 3.

Some of the main points in the theory of PNN evolution were already reviewed in 1982. Renzini and Schonberner gave an update, while Pottasch used a variety of distance determination options to compare an empirical luminosity versus T_{eff} plot with theoretical predictions. A number of objects near the galactic center seemed to depart substantially from the theoretical pattern, but the temperatures of these galactic center objects do not seem to be well determined. Using assumed initial mass functions and dM/dt rates for late AGB evolution, R.A. Shaw calculated some expected theoretical distributions of PNN in the logL - logT plane. These differed from the observations and suggested strong observational selection effects. Köppen emphasized the importance of linking PNN evolution with that of the PN itself and proposed some modifications or refinements of the Kwok two-stream model. In essence, dM/dt changes with time, perhaps increasing towards the end of AGB evolution as an OH-IR object, and possibly accompanied by Renzini's super wind.

PN with massive nuclei will evolve with a greater speed than "normal" PNN so that time-dependent effects should be readily observable. Such objects would attain high luminosities; yet, the upper limit of L(PN) seems rather sharply defined in other galaxies suggesting that there is no "high luminosity tail" in the distribution. We are not sure we have detected PNN with masses as high as 1.0 or even 0.9 m(sun). Tylenda suggests that NGC 2440, 6302, and 7027 may have large masses; NGC 6445, 6741, IC 2165, and M2-2 are other possible candidates.

Binary stars among PNN offer a number of unique opportunities to assess their masses and radii and to investigate possible effects of binarity upon nebular evolution. From his analysis of spectroscopic binaries, Mendez concluded that probably not more than 20% are close pairs, but if we allow for a range in masses and consider wider doubles, including eventual visual systems such as NGC 246, the percentage may be higher. A number of investigators have examined the possible role role of binarity in determining the shapes and forms of PN. J. Lutz and N.J. Lame secured monochromatic images of Abell 14, H3-75, and K1-2 with a CCD detector and concluded that the close binary character of their PNN may have influenced their structures.

In an excellent presentation, H.E. Bond noted that binary PNN are not freaks. Among the best known close binary systems are the nuclei of Abell 35, 41, 56, 63, and K1-2. Data for the best determined systems confirm that the stars are pre-white dwarfs and have masses of about 0.6 m(sun). Further clues to PNN masses are provided by the pulsations of the nucleus of K1-16. The pulsations involve a number of overtones so that evolutionary effects on the period are difficult to assess. Tutukov discussed the theory of close binary evolution. Can symbiotic stars evolve into PN? Can we acquire systems with a common envelope and can binary PNN evolve into cataclysmic variables?

What happens as PNN plummet to become white dwarfs (WD's)? J. Liebert reported that the low end of the PNN sequence and the high end of the WD region pose problems. The overlaps seem best defined by He-rich WD's and indicate a continuous sequence of stars from WC, OVI (or other He-rich, H-poor nuclei), through pulsating stars such as K1-16, field objects such as H1504+65 (which was discussed by Shipman), and finally the He-rich DO WD's, in decreasing steps of effective temperature. On the other hand, H-rich nuclei appear to evolve rapidly, but it is difficult to understand why hot DA stars are so rare. The PNN of NGC 7293 have appropriate values of (log g, T_{eff}) but this may be the only known example at this time.

Recently, a few very old, low-surface-brightness nebulae have been found around previously known faint blue stars. The best example is the hot DA, PG 0950+139, which is surrounded by a faint nebula discovered by Ellis, Grayson, and Bond. There is also a compact nebula (unresolved from the star) for which Liebert et al. find an electron density of the order of 1×10^6/cc, a mass of about 10^{-8} that of the sun and a radius of about 10 A.U. The star (log g = 7.5 T_{eff} = 70,000°K) has a cooling age of about 5×10^5 years; the faint nebula could have been ejected earlier.

Many excellent papers were well presented in poster sessions. It is unfortunate that the published volume can contain only abstracts which do not do justice to the wealth of high-quality data so often revealed.

In summary, new observational techniques and theoretical insights permit us to progress rapidly in the study of planetary nebulae. The importance of surveys is well recognized. We might have missed NGC 6751 with its engaging halo, which Ms. Chu has called to our attention. Many other examples exist. The Strasbourg catalogue is going to be of inestimable value to all of us working in this field.

At this epoch, however, a worthwhile undertaking might be to select a small sample of representative planetaries, preferably those whose distances could be established by the Kudritzki et al. methods. I would emphasize objects with well-established distances, for then we can express results in cgs, SI, or solar units. These objects should be observed intensively in the optical, IR, UV, and r.f. ranges with as high a spatial resolution as is practical. We could map images as observed in H, He I, He II, [N I], [N II], [O II], [O III], [Ne III], [Ne V], [S II], [S III], Ar III], [Ar IV], etc., and determine $\langle N_\epsilon \rangle$, $\langle T_\epsilon \rangle$, and velocity data point by point over the surface. Bits and pieces of such information exist already for many objects but the story must be put together more completely and the missing gaps filled in. By selecting candidates from young PN, such as NGC 6572, IC 4997, and BD+30 3639, to aged objects such as NGC 7293 and some Abell PN, we can provide challenging grist for the mills of the model builders and hydrodynamicists. We should also include "classical" PN such as NGC 40, 1535, 3242, 4361, 7009, 7662, and IC 418. We must not forget NGC 2392 with its intricate internal motions, for which Mendez et al. (1987) have deduced a PNN mass of 0.90 (which would imply rapid evolutionary changes) and an effective temperature of 47,000 K which is not consistent with its high excitation. In this PNN the UV flux from a wind must play a dominating role.

The record of firm accomplishments over the last half century is breathtaking. In 1937 I went to Harvard to work with Prof. Menzel. Very soon after my arrival he told me, "We are going to study planetary nebulae; they are very important."

List of references
(References to papers presented at this symposium are not listed.)

Adam, J., Köppen, J. 1985, Astron. Astrophys., 142, 461.
Aller, L.H. 1948, Ap. J., 108, 462.
Balick, B. 1987, Sky and Telescope, 73, 125.
Barker, T. 1978, Ap. J., 219, 914.
Chu, Y.-H., Kwitter, K.B., Kaler, J.B., and Jacoby, G. 1984, P.A.S.P., 96, 598.
Clegg, R.E.S., Harrington, J.P., Barlow, H.J., and Walsh, J.R. 1987, Ap. J., 314, 551.
Faundez-Abans, M., and Maciel, W.J. 1986, Astron. Astrophys., 158, 228.
Heap, S., and Augensen, H.J. 1987, Ap. J., 313, 268.

Kaler, J.B., 1970, Ap. J., 160, 887.
Mendez, R.H., Kudritzki, R.P., Herrero, A., Husfeld, D., and
    Groth, H.G. 1987, Astron. Astrophys. (in praess).
Peimbert, M., and Torres-Peimbert, S. 1983, I.A.U. Symposium No. 103:
    Planetary Nebulae, ed. Flower, D.R., Dordrecht, Reidel Publishers.
Reay, N.K., Pottasch, S.R., Atherton, P.D., and Taylor, K. 1984,
    Astron. Astrophys., 137, 113.
Shields, G.A., Aller, L.H., Keyes, C.D., and Czyzak, S.J. 1981,
    Ap. J., 248, 569.
Wilson, O.C., and Aller, L.H. 1954, Ap. J., 119, 243.

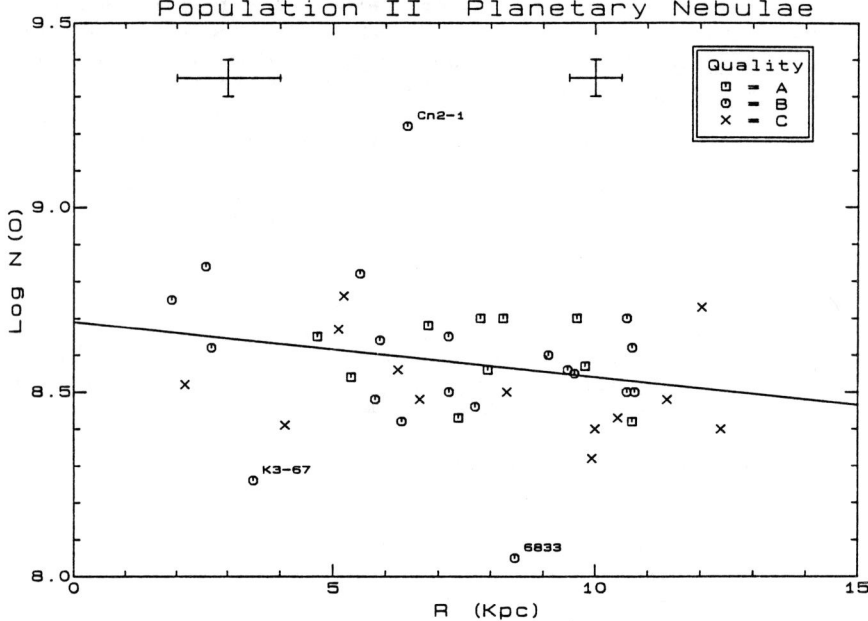

Figure 1. Oxygen/Hydrogen Ratio for Type II Planetary Nebulae.

We plot logN(O)/N(H) + 12 versus distance, R (in kpc), from the galactic center for population type II PN. Much of the scatter arises from observational errors in both distance estimates and abundances. The arrows give a rough estimate of the expected errors.

COMMENTS ON THE APPLICATIONS OF PLANETARY NEBULAE RESEARCH

Manuel Peimbert
Instituto de Astronomía
Universidad Nacional Autónoma de México
Apartado Postal 70-264, México 04510 D.F., México

ABSTRACT. A discussion is given about some of the implications that advances in the study of PN are having in other areas of research such as: atomic physics, stellar formation rates, stellar evolution models, dust production, chemical evolution of galaxies, pregalactic helium abundance, stellar dynamics and the extragalactic distance scale.

1. INTRODUCTION

Professor Aller has given us a summary of the reviews and poster papers of this Symposium on the structure and evolution of planetary nebulae and their central stars. In my summary of the excellent presentations of this Symposium I will concentrate on some of the results derived from PN that are relevant for other branches of astrophysics and of atomic physics. This is a biased summary since I will mainly cover recent results that are familiar to me.

2. PN AS ATOMIC PHYSICS LABORATORIES

Many of the review papers of the 1982 London Symposium on PN stress the relationship between atomic physics and PN (Flower 1983). In what follows I will mention some aspects of this relationship.

2.1 Comparison of Ionization Structure Models with Observations

A superb account of some of these aspects is presented in the review by Harrington (1988).
One of the most successful interactions between models and observations has occurred in the area of charge transfer reactions. Based on the comparison between ionization structure models and observations of NGC 7027 Pequignot et al. (1978, see also Pequignot 1980) concluded that the models were unable to account for several low and intermediate-excitation lines and proposed that charge transfer reactions with atomic hydrogen should be considered, in particular they suggested that the reaction

$$O^{+2} + H \rightarrow O^{+} + H^{+}$$

was fast, with $\beta(O^{+2}) \simeq 8 \times 10^{-10}$ cm^3 s^{-1}. This suggestion was confirmed by Butler et al. (1979)

who found that $\beta(O^{+2}) = 6 \times 10^{-10}$ cm^3 s^{-1}.

At the edge of extended objects of low degree of ionization, like NGC 6720 and NGC 7293 it has been found that the Ne^{++} region coincides with the O$^+$ region (Hawley and Miller 1977, 1978; Hawley 1978); this result is explained by the large value of β (O^{+2}) (Pequignot 1980).

2.2 Electron Densities

Since the pioneering paper by Osterbrock and Seaton (1957) on the electron densities in PN there has been an interplay between the observed line intensity ratios and the atomic parameters involved (e.g. Saraph and Seaton 1970).

Stanghellini and Kaler (1988) find that the densities derived from the [Cl III] line intensities are higher by a factor of 2.6 relative to the densities derived from the [S II] and [O II] line intensities, which probably indicates that the [Cl III] atomic parameters should be revised.

2.3 Abundances

Different lines of the same element should yield the same ionic abundance. This has not been the case for He$^+$ abundances determined from different He I lines nor for C^{++} abundances determined from the λ C II recombination line and the 1909 C III] collisionally excited line.

2.3.1 He$^+$ Abundances

The He I lines are affected by radiative transfer effects and collisions from the 2^{3}S metastable level. Cox and Daltabuit (1971) were the first to consider the effect of collisional excitation from the 2^{3}S level. Peimbert and Torres-Peimbert (1971) by comparing the abundances derived from the $\lambda\lambda$ 5876 and 4471 lines concluded that the theoretical predictions had overestimated the collisional effects by about a factor of three. Berrington et al. (1985), see also Ferland (1986), based on an 11 state ab initio computation obtained for λ 5876 results similar to those by Cox and Daltabuit. Peimbert and Torres-Peimbert (1987a) by comparing the abundances derived from the $\lambda\lambda$ 5876 and 6678 He I lines found again that the collisional cross sections had been overestimated by about factor of 3 or that the population of the 2^{3}S level was a factor of 3 smaller. Berrington and Kingston (1987), see also Clegg (1987), based on a 19 state ab initio computation found that the collisional effects were about a factor of 1.4 to 1.6 times smaller for $\lambda\lambda$ 5876 and 6678 than those derived from the 11 state computation. Clegg (1987) estimated that collisional ionization reduced the population of the 2^{3}S level by a few percent, and Clegg and Harrington (1988) have estimated that under certain conditions photoionization can reduce the 2^{3}S level population by as much as 20%. Peimbert and Torres-Peimbert (1987b) based on observations of $\lambda\lambda$ 5876, 6678 and 10830 still find that there is a difference of about 1.8 between the theoretical predictions and the observations; therefore an additional mechanism to depopulate the 2^{3}S level is needed, probably a charge exchange reaction that has not been considered. Higher quality observations of all the He I lines, in particular of λ 10830, are needed to advance in this problem.

2.3.2 C^{++} Abundances

The C^{++} abundance derived from λ 4267 is in general higher than that derived from λ 1909, the difference can reach in extreme cases an order of magnitude; nevertheless, it should be mentioned that there are objects that do not show this anomaly like IC 418, the Orion nebula and NGC 3918 (Torres-Peimbert et al. 1980; Clegg et al. 1987). General discussions of this problem have been given in the literature (e.g. Torres-Peimbert et al. 1980; Barker 1982; French 1983; Kaler 1986; Clegg 1988; see also references in these papers). Kaler (1986) has suggested that the discrepancy could be solved either by increasing the C$^+$ (λ 4267) effective recombination coefficient or by decreasing the target area of the $C^{++}(2s^2\ ^1S - 2s2p^3P)$ transition by about a factor of four. Alternatively, Peimbert (1983) has suggested that the discrepancy could by due to spatial variations in the electron temperature produced by inhomogeneities in the C/H abundance ratio; the regions with higher C/H abundance ratios would be at lower temperatures and would contribute preferentially to λ 4267 while the regions with lower C/H abundance ratios would be at higher temperatures and would contribute preferentially to λ (1909). Notice that the inhomogeneities could be due to carbon rich pockets embedded in a carbon poor medium, but could also be due to a C/O ratio decreasing outwards.

There is observational evidence in favor of non homogeneous objects. PN are produced by several mass loss stages; the outer layers of the nebula are the least affected by stellar nucleosynthesis showing mainly products of hydrogen burning. Alternatively, continuous mass loss reveals layers of the central star where helium burning has occurred.

There is evidence that many of the AGB stars evolve from oxygen to carbon stars, for example J-type carbon stars, have silicate dust in their envelopes (see the review by Knapp 1988 and references therein), therefore for these objects we would expect an increase of the C/O ratio from the outer layers inwards. Kudritzki and Méndez (1988) have discussed NGC 246 which has a peculiar PNN with a photosphere made mainly of He and C and with no trace of H (Husfeld 1986), but with normal nebular abundances (Heap 1975). Moreover, according to Méndez et al. (1986) 35% of all the spectroscopically well studied PNN belong to the extreme helium rich class.

Another non homogeneous object is NGC 40 where the central star is a Wolf-Rayet with N(He)/N(C) $\approx$ 15 and no hydrogen (Benvenuti et al. 1982); the inner layers of the nebula show very strong C IV 1550 lines, which apparently imply an overabundance of carbon, and the outer layers of the nebula show almost normal H,C, N, O abundances (Clegg et al. 1983). Moreover to explain the discrepancy between the temperature determined from the stellar nucleus and the much lower value inferred from the ionization of the nebula, Bianchi and Grewing (1987) have suggested the existence of a carbon curtain at the inner edge of the nebula.

Another extreme non homogeneous object is A30 which shows He and C rich condensations (Jacoby 1979; Hazard et al. 1980; Jacoby and Ford 1983; Peimbert 1983) with N(He)/N(H) = 9 $\pm$ 3, N(He)/N(C) = 13.2 $\pm$ 4, N(He)/N(O) = 132 $\pm$ 42 and N(He)/N(H) = 86 $\pm$ 29.

Barker (1982, 1983, 1984, 1985, 1986), has found that the $\lambda\lambda$ 4267,1909 discrepancy becomes largest closer to the central stars of NGC 6720, NGC 7009, NGC 6853, NGC 3242 and NGC 7662; a C/O ratio decreasing outwards in the parent star and carbon rich fast winds ejected in the later stages of evolution might also help to explain these results.

2.4 Non-Linearity of the Detectors

From observations of the 5007/4959 [O III], the 6583/6548 [N II], and the 4686/4541 He II line intensity ratios it is possible to check the linearity of a given detector; significant deviations from linearity have been found for IDS systems and for the ESA photon counting detector (*e.g.* Rosa 1985; Llebaria et al. 1986; Peimbert and Torres-Peimbert 1987a).

3. FORMATION RATES AND THE INITIAL MASS FUNCTION

To a first approximation the formation rates in the solar vicinity of: Miras, OH/IR stars, PN, white dwarfs and the main sequence turn off rate agree within a factor of 4! (Phillips 1988). This result is in agreement with the ideas about the genetic relation between Miras, OH/IR stars, PN and white dwarfs.Note that due to their larger scale height Phillips argues that not all the Miras become PN.

The PN formation rate depends on the fourth power of the distance scale adopted. Recent results by Barlow (1988), Kudritzky and Méndez (1988) and Mallik and Peimbert (1988) favor long distance scales, like that of Cudworth (1974), over short distance scales, like those of Cahn and Kaler (1971) and Daub (1982).

The white dwarf and the PN formation rate derived from the Cudworth (1974) or the Mallik and Peimbert (1988) distance scales are in good agreement but are smaller than the formation rates derived from Miras or OH/IR stars. The determination of distances was discussed by Lutz (1988) and Terzian (1988) and to advance in this problem we need more distance determinations like those that could be derived from second epoch studies of the Lick survey or of radio data and those derived from spectroscopic studies of the central stars. I also consider that the accuracy in the estimation of the Mira and OH/IR star formation rates should be improved.

4. CONSTRAINTS ON STELLAR EVOLUTION MODELS

Observations of PN provide strong constraints on the late stages of stellar evolution models.

The observed enrichment of He, C and N in the PN envelopes of the solar vicinity, the halo, the SMC and the LMC have to be explained by stellar evolution models. Modest agreement between the observations and the predictions by Renzini and Voli (1981) has been obtained (*e.g.* Aller 1983; Kaler 1983; Peimbert 1985; Perinoto 1987; and references in these papers). The computations by Renzini and Voli have been made for two initial compositions and considering two parameters: $\alpha =$ l/H, the ratio of the mixing length to the pressure scale height, and η which multiplied by the Reimers' rate (1975) gives the mass loss rate during the AGB phase. A new grid of models is needed for a larger range of α and η values and of chemical compositions.

Dufour (1988) finds an inverse correlation between O/H and stellar mass; the O/H and O/N ratios decrease in the sequence Type II PN → Type I PN → NGC 6164-5 → η Car. Apparently this is due to an increase in the O to N conversion with mass in the 1 to 40 $M_\odot$ range.

The observed increase of the C/O ratio in layers closer to the stellar center, see § 2.3.2, has to be explained by the models. In this respect, observations and predictions of C^{12}/C^{13} ratios as a function of distance to the PN center would be very valuable (see the discussion by Knapp 1988).

NGC 6302 is O rich, it shows OH and HCN molecules (*e.g.* Rodríguez 1988), and it is probably one of the most massive PN known. Maybe this is a case of a hydrodynamical ejection of the whole envelope that would correspond to an object with $3 \leq M/M_\odot \leq 8$ (Renzini 1988).

PN envelopes are formed by stellar mass loss in the form of winds, just after the AGB phase a hot wind is emitted that compresses the material previously ejected by the ordinary wind (*e.g.* Kwok *et al.* 1978; Kwok 1982, 1983, 1988; Kahn 1983, 1988). The formation of toroids and bipolar structures around objects like NGC 2346, NGC 2440 and NGC 6302 has to be explained by the stellar evolution process. To explain these type of configurations different mass loss stages have been proposed from PNN in binary systems: (*e.g.* Morris 1981; Livio 1982; Renzini 1983; Iben and Tutukov 1988) and from individual PNN with high rotational velocities (Heap 1982; Calvet and Peimbert 1983).

Another important constraint on stellar evolution models is the observed fraction of PNN in binary systems. Bond (1988)and Méndez (1988) have discussed the estimates of the fraction of PNN in close and wide binaries respectively. Alternatively Iben and Tutukov (1988) have estimated the theoretical fraction of close binaries as well as the expected morphology and chemical composition of the envelopes.

The average central star mass derived from the sample by Kudritzky and Méndez (1988) is 0.69 $M_\odot$ and it is higher than those derived from the samples by Aller *et al.* (1987), 0.60 $M_\odot$, and Monk *et al.* (1988), 0.58 $M_\odot$. It is not clear why the difference is so large. Pottasch (1988), Tylenda (1988) and Barlow (1988) have also discussed the masses of PNN.

Sizes for the nebulae and magnitudes for the central stars of PN in the Magellanic Clouds from the Space Telescope will provide strong observational constraints to stellar evolution models (*e.g.* Barlow 1988).

5. DUST PRODUCTION

It has been suggested that probably red giant stars are the main producers of interstellar dust in the Galaxy. Several authors have found that the M_{dust}/M_{gas} ratio decreases by two to three orders of magnitude with the size of PN (Nata and Panagia 1981; Pottasch et al. 1984; Pottasch 1987; Lenzuni et al. 1987), result that contradicts the idea that PN are an important source of dust for the ISM. Lenzuni (1987) argues that the bulk of ISM dust grains originates from the winds of carbon stars and AGB stars before they may produce a PN.

Alternatively, Mallik and Peimbert (1988) from a distance scale independent sample of PN and considering the effect that the filling factor has in the gaseous mass determination find that M_{dust}/M_{gas} does not vary substantially with the size of PN and that the average M_{dust}/M_{gas} value is equal to 6×10^{-3}. This value is similar to the value of 7×10^{-3} estimated by Savage and Mathis (1979) for diffuse clouds in the Galaxy.

It is well known that Fe, Si, Mg, and Al are depleted in gaseous envelopes of PN (e.g. Shields 1978, 1983; Pwa et al. 1986). Pwa et al. compute a mass fraction of 3.5×10^{-3} for these elements if the abundances were solar. From the assumption that about 15% of the oxygen is locked in silicate cores and about 5% ± 5% could be locked in polymer mantles (Meyer 1985) and if most of Fe, Si, Mg and Al were in dust grains we would expect a total M_{dust}/M_{gas} of about 5×10^{-3} for solar abundances in excellent agreement with the value derived by Mallik and Peimbert (1988). Therefore if most of the mass ejected by intermediate mass stars becomes part of the PN then according to Mallik and Peimbert, it is still possible that most of the dust grains originate in PN.

6. MODELS OF GALACTIC CHEMICAL EVOLUTION

Based on the determination of the chemical composition of PN in: the solar vicinity, the halo of the Galaxy and other galaxies, it has been found that PN are enriching the ISM with He, C and N (e.g. Kaler 1979; Peimbert and Serrano 1980; Torres-Peimbert et al. 1980; Aller and Czyzak 1983; Clegg 1988; Barlow 1988; and references in these articles). The observed values can be used as representative of the contribution of intermediate mass stars, those with $1 \leq M/M_\odot \leq 8$, to the enrichment of the ISM.

Tinsley (1978) considered that the solar C/O ratio and estimates of nucleosynthesis in massive stars are consistent with a major source of C arising from PN precursors. Sarmiento and Peimbert (1985), Mallik and Mallik (1985) and Matteucci (1986) found that most of the C in the ISM comes from intermediate mass stars.

Based on models of galactic chemical evolution an open discussion on the primary or secondary origin of N and on the relative importance of different stellar mass ranges in the N production is present in the literature (e.g. Edmunds and Pagel 1978; Peimbert and Serrano 1980; Serrano and Peimbert 1983; White and Audouze 1983; Matteucci 1986; Diaz and Tosi 1986; Forieri 1986; Pagel 1986, 1987a; Peimbert 1987). Peimbert (1987) finds that the N enrichment produced by PN of Types II and III is smaller in comparison to that produced by Type I PN, he also finds that the He enrichment produced by PN of Types II and III of the solar neighborhood is negligible in comparison to that produced by Type I PN.

The high $\Delta Y/\Delta Z$ values observed in the ISM are not easily explained from stellar evolution models and the observed IMF (see Peimbert 1986 and references therein). A very high relative contribution to the enrichment of the ISM by He-rich objects, like Type I PN and the Crab nebula (e.g. Davidson and Fesen 1985; Henry 1986; and references therein), is needed to be able to explain the observed $\Delta Y/\Delta Z$ values.

7. CHEMICAL COMPOSITION OF THE HALO

The four known Type IV PN permit to discuss the chemical composition of the ISM at the time the stars in the halo were formed (see the discussion by Clegg 1988 and references therein).

Dinerstein (1987, see also Garnett and Dinerstein 1988) suggested that NGC 2242 might be a fifth halo PN. Torres-Peimbert *et al.* (1988) found that NGC 2242 and NGC 4361 have very similar properties: a) very high degree of ionization, b) very optically thin, they only trap about 4% of the Lyman continuum photons, and c) very similar surface brightnesses and electron densities. The abundances derived by Torres-Peimbert *et al.* are presented in Table 1.

TABLE 1. Total Abundances, 12 + log N(X)/N(H)

	NGC 2242	NGC 4361	Orion	Sun
He	10.99	11.00	11.01	...
O	8.05	7.94	8.65	8.92
Ne	7.39	7.25	7.80	...
Ar	6.19	6.07	6.65	...

Other parameters for NGC 4361 are: distance from the plane 830 pc, mass of the central star 0.55 $M_\odot$, and ionized mass of the envelope, that corresponds to the total mass because the object is optically thin, 0.074 $M_\odot$; the distance from the plane for NGC 2242 is 870 pc (Kudritzky and Méndez 1988; Méndez *et al.* 1988; Torres-Peimbert *et al.* 1988). The two extremely low mass values for the central star and the envelope of NGC 4361 together with its distance from the plane, indicate that it is a population II object. Due to their location relative to the plane of the Galaxy NGC 2242 and NGC 4361 are not as extreme population II objects as the other four halo PN, on the other hand their He/H, O/H and Ne/H ratios are similar to those of DDDM-1 and BB-1. From the values of Table 1 and the previous discussion, it follows that these PN should be classified as population II objects, either of Type III or of Type IV.

The central stars of NGC 4361 and NGC 1360 are very similar (Kudritzky and Méndez 1988; Méndez *et al.* 1988; Mc Carthy 1988). Therefore NGC 1360 might also be a Type IV PN; this object is located 510 pc from the plane of the Galaxy and there are no abundance determinations for the gaseous envelope.

8. GRADIENTS IN THE GALAXY AND M31

PN present several advantages over H II regions for the study of chemical abundance gradients across the disc of the Galaxy: a) we have a considerably larger sample available, b) since PN show a larger scatter in the direction perpendicular to the plane, it is possible to observe them in the optical domain over a larger range of galactocentric distances, c) in general small PN have higher surface brightnesses than galactic H II regions observable in the optical domain; the disadvantages are: a) the initial He, C, N, and O could have been modified by the evolution of their parent stars, b) they might show non circular orbits, and c) there is a spread in the ages of the stellar progenitors.

Maciel and Faúndez-Abans (1985) have derived a well defined electron temperature gradient from Type II galactic PN given by $\sim 600 \pm 120$ K kpc^{-1}; this value is about 1.5 times larger than that derived from galactic H II regions, the difference could be real since there are many different factors affecting each gradient. Faúndez-Abans and Maciel (1986, 1987a) have derived abundance gradients for He, C, N, O, Ne, S, Cl and Ar relative to H; their results are similar to those derived for H II regions and imply that most of the PN in their sample show orbits of relative low excentricity and that their age spread is small compared with the age of the disc of the Galaxy.

By dividing Type II PN into two subtypes Type IIa with log (N/H) + 12 ≥ 8.0 and Type IIb with log (N/H) + 12 < 8.0 Faúndez-Abans and Maciel (1987b) were able to obtain a much better correlation for the Type IIb N/H gradient. This is probably due to two reasons: a) the smaller the N/H value the smaller the N contamination due to stellar evolution, and b) the N contamination might be due to two different processes, to the first dredge up that converts initial C into N and to a second source of N which could be due to O or to freshly made C, the second N source might not be correlated with the initial CNO abundances but with the mass or with the binary nature of the nucleus of the PN.

Ford (1983) determined abundance gradients of O/H and N/H for M31 that are considerably flatter than those for the Galaxy. Jacoby and Ford (1986) from the study of the chemical composition of three PN and an H II region in M31 found that M31 experienced considerably more enrichment of its interstellar medium during its early life than did the Galaxy, they also find that there is little evidence for a correlation between kinematics and metallicity in the halo of M31. More abundance determinations of PN in M31 should be carried out to advance in the discussion of the chemical evolution of M31.

9. PREGALACTIC HELIUM ABUNDANCE

A review on the relevance for cosmology and particle physics of the pregalactic or primordial helium abundance, Yp, has been given by Boesgaard and Steigman (1985). Clegg (1988) has discussed the derivation of Yp and from the four halo PN obtains that Yp = 0.23 ± 0.03. It should be mentioned that the observations of three of the four halo PN have not been corrected for the nonlinearity of the detector, the effect is very small, of the order of 0.01 in Y, but it would go in the direction of increasing Yp.

It would be very important to determine the C abundance of NGC 2242 and NGC 4361 to estimate the amount of He due to stellar evolution and to be able to determine Yp. It would also be very important to compute stellar evolution models for metal poor low mass stars to be able to predict the He enrichment due to stellar evolution. Peimbert (1983) and Clegg (1988) based their He enrichment estimates on models and observations of objects with higher masses and higher heavy element content than those of Type IV PN.

Maciel (1988a,1988b) has estimated Yp by combining Type IIb PN with galactic H II regions observed by Shaver et al. (1983) and with O poor extragalactic H II regions from the samples by Kunth and Sargent (1983), Pagel et al. (1986) and Pagel (1987b). He finds that Yp = 0.234 ± 0.004 (1σ) and that Δ Y/Δ Z = 3.5 ± 0.3 (1σ). The Yp value derive by Maciel depends mainly on the O poor H II regions, while the Δ Y/Δ Z ratio depends mainly on the PN. With respect to the PN there are three corrections that should be made to the results by Maciel: a) some of the observations should be corrected for the non linearity of the detectors which will go in the direction of increasing Δ Y/Δ Z, b) the collisional excitation from the 2^3S He^0 level (see § 2.3.1) should be taken into account, and c) even if the abundances of Type IIb PN are not considerably affected by stellar evolution, all theoretical models predict at least some stellar He enrichment, this effect will go in the direction of decreasing Δ Y/Δ Z. These three effects are small and will affect the Δ Y/Δ Z ratio by less than a factor of ~ 1.3.

10. STELLAR DYNAMICS OF THE GALAXY AND OTHER GALAXIES

PN have been used to study the stellar dynamics of the Galaxy and other galaxies. From the velocity and distance data for 252 PN at galactocentric distances from 4 to 19 kpc, Schneider and Terzian (1983) derived the galactic rotation curve. They find a rising rotation curve with increasing galactocentric distances similar to those found in most Sb and Sc galaxies (Rubin et al. 1980, 1982).

Masses, mass distributions and velocity dispersions have been obtained for galaxies as far away as a few megaparsecs. Barlow (1988) reviews the results for the SMC and Ford et al. (1988)

for other extragalactic systems.

11. THE EXTRAGALACTIC DISTANCE SCALE

Ford et al. (1988) have given us an excellent account on the use of PN as standard candles. Objects used as standard candles should have approximately the same absolute flux, therefore we should analyze if the brightest PN in different galaxies are expected to have similar absolute fluxes. To a first approximation a PN emission line flux will depend on two quantities: on the number of Lyman continuum photons produced by the central star and on the ability of the nebula to trap these photons, i.e. if the nebula is optically thick or optically thin to Lyman continuum radiation.

The distance scales by Cudworth (1974) for optically thin and optically thick objects indicate that the brightest PN in the solar vicinity are optically thin which would make them poor standard candles because the absolute line fluxes would be a function not only of the central star properties but of the gas distribution in the nebula as well. Alternatively Gathier (1987) and Mallik and Peimbert (1988) have found that the brightest solar vicinity PN are optically thick and that the overwhelming majority of solar vicinity PN are optically thick contrary to the results by Cudworth and others. These results indicate that the emission line fluxes of the brightest PN depend directly on the properties of the central stars and not on the fraction of photons absorbed by the nebular shells.

Another property that makes good standard candles of PN is that the interstellar extinction can be easily determined from the Balmer line decrement.

It is a pleasure to acknowledge fruitful discussions with many of the participants to the Symposium, in particular with D.C.V. Mallik and S. Torres-Peimbert.

REFERENCES

Aller, L. H. 1983, in D. R. Flower (ed.), *Planetary Nebulae, IAU Symp. No. 103*, Dordrecht: Reidel, p.1.
Aller, L. H. and Czyzak, S. J. 1983, *Astrophys. J. Suppl.* **51**, 211.
Aller, L. H., Keyes, C. D., Maran, S. P., Gull, T. R., Michalitsianos, A. G., and Stecher, T. P. 1987, *Astrophys. J.* **320**, 159.
Barker, T. 1982, *Astrophys. J.* **253**, 167.
Barker, T. 1983, *Astrophys. J.* **267**, 630.
Barker, T. 1984, *Astrophys. J.* **284**, 589.
Barker, T. 1985, *Astrophys. J.* **294**, 193.
Barker, T. 1986, *Astrophys. J.* **308**, 314.
Barlow, M. J. 1988, this volume.
Benvenuti, P., Perinotto, M., and Willis, A. J. 1982, in C. de Loore and A. J. Willis (eds.), *Wolf-Rayet Stars, IAU Symp. No. 99*, Dordrecht: Reidel, p. 453.
Berrington, K., Burke, P. G., Freitas, L. C. G., and Kingston, A. E. 1985, *J. Phys. B* **18**, 4135.
Berrington, K. and Kingston, A. E. 1987, *J. Phys. B* **20**, 6631.
Bianchi, L. and Grewing, M. 1987, *Astron. Astrophys.* **181**, 85.
Boesgaard, A. and Steigman, G. 1985, *Ann. Rev. Astron. Astrophys.* **23**, 319.
Bond, H. E. 1988, this volume.
Butler, S. E., Bender, C. F., and Dalgarno, A. 1979, *Astrophys. J.* **230**, L59.
Cahn, J. H. and Kaler, J. B. 1971, *Astrophys. J. Suppl.* **22**, 319.
Calvet, N. and Peimbert, M. 1983, *Rev. Mexicana Astron. Astrofis.* **5**, 319.
Clegg, R. E. S. 1987, *Mon. Not. R. astron. Soc.* **229**, 31p.
Clegg, R. E. S. 1988, this volume.
Clegg, R. E. S. and Harrington, J. P. 1988, abstract of contributed paper, this volume.
Clegg, R. E. S., Harrington, J. P., Barlow, M.J., and Walsh, J. R. 1987, *Astrophys. J.* **314**, 551.
Clegg, R. E. S., Seaton, M. J., Peimbert, M., and Torres-Peimbert, S. 1983, *Mon. Not. R. astron. Soc.*

205, 417.
Cox, D. P.and Daltabuit, E. 1971, *Astrophys. J.* **167**, 257.
Cudworth, K.M. 1974, *Astron. J.* **79**, 1384.
Daub, C. T. 1982, *Astrophys. J.* **260**, 612.
Davidson, K. and Fesen, R. A. 1985, *Ann. Rev. Astron. Astrophys.* **23**, 119.
Diaz,A. I. and Tosi, M. 1986, *Astron. Astrophys.* **158**, 60.
Dinerstein, H. L. 1987, comment made during the discussion of this Symposium.
Dufour,R. J. 1988, abstract of contributed paper, this volume.
Edmunds, M. G. and Pagel, B.E.J. 1978, *Mon. Not. R. astron. Soc.* **185**, 77p.
Faúndez-Abans, M. and Maciel, W. J. 1986, *Astron. Astrophys.* **158**, 228.
Faúndez-Abans, M. and Maciel, W. J. 1987a, *Astrophys. Space Sci.* **129**, 353.
Faúndez-Abans, M. and Maciel, W. J. 1987b, *Astron. Astrophys.* **183**, 324.
Ferland, G. J. 1986, *Astrophys. J.* **310**, L67.
Flower, D.R., ed., 1983, *Planetary Nebulae, IAU Symp. No. 103*, Dordrecht: Reidel.
Ford, H.C. 1983, in D.R. Flower (ed.), *Planetary Nebulae, IAU Symp. No. 103*, Dordrecht: Reidel, p 443.
Ford, H.C., Ciardullo, R., Jacoby, G.H., and Hui,X. 1988, this volume.
Forieri, C. 1986, in C. Chiosi and A. Renzini (eds.), *Spectral Evolution of Galaxies*, Dordrecht: Reidel, p. 473.
French, H. B. 1983, *Astrophys. J.* **273**, 214.
Garnett, D. R. and Dinerstein, H. L. 1988, *Astron. J.* **95**, 119.
Gathier, R.1987, *Astron. Astrophys. Suppl.* **71**, 245.
Harrington, J. P. 1988, this volume.
Hawley, S. A. 1978, *Publ. Astron. Soc. Pacific* **90**, 370.
Hawley, S. A. and Miller, J. S. 1977, *Astrophys. J.* **212**, 94.
Hawley, S. A. and Miller, J. S. 1978, *Publ. Astron. Soc. Pacific* **90**, 39.
Hazard, C., Terlevich, R., Morton, D. C., Sargent, W. L. W., and Ferland, G. 1980, *Nature* **285**, 463.
Heap, S. R. 1975, *Astrophys. J.* **196**, 195.
Heap, S. R. 1982, in C. W. H. de Loore and A. J. Willis (eds.), *Wolf-Rayet Stars, Observations, Physics and Evolution*, Dordrecht: Reidel, p. 423.
Henry, R. B. C. 1986, *Publ. Astron. Soc. Pacific* **98**, 1044.
Husfeld, D. 1986, Ph. D. thesis, Universität München.
Iben, I., Jr. and Tutukov, A. V. 1988, this volume.
Jacoby, G. H.1979, *Publ. Astron. Soc. Pacific* **91**, 754.
Jacoby, G. H. and Ford, H. C. 1983, *Astrophys. J.* **266**, 298.
Jacoby, G. H. and Ford, H. C. 1986, *Astrophys. J.* **304**, 490.
Kahn, F. 1983, in D. R. Flower (ed.), *Planetary Nebulae, IAU Symp. No. 103*, Dordrecht: Reidel, p. 305.
Kahn, F. 1988, this volume.
Kaler, J. B. 1979, *Astrophys. J.* **228**, 163.
Kaler, J. B. 1983, in D. R. Flower (ed.), *Planetary Nebulae, IAU Symp. No. 103*, Dordrecht: Reidel, p. 245.
Kaler, J. B. 1986, *Astrophys. J.* **308**, 337.
Knapp, G. 1988, this volume.
Kudritzki, R. P. and Méndez, R. H. 1988, this volume.
Kunth, D. and Sargent, W. L. W. 1983, *Astrophys. J.* **273**, 81.
Kwok, S. 1982, *Astrophys. J.* **258**, 280.
Kwok, S. 1983, in D. R. Flower (ed.), *Planetary Nebulae, IAU Symp. No. 103*, Dordrecht: Reidel, p. 293.
Kwok, S. 1988, this volume.
Kwok, S., Purton, C. R., and Fitzgerald, P. M. 1978, *Astrophys. J.* **219**, L125.

Lenzuni, P. 1987, in A. Preite-Martinez (ed.), *Planetary and Protoplanetary Nebulae: from IRAS to ISO*, Dordrecht: Reidel, p. 254.
Lenzuni, P., Natta, A., and Panagia, N. 1987, in A. Preite-Martinez (ed.), *Planetary and Proto- planetary Nebulae: from IRAS to ISO*, Dordrecht: Reidel, p. 249.
Livio, M. 1982, *Astron. Astrophys.* **105**, 37.
Llebaria, A., Nieto, J. L., and di Serego Alighieri, S. 1986, *Astron. Astrophys.* **168**, 389.
Lutz, J. 1988, this volume.
Maciel, W. J. 1988a, abstract of contributed paper, this volume.
Maciel, W. J. 1988b, *Astron. Astrophys.*, in press.
Maciel, W. J. and Faúndez-Abans, M. 1985, *Astron. Astrophys.* **149**, 365.
Mallik, D. C. V. and Mallik, S. V. 1985, *J. Astrophys. Astron.* **6**, 113.
Mallik, D. C. V. and Peimbert, M. 1988, *Rev. Mexicana Astron. Astrofis.*, submitted.
Matteucci, F. 1986, *Publ. Astron. Soc. Pacific* **98**, 973.
Mc Carthy, J. K. 1988, private communication.
Méndez, R. H. 1988, this volume.
Méndez, R. H., Kudritzki, R. P., Herrero, A., Husfeld, D., and Groth, H. G. 1988, *Astron. Astrophys.* **190**, 113.
Méndez, R. H., Miguel, C. H., Heber, U., and Kudritzki, R. P. 1986, in K. Hunger, D. Schönberner, and N. Kameswara (eds.), *Hydrogen Deficient Stars and Related Objects, IAU Coll. 87*, Dordrecht: Reidel p. 323.
Meyer, J. P. 1985, *Astrophys. J. Suppl.* **57**, 151.
Monk, D. J., Barlow, M. J., and Clegg, R. E. S. 1988, abstract of contributed paper, this volume.
Morris, M. 1981, *Astrophys. J.* **249**, 572.
Natta, A. and Panagia, N. 1981, *Astrophys. J.* **248**, 189.
Pagel, B. E. J. 1986, in E. Vangioni-Flan et al. (eds.), *Advances in Nuclear Astrophysics*, France: èditions Frontiéres, p. 53.
Pagel, B. E. J. 1987a, in G. Gilmore and B. Carswell (eds.), *The Galaxy*, Dordrecht: Reidel, p. 341.
Pagel, B. E. J. 1987b, unpublished.
Pagel, B. E. J., Terlevich, R. J., and Melnick, J. 1986, *Publ. Astron. Soc. Pacific* **98**, 1005.
Peimbert, M. 1983, in P. Shaver and D. Kunth (eds.), *Primordial Helium*, Garching: ESO, p. 267.
Peimbert, M. 1985, *Rev. Mexicana Astron. Astrofis.* **10**, 125.
Peimbert, M. 1986, *Publ. Astron. Soc. Pacific* **98**, 1057.
Peimbert, M. 1987, in A. Preite-Martinez (ed.), *Planetary and Protoplanetary Nebulae: from IRAS to ISO*, Dordrecht: Reidel, p. 91.
Peimbert, M. and Serrano, A. 1980, *Rev. Mexicana Astron. Astrofis.* **5**, 9.
Peimbert, M. and Torres-Peimbert, S. 1971, *Astrophys. J.* **168**, 413.
Peimbert, M. and Torres-Peimbert, S. 1987a, *Rev. Mexicana Astron. Astrofis.* **14**, 540.
Peimbert, M. and Torres-Peimbert, S. 1987b, *Rev. Mexicana Astron. Astrofis.* **15**, 117.
Pequignot, D. 1980, *Astron. Astrophys.* **81**, 356.
Pequignot, D., Aldrovandi, S. M. V., and Stasińska, G. 1978, *Astron. Astrophys.* **63**, 313.
Perinotto, M. 1987, in A. Preite-Martinez (ed.), *Planetary and Protoplanetary Nebulae: from IRAS to ISO*, Dordrecht: Reidel, p. 13.
Phillips, P. 1988, this volume.
Pottasch, S. R. 1987, in S. Kwok and S. R. Pottasch (eds.), *Late Stages of Stellar Evolution*, Dordrecht: Reidel, p. 355.
Pottasch, S. R. 1988, this volume.
Pottasch, S. R. et al. 1984, *Astron. Astrophys.* **138**, 10.
Pwa, R., Pottasch, S. R., and Mo, J. E. 1986, *Astron. Astrophys.* **164**, 184.
Reimers, D. 1975, *Mem. Soc. Roy. Sci. Liege, 6e Ser.* **8**, 369.
Renzini, A. 1983, in D. R. Flower (ed.), *Planetary Nebulae, IAU Symp. No. 103*, Dordrecht: Reidel, p. 267.

Renzini, A. 1988, this volume.
Renzini, A. and Voli, M. 1981, *Astron. Astrophys.* **94**, 175.
Rodríguez, L.F. 1988, this volume.
Rosa, M. 1985, *ESO Messenger* **39**, 15.
Rubin, V. C., Ford, W. K., and Thonnard, N. 1980, *Astrophys. J.* **238**, 471.
Rubin, V. C., Ford, W. K., Thonnard, N., and Burstein, D. 1982, *Astrophys. J.* **261**, 439.
Saraph, H.E. and Seaton, M.J. 1970, *Mon. Not. R. astron. Soc.* **148**, 367.
Sarmiento, A. and Peimbert, M. 1985, *Rev. Mexicana Astron. Astrofis.* **11**, 73.
Savage, B. D. and Mathis, J. S. 1979, *Ann. Rev. Astron. Astrophys.* **17**, 73.
Schneider, S. E. and Terzian, Y. 1983, *Astrophys. J.* **274**, L61.
Seaton, M. J. and Osterbrock, D. E. 1957, *Astrophys. J.* **125**, 66.
Serrano, A. and Peimbert, M. 1983, *Rev. Mexicana Astron. Astrofis.* **8**, 117.
Shaver, P. A., Mc Gee, R. X., Newton, L. M., Danks, A. C., and Pottasch, S. R. 1983, *Mon. Not. R. astron. Soc.* **204**, 53.
Shields, G. A. 1978, *Astrophys. J.* **219**, 565.
Shields, G. A. 1983, in D. R. Flower (ed.), *Planetary Nebulae, IAU Symp. No. 103*, Dordrecht: Reidel, p. 259.
Stanghellini, L. and Kaler, J. B. 1988, abstract of contributed paper, this volume.
Terzian, Y. 1988, this volume.
Tinsley, B.M. 1978, in Y. Terzian (ed.), *Planetary Nebulae, Observations and Theory, IAU Symp. No. 76*, Dordrecht: Reidel, p. 341.
Torres-Peimbert, S., Peimbert, M., and Daltabuit, E. 1980, *Astrophys. J.* **238**, 133.
Torres-Peimbert, S., Peimbert, M., and Peña, M. 1988, *Astron. J.*, submitted.
Tylenda, R. 1988, this volume.
White, S. D. and Audouze, J. 1983, *Mon. Not. R. astron. Soc.* **203**, 603.

AUTHOR INDEX

Acker, A.	39, 52, 309	Desai, J. N.	187, 200
Aitken, D. K.	203	Diesch, Chr.	182
Aller, L. H.	567, 306, 219	Dinerstein, H. L.	206, 214
Anandarao, B. G.	187, 189, 200	Dopita, M. A.	208, 352, 356
Andronov, I. L.	451	Dufour, R. J.	216
Apparao, K.M.V.	304	Echevarría, J.	302
Arévalo, V. M.	49	Eder, J.	448
Arkhipova, V. P.	50, 54, 57	Engels, D.	450
Aspin, C.	178	Escalante, V.	225
Atherton, P. D.	207	Esteban, C.	63
Azzopardi, M.	351	Falcón, L. H.	179
Balick, B.	83, 173, 181	Falomo, R.	444
Baluteau, J.-P.	59	Faulkner, D. J.	189
Banerjee, D. P. K.	187, 189	Faúndez-Abans, M.	222
Barlow, M. J.	197, 319, 354, 355	Feast, M. W.	167
Barnstedt, J.	182	Feibelman, W. A.	313, 456
Basart, J. P.	176	Feklistova, T. H.	218
Bässgen, G.	199, 227, 318	Ferrari-Toniolo, M.	57
Bässgen, M.	199, 227, 318	Fierro, J.	449
Bentley, A. F.	312	Fleming, T. A.	310
Bianchi, L.	182, 199, 307, 314, 318, 447	Ford, H.C.	335, 352, 357
Bignell, C.	60, 63, 210	Gathier, R.	302
Bond, H. E.	251, 310, 461	Geballe, T. R.	178, 445, 446
Butler, K.	317	Goodrich, R. W.	447
Campos, J.	169	Grauer, A. D.	310
Capellaro, E.	61, 172	Greenhouse, M. A.	170
Carr, J. S.	206	Grewing, M.	182, 199, 227, 241, 307, 314, 318
Cerrato, S.	199, 318		
Chatterjee, T. K.	169	Groth, H. G.	168
Chavarría-K., C.	49	Gruenwald, R. B.	224
Chu, Y.-H.	105, 183, 198	Guilloteau, S.	453
Ciardullo, R.	310, 335, 357	Gutiérrez-Moreno, A.	51, 53
Clegg, R. E. S.	139, 195, 197, 201, 204, 211, 223, 354, 355, 443	Habing, H. J.	359, 445, 446
		Hachisu, I.	541
Cortés, G.	51, 53	Harpaz, A.	454
Costero, R.	302	Harrington, J. P.	157, 201, 211
Cox, A. N.	311	Harvey, P. M.	206
Cristiani, S.	191	Hawkins, G.	186
Daub, C. T.	176	Hayward, T. L.	170

Healy, A. P.	205	Manchado, A.	63, 184, 194, 196, 220
Heap, S.	308	McCall, M.	228
Heathcote, S. R.	180, 192	McCarthy, J. K.	316
Henrichs, H.	456	McLean, I. S.	178
Henry, R. B. C.	305	Meakes, M.	461
Herrero, A.	168, 317	Meatheringham, S. J.	189, 208, 352, 356
Hoare, M. G.	202, 443	Méndez, R. H.	168, 261, 273, 317
Hoey, M. J.	177, 193	Mendoza, E. E.	49
Huggins, P. J.	205	Meyssonnier, N.	351
Hui, X.	335	Middlemass, D.	195, 217
Husfeld, D.	168	Monk, D. J.	197, 201, 354, 355
Hutsemékers, D.	317	Moreno, H.	51, 53
Iben, Jr., I.	505	Moreno, M. A.	171, 185
Icke, V.	181	Morgan, D. H.	356
Igumenshchev, I. V.	459	Moskalenko, E. I.	50
Jacoby, G. H.	183, 198, 303, 335, 357	Nikitin, A. A.	218
Jain, S. K.	187	Noriega-Crespo, A.	228
Jasniewicz, G.	309	Noskova, R. I.	57
Jiying, L.	62	O'Dell, C. R.	458
Kahn, F. D.	411	Olling, R.	63
Kaler, J. B.	220, 229, 313, 456	Omont, A.	453
Kato, M.	457, 541	Ortolani, S.	191
Kazes, I.	209	Pascoli, G.	455
Keyes, C. D.	219	Payne, P. W.	208
Kholtygin, A. F.	218	Payne, H. E.	205
Kinman, T. D.	167	Peimbert, M.	212, 577
Kirkpatrick, R. C.	175	Peña, M.	58, 353, 449
Knapp, G. R.	381	Pequignot, D.	59, 224
Kohoutek, L.	29	Perinotto, M.	293
Kolesnik, I. G.	460	Persi, P.	57
Köppen, J.	52, 523	Phillips, J. P.	194, 425
Kostyakova, E. B.	55, 56	Phillips, J. A.	205
Kudashkina, L. S.	451	Pilyugin, L. S.	460
Kudritzki, R. P.	168, 273, 317	Pismis, P.	185
Kwitter, K. B.	303	Pottasch, S. R.	60, 63, 63, 174, 184, 196, 196, 200, 210, 216, 220, 301, 302, 481
Kwok, S.	401, 452	Preite-Martinez, A.	9, 57
Lame, N. J.	173, 462	Preston, H. L.	181
Lasker, B. S.	167	Ratag, M.	216
Leene, A.	174	Reay, N. K.	207, 301
Lequeux, J.	351	Recillas, E.	307
Lester, D. F.	206	Renzini, A.	391
Lewis, B. M.	448, 450	Roche, P. F.	117, 178, 203
Liebert, J.	545	Rodríguez, L. F.	129
Livio, M.	461	Rosado, M.	171
López, J. A.	179	Roth, Ma.	317
Lozinskaya, T. A.	50	Roth, Mi.	179
Lucas, R.	453	Rubin, R. H.	221
Lutz, J. H.	65, 173, 462	Rudnitskij, G. M.	451
Maciel, W. J.	73, 213, 222	Ruiz, M. T.	179, 192, 353
Mallik, D. C. V.	187, 493	Sabbadin, F.	61, 172, 191, 444
Mampaso, A.	63, 184, 194, 196, 220		

Sahu, K. C.	196, 200	Thronson, Jr., H. A.	170
Sapar, A. A.	218	Torres, A. V.	308
Saurer, W.	168	Torres-Peimbert, S.	1, 58, 212
Schmidt-Voigt, M.	543	Turatto, M.	61, 172
Schönberner, D.	463	Tutukov, A. V.	459, 505
Seitter, W. C.	315	Tylenda, R.	531, 542
Shaw, R. E.	313, 456, 473	Ukita, N.	204, 216
Shchelkanova, A. Yu.	215	van der Veen, W. E.	359, 445, 446
Shibata, K.	188, 190	Viegas-Aldrovandi, S. M.	226
Shipman, H. L.	305, 555	Volk, K.	452
Shustov, B. M.	459	Walsh, J. R.	195, 204, 223, 443
Sitnik, T. G.	50	Walton, N. A.	207, 301
Smith, C. H.	203	Webster, B. L.	208, 352, 356
Smith, M. G.	178	Weidemann, V.	540
Spoelstra, T.	301	Weinberger, R.	93, 168
Stanghellini, L.	220	Weller, W. G.	180, 192
Starrfield, S.	311	Werner, M. W.	214
Stasińska, G.	542, 542	Whelan, D.	177
Stenholm, B.	52	Wood, P. R.	189, 358
Storey, J.W.V.	208	Xingchun, F.	62
Surdej, J.	317	Yesipov, V. F.	54
Szczerba, R.	539	Yongwei, H.	62
Tamura, S.	188, 190, 209	Yudin, B. F.	54
Tarafdar, S. P.	304	Zijlstra, A.	60, 63, 210, 216
te Lintel, P.	210, 359	Zuckerman, B.	186
Terzian, Y.	17, 205, 448		

OBJECT INDEX

This index follows the notation by each author, therefore a given object might appear under several designations. English characters are given precedence over numbers. Objects in constellations are ordered by their variable star names, greek characters are ordered by constellations. Page numbers refer to the title page of the article in which the object is mentioned.

AA Dor	273		456, 473
AFGL 2513	453	A 80	303
AG Car	216		
AG +30 1906.0	39	BB- 1	139, 577
Ap 3- 1	29	BD +30 3639	9, 17, 39, 83, 117,
ARO 502	39		139, 174, 178, 203, 206,
A 2	105, 172		229, 313, 444, 567
A 5	303	BD −12 133	39
A 7	174, 545	BD −12 134	39
A 9	303	BN 0808+11	29
A 11	29	BN 0950+13	29
A 13	171	BPM 25114	555
A 14	261, 462, 567	B1 3- 6	29
A 18	303, 304	B1 3- 11	29
A 21	9, 174, 304, 531	B1 B	29
A 24	171, 261		
A 30	83, 105, 117, 139, 183, 241,	Campbell's star	39
	261, 293, 304, 473, 577	Cen A	335
A 31	174, 304, 531	CIT 6	381, 453
A 32	29	CPD−56 8032	381
A 33	261, 304	Cn 1- 1	261, 493
A 35	39, 174, 251, 261, 309,	CRL 618	17, 29, 93, 381
	314, 461, 493, 567	CRL 2343	381
A 36	308, 309, 315	CRL 2688	381
A 41	251, 461, 493, 505, 567	CTS- 1	61
A 43	83		
A 45	303	DDDM- 1	139, 157, 202, 215, 577
A 46	251, 461, 493	DeHt 1	29
A 56	567	DeHt 2	39
A 58	139, 315	DHW 5	171
A 59	303	DS Dra	251
A 63	251, 312, 461, 493, 504, 567	DS 1	251, 493
A 72	83	DS 2	29
A 77	29	Dumbbell nebula	39, 105
A 78	105, 139, 184, 185, 241, 293, 315,	DWH 2	309

593

EGB 5	261, 273, 481	HBDS 1	29
EL 0103+73	29	Hb 5	194, 241
EL 0419+72	29	Hb 12	206, 401
EL 1647+64	29	HD 112313	251
ER 8	555	HD 161796	381
ESO-097-03	29	HD 184738	39
ESO-166-PN21	192	HEFE 1	29
ESO-180-05	29	He 1- 3	29
ESO-182-04	29	He 1- 4	303
ESO-212-08	29	He 1- 5	105
ESO-280-02	29	He 2- 30	179
ESO-289-19	29	He 2- 36	93, 261, 462, 493
ESO-308-08	29	He 2- 61	51
ESO-314-12	29	He 2- 99	229, 313
ESO-328-04	29	He 2- 104	39
ESO-328-40	29	He 2- 108	261, 273
ESO-367-03	29	He 2- 111	93, 105, 241
ESO-369-01	29	He 2- 131	65, 182, 191, 261, 481
ESO-390-05	29	He 2- 133	29
ESO-391-02	29	He 2- 138	51, 53, 261, 273, 531
ESO-392-05	39	He 2- 151	51, 53, 261, 273
ESO-426-13	29	He 2- 162	261, 273
ESO-429-04	29	He 2- 182	261, 273
ESO-429-17	29	He 2- 274	182
ESO-450-16	29	He 2- 442	54
ESO-451-03	29	He 2- 467	57
ESO-456-64	29	Helix nebula	105
ESO-456-73	29	HFG 1	251, 310, 461
ESO-515-19	29	HH 43C	447
ESO-520-30	29	HH 43N	447
ESO-522-29	29	HR 4049	453
Eskimo nebula	105, 458	HtDe 2	29
		HtDe 3	29
FEGU 248- 5	29	HtDe 4	29
FX Ser	453	HtDe 5	29
		HtDe 7	29
GCRV 11983	39	HtDe 8	29
GD 40	555	HtDe 10	29
GD 140	555	HtDe 11	29
GD 229	555	HtDe 12	29
GK Per	251	HtDe 13	29
GL 618	447	HtTr 1	29
Grw +70 8247	555	HtTr 2	29
GR 0155+10	29	HtTr 3	29
Gr 488	555	HtTr 4	29
GW Vir	251, 545	HtTr 5	29
Gum 60	39	HtTr 6	29
G0.9+1.3	129	HtTr 7	29
G195-19	555	HtTr 8	29
G240-72	555	HtTr 9	29
G349.2-0.2	129	HtTr 10	29

HtTr 11	29	IRC +50096	453
HtTr 12	29		
HtTr 13	29	JnEr 1	29
HtTr 14	29	Jn 1	531
Hu 1- 2	55, 139, 172, 241	J 320	83
Hu 2- 1	293	J 900	9
H 2- 1	261, 273		
H 2- 20	29	KV Vel	251
H 2- 30	29	KY 9296	39
H 3- 29	172	K 1- 2	251, 461, 462, 493, 567
H 3- 75	462, 567	K 1- 14	229
H 4- 1	139	K 1- 16	1, 251, 311, 531, 555, 567
H 1504+65	545, 555, 567	K 2- 2	303, 531
		K 2- 5	303
IC 289	83, 105	K 2- 13	29
IC 418	9, 17, 83, 117, 129, 139, 157,	K 3- 2	17
	182, 191, 197, 199, 202, 212,	K 3- 62	17
	217, 241, 261, 273, 293, 307,	K 3- 72	172
	309, 444, 481, 567, 577	K 3- 82	168
IC 1295	105	K 4- 13	29
IC 1297	105, 191	K 4- 14	29
IC 1454	83	K 4- 18	29
IC 1747	83	K648	139, 293, 391, 493
IC 2120	39, 60, 172		
IC 2149	241, 293	LB 3459	273
IC 2165	139, 199, 241, 531, 567	Leo Group	335
IC 2448	182, 199, 241, 261, 273, 481	LMC	1, 273, 319, 335, 352,
IC 2501	199		353, 354, 355, 357
IC 3568	55, 83, 105, 241,	LMC J 1	319
	293, 307, 308	LMC J 2	319
IC 4593	83, 105, 189, 293, 308	LMC J 3	319
IC 4634	191	LMC J 4	319
IC 4637	261, 273	LMC J 5	319
IC 4642	182, 304	LMC J 6	319
IC 4686	182	LMC J 8	319
IC 4776	182	LMC J 9	319
IC 4997	17, 56, 129, 139,	LMC J 10	319
	209, 217, 567	LMC J 11	319
IC 5117	17, 444	LMC J 12	319
IC 5217	83, 93	LMC J 13	319
IN Com	251	LMC J 18	319
IRAS 17516-2525	446	LMC J 19	319
IRAS 17581-1744	453	LMC J 20	319
IRAS 18095+2704	401	LMC J 23	319
IRAS 1912+172P09	29, 261, 493	LMC J 26	319
IRAS 19454+2920	401	LMC J 28	319
IRC +00509	381	LMC J 29	319
IRC +10011	381	LMC J 30	319
IRC +10216	381, 401	LMC J 33	319
IRC +10420	359, 381	LMC J 36	319
IRC +30374	453	LMC J 38	319

LMC J 40	319	M 4- 1	29
LMC J 41	319	M 4- 11	303
LMC N 66	353	M 4- 18	229, 308
LMC N 201	319	M27	39
LMC P 25	319	M31	1, 39, 335, 351, 357, 425
LMC P 40	1	M32	39, 335, 357
LMC WS 35	353	M71	391
LoTr 5	39, 229, 309, 493	M81	39, 335
Lo 2	29	M87	335
LP 701- 29	555		
LSE 125	261, 273	NGC 40	1, 83, 139, 229, 241, 293,
LS II +30 04	39		307, 308, 312, 567
LSS 1362	261	NGC 127	39
LSS 2018	251, 505	NGC 185	39, 357
LT 5	251, 261	NGC 205	39, 335, 357
L 795- 7	555	NGC 246	39, 261, 273, 304, 312, 567
		NGC 436	304
MaC 2- 1	29	NGC 650- 1	83, 105, 172, 261
MaC 2- 2	29	NGC 1360	157, 174, 261, 273,
MaC 2- 3	29		304, 309, 577
MaC 2- 4	29	NGC 1514	83, 261, 304, 462, 493
Magellanic Clouds	73, 139, 157,	NGC 1535	1, 83, 105, 157, 182,
	356, 481		191, 199, 229, 241, 261,
MA 2	29		273, 293, 308, 481, 567
MA 3	29	NGC 1545	105
MA 13	29	NGC 2022	83, 105, 198
MT Ser	251, 505	NGC 2149	83
MyCn 26	29	NGC 2242	62, 577
Mz 3	93, 180, 241	NGC 2346	83, 129, 139, 187, 204,
M 1- 1	188, 241		205, 251, 261, 381, 461,
M 1- 2	261		493, 505, 577
M 1- 7	172	NGC 2359	216
M 1- 8	172	NGC 2371- 2	83, 293, 307, 308, 456
M 1- 9	17	NGC 2392	1, 17, 93, 105, 181, 193,
M 1- 16	17		194, 241, 261, 273, 304,
M 1- 18	172		308, 458, 481, 531, 567
M 1- 26	17, 261, 273	NGC 2438	105, 198
M 1- 28	303	NGC 2440	1, 39, 83, 139, 180, 182,
M 1- 67	261		207, 214, 217, 229, 241, 306,
M 1- 78	216		481, 493, 531, 567, 577
M 1- 79	168	NGC 2451	555
M 1- 91	93, 447	NGC 2452	83, 199, 241
M 1- 92	381	NGC 2474- 75	29, 39
M 2- 2	105, 567	NGC 2516	493
M 2- 3	567	NGC 2610	83, 105
M 2- 9	83, 93, 178, 182, 194, 241, 447	NGC 2792	199
M 2- 52	172	NGC 2818	139, 493
M 2- 55	172	NGC 2867	199, 241, 306
M 2- 56	447	NGC 2899	179, 241
M 3- 30	456	NGC 3031	335
M 3- 44	53	NGC 3115	335

NGC 3132	105, 139, 199, 241, 261, 462, 493	NGC 6741	306, 531, 567
		NGC 6751	105, 198, 241
NGC 3211	1, 199	NGC 6772	93, 105
NGC 3242	17, 83, 105, 181, 182, 200, 225, 229, 241, 261, 273, 293, 304, 481, 567, 577	NGC 6781	83, 105
		NGC 6790	17, 129, 312
		NGC 6803	212
NGC 3377	335	NGC 6804	105
NGC 3379	335	NGC 6818	304
NGC 3384	335	NGC 6824	493
NGC 3568	83	NGC 6826	1, 9, 83, 93, 105, 139, 181, 195, 196, 198, 241, 261, 293, 307, 312, 567
NGC 3587	83		
NGC 3918	1, 65, 93, 117, 139, 157, 182, 199, 201, 225, 567, 577		
		NGC 6852	303
NGC 4071	39	NGC 6853	9, 174, 261, 304, 312, 577
NGC 4361	105, 157, 241, 261, 273, 308, 567, 577	NGC 6857	105
		NGC 6888	216
NGC 4486	335	NGC 6891	9, 55, 105, 182, 198, 261, 273, 293, 306, 481
NGC 5102	335		
NGC 5128	39, 335	NGC 6894	83, 105, 567
NGC 5189	172, 293, 307	NGC 6905	83, 194, 304, 307, 312
NGC 5315	65, 191	NGC 7008	83, 312, 493
NGC 5866	335	NGC 7009	49, 83, 105, 139, 175, 178, 181, 182, 191, 212, 229, 261, 273, 293, 312, 481, 567, 577
NGC 5882	182, 261		
NGC 5979	65		
NGC 6026	105	NGC 7026	83, 172, 194, 241
NGC 6058	105, 308, 493	NGC 7027	9, 17, 39, 83, 117, 129, 139, 157, 174, 176, 178, 203, 212, 224, 225, 229, 301, 306, 381, 481, 531, 567, 577
NGC 6153	189		
NGC 6164- 5	216		
NGC 6210	17, 49, 83, 105, 198, 293, 308, 312		
		NGC 7048	83
NGC 6302	17, 39, 93, 117, 129, 180, 205, 214, 241, 381, 493, 531, 567, 577	NGC 7139	83
		NGC 7293	9, 83, 93, 105, 129, 157, 174, 204, 223, 229, 241, 261, 273, 241, 261, 273, 304, 312, 381, 545, 567, 577
NGC 6309	83, 105, 198		
NGC 6369	105, 169, 481		
NGC 6397	273, 567	NGC 7354	83, 105, 241
NGC 6445	83, 531, 567	NGC 7635	216
NGC 6537	93, 241, 531, 567	NGC 7662	83, 105, 117, 178, 181, 198, 212, 241, 251, 567, 577
NGC 6543	9, 55, 83, 93, 105, 139, 181, 186, 195, 196, 198, 241, 293, 307, 308, 312, 567		
		NML Cyg	359, 381
		NML Tau	359
NGC 6565	65, 481, 567		
NGC 6567	65	OH 16.1-0.3	359
NGC 6572	17, 55, 117, 212, 225, 241, 293, 312, 444, 567	OH 19.2-1.0	359
		OH 20.7+0.1	359
		OH 21.5+0.5	359
NGC 6590	17	OH 26.5+0.6	359
NGC 6629	182, 261, 273, 493	OH 26.6+0.6	359
NGC 6644	306	OH 30.1-0.2	359
NGC 6720	55, 83, 105, 129, 170, 186, 198, 223, 381, 567, 577	OH 30.1-0.7	359
		OH 30.7+0.4	359
NGC 6729	455		

OH 32.0–0.5	359	PK 206–40.1	182, 191
OH 32.8–0.3	359	PK 208+33.1	241
OH 39.7+1.5	359	PK 215–24.1	182, 191, 307
OH 44.8–2.3	359	PK 223– 2.1	60
OH 127.9–0.0	359	PK 234+ 2.1	182, 241
OH 231.8+4.2	359	PK 261+32.1	182, 241
Orion nebula	577	PK 277– 3.1	179, 241
Owl nebula	105	PK 285–14.1	182
		PK 307– 3.1	307
PB 8	261	PK 309– 4.2	191
Pe 2– 3	29	PK 315– 0.1	241
PG 0108+101	545	PK 315–13.1	182, 191
PG 0122+200	251	PK 327+10.1	182
PG 0950+139	545, 567	PK 329+ 2 1	261
PG 1015+014	555	PK 331– 1.1	241
PG 1031+234	555	PK 334– 9.1	182
PG 1159–035	1, 251, 311, 545, 555	PK 345– 8.1	182, 191
PHL 932	261	PK 349+ 1.1	241
PK 0+12.1	191	PK 358–21.1	191
PK 1– 6.2	182	PK 359– 0.1	241
PK 2–13.1	182	Pleiades	493
PK 9–51.1	182	PLX 4591	39
PK 10+ 0.1	241	PN 4– 2.1	39
PK 10+18.1	182	PN 45– 2.1	567
PK 10+18.2	241	PN 49+ 1.1	39
PK 27– 9.1	182	PN 60– 3.1	39
PK 35– 0.1	60	PN 64+ 5.1	39
PK 36– 1.1	241	PN 136+ 5.1	310
PK 36–57.1	241	PN 164+ 3.1	39
PK 37–34.1	182, 191	PN 254+ 5.1	39
PK 45– 2.1	443, 567	PN 271+16.1	39
PK 54–12.1	182	PN 284– 5.1	39
PK 60– 3 1	39	PN 298– 4.1	39
PK 61– 9.1	307	PN 299– 1.1	39
PK 81–14.1	241	PP 40	29
PK 83+12.1	241, 307	Pu– 1	303
PK 86– 8.1	241	Pu– 2	303
PK 89+ 0.1	241	PW– 1	531
PK 96+29.1	241, 307	PZ Cas	381, 451
PK 106–17.1	241		
PK 107+ 2.1	241	R Cas	451
PK 118+ 2.1	60	R CrB	453
PK 120+ 9.1	307	RCW 43	179
PK 123+34.1	307	RCW 124	39
PK 130–11.1	241	Ring nebula	105
PK 132– 0 1	60	ROB 162	273
PK 169– 0.1	60	R Peg	451
PK 176+ 0.1	60	RS Vul	505
PK 189+19.1	307	R Tau	451
PK 195– 0.1	60	RT Vir	451
PK 197+17.1	241	RWT 152	29

RX Boo	451	Sp 1	261
RY Dra	381	Sp 1- 1	39
RY Gem	505	SP 2- 14	29
		Sp 4- 1	29
Sand 3	308	St 1- 1	39
SAO 163075	381	Sun	577
SAO 181201	314	SwSt 1	182, 293
Saturn neb	105	S 22	29, 50
SAWI 1	29	S 188	303
SAWI 2	29		
SAWI 3	29	Tc 1	191, 261, 273
SAWI 4	29	Th 1- 1	29
SAWI 5	29	Th 3- 11	29
SAWI 6	29	Th 3- 34	29
SAWI 7	29		
Sa 1- 1	39	U Her	451
Sa 2- 180	39	U Ori	455
Sa 3- 119	29	U Sge	505
Sa 4- 1	29	UU Sge	251, 505
S Cep	453		
Sh 1- 3	39	VERA 90	29
Sh 1- 89	93	VERA 104	29
Sh 2- 6	39	Virgo cluster	335
Sh 2- 68	29	VV 47	39, 545
Sh 2- 71	241	VV 94	39
Sh 2- 71	194	VV' 168	39
Sh 2- 216	29	VW Pyx	251
Sirius B	304, 391, 555	VY CMa	381
SMC	1, 273, 319, 351, 354, 355, 357, 577	Vy 2- 2	17, 129, 205, 209, 210, 381, 401, 443, 444, 567
SMC J 2	319	V471 Tau	251
SMC J 4	319	V477 Lyr	251
SMC J 6	319	V605 Aql	315
SMC J 7	319	V651 Mon	251, 261, 505
SMC J 9	319	V1016 Cyg	29
SMC J 10	319		
SMC J 12	319	WeDe 1	29
SMC J 13	319	We- 2	303
SMC J 15	319	We- 5	303
SMC J 16	319	Wray 16- 25	39
SMC J 17	319	WS 1	171
SMC J 18	319		
SMC J 22	319	XY Pup	505
SMC J 24	319		
SMC JP 34	319	Y-C 34	29
SMC L 302	319	Y-C 37	29
SMC N 2	1, 319	Y-C 39	29
SMC N 5	1	Y-C 40	29
SMC SP 32	319	Y-C 41	29
Sm 2	29	Y-C 42	29
Sm 3	29	Y-C 43	29

Y-C 44	29	38– 0.1	29
Y-C 45	29	40+ 7.1	29
Y-C 46	29	41– 0.1	29
Y-C 47	29	43+ 1.1	29
YM 29	50	51+ 2.1	29
		59– 1.1	29
Z Cyg	451	75+ 5.1	29
ZWG 204.005	62	75+35.1	29
ZZ Cet	545, 555	94+38.1	29
		97+ 3.1	29
ς Pup	273, 293	99– 8.1	29
τ Sco	273	124+10.1	29
υ Sgr	453	128– 4.1	29
ω Cen	391	136+ 5.1	29
40 Eri B	555	137+16.1	29
89 Her	381, 453	138+ 4.1	29
		148–48.1	29
0+ 2.2	29	149– 9.1	29
0– 6.1	29	156+12.1	29
1– 0.1	29	158+ 0.1	29
1– 3.3	29	164+31.1	29
1– 3.4	29	166– 6.1	29
1– 3.5	29	173+ 2.1	29
1– 3.6	29	192+ 7.1	29
1– 3.7	29	196–12.1	29
1– 3.8	29	197– 6.1	29
1– 3.9	29	203–18.1	29
2+ 1.1	29	204–13.1	29
3– 3.1	29	204–16.1	29
3– 4.10	29	205–26.1	29
6+ 1.1	29	211+18.1	29
9– 6.1	29	211+22.1	29
9– 8.1	29	214+31.1	29
11+17.1	29	218–10.1	29
11–14.1	29	219+ 7.1	29
14–25.1	29	221+ 4.1	29
22+ 4.1	29	221+ 5.2	29
23+ 1.1	29	221+46.1	29
23+ 4.1	29	223+ 4.1	29
28– 4.2	29	227+33.1	29
30+ 6.1	29	231– 8.1	29
31– 0.2	29	235+ 4.1	29
34–10.1	29	239–18.1	29
35– 2.1	29	241– 7.1	29
36+20.1	29	242– 3.1	29
36+21.1	29	245– 3.1	29
36– 1.2	29	247– 4.1	29
36– 2.1	29	247–21.1	29
37– 2.1	29	248–12.1	29
37– 3.1	29	249–22.1	29
38+14.1	29	251– 4.1	29

265+ 5.1		29	336- 2.1		29
266+ 2.1		29	336- 8.1		29
271- 8.1		29	341+17.1		29
273+ 6.1		29	341-15.1		29
274- 0.1		29	343+16.1		29
284-39.1		29	343- 0.1		29
292- 3.1		29	345+10.1		29
294+4.1		182	346+19.1		29
299- 4.1		29	347+ 7.1		29
308- 1.1		29	349-10.1		29
309+ 6.1		29	351-10.2		29
310+ 2.1		29	353+ 8.1		29
315+59.1		29	353-55.1		29
321- 3.1		29	355- 0.1		29
324- 1.1		29	356- 0.1		29
327+14.1		29	358+ 1.4		29
329+12.1		29	358+ 2.1		29
332-16.1		29	358- 2.2		29
332-16.2		29	358- 3.2		29
333- 4.1		29			
335+12.1		29	-22 3467		251
335- 3.1		29	0623+71		251
336- 1.1		29			

SUBJECT INDEX

Page numbers refer to the title page of the article in which the subject is mentioned.

abundance gradients	139
abundances	
– C/N/O	117, 139, 214, 216, 217
	222, 319, 481, 505
– Mg	217
– helium	139, 211, 212, 273, 493
	481, 577
– nebular	1, 9, 73, 139, 191
	192, 195, 196, 211, 212
	214, 216, 218, 219, 220
	221, 226, 229, 319, 493
	353, 354
– photospheric	555
accretion rate	523
AGB stars	174, 199, 305, 359, 381
	391, 401, 425, 454, 446
	473, 505, 523, 531, 541
	543, 567, 577
age - dynamical	319
age - nebular	356, 391, 481, 577
angular diameters	359
ansae	93
aperture synthesis	17
atmosphere blanketing	273
atmospheres NLTE models	168, 273, 305
	306, 319
atomic data	157, 577
atomic processes	211, 212, 224, 225
	226, 228
B nebulae	93
Baade's window	167
binaries - spectroscopic	261
binary stars	1, 94, 204, 251, 261
	309, 310, 312, 314, 359
	460, 461, 462, 493, 505
	541, 567
bipolar flows	1, 93

bipolar structures	83, 105, 129, 173, 180
	187, 188, 204, 208, 447
black body	273
Bowen fluorescence	157
C nebulae	93
C-rich envelope	117
C-rich stars	453
Carbon stars	117, 381
Carbon stars - J-type	381
cataclysmic variables	251
catalogues	39
cavities	411
charge exchange	228
circumstellar dust	261
circumstellar shells	9, 117, 401, 450, 453
collimated flows	181, 184
color-color diagrams	9, 359, 401
common envelope	251, 310, 461, 505
	541
condensations	93
data processing	177
direct images	1, 49, 83, 105, 168
	169, 172, 177, 461, 462
direct images - narrow band	171, 173
	180, 185, 193, 301, 302, 351
disk population	335
distance scale	73, 189, 319, 357, 425
	493, 540, 577
distances	17, 65, 73, 93, 189, 192
	196, 215, 216, 335, 359
	425, 481, 523, 567
– kinematic	17
– spectroscopic	168, 273
dredge-up episodes	139
dust	117, 139, 157, 197, 201

	202, 203, 381, 523
– C-rich	567
– carbide	117
– carbon	381
– composition	117, 139, 201
	202, 203, 217, 577
– distribution	117, 203
– emission	9, 445
– heating	9, 117, 174, 201, 203
– O-rich	567
– silica	117
– size	201
– temperatures	9
dust and gas envelopes	359
dust to gas ratio	202, 577
dynamical evolution	105
dynamical models	523
Eddington limit	273
emission line profiles	93, 105, 319, 335
energy balance method	9, 229
	306, 481
envelope acceleration	457
envelope ejection	391
evolution to pn	381
evolutionary time scales	481, 523
evolutionary tracks	401, 463, 473, 539
expansion velocities	1, 39, 65, 93, 182
	187, 189, 190, 191, 194
	196, 198, 204, 209, 356
	411, 425, 481, 567
extinction	
– distances	65, 481
– internal	117
– interstellar	308, 577
extragalactic distances	577
extragalactic pn	1, 39, 213, 319, 335
	351, 353, 354, 355, 356
	357
Fabry Perot interferometry	178, 189, 195
	197
far infrared lines	214
filling factor	65
galactic	
– chemical evolution	577
– rotation curve	577

– center	63, 139, 167, 219, 481
– distribution	73, 425, 493
– gradients	73, 219, 577
– structure	73
galaxy - masses	577
galaxy - velocity dispersion	577
galaxy kinematics	352
globules	223
H burning stars	463, 505
H deficient stars	391
H I line profiles	17, 65, 93, 129
	202, 216
H II regions - compact	216
halo pn	62, 215, 335, 577
halos	195, 196, 197, 198, 199
	202, 17, 105
Hβ flux	39, 319
He burning stars	463, 505
He I self absorption	211, 212
height distribution	493
high velocity flows	1, 93, 180, 181, 183
	184, 194
HR diagram	539, 542, 545, 555
	463, 473, 481, 493, 531
	139, 229, 319, 401, 425
hydrodynamical calculations	391, 457
	459, 505, 543
hydrodynamical shaping	83, 411
initial mass function	473, 577
instabilities	411
internal motions	93, 105, 319
ionization correction factor	219, 319
ionization structure and models	117, 577
infrared	
–color-color diagram	9, 58, 448
	449, 452
– emission	117, 201, 202
– energy distribution	57
– excess	9, 60, 210, 359
– features - SiC	117
– features - silicate	117
– features - unidentified	117
– fluxes	9, 117, 401, 523
– halos	9
– images	117, 178, 179, 208
– maps	9
– photometry	9, 39, 54, 58, 445

		449
– spectra		59, 443, 445, 446
IRAS		
– colors		9, 63
– images		186
– sources		9, 52, 64, 174, 210
		251, 315, 359, 381, 567
		445, 448, 450, 452
– spectra		9, 401
IUE spectra		1, 39, 139, 251, 273
		293, 308, 313, 314, 317
		319, 355
K-type stars		312
kinematical models		93
kinematics		1, 73, 180, 181, 183
		184, 192, 196, 200, 205
		319, 458, 493
knots		183
late type stars		52
line intensities		39
line profile fits		157, 229, 273
line profiles		93, 181, 182, 188, 189
		190, 198, 200, 204, 209
		241
long period variables		451
luminosities		359
m-s death rate		493
m-s turnoff mass		425, 493
M-type stars		21, 312, 381
magnetic braking		251
magnetic fields		17, 193, 455
magnitude-age diagram		481, 542
magnitude-radius relation		531
maser sources		359
mass loss		381, 567, 577
mass loss rates		1, 17, 129, 199, 207
		241, 261, 273, 293, 317
		359, 401, 411, 454, 493
		543
mass tranfer		505
mass-density diagram		319
mass-radius relation		65, 83, 319, 523
mass-semimajor axis relation		505
mass-T_{eff} diagram		463
Miras		359, 401, 425, 567, 577

model atmospheres	241, 261
molecules	129
– C_2H_2	446
– CN	17, 129
– CO	17, 129, 157, 204, 205
	216, 381, 401
– H_2	17, 129, 157, 206, 207
	208
– H_2O	450
– HCN	129, 204, 453
– MgS	117
– OH	17, 129, 205, 209, 210
	205, 446
morphological classification	105
morphology	1, 60, 65, 83, 105
	169, 172, 174, 179, 193
	411, 455, 458, 461, 462
	481, 567
– butterfly	1, 83
– elliptical	1, 83
– round	1, 83
– type I	105
– type II	105
multiple shells	1, 105, 458, 523, 567
multiple structures	173, 184, 193, 200
multiple winds model	411
nebular	
– density	1, 60, 83, 195, 214
	523, 577
– density structures	157, 175
	197, 199, 220, 221, 223
	353
– dimensions	39, 186, 577
– emission	9, 226
– evolution	356, 523, 545, 567
– excitation classification	319
– mass	1, 60, 83, 207, 319
	355
– mass distribution	473
– spectral evolution	17
– structure	39, 180, 205, 206, 208
	444, 567
– temperatures	10, 195, 223, 319
	353
– temperature structure	176
neutron stars	505
novae	315
nuclear timescales	391

O-rich envelope	117	– identification	171, 192, 301
O rich stars	453		303
O-type stars	216, 308, 309	– luminosity	307, 335
Of-type stars	229, 308, 309	– magnitudes	1, 229, 301, 302
OH/IR stars	129, 359, 209, 210, 391	– parameters	1, 39, 241, 293
	401, 425, 445, 448, 449		545
	450, 577	– photometry	309, 310
OH/IR stars - formation	73	polarization	17, 455
OH/IR stars - evolution rate	425	population I OB stars	293
optical depths	157, 523	population I stars	65, 391
orbital periods	261	population II stars	391
OVI-type stars	229, 251, 545, 311	post AGB evolution	391
		post AGB stars	359, 445, 457, 463, 473
		pregalactic helium to hydrogen	213, 577
P Cygni profiles	241, 273, 293, 318, 446	proper motions	65
photoionization models	157, 217, 219	proto-planetary nebulae	401, 443, 445
	223, 224, 227, 228, 443		446, 447, 448, 452, 455
photometry	55, 56, 57, 168, 215		458
photometry - narrow band	49, 50, 57	pulsating stars	251, 311, 545, 467
physical processes	1, 139, 157	pulsation analysis	311
planetary nebulae		pulsation modes	251
– birth rates	73, 251, 425, 473		
	493, 540, 545, 577		
– candidates	335	radial velocities	1, 39, 73, 335, 352, 381
– classification	73	radiation driven wind theory	273
– compact	17, 444	radiation transfer	157, 227
– formation	505	radio fluxes	17, 39, 60, 401, 481
– ionization	523	radio interferometric observations	39
– low excitation	52	radio maps	17, 176
– luminosity function	335	radio spectra	17
– misclassified	21, 52	reflection effect	251
– new	1, 21, 61, 62, 63	Roche lobe	505, 541
	64, 167, 335, 351		
– origin	523		
– positions	21, 61, 62	shaping mechanisms	83
– progenitor mass	319, 381	shell formation	391, 523
	425, 493	Shklovsky method	319, 481
– progenitors	117, 359, 451, 567	shock excited lines	129
– progenitors C-rich	359	shock heating	447
– progenitors O-rich	359	shocks	83, 129, 157, 180, 411
– surveys	335		447, 456, 460
– time evolution	531	space missions	65, 577
planetary nebulae central stars		spatial density	73
– absolute magnitude distribution		speckle interferometry	1, 319, 443
	545	spectroscopic data	17, 50, 51, 52, 53
– chemical composition	555		54, 56, 59, 168, 179
– classification	493		183, 184, 190, 191, 192
– cool	261		197, 215, 219, 220, 229
– evolution	293, 367, 463, 473		273, 309, 310, 317, 319
	531, 539		354, 444, 458
– helium rich	251, 273	standard candles	335, 357, 577

statistical method	481
stellar	
– CO core	505
– core mass	359, 401, 523
– envelope	359, 545
– flux distribution	157, 308, 314
– formation	505
– gravities	1, 273, 316, 481, 545
– mass	251, 273, 355, 454, 481
	531, 539, 340, 567, 577
– mass-luminosity relations	531
– populations	493
– rotation	555
– spectra	9, 229, 545
– structure and evolution models	
	261, 311, 425, 451, 542
	543, 577
– temperatures	1, 9, 229, 273
	307, 314, 316, 335, 481
Stoy temperatures	157, 319
super giant stars	452
superwind	391, 411, 458, 505, 523
	567
symbiotic stars	21, 51, 52, 57,
terminal velocity	1, 273
thermal pulses	391
thermal timescales	391
transition phase	210, 391
transition time	391, 463
Type I pn	139, 208, 211, 212, 213
	319, 353, 493
Type II pn	139, 213, 219, 222, 493
	577
Type III pn	139, 493
Type IV pn	139, 493, 577
UV spectroscopy	50

variability	1
– infrared flux	359
– photometric	55, 56, 261
– radial velocity	261
VLA data	17, 60, 129, 176, 210
	401, 443, 567
white dwarfs	251, 273, 355
– CO	505
– DA	391, 545, 555, 567
– birth rate	545
– evolution	555
– evolutionary lifetimes	545
– formation	73, 391
– formation rate	425, 577
– magnetic	555
– mass	540
– mass distribution	493, 555
– non DA	391, 545, 555
– progenitors	555
wind features	241
wind timescales	391
wind velocity	207, 241
winds	1, 65, 93, 157, 174
	194, 273, 293, 381, 401
	411, 473, 456, 457, 458
	523, 531, 567, 577
– interacting	83
WR features	545
WR-type stars	1, 229, 293, 355, 391
	216, 261, 307, 308, 313
	315, 319
X-ray observations	50, 241, 304
Zanstra luminosities	229
Zanstra temperatures	1, 9, 157, 229, 241
	301, 305, 306, 315, 319
	481, 523, 531, 567

QB 855.5 .I67 1987
International Astronomical
 Union. Symposium 1987 :
Planetary nebulae